PROBABILITY AND STOCHASTIC MODELING

PROBABILITY AND STOCHASTIC MODELING

VLADIMIR I. ROTAR

CRC Press is an imprint of the
Taylor & Francis Group, an **informa** business
A CHAPMAN & HALL BOOK

CRC Press
Taylor & Francis Group
6000 Broken Sound Parkway NW, Suite 300
Boca Raton, FL 33487-2742

First issued in paperback 2019

ISBN-13: 978-1-4398-7206-2 (hbk)
ISBN-13: 978-0-367-38094-6 (pbk)

Visit the Taylor & Francis Web site at
http://www.taylorandfrancis.com

and the CRC Press Web site at
http://www.crcpress.com

To the memory of my parents I.M. and R.F. Rotar

To my teacher Yu.V. Prokhorov

To Galya

To Igor

Preface

This book is intended as a first course in probability with an emphasis on stochastic modeling. Distinctive features of the book concern its contents and its format as well.

The Contents. The exposition is rigorous; with rare exceptions, all assertions are proven; almost every topic found in a traditional introductory probability course is covered. On the other hand, the book pays substantial attention to *stochastic modeling*, which is atypical for first-course textbooks on probability. Cases in point are Markov chains, birth–death processes (including queuing processes), reliability models, and other topics, both theoretical and applied.

We also consider a *number of concrete models* (for example, a model of financial markets, or the principal components scheme); sometimes, even with real-world data. The goal here is not to teach particular models or numerical methods but rather to help the student to better appreciate general concepts and theoretical results, and to demonstrate practical possibilities and restrictions of different approaches under consideration. The same concerns examples and exercises on *numerical calculation with the use of Excel.*

Besides the traditional material, we also pay attention to *topics usually skipped* (or almost skipped) in introductory courses, although in the author's opinion, they are becoming increasingly important. In particular, this concerns martingales, classification of dependency structures, and risk evaluation.

The format of the book. The material is presented in the form of *two nested "routes."* Route 1 contains the basic material designed for a one-semester course. This material is *self-contained* and has a moderate level of difficulty.

Route 2 contains all of Route 1, offers a more complete exposition, and is suited for a two-semester course or self-study. It is *slightly* more difficult but should be approachable for any reader familiar with primary concepts of calculus and linear algebra.

The routes are explicitly designated. To facilitate reading, we use *markers similar to road signs*. For the reader to have a general outline of the book's structure, we also provide the Table of Contents with the same signs.

Potential audience. Due to the features mentioned, the book is suitable for people who are *not planning to take a second course on stochastic models but rather wish to have one comprehensive course (and a book) on theory and applications.* For the same reason, the book may serve as a text for students with *majors in mathematics and statistics* as well as those *with majors in computer sciences, economics, finance, or physics.*

The prerequisites are two semesters of calculus and a course in linear algebra. There is no need for any specialized mathematical background outside of these areas. The book also contains an appendix with some general definitions and frequently used formulas.

Acknowledgments

My sincere thanks go to my colleagues for useful discussions of various scientific and pedagogical questions relevant to this book. They include

Rabi Bhattacharya, University of Arizona at Tuscon, USA
Richard Bradley, University of Indiana at Bloomington, USA
Alexander Bulinskiy, Moscow State University, Russia
Youri Davydov, University of Lille 1, France
John Elwin, San Diego State University, USA
Patrick Fitzsimmons, University of California at San Diego, USA
Oleg Izhvanov, General Atomics, San Diego, USA
Victor Korolev, Moscow State University, Russia
Jeffrey Liese, California Polytechnic State University, San Luis Obispo, USA
R. Duncan Luce, University of California at Irvine, USA
Donald Lutz, San Diego State University, USA
Alexander Novikov, University of Technology, Sydney, Australia
Yosef Rinott, Hebrew University, Jerusalem, Israel
Michael Sharp, University of California at San Diego, USA
Arkady Tempelman, The Pennsylvania State University, USA
Igor Ushakov, University of California at San Diego and Qualcomm, USA
Lee Van de Wetering, San Diego State University, USA
Hans Zwiesler, Ulm University, Germany

I really appreciate their help.

Sunil Nair was a wonderful editor. I felt very comfortable with all matters concerning the preparation of the book.

Additional thanks go to the editorial staff at Taylor & Francis, in particular, to Shashi Kumar for help in the preparation of the final file.

V. R.

Contents

Flag-signs in the margins designate a route: either Route 1 (which is entirely contained in Route 2) or Route 2; see the Preface and Introduction for a description of these routes. A new flag-sign is posted only at the moment of a route change.

Introduction

As mentioned in the preface, the material of the book is presented in the form of two nested "routes." Route 1 consists of the material designed for a one-semester course. Route 2 is intended for a broader and deeper study, and is better suited for a two-semester course.

Route 1 is self-contained.

The special "road signs" will help the reader to continue in the chosen route. For example, the sign

Route 1 $\Rightarrow$ *page 111*

indicates that the readers who chose Route 1 should advance to p.111 to continue the route. Below this sign, a small "flag" in the margin designates which route runs now (as in the margin here). In other words, if the reader does not switch to the page mentioned (as p.111 above), she/he will have entered Route 2.

2

If the reader goes to the page mentioned (as p.111 above), then in the margin of this page, she/he will see a sign confirming that this is the right place to move to, and showing a particular location on the page where Route 1 picks up (as in the margin here).

1,2

The corresponding flag-signs are also placed in the Table of Contents.

In the exercises, the problems belonging to Route 2 are marked by an asterisk *. However, if a whole section belongs to Route 2, in the exercises, we mark by * only the title of this section. In the exercises for chapters belonging entirely to Route 2, we naturally do not mark anything.

Occasionally, purely technical proofs or additional remarks are enclosed by the signs ► ◄. This material may be omitted in the first reading.

If we have not used a definition or fact recently or are using it for the first time, then the corresponding references are given. This is being done *just in case*. If the reader is already familiar with a referred item, it makes sense to ignore the reference and move ahead.

Certainly, when moving along Route 1, the reader is welcome to look around or venture into areas that are not included in the first route. However, the reader should not be discouraged if something seems difficult. Route 2 is indeed slightly more involved than Route 1, but only slightly, and requires just a more in-depth reading.

Another matter is that if you are taking a one-semester course, then it may be reasonable to postpone the material of Route 2—at least, most of this material—for a while, and return to it when you have more time and experience. Enticing topics you skipped on the way will await you.

More technical remarks. The symbol ■ indicates the end of a proof, while the symbol □ marks the end of an example or a series of examples.

The numbering of sections and formulas in each chapter is self-contained. The adopted system of references is clear from the following examples.

Section 2.3 is the third subsection of the second section of the *current* chapter.

The formula (2.3.4) is the fourth formula of the third subsection of the second section of the *current* chapter.

Example 2.3-4 is the fourth example from Section 2.3 of the *current* chapter.

In each chapter, theorems, propositions, and corollaries are being enumerated in a linear fashion through the whole chapter: the theorem that appears after Proposition 2 is Theorem 3, and the corollary following Theorem 3 is Corollary 4.

If we refer to a formula, section, example, etc., from another chapter, we write the number of the chapter to which we are referring in bold font. For instance, Section **1**.2.3 is the third subsection of the second section of the *first* chapter. Formula (**1**.2.3.4) is the formula (2.3.4) of the *first* chapter. Theorem **1**.2 is Theorem 2 of Chapter 1, etc.

The following *abbreviations* are used throughout the entire book:

a.s. – almost surely
c.d.f. – cumulative distribution function
CLT – central limit theorem
c.v. – coefficient of variation
d.f. – distribution function (the adjective "cumulative" is omitted)
EU – expected utility
EUM – expected utility maximization
FSD – first stochastic dominance
g.f. – generating function
i.i.d. – independent and identically distributed
l.-h.s. – left-hand side
LLN – law of large numbers
m.g.f. – moment generating function
p.d.f. – probability density function
RDEU – rank dependent expected utility
r.-h.s. – right-hand side
r.v. – random variable
r.vec. – random vector

Chapter 1

Basic Notions

1 SAMPLE SPACE AND EVENTS

1.1 Sample space

When building a model of any experiment, we first specify the set of *all* possible *non-decomposable* outcomes which may occur as a result of the experiment. We call such a set a *sample space*, or a *space of elementary outcomes*. A traditional notation for a sample space is Ω, and an element of this space (called an individual or *elementary* outcome) is denoted by ω. So, $\Omega = \{\omega\}$. Formally, Ω is just a set of elements of an arbitrary nature. However, for a particular experiment, each ω designates a possible outcome.

The number of elements in Ω will be denoted by $|\Omega|$.

EXAMPLE 1. In the case of *rolling a cubic die*, there are six possible outcomes. We denote them by ω_i, $i = 1, ..., 6$ and write $\Omega = \{\omega_1, ..., \omega_6\}$ and $|\Omega| = 6$. Note also that the outcome "an even number was rolled" is not elementary because it may be decomposed into "smaller" outcomes $\omega_2, \omega_4, \omega_6$.

EXAMPLE 2. Consider *tossing a coin*. Each toss can come up heads or tails. We can denote these outcomes by ω_1 and ω_2 and write $\Omega = \{\omega_1, \omega_2\}$ and $|\Omega| = 2$.

When *tossing a coin twice*, we may encounter one of four outcomes: *HH, HT, TH, TT*, where *H* stands for head and *T* for tail. It is worth emphasizing that *HT* and *TH* are different outcomes, so $|\Omega| = 4$.

Now, denote by N_n the number of outcomes in the case of n tosses. *For each* particular outcome of $n-1$ tosses, we have two possible outcomes for n tosses: either the last, nth, toss comes up heads or it comes up tails. Consequently, $N_n = 2N_{n-1}$. Then we can write $N_n = 2 \cdot 2 \cdot N_{n-2} = ... = 2^{n-1}N_1 = 2^n$ because $N_1 = 2$. Thus, $|\Omega| = 2^n$. □

Example 2 fits the following general scheme. Consider a *sequence of n trials* each of which may be either successful or not. In Example 2, we may call a success an appearance of a head; if we consider n newborns, then we may call a success a birth of a girl. In a quality control of n items, we talk about success if an item meets certain requirements.

Tossing a coin and a birth of a child are certainly different phenomena, but from a mathematical standpoint, they may be represented by the same model. Let us "mark" success by 1 and failure by 0. Then each particular outcome may be identified with a sequence of ones and zeros. For example, a sequence 1001 corresponds to four trials for which only the first and the last were successful.

As shown in Example 2, for n trials, Ω consists of 2^n outcomes (or sequences of ones and zeros of length n).

EXAMPLE 3. Consider an all-play-all tournament of 5 hockey teams where each team should play every other exactly once, and the rules exclude ties. Each game may be identified with a pair of teams. In Section 3.3, we will prove that one can select exactly $n(n-1)/2$ pairs from n elements, so there will be a total of $\frac{5\cdot4}{2} = 10$ games. Since ties are excluded, each game may have one of two results. For example, if the teams are numbered, we may define the result of a game to be a success if the winner's number is less than the number of the other team. Thus, we have 10 trials (games), and the total number of all possible outcomes of the tournament is $2^{10} = 1024$. If the number of teams is six, just one larger, then the number of games is $\frac{6\cdot5}{2} = 15$, and $|\Omega| = 2^{15} = 32,768$. Perhaps, the reader would not expect these numbers to be so large. □

In all examples above, the space Ω is finite; in symbols, $|\Omega| < \infty$. This is certainly not always the case.

EXAMPLE 4. Consider the number of cosmic particles registered by a device during a certain period of time. Such a number cannot be arbitrarily large. However, since we do not know the largest possible number, when modeling this phenomenon, it makes sense to identify the set of all outcomes with that of all non-negative integers and write $\Omega = \{\omega_0, \omega_1, \omega_2, ...\}$, where the outcome ω_i corresponds to i particles registered. (Here, it is convenient to start with ω_0 rather than with ω_1.) The space Ω is infinite but *countable*, that is, we can assign a natural number to each outcome. □

Sample spaces with a finite or countable number of outcomes are said to be *discrete*; all others are called *non-discrete*.

EXAMPLE 5. You come to a bus stop knowing that buses come each hour, but you do not know the schedule. The waiting time for the next bus may be equal to any number between 0 and 1, so we may identify Ω with an interval $[0,1]$. One cannot count all points from an interval, and hence the space Ω is non-discrete. This is also the case with the next three examples.

EXAMPLE 6. The error of a measurement can theoretically be any number, positive or negative, and Ω may be identified with the real line.

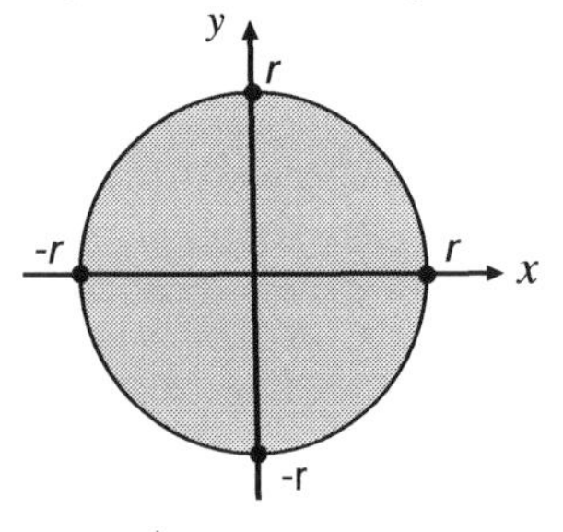

FIGURE 1.

EXAMPLE 7. John throws a dart at a circular target. Assuming that the dart always hits the target, we may identify the sample space with the set of points in an (x,y)-coordinate system, for which $x^2+y^2 \leq r^2$, where r is the radius of the target disk; see Fig. 1. So, each ω is a point (x,y), and $\Omega = \{(x,y) : x^2+y^2 \leq r^2\}$. □

Curves depicting the evolution of atmospheric pressure during different time periods or family trees of different families in a country may serve as examples of outcomes of a more complicated nature. We consider such spaces in Chapters 4, 12, 13, 16 and others devoted to random processes.

Route 1 ⇒ *page 5*

2

EXAMPLE 8 shows that one should be cautious when verifying whether a sample space is discrete or not. Consider an *infinite* sequence of trials marking again a success by 1 and a failure by 0. So, an outcome is an infinite sequence of ones and zeros. Let us write before any such a sequence zero and a point. For example, instead of the sequence 1001..., we write 0.1001.... The last expression may be viewed as a number from the interval $[0,1]$ written in binary notation. For example, in this notation, the number $0.1001... = 1\cdot\frac{1}{2}+0\cdot\frac{1}{4}+0\cdot\frac{1}{8}+1\cdot\frac{1}{16}+...$. We see that any sequence of ones and zeros corresponds to a unique number from $[0,1]$, and vice versa.

Thus, the sample space may be identified with $[0,1]$ and, hence, it is non-discrete. We do not say that such an identification is always useful in applications, but it clears up the situation. □

1.2 Events

By definition, given a particular sample space Ω, an *event* is a set of elements from Ω. Quite often, we may specify an event by a word description rather than listing all elements from the event.

EXAMPLE 1. For a family with two children, let us write the sample space as $\Omega = \{bb, bg, gb, gg\}$, where b stands for a boy and g for a girl. We could define an event A to be the set $A = \{bb, bg, gb\}$, but we may also write $A = \{\textit{at least one boy}\}$. □

We define standard operations on events $A, B, \ldots$.

- The *complement* A^c of A is the set of all elements from Ω *not* belonging to A.
- The *union* $A \cup B$ of events A and B is the set of all points from Ω belonging *either* to A *or* to B (or to both).
- The *intersection* $A \cap B$ is the set of all points from Ω belonging to both A *and* B. For $A \cap B$, we will usually use a *simpler notation* AB, and call it the *product* of A and B.

The above definitions are illustrated in what is called the *Venn diagrams* in Fig. 2ab. In each diagram, the rectangle represents a space Ω. In diagram (a), the shaded region represents a set A, and the non-shaded—the complement A^c. In diagram (b), AB^c is the set of all points from A that are not in B, A^cB is the set of all points from B that are not in A, the shaded region represents the union $A \cup B$, the strongly shaded region—the product AB.

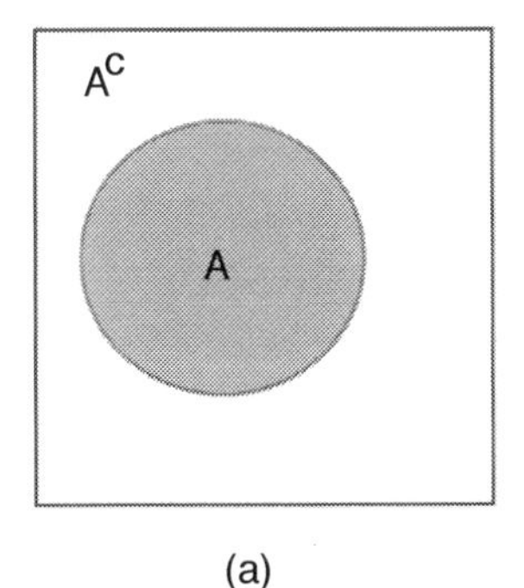

(a)

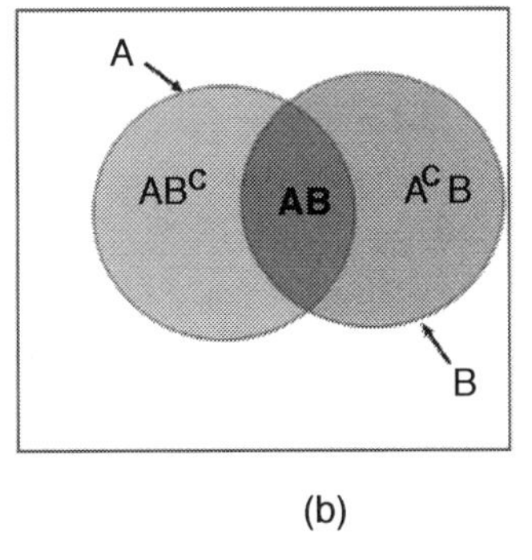

(b)

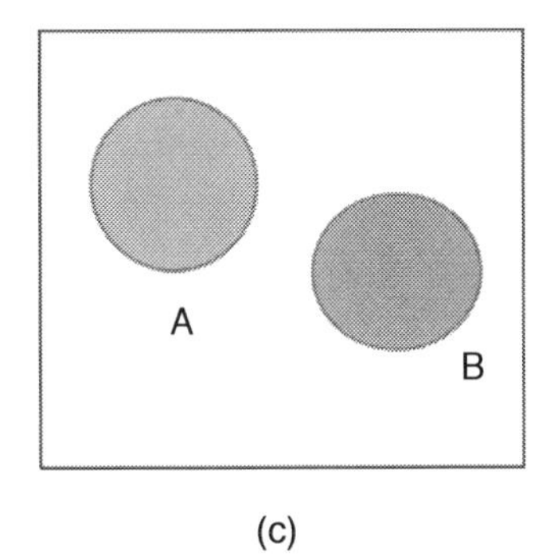

(c)

FIGURE 2.

EXAMPLE 2. In the situation of Example 1, let $A = \{at\ least\ one\ boy\}$ and $B = \{at\ least\ one\ girl\} = \{bg, gb, gg\}$. Then $AB = \{one\ girl\ and\ one\ boy\} = \{bg, gb\}$, and $A \cup B$ equals the whole space Ω.

EXAMPLE 3. There are exactly three contestants numbered 1,2,3, and each may take any place in a contest. In Section 3.2, we will show that n elements may be arranged (or permuted) in $n!$ ways. So, each outcome may be identified with one of $3! = 6$ permutations or arrangements of the numbers (contestants) 1,2,3. For example, the permutation (2,3,1) means that Contestant 2 took first place, Contestant 3 second, and Contestant 1 third. Let $A = \{Contestant\ 1\ took\ first\ place\}$ and $B = \{Contestant\ 2\ took\ second\ place\}$. Since $A = \{(1,2,3), (1,3,2)\}$ and $B = \{(1,2,3), (3,2,1)\}$, we have $AB = \{(1,2,3)\}$ and $A \cup B = \{(1,2,3), (1,3,2), (3,2,1)\}$.

EXAMPLE 4. We revisit Example 1.1-7. Let $A = \{(x,y) \in \Omega : x \geq 0\}$ and $B = \{(x,y) \in \Omega : y \geq 0\}$. The situation is illustrated in Fig. 3. The set A corresponds to the right half of the disk (x's are positive) and the set B to the top half of the disk (y's are positive). The whole shaded region corresponds to $A \cup B$, the strongly shaded to AB. □

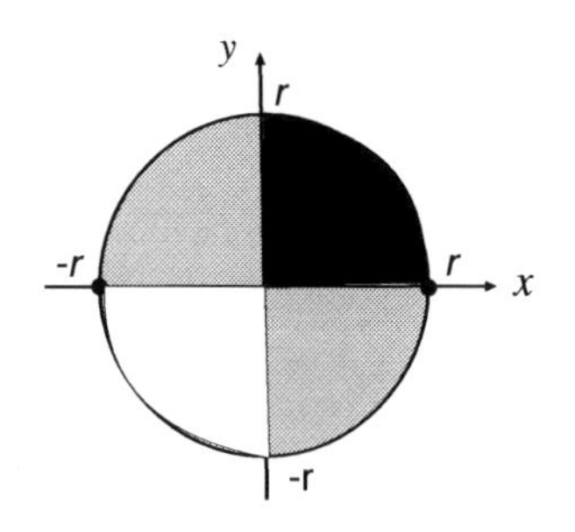

FIGURE 3.

The empty set, denoted by the symbol $\emptyset$, is also considered an event.

Two events, A and B, are said to be *disjoint* or *mutually exclusive* if they do not have common outcomes (or points when we use figures). Clearly, this is equivalent to the relation $AB = \emptyset$; see also Fig. 2c.

EXAMPLE 5. In Example 3, let $E = \{Contestant\ 2\ took\ first\ place\}$. Since first place cannot be occupied by Contestant 1 and Contestant 2 simultaneously, $AE = \emptyset$.

EXAMPLE 6. Without specifying a sample space, consider the events $A = \{The\ name\ of\ a\ man\ passing\ by\ is\ John\}$ and $B = \{It\ will\ be\ raining\ tomorrow\}$. Are these events disjoint? Sometimes people say "yes," but it is not true: A and B may occur simultaneously. □

Certainly, in the same fashion we can consider unions and intersections of an arbitrary number of events.

EXAMPLE 7. In Example 3, let $C = \{Contestant\ 3\ took\ third\ place\}$. Since $C = \{(1,2,3), (2,1,3)\}$, we have $A \cap B \cap C$ or simply $ABC = \{(1,2,3)\}$ and $A \cup B \cup C = \{(1,2,3), (1,3,2), (3,2,1), (2,1,3)\}$. □

The following proposition proves to be useful in many situations.

Proposition 1 *1) For any event A,*

$$(A^c)^c = A. \tag{1.2.1}$$

2) De Morgan's laws: For any events A and B,

$$(A \cup B)^c = A^c \cap B^c, \ (A \cap B)^c = A^c \cup B^c. \tag{1.2.2}$$

A *memorizing rule*: when distributing the complement sign, flip the operation symbol. A similar law is true for three or more events; see Exercise 6 for a formal statement.

Proof. Relation (1.2.1) is clear. Furthermore, let $\omega \in (A \cup B)^c$. Then ω is *not* contained in $A \cup B$, which means that $\omega \notin A$ *and* $\omega \notin B$. Consequently, $\omega \in A^c \cap B^c$. Going the other way is similar, and the first relation in (1.2.2) is proved.

Let $\omega \in A^c \cup B^c$. Then either $\omega \notin A$, or $\omega \notin B$. This means that ω cannot belong to A and B simultaneously. Hence, it belongs to $(A \cap B)^c$. Again, going the other way is similar. ■

Route 1 ⇒ *page 8*

2

1.3 Class of events[1]

To make the theory more rigorous, we should fix the class $\mathcal{A}$ of sets A of interest. For the theory we are building to be complete and non-contradictory, the class $\mathcal{A}$ should be sufficiently rich. To this end, we assume the following properties to be true:

(a) If $A \in \mathcal{A}$, then the complement A^c also belongs to $\mathcal{A}$.
(b) If events $A_1, A_2, \ldots$ are from $\mathcal{A}$, then their union $\cup_i A_i$ also belongs to $\mathcal{A}$.
(c) If events $A_1, A_2, \ldots$ are from $\mathcal{A}$, then their intersection $\cap_i A_i$ also belongs to $\mathcal{A}$.
(d) The empty set $\emptyset$ and the whole space Ω belong to $\mathcal{A}$.

It is noteworthy that Properties (a)-(b) imply Properties (c)-(d), but we list all of them for the completeness of the picture. For simplicity, we will show that (c)-(d) follow from (a)-(b) for the case of two events.

Let $A_1, A_2 \in \mathcal{A}$. Using Properties (a) and (b), we have $A_1^c, A_2^c \in \mathcal{A}$, and hence $A_1^c \cup A_2^c \in \mathcal{A}$. Consequently, $(A_1^c \cup A_2^c)^c \in \mathcal{A}$. By (1.2.2), $(A_1^c \cup A_2^c)^c = (A_1^c)^c \cap (A_2^c)^c = A_1 \cap A_2$. Thus, $A_1 \cap A_2 \in \mathcal{A}$. The proof for the general case is similar; see also Exercise 6.

To prove (d), consider any set $A \in \mathcal{A}$. We have $\emptyset = A \cap A^c \in \mathcal{A}$ by property (c), and $\Omega = A \cup A^c \in \mathcal{A}$ by property (b).

A class $\mathcal{A}$ with the above properties is called a σ-*field* or *sigma-field.*

If a space Ω is discrete, we can choose $\mathcal{A}$ to be the class of all subsets of Ω (though we do not have to do that). Clearly, in this case all the properties above hold.

The non-discrete case is more complicated, and the class $\mathcal{A}$ should not be too rich. For example, if the sample space is the real line, then it may be shown that it is impossible to define probabilities (which we are going to do in the next section) simultaneously for *all* sets. So, some sets should be excluded from consideration. These sets are those for which it is impossible to define the notion of length; we skip the formal definition. Fortunately, such sets are rather exotic, and the exclusion of them by no means prevents us from building models of real phenomena. (For detail, see an advanced textbook on Measure Theory or Probability, e.g., [10], [24], [42], [47].)

Thus, formally, in the general framework, we assume that two things are given: a sample space Ω and a class $\mathcal{A}$ satisfying Properties (a)–(d) above. Only sets from the class $\mathcal{A}$ are considered events. In Section 2.4, we discuss which $\mathcal{A}$ should be chosen for the theory not to be self-contradictory.

[1] This section may be skipped in the first reading.

1,2

2 PROBABILITIES

2.1 Probability measure

Consider a sample space Ω. Our next step is to assign to events A from Ω values $P(A)$ which will be referred to as the probabilities of the events A. In other words, we want to introduce a function $P(A)$ defined on events A and having natural, in this context, properties. These properties are called probability axioms and may be stated as follows.

(i) For any A,

$$0 \leq P(A) \leq 1. \tag{2.1.1}$$

(ii) $P(\Omega) = 1$.

(iii) *The additivity property*: for any *disjoint* events $A_1, A_2, ...$

$$P(\cup_i A_i) = \sum_i P(A_i). \tag{2.1.2}$$

We call a function $P(A)$ satisfying the properties (i)-(iii) a *probability distribution*, or a *probability measure* on Ω. A particular value of $P(A)$ is called the *probability of an event* A.

To illustrate (2.1.2), let us look at Fig. 2c. The events A and B there are disjoint, and in accordance with (2.1.2), the probability of their union should be equal to the sum of their probabilities. However, we should not expect the same in the situation in Fig. 2b. In this case, A and B are not disjoint, and the additivity property for these events may not hold. In Section 2.5, we consider this issue in detail.

Next, note that the above definition implies

$$P(\emptyset) = 0.$$

Indeed, $\emptyset \cap \Omega = \emptyset$, so the empty set $\emptyset$ and Ω are disjoint events. Hence, $1 = P(\Omega) = P(\emptyset \cup \Omega) = P(\emptyset) + P(\Omega) = P(\emptyset) + 1$. Thus, $1 = P(\emptyset) + 1$, which implies $P(\emptyset) = 0$.

It is also noteworthy that in Axiom (i), we might require just $P(A) \geq 0$. Indeed, $P(A) \leq P(A) + P(A^c) = P(A \cup A^c) = P(\Omega) = 1$, and we come to (2.1.1). We presented Axiom (i) in the above form for the completeness of the picture.

To some degree, the whole present book is concerned with the question of how to define and/or specify probability measures in particular problems. However, we are able to give first examples immediately.

2.2 Probabilities in discrete spaces

We start with a discrete sample space $\Omega = \{\omega_1, \omega_2, ...\}$. Suppose that somehow, proceeding from the nature of a particular problem under consideration, we have assigned to each ω_i a number p_i that in a certain sense reflects the likelihood of this outcome, i.e., to what extent it is likely to occur. Such an assignment may be presented as the array

$$\begin{array}{cccc} \omega_1 & \omega_2 & \omega_3 & \cdots \\ p_1 & p_2 & p_3 & \cdots \end{array} \tag{2.2.1}$$

Because of the axioms (i), (ii) and (iii), we can choose only p_i's such that for all i

$$p_i \geq 0, \text{ and } \sum_i p_i = 1. \tag{2.2.2}$$

We call p_i's *elementary probabilities.*

EXAMPLE 1. Consider tossing a coin. Let p_1, p_2 be the elementary probabilities corresponding to the two outcomes, ω_1, ω_2. If the coin is regular (symmetric), it is reasonable to set $p_1 = p_2$. The second equation for the two unknowns, p_1 and p_2, is $p_1 + p_2 = 1$. A solution to these two equations is $p_1 = p_2 = 1/2$.

EXAMPLE 2. In the case of rolling a die, we have 6 unknown elementary probabilities, $p_1, ..., p_6$ such that $p_1 + ... + p_6 = 1$. If the die is fair, we assume all probabilities are equal, which leads to all $p_i = 1/6$. □

These two examples fit a general scheme which we will call

The classical scheme. Suppose that a sample space Ω is finite, i.e., $|\Omega| < \infty$, and all outcomes are equally likely. Then, all elementary probabilities are identical, and their total sum equals one. Then, for all i, the elementary probability

$$p_i = \frac{1}{|\Omega|}. \tag{2.2.3}$$

Certainly, in general, elementary probabilities need not be identical.

EXAMPLE 3. Suppose that in an area, a sunny day is twice as likely as a non-sunny day. So, we have two outcomes: "sunny" and "non-sunny," and by assumption, $p_1 = 2p_2$. Then $1 = p_1 + p_2 = 3p_2$. So, $p_2 = \frac{1}{3}$, and $p_1 = \frac{2}{3}$.

EXAMPLE 4. Consider a "crooked" (or loaded) die that has been tampered with to make landing of some faces to be more likely than others. Suppose that for such a die, rolling two is twice as likely as rolling one, rolling three is three times as likely as rolling one, and so on. So, we can write $p_2 = 2p_1, p_3 = 3p_1, ..., p_6 = 6p_1$, and thus, $1 = p_1 + p_2 + ... + p_6 = p_1 + 2p_1 + 3p_1 + ... + 6p_1 = p_1(1 + 2 + ... + 6) = 21p_1$. Hence, $21p_1 = 1$, and $p_1 = \frac{1}{21}, p_2 = \frac{2}{21}, ..., p_1 = \frac{6}{21}$.

EXAMPLE 5. Consider an infinite discrete sample space $\Omega = \{\omega_0, \omega_1, \omega_3, ...\}$. Let $p_0 = \frac{1}{2}$, and each probability p_i be half the previous; that is, $p_i = \frac{1}{2}p_{i-1}$. Thus, $p_0 = \frac{1}{2}, p_1 = \frac{1}{4}, p_2 = \frac{1}{8}, ...$. Since the total (infinite) sum $\frac{1}{2} + \frac{1}{4} + \frac{1}{8} + ... = 1$ (see the Appendix, (2.2.11)), the probability distribution has been properly defined. For now, this example looks formal, but as we will see in Chapter 3, it may serve as a good model for many experiments. □

Now, it is natural to define $P(A)$, the probability of an event A, as the sum of all elementary probabilities corresponding to the outcomes from A; that is,

$$P(A) = \sum_{i:\omega_i \in A} p_i. \tag{2.2.4}$$

We also set, by definition, $P(\emptyset) = 0$. The reader is invited to verify that the function $P(A)$ defined above satisfies the axioms (i)–(iii) from Section 2.1.

EXAMPLE 6. In Example 4, $P(\textit{rolling an even number}) = p_2 + p_4 + p_6 = \frac{2}{21} + \frac{4}{21} + \frac{6}{21} = \frac{12}{21} = \frac{4}{7}$. □

Let us come back to

The classical scheme. Denote by $|A|$ the number of outcomes in A. Because the probability of each elementary outcome is $1/|\Omega|$, adding up all elementary probabilities of outcomes in A, we obtain

$$P(A) = |A| \cdot \frac{1}{|\Omega|} = \frac{|A|}{|\Omega|}. \tag{2.2.5}$$

EXAMPLE 7. There are exactly two participants in a contest that consists in answering questions. The first contestant chooses at random one question out of eight, and after that the second contestant chooses a question out of the seven remaining. (Regarding the general terminology, we say that an element is selected from a set of elements *at random*, if all elements are equally likely to be selected.) There are three easy and five difficult questions. Which contestant has an advantage: the first or the second? We will get an answer at the end of this example, but first consider a general scheme which is used for this type of problem.

Suppose an urn contains b black and r red balls, and set $n = b + r$.

(a) If a ball is drawn from the urn at random, what is the probability of $A = \{\textit{the ball is red}\}$, or shortly $P(\textit{red})$? There are n ways in total to select one ball from the n balls, and among these ways there are r ways to choose a red ball. In other words, $|\Omega| = n$, while $|A| = r$, and in accordance with (2.2.5), $P(\textit{red}) = r/n$. So, if there are three red balls and two black balls, the probability of drawing a red ball is $\frac{3}{3+2} = 0.6$.

(b) Now, let two balls be drawn at random, one after the other, without replacing the first ball back in the urn. What is the probability that both balls are red? Let us denote the event in hand as $A = \{\textit{red, red}\}$. There are n ways to draw the first ball, and *for each* of these n ways, there are $n - 1$ ways to draw the second ball. Consequently, there are $n(n-1)$ ways to draw two balls (see a general scheme in Section 3.1). In other words, $|\Omega| = n(n-1)$. Similarly, there are $r(r-1)$ ways to choose two red balls from the r red balls, that is, $|A| = r(r-1)$. Hence,

$$P(\textit{red, red}) = \frac{r(r-1)}{n(n-1)}.$$

So, in the above numerical example, the probability of drawing two red balls is $\frac{3 \cdot 2}{5 \cdot 4} = 0.3$.

(c) Two balls are drawn at random without replacement. What is the probability that the second ball is red (whatever the first is)? The answer, which perhaps looks somewhat unexpected, is that this probability is the same as it would be for the first ball: r/n. Thus, no contestant in the above problem has an advantage.

To make it more plausible, note that it is equally likely to select first a red ball and then a black one, or first, a black and then a red. Indeed, the number of combinations of the former type is rb, and of the latter type br. Since the sample space is the same as in Problem (b), $|\Omega|$ is still $n(n-1)$, and hence, the probabilities of both events are the same: $\frac{br}{n(n-1)}$.

Let us derive rigorously the probability that the second ball is red. There are $r(r-1)$ ways to select two red balls, and there are br ways to select a black ball first, and after that a red ball. Then, for the event A in hand, $|A| = r(r-1) + br = r(r+b-1) = r(n-1)$.

Hence,

$$P(\textit{the second ball is red}) = \frac{r(n-1)}{n(n-1)} = \frac{r}{n}.$$

(Another way of reasoning is to observe that there are r ways to select a red ball which will be the second, and *for each* such a selection, there are $n-1$ ways to select any ball out of the remaining $n-1$. So, again $|A| = r(n-1)$.)

After we consider a counting technique in Section 3, in Exercise 21, the reader will be invited to prove a similar result for an arbitrary draw. □

2.3 Examples of probability measures in non-discrete spaces

EXAMPLE 1. Let us return to the waiting-time problem in Example 1.1-5. Buses come each hour, and we have identified Ω with the interval $[0,1]$. Suppose that all possible values of the bus arrival time are equally likely. Rigorously, we will understand it as follows: for any subinterval Δ in $[0,1]$, the probability that the waiting time will fall within Δ does not depend on the location of Δ and is proportional to its length; see Fig. 4. Since the total length of $[0,1]$ is one, eventually we set $P(\Delta) = \ell(\Delta)$, where ℓ stands for length. For example, it is equally likely whether we will wait less than half an hour, or greater, or the waiting time will be between $\frac{1}{4}$ and $\frac{3}{4}$ of an hour: in all three cases, the probability is $1/2$. The probability distribution of this type is called *uniform*.

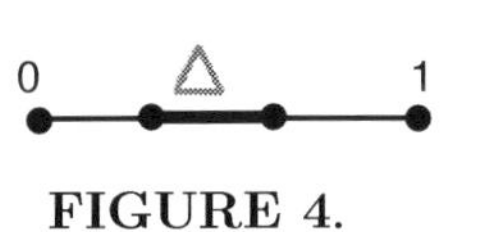

FIGURE 4.

The above construction is natural but it leads to a slightly unexpected conclusion: since a point is an interval of zero length, the probability of each point is zero. For example, the probability that we will wait *exactly* 30 min. ("not a microsecond less, not a microsecond more") is zero. However, the probability that the integer number of minutes will be 30, that is, the waiting time will be larger than or equal to 30 and less than 31 min., is equal to the length of a one-minute interval, i.e., $1/60$.

Now, let us observe that, since the probability of each point is zero, we cannot present $P(\Delta)$ as the sum of elementary probabilities as we did in (2.2.4): the probabilities of the points of which the interval Δ consists are equal to zero, and the sum of zeros is zero rather than the length of Δ as defined.

This is the main issue in defining probabilities in non-discrete spaces, and we see that it is reasonable to proceed not from the notion of an elementary probability but directly from the notion of the probability of an event. In our case, we set $P(\Delta) = \ell(\Delta)$, and proceeding from this, we define the probability of a set as its length.

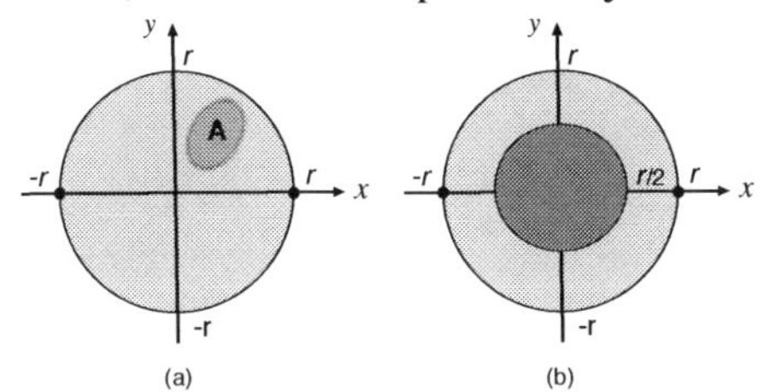

FIGURE 5.

EXAMPLE 2. Let us revisit Example 1.1-7, and assume that all points in the disk are equally likely to be hit.

Similar to Example 1, this means that the probability that the dart hits a region A in the disk (see Fig. 5a) does not depend on the location of the region and is proportional to its area. Since the total area is πr^2, we write $P(A) = \dfrac{s(A)}{\pi r^2}$, where $s(A)$ is the area of A. Such a probability distribution is called *uniform* on a set (in our case, it is a disk) in a plane. In particular, the probability that the distance from the point hit to the center is less than $r/2$ (see Fig. 5b; the interior disk corresponds to the event in question) equals $\dfrac{\pi(r/2)^2}{\pi r^2} = \dfrac{1}{4}$. □

Certainly, distributions on non-discrete spaces may differ from uniform ones; later we consider a large variety of non-uniform distributions.

2

Route 1 ⇒ page 13

Here is an example of a non-uniform distribution on a non-discrete space.

EXAMPLE 3. Suppose that the player in Example 2 is skillful enough, and it is more likely that he hits points that are closer to the center. To reflect this, we assign a weight $p(x,y)$ to each point (x,y). While the area of A is the usual integral $\int\int_A dxdy$, in this case, we set $P(A) = \int\int_A p(x,y)dxdy$, taking points with their weights. In Chapter 6, we consider the similar notion of a *density*, and the last term has the following analogy. A physical disk (say, metallic) may have a different density at different points, and parts of the disk where the density is higher, are heavier. Here, we deal not with physical weight but with a weight function measuring "likelihood."

Since $P(\Omega) = 1$, we must choose a weight function $p(x,y)$ such that $\int\int_\Omega p(x,y)dxdy = 1$. For example, let $p(x,y) = k/\sqrt{(x^2+y^2)}$, where k is a constant. Note that we have assigned higher weights to points that are closer to the center. Using the polar system of coordinates, it is not difficult to compute that $\int\int_{\sqrt{x^2+y^2} \leq r} \frac{k}{\sqrt{(x^2+y^2)}} dxdy = k \cdot 2\pi r$. So, k must be equal to $1/2\pi r$. Eventually, $P(A) = \frac{1}{2\pi r}\int\int_A \frac{1}{\sqrt{(x^2+y^2)}} dxdy$. For instance, if A is the interior disk in Fig. 5b, then $P(A) = \frac{1}{2\pi r}\int\int_{\sqrt{x^2+y^2} \leq r/2} \frac{1}{\sqrt{(x^2+y^2)}} dxdy = \frac{1}{2}$; compare to $\frac{1}{4}$ in Example 2. We again skip details of integration. □

2.4 Probability space[1]

In this short section, we clarify the notion of a probability measure $P(A)$ and make it more precise. As was noted in Section 1.3, when building a model of an experiment, we fix a class $\mathcal{A}$ of sets from the sample space Ω, satisfying certain properties, refer to only sets from this class as events, and deal only with these sets. In the discrete case, $\mathcal{A}$ may include all subsets of Ω; in the general case, it is not so.

Therefore, when defining a probability measure $P(A)$, we do it *only for sets from* $\mathcal{A}$, and we *presuppose that such a measure does exist.* In particular models, we should make sure that this is indeed true. As was already mentioned in Section 1.3, in non-discrete models it is impossible to define a probability measure on *all* sets from the sample space. So, we should exclude some (rather exotic) sets from consideration and deal with a class $\mathcal{A}$ for which the measure $P(A)$ is well defined.

In this book, for all models we consider, we take for granted that such a class $\mathcal{A}$ exists and all particular events of interest are from this class. On the other hand, it is noteworthy

[1]This section may be skipped in the first reading.

that in all models below, the very nature of these models and the way in which we construct concrete probability measures, make this assumption quite plausible.

Thus, a probability model of a random experiment consists of three components: a sample space Ω, a class of events $\mathcal{A}$, and a probability measure $P(A)$ defined on events from $\mathcal{A}$. The triple $\{\Omega, \mathcal{A}, P\}$ is called a *probability space*.

2.5 Some properties of probability measures

1,2

Proposition 2 *(a) For any event A,*

$$P(A^c) = 1 - P(A). \tag{2.5.1}$$

(b) If $A \subseteq B$ (all points from A are in B), then

$$P(A) \leq P(B). \tag{2.5.2}$$

(c) For any events $A_1, A_2, ...,$

$$P(\cup_i A_i) \leq \sum_i P(A_i), \tag{2.5.3}$$

where $\cup_i A_i$ is the union of A_i's.

A formal proof will be given later. Assertion (2.5.1) is understandable and may be illustrated by Fig. 2a; assertion (2.5.2)—by Fig. 5b (the probability of the strongly shaded region is less than that of the whole shaded region).

To clarify (2.5.3), consider the particular case where any probability $P(A)$ may be presented as the sum of the elementary probabilities of outcomes in A, as we did in (2.2.4). Consider two events, A and B, and look at Fig. 2b. We see that if we add up $P(A)$ and $P(B)$, then the probabilities of elementary outcomes from AB will be counted twice, so in the general case, $P(A \cup B) \neq P(A) + P(B)$ and we can only write that $P(A \cup B) \leq P(A) + P(B)$. As a matter of fact, the following is true.

Proposition 3 *(The addition rule). For any A and B,*

$$P(A \cup B) = P(A) + P(B) - P(AB). \tag{2.5.4}$$

EXAMPLE 1. In a country, 20% of people love classical music, 30% love jazz, and 10% love classical music *and* jazz. What is the probability that a person chosen at random loves either classical music *or* jazz (or both)? Denoting the corresponding events by C and J respectively, we have $P(C \cup J) = P(C) + P(J) - P(CJ) = 0.2 + 0.3 - 0.1 = 0.4$. $\square$.

Now, let us turn to proofs. It is convenient to begin with

Proof of Proposition 3. When going along the proof below, it is convenient to use also Fig. 2b. The events AB^c (the set of points from A but not from B), AB, and A^cB (the set of points from B but not from A) are disjoint, and the union of them is $A \cup B$. Hence, by

(2.1.2), $P(A \cup B) = P(A \cup A^c B) = P(A) + P(A^c B) = P(A) + P(A^c B) + P(AB) - P(AB) = P(A) + P(A^c B \cup AB) - P(AB) = P(A) + P(B) - P(AB)$. ■

Proof of Proposition 2. (a) Since A and A^c are disjoint, by (2.1.2), $P(A) + P(A^c) = P(\Omega) = 1$.

(b) The events A and $A^c B$ are disjoint, and because $A \subseteq B$, we have $B = A \cup A^c B$. Hence, by (2.1.2), $P(A) \leq P(A) + P(A^c B) = P(A \cup A^c B) = P(B)$.

(c) We prove it by induction. For one event, (2.5.3) is trivial, and for two events it follows from (2.5.4). Assume that for all $k \leq n$ and $n \geq 2$,

$$P(\cup_{i=1}^{k} A_i) \leq \sum_{i=1}^{k} P(A_i). \tag{2.5.5}$$

Then, using (2.5.5) consecutively for $k = 2$ and for $k = n$, we have

$$\begin{aligned} P\left(\cup_{i=1}^{n+1} A_i\right) &= P((\cup_{i=1}^{n} A_i) \cup A_{n+1}) \\ &\leq P(\cup_{i=1}^{n} A_i) + P(A_{n+1}) \leq \sum_{i=1}^{n} P(A_i) + P(A_{n+1}) = \sum_{i=1}^{n+1} P(A_i). \ ■ \end{aligned}$$

2

Route 1 ⇒ page 16

Next, we consider the counterpart of (2.5.4) for three or more events.

Proposition 4 *For any events $A_1, ..., A_n$,*

$$P(\cup_{i=1}^{n} A_i) = \sum_{i=1}^{n} P(A_i) - \sum_{i_1 < i_2} P(A_{i_1} A_{i_2}) + ... + (-1)^{k-1} \sum_{i_1 < i_2 < ... < i_k} P(A_{i_1} A_{i_2} \cdots A_{i_k}) + ... + (-1)^{n-1} P(A_1 A_2 \cdots A_n). \tag{2.5.6}$$

In $\sum_{i_1 < i_2 < ... < i_k}$ above, the summation is over all possible combinations of the numbers $i_1 < i_2 < ... < i_k$ taken from the set $\{1, ..., n\}$. In Exercise 15, the reader is invited to prove (2.5.6) by induction.

2.6 Continuity of probability measures

A sequence of events $A_1, A_2, ...$ is said to be *non-decreasing* if

$$A_1 \subseteq A_2 \subseteq ... \subseteq A_n \subseteq ...,$$

and *non-increasing* if

$$A_1 \supseteq A_2 \supseteq ... \supseteq A_n \supseteq ... .$$

We call both sequences *monotone*. Clearly,

$$A_n = \cup_{i=1}^{n} A_i \text{ for a non-decreasing sequence } \{A_n\}, \tag{2.6.1}$$

$$A_n = \cap_{i=1}^{n} A_i \text{ for a non-increasing sequence } \{A_n\}. \tag{2.6.2}$$

We define

$$\lim_{n\to\infty} A_n = \cup_{i=1}^{\infty} A_n \quad \text{for a non-decreasing sequence, and}$$
$$\lim_{n\to\infty} A_n = \cap_{i=1}^{\infty} A_n \quad \text{for a non-increasing sequence.}$$

Proposition 5 *(The continuity of $P(A)$). For any monotone sequence A_n,*

$$\lim_{n\to\infty} P(A_n) = P(\lim_{n\to\infty} A_n). \tag{2.6.3}$$

EXAMPLE 1. Consider a random number generator which we view here as a device or program that somehow chooses a point from the interval $[0,1]$. We assume the generator to be perfect in the sense that all points are equally likely to be chosen. This means that the probability that the point chosen falls within an interval Δ equals $\ell(\Delta)$, the length of Δ; see also Example 2.3-1 and Fig. 4. We have observed in this example that the probability assigned to a point should equal zero. Now, we show how to prove it rigorously.

Let us fix a point, say, $\frac{1}{2}$, and set the interval $\Delta_n = [\frac{1}{2} - \frac{1}{n+1}, \frac{1}{2} + \frac{1}{n+1}]$. So, $\Delta_1 = [0,1]$, $\Delta_2 = [\frac{1}{6}, \frac{5}{6}]$, $\Delta_3 = [\frac{1}{4}, \frac{3}{4}]$, and so on.

One may say, that the intervals Δ_n are "squeezing" to the point $1/2$ as $n \to \infty$. More rigorously, if we denote by $[1/2]$ the set that consists of only one point $1/2$, then $\cap_1^{\infty}\Delta_n = [1/2]$. Thus, since the sequence Δ_n is non-increasing, we have $\lim_{n\to\infty}\Delta_n = \cap_1^{\infty}\Delta_n = [1/2]$.

For simplicity, let us write $P(\Delta)$ for the probability that the point chosen is in Δ. Then $P(\Delta_n) = \ell(\Delta_n) = \frac{2}{n+1}$, and by Proposition 5,

$$P([1/2]) = P(\lim_{n\to\infty}\Delta_n) = \lim_{n\to\infty} P(\Delta_n) = \lim_{n\to\infty}\frac{2}{n+1} = 0. \ \square$$

Proof of Proposition 5. By the counterpart of rule (1.2.2) for many events (see also Exercise 6),

$$P(\cap_{i=1}^{n} A_n) = 1 - P((\cap_{i=1}^{n} A_n)^c) = 1 - P(\cup_{i=1}^{n} A_n^c). \tag{2.6.4}$$

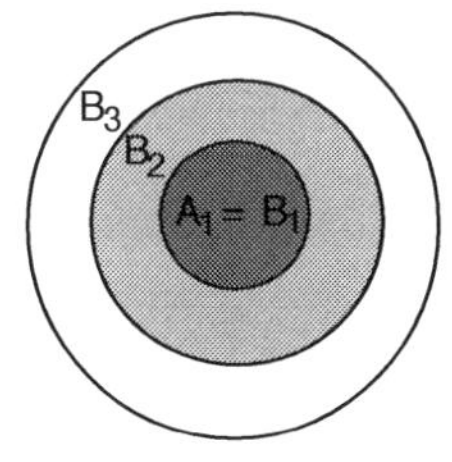

FIGURE 6.

Now, if a sequence $\{A_n\}$ is non-increasing, $\{A_n^c\}$ is non-decreasing; the reader is invited to prove it on her/his own. From this and (2.6.4), it follows that it suffices to prove the proposition for a non-decreasing sequence $\{A_n\}$.

Set $B_1 = A_1$, and for $i \geq 2$ set $B_i = A_i A_{i-1}^c$, the set of points from A_i that are not in A_{i-1}. See also Fig. 6, where we consider as A's concentric circles merely for simplicity. Since the sequence $\{A_n\}$ is non-decreasing, the events B_i are disjoint and $A_n = \cup_{i=1}^{n} B_i$. Hence, $\cup_{i=1}^{\infty} A_i = \cup_{i=1}^{\infty} B_i$. Therefore,

$$P(\lim A_n) = P(\cup_{i=1}^{\infty} A_i) = P(\cup_{i=1}^{\infty} B_i) = \sum_{i=1}^{\infty} P(B_i)$$
$$= \lim_{n\to\infty}\sum_{i=1}^{n} P(B_i) = \lim_{n\to\infty} P(\cup_{i=1}^{n} B_i) = \lim_{n\to\infty} P(A_n). \ \blacksquare$$

3 COUNTING TECHNIQUES AND PROBABILITIES IN THE CLASSICAL SCHEME

We consider here a finite sample space Ω, and suppose that all elementary outcomes are equally likely. As was shown in Section 2.2, in this case, for any event A,

$$P(A) = |A|/|\Omega|, \tag{3.0.1}$$

where $|A|$ and $|\Omega|$ are the numbers of elementary outcomes in A and Ω, respectively. Thus, in the classical scheme, the general approach to computing probabilities is simple. However, in concrete problems, computing (or counting) the numbers $|A|$ and $|\Omega|$ may turn out to be challenging. In this section, we systematize some counting techniques.

3.1 The basic counting principle

This principle is also called the *the multiplication rule of counting*. If a procedure can be carried out in m ways and, for *each* such a way, another procedure can be carried out in n ways, then together, there are mn ways of performing both procedures. To demonstrate this, denote by (i, j) the outcome for which the first procedure is performed in the ith way, and the second in the jth way. Since $i = 1,...,m$ and $j = 1,...,n$, all possible pairs may be presented as the rectangular array

$$\begin{matrix} (1,1) & \dots & (1,n) \\ \vdots & \ddots & \vdots \\ (m,1) & \dots & (m,n) \end{matrix}$$

We see that there are mn distinct pairs (i, j).

EXAMPLE 1. Each weekday, Donald has a lunch in one of two restaurants, and in each, he orders either meat, or fish, or salad. There are two procedures: choosing a restaurant, and choosing a meal. The number of possible outcomes is $2 \cdot 3 = 6$. □

Similarly, if there are k procedures, where the ith procedure may be performed in m_i ways regardless of how any of the other procedures was performed, then there are $m_1 \cdot ... \cdot m_k$ possible outcomes.

EXAMPLE 2. A license plate consists of a digit followed by three letters, followed by three more digits. If the first digit cannot be 0, then the total number of possible licence plates is equal to $9 \cdot 26 \cdot 26 \cdot 26 \cdot 10 \cdot 10 \cdot 10 = 158,184,000$.

EXAMPLE 3 (*A sequence of trials*) is a basic one. Consider n consecutive trials and assume that each separate trial may have one of m possible results (by contrast with two: success and failure that were considered in Section 1.1). Then the total number of outcomes for the series of trials is $\underbrace{m \cdot ... \cdot m}_{n \text{ times}} = m^n$. □

3.2 Permutations

In how many ways can we arrange or permute n distinct objects? The answer is $n!$, and to show it, we may apply the basic counting principle. The first place in an arrangement may be occupied by any of the n objects, the second place by any of the remaining $n-1$, the third by any of the remaining $n-2$, and so on. The last, nth, place may be occupied by the only one remaining object. So, the total number of outcomes is $n \cdot (n-1) \cdot \ldots \cdot 2 \cdot 1 = n!$.

EXAMPLE 1. Eight different problems are to be assigned to eight students, one problem to each. In how many ways can it be done? Each assignment corresponds to a particular arrangement of problems: once they are arranged in a particular order, you may give the first problem to the first student, the second problem to the second student, and so on. The total number of arrangements is $8! = 40,320$, which may seem higher than expected.

EXAMPLE 2. Three men and three women take six seats at a table in a singles party. The rule is that no woman should be a neighbor of a woman. Let m stand for a man, and w for a woman. Starting from the first seat at the table, we can consider *two* types of arrangements: either $mwmwmw$, or $wmwmwm$. We can arrange any of $3! = 6$ permutations of the men, and $3! = 6$ of the women. So, by the basic counting principle, the total number of outcomes is $2 \cdot 6 \cdot 6 = 72$.

Now, let the table be round. We are not interested where each person will sit, but rather who will be a right or left neighbor of whom. Let us label the men by m_1, m_2, m_3 and the women by w_1, w_2, w_3. It does not matter where the first man sits (so consider him occupying the "first" seat), but it matters whether the men will be arranged in the order m_1, m_2, m_3 or m_1, m_3, m_2. (For example, in the arrangement $m_1, w_1, m_2, w_2, m_3, w_3$, each woman sits between the same men as in $m_1, w_3, m_3, w_2, m_2, w_1$, but the left and right neighbors differ.) So, we have two ways to arrange men, and we still have $3! = 6$ ways to seat the women between the men. Eventually, the number of all outcomes is $2 \cdot 6 = 12$.

Now, the guests want to dance. What is the number of possible arrangements? To arrange pairs, it suffices to arrange the women in a particular order, and suggest the first woman to dance with m_1, and so on. Thus, there are $3! = 6$ possible outcomes.

EXAMPLE 3 (*the Monty Hall problem* [44]) is popular; the name comes from a game show host. You are on a game show. There are three doors: behind one door is a car; behind the others are goats. You pick a door and may get what is behind it. Suppose you picked door No.1, but you did not open it yet. The host, who knows what is behind the doors, opens another door, say, No.3, with a goat behind it, and offers you to switch to door No.2. Is it beneficial to switch?

Quite frequently, people answer that it does not matter whether to switch or not because the chances for the car to be behind No.1 or No.2 are the same. As a matter of fact, this was true only at the very beginning of the game, but once the host opened a door, additional information becomes available.

More precisely, if all outcomes are equally likely, the probability that the car is behind the door you have chosen is $\frac{1}{3}$. Indeed, there are $3! = 6$ outcomes in total, and among them there are $2! = 2$ outcomes for which the car is behind the door you have chosen. (There are two ways to put the two goats behind the two doors you did not pick.) So, the probability is $\frac{2}{6} = \frac{1}{3}$. Hence, the probability to pick a door hiding a goat is $\frac{2}{3}$.

Furthermore, the host is always able to open a door with a goat, and he would never open the door with the car because in this case the offer to switch will be meaningless. Obviously, if you picked the desired door (the probability is $\frac{1}{3}$), you should not switch, and if you picked a wrong door (the probability is $\frac{2}{3}$), you should. Consequently, though you do not know whether you chose the door with the car, *it is more likely that you are in the situation where you should switch* than where you should not. Thus, it is reasonable to switch. □

In connection with the last example, it is a good occasion to define the notion of the *odds ratio* of an event A, which is

$$\frac{P(A)}{P(A^c)} = \frac{P(A)}{1-P(A)}. \tag{3.2.1}$$

If the odds ratio is k, it is common to say that the "odds are k to 1" in favor of A. For instance, in the last example, for the switching strategy, the odds are 2 to 1 in favor of getting the car.

3.3 Partitions into groups

Suppose n distinct objects should be divided into k distinct groups with the given sizes $n_1, n_2, ..., n_k$, respectively. So, $n_1 + n_2 + ... + n_k = n$. Let N_n be the number of all possible divisions. We show that

$$N_n = \frac{n!}{n_1!n_2!\cdots n_k!}. \tag{3.3.1}$$

Indeed, let us imagine that we are arranging all objects in a particular order, and we are doing it in the following way. First, we divide the objects into the groups mentioned in one of the N_n ways. After that, we line up the objects that got into the first group in one of $n_1!$ possible ways. Next, after the elements of the first group, we line up the elements of the second group in one of $n_2!$ possible ways, and so on. By the basic counting principle, there are exactly $n_1!\cdot n_2!\cdots n_k!$ ways to line up the elements in this fashion. So, the total number of possible arrangements of this type is $N_n \cdot n_1! \cdot n_2! \cdot ... \cdot n_k!$.

On the other hand, each such an arrangement leads to one of the $n!$ permutations of the n objects, and vice versa, every permutation may be accomplished in one of the ways described. So, $n! = N_n \cdot n_1! \cdot n_2! \cdot ... \cdot n_k!$, which implies (3.3.1). ■

EXAMPLE 1. Each day in a week may be either sunny, or cloudy, or rainy. Assuming that all outcomes are equally likely (which is certainly not always the case), find the probability that there will be exactly three sunny, two cloudy, and two rainy days.

We should divide seven different days into the three corresponding groups, which in accordance with (3.3.1) may be done in $7!/(3!2!2!)$ ways. On the other hand, there are totally 3^7 possible outcomes; see Example 3.1-3. Then, by (3.0.1), the probability in hand is $\dfrac{7!}{3!\cdot 2!\cdot 2!}3^{-7} \approx 0.096$. □

3.4 Samples

3.4.1 Unordered samples or combinations

In the model of Section 3.3, consider the case of two groups and set $n_1 = k$. Then $n_2 = n-k$, and the number of partitions is equal to the quantity $\dfrac{n!}{k!(n-k)!}$ which is usually denoted by $\binom{n}{k}$. So,

$$\binom{n}{k} = \frac{n!}{k!(n-k)!}. \tag{3.4.1}$$

As a rule, the expression $\binom{n}{k}$ is read "n choose k." This is connected with the fact that the quantity (3.4.1) is also equal to the number of ways to select k objects from n distinct objects. Indeed, selecting k objects is equivalent to dividing the total group into two groups: of those k objects that we will select, and those $n-k$ that we will not. It is worth emphasizing that in this case, we do not consider the order in which the objects have been selected: each possible sample is characterized merely by the elements chosen.

For example, if you choose a bouquet of 4 from 10 available roses in a store, there are $\frac{10!}{4!6!} = 210$ ways to do that.

A sample defined above is called *unordered* or a *combination.*

Now, observe that $n! = [n(n-1)(n-2)\cdots(n-k+1)]\cdot(n-k)!$. Sometimes, it is convenient to cancel out $(n-k)!$ in the denominator and numerator of (3.4.1) and write

$$\binom{n}{k} = \frac{n(n-1)(n-2)\cdots(n-k+1)}{k!}. \tag{3.4.2}$$

In particular, when $k=2$, the number of pairs we can select from n objects equals

$$\binom{n}{2} = \frac{n(n-1)}{2}. \tag{3.4.3}$$

Note also that

$$\binom{n}{k} = \binom{n}{n-k}. \tag{3.4.4}$$

To prove this, it suffices to switch k and $n-k$ in (3.4.1) or to observe that in order to choose k objects from n, we may choose $n-k$ objects that will not be selected.

EXAMPLE 1. There are 2^n outcomes in the experiment of tossing a coin n times. Among them, there are $\binom{n}{k}$ outcomes resulting in k heads, since there are $\binom{n}{k}$ ways to choose k tosses out of n, for which the coin will land heads. If the coin is regular, all outcomes are equally likely, and in accordance with (3.0.1), calling a head a success and a toss a trial, we can write

$$P(k \text{ successes in } n \text{ trials}) = \binom{n}{k}\Big/2^n. \tag{3.4.5}$$

Certainly, we used the coin-example merely for illustration, and

Formula (3.4.5) *is true for any series of trials with equally likely outcomes.*

EXAMPLE 2. What is the probability of having three aces and two kings in a poker hand of five cards? There are $\binom{52}{5}$ different hands, which are assumed to be equally likely. There are $\binom{4}{3}$ ways to get three aces out of four, and $\binom{4}{2}$ ways to get two kings out of four. Therefore, by the basic counting principle, there are exactly $\binom{4}{3}\cdot\binom{4}{2}$ hands of the type mentioned. Eventually, the desired probability is $\dfrac{\binom{4}{3}\cdot\binom{4}{2}}{\binom{52}{5}} = \dfrac{4\cdot 6}{\binom{52}{5}} \approx .00001$ □

EXAMPLE 3 (*The hypergeometric distribution*). An urn has r red and b black balls; $r+b=n$. Suppose we select $k \leq n$ balls without replacement. Denote by p_m the probability that the sample will contain m red balls. If $m > r$ or $k-m > b$, then $p_m = 0$. To find p_m for $k-b \leq m \leq r$, note to there are exactly $\binom{r}{m}$ ways to select m red balls from the r red balls, and for each such a selection, there are $\binom{b}{k-m}$ ways to select the remaining $k-m$ balls from the b black balls. By the basic principle, the total number of samples with the property mentioned is $\binom{r}{m}\cdot\binom{b}{k-m}$. Thus, for $k-b \leq m \leq r$,

$$p_m = \frac{\binom{r}{m}\cdot\binom{b}{k-m}}{\binom{n}{k}}. \tag{3.4.6}$$

The collection of probabilities $\{p_m\}$ constitutes what is called the *hypergeometric probability distribution.*

EXAMPLE 4. As a good exercise, let us prove the binomial formula

$$(a+b)^n = \sum_{k=1}^{n} \binom{n}{k} a^k b^{n-k}. \tag{3.4.7}$$

Imagine that we have carried out the multiplication in $(a+b)^n = (a+b)\cdot\ldots\cdot(a+b)$ in the straightforward fashion but still did not combine like terms. Each term in the expansion obtained is a product of n factors, and each factor is either a or b. By the basic counting principle (see also Example 3.1-3), there will be exactly 2^n such products. In how many of them is the number of a's exactly equal to k (and, consequently, the number of b's is $(n-k)$)? The number of such products is equal to the number of ways of choosing k factors (which will be equal to a) from the total n factors. Thus, there are $\binom{n}{k}$ products equal to $a^k b^{n-k}$. After combining these like terms, we come to $\binom{n}{k}a^k b^{n-k}$. It remains to add up the terms of this type over all k. □

3.4.2 Ordered samples

Suppose we have an urn with n numbered balls, and we select consecutively, one after the other, k balls from the urn without replacement. Now, the order in which the balls have been selected matters, and such a sample is called *ordered.*

Since we select balls without replacement, there are n possible ways to select the first ball, there are $n-1$ ways to select the second ball, and so on. In the last, i.e., the kth draw, we select one of $n-(k-1) = n-k+1$ balls. Thus, the total number of samples is equal to the quantity

$$(n)_k = n(n-1)\cdots(n-k+2)(n-k+1). \tag{3.4.8}$$

There is an alternative way to get the same formula. In order to obtain all possible ordered samples of a size of k, we may first consider all possible unordered combinations of k balls out of n, and for each such a combination, consider all possible arrangements (or permutations) of the k balls chosen. Thus, the total number of ordered samples is equal to $k!\cdot\binom{n}{k}$, which leads to (3.4.8) by virtue of (3.4.2).

EXAMPLE 1. Two families, A and B, have decided to travel together. Family A has a car with 6 seats and consists of two parents, of whom only one drives, and three kids. Family B has a car with 5 seats, consists of two parents, both of whom drive, and one kid. If all ways of choosing drivers and seats for passengers are equally likely, what is the probability that each family will occur in its car?

First, we compute $|\Omega|$. There are $3 \cdot 2 = 6$ of ways of choosing drivers. (The sample is ordered because it matters which driver will drive which car.) Also, the families should choose 6 seats from the 9 passenger seats, and this may be done in $9 \cdot 8 \cdot ... \cdot 4$ ways. (The sample is ordered since it matters who will sit in which seat.) Thus, $|\Omega| = 6 \cdot 9 \cdot 8 \cdot ... \cdot 4$.

If the families travel in their own cars, family A should choose 4 seats out of 5, which may be done in $5 \cdot 4 \cdot 3 \cdot 2$ ways. Family B should choose a driver out of 2, and 2 seats from 4. The total number of ways to do that is $2 \cdot 4 \cdot 3$. Hence, the desired probability is

$$\frac{(5 \cdot 4 \cdot 3 \cdot 2) \cdot (2 \cdot 4 \cdot 3)}{6 \cdot 9 \cdot 8 \cdot ... \cdot 4} = \frac{5}{378}.$$

EXAMPLE 2 (*The birthday problem*) is very popular. What is the probability that no two people in a group of k people have a common birthday? We do not take into account leap years and assume that for each person chosen at random, all birthday dates are equally likely (though, in reality, it is not completely true). Certainly, for $k > 365$, the probability under discussion is zero. Let $k \leq 365$. We can view it as if k people consecutively select k balls from 365 numbered balls. Then, the total number of outcomes with different birthdays is equal to $365 \cdot 364 \cdot ... \cdot (365 - k + 1)$. On the other hand, there are 365^k total outcomes, and thus for the event A in hand,

$$P(A) = \frac{365 \cdot 364 \cdot 363 \cdot ... \cdot (365 - k + 1)}{365^k} = \frac{365}{365} \cdot \frac{364}{365} \cdot \frac{363}{365} \cdot ... \cdot \frac{365 - k + 1}{365}$$
$$= \left(1 - \frac{1}{365}\right)\left(1 - \frac{2}{365}\right) \cdot ... \cdot \left(1 - \frac{k-1}{365}\right). \tag{3.4.9}$$

Counterintuitively, this probability is small for a "moderate" k. For example, for k being equal to just 23, it is $\approx \frac{1}{2}$, and for $k = 55$, it is ≈ 0.01.

► To estimate the probability (3.4.9), we may apply the inequality $1 + x \leq e^x$ and the formula $1 + ... + k = \frac{k(k+1)}{2}$. (Regarding the inequality, graph $1 + x$ and e^x, and realize that the former graph is tangent to the latter at $x = 0$.) We have

$$P(A) = \left(1 - \frac{1}{365}\right)\left(1 - \frac{2}{365}\right) \cdot ... \cdot \left(1 - \frac{k-1}{365}\right)$$
$$\leq \exp\left\{-\frac{1}{365}\right\} \cdot \exp\left\{-\frac{2}{365}\right\} \cdot ... \cdot \exp\left\{-\frac{k-1}{365}\right\}$$
$$= \exp\left\{-\frac{1}{365}(1 + 2 + ... + (k-1))\right\} = \exp\left\{-\frac{1}{365} \cdot \frac{(k-1)k}{2}\right\} = \exp\left\{-\frac{(k-1)k}{730}\right\}.$$

For $k = 23$, the estimate above gives ≈ 0.4999; for $k = 50$, it is ≈ 0.03487; for $k = 100$ it is $\approx 0.000,001$. ◄ □

Route 1 ⇒ page 26

2

3.5 Sampling and occupancy problems

First, let us consider the selection procedure more systematically.

As above, suppose that an urn contains n balls, and k balls are drawn, one at a time. The sampling or selection may be carried out

1) *with replacement*; that is, each ball drawn is returned to the urn; or

2) *without replacement.*

The sample of k balls may be either

a) *ordered*; that is, samples containing the same balls but drawn in different orders are viewed as different; or

b) *unordered*; that is, we do not distinguish samples consisting of the same balls.

Consider all four cases that result from combinations of the above: 1a, 1b, 2a, 2b.

1a) We treat k draws as trials such that each may have one of n possible results (balls). In accordance with the basic counting principle, the total number of outcomes $|\Omega| = n^k$.

1b) The total number of outcomes is $\binom{n+k-1}{k}$, which will be proved at the end of this section.

2a) In Section 3.4.2, we have shown that the number of all outcomes is $(n)_k = n(n-1)\cdots(n-k+2)(n-k+1)$.

2b) As was shown in Section 3.4.1, the number of outcomes equals $\binom{n}{k} = \frac{n!}{k!(n-k)!}$.

Next, we show that the same model may be presented in terms of the *occupancy problem* that consists in distributing k objects (we will call them particles) into n groups; we will identify them with boxes. For example, employees are distributed between different divisions of a company, or nuclear particles (for example, electrons) occupy orbits with different energy levels, or the same particles occupy small parts of the phase space (the location of a particle and its impulse), etc.

Consider the following rules of distribution:

1) Each box may contain *any number of particles*;

2) Each box may contain *at most one particle.*

Also, particles may be viewed as

a) *distinguishable*;

b) *indistinguishable*.

From a modeling point of view, this is not a new problem because

> All four cases, 1a, 1b, 2a, 2b, may be put into one-to-one correspondence with those of the sampling problem above if we identify the ith particle with the ith draw, and the jth box with the jth ball.

Indeed, to *place* a particle into a box, we should *select* a box. If we are not allowed to select the *same* box for the next particle, every box can contain at most one particle. If a *sample of boxes* (which play the role of balls) is ordered, we distinguish which particle went to which box; that is, we distinguish particles. Dealing with non-ordered samples, we know only which boxes were chosen and how many times (if we do it with replacement); that is, we know only how many particles each box contains.

We present all possible formulas in

TABLE 1. Probabilities for the sampling and occupance problems: n balls/boxes and k draws/particles

	1) with replacement / any number of particles	2) without replacement / at most one particle
a) ordered samples / distinguishable particles	n^k	$n(n-1)\cdots(n-k+1),\ k \leq n$
b) unordered samples / indistinguishable particles	$\binom{n+k-1}{k}$	$\binom{n}{k},\ k \leq n$

EXAMPLE 1. Each day, the president of a company inspects one of the n divisions of the company. A division is selected randomly regardless of which divisions have been selected on previous days. Find the probabilities of the following events: (a) On the nth day, each division will have been inspected; (b) On the nth day, only the first division will have not been inspected; (c) On the nth day, exactly one division will have not been inspected.

We may view problems (a)–(c) as the distribution of n particles (inspections, or more precisely, inspection dates) into n boxes (divisions). Particles are distinguishable because it matters on which day a particular division will be inspected. Hence, $|\Omega| = n^n$.

In case (a), since during n days all n divisions were inspected (no box is empty), no division has been inspected twice (no box has two particles). The only point is that in which order the inspections have been carried out. So, $|A| = n!$, and $P(A) = n!/n^n$.

In case (b), all boxes except the first are not empty. This means that one box out of $(n-1)$ will have two particles. First, we choose a box into which we place 2 particles, which can be done in one of $(n-1)$ ways, and choose two particles from n which we will place into the box chosen. In accordance with (3.4.3), this may be done in $n(n-1)/2$ ways. It remains to distribute $(n-2)$ particles into $(n-2)$ boxes in a way that no box will be empty. Similar to case (a), this may be done in $(n-2)!$ ways. So, if we denote by A_1 the event in hand, then $|A_1| = (n-1) \cdot \frac{n(n-1)}{2} \cdot (n-2)! = (n-1)n!/2$, and

$$P(A_1) = \frac{(n-1) \cdot n!}{2n^n}. \tag{3.5.1}$$

Considering case (c), denote by A_i the same event as A_1 with the only exception that the ith box is the only empty box (rather than the first). The event in hand is $A = \cup_{i=1}^{n} A_i$. Clearly, the events A_i's are disjoint and, by symmetry, have the same probability as A_1. Consequently, by virtue of (3.5.1),

$$P(A) = \sum_{i=1}^{n} P(A_i) = nP(A_1) = \frac{n(n-1) \cdot n!}{2n^n}.$$

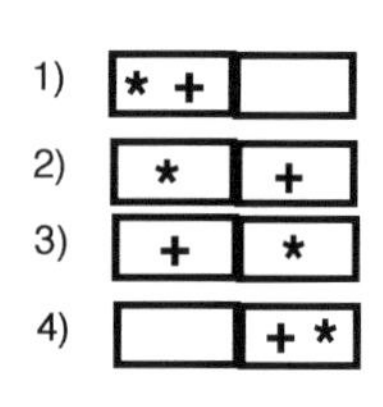

FIGURE 7.

EXAMPLE 2 demonstrates that the probabilities corresponding to different distributions of particles strongly depend on which conditions from those mentioned above we adopt. Consider two particles, two boxes, and first, Case 1a. Since the particles are distinguishable, we mark them by different symbols, say $+$ and $*$. The four possible ways of distribution are presented in Fig. 7. If all four outcomes are equally likely, the probability of each is $1/4$, and for instance, the probability that each box will have one particle is $1/2$. One should reason in this fashion when the particles are distinguishable, at least, in principle. For example, distributing "real" balls even if it is impossible to distinguish them visually.

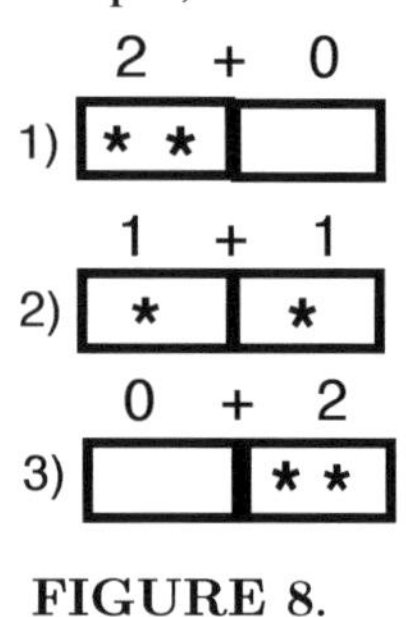

FIGURE 8.

Now, consider Case 1b where particles are, in principle, indistinguishable. In this case, we must mark them by one symbol, say $*$. For instance, we consider all possible ways to present the number 2 as the sum of two non-negative numbers: $2 = 2+0, 2 = 1+1, 2 = 0+2$. We may view it as if we distribute two "ones" (particles) between the two terms or summands (boxes).

In Case 1b, there are only three possible outcomes, which is shown in Fig. 8. The probability of each is $1/3$, and therefore, the probability that each box will receive one particle is also $1/3$.

A more serious example concerns nuclear particles. Some of them (photons, nuclei) exhibit behavior corresponding to Case 1b, while some (electrons, protons) to Case 2b. The former case concerns what is known as the Bose-Einstein statistics, the latter concerns the Fermi-Dirac statistics. □

It remains to prove the formula $\binom{n+k-1}{k}$ for the case 1b. We will deal with particles and boxes. Suppose that we have drawn $n+k-1$ vertical lines, then have selected k of them (which may be done in $\binom{n+k-1}{k}$ ways), and have replaced the lines chosen by the symbol $*$. For example, for $n = 5$ and $k = 4$, suppose that we had this: $||||||||$, and after a replacement this: $*||**|*|$.

We interpret it as if one (indistinguishable) particle went to the first box, none to the second box, two to the third, one to the fourth, and none to the fifth box.

Each of the $\binom{n+k-1}{k}$ procedures corresponds to one and only one distribution of particles, so the assertion is proved. ■

3.6 Matching

Suppose that n numbered objects have been randomly arranged in a way that all $n!$ arrangements are equally likely. If after the arrangement, the ith object still occupies the ith position, we call this a match in the ith position. A traditional example concerns n men at a party who left their hats in a cloakroom and when leaving, took hats at random. A match occurs if a man took his own hat. Or suppose that a deck of cards labeled $1, ..., n$ is shuffled, and the cards are dealt out one at a time. If the ith card dealt is the card labeled i, we say that a match has occurred.

Let $p_k(n)$ be the probability that exactly k matches will occur. In particular, $p_0(n)$ is the probability that there will be no matches. Set

$$S(m) = \sum_{k=0}^{m} \frac{(-1)^k}{k!},$$

and note that this sum represents the first $m+1$ terms of the Tailor expansion of e^x evaluated at $x = -1$; see the Appendix, (2.2.4). Hence, for large m, the quantity $S(m)$ is close to $\frac{1}{e} \approx 0.367879441$. The approximation is pretty good; for instance $S(5) \approx 0.36667$, and $S(10) \approx 0.367879464$; that is, $S(10)$ equals $1/e$ up to the seventh digit.

Proposition 6 *For any* $k = 0, ..., n$,

$$p_k(n) = \frac{1}{k!} S(n-k). \tag{3.6.1}$$

In particular, the probability that there will be no matches is

$$p_0(n) = S(n). \tag{3.6.2}$$

To make the result more plausible, let us derive from (3.6.1) the probability that there will be matches in all positions. We have $p_n(n) = \frac{1}{n!}S(0) = \frac{1}{n!}$ because $S(0) = 1$. This is what had to be expected: all matches occur only for one permutation: $1, 2, ..., n$.

In particular, from the proposition it follows that

$$P(no\ matches) \approx \frac{1}{e}, \tag{3.6.3}$$

and this approximation is accurate up to the seventh digit for $n \geq 10$. Furthermore,

$$P(exactly\ k\ matches) \approx \frac{e^{-1}}{k!}, \tag{3.6.4}$$

and for $n-k \geq 10$, this approximation is even more accurate because we divide the error of the approximation of $S(n-k)$ by $k!$.

Proof of Proposition 6. Denote by A_i the event that there is a match in the ith place. Then $\cup_{i=1}^{n} A_i$ is the event that there will be at least one match. To compute its probability, we will use (2.5.6). To compute $P(A_{i_1} A_{i_2} \cdots A_{i_k})$, let us imagine that we want to carry out an arrangement for which there are matches in the (fixed) places $i_1 < i_1 < ... < i_k$ (and perhaps in other places also). To this end, we should place the object i_1 in the i_1th place, the object i_2 in the i_2th place, and so on, finishing with the object i_k. This can be done only in one way. The remaining $n-k$ objects may be distributed between the remaining $n-k$ places in an arbitrary way; that is, in any of $(n-k)!$ ways. So, the total number of the mentioned arrangements is $(n-k)!$, or in other words, $|A_{i_1} A_{i_2} \cdots A_{i_k}| = (n-k)!$. Since the total number of all possible arrangements is $n!$, we have $P(A_{i_1} A_{i_2} \cdots A_{i_k}) = \frac{(n-k)!}{n!}$.

To compute the sum $\sum_{i_1 < i_2 < ... < i_k} P(A_{i_1} A_{i_2} \cdots A_{i_k})$ in (2.5.6), observe that this sum consists of $\binom{n}{k} = \frac{n!}{k!(n-k)!}$ terms (the number of ways in which we can choose k numbers $i_1 < i_2 < ... < i_k$ from the collection $\{1, 2, ..., n\}$), and each term equals $\frac{(n-k)!}{n!}$. Hence,

$$\sum_{i_1 < i_2 < ... < i_k} P(A_{i_1} A_{i_2} \cdots A_{i_k}) = \frac{n!}{k!(n-k)!} \cdot \frac{(n-k)!}{n!} = \frac{1}{k!}.$$

Substituting this into (2.5.6), we obtain

$$P(\cup_{i=1}^{n} A_i) = \sum_{k=1}^{n} (-1)^{k-1} \frac{1}{k!}.$$

Then

$$p_0(n) = 1 - P(\cup_{i=1}^{n} A_i) = 1 - \sum_{k=1}^{n} (-1)^{(k-1)} \frac{1}{k!} = 1 + \sum_{k=1}^{n} (-1)^{k} \frac{1}{k!} = \sum_{k=0}^{n} (-1)^{k} \frac{1}{k!} = S(n).$$

Thus, (3.6.2) has been proved.

To obtain (3.6.1), first, denote by $B(i_1,...,i_k)$ the event that the matches occurred in places $i_1 < ... < i_k$, and *only* in these places. Then for different collections $i_1,...,i_k$, these events are disjoint, and by symmetry have the same probability. That is, $P(B(i_1,...,i_k)) = P(B(1,...,k))$, the probability that matches occurred in the first k places and only in these places.

Note also that there are exactly $\binom{n}{k}$ events $B(i_1,...,i_k)$. Consequently,

$$p_k(n) = P(\cup_{i_1<...<i_k} B(i_1,...,i_k)) = \sum_{i_1<...<i_k} P(B(i_1,...,i_k)) = \binom{n}{k} P(B(1,...k)). \quad (3.6.5)$$

To compute $P(B(1,...k))$, denote by $f(r)$ the number of those permutations of the numbers $1,...,r$ that lead to no matches. Since $\frac{f(r)}{r!} = p_0(r)$, we have $f(r) = r! \cdot p_0(r)$. Next, observe that $B(1,...,k)$ consists of all permutations for which the first k positions are matches, and the remaining $(n-k)$ positions are no matches. Hence, the number of outcomes in $B(1,...,k)$ equals $f(n-k) = (n-k)! \cdot p_0(n-k)$. Eventually,

$$P(B(1,...k)) = \frac{f(n-k)}{n!} = \frac{(n-k)! \cdot p_0(n-k)}{n!},$$

and in view of (3.6.5),

$$p_k(n) = \frac{n!}{k!(n-k)!} \cdot \frac{(n-k)! \cdot p_0(n-k)}{n!} = \frac{1}{k!} p_0(n-k) = \frac{1}{k!} S(n-k). \blacksquare$$

4 EXERCISES

1. Mark each statement below true or false.

   (a) In any probability model, the number of all possible outcomes is finite.

   (b) In any probability model, all outcomes are equally likely.

   (c) $P(A \cup B) = P(A) + P(B)$ for any events A and B.

   (d) $P(A^c) = 1 - P(A)$ for any event A.

2. Describe the sample space for the situations below and compute $|\Omega|$.

(a) There are two political parties in a country. Each of n citizens either votes for a party, or does not vote at all, or uses his right to come and vote against both parties.

(b) A professor fills out an attendance list for a class of 20 students.

(c) In a children's party, ten different presents are prepared for ten children; one present for each.

(d) A fitness club is interested in the number of men and women visiting the club during a day.

3. Two dice are rolled. Let $A = \{the\ sum\ of\ the\ dice\ is\ odd\}$, $B = \{at\ least\ one\ die\ is\ even\}$, $C = \{at\ least\ one\ die\ is\ odd\}$. For each pair of these events, figure out whether the events in the pair are disjoint. Compare the events in each of the following pairs of events: A and $B \cup C$, A and BC, AB and C, AC and B, A^cB and A^cC, ABC and BC.

4. In Example 1.2-4, describe A^cB^c.

5. Prove the second relation in (1.2.2) replacing A by A^c and B by B^c in the first.

6. Similar to Proposition 1, show that for any events $A_1, A_2, \ldots$,

$$(\cup_i A_i)^c = \cap_i A_i^c, \ (\cap_i A_i)^c = \cup_i A_i^c,$$

where $\cup_i A_i$ and $\cap_i A_i$ stand for the union and intersection of all A_i's, respectively.

7. Explain why it is impossible to build a discrete model with an infinite number of outcomes, for which all outcomes are equally likely. (*Hint*: Suppose that this is true and add up all elementary probabilities.)

8. Suppose that A and B are disjoint, $P(A) = 0.4$ and $P(B) = 0.5$. Find the probabilities that (a) either A or B occur; (b) both A and B occur; (c) A occurs but B does not.

9. In an area, for two consecutive days, the probability that it is raining on both is 0.1, and the probability that there is no rain on any of these two days is 0.6. Also, it is equally likely whether the first day is rainy and the second is not, or vice versa. Find the probabilities of all possible elementary outcomes. Find the probability that there will be rain on at least one day.

10. A point (a,b) is chosen from the square $R = \{0 \leq a \leq 1, 0 \leq b \leq 1\}$. Suppose that the probability distribution on R is uniform; that is, the probability that a point (a,b) comes from a region in R, equals the area of this region. (a) Find the probabilities that $a \geq b$ and $2a \geq b$. (b) Consider the quadratic equation $x^2 + 2ax + b = 0$ with the coefficients (a,b). Find the probability that the equation has real solutions; has only one real solution.

11. Prove that $P(AB^c) = P(A) - P(AB)$, and draw an illustration picture.

12. In a city of Pleasant Corner, 20% of households have a pool, 60% have air conditioning, and 70% have at least one of these features. Find the probabilities of all possible combinations of these features. Illustrate it by the Venn diagram.

13. Let A and B be events, and let C be the event that exactly one of the events A or B occurs. Using operations on events, write a formula for C. Write a formula for $P(C)$ in terms of $P(A)$, $P(B)$, and $P(AB)$.

14. Let $P(A) = 0.9$ and $P(B) = 0.2$. Can $P(AB) = 0$? In general, if $P(A) + P(B) > 1$, can $P(AB) = 0$? Eventually, show that $P(AB) \geq P(A) + P(B) - 1$, and find to which value $P(AB)$ can be equal if $P(A) = 0.9$ and $P(B) = 0.2$.

15. (a) Write a counterpart of (2.5.4) for three events. (*Advice*: Start with $P(A \cup B \cup C) = P((A \cup B) \cup C)$, and use (2.5.4).) (b)* Prove (2.5.6) by induction.

16. How many seven-digit telephone numbers can be arranged if a telephone number does not begin with 0 or 1? Find the probabilities that a randomly selected number contains exactly three ones, exactly three fours?

17. In Example 3.2-3, what is the probability to get the car if you follow the switching-strategy?

18. Clearly, starting from zero, there are 100,000 numbers which may be written by five digits or less. Show that this also follows from the basic counting principle. How many numbers may be written using five digits in the binary system?

19. You mark five cells from fifty in a lottery ticket. Find the probability that you have guessed all five numbers; exactly three of them.

20. You select at random k numbers from the sequence $1, ..., n$. Show that the probability that you will choose a particular combination of numbers, say, $1, 2, ..., k$, is $1/\binom{n}{k}$. Suppose that now you select k numbers one at a time without replacement, and distinguish samples containing the same numbers but selected at different orders. That is, you consider ordered samples. Give an heuristic argument that the probability that the ordered sample you selected contains $1, 2, ..., k$ (perhaps, in a different order) is the same $1/\binom{n}{k}$. Prove it rigorously.

21. Similar to Example 2.2-7c, prove that the probability of selecting a red ball at any draw is the same as at the first draw.

22. In Bridge, fifty-two cards are dealt to four players, thirteen to each.

    (a) Write a formula for p_1, the probability that a particular player, say, the first, will get a whole suit (all thirteen cards will come from the same suit).

    (b) Write a formula for p_{12}, the probability that two particular players, say, the first and the second, will get whole suits each.

    (c) Write a formula for p_{123}, the probability that three particular players, say, the first, the second, and the third, will get whole suits each.

    (d) Explain that p_{1234}, the probability that all players will get whole suits, is equal to p_{123}.

    (e) Write a formula for the probability that at least one player will get a whole suit in terms of the above probabilities. (*Advice*: Formula (2.5.6) may help.)

23. There are ten pairs of shoes in a closet.

    (a) If you choose two shoes at random, what is the probability that it will be a pair?

    (b) You have chosen four shoes at random. Find the probability that among them, there will be at least one pair in two ways: directly and computing the probability of the complement of the event in hand.

24. Let $P(A_1) = P(A_2) = 1$. Show that $P(A_1 \cup A_2) = P(A_1 A_2) = 1$. (*Advice*: (2.5.4) may save time.)

25. Among 200 items, there are 10 defective. If you choose at random 20, what is the probability, that all will be non-defective?

26. Each week, on one of the weekdays, Joan receives a flyer advertisement from a particular store. It happened that in four of the last five weeks, the fliers came on Fridays. Given this, to what extent is it plausible that all weekdays are equally likely to be a day of receipt? (*Advice*: Suppose this is true and find the probability of the event occurring. Jump to a conclusion proceeding from the value of this probability.)

27. In a party of ten, each person shakes hands with each. What is the number of all handshakes?

28. Suppose n cards from a well shuffled deck of 52 are dealt out. If $5 \leq n \leq 52$, what is the probability that the first five cards are (a) spades, (b) red?

29. Seven apples, three oranges, and five lemons are randomly distributed into three boxes. No box can contain more than five fruits. Find the probability that (a) each box contains an orange; (b) exactly one box contains no oranges. (*Hint*: There are 15 fruits and 15 places in the boxes.)

30. A professor is preparing a final for n students. Each student will be given a theoretical question and a particular problem on calculations. The professor has prepared n theoretical questions and n calculation problems. How many combinations of a theoretical question and calculation problem can be arranged? What is the total number of possible outcomes for the final? Using software, say, Excel, compute the answer for $n = 2, 3, 5, 6, 8, 10$, and watch how it grows.

31. Suppose that for each member of a family of five people, all months are equally likely to be that of the birthday. Find the probabilities that (a) all five were born in the same month; (b) all five were born in different months; (c) two were born in the same month, two in another month, and one in a month different from the first two.

32. Each of one hundred students independently chooses one of five elective courses. Write a formula for the probability that each course will be chosen by twenty students.

33. An elevator in a ten-floor building leaves the first floor with six passengers. Assuming all possible outcomes to be equally likely, find the probabilities of the following events.

    (a) Three passengers will get off the elevator on the second floor, one passenger on the third floor, and two on the fifth.

    (b) Three passengers will get off on the second floor, and one on the third.

    (c) No two passengers will get off on the same floor.

    (d) Two passengers will get off on the same floor, and four more on another floor.

    (e) Two passengers will get off on the same floor, and three more on another floor.

34. A well-shuffled deck of 52 cards is dealt out. Find the probabilities of the following events.

    (a) The fourth card is a king.

    (b) Among the first five cards, there are cards from each suit.

    (c) There are k cards between the king and ace of spades.

35. John has five pairs of shoes. If he puts the shoes absolutely randomly into five shoe boxes, what is the probability that (a) each pair will go to the same box; (b) each box will have one left and one right shoe?

36. In Example 3.5-1, find the probabilities that (a) On the k-th day ($k < n$), there will be a division that will not have been inspected; (b) On the kth day ($k < n$), each division will have been inspected; (c) On the $(n+1)$th day, each division will have been inspected.

37. There are $n \geq 3$ pairs of socks in a drawer, and three pairs are black. Find the probability that two randomly selected socks are black. Does this probability get larger when n is increasing? For which n is this probability less than $1/5$?

38. In a country, there are only five first names for boys, and all five names are equally likely to be given. (a) Find the probability that four boys selected at random have different names. (b) Find the same probability for six boys.

39. Fifteen people are randomly seated in a row of thirty chairs. Write a formula for the probability that the people will occupy fifteen adjacent seats?

40. Ten women and nine men attend a lecture. Suppose that all orders in which they can leave the room after the lecture is over are equally likely. Find the probability that (a) all women will leave first, (b) the second person leaving the room will be a woman.

41. There are 4 roses and 5 lilies. Five flowers are randomly selected. Find the probability that the bouquet will contain 2 roses and 3 lilies.

42. Explain without any calculations why $\binom{n}{n} = 1$ and $\binom{n}{1} = \binom{n}{n-1} = n$.

43. (a) Which is larger: $\binom{100}{50}$ or $\binom{100}{51}$?

    (b) For a fixed n, consider $\binom{n}{k}$ as a function of k. Figure out for which k it is increasing and for which it is decreasing. Where does it attain its maximal value for an even n, for an odd n? In what sense is the picture symmetric? (*Advice*: Compare the values for k and $k+1$.)

    (c) If you toss a coin 100 times, which is more likely: getting 50 heads or 49 heads; 50 heads or 51 heads; 49 heads or 51 heads; 41 heads or 59 tails?

44. Using the binomial formula, prove that

$$\sum_{k=0}^{n}(-1)^k\binom{n}{k} = 0.$$

45. (*Pascal's triangle.*) Prove that

$$\binom{n}{k} = \binom{n-1}{k-1} + \binom{n-1}{k} \tag{4.1}$$

rigorously, and give a combinatorial interpretation. Proceeding from (4.1), show that all combinations $\binom{n}{k}$ can be arranged in the triangle below, which is constructed in the following way. We start with one in the first row as is shown below, and then, moving down, add two adjacent numbers and write the sum directly underneath. (An empty position is treated as zero.) In other words, each number in the Pascal's triangle is the sum of the two numbers above it. It helps counting combinations.

1
1 1
1 2 1
1 3 3 1
1 4 6 4 1
. . . .

46.* Consider the matching scheme of Section 3.6. Explain without calculations why $p_{n-1}(n)$ must be equal to zero, and why $\sum_{k=0}^{n} p_k(n)$ must be equal to one. Show that both assertions are consistent with (3.6.1). (Hint: For any a_k, b_m, one may write $\sum_{k=0}^{n} a_k \sum_{m=0}^{n-k} b_m = \sum_{j=0}^{n}\sum_{i=0}^{j} a_i b_{j-i}$.)

47.* Ten guests came to a diner party. The host had had plans where each guest would sit, but the guests ignored it and chose seats at random. Find the probability that at most two guests chose the seats that had been intended for them.

48.* Using software, compute $p_3(4), p_2(4), p_3(10), p_5(10)$, and $p_8(10)$ in the model of Section 3.6. Compare the results with what approximation (3.6.4) gives. Explain the result of the comparison, and provide a general rule of thumb for using (3.6.4).

Chapter 2

Independence. Conditional Probability

1 INDEPENDENCE

1.1 Definitions and examples

Consider a sample space Ω, events A, and a probability measure $P(A)$ defined on the events A.

Loosely speaking, we say that events A_1 and A_2 are independent if they occur "independently" from each other: the occurrence of each has no impact on the occurrence of the other.

Formally, events A_1 and A_2 are said to be *independent* if

$$P(A_1A_2) = P(A_1)P(A_2). \tag{1.1.1}$$

EXAMPLE 1. A card is drawn from a deck of 52. Let $A_1 = \{\text{an ace is drawn}\}$ and $A_2 = \{\text{a spade is drawn}\}$. As each suit has exactly one card of each value, our intuition tells us that the events A_1, A_2 should be independent, and this is true. Indeed, $P(A_1) = \frac{4}{52} = \frac{1}{13}$, $P(A_2) = \frac{13}{52} = \frac{1}{4}$. On the other hand, the event A_1A_2 consists of only one outcome: the ace of spades, and hence $P(A_1A_2) = \frac{1}{52}$. Since $P(A_1)P(A_2) = \frac{1}{13} \cdot \frac{1}{4} = \frac{1}{52}$, relation (1.1.1) holds.

EXAMPLE 2. An urn contains r red balls and b black balls. We draw two balls, one at a time, without replacement. Let $A_1 = \{\text{the first ball is red}\}$ and $A_2 = \{\text{the second ball is red}\}$. Now, our intuition tells us that these events should be dependent, and this is again true. In Example **1**.2.2-7, we have shown that $P(A_1) = P(A_2) = \frac{r}{n}$, where $n = r + b$. We have also shown there that the probability that both balls will be red, $P(A_1A_2) = \frac{r(r-1)}{n(n-1)}$. We see that the last expression is not equal to $P(A_1)P(A_2)$, and hence (1.1.1) does not hold.

EXAMPLE 3. Now, suppose it is *given* that A_1 and A_2 are independent, and $P(A_1) = \frac{1}{13}$, $P(A_2) = \frac{1}{4}$. (We took the same numbers as in Example 1 on purpose but so far we are not dealing with cards.) Find $P(A_1 \cup A_2)$. Without the independency condition we could not find this probability, but proceeding from this condition and (**1**.2.5.4), we can write

$$\begin{aligned} P(A_1 \cup A_2) &= P(A_1) + P(A_2) - P(A_1A_2) = P(A_1) + P(A_2) - P(A_1)P(A_2) \\ &= \frac{1}{13} + \frac{1}{4} - \frac{1}{13} \cdot \frac{1}{4} = \frac{16}{52} = \frac{4}{13}. \end{aligned}$$

Now, let us revisit Example 1 where $A_1 \cup A_2$ is the event that a card drawn from a deck is either an ace or a spade (or both). The reader may find $P(A_1 \cup A_2)$ by straightforward computing the number of outcomes in this event. Certainly, the answer should be the same as above: $\frac{4}{13}$. □

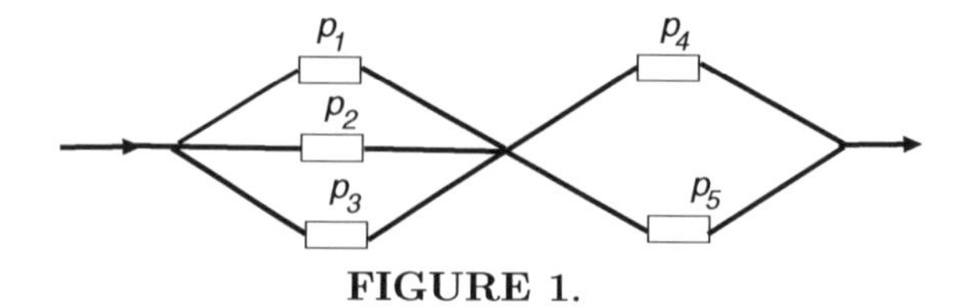

FIGURE 1.

We say that events A_1, A_2, A_3 are *mutually independent,* if they are *pairwise independent*, that is,

$$P(A_1A_2) = P(A_1)P(A_2), \;\; P(A_1A_3) = P(A_1)P(A_3), \; P(A_2A_3) = P(A_2)P(A_3), \qquad (1.1.2)$$

and additionally,

$$P(A_1A_2A_3) = P(A_1)P(A_2)P(A_3). \qquad (1.1.3)$$

The significance of this definition is based on the fact that three events, A_1, A_2, A_3, may be *pairwise* independent while (1.1.3) is not true.

EXAMPLE 4.[1] An urn contains four balls: red, blue, green, and a ball having stripes of all the three colors. One ball is drawn at random. Let the event $R = \{$the color red is present in the ball drawn$\}$, and the events B and G be similar events for blue and green.

Clearly, $P(R) = P(B) = P(G) = \frac{1}{2}$ since, for example, the red color is present in two balls. The event RB occurs only if the ball with all colors is drawn, and the same is true for RG, BG, and RBG. Consequently, $P(RB) = P(RG) = P(BG) = P(RBG) = \frac{1}{4}$. We see that $P(RB) = P(R)P(B)$ just because $\frac{1}{2} \cdot \frac{1}{2} = \frac{1}{4}$, and the same is true for the other two pairs of colors. However, $P(R)P(B)P(G) = \frac{1}{2} \cdot \frac{1}{2} \cdot \frac{1}{2} = \frac{1}{8}$, while $P(RBG) = \frac{1}{4}$. Thus, the events R, B, G are pairwise independent but are not mutually independent. □

One more example on mutual independence is given in Exercise 8. However, it is worth noting that situations similar to that in Example 4 are rare. Usually, in particular models, pairwise independent events turn out to be mutually independent as well.

An extension of the above definition to the case of n events may be stated as follows. We call $A_1, ..., A_n$ mutually independent if for any $k = 1, ..., n$, and any collection of events $A_{i_1} \cdots A_{i_k}$ selected from the original sequence $A_1, ..., A_n$,

$$P(A_{i_1} \cdots A_{i_k}) = P(A_{i_1}) \cdots P(A_{i_k}). \qquad (1.1.4)$$

See also Exercise 9.

From now on, when talking about more than two independent events, we will always mean mutual independence. This concerns, in particular,

EXAMPLE 5. A system consists of five components configured as shown in Fig. 1. Denote by p_i the probability that the component i works, and suppose that the components function *independently*.

[1]The example is similar to the well known example by *S.N. Bernstein*. The only difference is that in the latter, a regular tetrahedron is being tossed. We simplify the wording.

By virtue of independency, the probability that all components will work is $p_1 \cdot p_2 \cdot ... \cdot p_5$. Indeed, if $A_i = \{component\ i\ works\}$, then $P(A_1A_2 \cdots A_5) = P(A_1)P(A_2) \cdots P(A_5)$.

Now, suppose that for a signal to go through the system, *both* consecutive parts should work, but in each of them, it suffices that *at least one* parallel component works. Let us compute the probability that the signal will go through the system. The probability that component i will not work is $1 - p_i$. Then, in view of independency, the probability that *none* of the three parallel components of the first consecutive part will work, is $(1 - p_1)(1 - p_2)(1 - p_3)$, and hence the probability that *at least one* component will work is the *complement probability* $1 - (1 - p_1)(1 - p_2)(1 - p_3)$. This is the probability that the signal will go through the first part. For the second part, the similar probability is $1 - (1 - p_4)(1 - p_5)$.

Now, we want the signal to go through the first *and* the second part; that is, we are talking about the product of events. So, we should consider the product of the probabilities above: $(1 - (1 - p_1)(1 - p_2)(1 - p_3)) \cdot (1 - (1 - p_4)(1 - p_5))$. If all p_i's are equal to the same p, we will have $(1 - (1 - p)^3)(1 - (1 - p)^2)$. □

EXAMPLE 6 concerns a non-discrete space.

(a) We identify the sample space with the square in Fig. 2a, and assume that the probability distribution on this square is uniform. As we have already discussed in Section **1**.2.3, this means that the probability of any region in the square does not depend on the location of this region and is proportional to the area of the region. Since the total area of the square is 4, for any region (or event) A, the probability $P(A) = \frac{1}{4}s(A)$, where $s(A)$ is the area of A.

Let $A_1 = \{x_1 \geq 0.5\}$, $A_2 = \{x_2 \geq 0.5\}$ (see also Fig. 2a). Then $P(A_1) = \frac{1}{4}(0.5 \cdot 2) = \frac{1}{4}$, which is natural: A_1 constitutes one-fourth of the whole square. Similarly, $P(A_2) = \frac{1}{4}$.

As is easy to see, $P(A_1A_2) = \frac{1}{16}$ because the intersection of A_1A_2 constitutes one-sixteenth of the square; see again Fig. 2a. Thus, $P(A_1A_2) = P(A_1)P(A_2)$, and hence A_1 and A_2 are independent. Certainly, this had to be expected: if we picked a point at random from the square, the knowledge of the first coordinate would not give us any clue what the second coordinate could be.

(b) Now, consider the uniform distribution on the square in Fig. 2b. Since now the total area is equal to two, $P(A) = \frac{1}{2}s(A)$. Consider the same events $A_1 = \{x_1 \geq 0.5\}$ and $A_2 = \{x_2 \geq 0.5\}$ as above; how they look is shown in Fig. 2b. The intersection A_1A_2 contains only one point: $(0.5, 0.5)$. The area of a set consisting of one point is zero; so, $P(A_1A_2) = 0$.

Then, we do not need to compute $P(A_1)$ and $P(A_2)$; it suffices to observe that these

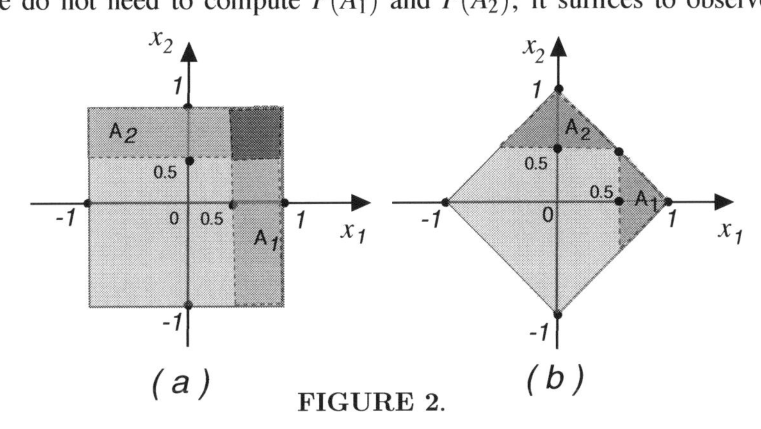

FIGURE 2.

probabilities are positive. Thus, $P(A_1A_2) \neq P(A_1)P(A_2)$, and consequently, A_1, A_2 are dependent events. This is also not surprising; in this case, once we know the first coordinate, we can say something about the second. For example, if $x_1 \geq 0.5$, then x_2 must be less than or equal to 0.5. □.

1.2 Independent trials

1.2.1 Binomial scheme

A sequence of *n independent* trials is being performed. Each trial results in a success with probability p and in a failure with probability $q = 1 - p$. Saying "independent," we mean that the events $A_1 = \{\textit{the first trial is successful}\}$, ..., $A_n = \{\textit{the n-th trial is successful}\}$ are mutually independent. Note also that the complement $A_i^c = \{\textit{the i-th trial results in a failure}\}$.

As in Section **1**.1, we mark by 1 a success, and by 0 a failure, and identify each outcome ω with a sequence of ones and zeros of a length of n.

Denote by $[\omega]$ the event consisting of one outcome ω. Such a set is also called a *singleton*.

Set, for example, $n = 5$, and consider $\omega = 10010$. Then $[\omega] = A_1A_2^cA_3^cA_4A_5^c$. All these events are independent (see also Exercise 2). Hence,

$$P([\omega]) = P(A_1)P(A_2^c)P(A_3^c)P(A_4)P(A_5^c) = p \cdot q \cdot q \cdot p \cdot q = p^2q^3.$$

Note that "2" above is the number of successes, and "3" is the number of failures.

Similarly, if the number of successes corresponding to an outcome ω is k (and, hence, the number of failures is $n - k$), then

$$P([\omega]) = p^kq^{n-k}. \tag{1.2.1}$$

How many outcomes are there with the same probability? As many as we can select k trials to be successful from the n trials; that is, $\binom{n}{k}$. Adding up the probabilities (1.2.1) for all such ω's, we arrive at the following:

If the event $B_{k,n} = \{\text{exactly } k \text{ successes in } n \text{ trials}\}$, then

$$P(B_{k,n}) = \binom{n}{k}p^kq^{n-k} = \frac{n!}{k!(n-k)!}p^kq^{n-k}. \tag{1.2.2}$$

The above probabilities are called *binomial* because they coincide with the terms in the binomial expansion of $(p+q)^n$. (See (**1**.3.4.7), and also Exercise 17.)

EXAMPLE 1. Suppose twenty-five percent of the customers entering a grocery store buy a dairy product. We randomly select 7 customers.

a) What is the probability that *exactly* 5 out of the 7 customers will buy a dairy product? The probabilities $p = 0.25$, $q = 0.75$, and the desired probability is

$$P(B_{5,7}) = \binom{7}{5}(0.25)^5(0.75)^2 = \frac{7!}{5!2!}(0.25)^5(0.75)^2 \approx 0.012.$$

b) Now, let us compute the probability that *at most* 5 customers bought a dairy product. Proceeding in a straightforward fashion, we may add up all probabilities $P(B_{k,7})$ for $k =$

$0, 1, ..., 5$, but this is tedious. Instead, we may compute the probability that the number of customers is larger than 5, that is, $P(B_{6,7}) + P(B_{7,7})$, which is shorter. After that we will subtract this probability from one. We have

$$P(B_{6,7}) + P(B_{7,7}) = \binom{7}{6}(0.25)^6(0.75)^1 + \binom{7}{7}(0.25)^7(0.75)^0$$
$$= \frac{7!}{6!1!}(0.25)^6(0.75)^1 + \frac{7!}{7!0!}(0.25)^7(0.75)^0 \approx 0.0013.$$

Then the desired probability is $\approx 1 - 0.0013 = 0.9987$. □

To find the probability that there will be no successes, we may set $k = 0$ in (1.2.2) and after simple calculations get

$$P(\textit{no successes in n trials}) = q^n. \tag{1.2.3}$$

However, we can also easily obtain it directly. The event in hand is equal to the intersection of the n independent events $A_1^c \cdots A_n^c$, and each of them has a probability of q. This directly implies (1.2.3).

EXAMPLE 1 revisited. If twenty-five percent of customers buy a dairy product, guess how likely it is that none of seven will do it. The probability of interest is $(0.75)^7 \approx 0.13$, which is not that small. For 10 customers it is $(0.75)^{10} \approx 0.056$, and for 20, it is ≈ 0.003. □

In conclusion, note that for such standard problems as above, it is reasonable to use software. In particular, the command for $P(B_{k,n})$ in Excel is

=BINOMDIST(k,n,p,FALSE),

and for the cumulative probability $P(\textit{the number of successes in n trials is less than or equal to k})$, it is

=BINOMDIST(k,n,p,TRUE) .

1.2.2 Binomial trees

The model below may describe different phenomena; we *interpret* it in terms of a stock market.

A simple model of the evolution of a stock price supposes that each day, the price either increases by a factor $u > 1$, or decreases by a factor $d < 1$. The symbols come after "up" and "down."

The situation is illustrated by the tree diagram in Fig. 3a where we consider three moments of time. The initial price is one. If the price rises, its new value is u, and if it rises again, its value becomes u^2. If the price first rises, and after that it drops, its value becomes ud, etc.

Each outcome corresponds to a path in the tree. There are 4 paths in this case. For short, we denote, for example, the path corresponding to two rises as *(up, up)*. Note that the paths *(up, down)* and *(down, up)* lead to the same price, and this is why the tree we constructed looks as it does in Fig. 3a.

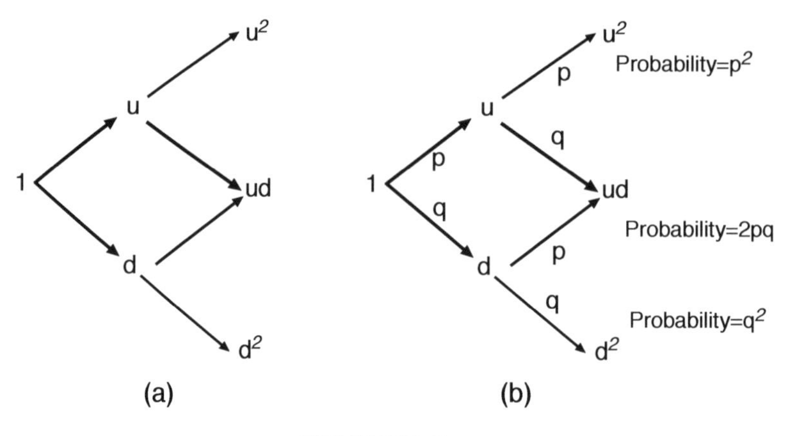

FIGURE 3.

Suppose that each time the price moves up or down with respective probabilities p and q, $p+q=1$, and that what happens on a particular day does not depend on what happened before. Then the probability of the path *(up, up)* is $p \cdot p = p^2$, for the path *(up, down)*, it is pq, and so on. Once we know the probabilities of the paths, we can compute the probabilities of all three possible prices at the end, which is shown in Fig. 3b. (We get $2pq$ because two paths lead to the value ud.)

In the case of n moments of time, we have a similar picture. Each day may be viewed as a trial; a success corresponds to moving up, a failure to moving down, and hence, there are 2^n possible outcomes (paths). The price on the last, nth, day depends only on the total number of successes (moves up): if there were k "ups," then the price equals $u^k d^{n-k}$. Since $k = 0, ..., n$, there are $n+1$ of such values. The probability of a value $u^k d^{n-k}$ is the probability $P(B_{k,n})$ in the binomial model above, and equals the binomial probability $\binom{n}{k} p^k q^{n-k}$.

Route 1 ⇒ page 37

2

1.2.3 Multinomial scheme

Suppose that each of n independent trials may end up with m different results labeled $1, ..., m$. Suppose also that the probability that result i occurs is the same for all trials and equals p_i with $p_1 + ... + p_m = 1$. For a particular outcome ω, denote by n_i the number of trials having result i. Clearly, $n_1 + ... + n_m = n$. Then the event $[\omega]$ (we call it a singleton) is the product of n *independent* events among which n_1 events have a probability of p_1, n_2 events have a probability of p_2, ... , and n_m events have a probability of p_m. Then

$$P([\omega]) = p_1^{n_1} \cdots p_m^{n_m}. \tag{1.2.4}$$

There are $n!/(n_1! \cdots n_m!)$ outcomes of this type, since this is the number of ways in which we can distribute n objects (in our case, trials) into m distinct groups (in our case, results) with given sizes $n_1, n_2, ..., n_m$; see Section 1.3.3. Hence,

$$P(n_1 \textit{ results of type } 1, \ldots, n_m \textit{ results of type } m) = \frac{n!}{n_1!n_2!\cdots n_m!} p_1^{n_1} \cdots p_m^{n_m}. \tag{1.2.5}$$

EXAMPLE 1. For a loaded die as in Example 1.2.2, $p_i = \frac{i}{21}$. If we roll the die 7 times, then

$$P(\text{one roll shows three, two rolls—five, four rolls—six})$$
$$= \frac{7!}{0!0!1!0!2!4!}\left(\frac{1}{21}\right)^0\left(\frac{2}{21}\right)^0\left(\frac{3}{21}\right)^1\left(\frac{4}{21}\right)^0\left(\frac{5}{21}\right)^2\left(\frac{6}{21}\right)^4$$
$$= \frac{7\cdot 6\cdot 5}{2}\cdot\frac{3\cdot 5^2\cdot 6^4}{21^7} \approx 0.0057. \ \square$$

1,2

2 CONDITIONING

2.1 Conditional probability

By definition, the conditional probability of event A given event B is

$$P(A\,|\,B) = \frac{P(AB)}{P(B)}, \tag{2.1.1}$$

provided $P(B) \neq 0$. The symbol $P(A\,|\,B)$ is read as "the probability of A given B."

From the definition of independence (1.1.1), it follows that

If A and B are independent, then $P(A\,|\,B) = P(A)$. (2.1.2)

Indeed, if $P(AB) = P(A)P(B)$, then the factor $P(B)$ in (2.1.1) cancels.

EXAMPLE 1. Suppose 4% of students have A's for all mathematics courses, and 10% for a probability course. Mary received an A for a probability course. What is the probability that she has A's for all mathematics courses? Keeping in mind that the event

$$\{\,A\textit{ for Mathematics}\} \subset \{\,A\textit{ for Probability}\},$$

we have

$$P(\,A\textit{ for Mathematics}\,|\,A\textit{ for Probability}) = \frac{P(\,A\textit{ for Mathematics AND A for Probability})}{P(A\textit{ for Probability})}$$

$$= \frac{P(\,A\textit{ for Mathematics})}{P(A\textit{ for Probability})} = \frac{0.04}{0.1} = 0.4.$$

EXAMPLE 2 is well known. You visit a family with two children, and you know that one child is a boy. For simplicity, consider the case of equally likely outcomes in the sample space $\{bb, bg, gb, gg\}$. What is the probability that the other child is also a boy?[1]

Often, people give the answer $1/2$, reasoning that it is equally likely that the other kid is a boy or girl. The correct answer is $1/3$, and apparently the mistake consists in unconscious identifying the outcomes bg and gb. To adjust our intuition, observe that since it is *given* that one child is a boy, we should consider only the outcomes bb, bg, gb. They are equally likely, and hence, the probability of interest is $1/3$. The formal proof should run as follows.

Let events $A = \{bb\}$, $B = \{bb, bg, gb\}$. Since $A \subset B$,

$$P(A\,|\,B) = \frac{P(AB)}{P(B)} = \frac{P(A)}{P(B)} = \frac{1/4}{3/4} = \frac{1}{3}. \; \square$$

EXAMPLE 3. Each day, the president of a company inspects one of the company's five divisions. The president chooses a division at random, and the choice does *not* depend on which divisions were inspected on previous days. Suppose a division has not been inspected for 30 consecutive days. What is the probability that it will be inspected tomorrow? The same as on the first day: $\frac{1}{5} = 0.2$, and this follows from (2.1.2). Sometimes, people mistakenly think that in such a situation, since the event has not happened for long time, it should happen soon, but this is not correct. *In the case of independency*, regardless of what happened before, each time "everything starts over as from the very beginning."

Another matter is that the probability that a division will not be inspected thirty days in a row is $(1-0.2)^{30} = (0.8)^{30} \approx 0.00124$; that is, rather small. So, if this event has indeed occurred, then it is reasonable to suppose that the very independency assumption may not be correct: for some reason, the president excluded or forgot about this division, and the division faces a non-routine situation. However, if we do know that the independence assumption holds, and just a rare event indeed occurred, the answer is $1/5$. $\square$

From definition (2.1.1), we immediately obtain what is called the
Multiplication Rule:

$$P(AB) = P(B)P(A\,|\,B). \tag{2.1.3}$$

EXAMPLE 4. Research shows that about 8 percent of males, and only 0.5 percent of females, are color blind in some way or another. Suppose that in a country, the population consists of 48% males and 52% females. What is the probability that a person chosen at random is a color blinded male? Setting, for short, $A = \{\textit{color blind}\}$, $B = \{\textit{male}\}$, we have $P(B) = 0.48$, $P(A\,|\,B) = 0.08$, and hence, $P(AB) = P(B)P(A\,|\,B) = 0.48 \cdot 0.08 = 0.0384$. A similar probability for a color blind female is $0.52 \cdot 0.005 = 0.0026$. $\square$.

2.2 Multiplication rule and probabilities on trees

Here, we consider a tree with a more general structure than in Section 1.2.2. Suppose that the price of a stock may evolve during three consecutive days as it is shown in Fig. 4a. (Certainly, this is just a study example.)

[1] One may find in the literature (perhaps, the first time it appeared in [39]) or on the Internet the following wording of this problem: "The king comes from a family of two children. What is the probability that he has a brother?"

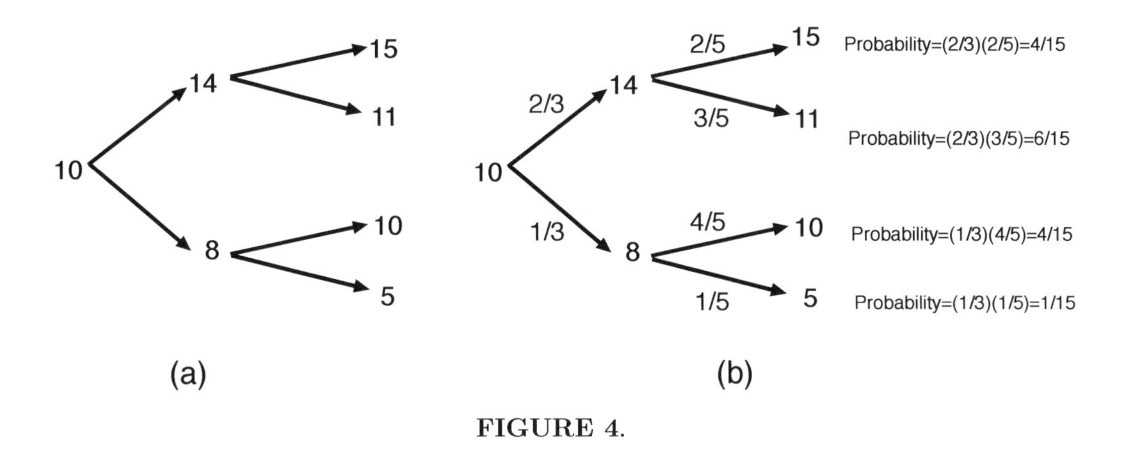

FIGURE 4.

The initial price is 10. At the end of the first day, the price may be either 14 or 8. If the price is 14, its next value may be either 15 or 11, while if the price is 8, then at the end of the second day, the price may be either 10 or 5.

Assume now that the probabilities of moving up and down depend on the previous history, and have the following values. At the initial moment, they are $2/3$ and $1/3$, as is shown in Fig. 4b. If the price has increased, then on the next day, these probabilities are $2/5$ and $3/5$, while if the price dropped, then they are $4/5$ and $1/5$; see again Fig. 4b.

It is worth emphasizing that these probabilities are *conditional*. For example, the probability $2/5$ is the probability that the price will rise *given* that it has risen on the first day. To compute the probability of each of four possible outcomes (paths), we can apply the multiplication rule. For example, let the event $C_1 = \{\textit{the price increased on the first day}\}$ and $C_2 = \{\textit{the price increased on the second day}\}$. Then the probability of the "top" path (the price increased on both days) is $P(C_1C_2) = P(C_1)P(C_2 \,|\, C_1) = \frac{2}{3} \cdot \frac{2}{5} = \frac{4}{15}$.

The probabilities of the other three paths are computed in Fig. 4b. All these probabilities are also those of the eventual values of the price.

2.3 The law of total probability

Consider an experiment with a sample space Ω, and two events, B_1 and B_2, that are *disjoint* and *together constitute the whole* Ω. That is, $B_1B_2 = \emptyset$, and $B_1 \cup B_2 = \Omega$. We call $\{B_1, B_2\}$ a *partition* of Ω. Assume $P(B_1) \neq 0$ and $P(B_2) \neq 0$.

One may view B_1 and B_2 as conditions under which the experiment may be performed. For example, if we talk about a runner's result in a competition to be held tomorrow, then B_1 may be the event that it will be raining tomorrow, and B_2 is the event that it will not be raining.

Next, we introduce an event A which may be viewed as the result of the experiment. Say, A is the event that the runner will beat his best personal time. The following formula, in spite of its simplicity, proves to be a very useful tool in solving a great many of problems.

The formula for total probability (or **the law of total probability**). For any A,

$$P(A) = P(A \,|\, B_1)P(B_1) + P(A \,|\, B_2)P(B_2). \tag{2.3.1}$$

Proof. Since B_1 and B_2 are disjoint, the events AB_1 and AB_2 are also disjoint, and since $B_2 = B_1^c$, we have $AB_1 \cup AB_2 = A$. Then, by virtue of (2.1.3),

$$P(A) = P(AB_1) + P(AB_2) = P(A \mid B_1)P(B_1) + P(A \mid B_2)P(B_2). \; \blacksquare$$

EXAMPLE 1. Suppose that students who do their homework have a 95% chance of passing a final exam, while the same figure for students who do not do the homework is 20%. Suppose that 85% of the students do the homework. What is the probability that a student chosen at random will pass the exam?

Let events $B_1 = \{$*the student does the homework*$\}$, $B_2 = \{$*the student does not do the homework*$\}$, $A = \{$*the student passes the exam*$\}$. We are given: $P(A \mid B_1) = 0.95$, $P(A \mid B_2) = 0.2$, $P(B_1) = 0.85$, and hence $P(B_2) = 0.15$. Then

$$P(A) = P(A \mid B_1)P(B_1) + P(A \mid B_2)P(B_2) = 0.95 \cdot 0.85 + 0.2 \cdot 0.15 = 0.8375.$$

EXAMPLE 2. There are two urns. The first contains 3 red and 2 black balls; the second 4 red and 5 black balls. You draw at random a ball from the first urn, put it into the second, and after that you draw a ball from the second urn, also at random. What is the probability that the ball drawn at the end is red?

Set $B_1 = \{$*the ball drawn from the first urn is red*$\}$, $B_2 = \{$*the ball drawn from the first urn is black*$\}$, $A = \{$*the ball drawn from the second urn is red*$\}$. Clearly, $P(B_1) = \frac{3}{5} = 0.6$, and hence $P(B_2) = 0.4$.

If B_1 occurs, then the second urn will have 5 red and 5 black balls. Consequently, $P(A \mid B_1) = \frac{5}{10} = 0.5$. Similarly, $P(A \mid B_2) = \frac{4}{10} = 0.4$. Thus,

$$P(A) = P(A \mid B_1)P(B_1) + P(A \mid B_2)P(B_2) = 0.5 \cdot 0.6 + 0.4 \cdot 0.4 = 0.46. \; \square$$

The reader can guess how the formula for total probability looks in the case of more than two conditions. Consider a *partition* of Ω into *n disjoint* events $B_1, ..., B_n$ such that $\cup_{i=1}^n B_i = \Omega$. Assume that $P(B_i) \neq 0$ for all i. Then for any event A,

$$P(A) = \sum_{i=1}^n P(A \mid B_i)P(B_i). \tag{2.3.2}$$

The proof is similar to the proof for two events.

EXAMPLE 3. We roll a die and then toss a coin as many times as the die shows. What is the probability of the event $A = \{$*the coin lands only heads*$\}$? Let $B_i = \{$*the die shows i*$\}$, $i = 1, ..., 6$. Clearly, $P(B_i) = \frac{1}{6}$ for all $i = 1, ..., 6$. If the coin is tossed i times, the probability that only heads will appear is $\left(\frac{1}{2}\right)^i$, which gives us $P(A \mid B_i)$. Then

$$P(A) = \sum_{i=1}^6 P(A \mid B_i)P(B_i) = \sum_{i=1}^6 \left(\frac{1}{2}\right)^i \cdot \frac{1}{6}$$
$$= \frac{1}{6}\sum_{i=1}^6 \left(\frac{1}{2}\right)^i = \frac{1}{6} \cdot \frac{1}{2}\sum_{i=1}^6 \left(\frac{1}{2}\right)^{i-1} = \frac{1}{12} \cdot \frac{1 - \left(\frac{1}{2}\right)^6}{1 - \frac{1}{2}} \approx 0.164.$$

(We used the formula for the sum of a geometric series; see the Appendix, (2.2.11).) $\square$

Next, we demonstrate the usefulness of the formula for total probability in a more sophisticated and important model.

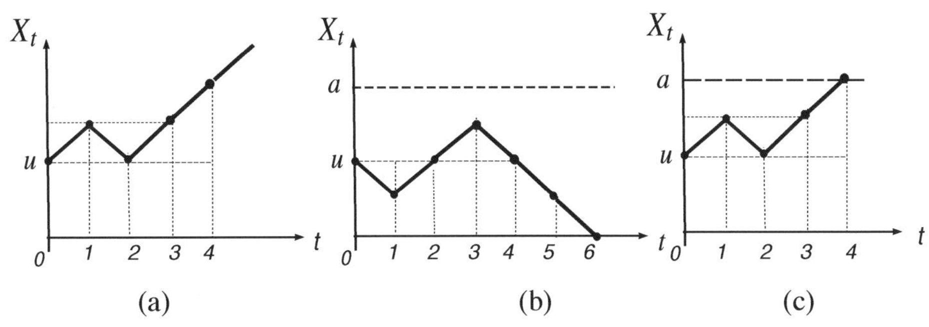

FIGURE 5. X_t is the total capital at time t. Diagram (b) corresponds to the case of ruin; Diagram (c)—to the case when the capital first reaches level a.

2.4 Random walk and ruin probability

A company has an initial capital u, and at consecutive moments of time, either gains or loses one unit of money with probabilities p and $q = 1 - p$, respectively. The result at each moment does not depend on what has happened before. A particular realization is shown in Fig. 5a. We connect the points by straight lines just to make the picture more illustrative.

The term "*random walk*" comes after another phenomenon which may be described by the same model. It concerns the motion of a particle which moves at each discrete moment of time either to the right or to the left by one unit of distance. For a particle immersed in a liquid, such a motion is connected with the bombardment by the molecules of the surrounding medium. (In fact, the particle is moving in a three-dimensional space, so we are talking about the projection of such a motion in a particular direction.)

Let us come back to the first interpretation. We assume u to be an integer, fix another integer $a \geq u$, and suppose that the process comes to a stop when the capital either reaches zero level (which we call *ruin*) or the level a, whichever comes first. For example, a company having an initial capital of u runs its business until the first moment when the company either runs out of money or its capital reaches a level a. See also Fig. 5bc.

The reader may also think about the simplest game of chance of two players tossing a coin (which may be non-symmetric). Each toss, one player pays to the other one unit depending on the result of the toss. In this case, a is the *total* initial capital of both players, while u and $a - u$ are the initial capitals of the first and second player, respectively. The game comes to a stop when one player runs out of money.

Let A_u be the event that the capital of the company dropped to zero level, and it happened before the capital reached the level a. Set $q_u = P(A_u)$ and call it a *ruin probability*. The index u indicates that this probability depends on the initial level u.

Let C_u be the event that the capital reached the a-level, and it happened before the capital dropped to zero. Set $p_u = P(C_u)$.

One can guess that $p_u + q_u = 1$, which is true though it should be proved. Theoretically, it may happen that the process will never reach either of the boundaries of the interval $[0, a]$ ("corridor" $[0, a]$ in Fig. 5bc). We will show later that the probability of such an event is zero.

To find q_u, first note that

$$q_0 = 1, \;\; q_a = 0. \tag{2.4.1}$$

(If $u = 0$, the company is already ruined; if $u = a$, the company already has the amount a.)

Let events $B_1 = \{$*at the first time, the company gains one*$\}$, and $B_2 = \{$*at the first time, the company loses one*$\}$. Now, observe that if at the first step the capital rises, the random walk starts over but from the level $u+1$, and the probability to be ruined under this condition becomes q_{u+1}. This means that $P(A_u \,|\, B_1) = q_{u+1}$. Similarly, $P(A_u \,|\, B_2) = q_{u-1}$.

Consequently, by the formula for total probability, for $u = 1, 2, ..., a-1$,

$$q_u = P(A_u \,|\, B_1)P(B_1) + P(A_u \,|\, B_2)P(B_2) = q_{u+1}p + q_{u-1}q. \tag{2.4.2}$$

This is an equation for q_u. At the end of this section, we show that together with boundary condition (2.4.1), a solution to (2.4.2) is unique. Then, if we guess what this solution is, it will suffice to check it by substitution. We distinguish two cases.

Case 1 (*Symmetric random walk*). Let $p = 1/2$. It is straightforward to verify that in this case,

$$q_u = \frac{a-u}{a} \tag{2.4.3}$$

satisfies both (2.4.1) and (2.4.2).

To find p_u, it suffices to observe that the distance from the initial point u to the level a, is $a-u$, and since the walk is symmetric, the probability of hitting a first, starting from u, equals the probability of hitting 0 first, starting from $a-u$. In other words,

$$p_u = q_{a-u} = u/a.$$

We see that in the symmetric case, the answer is simple: the probability of not being ruined is proportional to the initial capital. We also see that indeed $p_u + q_u = 1$, so the probability that the capital will be always within the corridor $[0, a]$, is zero.

Case 2 (*Non-symmetric random walk*). Now, let $p \neq 1/2$. If $p = 0$, then obviously $q_u = 1$, so we assume $p \neq 0$. For the non-symmetric walk, the solution is more complicated. First, set

$$r = (1-p)/p. \tag{2.4.4}$$

Not that this is the odds ratio of the event B_2 corresponding to losses. (See (**1**.3.2.1).) In other words, at each time moment, the odds are r to 1 in favor of losses, and hence, $\frac{1}{r}$ to 1 in favor of gains.

The reader is invited to verify by substitution (it will require simple algebra) that in this case, q_u satisfying (2.4.1) and (2.4.2) is

$$q_u = \frac{r^u - r^a}{1 - r^a}. \tag{2.4.5}$$

To find p_u, we follow the same logic as before, replacing in q_u the argument u by $a-u$. However, since now the walk is not symmetric, we should also replace p by $1-p$. If we do that in formula (2.4.4), the new r becomes $p/(1-p) = 1/r$. Thus,

$$p_u = \frac{(1/r)^{a-u} - (1/r)^a}{1 - (1/r)^a} = \frac{r^u - 1}{r^a - 1} = \frac{1 - r^u}{1 - r^a}. \tag{2.4.6}$$

The reader is invited to check on her/his own that again $p_u + q_u = 1$.

Analyzing the *ruin probability formula* (2.4.5), we come to some interesting observations. Let us follow the game interpretation.

EXAMPLE 1. Let u=\$9, the level a=\$10, and p=0.4. Then the probability to reach \$10 is $p_9 = (1-(3/2)^9)/(1-(3/2)^{10}) \approx 0.660$, and is larger than $\frac{1}{2}$.

We see that the size of the initial capital matters. Though the game is *non-favorable for the first player*, due to the fact that she/he is within one step of reaching the level a, the *probability of winning is higher than this probability for the second player.* □

Next, assume that the stakes are reduced in half (by a factor of two). How does this change the ruin probability?

If we adopt the new stake as a unit of money, the initial capital measured in the new units will be equal to $2u$, and the upper level—to $2a$. In the symmetric case, the ruin probability (2.4.3) will not change after such a substitution, but in the case $p \neq 1/2$, the new probability will be equal to

$$q_u^* = \frac{r^{2u} - r^{2a}}{1 - r^{2a}} = \frac{r^u + r^a}{1 + r^a} \cdot \frac{r^u - r^a}{1 - r^a} = \frac{r^u + r^a}{1 + r^a} \cdot q_u.$$

If $p < 1/2$ (and hence $r > 1$), the last fraction is greater than 1, and consequently, $q_u^* > q_u$.

If $p > 1/2$ (and hence $r < 1$), then $q_u^* < q_u$.

Thus, if a game is *non*-favorable for a player ($p < 1/2$), she/he should play higher stakes.

EXAMPLE 1 revisited. Suppose the stake is only 10¢. Then, in (2.4.5), we should set $a =$ 100, and $u = 90$. Then the new probability of reaching \$10 equals $(1-(3/2)^{90})/(1-(3/2)^{100})$ ≈ 0.0173. (Compare with 0.660 above.) □

Next, we set $p > \frac{1}{2}$, and suppose that the company continues its business no matter the level of the capital it reaches. This corresponds to the case of $a = \infty$.

Letting $a \to \infty$ in (2.4.5), and keeping in mind that $r < 1$, we get the simple formula

$$q_u = r^u. \tag{2.4.7}$$

EXAMPLE 2. Let $p = \frac{2}{3}$. Then $r = \frac{1}{2}$, and

$$q_u = \left(\frac{1}{2}\right)^u = 2^{-u}.$$

Clearly, if the initial capital u is small, the ruin probability is large. Suppose the company wants to know for which initial capital the ruin probability is not larger than 5%. This amounts to $2^{-u} \leq 0.05$, which implies $u \geq \log_2 20 \approx 4.32$. So, the choice of $u = 5$ will be more than enough. □

► It remains to prove that a solution to (2.4.1)–(2.4.2) is unique. For $p = 0$, the probability $q_u = 1$, so this case may be excluded. Then, from (2.4.2), it follows that

$$q_{u+1} = \frac{1}{p}(q_u - q q_{u-1}). \tag{2.4.8}$$

Since $q_0 = 1$, setting $u = 1$, we have $q_2 = \frac{1}{p}(q_1 - q)$. Then $q_3 = \frac{1}{p}(q_2 - qq_1) = \frac{1}{p}([\frac{1}{p}(q_1 - q)] - qq_1)$. We do not need to simplify it, but just observe that using recurrence relation (2.4.8) and proceeding in the same fashion, we may represent any q_u as a linear (!) function of q_1. In particular, this concerns q_a. Because $q_a = 0$, this gives a linear equation for unknown q_1. So, a solution q_u is unique. ■ ◄

2.5 Choosing the best object

The problem is known by different names including the secretary problem, the sultan's dowry problem, and the choosy bride problem. It may be stated as follows.

There are n distinct objects that may be compared by quality. In particular, if we had been able to inspect all objects and after that to select one, we would have been able to select the best. However, we are allowed to "screen" the objects only consecutively. When inspecting an object in a row, we are able to compare it with (and only with) those we have already seen, and make one of two decisions.

Either we select this object, and the screening is over, or we reject it and move to the next object. However, we cannot move back: once an object is rejected, it is lost. The objective is to maximize the probability of choosing the best object.

We assume that all $n!$ arrangements of objects are equally likely.

The problem we stated is a particular case of a general problem concerning optimal *stopping rules* in *sequential analysis*. In our case, the optimal strategy has the following structure. We screen the first r objects (we may call it learning), and then, keeping screening, we select the first object that turns out to be better than all the previous. Proving the optimality of such a strategy is beyond the scope of this book. We take it for granted but will find the optimal r.

In particular, we will obtain the remarkable result that for large n, the optimal

$$r \approx \frac{n}{e}; \tag{2.5.1}$$

that is, we should screen about 37% of the objects $\left(\frac{1}{e} \approx 0.37\right)$ but reject them and keep screening until we come to the object that is the best among those we have already screened. Let us consider it in more detail.

Note right away that if the best object is among the first r, following the strategy under consideration, we will miss the best object. However, this just means that even for the optimal strategy, the probability of choosing the best is less than one.

As has been supposed, the objects are presented in a sequence. Let us assign the number i to the object in the ith place. Let A be the event that the best object will be chosen, and B_i be the event that the best object is object i. Since all arrangement are equally likely, $P(B_i) = \frac{1}{n}$.

As has been noted,

$$P(A \mid B_i) = 0 \text{ for } i \le r. \tag{2.5.2}$$

Suppose that the best object is in a place $i > r$. Consider the first $(i-1)$ objects (that is, those that are presented before the best), and denote by k the number of the object that is the best among these $i-1$. If $k > r$ (object k is between objects r and i), then following our strategy, we will not screen more than k objects and choose either the object k or an object before object k. In other words, we will finish screening before arriving at the ith place, and the best object will not be chosen.

If $k \le r$, following our strategy, we will reject all $(i-1)$ objects and choose the best when we are presented with the ith (and best) object.

Furthermore, all arrangements are equally likely, and this concerns the first $(i-1)$ objects as well. Therefore, the best object among the first $(i-1)$ is equally likely to be in any of $(i-1)$ places. Eventually, we conclude that

$$P(A\,|\,B_i) = \frac{r}{i-1} \text{ for } i > r. \tag{2.5.3}$$

Then, by (2.5.2)–(2.5.3) and the formula for total probability,

$$P(A) = \sum_{i=1}^{n} P(A\,|\,B_i)P(B_i) = \sum_{i=r+1}^{n} \frac{r}{i-1} \cdot \frac{1}{n} = \frac{r}{n} \sum_{i=r+1}^{n} \frac{1}{i-1}. \tag{2.5.4}$$

We have found $P(A)$, and it remains to find r at which $P(A)$ attains its maximum. For a particular n, it is not difficult to find a numerical solution; one can do it even in Excel or write a program to find it for any n. To find an analytical approximation, we approximate the r.-h.s. of (2.5.4) by

$$\frac{r}{n} \int_{r+1}^{n} \frac{1}{x-1} dx = \frac{r}{n} \ln\left(\frac{n-1}{r}\right) \approx \frac{r}{n} \ln\left(\frac{n}{r}\right)$$

for large n. Differentiating the last expression as a function of r, it is not difficult to find that the maximum is attained at $r = n/e$, so we have arrived at (2.5.1).

2.6 Bayes' formula

Suppose Mr. K had a routine medical test performed to determine if he had a certain disease. The result of the test turned out to be positive (which is bad). However, the test is not perfect. For 5% of healthy people, the results occur to be positive, and for 15% of people who actually have this disease, the test produces a negative result. It is also known that for the entire population Mr. K belongs to, 10% of people have this particular disease.

What can we do about it? Since the test result is positive, we expect that the probability that Mr. K has the disease is greater than 10%, the actual probability for a randomly selected person from the whole population. But what is the exact probability? Let us state the problem rigorously.

Consider an experiment with a sample space Ω and a partition into two disjoint events B_1 and B_2. We view these events as *hypotheses* regarding the experiment. In this context, the probabilities $P(B_1)$ and $P(B_2)$ are called *prior probabilities* (prior to the experiment). We assume them to be positive.

In our example, $B_1 = \{$*Mr. K has the disease*$\}$ and $B_2 = \{$*Mr. K does not have the disease*$\}$. Prior to the test, we should set $P(B_1) = 0.1$ and $P(B_2) = 0.9$.

Next, consider an event A which we view as a result of the experiment. In our example, $A = \{$*the test is positive*$\}$. We are interested in $P(B_1\,|\,A)$, the probability that Mr. K has the disease given that the test has shown a positive result.

The probabilities $P(B_1 \,|\, A)$ and $P(B_2 \,|\, A)$ are called *posterior* (after the experiment). By the multiplication rule and the formula for total probability,

$$P(B_1 \,|\, A) = \frac{P(AB_1)}{P(A)} = \frac{P(A \,|\, B_1)P(B_1)}{P(A)} \tag{2.6.1}$$

$$= \frac{P(A \,|\, B_1)P(B_1)}{P(A \,|\, B_1)P(B_1) + P(A \,|\, B_2)P(B_2)}, \tag{2.6.2}$$

and a similar formula is true for $P(B_2 \,|\, A)$.

The presentation (2.6.2) known as *Bayes' formula* after English mathematician Thomas Bayes, can be applied to a great many of problems. If we already know $P(A)$, it suffices to use (2.6.1).

EXAMPLE 1. Let us return to our Mr. K. We know that the test is positive only for 85% of people having the disease, so $P(A \,|\, B_1) = 0.85$. On the other hand, $P(A \,|\, B_2) = 0.05$. Hence, by (2.6.2),

$$P(B_1 \,|\, A) = \frac{0.85 \cdot 0.1}{0.85 \cdot 0.1 + 0.05 \cdot 0.9} \approx 0.654,$$

which is much larger than the prior probability 0.1. □

The Bayes formula for many events $B_1, ..., B_n$ is similar: for any $i = 1, ..., n$,

$$P(B_i \,|\, A) = \frac{P(A \,|\, B_i)P(B_i)}{\sum_{k=1}^{n} P(A \,|\, B_k)P(B_k)}. \tag{2.6.3}$$

The next model is of a general nature.

2.7 The two-armed bandit problem

Sometimes, a slot machine is called an one-armed bandit since it has one handle (nowadays it is usually a button). A two-armed bandit is a slot machine with two handles (which do not exist in reality). The chances of winning for one handle is larger than for the other, but we do not know for which. Trying the handles, we may estimate the chances for each handle to be luckier of the two. Such a model may be used in many realistic applied problems; for example, in allocating resources among competing projects whose quality is being compared during consecutive trials.

Let p_1 be the probability of winning for the better handle, and p_2 be the same probability for the other handle; $p_1 > p_2$. We face two hypotheses: $B_1 = \{\textit{the best handle is the left one}\}$ and $B_2 = \{\textit{the best handle is the right one}\}$. Suppose that at the beginning, the hypotheses are equally likely to be true, so the prior probabilities $P(B_1) = P(B_2) = \frac{1}{2}$.

Optimal behavior in this problem may be defined in different ways. We may try to figure out which hypothesis is true, or maximize the number of wins. (Though we should specify what the latter means since the number of wins is random.) Not going deeper into it, we specify a strategy which seems quite natural, and indeed fits a number of reasonable criteria in a certain sense.

Since the prior probabilities are equal to each other, at the first moment, it makes no difference which handle to pull. So, we pull whichever we like. Let A be the event corresponding to the result of the first play. Then we compute the posterior probabilities $P(B_1 \,|\, A)$ and

$P(B_2 \mid A)$, and choose the handle corresponding to the largest of the posterior probabilities. After that, we pull the handle chosen, and recalculate the posterior probabilities $P(B_1 \mid A)$ and $P(B_2 \mid A)$, *where now A corresponds to the results of the two plays.* After that, we again choose the handle for which the posterior probability is larger, and continue in the same fashion.

Certainly, since $P(B_1 \mid A) + P(B_2 \mid A) = 1$, it suffices to compute $P(B_1 \mid A)$, and choose the left handle if $P(B_1 \mid A) > \frac{1}{2}$. If the posterior probabilities are the same, it does not matter which handle to pull.

Assume that, at the first time, the player pulled the left handle and won. Since $P(A \mid B_1) = p_1$ and $P(A \mid B_2) = p_2$, by (2.6.2),

$$P(B_1 \mid A) = \frac{p_1 \cdot \frac{1}{2}}{p_1 \cdot \frac{1}{2} + p_2 \cdot \frac{1}{2}} = \frac{p_1}{p_1 + p_2}.$$

Since $p_1 > p_2$, the probability $P(B_1 \mid A) > \frac{1}{2}$, and then the player should stick to the left handle, which is quite natural: he won.

Suppose that at the second moment, the player lost. Then $P(A \mid B_i) = p_i(1 - p_i)$, and

$$P(B_1 \mid A) = \frac{p_1(1-p_1) \cdot \frac{1}{2}}{p_1(1-p_1) \cdot \frac{1}{2} + p_2(1-p_2) \cdot \frac{1}{2}} = \frac{p_1(1-p_1)}{p_1(1-p_1) + p_2(1-p_2)}.$$

The last expression is less than $\frac{1}{2}$ (the player switches to the other handle) if and only if $p_1(1-p_1) < p_2(1-p_2)$. This is equivalent to $(p_1 - p_2)(p_1 + p_2 - 1) > 0$. Since $p_1 > p_2$, this is equivalent to $p_1 + p_2 > 1$.

This result should also seem natural. For example, assume p_1 to be large, say, equal to 0.99, and p_2 to be moderate, say, 0.4. Then $p_1 + p_2 > 1$. On the other hand, if the player picked the wrong handle, he still can win: the probability 0.4 is not small. However, if he lost at the second time, it would be unlikely that he is dealing with the best handle: the probability of this event for the best handle is $0.99 \cdot 0.01 = 0.0099$. So, it is indeed reasonable to switch handles.

We continue this discussion in Exercise 41.

Route 1 ⇒ page 50

2

3 THE BOREL–CANTELLI THEOREM

This section is concerned with infinite sequences of events.

Consider a sequence of events $A_1, A_2, \ldots$ and the event A that an infinite number of the events $A_1, A_2, \ldots$ will occur. People also say that in this case, the events A_n occur infinitely often. Our object of interest is $P(A)$.

EXAMPLE 1. Consider a sequence of independent trials with a probability of success $p > 0$. Let A_n be the event that the nth trial is successful. What is the probability that if we run these trials an infinite number of times, then there will be an infinite number of successes? In other words, what is the probability that there will be no final success after which there will be no further successful trials?

Our intuition tells us that this probability should be equal to one, and this is true. The probability that there will be no successes during n trials is $(1-p)^n \to 0$ as $n \to \infty$. Hence, the probability that a success will never happen is zero. However, since the trials are independent, once a success has occurred, the process starts over, and the second success also occurs with probability one, and so on: the sequence of successes is infinite.

Now, suppose that the probability of success on the nth trial depends on n, and denote this probability by p_n. If $p_n = 0$ for n greater than some n_0, then with probability one, the number of successes will not exceed n_0. In this case, only a finite number of events A_n will occur.

Then, one may conjecture that if all p_n are positive but $p_n \to 0$ as $n \to \infty$ sufficiently fast, then the number of successes can be finite. In other words, at some random moment, a last success will occur, and there will be no further successes. □

Let us return to the general scheme.

Theorem 1 *(Borel-Cantelli). Set $p_n = P(A_n)$.*

1. *If the series $\sum_{n=1}^{\infty} p_n < \infty$, then $P(A) = 0$.*

2. *If the series $\sum_{n=1}^{\infty} p_n = \infty$, and the events $A_1, A_2, \ldots$ are independent, then $P(A) = 1$.*

Corollary 2 *If $A_1, A_2, \ldots$ are independent, then $P(A)$ is either 0 or 1.*

EXAMPLE 1 revisited. In this example, p_n is the probability of success in the nth trial. If $p_n = p > 0$, the series $\sum_{n=1}^{\infty} p_n$ clearly diverges, and the theorem confirms that the number of successes in the infinite sequence of trials is infinite. The same is true if $p_n = 1/n$ because the harmonic series $\sum_{n=0}^{\infty} \frac{1}{n} = \infty$. However, if $p_n = 1/n^2$, the series $\sum_{n=1}^{\infty} p_n$ converges, and hence $P(A) = 0$. This means that with probability one, the number of successes is finite.

EXAMPLE 2. (The idea of this example may be found in [28]). Three games of heads and tails are played simultaneously and independently, or in other words, three regular coins are being tossed. For a particular game, we define a tie at an even moment $2n$ if the numbers of heads and tails at the end of the $2n$ tosses are the same, that is, are equal to n. Let A_n be the event that at the moment $2n$, all three games are tied. We show that it may happen only finitely many times.

For one game, in accordance with (1.2.2), the probability of a tie at a moment $2n$ is equal to $u_{2n} = \binom{2n}{n}\left(\frac{1}{2}\right)^n\left(\frac{1}{2}\right)^n = \frac{(2n)!}{n!n!}2^{-2n}$. We apply Stirling's formula

$$n! \sim \sqrt{2\pi n}\, n^n e^{-n}. \tag{3.1}$$

(As usual, the notation $a_n \sim b_n$ means that $\frac{a_n}{b_n} \to 1$. A proof of this formula may be found in many books on Advanced Calculus. Less traditional proofs are contained, e.g., in [13,

II.9] and in [39, 4.3].) Substituting it in the formula for u_{2n}, we have

$$u_{2n} \sim \frac{\sqrt{2\pi 2n}(2n)^{2n}e^{-2n}}{(\sqrt{2\pi n}(n)^n e^{-n})^2} \cdot 2^{-2n} = \frac{1}{\sqrt{\pi n}}. \tag{3.2}$$

Hence,

$$P(A_n) = u_{2n}^3 \sim \frac{1}{\pi^{3/2} n^{3/2}}.$$

Consequently, the series $\sum_{n=1}^{\infty} P(A_n)$ converges. See also Exercise 44. □

Proof of Theorem 1. Let $B_n = \bigcup_{k=n}^{\infty} A_k$. The sequence B_n is decreasing: $B_1 \supseteq B_2 \supseteq \ldots$. Let us realize that

$$A = \bigcap_{n=1}^{\infty} B_n = \bigcap_{n=1}^{\infty} \bigcup_{k=n}^{\infty} A_k. \tag{3.3}$$

Indeed, if an elementary outcome ω belongs to an infinite number of A_k's, then $\omega \in B_n$ for *any* n. Then ω belongs to the intersection $\bigcap_{n=1}^{\infty} B_n$.

Conversely, suppose that ω belongs only to a finite number of A_k's. Then there exists n_0 such that $\omega \notin B_n$ for all $n > n_0$. Then $\omega \notin \bigcap_{n=1}^{\infty} B_n$.

By (3.3) and the continuity property of probability measures (see Section 1.2.6),

$$P(A) = P(\lim_{n\to\infty} B_n) = \lim_{n\to\infty} P(B_n) = \lim_{n\to\infty} P\left(\bigcup_{k=n}^{\infty} A_k\right). \tag{3.4}$$

By (3.4),

$$P(A) \le \lim_{n\to\infty} \sum_{k=n}^{\infty} P(A_k) = \lim_{n\to\infty} \sum_{k=n}^{\infty} p_k = 0$$

if the series $\sum_{k=1}^{\infty} p_k$ converges. Since any probability is non-negative, this implies $P(A) = 0$.

Now, suppose that the A_k's are independent events and the series $\sum_{k=1}^{\infty} p_k$ diverges. For $m > n$, using De Morgan's law (Proposition 1.1) and the independence assumption, we have

$$P\left(\bigcup_{k=n}^{m} A_k\right) = 1 - P\left(\left(\bigcup_{k=n}^{m} A_k\right)^c\right) = 1 - P\left(\bigcap_{k=n}^{m} A_k^c\right) = 1 - \prod_{k=n}^{m} P(A_k^c) = 1 - \prod_{k=n}^{m}(1 - p_k).$$

From this, using the inequality $1 + x \le e^x$, we get

$$P\left(\bigcup_{k=n}^{m} A_k\right) \ge 1 - \prod_{k=n}^{m} \exp\{-p_k\} = 1 - \exp\left\{-\sum_{k=n}^{m} p_k\right\} \to 1$$

as $m \to \infty$ because the series $\sum_{k=1}^{\infty} p_k$ diverges and in limit we have the expression of the form $1 - e^{-\infty}$. Thus, for all n,

$$P(B_n) = P\left(\bigcup_{k=n}^{\infty} A_k\right) = \lim_{m\to\infty} P\left(\bigcup_{k=n}^{m} A_k\right) \ge 1.$$

Since any probability is not larger than one, the above probability must equal exactly one. Thus, $P(B_n) = 1$ for any n. Then, by (3.4), $P(A) = \lim_{n\to\infty} P(B_n) = 1$. ■

4 EXERCISES

1. In the city of Pleasant Corner, 20% of houses have a pool, 60%—air conditioning. Suppose that for each citizen of Pleasant Corner, the decisions whether to have a pool and whether to have air conditioning are independent. Find the probability that a citizen chosen at random has either a pool or air conditioning (or both).

2. Prove that, if A and B are independent, the same is true for A^c and B, A^c and B^c.

3. (a) Consider two singletons, $[\omega_1]$ and $[\omega_2]$; that is, events containing only one outcome each. Are they independent? (*Hint*: An answer that seems natural, may be not true in one special case. The same concerns the next question.)

   (b) Are disjoint events independent?

   (c) Let $P(A_1) = P(A_2) = 1$. Are A_1, A_2 independent? (*Advice*: Revisit Exercise **1**-24.)

4. Two cards are selected at random from a deck of 52. Guess whether the events $A_1 = \{$*at least one card is a king*$\}$, $A_2 = \{$*at least one card is an ace*$\}$ are independent. Justify your guess rigorously.

5. Two dice are rolled. Let X and Y be the numbers appeared; and events
   $A = \{X \text{ is even}\}$, $B = \{X+Y \text{ is even}\}$,
   $\overline{A} = \{X \text{ is divided by } 4\}$, $\overline{B} = \{X+Y \text{ is divided by } 4\}$.
   Check for independence the pairs A, B and $\overline{A}, \overline{B}$.

6. (a) Consider the tree in Fig. 4b, Section 2.2. Are the events $A = \{$*on the second day, the price is larger than 9*$\}$, $B = \{$*on the second day, the price is smaller than 12*$\}$ independent?

   (b) In general, if $\Omega = \{\omega_1, \omega_2, \omega_3, \omega_4\}$ with given elementary probabilities, in which case are the events $A = \{\omega_1, \omega_2, \omega_3\}$ and $B = \{\omega_2, \omega_3, \omega_4\}$ independent?

7. (a) Consider two trials. We do *not* impose any independence condition but assume that all outcomes are equally likely. Our intuition tells us that in this case, trials are independent in the sense that, for example, the events $A_1 = \{$ *the first trial is successful*$\}$ and $A_2 = \{$ *the second trial is successful*$\}$ are independent. Proceeding from the definition of independence, show that this is indeed true.

   (b) Do the same for n trials.

8. Two dice were rolled. Let $A_1 = \{$*the first die rolled an even number*$\}$, $A_2 = \{$*the second die rolled an odd number*$\}$, $A_3 = \{$*the sum of the results is odd*$\}$. Show that these events are pairwise independent but are mutually dependent.

9. Make sure that definition (1.1.4) for $n = 3$ indeed leads to the definition of mutual independence for three events.

10. A system consists of components configured as shown in Fig. 6a; p_i is the probability that component i works; the components function independently. Find the probability that a signal will go through.

11. To be hired by a company, an applicant should pass two tests. The first test contains 10 questions, and for each question, the probability of giving a correct answer is p_1. The second test contains 20 questions, and for each question of this test, the probability of a correct answer is p_2. Whether an answer is correct does not depend on the answers to other questions. To pass each test one should answer correctly at least a half of questions. (a) Find the probability that the applicant will pass both tests. (*Hint*: The formula you should get may contain the symbol $\sum$.) (b) Arrange a file in Excel for computing the answer for given p_1 and p_2.

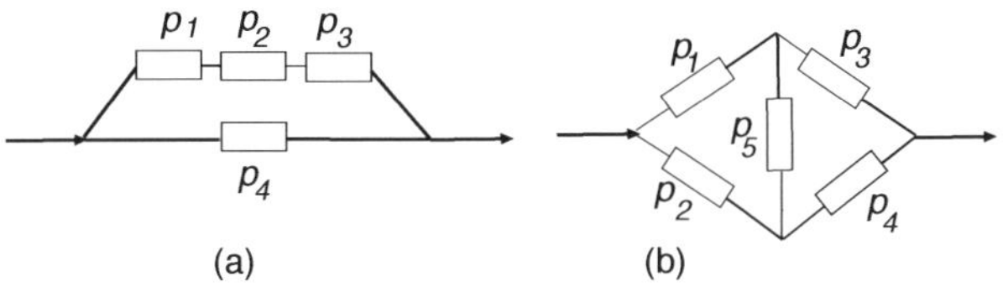

(a) (b)

FIGURE 6.

12. Derive (**1**.3.4.5) from (1.2.2).

13. 9% of people who are given a particular drug experience a side effect. Find the probability that at least two of fifteen people selected at random will have side effects.

14. Four dice are rolled five times. Write an expression for the probability that exactly three times all four dice will show six.

15. We toss a coin 100 times and are interested in the probability that there will be at most 50 heads. Explain why this probability is larger than 0.5, while if you toss 101 times, it is exactly 0.5. Compute these probabilities using Excel or another software. (*Advice*: Look over Exercise 1.43.)

16. In a university, 55% of students are females. Consider the probability that in a class of 100 students there will be at most 55 females. Do you expect that this probability is 0.55? Check your guess using software. (*Comments*: The fact that the "probability of a female" is 0.55, means that *on the average* there will be 55 females in the class. We see that this does not mean that the probability of having 55 females or less should be 0.55. We discuss it in detail in Section **3**.4.1, and for now, let us just keep it in mind.)

17. Explain why for a fixed n, the events $B_{k,n}$ in Section 1.2.1 are disjoint, and why the total sum of all probabilities $P(B_{k,n})$ must be one. Give a heuristic explanation and show it rigorously using the binomial formula (**1**.3.4.7).

18. You are rolling a die. One face of the die is painted green, another red, and the rest of the faces are black. Find the probabilities of the following events.

    (a) In the first three rolls, the die lands with black faces up, and in the next two with green faces up.

    (b) In five rolls, the die shows a black face exactly 3 times, and a green face two times. Realize the difference between this question and the previous.

    (c) The die will show a black face for the first time in the sixth roll.

    (d) The die will show a black face after the fifth roll. Realize the difference between this question and the previous.

    (e) There will be either green or red faces in the first five rolls.

    (f) The first run of one color of a length of 3 will happen in the fifth roll. (*Hint*: This means, in particular, that there will be a change of color in the sixth roll.)

19.* You roll a regular die six times. Write formulas for the probabilities of the following events.

    (a) The first three rolls show "six", the fourth and fifth "five", and the sixth "four".

    (b) The die will roll "six" three times, "five" two times, and "four" one time.

20. For a family having two children, assuming that all outcomes are equally likely, find the probability that there is at least one boy given that there is at least one girl.

21. Two dice are rolled. Given that the sum is divided by 3, find the probability of two threes.

22. In a game of bridge, you did not get spades. Write a formula for the probability that your partner does not have spades too. (Bridge is played with a standard 52-card deck by two pairs of partners; each player gets 13 cards.)

23. Find $P(A_1 \,|\, A_2)$ in Example 1.1-6 for both cases. Try to minimize calculations.
    Now, let $A_1 = \{x_1 \geq 0\}$, $A_2 = \{x_2 \geq 0\}$. Figure out whether A_1 and A_2 are independent for both cases.

24. Show that if $A \subseteq B$, then $P(A\,|\,B) = P(A)/P(B)$. Find $P(A\,|\,B)$ for $A \supseteq B$, and for disjoint A and B. Explain why the last two answers are obvious.

25. Consider the binomial tree in Section 1.2.2. Suppose that the probabilities of "moving up and down" on the first day equal $1/2$, but on the second day, the price moves up with a probability of p_1 if on the first day it moved up, and with a probability of p_2 if on the first day it dropped. (a) Find the probability that on the second day the price will equal ud. (b) *Given* that the price took on the value ud, find the probability that it moved up on the first day.

26. For the tree in Fig. 4b, Section 2.2, find the probability that the price increased on the first day given that it did not exceed 14 on the second day.

In Problems 27–37 below, the reader should be ready that some of them may concern either the formula for total probability, or the Bayes formula, or both formulas.

27. In a multiple choice exam, a question has two answers, only one of which is correct. Suppose 80% of students know the answer, and those who do not know choose an answer at random. (a) What is the probability that a student chosen at random will answer the question correctly? (b) Suppose a student has answered the question correctly. What is the probability that she/he was just guessing?

28. In a region, a hiker may come across a rattlesnake. For a mountain area and a randomly chosen day, the probability of this event is 0.02; for valleys, it is 0.01. Joan, when choosing one from these two areas, does it at random. (a) What is the probability that Joan will see a rattlesnake on a randomly chosen day? (b) Suppose that Joan has seen a rattlesnake. We are interested in the probability that it happened in the mountains. Explain why the answer $2/3$ sounds very plausible. Provide rigorous calculations.

29. (*Pòlya's scheme.*) An urn has b black and r red balls. A ball is drawn and put back into the urn together with c balls of the same color. (So, now there are $b+r+c$ balls in the urn.) After that again a ball is drawn.
    (a) Do you expect that the probability that the second ball is red, depends on c?
    (b) Prove that this probability is the same as for the first ball: $r/(b+r)$.
    (c) Consider the probability that the first ball is red given that the second is red. Do you expect that this probability depends on c? Find it.

30. A systems consist of components configured as shown in Fig. 6b; p_i is the probability that component i works; the components function independently. Find the probability that a signal will go through.[1] (*Advice*: Condition on the fifth component and apply the formula for total probability.)

[1]The nice idea to consider a circuit of the second type may be found in [39].

31. Two friends are fishing on a river and their time is limited. For each, the probability of catching a fish is $p < 1$. The probability that somebody will catch two fish is zero and the results of fishing for each are independent. The two friends agree to the following. If each catches a fish, each will bring one home. If only one is lucky, they will toss a coin to determine who will come home with the only fish. Does their agreement increase the likelihood for each to bring home a fish, or should the friends just arrange a joint dinner with what they caught? First guess the answer, then verify it by computing the probability of bringing home a fish for each friend separately

32. A transmitter is receiving a digital text, i.e., a sequence of digits. Suppose that each digit in the sequence is equally likely to be any number from 0 to 9, and the digits are appearing independently from each other. Each digit is transmitted correctly with probability p; and with probability $q = 1 - p$, it is replaced by another digit with equal probabilities due to noise. Suppose an output sequence is "$0, 1$." What is the probability that, as a matter of fact, it was "$0, 0$"? (*Hint*: The problem is on the conditional probability $P(B|A)$ and the formula for total probability. Realize what is A, and what is B.)

33. Consider the Bayes framework of Section 2.6. Suppose that an event A concerns the result of the experiment, and we know that $P(AB_1) > P(AB_2)$. Given A, which hypothesis is more probable: B_1 or B_2? Justify the answer.

34. On the average, out of 100 people who are taking a professional exam, twelve took a special tutoring course that increases the percentage of passing the exam from 70% to 90%. Ms. K has passed the exam. What is the probability that she had taken the tutoring course?

35. There are two groups of students. The first consists of ten males and eleven females; the second consists of five males and six females. Two representatives, one from each group, were chosen at random. After that, one more student was chosen for another job at random from the whole group of the thirty students remaining.

    (a) Find the probability that the last student chosen is a female.

    (b) Given that the last student chosen is a female, find the probability that the first two students chosen are also females.

36. A country is divided into three provinces with 40%, 35%, and 25% of the total population. Due to climate conditions, in the first province, 55% of the people are interested in gardening, while in the other provinces this percentage is equal to 30% and 20%, respectively. A representative of a company selling books on gardening receives a call. What is the probability that it comes from the first province? Compare your answer with the corresponding prior probability. Explain the result of the comparison.

37. A company is advertising a product it produces. The management places an advertisement in newspapers, and a fancy advertisement on TV. The following average figures have been observed. Among 50 potential buyers, one sees an ad in a newspaper and only there, and five see the ad on TV (and perhaps in a newspaper also). Among 10 customers who have seen the ad only in a newspaper, there is only one who actually purchases the product, and among 10 who have seen the TV ad, there are three who purchase the product. At last, among 30 potential buyers who did not see any ad, only one will purchase the product.

    Let Mr. Smith be a potential customer.

    (a) What is the probability that Mr. Smith will buy the product?

    (b) What is the probability that Mr. Smith saw an ad and bought the product?

(c) Given that Mr. Smith has bought the product, do you expect the probability that he has seen the ad on TV to be equal to, or less than, or larger than 0.1? Find this probability.

38. In Example 2.4-1, the probability $p_9 \approx 0.66$ and is larger than 0.4, the probability to "move up." Explain this from a heuristic point of view.

39. Give a heuristic explanation why the ruin probability in (2.4.5) must be smaller than in (2.4.7). Double check it using simple algebra.

40. Problems below concern the model of Section 2.4 with $a = \infty$.

(a) For a separate moment of time, the odds are 3 to 1 in favor of gains. Which initial capital should company start with for the ruin probability to be less than 0.01?

(b) Show that $q_1 \geq q$ and explain, from a heuristic point of view, that it must be expected.

(c) Mark starts a risky business in which, on each day, he will make a profit of $500 with a probability of $p = 0.9$, but may suffer a loss of $500 with a probability of $q = 0.1$. Mark takes a risk starting with zero initial capital, and hopes that he will not lose the first time and will be always able to cover losses in the future from the capital he will earn. Find the probability that Mark's hope will come true. For which p (perhaps not equal to 0.9) is this probability larger then 0.95?

41. Suppose that in the situation of the two-armed bandit problem in Section 2.7, you follow the strategy suggested there.

(a) Show that (i) if you lose at the first time, then you will switch; (ii) if you win at the first time, and $p_1 < \frac{1}{2}$, then you will never switch at the second time.

(b) Let $p_1 = 0.3$, $p_2 = 0.2$. Suppose you won at the first time, and after that began to lose in each play. At which moment will you switch?

(c) For the previous problem, arrange a file in Excel, or write a program using another software, allowing you to find the moment of switch for any p_1, p_2.

42.* For a sequence of events A_n such that $P(A_n) \to 0$, should $P(A_n \text{ occur infinitely often}) = 0$?

43.* Consider the situation of Example 3-1. Let A_n be the event that the nth trial is successful, B_n be the event that the first success occurred in the nth trial, and A be defined as in Section 3.

(a) Find $P(B_n)$ in terms of $p_n = P(A_n)$.

(b) Suppose $p_n = 1 - e^{-1/n}$. What is the order of the convergence $p_n \to 0$? What is $P(A)$? Write an expression for $P(B_n)$. Estimate the order of this probability for large n recalling that $\sum_{k=1}^{n} \frac{1}{k} - \ln n \to \gamma$, where the constant γ (called the Euler-Mascheroni constant) approximately equals 0.58.

(c) Let $p_n = 1 - e^{-1/n^2}$. What is the order of the convergence $p_n \to 0$? Find $P(A)$? Find the probability that there will be no successes at all, using the fact that $\sum_{n=1}^{\infty} \frac{1}{n^2} = \frac{\pi^2}{6}$.

44.* Regarding Example 3-2, can we apply the Borel-Cantelli theorem for one or two games of heads and tails?

45.* Consider an infinite sequence of trials with probabilities of success perhaps depending on the numbers of the trials. (a) Can the number of successes be finite with probability 0.5 if the trials are independent? Justify the answer. (b) Give an example of a sequence of dependent trials for which the same probability is 0.5. (*Advice*: For example, you may assume that the chances for successes in the trials following the first, depend on the result of the first trial.)

Chapter 3

Discrete Random Variables

1,2

1 RANDOM VARIABLES AND VECTORS

1.1 Random variables and their distributions

Loosely put, a random variable (r.v.) is a variable whose values are determined "by chance." More precisely, the value of a r.v. depends on which elementary outcome has occurred when the experiment under consideration has been performed.

For example, when tossing a coin twice, we may be interested merely in the number of heads, no matter in which order they appeared. Possible results are given in the table below.

Outcomes	The number of heads
$\omega_1 = \{HH\}$	2
$\omega_2 = \{HT\}$	1
$\omega_3 = \{TH\}$	1
$\omega_4 = \{TT\}$	0

So, while there are four outcomes, the number of heads may take on three values: either 0, or 1, or 2. We also see that the number of heads is completely determined by the outcome occurred.

In the general case, when considering an experiment with a sample space $\Omega = \{\omega\}$, *we define a random variable as a function* $X(\omega)$ *on the space* Ω.

EXAMPLE 1. In the case of rolling two dice, the sample space $\Omega = \{\omega\}$ consists of 36 outcomes. The r.v. $X(\omega)$, the sum of the numbers on the dice, may assume all integer values from 2 to 12.

EXAMPLE 2. We choose at random a point from the disk $\{(x,y) : x^2+y^2 \leq 1\}$; see Example **1**.1.1-7. An elementary outcome ω may be identified with a point (x,y) from the disk. The distance from the point chosen to the center of the disk is the r.v. $X(\omega) = \sqrt{x^2+y^2}$ which may take on any value from the interval $[0,1]$. □

As a rule, to simplify the notation, we will write X for $X(\omega)$.

A r.v. is said to be *discrete* if it can take on only a finite or countable number of possible values. If it is not the case, we call the r.v. *non-discrete*. Another term in use is a *continuous r.v.*, but it is slightly misleading; see Example 6 below.

EXAMPLE 3. The number of successes in n trials is a discrete r.v. assuming $n+1$ values: $0, 1, \ldots, n$.

EXAMPLE 4. The number of "stars falling from the sky" during a period of time is a discrete r.v. taking on values $0, 1, 2, \ldots$. Certainly, this number cannot be arbitrarily large, but when modeling this phenomenon, it makes sense to view the set of all possible values as that of all non-negative integers: $0, 1, 2, \ldots$. This set is infinite but countable.

EXAMPLE 5. As was told, the r.v. in Example 2 may assume any value from the interval $[0,1]$. We cannot count all points from an interval, and hence this r.v. is non-discrete.

EXAMPLE 6. Michael usually goes to school on weekdays, but sometimes he misses it. The trip to school takes from twenty to twenty-five minutes. Let X be the (random) time it takes to come to school on a randomly chosen day. The set of all possible values of X consists of the number 0 and the interval $[20,25]$. So, the r.v. X is non-discrete but because 0 is separated from the interval $[20,25]$, it is not reasonable to call X continuous. □

In this chapter, we consider discrete r.v.'s. Let a (discrete) r.v. $X = X(\omega)$ take on values $x_1, x_2, \ldots$. Suppose that the probability measure $P(A)$ on sets from the sample space Ω is known. Then we can—at least, theoretically—compute the probabilities of the events $\{X = x_i\}$ for all i. Set $f_i = P(X = x_i)$.

Clearly, the total sum of all probabilities f_i,

$$\sum_i f_i = 1, \tag{1.1.1}$$

where the summation is over all i. The set of all values along with the respective probabilities:

$$\begin{matrix} x_1 & x_2 & x_3 & \ldots \\ f_1 & f_2 & f_3 & \ldots \end{matrix} \tag{1.1.2}$$

specifies the *probability distribution of the r.v. X.*

Another way to describe the same distribution is to consider the function $p(x) = P(X = x)$ for all x's. For $p(x)$ so defined, $p(x_i) = f_i$ for all i, and $p(x) = 0$ for any x that is not equal to some x_i. The function $p(x)$ is usually called a *probability mass function.* (We may view (1.1.2) as if we had distributed unit mass, see (1.1.1), in a way that mass f_i is located at point x_i.)

EXAMPLE 7. Consider tossing a fair coin two times. Suppose that on each toss that comes up heads, we win \$1, and if a toss comes up tails we lose \$1. Let X be the total winning. Clearly, X takes on three values: $-2, 0, 2$ with probabilities $1/4$ (the outcome TT), $1/2$ (the outcomes HT and TH), and $1/4$ (the outcome HH), respectively. Hence, we can represent the distribution of X as the array

$$\begin{matrix} -2 & 0 & 2 \\ 1/4 & 1/2 & 1/4. \end{matrix}$$

EXAMPLE 8. A special insurance plan pays \$150 to passengers of an airline in the case of a flight delay larger than 3 hours. If a delay exceeds 12 hours, the plan pays an additional \$300.

Suppose that 95% of flights run without a delay larger than 3 hours. For the *rest* of flights, 90% of delays do not exceed 12 hours, and 10% do.

To find the distribution of the r.v. X of the payment to a separate insured, we observe that the insurance pays nothing with a probability of 0.95, pays \$150 with a probability of

$0.05 \cdot 0.9 = 0.045$, and pays $150+300 = 450$ dollars with a probability of $0.05 \cdot 0.1 = 0.005$. Hence, the distribution of X may be represented by the array

$$\begin{matrix} 0 & 150 & 450 \\ 0.95 & 0.045 & 0.005. \end{matrix} \quad \square$$

In the next examples, we consider only probabilities f_i's not providing arrays similar to those above.

If a r.v. X takes on values $x_1, x_2, \ldots$, we will often write it in short as $X = x_1, x_2, \ldots$.

EXAMPLE 9 (*The Poisson distribution*). Let $X = 0, 1, 2, \ldots$. Frequently, such a r.v. is associated with the number of "arrivals at a system." For example, it may be the number of e-mail messages received by a web server, or the number of falling stars from Example 4, or the number of spelling mistakes in a randomly chosen page. Assume that

$$f_i = P(X = i) = e^{-\lambda} \frac{\lambda^i}{i!}, \quad i = 0, 1, \ldots, \tag{1.1.3}$$

where λ is a positive parameter. We will see in Section 2.1 that λ may be interpreted as the mean value of arrivals.

For now, we introduce this distribution formally. In Section 4.5 and further, we will see why this distribution plays an extremely important role in theory and applications.

To check the validity of (1.1.1), we use the Taylor expansion of the exponential function (the Appendix , (2.2.4)):

$$\sum_{i=0}^{\infty} f_i = \sum_{i=0}^{\infty} e^{-\lambda} \frac{\lambda^i}{i!} = e^{-\lambda} \sum_{i=0}^{\infty} \frac{\lambda^i}{i!} = e^{-\lambda} \cdot e^{\lambda} = 1.$$

EXAMPLE 10 (*The binomial distribution*). In this and the next example, we deal with a sequence of n independent trials with the probability of success in a separate trial equal to p. Let X be the number of successes. In accordance with what we got in Section **2**.1.2.1,

$$f_k - P(X = k) = \binom{n}{k} p^k q^{n-k}, \tag{1.1.4}$$

where $q = 1 - p$, and

$$\binom{n}{k} = \frac{n!}{k!(n-k)!} = \frac{n(n-1)\cdots(n-k+1)}{k!}.$$

Assume, for instance, that for a car owner, the probability that during a year she/he will face a car body damage (perhaps, minor), is 0.2. What is the probability that during ten years, there will be *at least* nine "lucky" years? Suppose that whether a damage occurs in a year does not depend on what happened in the other years. Then the number of lucky years is a binomial r.v. with parameters $n = 10$ and $p = 0.8$, and

$$P(X \geq 9) = \binom{10}{9} (0.8)^9 (0.2)^1 + \binom{10}{10} (0.8)^{10} (0.2)^0 = 10(0.8)^9 (0.2) + 1 \cdot (0.8)^{10} \cdot 1 \approx 0.38.$$

EXAMPLE 11 (*The geometric distribution*). For the same scheme, we consider the r.v. N, the number of the *first* successful trial. This number is equal to some k if the kth trial is successful (the probability is p), but the first $(k-1)$ are not. These events are independent, and the probability of the latter is q^{k-1}, where again $q = 1-p$. (See Section **2**.1.2.1 for detail.) So,

$$f_k = P(N = k) = pq^{k-1}. \tag{1.1.5}$$

On the way, note that

$$P(N > k) = q^k, \tag{1.1.6}$$

since this is the probability that there will be no success during the first k trials.

For instance, in the previous example with car damage, let us view damage as "success". Then, (1.1.6) will imply that the probability that there will be no damage during ten years is $(0.8)^{10} \approx 0.107$.

EXAMPLE 12 (*The hypergeometric distribution*). This distribution is described in Example **1**.3.4.1-3.

EXAMPLE 13 (*The doubling strategy*). A player plays a game of chance consisting in a sequence of independent bets with probability $p > 0$ of winning at each bet. Having started with a stake of one, the player plays until the first win, *doubling* her/his bet after each loss. After the first win, the player quits. In casinos, such a strategy is also called a "negative progression."

If the first success happens in the first play, the player gets one unit of money and quits. If he loses, he bets 2, and if the second try is successful, the player quits with the profit equal to $2-1=1$. If the first success happens at the third try, the player gets 4 (since he bet 4), and the profit is $4-1-2=1$. In general, if the first success happens at the $(k+1)$th try, the player's profit will be $2^k-(1+2+4+...+2^{k-1})=1$. (See the formula for a geometric series in the Appendix , (2.2.11).)

The r.v. N, the moment of the first win, has a geometric distribution, and $P(N > n) = q^n$, where $q = 1-p$. If $p > 0$, then $0 \le q < 1$, and $P(N > n) \to 0$ as $n \to \infty$; that is, the probability that a success will never happen is zero, so the game will come to the end with unit probability.

Thus, the doubling strategy allows a player to get one (unit of money equal to the first bet) with probability one, i.e., without any risk to lose money. It is worth noting, however, that this presupposes that the player should, at least theoretically, have an infinite initial capital.

Consider a particular game of roulette. Suppose that a player always bets on red and in the casino, the minimal bet is \$1 and the maximal bet is \$200. The player starts with a stake of \$1. What is the probability that the player will fail to use the doubling strategy?

At each play, the probability of success is $p = 9/19$ (there are 18 red, 18 black, and 2 green cells), and $q = 10/19$. The highest n for which 2^{n-1}, the bet at the nth play, does not exceed 200, is $n = 8$, and $P(N > 8) = (10/19)^8 \approx 0.006$.

If N exceeds 8, the player will not be able to apply the doubling strategy. Assume that he will quit in this case; that is, after the 8th play if it turns out to be non-successful. Then he will lose $1+2+...+2^7 = 2^8-1 = 255$ (dollars) but it may happen with a small probability of 0.006.

So, the random profit X equals 1 with probability 0.994, and -255 with probability 0.006. This is up to the player whether to play such a game.

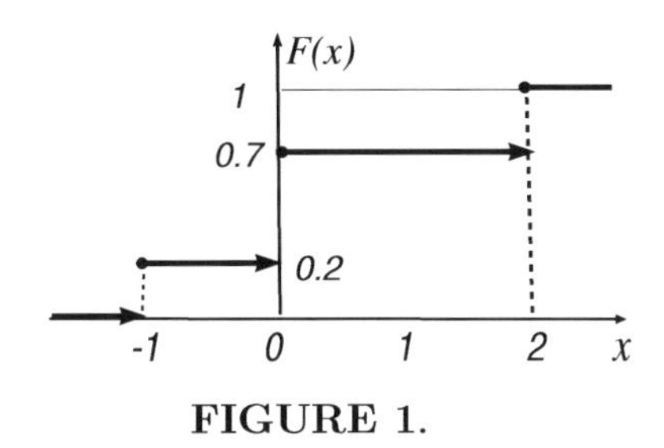

FIGURE 1.

We revisit this problem in Section 4.5.1. The reader who will choose to read Chapter 15 on martingales, will see that this "not-very-serious" problem may serve as a good illustration of an important theoretical issue (see Section **15**.2.1). □

Next, we introduce the notion of the *cumulative distribution function* (c.d.f.) of a r.v. X. By definition, it is the function

$$F(x) = P(X \leq x). \tag{1.1.7}$$

For a discrete r.v. X, to find $P(X \leq x)$, we should add up all f_i for which $x_i \leq x$; that is,

$$F(x) = P(X \leq x) = \sum_{i: x_i \leq x} f_i. \tag{1.1.8}$$

Thus, the c.d.f. is a step-function; it jumps at points x_i, and the size of the corresponding jump is f_i. We clarify it by

EXAMPLE 14. Let a r.v. X take on values -1, 0, and 2 with respective probabilities 0.2, 0.5, and 0.3.

For any $x < -1$, the c.d.f. $F(x) = P(X \leq x) = 0$ since X does not assume values less than -1; see Fig. 1.

For $x = -1$, we have $F(-1) = P(X \leq -1) = P(X = -1) = 0.2$, and this corresponds to the jump at point -1 in Fig. 1.

For $-1 \leq x < 0$, we still have $F(x) = P(X \leq x) = P(X = -1) = 0.2$; the function is constant between -1 and 2; see Fig. 1.

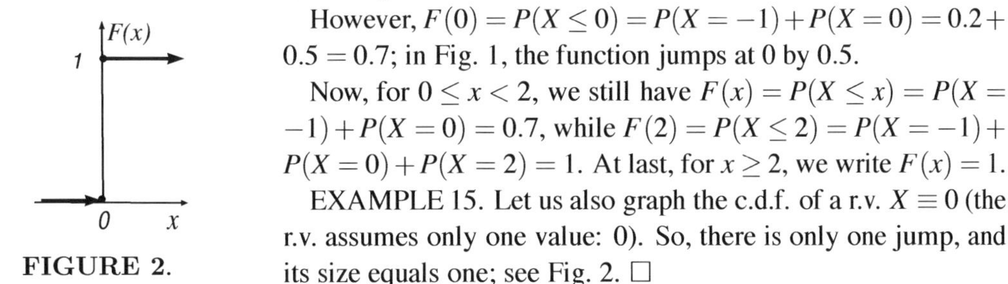

FIGURE 2.

However, $F(0) = P(X \leq 0) = P(X = -1) + P(X = 0) = 0.2 + 0.5 = 0.7$; in Fig. 1, the function jumps at 0 by 0.5.

Now, for $0 \leq x < 2$, we still have $F(x) = P(X \leq x) = P(X = -1) + P(X = 0) = 0.7$, while $F(2) = P(X \leq 2) = P(X = -1) + P(X = 0) + P(X = 2) = 1$. At last, for $x \geq 2$, we write $F(x) = 1$.

EXAMPLE 15. Let us also graph the c.d.f. of a r.v. $X \equiv 0$ (the r.v. assumes only one value: 0). So, there is only one jump, and its size equals one; see Fig. 2. □

Thus, at each point x_i, the c.d.f. $F(x)$ jumps up by the probability $f_i = P(X = x_i)$, and for points between neighbor x_i's, the c.d.f. is constant. In the general case, a typical graph looks as in Fig. 3.

From now on, when talking about a cumulative distribution function, we skip the adjective "cumulative" and use a shorter abbreviation d.f.

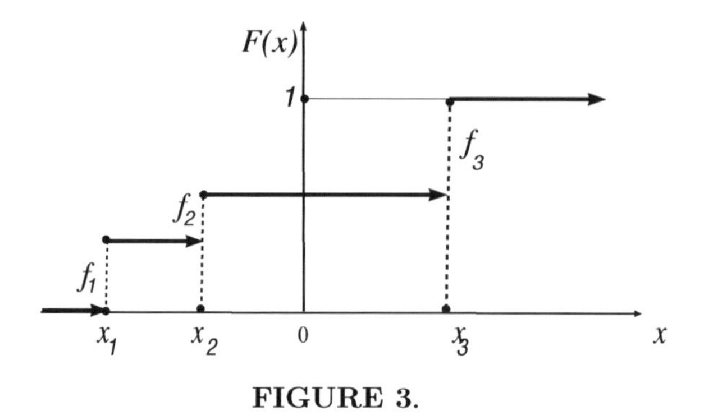

FIGURE 3.

1.2 Random vectors and joint distributions

1.2.1 Joint and marginal distributions

We call a vector $\mathbf{X} = (X_1, ..., X_k)$, where X_i's are r.v.'s, a k-dimensional *random vector* (r.vec.). For simplicity, we restrict ourselves to the case $k = 2$; so, $\mathbf{X} = (X_1, X_2)$. The results below may be easily carried over from the two dimensional to the multidimensional case.

Let the r.v. X_k, $k = 1,2$, takes on values $x_{k1}, x_{k2}, ...$ and $f_{ij} = P(X_1 = x_{1i}, X_2 = x_{2j})$. We say that the probabilities f_{ij} specify the *joint distribution* of (X_1, X_2), and call the f_{ij}'s *joint probabilities*. The definition is illustrated in Fig. 4.

Clearly, the total sum

$$\sum_i \sum_j f_{ij} = 1.$$

Probabilities $f_i^{(1)} = P(X_1 = x_{1i})$ and $f_j^{(2)} = P(X_2 = x_{2j})$ are called *marginal*. The collections of these probabilities: $f^{(1)} = \left(f_1^{(1)}, f_2^{(1)}, ...\right)$ and $f^{(2)} = \left(f_1^{(2)}, f_2^{(2)}, ...\right)$, specify the probability distributions of the r.v.'s X_1 and X_2 *separately*, and are called *marginal distributions*.

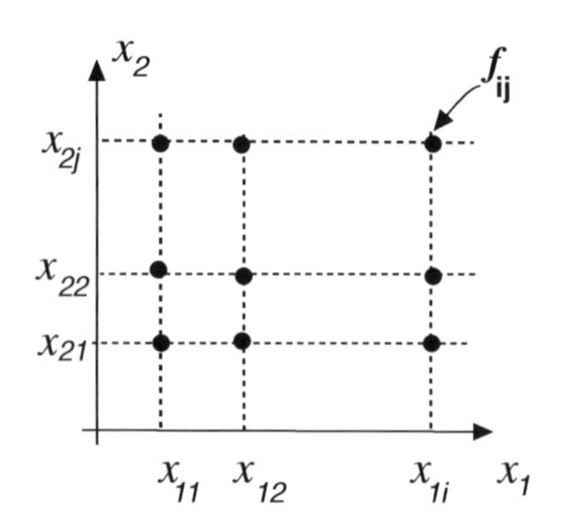

FIGURE 4. The bold-font points correspond to possible values of the vector **X**. The arrow shows to which point the probability f_{ij} corresponds.

If we know the joint distribution, we can find the marginal distributions. Indeed,

$$f_i^{(1)} = P(X_1 = x_{1i}) = \sum_j P(X_1 = x_{1i}, X_2 = x_{2j}) = \sum_j f_{ij}.$$

Appealing to Fig. 4, we may say that the probability $f_i^{(1)}$ may be found by adding up all probabilities corresponding to the points that lie in the vertical line with the first coordinate equal to x_{1i}.

Similarly, to find $f_j^{(2)}$, we add up all probabilities corresponding to the points that lie in the horizontal line with the second coordinate equal to x_{2j}. Thus,

$$f_i^{(1)} = \sum_j f_{ij}, \quad f_j^{(2)} = \sum_i f_{ij}. \tag{1.2.1}$$

It is worthwhile to emphasize that knowing only marginal probabilities, one cannot find joint probabilities. The latter probabilities reflect how the r.v.'s X_1 and X_2 depend on each other, and to find the joint distributions, one should know the structure of this dependency. Let us turn to examples.

EXAMPLE 1. A store has two working checkout counters, and X_1 and X_2 are the (random) numbers of customers at the first and second counter, respectively. Suppose that these numbers are not larger than three; the joint probabilities are given in the table below.

Values of X_1 $\Rightarrow$ / Values of X_2 $\Downarrow$	0	1	2	3
0	0.1	0.1	0	0
1	0.1	0.2	0.15	0
2	0	0.05	0.15	0.08
3	0	0	0.02	0.05

For example, f_{32}, the probability that there will be three customers at the first counter, and two at the second, is 0.08. The distribution is not symmetric: for example, f_{23}, the probability that there will be two customers at the first counter, and three at the second, is 0.02 and is smaller than f_{32}. It may reflect the fact that the first counter is closer to more popular departments. Some probabilities, say, f_{02}, equal zero, since if, for instance, there are no customers at the first counter, but there are two at the second, one customer from the latter will move to the first counter. (Actually, in reality, it may be not so: people are not always observant.)

How to find marginal probabilities? For instance, to find $f_2^{(1)}$, the probability that there will be two customers at the first counter, we should add up all probabilities in the column that corresponds to $X_1 = 2$, which amounts to $0 + 0.15 + 0.15 + 0.02 = 0.32$. Providing all similar calculations, we get the following modified table including marginal probabilities. (The sum of all marginal probabilities equals one, which is reflected in the very right bottom cell.)

Values of X_1 ⇒ / Values of X_2 ⇓	0	1	2	3	Marginal probabilities for X_2
0	0.1	0.1	0	0	0.2
1	0.1	0.2	0.15	0	0.45
2	0	0.05	0.15	0.08	0.28
3	0	0	0.02	0.05	0.07
Marginal probabilities for X_1	0.2	0.35	0.32	0.13	1

EXAMPLE 2. Assume that the children in a random pair of twins are of the same sex with probability 0.64, and that the probability that a boy will be born is 0.51. (Both numbers are close to real; for biological reasons, the probability of a boy to be born is slightly larger than that for a girl.) How to calculate the probabilities of all four possible variants of twins' pairs? Let a r.v. $X_1 = 0$ if the first twin is a boy, and let $X_1 = 1$ if it is a girl. Denote by X_2 the corresponding r.v. for the second twin. Then, for example, f_{01} is the probability that the first twin is a boy, and the second is a girl.

By assumption, $f_{00} + f_{11} = 0.64$, $f_{00} + f_{01}$ (the probability that first twin is a boy) $= 0.51$. Similarly, $f_{00} + f_{10} = 0.51$. To have the fourth equation for our four unknowns, recall that the total sum $f_{00} + f_{11} + f_{01} + f_{10} = 1$. Solving this system of equations, we get $f_{00} = 0.33$, $f_{01} = f_{10} = 0.18$, and $f_{11} = 0.31$. □

1.2.2 Independent r.v.'s

R.v.'s X_1 and X_2 are said to be *independent* if for *any* two sets D_1 and D_2 of real numbers, the events $\{X_1 \in D_1\}$ and $\{X_2 \in D_2\}$ are independent. (The event $\{X_1 \in D_1\}$ is the event that the r.v. X_1 assumed a value from D_1.)

By definition of independent events, this means that

$$P(X_1 \in D_1,\ X_2 \in D_2) = P(X_1 \in D_1)P(X_2 \in D_2). \tag{1.2.2}$$

In the multidimensional case, r.v.'s $X_1, ..., X_n$ are said to be *mutually independent* if for *any* sets $D_1, ..., D_n$ of real numbers, the events $\{X_1 \in D_1\}, ..., \{X_n \in D_n\}$ are mutually independent (see Section **2**.1.1.)

We restrict ourselves to the two-dimensional case.

Proposition 1 *Discrete r.v.'s X_1, X_2 are independent if and only if*

$$f_{ij} = f_i^{(1)} f_j^{(2)} \text{ for all } i, j. \tag{1.2.3}$$

A similar formula is true in the multidimensional case; we skip details.

First, we consider examples. In the first two, before applying Proposition 1, we try to guess an answer proceeding from *an heuristic understanding of independence: a r.v. X_2 does not depend on X_1 if any knowledge about the value that X_1 has assumed does not give us any information regarding the likelihood of possible values of X_2.*

It is also worth noting that in order to make sure that r.v.'s are independent, we must check (1.2.3) for *all* i and j, while to reject independence, it suffices to point out only *one* pair (i, j) for which (1.2.3) is not true.

EXAMPLE 1. Let a r.vec. $\mathbf{X} = (X_1, X_2)$ take on four vector-values corresponding to four points in Fig. 5a with equal probabilities $1/4$. We see that whatever value X_1 assumes: -1

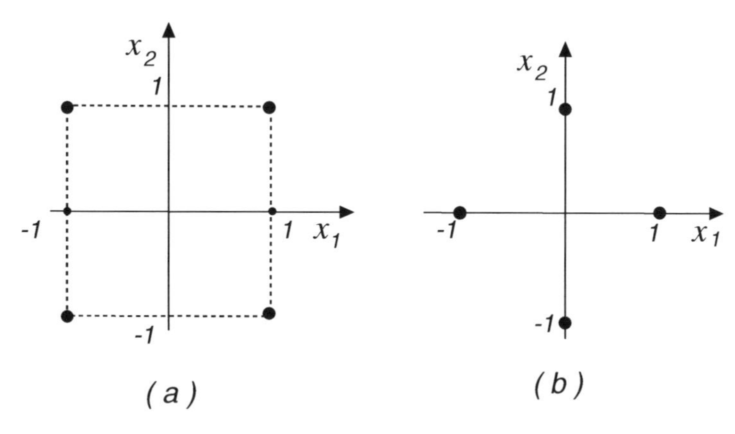

FIGURE 5.

or 1, the possible values of X_2 are equally likely. So, one should guess that X_1 and X_2 are independent.

To check this using Proposition 1, it is convenient to set $f_{ij} = P(X_1 = i, X_2 = j)$ for $i = \pm 1, j = \pm 1$. We have $f_1^{(1)} = P(X_1 = 1) = \frac{1}{4} + \frac{1}{4} = \frac{1}{2}$, and similarly, all other marginal probabilities $f_{-1}^{(1)} = f_1^{(2)} = f_{-1}^{(2)} = \frac{1}{2}$. Hence, $f_{ij} = \frac{1}{4} = \frac{1}{2} \cdot \frac{1}{2} = f_i^{(1)} f_j^{(2)}$ for all $i = \pm 1, j = \pm 1$.

Now let **X** take on values corresponding to the four points in Fig. 5b with equal probabilities. In this case, X_1 and X_2 take on values $-1, 0, +1$ with probabilities $\frac{1}{4}, \frac{1}{2}, \frac{1}{4}$. We see also that if $X_1 = 1$, then X_2 may take on only one value: 1, while if $X_1 = 0$, then X_2 may assume either 1 or -1. This observation guarantees that X_1 and X_2 are dependent.

Certainly, the same follows from Proposition 1. For example, $f_{11} = P(X_1 = 1, X_2 = 1) = 0$, while $f_1^{(1)} f_1^{(2)} = \frac{1}{4} \cdot \frac{1}{4} \neq 0$.

EXAMPLE 2. Let us revisit Example 1.2.1-2. The probability that the first twin is a boy, $f_0^{(1)}$, is 0.51. The same concerns the second twin: $f_0^{(2)} = 0.51$, while we have calculated that the probability that both are boys, f_{00}, equals 0.33. Since $0.51 \cdot 0.51 \neq 0.33$, X_1, X_2 are dependent.

EXAMPLE 3. An instructor is preparing a test-problem for students on checking independence; say, for simple r.v.'s X_1, X_2 assuming two values: 0 and 1. Particular details about the nature of X_1, X_2 are not essential. Assume that the instructor wants the r.v.'s to be independent (which the students will not know). Then, in the table for probabilities, she/he may first fill in cells for marginal probabilities, as in the first table below. After that the instructor fills out all cells for all joint probabilities just multiplying the corresponding marginal probabilities; see the second table.

$X_1 \Rightarrow$ $X_2 \Downarrow$	0	1	Marg.prob.
0			0.3
1			0.7
Marg.prob.	0.2	0.8	

$X_1 \Rightarrow$ $X_2 \Downarrow$	0	1	Marg.prob.
0	0.06	0.24	0.3
1	0.14	0.56	0.7
Marg.prob.	0.2	0.8	

After that the instructor may remove the figures for marginal probabilities and leave only the joint ones.

If the instructor wants to construct dependent r.v.'s, it suffices to fill out cells for joint probabilities in an arbitrary way: it would be a rare occasion if (1.2.3) turns out to be true for all i, j. However, it should be double checked, and if (1.2.3) is true for all i, j, it will be enough to slightly change the probabilities.

EXAMPLE 4. Suppose r.v.'s X_1 and X_2 both depend on a r.v. X_3. (More precisely, X_1 and X_3 are dependent, and the same is true for X_2 and X_3.) Can X_1 and X_2 be independent? One may think that they cannot, but this is not true. For instance, consider r.v.'s X_1 and X_2 about which *it is given that they are independent* and assume more than one value. Set $X_3 = X_1 + X_2$. Clearly, X_1 and X_3 are dependent, and so are X_2 and X_3. (If each X_i assumes only one value, then all three r.v.'s are independent (see also Exercise 12).) □

Proof of Proposition 1. First, let D_1, D_2 in (1.2.2) be *one point* sets $D_1 = [x_{1i}], D_2 = [x_{2j}]$. If X_1, X_2 are independent, then from (1.2.2) it follows that $f_{ij} = P(X = x_{1i}, X_2 = x_{2j}) = P(X_1 = x_{1i})P(X_2 = x_{2j}) = f_i^{(1)} f_j^{(2)}$, and we have arrived at (1.2.3).

Let now D_1, D_2 be arbitrarily, and (1.2.3) be true. Then

$$P(X_1 \in D_1,\ X_2 \in D_2) = \sum_{i:\, x_{1i} \in D_1} \sum_{j:\, x_{2j} \in D_2} f_{ij} = \sum_{i:\, x_{1i} \in D_1} \sum_{j:\, x_{2j} \in D_2} f_i^{(1)} f_j^{(2)}$$

$$= \left(\sum_{i:\, x_{1i} \in D_1} f_i^{(1)} \right) \left(\sum_{j:\, x_{2j} \in D_2} f_j^{(2)} \right) = P(X_1 \in D_1) P(X_2 \in D_2). \ \blacksquare$$

1.2.3 Convolution

Next, we derive a useful formula for the distribution of the sum $S = X_1 + X_2$ of two independent non-negative *integer-valued* r.v.'s X_1, X_2. (That is, X_1, X_2, and S may assume only values $0, 1, 2, ...$)

Set $f_i^{(1)} = P(X_1 = i)$, $f_j^{(2)} = P(X_2 = j)$, and $f_n = P(S = n)$. Then

$$f_n = \sum_{k=0}^{n} f_k^{(1)} f_{n-k}^{(2)}. \tag{1.2.4}$$

Proof is simple. The r.v. S is equal to n if and only if the r.v. X_1 is equal to *some* k and X_2 to $n-k$. The corresponding probability is $P(X_1 = k, X_2 = n-k) = P(X_1 = k)P(X_2 = n-k) = f_k^{(1)} f_{n-k}^{(2)}$ because X_1 and X_2 are independent. Summing over all k from 0 to n leads to (1.2.4). ■

The operation (1.2.4) is called *convolution*, and the formula itself—the *convolution formula*. We will also write it in the following compact form:

$$f = f^{(1)} * f^{(2)}, \tag{1.2.5}$$

where $f, f^{(1)}, f^{(2)}$ are the *collections* of the probabilities specifying the distributions of the r.v.'s S, X_1, and X_2, respectively:

$$f = (f_1, f_2, ...); \ \ f^{(1)} = (f_1^{(1)}, f_2^{(1)}, ...); \ \ f^{(2)} = (f_1^{(2)}, f_2^{(2)}, ...).$$

The symbol $*$ itself denotes the *convolution operation*.

EXAMPLE 1. Let independent r.v.'s

$$X_1 = \begin{cases} 1 \text{ with probability } p_1 = \frac{1}{3} \\ 0 \text{ with probability } q_1 = \frac{2}{3} \end{cases}, \quad X_2 = \begin{cases} 1 \text{ with probability } p_2 = \frac{1}{2} \\ 0 \text{ with probability } q_2 = \frac{1}{2} \end{cases}.$$

Clearly, S takes on values $0,1,2$. The problem is very simple and can be solved in a straightforward fashion, but we use this example to demonstrate how (1.2.4) works. We have

$$f_0 = f_0^{(1)} f_0^{(2)} = q_1 q_2 = \frac{2}{3} \cdot \frac{1}{2} = \frac{1}{3},$$

$$f_1 = \sum_{k=0}^{1} f_k^{(1)} f_{n-k}^{(2)} = f_0^{(1)} f_1^{(2)} + f_1^{(1)} f_0^{(2)} = q_1 p_2 + p_1 q_2 = \frac{2}{3} \cdot \frac{1}{2} + \frac{1}{3} \cdot \frac{1}{2} = \frac{1}{2},$$

$$f_2 = \sum_{k=0}^{2} f_k^{(1)} f_{n-k}^{(2)} = f_0^{(1)} f_2^{(2)} + f_1^{(1)} f_1^{(2)} + f_2^{(1)} f_0^{(2)} = q_1 \cdot 0 + p_1 p_2 + 0 \cdot q_2 = p_1 p_2 = \frac{1}{3} \cdot \frac{1}{2} = \frac{1}{6}. \quad \square$$

EXAMPLE 2. Consider an unlimited sequence of independent trials with the probability of success in an individual trial equal to p. In Example 1.1-10, we derived the probability distribution of the r.v. N, the number of the first successful trial. Let N_2 be the number of the second successful trial, and $N' = N_2 - N$, the number of trials one should perform after the first success and up to the second. Since the trials are independent, after the first success everything starts over as from the very beginning. So, N' does not depend on N and has the same distribution as N. Thus, in view of (1.1.5), $P(N = k) = P(N' = k) = pq^{k-1}$, where $q = 1 - p$.

Since $N_2 = N + N'$, and $P(N = 0) = P(N' = 0) = 0$, by the convolution formula,

$$P(N_2 = n) = \sum_{k=0}^{n} P(N = k)P(N' = n-k) = \sum_{k=1}^{n-1} pq^{k-1} pq^{n-k-1} = \sum_{k=1}^{n-1} p^2 q^{n-2} = (n-1)p^2 q^{n-2}.$$

Thus, we have derived the distribution of the number of the second successful trial; in short, the distribution of the moment of the second success. $\square$

1.2.4 Conditional distributions

For r.v.'s $X_1 = x_{11}, x_{12}, \ldots$ and $X_2 = x_{21}, x_{22}, \ldots$, set

$$f_{j|i} = P\left(X_2 = x_{2j} \,\middle|\, X_1 = x_{1i}\right),$$

the probability that X_2 assumed the jth value *given* that X_1 assumed the ith value. In the notation of Section 1.2.1,

$$f_{j|i} = \frac{P(X_1 = x_{1i},\ X_2 = x_{2j})}{P(X_i = x_{1i})} = \frac{f_{ij}}{f_i^{(1)}},$$

provided $f_i^{(1)} \neq 0$. For any fixed i, the total sum of all conditional probabilities

$$\sum_j f_{j|i} = 1.$$

Indeed, by (1.2.1), $\sum_j f_{j|i} = \sum_j \frac{f_{ij}}{f_i^{(1)}} = \frac{1}{f_i^{(1)}} \sum_j f_{ij} = \frac{f_i^{(1)}}{f_i^{(1)}} = 1$.

Thus, given $X_1 = x_{1i}$, we have assigned to *each* value x_{2j} its *conditional* probability $f_{j|i}$. These conditional probabilities determine a probability distribution on the set $\{x_{21}, x_{22}, x_{23}, ...\}$. We call this distribution the *conditional* distribution of X_2 given $X_1 = i$.

EXAMPLE 1. Consider the problem on twins from Examples 1.2.1-2 and 1.2.2-2. What is the probability that a twin is a boy given that the other twin is a boy? In our notation, it is

$$f_{0|0} = \frac{f_{00}}{f_0^{(1)}} = \frac{0.33}{0.51} = \frac{33}{51} \approx 0.647,$$

and the probability that the twin is a girl given that the other twin is a boy, is

$$f_{1|0} = \frac{f_{01}}{f_0^{(1)}} = \frac{0.18}{0.51} = \frac{18}{51} \approx 0.353.$$

As must be expected, the sum of these probabilities equals one.

EXAMPLE 2. We revisit Example 1.2.3-2. Suppose that we know that the second success has occurred at the tenth trial, but we do not know when the first success has occurred. Given the information we possess, the first success may occur only at one of the first nine trials. Are these nine trials equally likely to be successful, or for example, is the first success is less likely to occur near the beginning than, say, near the middle (fifth) trial? In the general case, for $k < n$, consider

$$P(N=k\,|\,N_2=n) = \frac{P(N=k,\, N_2=n)}{P(N_2=n)} = \frac{P(N=k,\, N'=n-k)}{P(N_2=n)} = \frac{P(N=k)P(N'=n-k)}{P(N_2=n)},$$

because N, N' are independent. Inserting results from Examples 1.2.2-5 and Example 1.2.3-2, we have

$$P(N=k\,|\,N_2=n) = \frac{pq^{k-1}pq^{n-k-1}}{(n-1)p^2q^{n-2}} = \frac{1}{n-1}.$$

Thus, all $n-1$ trials before the second success are *equally likely* to have been the first success. In particular, for $n = 10$, all conditional probabilities above equal $1/9$. □

---

2 EXPECTED VALUE

2.1 Definitions and properties

If a r.v. X is discrete and takes on values $x_1, x_2, ...$ with probabilities $f_1, f_2, ...$, respectively, then its *expected value* (or *mean value*, or simply *mean*) is defined by

$$E\{X\} = \sum_i x_i f_i, \tag{2.1.1}$$

provided the series (2.1.1) converges. The summation is over all possible i.

The case of an infinite expectation is considered in Section 2.3.

It may be said that $E\{X\}$ is the weighted average of the values x_i, where the role of weights is played by the probabilities f_i.

In Section 5.2, we provide a detailed and rigorous explanation as to why this characteristic may be indeed interpreted as the mean value of the r.v. In short, assume that we perform many independent replicas of an experiment, obtaining different independent values of the same r.v. X. Then, in a certain reasonable sense, the average of these values will be close to the quantity defined in (2.1.1).

It is interesting that the concept of expectation is analogous to that of a center of mass (or gravity) in Mechanics. If we distribute a unit mass in a way that mass f_i is located at point x_i in the real line, then the location of the center of mass will be given by (2.1.1).

EXAMPLE 1. Let us revisit Example 1.1-8 on an insurance against flight delays. We have obtained there the following distribution of the r.v. X, the payment to a separate client:

payment:	0	150	450
probability:	0.95	0.045	0.005

Hence, $E\{X\} = 0 \cdot 0.95 + 150 \cdot 0.045 + 450 \cdot 0.005 = 9$ dollars. Without going deeply into it, note that the premium in any insurance policy should be (at least slightly) larger than the mean payment: otherwise the insurance company will not be able to function. So, in our case, the premium should be—at least, a bit—larger than \$9.

EXAMPLE 2. If a r.v. X takes on just one value a, then $E\{X\} = a \cdot 1 = a$. If

$$X = \begin{cases} a \text{ with probability } p, \\ 0 \text{ with probability } 1-p, \end{cases} \tag{2.1.2}$$

then $E\{X\} = 0 \cdot (1-p) + a \cdot p = ap$. An obvious but important corollary from the last assertion may be stated as follows. Let A be an event, and the r.v.

$$I_A = \begin{cases} 1 & \text{if } A \text{ has occurred}, \\ 0 & \text{if } A \text{ has not occurred}. \end{cases} \tag{2.1.3}$$

Such a r.v. is called the *indicator* of an event A: it indicates whether A has occurred. Since A occurs with probability $P(A)$,

$$E\{I_A\} = P(A). \tag{2.1.4}$$

EXAMPLE 3 (*The geometric distribution*). Consider the r.v. N, the moment of the first success in a sequence of independent trials with the probability of success equal to p (Example 1.1-10). Since $P(N=k) = pq^{k-1}$, where $q = 1-p$,

$$E\{N\} = \sum_{k=1}^{\infty} k \cdot P(N=k) = \sum_{k=1}^{\infty} k \cdot pq^{k-1} = p \sum_{k=1}^{\infty} k \cdot q^{k-1}.$$

To compute the last sum, set $S(q) = \sum_{k=0}^{\infty} q^k = \frac{1}{1-q}$, provided $q < 1$, as a geometric series; see the Appendix , (2.2.11). Then $\sum_{k=1}^{\infty} k \cdot q^{k-1} = (\sum_{k=0}^{\infty} q^k)' = S'(q) = \frac{1}{(1-q)^2}$, and

$$E\{N\} = p \cdot \frac{1}{(1-q)^2} = \frac{1}{p}.$$

For instance, suppose that for a car owner, the probability of suffering a car body damage during a year is $p = 0.2$, and the results for different years are independent. What is the expected number of consecutive years without having car body damage? The answer is "four," since $\frac{1}{p} = 5$, and *on the average*, the first damage will occur in the fifth year.

EXAMPLE 4 (*The Poisson distribution*). Let X be a Poisson r.v. with a parameter λ; see Example 1.1-9. We show that this parameter is equal to the mean value of X. Indeed, because $P(X = k) = e^{-\lambda}\lambda^k/k!$,

$$E\{X\} = \sum_{k=0}^{\infty} k \cdot P(X = k) = \sum_{k=0}^{\infty} ke^{-\lambda}\frac{\lambda^k}{k!} = e^{-\lambda}\sum_{k=0}^{\infty} k\frac{\lambda^k}{(k-1)!k} = \lambda e^{-\lambda}\sum_{k=1}^{\infty}\frac{\lambda^{k-1}}{(k-1)!}.$$

Setting in the last sum $k - 1 = m$, we see that this sum equals $\sum_{m=0}^{\infty}\frac{\lambda^m}{m!} = e^{\lambda}$ (the Appendix, (2.2.4)). Substituting it, we have

$$E\{X\} = \lambda e^{-\lambda}e^{\lambda} = \lambda. \quad \square$$

Next, we list some properties of expectation, proofs of which are given at the end of this section.

1. If a r.v. $Y = u(X)$, where $u(x)$ is a function, then

$$E\{Y\} = E\{u(X)\} = \sum_i u(x_i)f_i, \tag{2.1.5}$$

where x_i's and f_i's are the same as in (2.1.1). The formula above is not obvious since, while in the definition (2.1.1) values x_i's are assumed to be *distinct*, in (2.1.5) values $u(x_i)$ may coincide.

Similarly, if $X_1 = x_{11}, x_{12}, ...$ and $X_2 = x_{21}, x_{22}, ...$ with joint probabilities f_{ij}, and $Y = u(X_1, X_2)$ for a function $u(x_1, x_2)$, then

$$E\{Y\} = E\{u(X_1, X_2)\} = \sum_i\sum_j u(x_{1i}, x_{2j})f_{ij}. \tag{2.1.6}$$

2. $E\{cX\} = cE\{X\}$ for any constant c.

3. For any r.v.'s $X_1, ..., X_n$,

$$E\{X_1 + ... + X_n\} = E\{X_1\} + ... + E\{X_n\}. \tag{2.1.7}$$

In particular, if $E\{X_i\}$ is the same for all i's and equals m, then for the sum $S_n = X_1 + ... + X_n$,

$$E\{S_n\} = \underbrace{m + ... + m}_{n \text{ times}} = mn. \tag{2.1.8}$$

Properties 2 and 3 imply the following two corollaries.

4. For any constants a and b, it is true that $E\{a + bX\} = E\{a\} + E\{bX\} = a + bE\{X\}$ (since the expectation of a constant is the constant itself).

5. If $E\{X\} = m$, and the r.v. $X' = X - m$, then $E\{X'\} = E\{X - m\} = E\{X\} - m = m - m = 0$. We call the r.v. X' *centered*, and the very operation—*centering*.

6. If $P(X_1 \geq X_2) = 1$, then $E\{X_1\} \geq E\{X_2\}$.

EXAMPLE 5. Let X be an air temperature measured in Fahrenheit, and Y be the same temperature measured in Celsius. As is known, $Y = 32 + 1.8X$. If we are talking about a future temperature, say, the highest temperature to be reached tomorrow, X is a r.v. If its *mean* value (on some day and in some area) is 25°C, then, by Property 4, the *same mean* temperature in Fahrenheit is $32 + 1.8 \cdot 25 = 77$.

EXAMPLE 6. Consider the *binomial distribution* of X, the number of successes in n independent trials; see (1.1.4). We can find $E\{X\}$ formally, proceeding directly from (2.1.1) and writing $E\{X\} = \sum_{k=1}^{n} k \cdot \binom{n}{k} p^k q^{n-k}$. The last sum is tractable but requires some calculations. The following presentation, which we will appeal to repeatedly, helps to simplify the problem. Let

$$X_k = \begin{cases} 1 & \text{if the } k\text{th trial is successful, and} \\ 0 & \text{otherwise.} \end{cases}$$

(i.e., the X_k's "count" successes.) Then,

$$X = X_1 + X_2 + ... + X_n. \tag{2.1.9}$$

If the probability of success is p, then $P(X_k = 1) = p$, and $E\{X_k\} = p$ for each k. Hence, by Property 3,

$$E\{X\} = E\{X_1\} + ... + E\{X_n\} = \underbrace{p + ... + p}_{n \text{ times}} = np. \tag{2.1.10}$$

Note that (2.1.10) is true for dependent trials as well, since Property 3 does not require the independence of the X's.

For instance, in Demography, when exploring surviving rates, people proceed from a randomly chosen group of n newborns; usually, $n = 100{,}000$. Denote by $s(x)$ the survival probability; that is, the probability that a randomly chosen newborn will survive to age x; and let $L_n(x)$ be the group's number of survivors to age x.

We may view the newborns as "trials," and survivorship—as success. Then $E\{L_n(x)\} = ns(x)$, regardless of whether the lifetimes of separate newborns are independent or not. For example, "Life Table for the total population: United States, 2007" [3] gives the estimate $s(50) \approx 0.9357$, and hence, on the average, there will be 93,570 survivors to age 50.

If the newborn lifetimes are independent, we may say more: $L_n(x)$ is binomial r.v. with parameters n and $p = s(x)$.

EXAMPLE 7. Consider now an unlimited sequence of independent trials, and denote by N_ν the moment of the νth success (more precisely, the number of the νth successful trial). How can we compute $E\{N_\nu\}$? Let τ_m be the number of trials between the $(m-1)$th and the mth success including the latter. In Example 1.2.3-2, we have observed that the τ_m's are independent r.v.'s each having the same geometric distribution with parameter p. It is also clear that

$$N_\nu = \tau_1 + ... + \tau_\nu. \tag{2.1.11}$$

From Example 3, it follows that $E\{\tau_m\} = 1/p$. Hence, $E\{N_\nu\} = \nu/p$. Note that here the independence of trials is essential: we used this assumption when showing that $E\{\tau_i\} = 1/p$. □

We have seen that for any r.v.'s X_1, X_2, it is true that $E\{X_1 + X_2\} = E\{X_1\} + E\{X_2\}$.

Is it true that $E\{X_1X_2\} = E\{X_1\}E\{X_2\}$?
In the general case, the answer is negative.

To show this, suppose X_1, X_2 equal the same r.v. X. In this particular case, the question is equivalent to the question whether $E\{X^2\} = (E\{X\})^2$. In Section 3, we will see that it is true only if X takes on just one value with probability one. Here, we restrict ourselves to a particular example.

EXAMPLE 8. Suppose X assumes values 1 or 0 with probabilities p and $1-p$, respectively. Then $E\{X\} = p$. On the other hand, X^2 assumes the values $1^2 = 1$ and $0^2 = 0$; that is, the same values as X, and with the same probabilities. Hence, $E\{X^2\} = p$, while $(E\{X\})^2 = p^2$. But $p = p^2$ only if $p = 0$ or 1, which is equivalent to the case where with probability one, X assumes one value: either 0 (if $p = 0$) or 1 (if $p = 1$). □

The case $X_1 = X_2$ is certainly extreme: the r.v.'s X_1, X_2 are equal to each other, and hence are "strongly dependent." The next assertion deals, in a sense, with the opposite case.

Proposition 2 *If r.v.'s X_1, X_2 are independent,*

$$E\{X_1X_2\} = E\{X_1\}E\{X_2\}. \tag{2.1.12}$$

A proof will be given at the end of the section.

EXAMPLE 9. We revisit the model of the stock market from Section **2**.1.2.2. It presupposes that each day the price of a stock is multiplied by a r.v.

$$X = \begin{cases} u & \text{with probability } p, \\ d & \text{with probability } q = 1-p. \end{cases}$$

This is what we will have on the next day if we invest one unit of money on the current day. We will call such a variable a *return* (per unit of money). Let X_i be the return on the ith day. Then the return for n days is the r.v. $Y_n = X_1 \cdot \ldots \cdot X_n$. In a stable market, we may assume that the daily returns are mutually independent, and extending (2.1.12) to the case of many r.v.'s, we may write that

$$E\{Y_n\} = E\{X_1\} \cdot \ldots \cdot E\{X_n\} = (up + dq)^n.$$

Assume that each weekday (stock exchanges do not work on weekends), the price either increases or drops by 1% with probabilities 0.51 and 0.49, respectively. Then, if we take 22 as the number of weekdays in a month, the mean monthly growth factor equals $(1.01 \cdot 0.51 + 0.99 \cdot 0.49)^{22} \approx 1.0044$, which is equivalent to a 0.44% mean monthly growth. For one year (consisting of about 250 working days), we write $(1.01 \cdot 0.51 + 0.99 \cdot 0.49)^{250} \approx 1.051$, which amounts to a 5.1% mean annual growth.

EXAMPLE 10. A company is planning to produce a new product in a year. The future price X the company will be able to charge depends on the market, and hence, is random. The number of sales, N, during say the first month, is random too. If we may assume that the price does not depend on the demand, we can write that the mean income for the first month is equal to $E\{XN\} = E\{X\}E\{N\}$. However, the price usually does depend on the demand, and to compute the mean future income, one needs to know the structure of this dependency. □

Proofs of Properties 1–6.

1) As was mentioned, the values $u(x_1), u(x_2), \ldots$ may coincide. Denote by $y_1, y_2, \ldots$ *all distinct* values of the r.v. $Y = u(X)$, and by $f_{Y1}, f_{Y2}, \ldots$ respective probabilities. To find $P(Y = y_k)$, one should add up all probabilities f_i corresponding to those i for which $u(x_i) = y_k$. So,

$$f_{Yk} = P(Y = y_k) = \sum_{i:\, x_i = y_k} f_i.$$

Then, by definition,

$$E\{Y\} = \sum_k y_k f_{Yk} = \sum_k y_k \sum_{i:\, u(x_i)=y_k} f_i = \sum_k \sum_{i:\, u(x_i)=y_k} u(x_i) f_i = \sum_i u(x_i) f_i, \tag{2.1.13}$$

and we have arrived at (2.1.5). The proof of (2.1.6) is similar: to find $P(Y = y_k)$, one should add up all joint probabilities f_{ij} corresponding to those i, j for which $u(x_{1i}, x_{2j}) = y_k$.

2) This is the particular case of $u(x) = cx$. By (2.1.5), $E\{cX\} = \sum_i cx_i \cdot f_i = c\sum_i x_i \cdot f_i = cE\{X\}$.

3) It suffices to prove (2.1.7) for $n = 2$. We keep the same notation and, as usual, denote by $f_i^{(1)}$, $f_j^{(2)}$ marginal probabilities. By virtue of (2.1.6),

$$\begin{aligned} E\{X_1 + X_2\} &= \sum_i \sum_j (x_{1i} + x_{2j}) f_{ij} = \sum_i x_{1i} \sum_j f_{ij} + \sum_j x_{2j} \sum_i f_{ij} = \sum_i x_{1i} f_i^{(1)} + \sum_j x_{2j} f_j^{(2)} \\ &= E\{X_1\} + E\{X_2\}. \end{aligned}$$

4)-5) We have already shown that these properties follow from Properties 2-3.

6) Let $Z = X_1 - X_2$, and $z_1, z_2, \ldots$ be all *distinct* values of Z with respective probabilities f_{Zi}. We do not have to consider zero probabilities, so without loss of generality we assume $f_{Zi} > 0$. By assumption, $Z \geq 0$ with probability one. So, all z_i's are non-negative since otherwise Z would be negative with a positive probability. Hence, $E\{Z\} = \sum_i z_i f_{Zi} \geq 0$. Then, $E\{X_1\} - E\{X_2\} = E\{X_1 - X_2\} = E\{Z\} \geq 0$, and hence $E\{X_1\} \geq E\{X_2\}$. ■

Proof of Proposition 2. We keep all above notations. Since X_1 and X_2 are independent, the joint probabilities $f_{ij} = f_i^{(1)} f_j^{(2)}$. Consequently, by (2.1.6),

$$\begin{aligned} E\{X_1 X_2\} &= \sum_i \sum_j x_{1i} x_{2j} f_{ij} = \sum_i \sum_j x_{1i} x_{2j} f_i^{(1)} f_j^{(2)} = \left(\sum_i x_{1i} f_i^{(1)}\right)\left(\sum_j x_{2j} f_j^{(2)}\right) \\ &= E\{X_1\}E\{X_2\}. \ \blacksquare \end{aligned}$$

Route 1 ⇒ page 74

2

2.2 A useful formula for the expectation of a non-negative integer-valued r.v.

Proposition 3 *Let a r.v.* $N = 0, 1, 2, \ldots$. *Then*

$$E\{N\} = \sum_{n=0}^{\infty} P(N > n). \tag{2.2.1}$$

The r.-h.s. of (2.2.1) may be also written in the form $\sum_{k=1}^{\infty} P(N \geq k)$ since $P(N > n) = P(N \geq n+1)$, and one may set $k = n+1$.

Proof. Below, we use the fact that $m = \sum_{k=1}^{m} 1$ (we add up m ones), and in the fourth step, we interchange the order of summation. So,

$$E\{N\} = \sum_{m=0}^{\infty} mP(N=m) = \sum_{m=1}^{\infty} mP(N=m) = \sum_{m=1}^{\infty}\left(\sum_{k=1}^{m} 1\right) P(N=m)$$

$$= \sum_{k=1}^{\infty}\sum_{m=k}^{\infty} 1 \cdot P(N=m) = \sum_{k=1}^{\infty}\left(\sum_{m=k}^{\infty} P(N=m)\right) = \sum_{k=1}^{\infty} P(N \geq k) = \sum_{n=0}^{\infty} P(N > n). \quad \blacksquare$$

We apply this proposition repeatedly; for the first time, in the next section.

2.3 Infinite expected values. Do they occur in models of real phenomena?

Formally speaking, the series (2.1.1) may diverge. From a purely mathematical point of view, it is not difficult to construct a r.v. X for which $E\{X\} = \infty$. The first example below is well known and connected with the so-called Saint Petersburg paradox. The paradox itself will be considered later, in Section **18**.2, where we touch on the expected utility theory.

EXAMPLE 1. Consider a game of chance consisting in tossing a regular coin until it comes up heads. Suppose that if the first head appears right away at the first toss, the payment equals 2, say, dollars. If the first head appears at the second toss, the payment equals 4, and so on; namely, if a head appears for the first time on the kth toss, the payment equals 2^k. Thus, the payment in this case is a r.v. X taking on values $2, 4, 8, \ldots, 2^k, \ldots$ with probabilities $\frac{1}{2}, \frac{1}{4}, \frac{1}{8}, \ldots, \frac{1}{2^k}, \ldots$, respectively, and $E\{X\} = 2 \cdot \frac{1}{2} + 4 \cdot \frac{1}{4} + 8 \cdot \frac{1}{8} + \ldots = 1 + 1 + 1 + \ldots = \infty$. $\square$

This example is somewhat contrived, and the question is whether we can encounter infinite expectations in models of real experiments with practical importance. The answer is positive, and it happens more often that one might expect.

EXAMPLE 2 (*Record value*). Consider an infinite sequence of independent and identically distributed (i.i.d.) r.v.'s X_0, $X_1, X_2, \ldots$. Let N be the index of the first X_i among $X_1, X_2, \ldots$ which is larger than X_0. It may be said that N is the moment of the new record value. For example, suppose that the X_i's are the daily numbers of customers of a store. Then N is the number of the first day where the daily number of customers exceeds the initial number on day 0.

Surprisingly, it turns out that $E\{N\} = \infty$.

To show this, consider the events $A_{in} = \{X_i \geq X_0, X_i \geq X_1, ..., X_i \geq X_n\}$, $i = 0, 1, ..., n$. So, if A_{in} occurs, then among the $n+1$ r.v.'s $X_0, X_1, ..., X_n$, the ith r.v. assumes a value that is no less than the value of each of the other n r.v.'s. Then the event $\{N > n\} = A_{0n}$.

Furthermore, because the X's are i.i.d., by the symmetry argument, among the r.v.'s $X_0, X_1, ..., X_n$, each has an equal chance to take on a value not smaller than the values of the others. Hence, $P(A_{in})$ is the same for all i, and, consequently, is equal to $P(A_{0n})$. On the other hand, at least one A_{in} should occur, which means that the union $\cup_{i=0}^{n} A_{in}$ should have unit probability. Thus,

$$1 = P(\cup_{i=0}^{n} A_{in}) \leq \sum_{i=0}^{n} P(A_{in}) = (n+1)P(A_{0n}).$$

(In the last equality, we used the fact that $P(A_{in}) = P(A_{0n})$ for all $i = 0, ..., n$.) Consequently, $P(N > n) = P(A_{0n}) \geq 1/(n+1)$ for all $n = 0, 1, ...$ By (2.2.1),

$$E\{N\} = \sum_{n=0}^{\infty} P(N > n) \geq \sum_{n=0}^{\infty} \frac{1}{n+1}.$$

As we remember from Calculus, the last series (called harmonic) diverges; that is, its sum is infinite.

It is worth emphasizing three things. First, we did not use the fact that the X's were discrete, so the result we obtained is true for any i.i.d. r.v.'s X_i. We will touch on the case of non-discrete r.v.'s in more detail in Exercise **6**.37.

Secondly, if the X_i's assume a finite number of values, the result is trivial. In this case, with a positive probability, X_0 assumes the maximal of all possible values, and in this case, the new record value will never appear: $N = \infty$. However, if the X_i's do not have a maximal value (for example, if they have the Poisson distribution), the result is not trivial at all, since in this case $P(N < \infty) = 1$.

Indeed, assume that the X's take on values $x_1, x_2, ...$, and X_0 has assumed a particular value x_k. If there is no maximal value among $x_1, x_2, ...$, then for any x_k, the probability $p_k = P(X_i > x_k) > 0$. Then, we may view the sequence $X_1, X_2, ...$ as a sequence of trials where success is the appearance of a value larger than x_k. Hence, *given* $X_0 = x_k$, the r.v. N has the geometric distribution with parameter p_k. In particular, $P(N < \infty \,|\, X = x_k) = 1$. Then, by the formula for total probability,

$$P(N < \infty) = \sum_{k} P(N < \infty \,|\, X = x_k) P(X = x_k) = \sum_{k} 1 \cdot P(X = x_k) = 1.$$

Thirdly, the fact that $E\{N\} = \infty$ does not mean that one should wait for the new record value infinitely long: we know that $P(N < \infty) = 1$. However, the mean value of the time it takes to face a new record value is infinite, which may be interpreted as follows.

Suppose we perform n realizations of the sequence $X_0, X_1, ...$; say, we consider n similar stores functioning independently. Let N_i be the waiting time for the record value for the ith store. Then the average waiting time $\frac{1}{n}(N_1 + ... + N_n)$ will be unlimitedly large for large n; more precisely, with probability one it will tend to infinity as $n \to \infty$. Rigorously, it follows from the law of large numbers we consider in detail in Section 5.2.

The same interpretation applies to

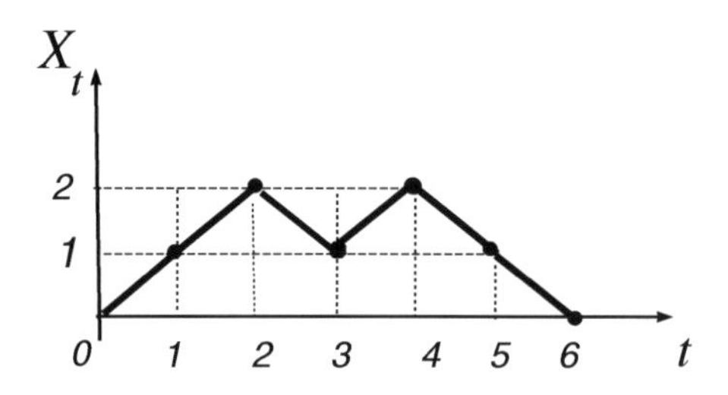

FIGURE 6. X_t is the position of the random walk at time t. In this particular realization, X_t has revisited zero at the first time at the moment $t = 6$.

EXAMPLE 3 (*"Revisiting zero"*). Consider a random walk as described in Section **2**.2.4. Assume that the process starts from the origin (zero), and at each step moves up or down with equal probabilities $1/2$; so, the walk is symmetric. Denote by τ the moment of the first return to zero (that is, the first moment when the numbers of moves up and down become equal). The situation is illustrated in Fig. 6.

For example, we toss a regular coin, and τ is the first moment when the number of heads is equal to the number of tails.

Again, contrary to intuition, $E\{\tau\} = \infty$. We will prove this by two methods in Sections **4**.4.1 and **15**.2.2. □

2.4 Jensen's inequality

Proposition 4 *Let X be a r.v. with a finite expectation, and let $u(x)$ be a concave function defined on a (finite or infinite) interval including all values of X. Then*

$$E\{u(X)\} \leq u(E\{X\}). \tag{2.4.1}$$

If u is convex (concave upward), then

$$E\{u(X)\} \geq u(E\{X\}). \tag{2.4.2}$$

Proof. Set $m = E\{X\}$, and denote by I the interval mentioned in the statement. If X takes on *only one value* corresponding to one of the endpoints of I, Jensen's inequality becomes an equality, and the proof is trivial. Assume that this is not the case. Then m is an interior point of I, and by Proposition 1 in the Appendix , there exists a number c such that

$$u(x) - u(m) \leq c(x - m) \quad \text{for any } x \in I. \tag{2.4.3}$$

Setting $x = X$, we write $u(X) - u(m) \leq c(X - m)$. Computing the expectations of both sides, we have $E\{u(X)\} - u(m) \leq c(m - m) = 0$, which amounts to (2.4.1).

To prove (2.4.2), it suffices to consider the function $-u(x)$ which is concave if $u(x)$ is convex. ■

Inequalities (2.4.1)–(2.4.2) have proved to be useful in a great many applied and theoretical problems, and we will repeatedly come back to these inequalities in next chapters. For now, we consider one example.

EXAMPLE 1 (*Random interest rate*). Assume that we have deposited an amount C_0 into a bank account, and the bank credits (or compounds) interest continuously (for details, see the Appendix , Section 3). If the annual interest rate does not change in time and equals δ, then at time t, the capital is given by

$$C_t = C_0 e^{\delta t}. \tag{2.4.4}$$

(See again the Appendix , Section 3.)

In general, the rate δ depends on the market and, consequently, is random. An interesting question is whether we underestimate or overestimate the expected income $E\{C_t\}$ if we replace the random rate δ by its mean value $E\{\delta\}$.

The answer immediately follows from Jensen's inequality. Since e^x is a convex function,

$$E\{C_t\} = C_0 E\left\{e^{\delta t}\right\} \geq C_0 e^{E\{\delta t\}} = C_0 e^{E\{\delta\}t}.$$

Thus, when replacing the rate δ by its mean $E\{\delta\}$, we underestimate the mean effective growth.

For instance, assume that for a "good" year, the annual rate is 8%, while for a "bad" year, it is 2%. Suppose that both scenarios are equally likely. In this case, if the initial capital $C_0 = 1$, then the expected capital at the end of a year is

$$E\{C_1\} = \frac{1}{2}e^{0.02} + \frac{1}{2}e^{0.08} \approx 1.0517, \text{ while } \ e^{E\{\delta\}} = e^{0.05} \approx 1.0513.$$

The difference is not large but may be significant for large investment. □

---

3 VARIANCE AND OTHER MOMENTS OF HIGHER ORDERS. INEQUALITIES FOR DEVIATIONS

3.1 Definitions and examples

We now know how to compute the mean value of a r.v. Our next goal is to find a suitable characteristic of the dispersion or variation of the values of r.v.'s.

EXAMPLE 1. Suppose you take a train and do not know whether the exit you need in the destination station is at the head or tail of the train. Compare two strategies. In the first strategy, you take either the first or the last car, whichever is closer. For the second strategy, you take the middle car. Let X and Y be the distance you will have to walk upon arrival following the first and second strategies, respectively. If both possible locations of the exit are equally likely,

$$X = \begin{cases} l \text{ with probability } 1/2 \\ 0 \text{ with probability } 1/2 \end{cases}, \qquad Y = l/2,$$

where l is the length of the destination station.

Thus, the mean values of both r.v.'s are the same, $l/2$, but the latter variable is certain, while the values of the former are different. □

We want to define a characteristic that would measure how far values of a r.v. X are from its mean *on the average*. The first thought is to consider the deviation $X' = X - m$, where $m = E\{X\}$, and calculate its mean value. However, as we saw in Section 2.1, $E\{X'\} = 0$, which reflects just the fact that *on the average* the r.v. deviates from the mean to the "right" as much as to the "left."

The quantity $E\{|X-m|\}$ is more informative, and is sometimes used in practice, but a much more common and more convenient characteristic is *variance* which is defined as follows:

$$Var\{X\} = E\{(X - E\{X\})^2\}. \tag{3.1.1}$$

In other terms, if $X = x_1, x_2, ...$ with probabilities $f_1, f_2, ...$, and $E\{X\} = m$, then

$$Var\{X\} = (x_1 - m)^2 f_1 + (x_2 - m)^2 f_2 + \dots . \tag{3.1.2}$$

EXAMPLE 1 revisited. Following (3.1.2), we have

$$Var\{X\} = \left(l - \frac{l}{2}\right)^2 \cdot \frac{1}{2} + \left(0 - \frac{l}{2}\right)^2 \cdot \frac{1}{2} = \frac{l^2}{4}, \quad \text{and}$$

$$Var\{Y\} = \left(\frac{l}{2} - \frac{l}{2}\right)^2 \cdot 1 = 0. \quad \square$$

Regarding the latter variance, in the general case, we have the same: if a r.v. X assumes one value, say a, with probability one, then $Var\{X\} = (a-a)^2 \cdot 1 = 0$.

The converse assertion is also true:

> If $Var\{X\} = 0$, then with probability one, X takes on only one value (which coincides with its expectation). (3.1.3)

Indeed, (3.1.2) is a sum of non-negative values. Thus, if the sum equals zero, so does each term. But if $(x_i - m)^2 f_i = 0$, then either $f_i = 0$, or $x_i = m$. ■

As a rule, when computing variances, it is convenient to use the following formula:

$$Var\{X\} = E\{X^2\} - (E\{X\})^2. \tag{3.1.4}$$

The proof is straightforward: if $E\{X\} = m$, then $Var\{X\} = E\{(X-m)^2\} = E\{X^2 - 2mX + m^2\} = E\{X^2\} - 2mE\{X\} + m^2 = E\{X^2\} - 2mm + m^2 = E\{X^2\} - m^2$. ■

First of all, note that since $Var\{X\} \geq 0$, from (3.1.4) it immediately follows that

$$E\{X^2\} \geq (E\{X\})^2, \tag{3.1.5}$$

and from (3.1.4) and (3.1.3) it follows that $E\{X^2\} = (E\{X\})^2$ if and only if X assumes a value with probability one.

Let us consider particular examples of computing variances.

EXAMPLE 2. Let us revisit Examples 2.1-1 and 1.1-8 on flight delays. We have already computed that $E\{X\}=9$. Then, since the distribution of X is given by

payment:	0	150	450
probability:	0.95	0.045	0.005

we have $E\{X^2\}=0^2\cdot 0.95+150^2\cdot 0.045+450^2\cdot 0.005=2025$. Hence, $Var\{X\}=2025-9^2=1944$. □

Now, let us observe that when computing variances, we square r.v.'s, and if for instance, a r.v. is measured in inches, the variance is measured in (inches)^2 (in square). To return to the original measurement units, we take the square root of the variance.

The quantity $\sigma_X=\sqrt{Var\{X\}}$ is called the *standard deviation* of X, and may be viewed as an adjusted, i.e., consistent with measurement units, dispersion characteristic. When it cannot cause misunderstanding, we omit the index X in σ_X.

EXAMPLE 1 one more time revisited. The standard deviation $\sigma_X=\sqrt{l^2/4}=l/2$, which is pretty much consistent with the statement of the problem: the r.v. deviates from its mean by $l/2$ to the left or to the right with equal probabilities.

EXAMPLE 2 revisited. In this case, the standard deviation $\sigma=\sqrt{1944}\approx 44.091$. □

Let us compute the variances of three more distributions.

EXAMPLE 3. Let X equal 1 or 0 with probabilities p and $q=1-p$, respectively. Then $E\{X\}=p$, $E\{X^2\}=1^2\cdot p+0^2\cdot q=p$, and by (3.1.4), $Var\{X\}=p-p^2=p(1-p)=pq$.

EXAMPLE 4. For a *Poisson r.v.* X with a parameter λ, we first compute

$$E\{X(X-1)\}=\sum_{k=0}^{\infty}k(k-1)e^{-\lambda}\frac{\lambda^k}{k!}=\sum_{k=2}^{\infty}k(k-1)e^{-\lambda}\frac{\lambda^k}{k!}=\lambda^2e^{-\lambda}\sum_{k=2}^{\infty}\frac{\lambda^{k-2}}{(k-2)!}$$
$$=\lambda^2e^{-\lambda}\sum_{m=0}^{\infty}\frac{\lambda^m}{m!}=\lambda^2e^{-\lambda}e^{\lambda}=\lambda^2$$

(we set $m=k-2$ in the fourth step). Then, since $E\{X\}=\lambda$, we have $Var\{X\}=E\{X^2\}-(E\{X\})^2=E\{X(X-1)\}+E\{X\}-(E\{X\})^2=\lambda^2+\lambda-\lambda^2=\lambda$.

EXAMPLE 5. For a r.v. N having a *geometric distribution* with a parameter p, as in Example 2.1-3, we use the function $S(q)=\sum_{k=0}^{\infty}q^k=\frac{1}{1-q}$. As above, we consider first

$$E\{N(N-1)\}=\sum_{k=1}^{\infty}k(k-1)pq^{k-1}=pq\sum_{k=1}^{\infty}k(k-1)q^{k-2}=pqS''(q)=pq\frac{2}{(1-q)^3}=\frac{2q}{p^2}.$$

Then, since $E\{N\}=1/p$, similar to the previous example,

$$Var\{N\}=E\{N(N-1)\}+E\{N\}-(E\{N\})^2=\frac{2q}{p^2}+\frac{1}{p}-\left(\frac{1}{p}\right)^2=\frac{q}{p^2}.\quad\square$$

In conclusion, we generalize the above definitions. For a r.v. X, the quantities $m_k=E\{X^k\}$ for a natural k and $\overline{m}_k=E\{|X|^k\}$ for any $k\geq 0$ are called the kth *moment* and the

kth *absolute moment* of the r.v., respectively. So, the expectation $E\{X\}$ is the first moment m_1.

Set $m = m_1$. The quantities $\mu_k = E\{(X-m)^k\}$ and $\bar{\mu}_k = E\{|X-m|^k\}$ are called the kth *central moment* and the kth *absolute central moment* of X, respectively. So, variance is the second central moment $\bar{\mu}_2$. We will use moment characteristics frequently.

3.2 Properties of variance

Some of the properties are not trivial and important, and *the reader is recommended to read this section carefully.* First, for any constant c,

$$Var\{X+c\} = Var\{X\}, \quad Var\{cX\} = c^2 Var\{X\}. \tag{3.2.1}$$

We may unify this as

$$Var\{a+bX\} = b^2 Var\{X\} \tag{3.2.2}$$

for any constants a, b. To verify this, recall that $E\{a+bX\} = a+bE\{X\}$. Hence, by (3.1.1), $Var\{a+bX\} = E\{(a+bX-E\{a+bX\})^2\} = E\{b^2(X-E\{X\})^2\} = b^2E\{(X-E\{X\})^2\} = b^2 Var\{X\}$.

Next, consider summation. We know that $E\{X_1+X_2\} = E\{X_1\}+E\{X_2\}$ for any r.v.'s X_1, X_2.

Is it true that $Var\{X_1+X_2\} = Var\{X_1\}+Var\{X_2\}$ for any r.v.'s X_1, X_2? In general, the answer is "no."

For example, let X_1 and X_2 be equal to the same r.v. X. Then, by (3.2.1), $Var\{X_1+X_2\} = Var\{2X\} = 4Var\{X\}$, while $Var\{X_1\}+Var\{X_2\} = 2Var\{X\}$.

In this example, the r.v.'s X_1, X_2 are "strongly dependent."

Proposition 5 *For any independent r.v.'s X_1, ..., X_n,*

$$Var\{X_1+...+X_n\} = Var\{X_1\}+...+Var\{X_n\}. \tag{3.2.3}$$

It is worth noting that independence in the above proposition can be relaxed to just pairwise independence (see Section 1.2.2); we will see this when proving Proposition 5 at the end of this section.

Corollary 6 *Let $X_1,...,X_n$ be independent and identically distributed with $E\{X_i\} = m$, and $Var\{X_i\} = \sigma^2$ for $i = 1,..,n$. Set $S_n = X_1+...+X_n$. Then*

$$E\{S_n\} = mn, \text{ and } Var\{S_n\} = \underbrace{\sigma^2+...+\sigma^2}_{n \text{ times}} = \sigma^2 n. \tag{3.2.4}$$

(For the first formula, see (2.1.8).)

The last relation leads to an unexpected consequence. From (3.2.4), it follows that the standard deviation of the sum S_n is $\sigma\sqrt{n}$. That is, the standard deviation, which we view as a measure of dispersion, grows proportionally to $\sqrt{n}$ rather than to n. For example, if an investor invests one unit of money in each of nine independent enterprises, and the random results have the same probability distributions, then the "uncertainty" of the total investment is only three times as large as the uncertainty of one separate investment (rather than nine times as large, as might be expected).

Corollary 7 *In the situation of Corollary 6, set*

$$\overline{X}_n = \frac{X_1 + ... + X_n}{n},$$

the average result. Then

$$E\{\overline{X}_n\} = m, \quad and \quad Var\{\overline{X}_n\} = \frac{\sigma^2}{n}. \tag{3.2.5}$$

Thus, the expected value of the average is the same as that of the X's, while $Var\{\overline{X}_n\}$ is decreasing when n is increasing. For arbitrarily large n, the variance vanishes, and hence, in a sense, so does the "uncertainty."

Proof of Corollary *7* is short. By (3.2.1) and (3.2.4),

$$E\{\overline{X}_n\} = \frac{E\{S_n\}}{n} = \frac{mn}{n} = m, \quad \text{and } Var\{\overline{X}_n\} = \frac{Var\{S_n\}}{n^2} = \frac{\sigma^2 n}{n^2} = \frac{\sigma^2}{n}. \blacksquare$$

EXAMPLE 1 (*Redistribution of risk*). At first glance, this example may seem like a small miracle. Two investors, Ann and David, expect random incomes amounting to random variables X_1 and X_2, respectively. We do not exclude the case when the X's may take on negative values, which corresponds to losses. For simplicity, suppose X_1 and X_2 are independent with the same probability distribution. Then X_1 and X_2 have the same expected value $m = E\{X_i\}$ and variance $\sigma^2 = Var\{X_i\}$.

Assume that Ann and David evaluate the riskiness of their investments by the variance of income, and being risk averse, they want to reduce the riskiness of their future incomes. To this end, Ann and David decide to divide the total income into equal shares, so each will have the random income

$$Y = \frac{1}{2}(X_1 + X_2).$$

Then for both Ann and David, in accordance with (3.2.5), the expected value of the new income will be

$$E\{Y\} = m,$$

that is, the same as before sharing the risk. On the other hand, by virtue of (3.2.5), the variance

$$Var\{Y\} = \frac{1}{2}\sigma^2,$$

and is half as large.

Although this result is easy to prove, this is a key fundamental fact. And it is indeed quite astonishing. The riskiness of the system as a whole did not change, the r.v.'s X_1 and X_2 remained as they were, but the level of risk faced by each participant has decreased.

Consider n participants of a mutual risk exchange, and denote their random incomes by $X_1, ..., X_n$. Assume again that the X's are independent and identically distributed, and set $m = E\{X_i\}$ and $\sigma^2 = Var\{X_i\}$. If the participants divide their total income into equal shares, then the income for each is $Y = \frac{X_1 + ... + X_n}{n}$.

In this case,

$$E\{Y\} = m, \quad \text{while} \quad Var\{Y\} = \frac{\sigma^2}{n},$$

and for large n, the variance is close to zero. Thus, for a large number of participants, the risk of each separate participant may be reduced nearly to zero.

The phenomenon we observed in its simplest form is called *redistribution of risk*. It is at the heart of most stabilization financial mechanisms; for example, insurance. People use insurance because they can redistribute the risk, making it small for each if the number of participants is large. Insurance companies play the role of organizers of such a redistribution. Of course, they do it for profit, although there are non-profit organizations of mutual insurance.

Similarly, the same argument applies to the optimization of investment portfolios, which leads to the principle "Do not keep all of your eggs in one basket." The reader may consider this in more detail when solving Exercise 60. □

EXAMPLE 2 (*A non-rare student mistake*). So, for *independent* r.v.'s X_1 and X_2, we have $Var\{X_1 + X_2\} = Var\{X_1\} + Var\{X_2\}$. Sometimes, for $Var\{X_1 - X_2\}$ people automatically conclude that it is $Var\{X_1\} - Var\{X_2\}$. However, this is not true. For example, if $Var\{X_1\} < Var\{X_2\}$, then $Var\{X_1\} - Var\{X_2\}$ is negative, and cannot be equal to $Var\{X_1 - X_2\}$ which is non-negative . As a matter of fact,

$$\begin{aligned} Var\{X_1 - X_2\} &= Var\{X_1 + (-X_2)\} = Var\{X_1\} + Var\{-X_2\} = Var\{X_1\} + (-1)^2 Var\{X_2\} \\ &= Var\{X_1\} + Var\{X_2\}. \end{aligned}$$

If your income is a r.v. X_1, and the expenses amount to a r.v. X_2, then the profit is $X_1 - X_2$. However, the uncertainty of income is not compensated by that of expenses.

EXAMPLE 3 (*The binomial distribution*). Consider a sequence of n independent trials with a success probability of p. Let X be the number of successes. We use representation (2.1.9) from Example 2.1-6. From Example 3.1-3 it follows that $Var\{X_i\} = pq$, and hence by (3.2.4), $Var\{X\} = npq$. □

Proof of Proposition 5. We consider the case $n = 2$: the proof for $n > 2$ is similar. Set $m_i = E\{X_i\}$ and recall that $E\{X_i - m_i\} = 0$, $i = 1, 2$. By Proposition 2,

$$\begin{aligned} Var\{X_1 + X_2\} &= E\{(X_1 + X_2 - (m_1 + m_2))^2\} = E\{(X_1 - m_1 + X_2 - m_2)^2\} \\ &= E\{(X_1 - m_1)^2\} + E\{(X_2 - m_2)^2\} + 2E\{(X_1 - m_1)(X_2 - m_2)\} \\ &= E\{(X_1 - m_1)^2\} + E\{(X_2 - m_2)^2\} + 2E\{(X_1 - m_1)\}E\{(X_2 - m_2)\} \\ &= E\{(X_1 - m_1)^2\} + E\{(X_2 - m_2)^2\} + 0 = Var\{X_1\} + Var\{X_2\} \quad \blacksquare \end{aligned}$$

3.3 Inequalities for deviations

This section concerns the probability that a r.v. will assume a "large" value. More precisely, for any r.v. X, the probability $P(X \geq x) \to 0$ as $x \to \infty$. Our goal is to estimate how quickly this probability is vanishing. For large x, the probability $P(X \geq x)$ is often called a *tail* of the distribution. Similarly, one can consider $P(X \leq x)$ as $x \to -\infty$. The latter is referred to as a left tail, and the former (in this context)—as a right tail.

Proposition 8 *(An inequality for deviations). Let a function $u(x)$ be non-negative and non-decreasing. Then for any x and any r.v. X,*

$$P(X \geq x) \leq \frac{E\{u(X)\}}{u(x)}, \tag{3.3.1}$$

provided that the r.-h.s. of (3.3.1) is finite.

EXAMPLE 1. Suppose that the quantity $K = E\{e^X\}$ is finite. This is not always true, and we will discuss this issue at the end of this example. Setting $u(x) = e^x$, we obtain that

$$P(X \geq x) \leq \frac{E\{e^X\}}{e^x} = Ke^{-x}. \tag{3.3.2}$$

So, the tail $P(X \geq x)$ converges to zero as $x \to \infty$ no slower than the exponential function.

For instance, if X is a Poisson r.v. with parameter $\lambda = 1$ (see Example 1.1-1), using the Taylor expansion for e^x (the Appendix , (2.2.4)), we get

$$K = \sum_{i=0}^{\infty} e^i \cdot e^{-1}\frac{1}{i!} = e^{-1}\sum_{i=0}^{\infty}\frac{e^i}{i!} = e^{-1}e^e = e^{e-1} \approx 5.57.$$

We consider bounds as (3.3.2) in much more detail in Chapter 8.

To see that K is not always finite, consider, for example, $X = 1, 2, ...,$ and $f_i = P(X = i) = c/i^2$ where the constant c is defined so that $\sum_{i=1}^{\infty} f_i = 1$. (It is known that $c = 6/\pi^2$, but here we do need this fact.) We have $E\{e^X\} = \sum_{i=1}^{\infty} e^i f_i = c\sum_{i=1}^{\infty} e^i \frac{1}{i^2} = \infty$. In this case, we should not expect that $P(X > x)$ is vanishing exponentially. Indeed, let $[x]$ be the integer part of x. Then

$$P(X > x) = c\sum_{[x]+1}^{\infty}\frac{1}{i^2} > c\int_{[x]+1}^{\infty}\frac{1}{(y+1)^2}dy = \frac{c}{[x]+2} \sim \frac{c}{x} \quad \text{as } x \to \infty. \ \square$$

Next, let us replace X in (3.3.1) by $|X|$ and set $u(x) = x^k$ for $x \geq 0$ and $k > 0$. Since $|X|$ is non-negative, it does not matter what $u(x)$ is equal to for $x < 0$, so we may set $u(x) = 0$ for negative x's. We come to

Corollary 9 *(Markov's inequality). For any $x > 0$,*

$$P(|X| \geq x) \leq \frac{E\{|X|^k\}}{x^k} = \frac{\overline{m}_k}{x^k}, \tag{3.3.3}$$

provided that the absolute kth moment $\overline{m}_k$ is finite (see the end of Section 3.1).

Hence, if $\overline{m}_k$ is finite, the tail $P(|X| > x)$ is vanishing at least as fast as the power function $1/x^k$.

Let $m = E\{X\}$. Setting $k = 2$ in (3.3.3) and replacing there X by $X - m$, we come to

Corollary 10 *(Chebyshev's inequality). For any* $x > 0$,

$$P(|X - m| \geq x) \leq \frac{E\{(X-m)^2\}}{x^2} = \frac{Var\{X\}}{x^2}. \tag{3.3.4}$$

Proof of Proposition 8. Using the fact that $u(x)$ is non-negative and that it is non-decreasing consecutively in the inequalities below, we have

$$\begin{aligned} E\{u(X)\} &= \sum_j u(x_j) f_j \geq \sum_{j:\, x_j \geq x} u(x_j) f_j \\ &\geq \sum_{j:\, x_j > x} u(x) f_j = u(x) \sum_{j:\, x_j \geq x} f_j = u(x) P(X \geq x). \ \blacksquare \end{aligned}$$

EXAMPLE 2. What is the probability that X will fall within $\pm k$ standard deviations from its expected value? In other words, what is $P(|X - m| < k\sigma)$, where $m = E\{X\}$, and $\sigma^2 = Var\{X\}$? For example, if $m = 5$ and $\sigma = 2$, what is $P(|X - 5| < k \cdot 2)$?

From (3.3.4), it follows that $P(|X - m| \geq k\sigma) \leq \dfrac{Var\{X\}}{(k\sigma)^2} = \dfrac{\sigma^2}{k^2\sigma^2} = \dfrac{1}{k^2}$, and hence,

$$P(|X - m| < k\sigma) \geq 1 - \frac{1}{k^2}. \tag{3.3.5}$$

For example, for $k = 2$ and for *any (!)* r.v., the probability that its value will fall within two standard deviations from its expected value is no less than $1 - \frac{1}{4} = 0.75$. $\square$

It is noteworthy that since (3.3.5) is a universal bound (valid for *any* r.v.), it may be not sharp, and in particular situations the probability in hand may be much larger.

EXAMPLE 3. Let N have a geometric distribution with $p = 1/2$. Then $q = 1/2$, and as we know from Examples 2.1-3 and 3.1-5, $m = E\{N\} = \frac{1}{p} = 2$, and $\sigma^2 = Var\{N\} = \frac{q}{p^2} = 2$. Hence, $\sigma = \sqrt{2}$. Remembering that $N = 1, 2, \ldots$, we have $P(|N - m| < 2\sigma) = P(2 - 2\sqrt{2} < N < 2 + 2\sqrt{2}) = P(1 \leq N \leq 4)$. For $p = 1/2$, the probability $P(N = k) = 1/2^k$ for all k. Consequently, $P(1 \leq N \leq 4) = \frac{1}{2} + \frac{1}{4} + \frac{1}{8} + \frac{1}{16} > 0.937$, which is substantially larger than 0.75. $\square$

Note also that nowadays, at least when it concerns well known distributions, bounds like those above have rather theoretical significance. If we know that a tail $P(X \geq x)$ is vanishing like e^{-x}, it can be useful for many purposes. However, for particular numerical calculations, we can use good software, and calculate exact probabilities (rather than obtain just bounds for them) for a large number of basic distributions. In Section 4, we consider, in particular, some Excel commands.

4 SOME BASIC DISTRIBUTIONS

In this section, we list some of distributions playing an important role in theory and applications. Most have already been introduced above. We summarize it and go into it somewhat deeper. A summarizing table is given in the Appendix , Tables, Table 1.

4.1 The binomial distribution

Let $n \geq 1$ be an integer and $p \in [0,1]$. The *binomial distribution* with parameters n and p is the distribution of a r.v. X taking on values $0, 1, ..., n$, and such that

$$f_k = P(X = k) = \binom{n}{k} p^k q^{n-k}, \tag{4.1.1}$$

where $q = 1 - p,\ \ k = 0, 1, ..., n$, and

$$\binom{n}{k} = \frac{n!}{k!(n-k)!} = \frac{n(n-1)\cdots(n-k+1)}{k!}, \tag{4.1.2}$$

the number of ways to choose k objects from n distinct objects. (The last representation in (4.1.2) makes sense for $k > 0$. For $k = 0$, we just set $\binom{n}{0} = 1$, which also follows from the first representation in (4.1.2).)

The binomial distribution with parameters n and p usually appears in applications as the distribution of the number of successes in the sequence of n independent trials, if the probability of success in a separate trial is equal to p; see Example 1.1-10. Adopting this interpretation, we can represent X as

$$X = X_1 + ... + X_n, \tag{4.1.3}$$

where $X_i = 1$ if the ith trial is successful, and $X_i = 0$ otherwise. (See also Example 2.1-6.) The r.v. X_i's defined above are called *Bernoulli's variables.* We have assumed that $P(X_i = 1) = p$, and since the trials are independent, the r.v.'s X_i are independent. In Examples 2.1-6 and 3.2-3, proceeding from representation (4.1.3), we have shown that

$$E\{X\} = np, \text{ and } Var\{X\} = npq.$$

The ***Excel command*** BINOMDIST provides calculations for the cumulative probability $P(X \leq k)$ and for $P(X = k)$. The respective commands are
=BINOMDIST(k, n, p,TRUE), and =BINOMDIST(k, n, p,FALSE).

4.2 The multinomial distribution

The next distribution is multivariate, and represents a natural generalization of the binomial distribution. Assume that each of n independent trials may have any of l possible results with respective probabilities $p_1, ..., p_l$ such that $\sum_{i=1}^{l} p_i = 1$. Denote by K_i the number of trials with result i, and consider the *joint distribution* of the r.v.'s $K_1, ..., K_l$. Let

$m_1,...,m_l$ be non-negative integers such that $m_1+...+m_l=n$. Then for any such integers,

$$P(K_1=m_1,...,K_l=m_l)=\frac{n!}{m_1!\cdot...\cdot m_l!}p_1^{m_1}\cdot...\cdot p_l^{m_l}. \qquad (4.2.1)$$

To sketch a proof, observe that, for n trials, there are exactly $n!/(m_1!\cdot...\cdot m_l!)$ outcomes having exactly m_1 results of the first type, m_2 results of the second type, and so on; see Section **1**.3.3. Since the trials are independent, the probability of each outcome mentioned is $p_1^{m_1}\cdot...\cdot p_l^{m_l}$. This implies (4.2.1).

It is worthwhile to emphasize that the marginal distributions, i.e., the distributions of K_i's are binomial. For example,

$$P(K_1=k)=\binom{n}{k}p_1^k(1-p_1)^{n-k}. \qquad (4.2.2)$$

Formally, (4.2.2) follows from (4.2.1) if we set $m_1=k$ and add up all probabilities (4.2.1) over all possible values of $m_2,...,m_l$. However, we can justify (4.2.2) without calculations if we consider n independent trials above, and call it a success if a trial ends up with the first result, and call it a failure otherwise. Then K_1 is the number of successful trials and has the binomial distribution with parameters n and p_1.

4.3 The geometric distribution

In the literature, two closely related distributions are called *geometric*. First, this is the distribution of a r.v. N taking values $1,2,...$ and such that

$$f_k=P(N=k)=pq^{k-1}, \qquad (4.3.1)$$

where the parameter $p\in[0,1]$, and $q=1-p$.

A classical example is the distribution of the number N of the first successful trial in a sequence of independent trials with the same probability of success p. In this case, the event $\{N>k\}$ occurs if the first k trials are not successful, which implies that

$$P(N>k)=q^k. \qquad (4.3.2)$$

We have computed in Examples 2.1-3 and 3.1-5 that

$$E\{N\}=\frac{1}{p},\ Var\{N\}=\frac{q}{p^2}. \qquad (4.3.3)$$

The geometric distribution has the following property:

$$P(N>m+k\,|\,N>k)=P(N>m) \qquad (4.3.4)$$

for any integers m and k. This may be clarified as follows.

Assume that we have already performed k trials, and there was no success in these trials (condition $N>k$). What is the probability that during the *next* m trials there will be no success either? The property (4.3.4) says that the past history has no effect on how long we will wait for a success after k trials: everything starts over "as from the very beginning," and the probability that it will happen after an additional m trials does not depend on k.

Such a property is called the *memoryless,* or the *lack of memory,* property.

With the use of the trial interpretation, (4.3.4) immediately follows from the independency of the trials: after each trial, the process starts over. Nevertheless, let us carry out a formal proof:

$$P(N > m+k \,|\, N > k) = \frac{P(N > m+k \text{ and } N > k)}{P(N > k)} = \frac{P(N > m+k)}{P(N > k)}$$

$$= \frac{q^{m+k}}{q^k} = q^m = P(N > m).$$

It makes also sense to emphasize that above we are dealing with integer m and k. For non-integer m and k, (4.3.4) may be not true. Say, $P(N > 2.5 + 2.5 \,|\, N > 2.5) = \frac{P(N>5)}{P(N>2.5)} = \frac{P(N>5)}{P(N>2)} = \frac{q^5}{q^2} = q^3$ while $P(N > 2.5) = q^2$.

Often, people also call "geometric" the distribution of the r.v. $K = N - 1$ which, naturally, assumes values $0, 1, 2, \ldots$. In other words, K is the number of failures before the first success. In this case, we have

$$P(K = k) = P(N = k+1) = pq^k,\ k = 0, 1, 2, \ldots. \tag{4.3.5}$$

It follows from (4.3.2) that

$$P(K > k) = q^{k+1}. \tag{4.3.6}$$

For the mean and variance, we have

$$E\{K\} = E\{N\} - 1 = \frac{1}{p} - 1 = \frac{q}{p}, \text{ and } Var\{N\} = Var\{K\} = \frac{q}{p^2}. \tag{4.3.7}$$

Note also, that for the r.v. K, (4.3.4) should be slightly changed; the reader is invited to consider this issue in Exercise 68.

The ***Excel command*** =NEGBINOMDIST($k, 1, p$) provides calculations for $P(K = k)$.

4.4 The negative binomial distribution

Consider again a sequence of independent trials with a probability p of being a success. As above we set $q = 1 - p$. Suppose that the trials are performed until a total of ν successes is accumulated, and let N_ν be the number of the trials required for this. In other words, N_ν is the number of the trial in which the νth success occurs. Then N_ν may assume values ν, $\nu + 1, \ldots$. The νth success can occur in the mth trial if and only if the mth trial is successful, and among the previous $m - 1$ trials, there are exactly $\nu - 1$ successes. Then, as it follows from (4.1.1),

$$P(N_\nu = m) = p\binom{m-1}{\nu-1}p^{\nu-1}q^{m-\nu} = \binom{m-1}{\nu-1}p^\nu q^{m-\nu},\ m = \nu, \nu+1, \ldots . \tag{4.4.1}$$

(For $\nu = 2$, we have obtained the same in Example 1.2.3-2 using convolution. Here, we applied a combinatorial technique.)

As in Example 2.1-7, denote by τ_i the number of trials after the $(i-1)$th success it takes until the ith success occurs. For $i = 1$, we set $\tau_1 = N_1$. Clearly, $N_\nu = \tau_1 + \tau_2 + ... + \tau_\nu$.

As was shown in Example 2.1-7, the r.v.'s τ_i are independent and have the geometric distribution. Then, in view of (4.3.3),

$$E\{N_\nu\} = \frac{\nu}{p},\ Var\{N_\nu\} = \frac{\nu q}{p^2}. \tag{4.4.2}$$

EXAMPLE 1 (*The Banach match problem*). A mathematician carries two match boxes with n matches in each. One box is in his left pocket, and the other is in his right pocket. When the mathematician needs a match, he chooses a box at random. Consider the moment when the mathematician takes a box, and it turns out to be empty. What is the probability that at this moment the other box contains exactly k matches?

We view a selection of a match box as a trial. Let A denote the event that the box which turns out to be empty is the right box, and at this moment there are k matches in the left box. This event occurs if and only if the $(n+1)$th choice of the right box (when it is, as matter of fact, empty) is made at the $(n+1+n-k)$th trial. (All n matches were taken from the right box, and $n-k$ from the left.)

Let us call the choice of the right box success, and let N_ν be the moment of the νth success. Then, by (4.4.1) with $p = 1/2$,

$$P(A) = (N_{n+1} = 2n+1-k) = \binom{2n-k}{n}\left(\frac{1}{2}\right)^{2n-k+1}.$$

Since the matchboxes are equally likely to be found empty, the probability in hand is

$$2P(A) = \binom{2n-k}{n}\left(\frac{1}{2}\right)^{2n-k}. \square$$

As in the case of the geometric distribution, we consider an alternative definition of the negative binomial distribution. This is the distribution of the r.v. $K_\nu = N_\nu - \nu$, which takes on values $0, 1, 2, ...$ In other words, K_ν is the number of failures at the moment of the νth success.

Since $P(K_\nu = m) = P(N_\nu = m+\nu)$, it follows from (4.4.1) that

$$P(K_\nu = m) = \binom{\nu+m-1}{\nu-1}p^\nu q^m,\ m = 0, 1, 2, ...,$$

or, due to the formula $\binom{n}{k} = \binom{n}{n-k}$,

$$P(K_\nu = m) = \binom{\nu+m-1}{m}p^\nu q^m,\ m = 0, 1, 2, ... \tag{4.4.3}$$

From (4.4.2), we get that

$$E\{K_\nu\} = \frac{\nu}{p} - \nu = \frac{\nu q}{p},\ Var\{K_\nu\} = \frac{\nu q}{p^2}. \tag{4.4.4}$$

The ***Excel command*** '=NEGBINOMDIST(m, ν, p)' provides calculations for $P(K_\nu = m)$, where p is the above parameter of the distribution.

Route 1 ⇒ page 87

Distribution (4.4.3) appears in many applications *including those which are not relevant to a sequence of trials and the numbers of successes*. Moreover, in some applications, the parameter ν in (4.4.3) is positive but not necessarily an integer. The last instance requires clarification.

2

First, let us observe that the quantity $\binom{r}{k}$ may be defined for any real number r and integer $k \geq 0$ if we adopt as a definition the last expression in (4.1.2), setting

$$\binom{r}{0} = 0, \text{ and } \binom{r}{k} = \frac{r(r-1)\cdots(r-k+1)}{k!} \text{ for } k = 1,2,\dots . \tag{4.4.5}$$

Consider a r.v. K_ν, not necessarily connected with a sequence of trials, whose distribution is formally defined by (4.4.3). The parameter ν in (4.4.3) is positive, and the coefficients $\binom{\nu+m-1}{m}$ are defined in accordance with (4.4.5). Since $\nu > 0$, all these coefficients are positive, and hence all probabilities in (4.4.3) are also positive. Then, to show that the distribution is well defined, it remains to prove that the sum of these probabilities equals one. To this end, we use the Taylor expansion of the function $(1-x)^{-\alpha}$ (see the Appendix, (2.2.9)), which gives

$$\sum_{m=0}^{\infty} P(K_\nu = m) = p^\nu \sum_{m=0}^{\infty} \binom{\nu+m-1}{m} q^m = p^\nu(1-q)^{-\nu} = p^\nu p^{-\nu} = 1.$$

The distribution (4.4.3) is called negative binomial with parameters p and ν. If ν is an integer, the distribution of the r.v. $K_\nu + \nu$ coincides with the distribution of the νth success. If $\nu = 1$, this is the geometric distribution with parameter p.

Formulas (4.4.4) remain true in the general case of an arbitrary positive ν. We skip here particular calculations. Another and easier way to show this, is to use moment generation functions; see Section **8**.1.2.

4.5 The Poisson distribution and theorem

1,2

4.5.1 Poisson approximation

As was introduced in Example 1.1-9, the *Poisson distribution* is that of a non-negative integer valued r.v. X such that

$$P(X = k) = e^{-\lambda}\lambda^k/k! \text{ for } k = 0,1,\dots , \tag{4.5.1}$$

where λ is a positive parameter. We have shown in Examples 2.1-4 and 3.1-4 that

$$E\{X\} = \lambda, \ \ Var\{X\} = \lambda, \tag{4.5.2}$$

so the parameter λ has an explicit sense: it is a mean value. The fact that the mean and variance are equal to each other is not accidental and may be explained by properties of the Poisson process we consider in Chapter 13.

There are many programs computing Poisson probabilities. The respective ***Excel commands*** for $P(X \leq k)$ and $P(X = k)$ are

=POISSON(k, λ, TRUE) and =POISSON(k, λ, FALSE).

There are at least two explanations why this distribution plays a very important role in theory and has a wide range of applications in diverse areas. First, the Poisson distribution appears in a key model of the flow of events or "arrivals," like occurring from time to time breakdowns of a system (say, electrical breakdowns in power lines), e-mail messages arriving at a mail account, customers making purchases, etc., etc. In Chapter 13, we consider in detail how some natural assumptions on the evolution of such processes lead to the Poisson distribution for the number of arrivals during any given period of time. See also a somewhat heuristic reasoning in Section 4.5.3 in this chapter.

Another explanation is connected with Poisson's theorem below which says that the Poisson distribution may be used as a good approximation for the binomial one with parameters (n, p) if n is "large" and p is "small" in a way that np is of a "moderate" size.

EXAMPLE 1. There are $n = 100$ students enrolled in a class. Assume that each student does not show up in a lecture with probability $p = 0.04$, and whether she/he attends the lecture does not depend on the behavior of other students. Let X be the number of students who missed a randomly selected lecture. This is a binomial r.v., and its mean equals $np = 100 \cdot 0.04 = 4$. So, n is "large," p is "small," whereas the mean number of students who will not attend the lecture, is neither large nor small, and assumes a "moderate" value of 4. We call such a situation as that of *rare events*. □

Consider a general scheme of n independent trials with a success probability p, and denote by X the number of successes. The r.v. X is binomial, $E\{X\} = np$, and

$$P(X = k) = \binom{n}{k} p^k (1-p)^{n-k}. \tag{4.5.3}$$

To state the theorem below rigorously, assume that the probability p depends on n, and

$$p = p_n = \frac{\lambda}{n} + o\left(\frac{1}{n}\right), \tag{4.5.4}$$

where λ is a positive number, and $o(1/n)$ is a function, or a remainder, which is essentially smaller than $1/n$ for large n. Rigorously, $n \cdot o(1/n) \to 0$ as $n \to \infty$.

The reader may just view this term as negligible when $n \to \infty$, or even set it equal to 0 *in the first reading.*

What is important to understand is that

$$E\{X\} = np_n = \lambda + n \cdot o(1/n) \to \lambda \text{ as } n \to \infty, \tag{4.5.5}$$

by the assumption on $o(1/n)$. Thus, $E\{X\} \approx \lambda$ for large n.

Note also that the r.v. X in this framework depends on n, so we write $X = X_n$.

Theorem 11 *(The Poisson theorem). For any k,*

$$P(X_n = k) \to e^{-\lambda}\frac{\lambda^k}{k!} \quad \text{as } n \to \infty. \tag{4.5.6}$$

We prove it in Section 4.5.2, but first, let us consider how to apply the approximation the Poisson theorem suggests, to particular problems.

EXAMPLE 1 revisited. Let $\lambda = np = 4$. Applying the Poisson approximation, we write

$$P(X = k) \approx e^{-\lambda}\frac{\lambda^k}{k!} = e^{-4}\frac{4^k}{k!}.$$

For example, the probability that at most one student will miss the lecture is

$$P(X = 0) + P(X = 1) \approx e^{-\lambda} + e^{-\lambda}\lambda = e^{-4} + 4e^{-4} \approx 0.092.$$

For the respective binomial probability, Excel gives ≈ 0.087, so the approximation is not so bad.

EXAMPLE 2. A gambler plays a game of chance that for each play, pays \$100 with probability 0.01, and pays nothing with probability 0.99. For simplicity, suppose that the participation in each play costs \$1. The gambler is going to play 100 times, whatever will happen during these plays. What are the odds?

In our problem, $n = 100$, $p = 0.01$, and $\lambda = np = 1$. Denote by X the number of wins. The gambler will lose \$100 with the probability $P(X = 0)$; there will be no gain or loss with the probability $P(X = 1)$; and the gambler will gain some money with the probability $P(X \geq 2) = 1 - P(X = 0) - P(X = 1)$. The Poisson approximation implies $P(X = 0) \approx e^{-\lambda} = \frac{1}{e}$, $P(X = 1) \approx e^{-\lambda}\lambda = \frac{1}{e}$, and hence $P(X \geq 2) \approx 1 - \frac{2}{e}$. Approximate numerical values are 0.37, 0.37, 0.26. For a risk lover, it is not extremely bad, so the decision is up to the gambler.

EXAMPLE 3 shows, first, that the Poisson approximation does not work if p is not small, and second, that the accuracy of approximation may be pretty high when p is small.

Set $n = 30$, and, initially, $p = 0.5$. Then $\lambda = 15$. The Excel worksheet in Fig. 7a, for $k = 0, ..., 30$ presented in Column A, shows the binomial (the r.-h.s. of (4.5.3)) and Poisson (the r.-h.s. of (4.5.6)) probabilities in columns B and C, respectively. The corresponding graphs are in the chart. The values of the cumulative probabilities $P(X \leq k)$ are given in columns E and F, and the difference between them—in column G. (Recall that $P(X \leq k)$ is called a (cumulative) distribution function (d.f), which explains the titles of these columns.)

We see that the distributions are not close: it is enough to take a look at the graphs in the chart. The maximal difference in absolute value between the d.f.'s is 0.0868 (for $k = 12$), which is large. This is not surprising since $p = 0.5$ is not "small."

Now, let n be still 30, but $p = 0.1$. Then $\lambda = 3$. The result given in Fig. 7b shows that now the distributions are fairly close: the graphs practically coincide, and the maximal difference between the d.f.'s is 0.0107 (for $k = 5$), which is not bad at all. We see that a good approximation can appear for relatively "moderate" n.

In the literature, one can find rigorous and pretty sharp estimates of the accuracy of the Poisson approximation. See, for example, the monograph [5] and references therein.

EXAMPLE 4. Let us revisit Example 1.1-13, and consider the particular problem described in this example, concerning the doubling strategy in a roulette game with a minimal bet of \$1 and a maximal bet of \$200. We have obtained there that the gambler fails to apply the doubling strategy with a probability of 0.006.

Suppose that a professor, when teaching Probability, considered this problem in his class of 100 students, and after the lecture, all students rushed to a casino (with rules described

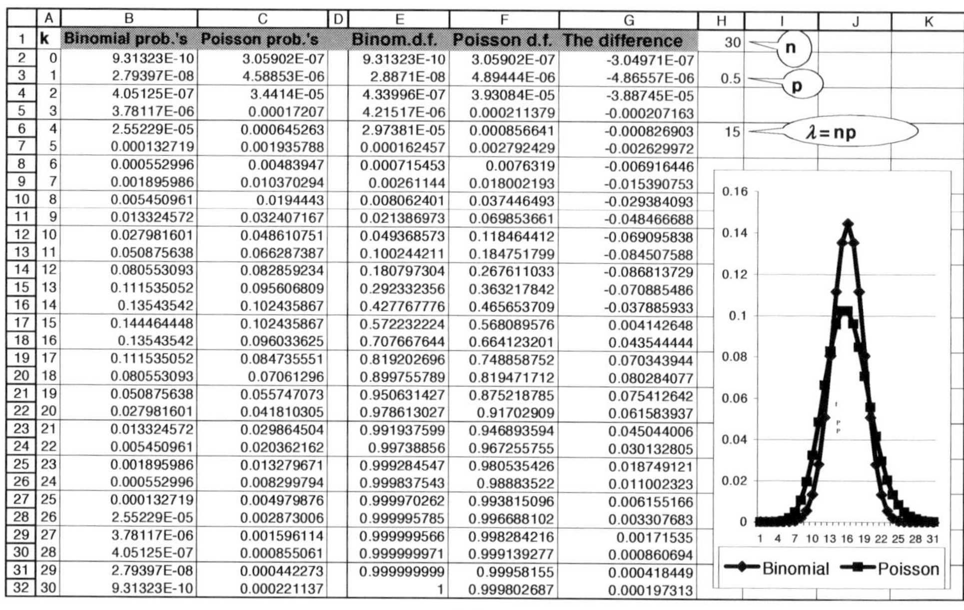

	A	B	C	D	E	F	G	H	I
1	k	Binomial prob.'s	Poisson prob.'s		Binom.d.f.	Poisson d.f.	The difference		
2	0	9.31323E-10	3.05902E-07		9.31323E-10	3.05902E-07	-3.04971E-07	30	n
3	1	2.79397E-08	4.58853E-06		2.8871E-08	4.89444E-06	-4.86557E-06	0.5	p
4	2	4.05125E-07	3.4414E-05		4.33996E-07	3.93084E-05	-3.88745E-05		
5	3	3.78117E-06	0.00017207		4.21517E-06	0.000211379	-0.000207163		
6	4	2.55229E-05	0.000645263		2.97381E-05	0.000856641	-0.000826903	15	$\lambda = np$
7	5	0.000132719	0.001935788		0.000162457	0.002792429	-0.002629972		
8	6	0.000552996	0.00483947		0.000715453	0.0076319	-0.006916446		
9	7	0.001895986	0.010370294		0.00261144	0.018002193	-0.015390753		
10	8	0.005450961	0.0194443		0.008062401	0.037446493	-0.029384093		
11	9	0.013324572	0.032407167		0.021386973	0.069853661	-0.048466688		
12	10	0.027981601	0.048610751		0.049368573	0.118464412	-0.069095838		
13	11	0.050875638	0.066287387		0.100244211	0.184751799	-0.084507588		
14	12	0.080553093	0.082859234		0.180797304	0.267611033	-0.086813729		
15	13	0.111535052	0.095606809		0.292332356	0.363217842	-0.070885486		
16	14	0.13543542	0.102435867		0.427767776	0.465653709	-0.037885933		
17	15	0.144464448	0.102435867		0.572232224	0.568089576	0.004142648		
18	16	0.13543542	0.096033625		0.707667644	0.664123201	0.043544444		
19	17	0.111535052	0.084735551		0.819202696	0.748858752	0.070343944		
20	18	0.080553093	0.07061296		0.899755789	0.819471712	0.080284077		
21	19	0.050875638	0.055747073		0.950631427	0.875218785	0.075412642		
22	20	0.027981601	0.041810305		0.978613027	0.91702909	0.061583937		
23	21	0.013324572	0.029864504		0.991937599	0.946893594	0.045044006		
24	22	0.005450961	0.020362162		0.99738856	0.967255755	0.030132805		
25	23	0.001895986	0.013279671		0.999284547	0.980535426	0.018749121		
26	24	0.000552996	0.008299794		0.999837543	0.98883522	0.011002323		
27	25	0.000132719	0.004979876		0.999970262	0.993815096	0.006155166		
28	26	2.55229E-05	0.002873006		0.999995785	0.996688102	0.003307683		
29	27	3.78117E-06	0.001596114		0.999999566	0.998284216	0.00171535		
30	28	4.05125E-07	0.000855061		0.999999971	0.999139277	0.000860694		
31	29	2.79397E-08	0.000442273		0.999999999	0.99958155	0.000418449		
32	30	9.31323E-10	0.000221137		1	0.999802687	0.000197313		

(a)

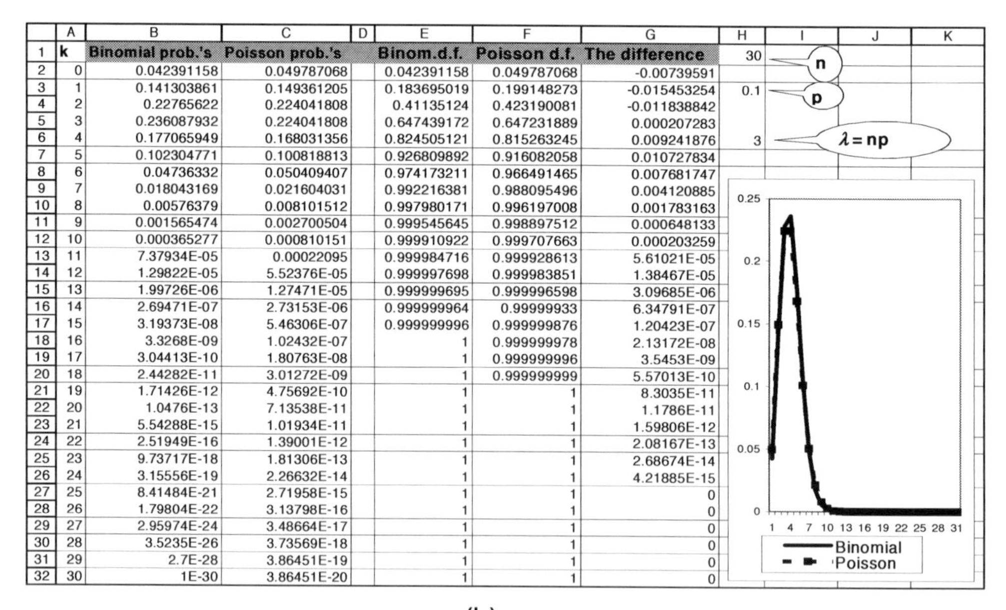

	A	B	C	D	E	F	G	H	I
1	k	Binomial prob.'s	Poisson prob.'s		Binom.d.f.	Poisson d.f.	The difference		
2	0	0.042391158	0.049787068		0.042391158	0.049787068	-0.00739591	30	n
3	1	0.141303861	0.149361205		0.183695019	0.199148273	-0.015453254	0.1	p
4	2	0.22765622	0.224041808		0.41135124	0.423190081	-0.011838842		
5	3	0.236087932	0.224041808		0.647439172	0.647231889	0.000207283		
6	4	0.177065949	0.168031356		0.824505121	0.815263245	0.009241876	3	$\lambda = np$
7	5	0.102304771	0.100818813		0.926809892	0.916082058	0.010727834		
8	6	0.04736332	0.050409407		0.974173211	0.966491465	0.007681747		
9	7	0.018043169	0.021604031		0.992216381	0.988095496	0.004120885		
10	8	0.00576379	0.008101512		0.997980171	0.996197008	0.001783163		
11	9	0.001565474	0.002700504		0.999545645	0.998897512	0.000648133		
12	10	0.000365277	0.000810151		0.999910922	0.999707663	0.000203259		
13	11	7.37934E-05	0.00022095		0.999984716	0.999928613	5.61021E-05		
14	12	1.29822E-05	5.52376E-05		0.999997698	0.999983851	1.38467E-05		
15	13	1.99726E-06	1.27471E-05		0.999999695	0.999996598	3.09685E-06		
16	14	2.69471E-07	2.73153E-06		0.999999964	0.99999933	6.34791E-07		
17	15	3.19373E-08	5.46306E-07		0.999999996	0.999999876	1.20423E-07		
18	16	3.3268E-09	1.02432E-07		1	0.999999978	2.13172E-08		
19	17	3.04413E-10	1.80763E-08		1	0.999999996	3.5453E-09		
20	18	2.44282E-11	3.01272E-09		1	0.999999999	5.57013E-10		
21	19	1.71426E-12	4.75692E-10		1	1	8.3035E-11		
22	20	1.0476E-13	7.13538E-11		1	1	1.1786E-11		
23	21	5.54288E-15	1.01934E-11		1	1	1.59806E-12		
24	22	2.51949E-16	1.39001E-12		1	1	2.08167E-13		
25	23	9.73717E-18	1.81306E-13		1	1	2.68674E-14		
26	24	3.15556E-19	2.26632E-14		1	1	4.21885E-15		
27	25	8.41484E-21	2.71958E-15		1	1	0		
28	26	1.79804E-22	3.13798E-16		1	1	0		
29	27	2.95974E-24	3.48664E-17		1	1	0		
30	28	3.5235E-26	3.73569E-18		1	1	0		
31	29	2.7E-28	3.86451E-19		1	1	0		
32	30	1E-30	3.86451E-20		1	1	0		

(b)

FIGURE 7. The accuracy of the Poisson approximation.

above) to apply the doubling strategy. What is the probability that at least one of them will lose?

A good idea is to apply the Poisson approximation. Let K be the number of those students out of 100 who will not be able to use the doubling strategy. Since $\lambda = 100 \cdot 0.006 = 0.6$, we have $P(K \geq 1) = 1 - P(K = 0) \approx 1 - e^{-\lambda} = 1 - e^{-0.6} \approx 0.45$.

This is not a small probability, and it will not be superfluous if the professor will call students' attention to this fact in class. □

4.5.2 Proof of the Poisson theorem

As was told above, the term $o(1/n)$ is negligible with respect to $1/n$ as $n \to \infty$. In the first reading, the reader may just set below this expression equal to 0. We have

$$
\begin{aligned}
P(X_n = k) &= \binom{n}{k} p_n^k (1 - p_n)^{n-k} \\
&= \frac{n(n-1)...(n-k+1)}{k!} \left(\frac{\lambda}{n} + o\left(\frac{1}{n}\right)\right)^k \left(1 - \frac{\lambda}{n} - o\left(\frac{1}{n}\right)\right)^{n-k} \\
&= \frac{n(n-1)...(n-k+1)}{k!} \cdot \frac{(\lambda + n\, o(1/n))^k}{n^k} \cdot \left(1 - \frac{\lambda}{n} - o\left(\frac{1}{n}\right)\right)^n \cdot \left(1 - \frac{\lambda}{n} - o\left(\frac{1}{n}\right)\right)^{-k} \\
&= \left(1 - \frac{\lambda}{n} - o\left(\frac{1}{n}\right)\right)^n \cdot \frac{(\lambda + n\, o(1/n))^k}{k!} \cdot \frac{n(n-1)...(n-k+1)}{n^k} \cdot \left(1 - \frac{\lambda}{n} - o\left(\frac{1}{n}\right)\right)^{-k} \\
&= \left(1 - \frac{\lambda}{n} - o\left(\frac{1}{n}\right)\right)^n \cdot \frac{(\lambda + n\, o(1/n))^k}{k!} \\
&\cdot \left[1\left(1 - \frac{1}{n}\right) \cdot ... \cdot \left(1 - \frac{k-1}{n}\right)\right] \cdot \left(1 - \frac{\lambda}{n} - o\left(\frac{1}{n}\right)\right)^{-k}.
\end{aligned}
$$

The first factor converges to $e^{-\lambda}$, the second—to $\lambda^k/k!$, the term in the brackets [..] and the fourth factor converge to 1. This implies (4.5.6). ■

Route 1 ⇒ page 92

2

4.5.3 Modeling a process of arrivals

Reasoning for now somewhat heuristically, we show why a Poisson r.v. proves to be a good model for the number of arrivals in a great many of problems concerning flows of events like customers entering a store on a particular day, traffic accidents, computer breakdowns, etc. This section differs from what we considered before in that now we explore the process of arrivals in its dynamics, in time.

Consider a unit *time* interval $[0, 1)$; it is convenient not to include the right endpoint. Denote by X the number of arrivals during this period. Let us divide the interval into n equal time subintervals $[\frac{k-1}{n}, \frac{k}{n})$, $k = 1, ..., n$, and set

$$
X_{kn} = \begin{cases} 1 \text{ if at least one arrival occurs in } [\frac{k-1}{n}, \frac{k}{n}), \\ 0 \text{ otherwise.} \end{cases}
$$

We endowed these r.v.'s by the second index n because they depend on into how many subintervals we divided the original time interval. Then the total sum $S_n = \sum_{k=1}^{n} X_{kn}$ is equal to the number of subintervals "with arrivals."

We make the following three assumptions.

A. For each n, the r.v.'s X_{kn} are independent. This is a strong requirement: it means that whether an arrival occurred in a particular subinterval does not depend on when and how many arrivals occurred outside of this subinterval. One can call it the memory-less property.

B. For large n (and hence for small subintervals), rigorously as $n \to \infty$, the probability $P(X_{kn} = 1)$ is proportional to the length of the subinterval. This is a natural requirement, and denoting the proportionality coefficient by λ, we adopt the presentation

$$P(X_{kn} = 1) = \lambda \cdot \frac{1}{n} = \frac{\lambda}{n}.$$

(Rigorously speaking, the above condition is true asymptotically, for large n, but as was said, we reason somewhat heuristically.)

C. For large n, rigorously as $n \to \infty$, the probability $P(X_{kn} \geq 2)$ is negligible in comparison with $P(X_{kn} = 1)$. This is also a weak requirement. It means that in a very small period of time, we expect at most one arrival.

In view of Properties A and B, and by the Poisson theorem, the distribution of S_n converges to the Poisson distribution with the parameter λ.

On the other hand, in view of Property C, $S_n \approx X$ for large n, or in other words, $S_n \to X$, as $n \to \infty$, in a reasonable proper sense.

To specify in which sense $S_n \to X$, we should state Property C more rigorously and carry out some calculations, which we skip here. Our goal is to provide heuristic arguments in favor of the Poisson model. We will consider this problem in a rigorous fashion in Chapter 13.

1,2

4.5.4 Summation property

EXAMPLE 1. A police department provides daily reports about traffic accidents in two areas. Denote by X_1 and X_2 the random numbers of accidents in the first and second area, respectively, on a randomly chosen day. We assume X_1 and X_2 to be independent and have the Poisson distributions with respective parameters λ_1 and λ_2. Let $X = X_1 + X_2$, the total number of accidents. Clearly, the mean $E\{X\} = E\{X_1\} + E\{X_2\} = \lambda_1 + \lambda_2$, but will the distribution of X be still Poisson? As it follows from the next proposition, the answer is positive. □

Proposition 12 *Let X_1 and X_2 be independent Poisson r.v.'s with parameters λ_1 and λ_2, respectively. Then the r.v. $X = X_1 + X_2$ is a Poisson r.v. with parameter $\lambda_1 + \lambda_2$.*

Proof. We make use of convolution formula (1.2.4). For $f_n = P(X = n)$, we have

$$f_n = \sum_{k=0}^{n} f_k^{(1)} f_{n-k}^{(2)} = \sum_{k=0}^{n} e^{-\lambda_1} \frac{\lambda_1^k}{k!} e^{-\lambda_2} \frac{\lambda_2^{n-k}}{(n-k)!}$$

$$= e^{-(\lambda_1+\lambda_2)} \frac{1}{n!} \sum_{k=0}^{n} \frac{n!}{k!(n-k)!} \lambda_1^k \lambda_2^{n-k} = e^{-(\lambda_1+\lambda_2)} \frac{1}{n!} \sum_{k=0}^{n} \binom{n}{k} \lambda_1^k \lambda_2^{n-k}$$

$$= e^{-(\lambda_1+\lambda_2)} \frac{(\lambda_1 + \lambda_2)^n}{n!},$$

because the last sum above is the binomial expansion of $(\lambda_1 + \lambda_2)^n$. (See, for example, (**1**.3.4.7).) ■

Thus, in the Poisson case, the sum of independent r.v.'s inherits the distribution of separate terms. The proof above is technical, but the very fact is important for models under discussion. It is also worth emphasizing that such a property is rather rare and may be considered a special property of the Poisson model.

4.5.5 Marked Poisson r.v.'s

Consider two independent Poisson r.v.'s, N_1 and N_2, with respective means λ_1 and λ_2.

For example, a company receives each day customers of two types; the daily numbers of customers of each type are independent and have Poisson distributions. What is the distribution of the number of customers of, say, the first type given the total number of customers? We will see that this distribution is binomial, and this may be considered a distinctive feature of the Poisson distribution.

Set $N = N_1 + N_2$.

Proposition 13 *Let* $p_1 = \dfrac{\lambda_1}{\lambda_1 + \lambda_2}$ *and* $p_2 = \dfrac{\lambda_2}{\lambda_1 + \lambda_2}$. *(So,* $p_1 + p_2 = 1$.*) Then for any* $n = 1, 2, \ldots$, *and* $k = 0, 1, \ldots, n$,

$$P(N_1 = k \,|\, N = n) = \binom{n}{k} p_1^k p_2^{n-k}. \tag{4.5.7}$$

Thus, given the total number of customers $N = n$, the distribution of the number of the customers of the first type, N_1, is binomial with parameters (p_1, n). Clearly, N_2 has the binomial distribution with parameters (p_2, n). (We will see that in this context, it is more convenient to use notations p_1, p_2 rather than traditional p and $q = 1 - p$.)

The next fact may be considered converse to the first.

Let N be the random number of some objects, and suppose that N is Poisson r.v. with a mean of λ. Each object, independently of the other objects and of the number of the objects, may belong to one of two types. For each object, the probability of belonging to the first type is p_1; to the second type, it is p_2. So, $p_1 + p_2 = 1$.

For example, a company offers two products, for \$100 and \$150. Each day, the company deals with a Poisson r.v. of customers, and the probability that a current customer is from those who buy the first product is a given p_1.

Coming back to the general wording, denote by N_1 and N_2 the numbers of objects of the first and second type, respectively. Clearly, $N_1 + N_2 = N$.

Proposition 14 *Let $\lambda_1 = p_1\lambda$ and $\lambda_2 = p_2\lambda$. Then the r.v.'s N_1 and N_2 are independent and have the Poisson distribution with respective parameters λ_1 and λ_2.*

The fact that the r.v.'s are Poisson is not very surprising; the fact that they are independent is less expected. If N is not a Poisson r.v., this may be not true, and the property established may be also viewed as a special property of the Poisson distribution. The r.v.'s N_1, N_2 are called *marked Poisson r.v.'s*: they count only "marked" objects.

EXAMPLE 1. Let us come back to the company that sells two products for the prices \$100 and \$150. Assume that the number of customers during a day is a Poisson r.v. N with a mean of 40, and on the average, 75% of customers buy the cheaper product. The object of study is the total daily income. If we had been solving the problem in a straightforward fashion, we would have introduced the r.v.'s

$$X_j = \begin{cases} 100 \text{ with probability } 3/4, \\ 150 \text{ with probability } 1/4; \end{cases}$$

the payment of the jth customer, and would have considered $X_1 + ... + X_N$, the sum where not only the separate terms are random, but the number of terms is random also.

This is a complex object. We will repeatedly consider it from different points of view—at the first time, in Section 6.2—and will see that it is not very easy to tackle it. However, in the case where N is a Poisson r.v. and X's assume a moderate number of values (in our case, just two), the problem may be essentially simplified.

Let N_1 and N_2 be the number of customers who will buy the first and the second product, respectively. By Proposition 14, N_1, N_2 are independent Poisson r.v.'s with parameters $\lambda_1 = \frac{3}{4}\cdot 40 = 30$ and $\lambda_2 = \frac{1}{4}\cdot 40 = 10$, respectively. The total income is the r.v.

$$S = \$100\cdot N_1 + \$150\cdot N_2.$$

Thus, the sum of, on the average, 40 r.v.'s, has been reduced to the sum of only two (!) r.v.'s. Such a sum is easily tractable. The first characteristics may be written immediately:

$$\begin{aligned} E\{S\} &= 100E\{N_1\} + 150E\{N_2\} = 100\lambda_1 + 150\lambda_2 = 4,500; \\ Var\{S\} &= 100^2 Var\{N_1\} + 150^2 Var\{N_2\} = 100^2\lambda_1 + 150^2\lambda_2 = 525,000; \\ \sigma_S &= \sqrt{Var\{S\}} \approx 725. \end{aligned}$$

Regarding the probabilities of separate values, they cannot be expressed in simple formulas, but are easily computed numerically since we deal with just two variables. See also Exercise 77. □

Proof of Proposition 13. By Proposition 12, the r.v. N has the Poisson distribution with parameter $\lambda = \lambda_1 + \lambda_2$. Since N_1, N_2 are independent,

$$\begin{aligned} P(N_1 = k \,|\, N = n) &= \frac{P(N_1 = k, N_1 + N_2 = n)}{P(N = n)} = \frac{P(N_1 = k,\, N_2 = n-k)}{P(N = n)} \\ &= \frac{P(N_1 = k)P(N_2 = n-k)}{P(N = n)} \\ &= \frac{\exp\{-\lambda_1\}\lambda_1^k}{k!}\cdot\frac{\exp\{-\lambda_2\}\lambda_2^{n-k}}{(n-k)!} \Big/ \frac{\exp\{-(\lambda_1+\lambda_2)\}(\lambda_1+\lambda_2)^n}{n!}. \end{aligned}$$

The exponential terms cancel out, and we get that

$$P(N_1 = k \,|\, N = n) = \frac{n!}{k!(n-k)!} \cdot \frac{\lambda_1^k \lambda_2^{n-k}}{(\lambda_1+\lambda_2)^n} = \binom{n}{k}\left(\frac{\lambda_1}{\lambda}\right)^k \left(\frac{\lambda_2}{\lambda}\right)^{n-k} = \binom{n}{k} p_1^k p_2^{n-k}. \blacksquare$$

Proof of Proposition 14. For $n \neq 0$, given $N = n$, the n objects may be identified with n independent trials, each of which may be successful (the object is of the first type) with probability p_1. Hence, for any n, and any $k \leq n$,

$$P(N_1 = k \,|\, N = n) = \binom{n}{k} p_1^k p_2^{n-k}. \tag{4.5.8}$$

For $n = 0$, (4.5.8) is also true because $P(N_1 = 0 \,|\, N = 0) = 1$; and by convention, $\frac{0!}{0!0!} p_1^0 p_2^0 = 1$ also. By the multiplication rule,

$$\begin{aligned} &P(N_1 = k, N_2 = m) = P(N_1 = k, N_2 = m, N = m+k) = P(N_1 = k, N = m+k) \\ &= P(N_1 = k \,|\, N = m+k) P(N = m+k) \\ &= \frac{(k+m)!}{k!m!} p_1^k p_2^m \, e^{-\lambda} \frac{\lambda^{k+m}}{(k+m)!} = e^{-p_1\lambda} \frac{(p_1\lambda)^k}{k!} e^{-p_2\lambda} \frac{(p_2\lambda)^m}{m!} = e^{-\lambda_1} \frac{\lambda_1^k}{k!} e^{-\lambda_2} \frac{\lambda_2^m}{m!}. \end{aligned}$$

This is the *product* of Poisson probabilities, so N_1, N_2 are Poisson and independent by Proposition 1. ■

In Exercise 78, we carry it over to the case of many Poisson r.v.'s.

---

5 CONVERGENCE OF RANDOM VARIABLES. THE LAW OF LARGE NUMBERS

5.1 Convergence of r.v.'s

First, we define the very notion of convergence of a sequence of r.v.'s X_n to a r.v. X. To this end, we should recall that the r.v.'s X_n and X are functions on a sample space $\Omega = \{\omega\}$; that is, $X_n = X_n(\omega)$ and $X = X(\omega)$. For this reason, we cannot define a limit as that of a sequence of numbers, since functions are more complex objects than numbers.

We consider two definitions. All limits below are those as $n \to \infty$; as a rule, we skip stating it explicitly.

We say that a sequence of r.v.'s $X_n = X_n(\omega)$ converges to a r.v. $X = X(\omega)$ *almost surely*, or *with probability one*, if

$$P(X_n \to X) = 1.$$

More precisely, this means that the set of all ω's for which $X_n(\omega) \to X(\omega)$ has probability one.

In this case, we will also write

$$X_n \stackrel{\text{a.s.}}{\to} X. \tag{5.1.1}$$

In many models, such convergence either does not take place, or if it does, it is difficult to prove. On the other hand, in many applications, it is sufficient to consider a weaker type of convergence when we just require $X_n - X$ to be small for large n with a probability close to one. Let us translate this heuristic definition into mathematical terms.

Saying that $X_n - X$ is small, we mean that $|X_n - X|$ is less than a sufficiently small positive number ε. So, we want $|X_n - X|$ to be less than ε for large n with a large probability.

Saying that a probability is large, we mean that it is close to one, and saying "for large n," we mean that $n \to \infty$. We are ready to give a formal definition.

A sequence of r.v.'s X_n converges to a r.v. X *in probability* if for any, arbitrary small, $\varepsilon > 0$,

$$P(|X_n - X| < \varepsilon) \to 1 \text{ as } n \to \infty. \tag{5.1.2}$$

In this case, we write

$$X_n \xrightarrow{P} X. \tag{5.1.3}$$

Note that (5.1.2) is equivalent to the relation

$$P(|X_n - X| \geq \varepsilon) \to 0 \text{ as } n \to \infty. \tag{5.1.4}$$

From a heuristic point of view, the fact that almost sure convergence implies that in probability, is clear. Indeed, let $P(X_n \to X) = 1$. This means that with probability one, the difference $|X_n - X|$ is vanishing as $n \to \infty$. This implies that for any positive number ε, the probability that $|X_n - X|$ will be larger than ε should be small for large n. A formal proof is postponed to Section 5.4. In the same section, we also give an example where convergence in probability does take place while almost sure convergence does not.

5.2 The law of large numbers (LLN)

Let $X_1, X_2, ...$ be a sequence of independent identically distributed (i.i.d.) r.v.'s. Let $S_n = X_1 + ... + X_n$, and $\overline{X}_n = S_n/n$. Set $m = E\{X_i\}$. It does not depend on i since X's are identically distributed.

The LLN says that though for each particular n, the average $\overline{X}_n$ is random, it is approaching a certain number, namely the mean m as n gets larger. To define it rigorously, we proceed from the definitions of convergence from the previous section. Note also that when we are talking about a certain number, we may view it as a r.v. assuming only one value: this number.

Theorem 15 *(The strong LLN).*

(a) Suppose that m is finite. Then

$$P\left(\overline{X}_n \to m\right) = 1. \tag{5.2.1}$$

(b) If for a number c,

$$P\left(\overline{X}_n \to c\right) = 1, \tag{5.2.2}$$

then m is finite, and $c = m$.

Another way of presenting relation (5.2.1) is

$$\overline{X}_n \overset{\text{a.s.}}{\to} m.$$

Since the *almost sure* convergence implies the convergence in probability, we may state the following

Corollary 16 *(The weak LLN). Suppose that m is finite. Then for any* $\varepsilon > 0$,

$$P\left(|\overline{X}_n - m| < \varepsilon\right) \to 1 \quad as \quad n \to \infty. \tag{5.2.3}$$

Another way of presenting relation (5.2.3) is

$$\overline{X}_n \overset{P}{\to} m.$$

Let us consider the particular but important case of a sequence of independent trials. Let p be the probability of success, and ν_n be the proportion (or frequency) of successes in n trials. The r.v. ν_n may be presented as $\frac{1}{n}(X_1 + ... + X_n)$, where the r.v.'s X_j are independent, $X_j = 1$ if the jth trial is successful, and $X_j = 0$ otherwise (See Section 4.1). Since $E\{X_j\} = p$, we arrive at

Corollary 17 *(The LLN for the binomial scheme (Bernoulli's theorem)). In the scheme of independent trials described above, with probability one,*

$$\nu_n \to p. \tag{5.2.4}$$

We will comment on this result at the end of this section.

In full, Theorem 15 will be proved in Section **15**.4.7; under an additional assumption, we prove it in the next subsection. Nevertheless, we are able right away to give a short *direct*

Proof of (5.2.3) assuming additionally that the variance $\sigma^2 = Var\{X_i\}$ is finite. By (3.2.5), $E\{\overline{X}_n\} = m$, and $Var\{\overline{X}_n\} = \sigma^2/n$. Then, by Chebyshev's inequality (3.3.4), for any $\varepsilon > 0$,

$$P\left(|\overline{X}_n - m| \geq \varepsilon\right) \leq \frac{1}{\varepsilon^2} Var\{\overline{X}_n\} = \frac{1}{\varepsilon^2}\frac{\sigma^2}{n} \to 0, \quad \text{as} \quad n \to \infty. \blacksquare \tag{5.2.5}$$

EXAMPLE 1. Suppose that, from past experience, it is known that the test score of a randomly chosen student taking a Probability exam is a random variable with a mean of 75 and a standard deviation of 5. First, consider the score X of one student chosen at random. What is the probability that this score will fall between 65 and 85? By Chebyshev's inequality (3.3.4),

$$P(65 < X < 85) = P(|X - 75| < 10) = 1 - P(|X - 75| \geq 10) \geq 1 - \frac{25}{10^2} = 0.75.$$

Now, let n students take the exam, X_i be the score of the ith student, and the average score $\overline{X}_n = (X_1 + ... + X_n)/n$. By (3.2.5), $E\{\overline{X}_n\} = 75$, and the LLN tells that for large n, the average score $\overline{X}_n$ should be much closer to 75 than a separate score X_i. In particular, since by (3.2.5) $Var\{\overline{X}_n\} = 25/n$,

$$P(65 < \overline{X}_n < 85) = P(|\overline{X}_n - 75| < 10) = 1 - P(|\overline{X}_n - 75| \geq 10) \geq 1 - \frac{1}{10^2}\frac{25}{n} = 1 - \frac{1}{4n}.$$

For a class of just $n = 25$ people this gives $P(65 < \overline{X}_n < 85) \geq 0.99$ in contrast with 0.75 for one student.

For a narrower range, say, $[70, 80]$, we have

$$P(70 < \overline{X}_n < 80) = P(|\overline{X}_n - 75| < 5) = 1 - P(|\overline{X}_n - 75| \geq 5) \geq 1 - \frac{1}{5^2}\frac{25}{n} = 1 - \frac{1}{n},$$

and for a class of 25 people, this probability is not smaller than 0.96. □

The LLN is a mathematical theorem but *it may be viewed as a fundamental law of nature.* First of all, due to this law, the random behavior of a great many of real processes in the long run exhibits a sort of stability. Consider, for instance, the consecutive values of daily income of a company in the long run. These values—denote them by $X_1, X_2, ...$—may be essentially random, uncertain. However, if the X's are i.i.d., the average income per day for n days, that is, $\overline{X}_n = \frac{1}{n}(X_1 + ... + X_n)$, for large n, in the long run, is practically certain, being close to the non-random value equal to the expected value of the X's.

Not of less importance is that the LLN allows us to estimate the mean values of r.v.'s. Suppose, for example, that we want to estimate the mean of the highest August temperature in a particular area. The only thing we can do is to review such a temperature, say, in the last 100 years and compute the average. Our intuition tells us that the average will be close to the mean value (though, perhaps, not exactly equal to), and the LLN confirms it and explains in which sense it is true.

It is also important to emphasize that, in a certain sense, *the LLN may serve as a justification of the whole theory we are building.* Consider an experiment which may result in a certain event A. In people's mind, the probability of such an event is often connected with the frequency of the occurrences of this event in a long sequence of independent replicas of the same experiment. For example, when rolling a die, we believe that in the long run, the proportion of, say, "ones" will be close to $1/6$, and we connect this number with the probability of rolling "one."

Corollary 17 says that our intuition is consistent with the theory (or the theory is consistent with our intuition): the proportion of occurrences of A in a series of independent replicas converges to the probability of A in a certain reasonable sense.

5.3 On a proof of the strong LLN

A complete proof of the strong LLN will be given in Section **15**.4.7. Here, we prove the first part of Theorem 15 assuming additionally that the quantity $\mu = E\{X^4\}$, the fourth moment, is finite. Also, without loss of generality, we set $m = 0$. If it is not so, we can consider the r.v.'s $X_i - m$ (whose expectations are zeros), and prove that $\frac{1}{n}\sum_{i=1}^{n}(X_i - m) \stackrel{\text{a.s.}}{\to} 0$. This will imply $\frac{1}{n}\sum_{i=1}^{n} X_i \stackrel{\text{a.s.}}{\to} m$.

Next, we set $S_n = X_1 + .. + X_n$, and consider $E\{S_n^4\} = E\{(X_1 + .. + X_n)^4\}$. Let us observe that the expansion of $(X_1 + .. + X_n)^4$ consists of terms X_i^4, $X_i^2X_j^2$, $X_i^3X_j$, $X_i^2X_jX_k$, and $X_iX_jX_kX_l$, where i, j, k, l are distinct.

The expectations of the terms of the last three types equal zero. Indeed, for example, in view of independency, $E\{X_i^3X_j\} = E\{X_i^3\}E\{X_j\} = E\{X_i^3\}m = 0$, since we assumed $m = 0$. The terms of the last two types are treated similarly.

Consider the terms of the first two types. There are n terms of the type X_i^4.

Also, there are $\binom{n}{2}$ possible combinations of distinct i, j. For particular i, j, there are $\binom{4}{2} = 6$ terms that equal $X_i^2 X_j^2$ because $(X_1 + .. + X_n)^4$ is the product of four factors: from two of them X_i comes, the rest give X_j.

Since $m = 0$, we have $E\{X_i^2\} = \sigma^2$, and $E\{X_i^2 X_j^2\} = E\{X_i^2\}E\{X_j^2\} = \sigma^4$. Thus, $E\{S_n^4\} = n\mu + 6\binom{n}{2}\sigma^4 = n\mu + 3n(n-1)\sigma^4$.

This implies $E\left\{\sum_{n=1}^{\infty} \frac{S_n^4}{n^4}\right\} = \sum_{n=1}^{\infty} \frac{E\{S_n^4\}}{n^4} = \sum_{n=1}^{\infty} \frac{n\mu + 3n(n-1)\sigma^4}{n^4} < \infty$, because, in the nth term, the numerator has an order of n^2, while the order of the denominator is n^4. Consequently, with probability one, the r.v. $\sum_{n=1}^{\infty} \frac{S_n^4}{n^4} < \infty$; that is, with probability one it takes only finite values. Indeed, if with a positive probability the sum $\sum_{n=1}^{\infty} \frac{S_n^4}{n^4}$ had been infinite, then its expected value would have been also infinite. If a series converges, its terms converge to zero. Thus, with probability one, $\frac{S_n^4}{n^4} = \left(\frac{S_n}{n}\right)^4 \to 0$, and hence, so does $\frac{S_n}{n}$. ■

Route 1 ⇒ page 101

5.4 More on convergence almost surely and in probability

2

The facts we consider here have their intrinsic values and are useful for future references.

First, let us consider a corollary from Borel-Cantelli's theorem (Theorem **2**.1), which, above else, illustrates the difference between the two types of convergence we have defined.

Proposition 18 *Let r.v.'s Y_n be independent, and $Y_n \xrightarrow{a.s.} 0$. Then, for any $\varepsilon > 0$,*

$$\sum_{n=1}^{\infty} P(|Y_n| > \varepsilon) < \infty. \tag{5.4.1}$$

Thus, by Proposition 18, the almost sure convergence of independent r.v.'s Y_n to zero implies the convergence of series (5.4.1). Then, the separate terms in (5.4.1) should converge to zero: $P(|Y_n| > \varepsilon) \to 0$, which is equivalent to the convergence in probability. On the other hand, the convergence of the terms to zero of a series does not imply the convergence of the series itself, so almost sure convergence is a stronger property.

EXAMPLE 1. Let ξ_n be i.i.d. r.v.'s assuming values $1, 2, \ldots$, and (omitting the index n) let $P(\xi > k) = 1/(1+k)$ for any integer $k \geq 0$. (The reader may observe that in this case, $P(\xi = k) = P(\xi > k-1) - P(\xi > k) = \frac{1}{k} - \frac{1}{k+1} = \frac{1}{k(k+1)}$ for $k \geq 1$, but we will not use this fact.)

Let $Y_n = \frac{1}{n}\xi_n$. Denoting by $[a]$ the integer part of a, we have $P(|Y_n| > \varepsilon) = P(\xi_n > \varepsilon n) = \frac{1}{1+[\varepsilon n]} \leq \frac{1}{1+\varepsilon n - 1} = \frac{1}{\varepsilon n} \to 0$ for any $\varepsilon > 0$. Hence, $Y_n \xrightarrow{P} 0$.

On the other hand, $Y_n \not\to 0$ almost surely. Indeed, if it were true, then by Proposition

18, the series $\sum_{n=1}^{\infty} P(|Y_n| > \varepsilon)$ would have converged, but this is not the case. As we saw, $P(|Y_n| > \varepsilon) = \dfrac{1}{1+[\varepsilon n]} \geq \dfrac{1}{1+\varepsilon n} \geq \dfrac{1}{1+n}$, provided $0 < \varepsilon \leq 1$. As is known, the series $\sum_{n=1}^{\infty} \dfrac{1}{1+n} = \infty$. □

Proof of Proposition 18 is short. First, as any r.v., $Y_n = Y_n(\omega)$. Let us fix $\varepsilon > 0$ and set $A_n = \{|Y_n(\omega)| > \varepsilon\}$. Let A be defined as in Section **2**.3. If for some ω, we have $Y_n(\omega) \to 0$, then this ω may belong to only a finite number of the events A_n. Hence, the event $\{Y_n(\omega) \to 0\} \subseteq A^c$. Consequently, if $P(Y_n \to 0) = 1$, then $P(A^c) = 1$, and hence, $P(A) = 0$. By Borel-Cantelli's Theorem **2**.1, this implies (5.4.1). ■

Proposition 18 deals with independent r.v.'s. The following lemma clarifies the difference between the above two types of convergence in the general case. The reader who is not familiar with the notion of *supremum* may replace the symbol sup below by max; this will not restrict the understanding of the essence of the matter.

Lemma 19 *Let $Y, Y_1, Y_2, \ldots$ be r.v.'s, and $\widetilde{Y}_n = \sup_{k \geq n} |Y_k - Y|$. Then $Y_n \xrightarrow{a.s.} Y$ if and only if $\widetilde{Y}_n \xrightarrow{P} 0$.*

In particular, from Lemma 19, it follows that almost sure convergence implies that in probability. Indeed, if $Y_n \xrightarrow{\text{a.s.}} Y$, then $\widetilde{Y}_n \xrightarrow{P} 0$. But $\widetilde{Y}_n \geq |Y_n - Y|$. Consequently, $Y_n - Y \xrightarrow{P} 0$.

Proof of Lemma 19 . First, note that by the Calculus definition of a limit of a numerical sequence, for each ω, the sequence $Y_n(\omega) \to Y(\omega)$ if and only if $\widetilde{Y}_n(\omega) \to 0$.

Let the event $C = \{\widetilde{Y}_n \not\to 0\}$. We will figure out when $P(C) = 0$.

Set $C_{mn} = \{\widetilde{Y}_n > 1/m\}$, where n, m are positive integers. Note that $\widetilde{Y}_{n+1} \leq \widetilde{Y}_n$, and hence the sequence C_{mn} is non-increasing in n. Let us show that

$$C = \bigcup_{m=1}^{\infty} \bigcap_{n=1}^{\infty} C_{mn}. \tag{5.4.2}$$

If $\omega \in C$, then there *exists* an m such that $\omega \in C_{mn}$ for *all* n. Indeed, suppose that it is not so. Then, whatever m is, for some n_0, the r.v. $\widetilde{Y}_{n_0} \leq 1/m$. But, since the sequence $\{\widetilde{Y}_n\}$ is non-increasing, the same will be true for all $n \geq n_0$. Eventually, it would mean that $Y_n(\omega)$ converges to zero, which contradicts the choice of ω.

If $\omega \notin C$, then for any m, the outcome $\omega \notin C_{mn}$ starting from some n, and hence ω does not belong to the r.-h.s. of (5.4.2).

From (5.4.2), it follows that $P(C) = 0$ if and only if $P(\bigcap_{n=1}^{\infty} C_{mn}) = 0$ for all m. By the continuity property and the fact that C_{mn} are non-increasing in n, the last relation is equivalent to the relation $\lim_{n\to\infty} P(C_{mn}) = 0$.

Thus, $P(C) = 0$ if and only if $P(\widetilde{Y}_n > 1/m) \to 0$ as $n \to \infty$ for any m. This is equivalent to the assertion that $P(\widetilde{Y}_n > \varepsilon) \to 0$ as $n \to \infty$ for any $\varepsilon > 0$. In its turn, this means that $\widetilde{Y}_n \xrightarrow{P} 0$. ■

6 CONDITIONAL EXPECTATION

6.1 Conditional expectation given a random variable

Consider r.v.'s X and Y. Our immediate goal is to define the quantity which we will denote by $E\{Y\,|\,X=x\}$, and which we will understand exactly as it sounds: the mean value of Y given that X took on a particular value x.

Suppose X takes on values $x_1, x_2, \dots$, and Y —values $y_1, y_2, \dots$. Set $f_{ij} = P(X = x_i, Y = y_j)$, a joint probability; $f_i^{(1)} = P(X = x_i)$, $f_j^{(2)} = P(Y = y_i)$, marginal probabilities; and $f^{(1)} = \left(f_1^{(1)}, f_2^{(1)}, \dots\right)$ and $f^{(2)} = \left(f_1^{(2)}, f_2^{(2)}, \dots\right)$.

(Here, we prefer the notation (X,Y) to (X_1, X_2) that we used in Section 1.2.)

In a usual way, we define conditional probabilities

$$f_{j|i} = P(Y = y_j\,|\,X = x_i) = \frac{P(X = x_i,\, Y = y_j)}{P(X = x_i)} = \frac{f_{ij}}{f_i^{(1)}}. \tag{6.1.1}$$

In the notation above, the order in which we consider X and Y is essential.

Sometimes, we will omit the index i of x_i, writing just x but keeping in mind that we consider only x's which coincide with one of the values of X. In particular, this means that for such x's, $P(X = x) \neq 0$. The same concerns the index j in y_j.

We define the conditional expectation as

$$E\{Y\,|\,X = x\} = \sum_j y_j P(Y = y_j\,|\,X = x). \tag{6.1.2}$$

For $E\{Y\,|\,X = x\}$ so defined, we will use the notation $m_{Y|X}(x)$. The function $m_{Y|X}(x)$ is often called a *regression function* of Y on X. When it does not cause misunderstanding, we omit the index $Y|X$ in $m_{Y|X}(x)$, writing just $m(x)$.

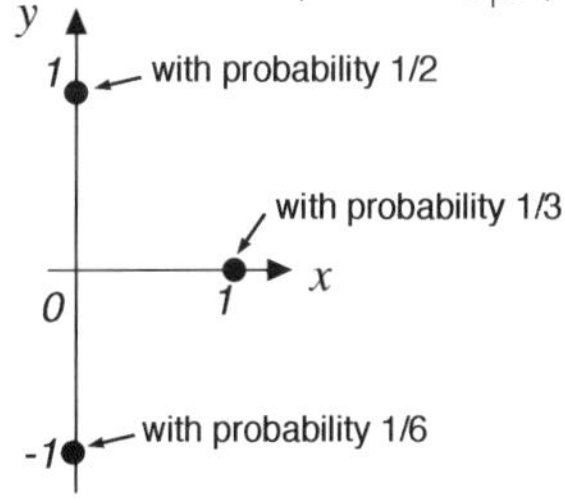

FIGURE 8.

EXAMPLE 1. Let a r.vec. (X,Y) take on vector-values $(0,1)$, $(1,0)$, $(0,-1)$ with probabilities $\frac{1}{2}, \frac{1}{3}, \frac{1}{6}$, as is shown in Fig. 8. If $X = 1$, then Y takes on only one value 0, so $P(Y = 0\,|\,X = 1) = 1$, and

$$m(1) = E\{Y\,|\,X = 1\} = 0.$$

In accordance with (6.1.2), $m(0) = E\{Y\,|\,X = 0\}$ $= 1 \cdot P(Y = 1\,|\,X = 0) + 0 \cdot P(Y = 0\,|\,X = 0)$ $+(-1) \cdot P(Y = -1\,|\,X = 0) = 1 \cdot \frac{1/2}{(1/2)+(1/6)} + 0 \cdot 0 + (-1) \cdot \frac{1/6}{(1/2)+(1/6)} = \frac{1}{2}$. $\square$

EXAMPLE 2. Let N_1 and N_2 be independent Poisson r.v.'s with parameters λ_1 and λ_2, respectively. Find $E\{N_1 \mid N_1 + N_2 = n\}$.

(Thus, N_1 plays the role of Y, and $N_1 + N_2$ plays the role of X. We also replaced x by n.)

For example, N_1 and N_2 are the numbers of men and women respectively, who have entered a store during a day. Assume that we know how many people totally have come. What is the mean number of men, *given* this information?

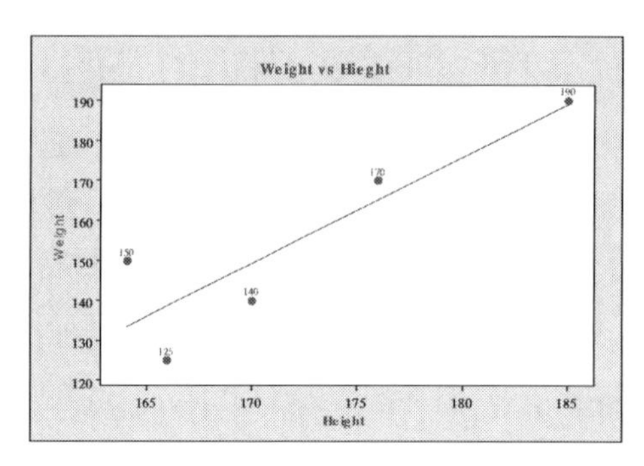

FIGURE 9. Scatter plot of weight (lb) vs height (cm) for a sample of 5 adults.

In Proposition 13, Section 4.5.5, we proved that the conditional distribution of N_1 given $N_1+N_2=n$ is the binomial distribution with parameters n and $p=\lambda_1/(\lambda_1+\lambda_2)$. Since the mean value of the binomial distribution is np,

$$E\{N_1\,|\,N_1+N_2=n\}=\frac{n\lambda_1}{\lambda_1+\lambda_2}.\ \square \tag{6.1.3}$$

Loosely speaking, the regression function $m(x)$ represents the dependence Y on X *on the average*. To estimate $m(x)$ in a concrete case, one may proceed from data concerning particular values of a random vector (X,Y). Such data are presented by a collection of pairs of numbers $(x_1,y_1),(x_2,y_2),\ldots$. Placing these points in a (x,y)-plane (it is called a *scatter plot*), we get a picture depicting the pattern of dependency.

For example, Fig. 9 represents the dependence between the human weight and height. Each point corresponds to a particular person; the x-coordinate equals the height of this person, and the y-coordinate—the weight. This is a study example; we considered just five adult males.

We can draw a "well fitting curve" presenting the pattern of dependency, and which may be considered an estimate of the regression function. Statistics suggests precise algorithms to do that, but here we skip the details. In Fig. 9, the straight line fits relatively well, and we can suppose that the regression is linear.

Certainly, this is not always the case. Fig. 10[1] represents the dependence of the life expectancy (the mean lifetime) in a country on the country's wealth measured as the gross domestic product (GDP) per person. (GDP is the total value of the goods and services produced in a country.)

Each point in the scatter-plot in Fig. 10 corresponds to a country; the x-coordinate is its GDP per person, and the y-coordinate is the mean lifetime in this country. We expect people in wealthier countries to live longer, and the data confirm it, but the regression is not linear: for wealthy countries, the life expectancy is practically the same.

[1]Presented with the permission of professor Lane Kenworthy, University of Arizona.

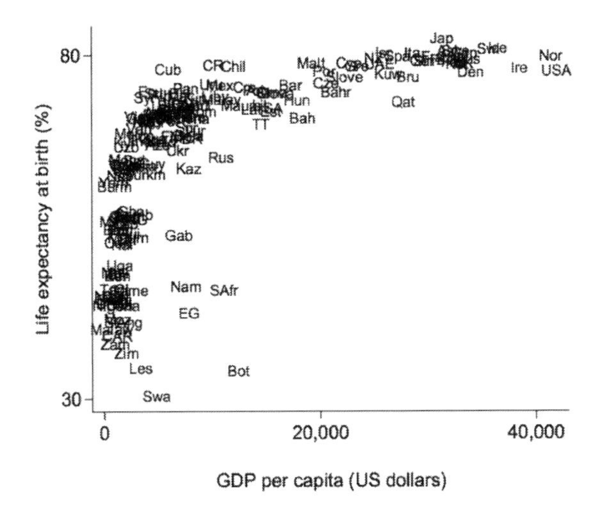

Note: Data are for 2005. 171 countries. For GDP per capita, currencies converted into U.S. dollars using purchasing power parities (PPPs).

Source: United Nations Development Programme (UNDP), *Human Development Report*, various years.

FIGURE 10. Life expectancy versus GDP.

Certainly, the regression function does not have to be increasing. For example, this concerns the dependence of gas mileage on car weight.

We return to general properties of the regression function $m(x) = m_{Y|X}(x)$. This is the mean value of Y given $X = x$. Since X is a random variable, its values x are different, random. To reflect this circumstance, let us replace the argument x in $m(x)$ by the r.v. X itself; that is, let us consider the r.v. $m(X)$. This is a function of X, and its significance is the same as above: it is the conditional mean value of Y given a value of X. However, since X is random, the conditional mean value of Y given X is also random, and the value of $m(X)$ depends on which value X has assumed.

The r.v. $m(X)$ has a special notation, namely $E\{Y\,|\,X\}$, and is called the *conditional expectation* of Y given X. It is important to keep in mind that this is a r.v., and since $E\{Y\,|\,X\}$ is a function of X, its value is completely determined by the value of X.

EXAMPLE 3. In the situation of Example 1, X takes on two values: 0 and 1 with probabilities $\frac{2}{3}$ and $\frac{1}{3}$, respectively; see also Fig. 9. Hence, $m(X)$ takes on the values $m(0) = \frac{1}{2}$ and $m(1) = 0$ with the above probabilities. Thus, $E\{Y\,|\,X\}$ is a r.v. taking on values $\frac{1}{2}$ and 0 with respective probabilities $\frac{2}{3}$ and $\frac{1}{3}$. $\square$

EXAMPLE 4. Let us return to Example 2 and set $N = N_1 + N_2$, $\lambda = \lambda_1 + \lambda_2$. By virtue of (6.1.3), $m_{N_1\,|\,N}(n) = \dfrac{\lambda_1}{\lambda}n$, and hence $E\{N_1\,|\,N\} = \frac{\lambda_1}{\lambda}N$. $\square$

6.2 The formula for total expectation and other properties of conditional expectation

Below, we assume all expectations to be finite.

1. (*The formula for total expectation.*) The main and extremely important property of conditional expectation is that for any X and Y,

$$E\{E\{Y\,|\,X\}\} = E\{Y\}. \qquad (6.2.1)$$

Putting it in words,

If we "condition Y on X" and after that,
compute the expected value of the conditional expectation,
then we come to the unconditional expectation $E\{Y\}$.

We call (6.2.1) the *formula for total expectation* or the *law of total expectation.*

Proof is straightforward. By (6.1.1) and definition (6.1.2),

$$E\{E\{Y\,|\,X\}\} = E\{m(X)\} = \sum_i m(x_i)f_i^{(1)} = \sum_i \left(\sum_j y_j P(Y = y_j\,|\,X = x_i)\right) f_i^{(1)}$$

$$= \sum_i \left(\sum_j y_j \frac{f_{ij}}{f_i^{(1)}}\right) f_i^{(1)} = \sum_i\sum_j y_j f_{ij} = \sum_j y_j \left(\sum_i f_{ij}\right) = \sum_j y_j f_j^{(2)} = E\{Y\}. \blacksquare$$

For future references, we also present another form of the formula for total expectation. This form is less compact than (6.2.1) but prepares us for concrete calculations. Let X assume values $x_1, x_2, ...$. Then the r.v. $E\{Y\,|\,X\}$ takes on values $E\{Y\,|\,X = x_1\}, E\{Y\,|\,X = x_2\}, ...$, and

$$E\{Y\} = \sum_i E\{Y\,|\,X = x_i\}P(X = x_i). \qquad (6.2.2)$$

The next properties are straightforward and quite plausible from a heuristic point of view. At the end of the section, we prove the second property; the others are proved similarly.

2. For any number c and r.v.'s X, Y, Y_1, and Y_2,

$$E\{cY\,|\,X\} = cE\{Y\,|\,X\} \quad \text{and} \quad E\{Y_1 + Y_2\,|\,X\} = E\{Y_1|X\} + E\{Y_2\,|\,X\}. \qquad (6.2.3)$$

3. If r.v.'s X and Y are independent, then $E\{Y\,|\,X\}$ is not random and equals $E\{Y\}$.

   (Loosely speaking , if Y does not depend on X, then values of X do not have any impact on the mean value of Y.)

4. First, $E\{X\,|\,X\} = X$, which is quite understandable: if X is *given*, then we treat X as a non-random variable (equal to X), and its mean value is X. Similarly, for any function $g(x)$,
$$E\{g(X)\,|\,X\} = g(X), \tag{6.2.4}$$
(given X, the mean value of $g(X)$ is $g(X)$).

5. Consider now $Y = g(X)Z$, where Z is another r.v. and $g(x)$ is a function. Then
$$E\{g(X)Z\,|\,X\} = g(X)E\{Z\,|\,X\}. \tag{6.2.5}$$
From a heuristic point of view, it is also understandable. When conditioning on X, we view X as a constant, and hence $g(X)$ may be brought outside of the conditional expectation.

We turn to examples.

EXAMPLE 1 (*A sum of random number of random variables*). The scheme of this example serves as a model in a wide variety of applied problems. Consider a r.v.
$$S = \sum_{i=1}^{N} X_i, \tag{6.2.6}$$
where not only X's are r.v.'s, but so is the number of them, N.

For instance, N is the number of cars that entered a national park during a day, and X_i is the number of passengers in the ith car. Then S in (6.2.6) is the total number of visitors on this day. In another example, N may be the number of traffic accidents during a month in a particular area, and X_i is the number of injured people in the ith accident. Then the total number of injuries is given by (6.2.6).

In the general case, the r.v. N in such examples is random as well as X_i's.

Assume all X's to be independent of each other, and independent of N. Suppose X's are identically distributed, and $E\{X_i\} = m$.

We show that, for the scheme (6.2.6),
$$E\{S\} = mE\{N\}. \tag{6.2.7}$$
For instance, if on the average 260 accidents occur, and there are 6 injuries per 100 accidents on the average, then the mean total number of injuries is $0.06 \cdot 260 = 15.6$.

Proof. We use (6.2.1); S stands for Y, and N for X. Conditioning S on N, we write $E\{S\} = E\{E\{S\,|\,N\}\}$. Given $N = n$, the conditional expectation $E\{S\,|\,N = n\} =$ $E\{\sum_{i=1}^{N} X_i\,|\,N = n\} = E\{\sum_{i=1}^{n} X_i\,|\,N = n\} = E\{\sum_{i=1}^{n} X_i\}$ because X's do not depend on N. But $E\{\sum_{i=1}^{n} X_i\} = mn$, and hence, $E\{S\,|\,N = n\} = mn$. Consequently, $E\{S\,|\,N\} = mN$, and $E\{S\} = E\{E\{S\,|\,N\}\} = E\{mN\} = mE\{N\}$. ■ □

EXAMPLE 2 (*The "beta" of a security*). In Finance, the "beta" (β) of a security (say, a stock or a portfolio of stocks) is a characteristic describing the relation of the security's return with that of the financial market as a whole. In other words, β shows how the return of a security depends on the situation in the market on the average.

Let Y be the (future) return of a security; that is, the income per \$1 investment. Let X be the future value of a market index, i.e., a global characteristic monitoring either the value of the market as a whole, or an essential part of it. (Typical examples are Dow Jones, or S&P indices.) Not going deeply into it, note that such characteristics may be adjusted with respect to a \$1 investment, and viewed as the return of the market (or the part of the market which the index is monitoring). We adopt the following simple model that sometimes works well: $Y = \xi X + \varepsilon$, where coefficients ξ and ε are *random* but do not depend on X.

Loosely put, ε characterizes the "random factors" that are relevant to the security and are not associated with the market. The random coefficient ξ reflects the impact of the global market situation on the value of the security.

In view of Properties 2-4, $E\{Y\,|\,X\} = E\{\xi X + \varepsilon\,|\,X\} = E\{\xi X\,|\,X\} + E\{\varepsilon\,|\,X\} = XE\{\xi|X\} + E\{\varepsilon\,|\,X\} = XE\{\xi\} + E\{\varepsilon\}$.

The mean value $E\{\xi\}$ is denoted by β and called "beta." We set $a = E\{\varepsilon\}$, and eventually write $E\{Y\,|\,X\} = \beta X + a$.

Note that β may be negative. For example, as a rule, the price of gold is growing when the market is dropping.

By definition, the market itself has a beta of 1. Indeed, in this case $Y = X$, and hence $\xi = 1$, $\varepsilon = 0$.

If the variation of the stock return is larger on the average than that of the market, the absolute value of beta is greater than 1, whereas for a stock whose returns vary less than the market's returns, $|\beta| < 1$. The reader may find particular values of β for many stocks in Internet sites concerning the stock market.

EXAMPLE 3. Let X_1, X_2 be independent and identically distributed. Find $E\{X_1\,|\,X_1 + X_2\}$. In other words, what is the mean value of one term given the value of the sum? By the symmetry argument, we can guess that the answer is simple:

$$E\{X_1\,|\,X_1 + X_2\} = \frac{1}{2}(X_1 + X_2).$$

This is, indeed, true, and may be shown as follows. Since X_1, X_2 are independent and have the same distribution, $E\{X_1\,|\,X_1 + X_2\} = E\{X_2\,|\,X_1 + X_2\}$ by symmetry. Then, using the properties above, we have $E\{X_1\,|\,X_1 + X_2\} = \frac{1}{2}(E\{X_1\,|\,X_1 + X_2\} + E\{X_2\,|\,X_1 + X_2\}) = \frac{1}{2}E\{X_1 + X_2\,|\,X_1 + X_2\} = \frac{1}{2}(X_1 + X_2)$. □

We give a

Proof of the second part of Property 2. Other properties are proved similarly. Let y_{1i}, y_{2j} denote the values of Y_1, Y_2, respectively. Then, following (6.1.2), we write

$$E\{Y_1 + Y_2\,|\,X = x\} = \sum_i \sum_j (y_{1i} + y_{2j}) P(Y_1 = y_{1i}, Y_2 = y_{2j}\,|\,X = x)$$

$$= \frac{1}{P(X = x)} \sum_i \sum_j (y_{1i} + y_{2j}) P(Y_1 = y_{1i}, Y_2 = y_{2j}, X = x)$$

$$= \frac{1}{P(X = x)} \left(\sum_i y_{1i} \sum_j P(Y_1 = y_{1i}, Y_2 = y_{2j}, X = x) + \sum_j y_{2j} \sum_i P(Y_1 = y_{1i}, Y_2 = y_{2j}, X = x) \right)$$

$$= \frac{1}{P(X = x)} \left(\sum_i y_{1i} P(Y_1 = y_{1i}, X = x) + \sum_j y_{2j} P(Y_2 = y_{2j}, X = x) \right)$$

$$= \sum_i y_{1i} P(Y_1 = y_{1i} \,|\, X = x) + \sum_j y_{2j} P(Y_2 = y_{2j} \,|\, X = x) = E\{Y_1 \,|\, X = x\} + E\{Y_2 \,|\, X = x\}.$$

It remains to replace x by X. ■

Route 1 ⇒ *page 108*

2

6.3 A formula for variance

Let us consider a counterpart of (6.2.1) for variances. We show that

$$Var\{Y\} = E\{Var\{Y \,|\, X\}\} + Var\{E\{Y \,|\, X\}\}, \tag{6.3.1}$$

where $Var\{Y \,|\, X\}$ is the variance of Y with respect to the conditional distribution of Y given X. In particular, we can write that $Var\{Y \,|\, X\} = E\{Y^2 \,|\, X\} - (E\{Y \,|\, X\})^2$.

To memorize (6.3.1), notice that in the terms of this formula, the order of the operations $E\{\cdot\}$ and $Var\{\cdot\}$ alternates. To prove (6.3.1), we make use of (6.2.1) in the following way:

$$\begin{aligned} Var\{Y\} &= E\{Y^2\} - (E\{Y\})^2 = E\{E\{Y^2 \,|\, X\}\} - (E\{E\{Y \,|\, X\}\})^2 \\ &= E\left\{E\{Y^2 \,|\, X\} - (E\{Y \,|\, X\})^2\right\} + E\left\{(E\{Y \,|\, X\})^2\right\} - (E\{E\{Y \,|\, X\}\})^2 \\ &= E\left\{Var\left\{Y \,|\, X\right\}\right\} + Var\left\{E\{Y \,|\, X\}\right\}. \;\blacksquare \end{aligned}$$

EXAMPLE 1. Consider again the sum of a random number of r.v.'s in (6.2.6), assuming all X's to be identically distributed, independent of each other, and independent of N. Set $E\{X_i\} = m$, and $Var\{X_i\} = \sigma^2$.

By (6.3.1), $Var\{S\} = E\{Var\{S \,|\, N\}\} + Var\{E\{S \,|\, N\}\}$. Reasoning similar to the proof of (6.2.7), we can write that, given N, the conditional variance $Var\{S \,|\, N\} = \sigma^2 N$. When proving (6.2.7), we have also shown that $E\{S \,|\, N\} = mN$. So, $Var\{S\} = E\{\sigma^2 N\} + Var\{mN\} = \sigma^2 E\{N\} + m^2 Var\{N\}$.

Thus, together with (6.2.7), we have arrived at

$$E\{S\} = mE\{N\}, \quad Var\{S\} = \sigma^2 E\{N\} + m^2 Var\{N\}. \tag{6.3.2}$$

Suppose now that N is a Poisson r.v. with parameter λ. Then, because in this case, $E\{N\} = Var\{N\} = \lambda$,

$$E\{S\} = m\lambda, \quad Var\{S\} = (\sigma^2 + m^2)\lambda. \tag{6.3.3}$$

Let, for instance, for an auto insurance portfolio, the number of claims being received by the claim department during a day is a Poisson r.v. N with $\lambda = 12$. Suppose the sizes of claims, X_i, satisfy conditions imposed above, the mean size of a claim is \$2500, and the standard deviation is \$500. The company wants to know the mean and standard deviation of the aggregate claim S.

By (6.3.3), $E\{S\} = 2500 \cdot 12 = 30{,}000$, and the standard deviation $\sigma_S = \sqrt{Var\{S\}} = \sqrt{(500^2 + 2500^2)12} \approx 8{,}831$. Loosely put, on the average, the aggregate claim amounts to \$30,000 plus-minus, on the average, \$8,800. □

In conclusion, we discuss one more basic and important for applications property of conditional expectation.

6.4 Conditional expectation and least squares prediction

An important problem of Probability Theory is to predict the value of a future observation Y given a present related observation X. Say, we want to predict the tomorrow value of a market index or the tomorrow highest air temperature given the today values of these characteristics. Since we know only X, any *predictor* is a function $g(X)$. The quality of prediction is usually measured by $d(g) = E\{(Y - g(X))^2\}$, which is referred to as the *mean square criterion*. If the prediction is perfect; that is, $Y = g(X)$, then $d(g) = 0$. Usually, this is unachievable, and we are looking for $g(x)$ minimizing $d(g)$.

The choice of the above criterion may be justified by two facts. First, $d(g)$ inherits features of the usual Euclidean distance, where we square the differences between the coordinates of the vectors and sum them up.

Second, it is natural to think that a good predictor should be perfect on the average: $E\{Y\} = E\{g(X)\}$; such a predictor is called *unbiased*. On the other hand, if a predictor $g(X)$ is unbiased, then $E\{Y - g(X)\} = 0$, and hence, $d(g) = Var\{Y - g(X)\}$. Thus, in this case, we are minimizing the variance of the difference $Y - g(X)$.

Theorem 20 *The least square predictor is $E\{Y\,|\,X\}$; that is the minimum of $d(g)$ is attained at the conditional expectation $m(x) = E\{Y\,|\,X = x\}$.*

First, note that $E\{E\{Y\,|\,X\}\} = E\{Y\}$, so the best predictor is unbiased.

Proof of Theorem 20. For any r.v. Z with a mean of m, and for any constant c, we have $E\{(Z-c)^2\} = E\{(Z-m-(c-m))^2\} = E\{(Z-m)^2\} - 2(c-m))E\{Z-m\} + (c-m)^2 = E\{(Z-m)^2\} + (c-m)^2$. Hence, the minimum of $E\{(Z-c)^2\}$ with respect to c is attained at $c = m$.

Now, $E\{(Y-g(X))^2\} = E\{E\{(Y-g(X))^2\,|\,X\}\}$. In the conditional expectation $E\{(Y - g(X))^2\,|\,X\}$, we can view X, and hence $g(X)$ also, as a constant. So, $g(X)$ plays the role of the c above. Consequently, for any value of X, the minimum $E\{(Y-g(X))^2\,|\,X\}$ is attained when $g(X) = E\{Y\,|\,X\}$. ■

1,2

7 EXERCISES

The use of software for solving problems below is recommended.

Section 1

1. "Hickory, dickory, dock,

   The mouse ran up the clock.

   The clock struck one,

   The mouse ran down,

   Hickory, dickory, dock."

   Suppose a word is chosen at random from this famous nursery rhyme. Write the probability distribution of the (random) number of letters in this word.

2. Let X be the product of the numbers on two dice rolled. (a) Does X assume all values between 1 and 36? (b) Trying to provide all calculations in mind (they are easy), find $P(X \leq 36)$, $P(X < 36)$, $P(X \leq 30)$, $P(X < 30)$, $P(X = 1)$. (c) Find the probability that X is a prime number. (By definition, one is not a prime.)

3. You draw two cards from a deck. If they are of different color, you win nothing. If both are black, you win \$5. If both are red, you win \$10, and additional \$15 if the two cards are a red ace and king. Write the probability distribution for the amount you will win.

4. A bag contains 5 red marbles and 15 blue marbles. Two marbles are drawn from the bag without replacement. Write the distribution of the number of red marbles selected.

5. (a) Which distribution below would you call symmetric, and which "skewed"?

$$\begin{array}{ccccc} 1 & 2 & 3 & 4 & 5 \\ 0.1 & 0.2 & 0.4 & 0.2 & 0.1 \end{array} \quad \text{and} \quad \begin{array}{ccccc} 1 & 2 & 3 & 4 & 10 \\ 0.1 & 0.2 & 0.4 & 0.2 & 0.1 \end{array}.$$

   (b) Rigorously speaking, the distribution of a discrete r.v. X is said to be *symmetric* if for a number s called a *center of symmetry*, $P(X = s + x) = P(X = s - x)$ for any $x > 0$. Explain the sense of this definition, and illustrate it by the example above. Explain why we may consider all x's in this definition. (*Hint*: If $m + x$ is not equal to some x_i, then $P(X = x + s) = 0$.)

   (c) Does the center of symmetry have to equal one of the values of the r.v.?

   (d) Compare the two following distributions:

$$\begin{array}{ccccc} 1 & 2 & 3 & 4 & 5 \\ 0.1 & 0.2 & 0.4 & 0.2 & 0.1 \end{array} \quad \text{and} \quad \begin{array}{ccccc} 1 & 2 & 3 & 4 & 5 \\ 0.2 & 0.1 & 0.4 & 0.1 & 0.2 \end{array}.$$

   What is the difference? Are both distributions symmetric? For which distribution does the "dispersion" or "variation" of values seem to you larger?

6. There are six items in a row. You mark at random two of them. Find the distribution of the number of items between the marked items.

7. The distribution of the number of telephone calls a manager is receiving during an hour is well approximated by the Poisson distribution with parameter $\lambda = 4$. Calculate the probability that during an hour, there will be at least 3 calls.

8. Let a r.v.

$$X = \begin{cases} -3.2 & \text{with a probability of } 1/7 \\ -2 & \text{with a probability of } 2/7 \\ 0 & \text{with a probability of } 1/14 \\ 5 & \text{with a probability of } 1/2 \end{cases}$$

   Graph the (cumulative) distribution function of X. If you show this graph to somebody who is taking a course in Probability, how would she/he compute $P(-3 \leq X \leq 3)$ just looking at the graph?

9. Let r.v.'s X_1 and X_2 be independent and both have the geometric distribution with parameter $p = 1/2$. Find $P(X_1 = X_2)$.

10. Let r.v.'s X_1, X_2 be the numbers on two dice rolled. How does the table of joint and marginal probabilities look in this case? Realize how this simple table may be of help for finding the distribution of $X_1 + X_2$, the sum of the numbers on the dice. Find this distribution using the table.

11. The two tables below give two different joint probability distributions of a random vector (X_1, X_2):

$X_1 \Rightarrow$ / $X_2 \Downarrow$	-1	0	1
-1	1/14	1/21	1/42
0	1/7	2/21	1/21
1	2/7	4/21	2/21

$X_1 \Rightarrow$ / $X_2 \Downarrow$	-1	0	1
-1	1/5	1/5	0
0	0	1/5	1/5
1	0	0	1/5

For both cases, find the marginal probabilities; figure out whether the random variables are dependent; find $P(X_1 = 0 \mid X_2 = 0)$, compare it with $P(X_1 = 0)$, and explain the result of the comparison.

12. Are r.v.'s $X_1 \equiv 3$ and $X_2 \equiv 5$ independent? ($X \equiv a$ means that X takes on only one value a.)

13. Let r.v.'s X_1 and X_2 be i.i.d. Show that $P(X_1 > X_2) = P(X_2 > X_1)$.

14. Let r.v.'s X_1 and X_2 assume the same values, and their joint distribution is symmetric in the sense that $f_{ij} = f_{ji}$. (a) How will the table of joint probabilities look in this case? (b) Do X_1, X_2 have the same marginal distributions? (c) Can X_1, X_2 be dependent? (d) Show that $P(X_1 > X_2) = P(X_2 > X_1)$.

15. Show without any calculations and using no particular formulas that the sum of two *independent* binomial r.v.'s with parameters (n_1, p) and (n_2, p) respectively, is a binomial r.v. with parameters $(n_1 + n_2, p)$.

16. Consider the game from Example 1.1-7, and suppose that the coin was tossed n times. Let W_n be the total winnings of the first player. (a) Which values can W_n assume? (b) Find $P(W_n = 0)$, (c) Find $P(W_n = k)$ for all $k = 0, \pm 1, \pm 2, ...$

17. A husband and wife have two cars. They can purchase special auto insurance policies for each car separately, such that each policy covers the loss only for the first accident occurred in the same year. Suppose that the numbers of accidents connected with each car are independent r.v.'s N_1, N_2 with the same distribution. The couple have also an option to buy a joint insurance for both cars which covers two accidents a year. If the premium for the joint insurance is double the premium of the individual policy, what decision should the couple make? Describe the distribution of the number of accidents which the insurance will cover for both cases and provide comparison. (*Advice*: Observe that these distributions are completely specified by $p_0 = P(N_1 = 0) = P(N_2 = 0)$, and $p_1 = P(N_1 = 1) = P(N_2 = 1)$.)

18. Regarding the convolution formula (1.2.5), is it true that $f^{(1)} * f^{(2)} = f^{(2)} * f^{(1)}$?

19. Find the distribution of the sum $X_1 + X_2$ of independent r.v.'s

$$X_1 = \begin{cases} 1 \text{ with probability } p_1 = \frac{1}{3} \\ 2 \text{ with probability } q_1 = \frac{2}{3} \end{cases}, \quad X_2 = \begin{cases} 2 \text{ with probability } p_2 = \frac{1}{5} \\ 3 \text{ with probability } q_2 = \frac{4}{5} \end{cases}.$$

20. (a) Two independent r.v.'s, X_1 and X_2, take on values 1,2 with *equal* probabilities. Without calculations, guess whether all possible values of the sum $X_1 + X_2$ are also equally likely. Give a heuristic explanation. Using the convolution formula, find all probabilities $f_n = P(X_1 + X_2 = n)$. Graph f_n as a function of n. (You may consider only integer n, or connect neighbor points by straight lines.)

(b) Do the same for X_1 and X_2 taking on *equiprobable* values $1, ..., r$ for a natural r.

(c) Connect this problem with Exercise 10 and compare the results.

21. A husband and wife played the same slot machine together. The wife started to play first, played until the first win and yielded her place to the husband. Then the husband played until his first win and after that both quit. If both together played 11 times, what is the probability that the wife won at the third play?

22. We revisit Example 1.2.1-1. (a) Just looking at the table of joint probabilities and providing all calculations in mind, find $P(X_1 = 3 \,|\, X_2 = 0)$. (b) Just looking at the table of joint probabilities and providing all calculations in mind, write the conditional distribution of X_2 given $X_1 = 0$; the conditional distribution of X_1 given $X_2 = 3$. (c) Guess whether it is true that $P(X_2 = 1 \,|\, X_1 = 1) = P(X_2 = 1)$, and that $P(X_1 = 1 \,|\, X_2 = 1) = P(X_1 = 1)$. Compute these probabilities and check your guess.

23. In Exercise 20b, for $r = 3$, find the distribution of X_1 given $X_1 + X_2 = 4$ in two ways: using formulas for conditional probabilities, and considering different outcomes given $X_1 + X_2 = 4$.

Section 2

24. Find the mean number of letters in the word taken at random from the rhyme in Exercise 1. Compare it with the number 5.1 that in some sources is referred to as the mean length of a word in an English text. Do you find the difference from what you got significant?

    Note that in our example, as well as when the mean 5.1 was being computed, each word was counted as many times as it was used. Since the most repeated words, like articles (a, the), pronouns (I, me, ...) are short, the mean is not as large as it would be if we had counted repeated words one time. Compute the mean for our verse in this way. Explain why it is small in comparison with the mean length of words in the English language.

25. (a) Explain without any calculations why the mean number on a die rolled should be equal to 3.5. Verify it proceeding from definition (2.1.1).

    (b) Let us revisit Exercise 5a. Say, without calculations, in which case the mean is larger.

    (c) Show that for any symmetric distribution (see a definition in Exercise 5), the expected value equals the center of symmetry.

26. You play a game of chance with the probability of winning at each play equal to $1/3$. You bet \$1 and quit if you win. If you lose, you bet \$2 and quit whatever happens. Find the expected value of your profit. Will this figure change, if you first bet \$2, and then \$1 (if you lose at the first time)? Guess and then check your guess.

27. We run a sequence of trials for which the probabilities of success are different. Namely, for the first trial, it is $1/2$, for the second, it is $1/4$, and so on: for the kth trial, it is $1/2^k$. Find the expected number of successes in n trials. Consider the case $n = \infty$.

28. There are four closed boxes, one of which contains a prize and the others are empty. A player can consecutively check boxes until she/he finds the prize, and in this case, she/he pays \$1 for each check (including the last box if the prize is there). The player may also divide the boxes into two equal parts and ask the host of the game in which group the box with the prize is. After the answer, the player asks the host which box among the two contains the prize. Each such a question also costs \$1. Let X and Y be the total payment for the first and second strategy, respectively. Which strategy is better on the average? Compute $E\{X\}$ and $E\{Y\}$. With what probability will the strategy that is worse on the average, lead to a better result?

29. Solve problem of Exercise 28 for 2^n, $n > 1$, boxes. Does the above probability increase or decrease as n is increasing? Find the limit as $n \to \infty$.

30. Let a r.v. $X = 1,2,3,4,5$, and $P(X = 3) = 0.2$. Assign probabilities to the other values in a way that it would be clear without any calculations that $E\{X\} = 3$. Do not make all probabilities you assign the same.

31. In Example 1.2.1-1, find the expected value of the number of customers at the first and second counter separately.

32. There are two classes of 20 and 30 students. No student takes both classes. A student is randomly selected out of these 50 students. Let X be the number of students in the class this student is taking. Now, a class is randomly chosen (out of two). Denote by Y the number of students in this class. Which do you think is larger: $E\{X\}$ or $E\{Y\}$? Explain your guess. Find both expectations and compare.

33. A company is going to put a new computer in the market in one year. A company's analyst presupposes that the future market price for such a computer may assume three values: $\$900, \$1000, \$1100$ with equal probabilities. The analyst also thinks that on the average, $60,000$ people will buy the computer. Make an assumption that will allow the analyst to find the expected value of the company's income. Find it. We continue in Exercise 51.

34. You roll a die until the first moment when "6" appears, but not more than ten times. Find the expected number of rolls. (Use of software is recommended.)

35. A r.v. X assumes values $1, 2, 4, \ldots$ with respective probabilities $\frac{2}{3}, \frac{2}{9}, \frac{2}{27}, \ldots$. Find $E\{X\}$.

36. A r.v. X has the Poisson distribution with parameter λ. Find $E\{1/(1+X)\}$.

37. Find the expected value of the product of the numbers on two dice rolled.

38. Is the independence condition necessary for $E\{X_1X_2\} = E\{X_1\}E\{X_2\}$? (*Advice*: You may consider, for instance, Example 1.2.2-1.)

39. Let X be the number of successes in a sequence of n trials with the probability of success equal to p. Show that the relation $E\{X\} = np$ is always true whether trials are independent or not.

40. Compute the expected value for the geometric distribution using formula (2.2.1) and compare your calculation with those in Example 2.1-3.

41. Does the fact that $E\{X\}$ has a "moderate" value, mean that X cannot assume "big" values? Let us consider a r.v.

$$X = \begin{cases} n \text{ with probability } 1/n \\ 0 \text{ with probability } 1 - 1/n \end{cases}$$

for large n. Such a r.v. may illustrate a possible future loss connected with a possibility of a catastrophic event, say, an earthquake. The probability of the event is small, but the damage, if the event occurs, is large. Compute $E\{X\}$ and comment on to what extent in this case the mean value reflects the "behavior" of the r.v.

42. (*The two envelopes problem.*) In a show, you are given two closed envelopes containing positive sums of money. It is known that one envelope contains twice as much as the other. You may choose one envelope and keep the money it contains. You pick one envelope at random, open it and see an amount of c. At this moment, the host offers you to switch to the other envelope. The common sense tells that whether you switch or not, on the average you will have the same. Nevertheless, you can reason as follows.

With equal probabilities, the other envelope contains either $2c$ or $\frac{c}{2}$. Hence, if you switch, then the mean amount you will get is $2c \cdot \frac{1}{2} + \frac{c}{2} \cdot \frac{1}{2} = \frac{5c}{4} > c$. So, you should switch.

Moreover, this conclusion does not depend on c. Hence, you should switch even not opening the envelope. Then, you can repeat the same argument with respect to the envelope you have switched to, and consequently, you should switch back. So, we have come to the absurdity that you should swap envelopes infinitely long. Where is a mistake in the above reasoning? (*Hint*: The quantity c above is a r.v. rather than a number. Denote by z the smaller amount (which is a number) and provide proper calculations of the mean value of the amount you will get in both cases: when you switch and when you do not.)

43.* The time X it takes for Michael to drive to school varies from day to day depending on weather and traffic conditions. Michael wishes to estimate the *mean* speed at which he drives. To this end, he estimates the mean time $E\{X\}$ computing the average time for a significant number of trips, and divides the distance to school by the estimate obtained. Another way is to compute the average speed in each trip (that is, to divide the distance by the time for each trip), and after that to compute the average speed for a number of trips. Which way is more precise? Is the mean speed over or underestimated when the worst way is used?

44.* Consider the r.v. X from Exercise 41 replacing 0 by a "small" number ε. Compare $1/E\{X\}$ and $E\{1/X\}$. Which theorem does this example illustrate?

45.* It is known that for any positive r.v. X, the function $n(s) = (E\{X^s\})^{1/s}$ is non-decreasing in s. Prove it, using Jensen's inequality. (*Hint:* We should prove that, if $s < t$, then $n(s) \leq n(t)$, which is equivalent to $E\{X^s\} \leq (E\{X^t\})^{s/t}$. Observe also that $X^s = (X^t)^{s/t}$.)

46.* Let us come back to Example 2.3-2. Let $\widetilde{N}$ be the index of the first X_i whose value is greater than or equal to X_0. Show that in the case of a finite number of the values of X's, the mean $E\{\widetilde{N}\} < \infty$. Find it, if $X_i = 0, 1, 2$ with respective probabilities $\frac{1}{2}, \frac{1}{3}, \frac{1}{6}$.

Section 3

47. Let $X = \pm 1$ with equal probabilities. Do we need to calculate $Var\{X\}$, or is the result obvious?

48. (a) Find the variance of the number on a die rolled.

    (b) Find the variance of the product of the numbers two dice rolled.

49. Find the mean and variance for the distribution in Exercise 6.

50. In which of two cases in Exercise 5d do you expect a larger variance? Check your guess computing the variances.

51. We revisit Exercise 33. Suppose that the number of potential buyers is a Poisson r.v. with a mean of 60,000, and it does not depend on the price of the product. Find the standard deviation of the company's income. (*Advice*: Take \$100 as a monetary unit.)

52. Find the variance in the problem of Exercise 27 for independent trials.

53. For any distribution with mean $m \neq 0$ and standard deviation σ, the ratio σ/m is called a *coefficient of variation* (c.v.). (a) How would you interpret the situation when the c.v. of one r.v. is larger than that of another? (b) Will the c.v. change if we multiply the r.v. by a number?

54. Which is larger: $E\{X^2\}$ or $(E\{|X|\})^2$? (Unlike in (3.1.5), we consider $|X|$.)

55.* Compare the problem in Exercise 54 with that in Exercise 45.

56. 8 cards are drawn from a 52-card deck. Find the expected number of spades. Does the answer depend on whether the cards were drawn with or without replacement? Can we reason in the same way when computing the variance?

57. 100 dice are rolled. What is the expected value and the variance of the sum of the faces shown?

58. Compute the standard deviation of the r.v. from Exercise 41, compare it with the mean and comment on the situation with large n.

59. Let $E\{X\} = 0$ and $Var\{X\} = 1$. Does it mean that X cannot assume "big" values? Consider

$$X = \begin{cases} n \text{ with probability } \frac{1}{2n^2} \\ 0 \text{ with probability } 1 - \frac{1}{n^2} \\ -n \text{ with probability } \frac{1}{2n^2}. \end{cases}$$

Compute $E\{X\}$ and $Var\{X\}$, and comment on the situation. Compute the third moment m_3 and the third absolute moment $\overline{m}_3$. Consider their behavior for large n.

60. There are n assets with random *returns* $X_1, ..., X_n$. The term "return" means that, if you invest \$1 in, say, the first asset, you will get an income of X_1 dollars; see also Example 2.1-9. Note that a return may be less than one.

    Assume that $X_1, ..., X_n$ are i.i.d., $E\{X_i\} = m$, $Var\{X_i\} = \sigma^2$. Consider two strategies of investing n units of money: either investing the whole sum into one asset, for example, into the first, or distributing the investment sum equally between n assets. For both strategies, compute (a) the expected total return, i.e., the return per unit of money; and (b) the standard deviation of the return. Proceeding from what you got, compare the two strategies.

61. Let $X_1, ..., X_n$ in Exercise 60 be independent, all expectations $E\{X_i\} = m$, but the variances be different, and $Var\{X_i\} = \sigma_i^2$. How would you distribute the money between the assets adopting the variance as a measure of riskiness? Would it influence the expected return of your investment? Find the minimal variance if the total investment equals one.

62. Let X be a Poisson r.v. with $\lambda = 2$, and Y be a geometric r.v. with the parameter $p = 0.3$. Assume X and Y to be independent. Find $E\{2X + 3Y\}$, $E\{\frac{X}{2} + \frac{Y}{3}\}$, $E\{2X + 3Y + 10\}$, $E\{\frac{X}{2} + \frac{Y}{3} + 10\}$, $Var\{2X + 3Y\}$, $Var\{\frac{X}{2} + \frac{Y}{3}\}$, $Var\{2X - 3Y\}$, $Var\{\frac{X}{2} - \frac{Y}{3}\}$, $Var\{2X + 3Y + 10\}$, $Var\{\frac{X}{2} - \frac{Y}{3} + 10\}$. Where did you use the independence assumption?

Section 4

63. Standard items produced in a factory go through quality control. The probability that an item taken at random does not meet certain requirements is 0.005. Suppose that the first 100 consecutive items have been found conforming. What is the probability that the total number of consecutive items before the first non-conforming will be at least 150? What assumption have you made before calculating the answer?

64. Mr. Smith bought a number of shares of a particular stock. Assume that each working day, the price per share drops with a probability of 0.4 and rises with a probability of 0.6. Mr. Smith has decided to sell all shares right away after the third drop. How many days on the average will Mr. Smith keep the shares?

65. Consider the r.v.'s N and K from Section 4.3. Show that the conditional distribution of K given $K > 0$ coincides with the distribution of N. Show that this is a special property of the geometric distribution.

66. In some world chess championships, ties were not counted, and the competition lasted until a participant got six points. Assuming that the results of games are independent, and the participants are of equal strength, find the probability that at the end of the championship, a participant will have $k = 0, 1, ..., 5$ points.

67. For a sequence of independent trials with the probability of success equal to p, what is the probability that there will be exactly five failures before the eighth success? How is this question connected with the negative binomial distribution?

68. Is (4.3.4) true for the r.v. K from Section 4.3? If not, write an appropriate counterpart.

69. Let X be a Poisson r.v. with parameter λ, and $p(\lambda) = P(X \text{ is even})$. Prove that $p(\lambda) = \frac{1}{2}(1 + e^{-2\lambda})$. Find the probability that X is odd. (*Advice*: Using Taylor's expansions, observe that $\frac{1}{2}(e^{\lambda} + e^{-\lambda}) = \sum_{k=0}^{\infty} \frac{\lambda^{2k}}{(2k)!}$.)

70. Assume that a medicine has a side effect which happens on the average with one patient out of 750. Suppose that in a hospital during a certain period, 1500 patients took this medicine. *Find* the expected value and the variance of the number of the patients who had the side effect, and *estimate*, using the Poisson approximation, the probability that at most 4 patients experienced the side effect. You are recommended to use software.

71. For an auto insurance portfolio, on the average, 10% of customers file claims for the coverage of their losses. Use the Poisson approximation to estimate the probability that at most 3 out of 25 customers will file claims.

72. Compute the coefficient of variation (see Exercise 53) for the geometric, binomial, and Poisson distribution. Describe the behavior of the ore the behavior of the coefficients of variation when in the second case n is increasing; and in the third case λ is increasing.

73. Suppose that, in a coastal area, during the hurricane season from August through November, hurricanes hit the coast with a monthly rate of 1.15, and outside the hurricane season the rate is 0.3 per month. The monthly number of hurricanes are independent Poisson r.v.'s. Find the probability that there will be not less than 7 hurricanes during a year.

74. In the situation of Exercise 17, assume that N_1, N_2 are independent Poisson r.v.'s with the same λ. Find the distribution of the total number of the claims covered for the case of the two separate insurances and for the case of the joint insurance. Compare these distributions.

75. Let Z_n be a Poisson r.v. with parameter λ equal to an integer n. Show that Z_n may be represented as $Y_1 + ... + Y_n$, where the Y's are independent Poisson r.v.'s with parameter $\lambda = 1$.

76. Suppose that in the scheme of Section 4.5.1, $p_n = \lambda/n$ (that is, we simplify the problem neglecting the remainder $o(1/n)$ in (4.5.4)). (a) Is $E\{X_n\}$ equal to the mean of the limiting Poisson distribution? (b) Is $Var\{X_n\}$ equal to the variance of the limiting Poisson distribution? What happens when $n \to \infty$?

77. Is the total income S in Example 4.5.5-1 a Poisson r.v.? Is it a linear combination of Poisson r.v.'s? Suppose $E\{N\} = 4$ (rather than 40) and find $P(S = 100)$, $P(S = 200)$, $P(S = 350)$.

78. (a) State generalizations of Propositions 13 and 14 for the case of k independent Poisson r.v.'s for the former proposition, and for the case of k types for the latter. (b) Prove the generalizations you stated.

79. The daily number of traffic accidents is a Poisson r.v. with $\lambda = 30$, and the probability that an accident causes a serious damage is $p = 0.3$. The outcomes of different accidents are independent. Find the probability that (a) the number of accidents with serious damages will exceed 8; (b) the number of accidents with serious damages exceeded 8 given that the number of accidents without serious damages equals 20; (c) the number of accidents with serious damages exceeded 8 given that the total number of accidents equals 20.

Sections 5 and 6

80. Each day, a manager of a store determines the daily number of customers. The numbers for different days are independent with the same Poisson distribution. Assume that the mean number of customers per day is 36. The manager does not know this, but she/he has data for the last n days, and can compute $\overline{X}_n$, the average number of customers per day. What must $\overline{X}_n$ be close to for large n? In what sense? Let $n = 100$. Estimate the probability that $\overline{X}_n$ falls within 4 from its expected value? What would it mean for the manager?

81. Bob wishes to check whether the total sum in his credit card statement is plausible. He adds up all figures concerning the purchases made, rounding each number to the nearest integer. Let S be the cumulative error after summation. Suppose there were 100 purchases, and the fraction part of the cost of each purchase is equally likely to be any number from $0, 0.01, ..., 0.99$. (The prices are often ending with 99c or 98c, but after taxation the distribution becomes more or less uniform.) If the fraction part is 0.5, Bob rounds to the nearest smaller or larger integer with equal probabilities. Find the mean and standard deviation of S. What would you consider the accuracy of the approximation? Estimate the probability that the absolute value of the error will exceed \$10; \$15.

82.* Let S_n be the number of successes in n independent trials with the probability of success in each trial equal to p. Let $\xi_n = \frac{1}{n}\max\{S_n, n - S_n\}$. Which is larger: $E\{\xi_n\}$ or $\max\{p, 1-p\}$? Find $\lim_{n\to\infty}\xi_n$. In what sense does ξ_n converge to its limit? (*Advice*: Use Jensen's inequity.)

83.* In Lemma 19, which is larger: $\widetilde{Y}_n$ or $|Y_n - Y|$? Does the convergence $Y_n(\omega) - Y(\omega) \to 0$ imply $\widetilde{Y}_n(\omega) \to 0$?

84. Suppose that in a sequence of n independent trials, the common probability of success is a random variable; say, it depends on conditions under which the trials are being performed. (For example, in the quality control scheme, the share of defected items may depend on conditions under which the particular batch of items had been manufactured, and may vary from day to day. However, we assume that, given the probability of success (for example, we know it for a particular batch), the trials are independent.) In the general scheme, denote by Θ the (random) probability of success. Find the expected value of the number of successes in the following situations. (a) $\Theta = 0.1$ or 0.2 with probabilities 0.3 and 0.7, respectively. (b) $\Theta = \frac{1}{2}, \frac{1}{4}, \frac{1}{8}, ...$ with respective probabilities $\frac{2}{3}, \frac{2}{9}, \frac{2}{27}, ...$ (c) For an integer m, the r.v. $\Theta = \frac{1}{m}, \frac{2}{m}, \frac{3}{m}, ..., \frac{m-1}{m}, 1$ with equal probabilities.

85. In Exercise 84, set $n = \infty$ and find the expected value of the number of the first success.

86. A gambler is playing a slot machine. The results of plays are independent, the probability of winning in each play is p. Denote by N_k the number of the play when the kth win occurred, and by τ_k the number of plays between the $(m-1)$th and the mth including the latter. (a) Find the *conditional* distribution and expectation of τ_3 given $N_3 = n$. (b) Consider the case of an arbitrary k. (*Advice*: Look over Example 1.2.4-2 and formula (2.1.11).)

87. Let us come back to Example 2.3-2. Write a formula for $P(N = k)$ in the case where X's have a geometric distribution. Demonstrate directly, using the formula for total expectation, that $E\{N\} = \infty$. (*Advice*: In both cases, condition on X_0.)

88. In an area, the number of days during a season when you can watch shooting (falling) stars is a Poisson r.v. with a parameter λ. The number of falling stars on each of such days is also a Poison r.v. with a parameter μ. The numbers of stars on different days are independent, and do not depend on the number of the days with falling stars. (a) Find the expected value of the total number of stars that can be watched during the season. (b) Find the variance of the same r.v. (c) Show that the distribution of the total number of stars is not Poisson. (Sometimes, the distribution in hand is called *Poisson-Poisson*.)

Route 1 $\Rightarrow$ *page 131*

Chapter 4

Generating Functions. Branching Processes. Random Walk Revisited

2

In previous chapters, we used methods connected with straightforward calculations of probabilities. Next, we introduce one of analytical methods of Probability Theory, based on what is called generating functions. It will allow us to solve several important problems, including the extinction problem and some sophisticated questions concerning random walk.

1 BRANCHING PROCESSES

Consider people, or organisms, or something else—we will use the term "particles"—that are able to produce offsprings. We distinguish the initial group of particles, and denote the size of this group by X_0. If $X_0 = 1$, we have just one "progenitor."

Each of X_0 particles produces a random number of offsprings, and all these offsprings constitute the first generation. The particles of the first generation produce their offsprings, they constitute the second generation, and the propagation process is continuing in the same fashion.

A nucleus, when it splits up due to a random collision with a neutron, yields a random number of neutrons. These new neutrons, when moving, may hit other nuclei, which leads to the next generation of neutrons, and so on: we are dealing with what is called a nuclear chain reaction.

Other examples are connected with the inheritance of family names, cell divisions, electron multipliers, etc., etc.

Merely for convenience, we assume that when a particle has produced its offsprings, it "dies" or, at least, is not counted anymore.

Assume that all particles act independently of each other, and the probability that a particle produces k offsprings is the same for all particles. We denote this probability by f_k, $k = 0, 1, 2, ...$, and the population size of the t-th generation by X_t , viewing $t = 0, 1, ..$ as a time moment.

Our goal is to calculate the *extinction probability*; that is, the probability that at some (random) moment, the size of the population of the particles will be reduced to zero. More rigorously, let us set $u_t = P(X_t = 0)$ and observe that $u_t \leq u_{t+1}$ for any t. (Once $X_t = 0$, there will be no particles in the next generations.) So, the sequence $\{u_t\}$ is non-decreasing, all $u_t \leq 1$, and hence, there exists a number $u = \lim_{t\to\infty} u_t$ which we will call an extinction probability.

Suppose, for a while, that $X_0 = 1$, and consider the events $A = \{\textit{the population will distinct}\}$ and $B_k = \{\textit{the original particle has produced k offsprings}\},\ k = 0, 1, 2, \dots$.

Since particles act independently, once a particle has been "born," the extinction probability for the part of the population generated only by this particle is the same u. If the original particle has produced exactly k offsprings, the whole population will disappear if and only if all "branches" corresponding to the k original offsprings disappear. Hence, the extinction probability for the whole population is u^k. Eventually, by the formula for total probability,

$$u = P(A) = \sum_{k=0}^{\infty} P(A\,|\,B_k)P(B_k) = \sum_{k=0}^{\infty} u^k f_k = \sum_{k=0}^{\infty} f_k u^k. \tag{1.1}$$

Let us define the function

$$G(z) = \sum_{k=0}^{\infty} f_k z^k, \tag{1.2}$$

and rewrite (1.1) as $u = G(u)$. Thus, u is one of solutions to the equation

$$z = G(z). \tag{1.3}$$

EXAMPLE 1. Suppose $f_k = 0$ for all $k \neq 0$ or 2. For example, a cell either splits into two cells, or does not and dies. Then $G(z) = f_0 + f_2 z^2$, and (1.3) is equivalent to

$$z = f_0 + f_2 z^2.$$

Using below the fact that in our case $f_0 + f_2 = 1$, for a solution to this quadratic equation, we have

$$z = \frac{1 \pm \sqrt{1 - 4f_0 f_2}}{2f_2} = \frac{1 \pm \sqrt{(f_0 + f_2)^2 - 4f_0 f_2}}{2f_2} = \frac{1 \pm \sqrt{(f_0 - f_2)^2}}{2f_2} = \frac{1 \pm (f_0 - f_2)}{2f_2}.$$

Since $f_0 + f_2 = 1$, in the "plus" case, $z = \frac{1}{2f_2}(1 + f_0 - f_2) = \frac{1}{2f_2}(2f_0) = f_0/f_2$. Similarly, the second root is $\frac{1}{2f_2}(2f_2) = 1$. We should choose the solution that is less than or equal to one. If $f_0 \geq f_2$ (which is equivalent to $f_0 \geq 0.5$), the extinction probability $u = 1$. If $f_0 < f_2$, we choose the smaller solution, so $u = f_0/f_2$. We will justify this choice in Section 3.

EXAMPLE 2. Suppose that the number of offsprings for a particular particle is Poisson with parameter λ. Then, using the expansion formula for the exponential function (see the Appendix, (2.2.4)), we have

$$G(z) = \sum_{k=0}^{\infty} e^{-\lambda} \frac{\lambda^k}{k!} z^k = e^{-\lambda} \sum_{k=0}^{\infty} \frac{(\lambda z)^k}{k!} = e^{-\lambda} e^{\lambda z} = e^{\lambda(z-1)}. \tag{1.4}$$

Thus, the equation (1.3) becomes

$$z = e^{\lambda(z-1)}.$$

Though it is impossible to solve this equation analytically, a numerical solution for any particular λ may be obtained even by a graphic calculator by graphing $G(z) = e^{\lambda(z-1)}$ and the function $f(z) = z$, and tracing the intersections.

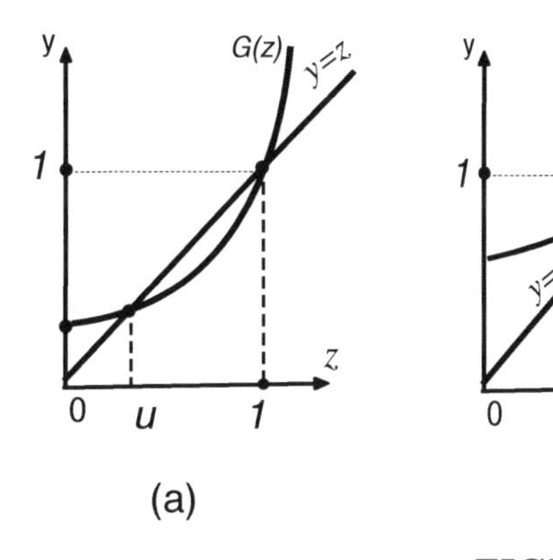

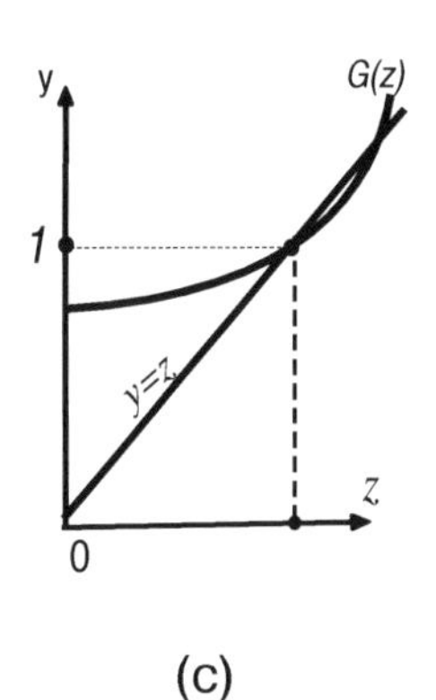

FIGURE 1.

Let us consider this in more detail. The values $G(0) = e^{-\lambda} < 1$, $G(1) = 1$, and the derivative $G'(1) = \lambda$. In Fig. 1a, we see the graph of $G(z)$ together with the function $f(z) = z$ for $\lambda > 1$. The function $G(z)$ is convex, and therefore may intersect the line $y = z$ only at two points. One point is $(1,1)$. Since the derivative $G'(1) > 1$, for the other intersection point, $z < 1$. This is the extinction probability u. Again, we will justify this choice in Section 3.

Fig. 1b illustrates the case $\lambda = 1$. The line $y = z$ is tangent to $G(z)$ at point $z = 1$ and the only solution to (1.3) is $z = 1$. So, the extinction probability equals one.

Fig. 1c corresponds to the case $\lambda < 1$. Now, $G'(1) < 1$, and for the intersection point different from $(1,1)$, we have $z > 1$. Consequently, the extinction probability is also equal to one.

The fact that the answers in the cases $\lambda > 1$ and $\lambda \leq 1$ are different looks natural if we recollect that λ is the mean value of the offsprings for each particle. The table below contains approximate values of u for some λ's that are larger than one.

λ	1.1	2	3	4
u	0.824	0.203	0.059	0.002

□

The function $G(z)$ above is called a *probability generating function* or simply a *generating function*. It proved to be a helpful tool in many problems, and in the next section, we consider it more systematically.

---

2 GENERATING FUNCTIONS

Let $a = \{a_0, a_1, a_2, ...\}$ be a numerical sequence. If the function

$$G_a(z) = a_0 + a_1 z + a_2 z^2 + ... = \sum_{k=0}^{\infty} a_k z^k \tag{2.1}$$

is finite on an interval $[0, z_0]$, we call it the generating function (g.f.) of the sequence a.

Consider a *non-negative integer-valued* r.v. X, set $f_k = P(X = k)$, and for non-negative z's, define

$$G_X(z) = E\{z^X\} = \sum_{k=0}^{\infty} f_k z^k. \tag{2.2}$$

By convention, we set the first term in (2.2), that is, $f_0 z^0 = f_0$ even if $z = 0$. The function $G_X(z)$ is called the generating function of the r.v. X. We see that $G_X(z)$ coincides with the g.f. of the sequence $f = \{f_0, f_1, f_2, ...\}$ that specifies the probability distribution of X. Since $\sum_k f_k = 1$, the function $G_X(z)$ is finite if $0 \leq z \leq 1$. For other z's, the g.f. may be finite, and may be not.

In Section 1, we have considered two examples of a g.f. Let us consider two more.

EXAMPLE 1. Let $X = 1$ or 0 with respective probabilities p and q. Then

$$G_X(z) = qz^0 + pz^1 = q + pz. \tag{2.3}$$

EXAMPLE 2. Suppose that the distribution of X is the second version of the geometric distribution (see Section **3**.4.3); namely, $f_k = pq^k,\ \ k = 0, 1, .., $ where $q = 1 - p$, and $0 \leq p \leq 1$. Then, by the formula for a geometric series (see the Appendix, (2.2.11)),

$$G_X(z) = \sum_{k=0}^{\infty} pq^k z^k = p \sum_{k=0}^{\infty} (qz)^k = \frac{p}{1 - qz}. \ \square \tag{2.4}$$

Next, we list some properties of the g.f.'s. of r.v.'s, skipping the index X in $G_X(z)$. From now on, we assume $E\{X^2\} < \infty$ for all X's under consideration.

A. $G(0) = f_0 = P(X = 0)$, and $G(1) = \sum_k f_k = 1$.

B. Differentiating (2.2) in z, we have

$$G'(z) = \frac{d}{dz}(f_0 + f_1 z + f_2 z^2 + f_3 z^3 ...) = f_1 + 2f_2 z + 3f_3 z^2 + ... = \sum_{k=1}^{\infty} k f_k z^{k-1}.$$

Hence,

$$G'(1) = \sum_{k=1}^{\infty} k f_k = \sum_{k=0}^{\infty} k f_k = E\{X\}. \tag{2.5}$$

C. Differentiating $G(z)$ one more time, we have

$$G''(z) = 2f_2 + 3 \cdot 2 \cdot f_3 z + ... \tag{2.6}$$

We see that $G''(z) \geq 0$, and hence, the *function* $G(z)$ *is convex*.

Properties A–C show, in particular, that the graphs we built in Fig. 1 for particular g.f.'s, as a matter of fact, are typical. We consider it in detail in Section 3.

D.

For independent r.v. X_1 and X_2,
$$G_{X_1+X_2}(z) = G_{X_1}(z) G_{X_2}(z), \tag{2.7}$$
and a similar representation is true for any number of mutually independent r.v.'s.

It suffices to consider two r.v.'s. Since X_1 and X_2 are independent, so are z^{X_1} and z^{X_2}, and

$$G_{X_1+X_2}(z) = E\{z^{X_1+X_2}\} = E\{z^{X_1}\}E\{z^{X_2}\} = G_{X_1}(z)G_{X_2}(z).$$

EXAMPLE 3. Let X have the binomial distribution with parameters (n,p). Let us use the representation (see, for example, Section **2**.4.1)

$$X = X_1 + ... + X_n, \tag{2.8}$$

where X_i's are independent, and $X_i = 1$ or 0 with respective probabilities p and $q = 1-p$. Combining it with (2.3) and (2.7), we have

$$G_X(z) = (q+pz)^n. \ \square \tag{2.9}$$

Now, let us recall that the distribution of $X_1 + X_2$ is the convolution of the distributions of X_1 and X_2; see Section **3**.1.2.3. For future purposes, we need the following generalization of (2.7).

We call the *convolution* of two numerical sequences $a = \{a_0, a_1, a_2, ...\}$ and $b = \{b_0, b_1, b_2, ...\}$ the sequence $c = \{c_0, c_1, c_2, ...\}$ such that

$$c_n = \sum_{k=0}^{n} a_k b_{n-k},\ n = 0, 1, 2, ...;$$

compare with (**3**.1.2.4). As in Section **3**.1.2.3, we present it as

$$c = a * b.$$

We prove that, similar to the case of r.v.'s,

$$G_{a*b}(z) = G_a(z)G_b(z). \tag{2.10}$$

Indeed, going in (2.10) from the right to the left, we have

$$\left(\sum_{k=0}^{\infty} a_k z^k\right)\left(\sum_{j=0}^{\infty} b_j z^j\right) = \sum_{n=0}^{\infty}\left(\sum_{k,j:k+j=n} a_k b_j\right) z^n = \sum_{n=0}^{\infty}\left(\sum_{k=0}^{n} a_k b_{n-k}\right) z^n = \sum_{n=0}^{\infty} c_n z^n. \ \blacksquare$$

In conclusion, we state without a proof

Theorem 1 *(The uniqueness theorem.) Different probability distributions have different g.f.'s.*

In particular, this means that if, when computing the g.f. of a r.v., we arrive at the g.f. of a familiar distribution (say, to the function $e^{\lambda(z-1)}$), we may be sure that the distribution of the r.v. in hand is exactly this distribution (in our example, the Poisson distribution with parameter λ).

3 BRANCHING PROCESSES REVISITED

We come back to branching processes, and use the notation of Section 1.

Let $X_0 = 1$. Denote by ξ the number of offsprings of a particle. We have assumed that the distribution of ξ is the same for all particles. Set, as in Section 1, $f_k = P(\xi = k)$, and $\mu = E\{\xi\}$. The equation (1.3) may be rewritten as

$$z = G_\xi(z). \tag{3.1}$$

As was already mentioned, Properties A–C show that the graphs in Fig. 1 are typical.

Fig. 1a illustrates the case $G'_\xi(1) = E\{\xi\} = \mu > 1$. There are two solutions that do not exceed one. We choose the smaller, so the extinction probability $u < 1$. A rigorous proof of why we should not select $z = 1$ as a solution, will be given a bit later.

Fig. 1c corresponds to the case $G'_\xi(1) = \mu < 1$. Here, the only solution which is less than or equal to one, is one. So, the extinction probability $u = 1$.

Let $G'_\xi(1) = \mu = 1$. First, suppose that $f_k > 0$ at least for one $k \geq 2$. Then, by (2.6), $G''(1) = 2f_2 + 3\cdot 2\cdot f_3 z + ... > 0$ for $z > 0$. This means that $G'_\xi(z)$ is growing, and the graph looks as in Fig. 1b. So, (3.1) has only one solution, and it is $u = 1$. Thus, though each particle produces on the average one offspring, the population still dies out with probability one.

It remains to consider the case $f_k = 0$ for all $k \geq 2$ and $\mu = 1$. This means that $0\cdot f_0 + 1\cdot f_1 = 1$, that is, $f_1 = 1$. This is a trivial case: the number of offsprings is not random and equals one, and hence, $u = 0$. Note also, that in this case, $G_\xi(z) = 1\cdot z = z$, and its graph coincides with the line $y = z$.

Next, we consider the mean value and variance of the size of the tth population. Let again $X_0 = 1$, and $\sigma^2 = Var\{\xi\}$.

Proposition 2 *For any $t = 1, 2, ...$,*

$$E\{X_t\} = \mu^t, \tag{3.2}$$

$$Var\{X_t\} = \begin{cases} \sigma^2\mu^{t-1}\left(\frac{\mu^t - 1}{\mu - 1}\right) & \text{if } \mu \neq 1, \\ \sigma^2 t & \text{if } \mu = 1. \end{cases} \tag{3.3}$$

We see, in particular, that if $\mu > 1$, then $E\{X_t\} \to \infty$ as $t \to \infty$; the population is unlimitedly growing. So, we were right when choosing in this case, as a solution to (3.1), the smaller value $u < 1$. Indeed, since the sequence of events $\{X_t = 0\}$ is non-decreasing, the fact that $u = 1$ implies that $P(\cup_{t=1}^{\infty}\{X_t = 0\}) = 1$, which means that with unit probability one of events $\{X_t = 0\}$ will occur. That is, at a (random) moment, the population will be reduced to zero. Clearly, in this case, the mean size of the population cannot converge to infinity.

If $\mu < 1$, then $E\{X_t\} \to 0$ as $t \to \infty$, which is consistent with the fact that in this case, the population disappears with probability one.

EXAMPLE 1. Let us revisit Example 1-1 and suppose that $f_0 < f_2$. Together with $f_0 + f_2 = 1$, this implies $f_2 > \frac{1}{2}$. Hence, $\mu = 2f_2 > 1$, and this is why we have chosen in Example 1-1 the smaller root. $\square$.

EXAMPLE 2. In Example 1-2, $G'(1) = \lambda$, which is the mean of the Poisson distribution (and equals μ in the terms of this section). If $\lambda > 1$, we choose the smaller solution. □

Proof of Proposition 2. Denote by ξ_{kt} the number of offsprings of the kth particle of the tth generation. Summing up the numbers of the offsprings of the particles in a generation, we get the size of the next generation. So,

$$X_{t+1} = \sum_{k=1}^{X_t} \xi_{kt}. \tag{3.4}$$

Therefore, by (**3**.6.2.7), $E\{X_{t+1}\} = \mu E\{X_t\}$. Since $E\{X_1\} = \mu$, (3.2) follows by induction.

Now, by virtue of (**3**.6.3.2), the relation (3.4) yields

$$Var\{X_{t+1}\} = \sigma^2 E\{X_t\} + \mu^2 Var\{X_t\} = \mu^t \sigma^2 + \mu^2 Var\{X_t\}.$$

Since $Var\{X_1\} = \sigma^2$, and this is consistent with (3.3) for $t = 1$, we can apply induction. Assuming that (3.3) is true, for $\mu \neq 1$, doing simple algebra, we get

$$Var\{X_{t+1}\} = \sigma^2 \mu^t + \mu^2 \cdot \sigma^2 \mu^{t-1} \left(\frac{\mu^t - 1}{\mu - 1} \right) = \frac{\sigma^2}{\mu - 1} \left(\mu^t \left(\mu^{t+1} - 1 \right) \right).$$

This leads to (3.3) with replacement t by $t+1$. The case $\mu = 1$ is similar and easier. ■

In conclusion, consider the case where $X_0 > 1$ with a positive probability. Let A be the event that the population will get extinct. Then $P(A \mid X_0 = k) = u^k$, where u is the extinction probability for the population generated by one of the original particles. Then the extinction probability for the whole population is $\sum_{k=0}^{\infty} u^k P(X_0 = k) = E\{u^{X_0}\} = G_{X_0}(u)$.

---

4 MORE ON RANDOM WALK

In this section, we will use the method of generating functions to explore some sophisticated questions concerning random walk.

4.1 Revisiting the initial state

Let $X_0 = 0$, and $X_t = \xi_1 + ... + \xi_t$, for $t = 1, 2, ...$, where the r.v.'s ξ_i are independent and all take on the values 1 or -1 with respective probabilities p and q. We will adopt the game interpretation.

Suppose there are two players; at each play, the first player wins one (unit of money) with probability p, and loses one with probability $q = 1 - p$. The gains-and-losses of the second player are opposite. If, merely for convenience, we view t as time, then the r.v. X_t is what the first player will get by time t. We do not involve initial capital, and just explore the gain-loss process X_t.

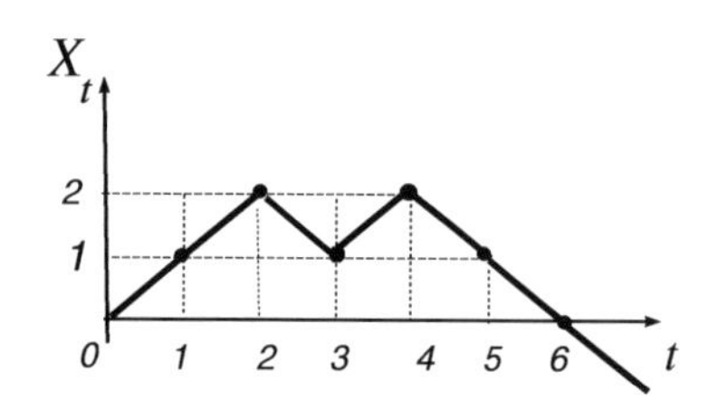

FIGURE 2. X_t has revisited zero at $\tau = 6$.

The initial balance is zero, and our first object of study is the random moment τ when the process revisits zero at the first time; see also Fig. 2. In other words, this is the moment when the players are again even. Formally, $\tau = \min\{t : X_t = 0, t > 0\}$. Set $r_n = P(\tau = n)$, and $\overline{q} = \sum_{n=1}^{\infty} r_n$. The quantity $\overline{q}$ is the probability that the process will revisit zero at some time; that is, will ever come back to the initial state.

If $\overline{q} < 1$, with the positive probability $\overline{p} = 1 - \overline{q}$, the process will never come back to zero, which means that one of the players will always have a positive profit.

We set $r_0 = 0$, and consider the g.f. of the sequence $r = \{r_0, r_1, r_2, ...\}$, i.e.,

$$G_r(z) = \sum_{n=0}^{\infty} r_n z^n. \tag{4.1.1}$$

Note that $G_r(1) = \sum_{n=0}^{\infty} r_n = \overline{q}$.

Theorem 3 *For* $0 \leq z \leq 1$,

$$G_r(z) = 1 - \sqrt{1 - 4pqz^2}. \tag{4.1.2}$$

A proof will be given at the end of this section.

Corollary 4 *The probability*

$$\overline{q} = 1 - |p - q| = \begin{cases} 2q \text{ if } p > q, \\ 1 \text{ if } p = q, \\ 2p \text{ if } p < q. \end{cases} . \tag{4.1.3}$$

(Say, if $p > q$, then $1 - |p - q| = 1 - p + q = 2q$.)

To obtain (4.1.3), it suffices to write

$$\overline{q} = G(1) = 1 - \sqrt{1 - 4pq} = 1 - \sqrt{(p+q)^2 - 4pq} = 1 - \sqrt{(p-q)^2} = 1 - |p - q|.$$

If $p = 0$ or 1, the problem is trivial: either with probability one $\xi_k \equiv 1$, or with the same probability $\xi_k \equiv -1$. Assume that $0 < p < 1$, and hence the same is true for q.

We distinguish three cases.

Case 1: $p > q > 0$. In this case, $q < \frac{1}{2}$, and $\overline{q} = 2q < 1$. The process will evolve as follows.

Starting from zero, with probability $\overline{p} = 1 - \overline{q} = p - q > 0$, the process will never come back to zero. With probability $\overline{q}$, the process will revisit the initial point zero, and "will start over as from the beginning." After that, with the same probability $\overline{q}$, the process will again revisit zero, and so on, up to the moment when the revisit will turn out to be the last. To show that it will happen with probability one, denote by N the number of revisits. The probability $P(N > k) = \overline{q}^{k+1}$. So, we are dealing with a *geometric distribution*, and $P(N > k) \to 0$, as $k \to \infty$, because $\overline{q} < 1$.

It remains to realize that, after the last revisit, X_t will be always positive. Indeed, $E\{\xi_k\} = p-q$, and the mean $E\{X_t\} = t(p-q) \to \infty$ as $t \to \infty$, for $p > q$. However, this would have been impossible if after a finite moment of time, X_t had become negative and never had come back to zero.

Case 2: $p < q$. It is similar: we may just switch the players considering the second as the first player in Case 1.

Case 3: $p = q = \frac{1}{2}$. Now, $\overline{q} = 1$, and the process will revisit zero infinitely many times: "it always comes back." Then $\sum_{n=1}^{\infty} r_n = 1$, and τ is a r.v. with the distribution determined by the sequence r. The g.f. $G_r = 1 - \sqrt{1-z^2}$, and as is easy to verify $G'(1) = \infty$. This, along with (2.5), immediately implies

Corollary 5 *In the case* $p = q = \frac{1}{2}$,

$$E\{\tau\} = \infty. \tag{4.1.4}$$

This non-trivial fact has already been commented on in Section **3**.2.3.

Next, we will find the precise formulas for the probabilities r_k. First, note that the process may revisit zero only at even moments of time. So, $r_{2k+1} = 0$, and it suffices to consider r_{2k}.

By Taylor's expansion (2.2.9) in the Appendix,

$$G_r(z) = 1 - \sum_{k=0}^{\infty} \binom{1/2}{k} (-4pqz^2)^k = -\sum_{k=1}^{\infty} (-4)^k \binom{1/2}{k} (pq)^k z^{2k}.$$

Then, by the definition of $G_r(z)$, for $k = 1, 2, ...$,

$$r_{2k} = -(-4)^k \binom{1/2}{k} (pq)^k. \tag{4.1.5}$$

Using the definition of binomial coefficients in the Appendix, (2.2.10), we can write

$$\begin{aligned}
\binom{1/2}{k} &= \frac{\frac{1}{2}(\frac{1}{2}-1)\cdots(\frac{1}{2}-k+1)}{k!} = \frac{(-1)^{k-1}(2k-3)(2k-5)\cdots 3\cdot 1}{k!2^k} \\
&= \frac{(-1)^{k+1}(2k-3)!}{k!2^k 2\cdot 4\cdots 2(k-2)} = \frac{(-1)^{k+1}(2k-3)!}{2^k\cdot 2^{k-2}(k-2)!k!} \\
&= \frac{(-1)^{k+1}}{4^k\cdot 2^{-2}}\cdot\frac{(k-1)}{2(k-1)k}\cdot\frac{(2k-2)!}{(k-1)!(k-1)!} = (-1)^{k+1}4^{-k}\frac{2}{k}\binom{2(k-1)}{k-1}.
\end{aligned} \tag{4.1.6}$$

Substituting this into (4.1.5), we arrive at

Corollary 6 *For all* $k = 1, 2, ...$,

$$r_{2k} = \frac{2}{k}\binom{2(k-1)}{k-1}(pq)^k. \tag{4.1.7}$$

To simplify the last expression, we use the Stirling formula (see comments on p.49)

$$n! \sim \sqrt{2\pi n}\, n^n e^{-n}. \tag{4.1.8}$$

Substituting it to the expression for the binomial coefficients in (4.1.7), after simple algebra, one can get that

$$r_{2k} \sim \frac{1}{2\sqrt{\pi}} \frac{1}{k\sqrt{k}} (4pq)^k. \tag{4.1.9}$$

Proof of Theorem 3. Let $u_n = P(X_n = 0)$. Below, we use the fact that ξ_j's are i.i.d., $r_0 = 0$, and that $\sum_{j=k+1}^{n} \xi_j$ has the same distribution as X_{n-k} . For $n = 1, 2, ...$, we have

$$u_n = P(X_n = 0) = \sum_{k=0}^{n} P(\tau = k, X_n = 0) = \sum_{k=1}^{n} P\left(\tau = k,\ \sum_{j=k+1}^{n} \xi_j = 0\right)$$

$$= \sum_{k=1}^{n} P(\tau = k) P\left(\sum_{j=k+1}^{n} \xi_j = 0\right) = \sum_{k=1}^{n} P(\tau = k) P(X_{n-k} = 0) = \sum_{k=0}^{n} r_k u_{n-k}. \tag{4.1.10}$$

Let the sequence $u = \{u_0, u_1, u_2, ...\}$ and the sequence $c = \{0, u_1, u_2, ...\}$. Then $c_0 = 0 = r_0 u_0$, which together with (4.1.10) implies $c = r * u$. Hence, by (2.10),

$$G_u(z) = 1 + \sum_{k=1}^{\infty} u_k z^k = 1 + \sum_{k=0}^{\infty} c_k z^k = 1 + G_c(z) = 1 + G_r(z) G_u(z).$$

From this, it follows that

$$G_r(z) = 1 - 1/G_u(z), \tag{4.1.11}$$

and it remains to find $G_u(z)$.

Since X_t may be equal to zero only for even t's, the probability $u_{2k+1} = 0$.

Furthermore, $u_{2k} = \binom{2k}{k} p^k q^k$. Decomposing $(2k)!$ into the products $1 \cdot 3 \cdots (2k-1)$ and $2 \cdot 4 \cdots 2k$, similar to(4.1.6), it is straightforward to verify that $\binom{2k}{k} = (-4)^k \binom{-1/2}{k}$. Then

$$G_u(z) = \sum_{k=0}^{\infty} u_{2k} z^{2k} = \sum_{k=0}^{\infty} \binom{-1/2}{k} (-4pqz^2)^k = (1 - 4pqz^2)^{-1/2},$$

by the Appendix, (2.2.9). It remains to substitute the last expression into (4.1.11). ■

4.2 Symmetric random walk: Ties and leadership

In this section, we consider the particular case $p = q = 1/2$. We will see that even in this case, the random walk turns out to be a much deeper phenomenon that one may expect, and some results look amazing.

4.2.1 The number of ties

First, let us count the number of revisits to zero. Following the game interpretation, we may call it also the number of ties because at these moments the players are even. Let Y_n be the number of revisits during n plays. It looks quite plausible that the mean number of revisits during, say, 400 plays should be, at least approximately, four times as large as this number for 100 plays. Surprisingly, it is not true: on the average, it will be only twice as large. More precisely, $E\{Y_n\}$ has an order of $\sqrt{n}$ rather than n, which is reflected in

Theorem 7 *As $n \to \infty$,*

$$\frac{E\{Y_n\}}{\sqrt{n}} \to \sqrt{\frac{2}{\pi}}. \tag{4.2.1}$$

EXAMPLE 1. Viewing 100 and 400 as "large" numbers, and using approximation (4.2.1), we have $E\{Y_{100}\} \approx \sqrt{\frac{2}{\pi}} \cdot \sqrt{100} \approx 7.98 \approx 8$, while $E\{Y_{400}\} \approx \sqrt{\frac{2}{\pi}} \cdot \sqrt{400} \approx 15.63 \approx 16$. □

Proof of Theorem 7 is relatively short. Since revisits to zero may occur only at even moments of time, $Y_{2n+1} = Y_{2n}$, and it suffices to consider only Y_{2n}. Let the r.v. $I_{2k} = 1$ if $X_{2k} = 0$, and $I_{2k} = 0$ otherwise. One may say that I_{2k}'s counts revisits. Clearly,

$$Y_{2n} = \sum_{k=1}^{n} I_{2k}.$$

and $E\{I_{2k}\} = P(X_{2k} = 0) = u_{2k}$. Then

$$E\{Y_{2n}\} = \sum_{k=1}^{n} u_{2k}.$$

Furthermore, for $p = q = 1/2$,

$$u_{2k} = \binom{2k}{k} 4^{-k} \sim \frac{1}{\sqrt{\pi k}}, \tag{4.2.2}$$

which has been already shown in Section **2**.3; see (**2**.3.2). Thus,

$$E\{Y_{2n}\} \sim \frac{1}{\sqrt{\pi}} \sum_{k=1}^{n} \frac{1}{\sqrt{k}}.$$

The last sum $\sum_{k=1}^{n} \frac{1}{\sqrt{k}} \sim \int_1^n \frac{1}{\sqrt{x}} dx = 2\sqrt{n}$, which implies

$$E\{Y_{2n}\} \sim \frac{2\sqrt{n}}{\sqrt{\pi}} = \sqrt{\frac{2}{\pi}} \cdot \sqrt{2n}.$$

Together with the fact that $Y_{2n+1} = Y_{2n}$, this implies (4.2.1). ■

4.2.2 The number of leadership intervals. The arcsine law

We call an interval $[t, t+1]$ an interval of the leadership of the first player if either X_t or X_{t+1} (or both) are positive. For example, in Fig. 2, the periods from $t = 0$ to $t = 6$ are those of leadership, and at $t = 6$, the leadership switches to the second player.

We are interested in which is more probable: a frequent change of leadership or a long leadership of one of the players. For example, which probability is larger: that one of the players will be a leader during 90% of time, or the proportion of the leadership time for each will lie between 45% and 55%? We will see that while the former probability approximately equals 0.41, the latter is about 0.064. Perhaps, this also looks unexpected.

Let Z_n be the number of leadership intervals of the first player for n plays, and $K_n = Z_n/n$.

Theorem 8 *For $0 \le a \le b \le 1$, as $n \to \infty$,*

$$P(a \le K_n \le b) \to \frac{1}{\pi}\int_a^b \frac{dx}{\sqrt{x(1-x)}} = \frac{2}{\pi}\left(\arcsin\sqrt{b} - \arcsin\sqrt{a}\right). \tag{4.2.3}$$

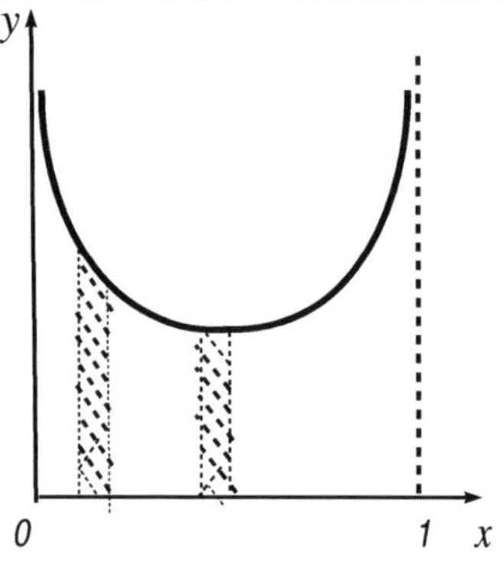

FIGURE 3.

The last equation in (4.2.3) may be checked by differentiation. The graph of the integrand $1/\sqrt{x(1-x)}$ in (4.2.3) is sketched in Fig. 3. We see that the closer a subinterval to an end point of $[0,1]$, the larger the area over this subinterval and under the curve.

The reader is invited to calculate on her/his own that approximation (4.2.3) indeed implies that $P(0.45 \le K_n \le 0.55) \approx 0.0638$ and $P(K_n \le 0.1 \text{ or } K_n \ge 0.9) \approx 0.410$ for large n.

Theorem 8 is a corollary from a more precise Theorem 9 below. Before stating it, note that, as is easy to realize, Z_{2n} may assume only even values.

Theorem 9 *For all $k = 0, 1, ..., n$,*

$$P(Z_{2n} = 2k) = u_{2k}u_{2n-2k}. \tag{4.2.4}$$

First, we derive the former theorem from the latter.

Proof of Theorem 8. We restrict ourselves to even moments of time. By (4.2.4),

$$P\left(a \le \frac{Z_{2n}}{2n} \le b\right) = \sum_{a\le(k/n)\le b} P(Z_{2n} = 2k) = \sum_{a\le(k/n)\le b} u_{2k}u_{2n-2k}.$$

Using (4.2.2), we approximate the last sum by

$$\sum_{a\le(k/n)\le b} \frac{1}{\pi\sqrt{k(n-k)}} = \frac{1}{\pi}\sum_{a\le(k/n)\le b} \frac{1}{\sqrt{(k/n)(1-(k/n))}}\cdot\frac{1}{n}.$$

The sum above is an integral sum for $\int_a^b \frac{dx}{\sqrt{x(1-x)}}dx$, and converges to this integral as $n \to \infty$. ■

Proof of Theorem 9 is somewhat lengthy, and we will present only a sketch. By convention, we set $Z_0 = 0$. Then for $n = 0$, (4.2.4) is obvious since $u_0 = P(X_0 = 0) = 1$.

Consider the following events: $B_m = \{\tau = 2m\}$, where τ is the moment of the first revisit to zero; $C_m^+ = \{X_i > 0 \text{ for all } i = 1, .., 2m-1\}$; and $C_m^- = \{X_i < 0 \text{ for all } i = 1, .., 2m-1\}$.

By the formula for total probability, for $k = 0, ..., n$,

$$P(Z_{2n} = 2k) = \sum_{m=1}^{\infty} P(Z_{2n} = 2k \,|\, B_m C_m^+) P(B_m C_m^+) + \sum_{m=1}^{\infty} P(Z_{2n} = 2k \,|\, B_m C_m^-) P(B_m C_m^-). \tag{4.2.5}$$

The reader is invited to realize that, for $k = 1, 2, ..., n-1$,

$$P(Z_{2n} = 2k \,|\, B_m C_m^+) = \begin{cases} 0 & \text{for } m \geq k+1, \\ P(Z_{2n-2m} = 2(k-m)) & \text{otherwise.} \end{cases}$$

Similarly,

$$P(Z_{2n} = 2k \,|\, B_m C_m^-) = \begin{cases} 0 & \text{for } m \geq n-k+1 \\ P(Z_{2n-2m} = 2k) & \text{otherwise.} \end{cases}$$

Also, by symmetry, $P(B_m C_m^+) = P(B_m C_m^-) = r_{2m}/2$, where as above, $r_m = P(\tau = m)$. Substituting this to (4.2.5), for $k = 1, ..., n-1$, we get

$$P(Z_{2n} = 2k) = \sum_{m=1}^{k} P(Z_{2n-2m} = 2(k-m)) \frac{r_{2m}}{2} + \sum_{m=1}^{n-k} P(Z_{2n-2m} = 2k) \frac{r_{2m}}{2}. \tag{4.2.6}$$

Next, we apply induction. As was noted, (4.2.4) is true for $n = 0$. Suppose that (4.2.4) holds if we replace n by $1, 2, ..., n-1$. Let $0 < k < n$. Using (4.1.10) in the second step below, we have

$$\begin{aligned} P(Z_{2n} = 2k) &= \frac{1}{2} \sum_{m=1}^{k} u_{2k-2m} u_{2n-2k} r_{2m} + \frac{1}{2} \sum_{m=1}^{n-k} u_{2k} u_{2n-2m-2k} r_{2m} \\ &= \frac{1}{2} u_{2n-2k} \sum_{m=1}^{k} u_{2k-2m} r_{2m} + \frac{1}{2} u_{2k} \sum_{m=1}^{n-k} u_{2n-2m-2k} r_{2m} \\ &= \frac{1}{2} u_{2n-2k} u_{2k} + \frac{1}{2} u_{2k} u_{2n-2k} = u_{2k} u_{2n-2k}. \end{aligned}$$

Now, we set $k = n$ and observe that, if $m > n$, then $P(Z_{2n} = 2n \,|\, B_m C_m^-) = 0$ and $P(Z_{2n} = 2n \,|\, B_m C_m^+) = 1$. From this and (4.2.5), it follows that

$$P(Z_{2n} = 2n) = \sum_{m=1}^{n} P(Z_{2n-2m} = 2n - 2m) \frac{r_{2m}}{2} + \sum_{m=n+1}^{\infty} \frac{r_{2m}}{2}. \tag{4.2.7}$$

Also, using (4.1.7), it is straightforward to verify that $r_{2m} = u_{2m-2} - u_{2m}$, which implies $\sum_{m=n+1}^{\infty} r_{2m} = u_{2n}$. Thus,

$$P(Z_{2n} = 2n) = \frac{1}{2} \left(\sum_{m=1}^{n} P(Z_{2n-2m} = 2n - 2m) r_{2m} + u_{2n} \right). \tag{4.2.8}$$

It remains to imply induction and again (4.1.10), which leads to $P(Z_{2n} = 2n) = u_{2n}$. The case $k = 0$ is considered similarly. ■

More detailed proof of both theorems above may be found, e.g., in [13], [42].

5 EXERCISES

1. Find the g.f. of the first version of the negative binomial distribution using (**3**.2.1.11).

2. Show that, if $G(z)$ is the g.f. of a r.v. X, then $G''(1) = E\{X(X-1)\}$. How to find $Var\{X\}$ using the g.f. $G(z)$? Do that for the second version of the geometric distribution.

3. Find the mean and variance of the binomial distributions using g.f.'s.

4. Show that the sum of Poisson independent r.v.'s is a Poisson r.v. using Theorem 1 and (2.7).

5. For the case $f_0 + f_2 = 1$, graph the extinction probability as a function of f_0.

6. Find the extinction probability in the case where ξ, the number of offsprings of a particle, has the second version of the geometric distribution as in Example 2-2.

7. Show that for the case $f_k = 0,\ k > 2$, the extinction probability is the same as in Example 1-1. Connect the condition $f_2 > f_0$ with the condition $\mu > 1$. (*Advice*: When solving (1.3), use the fact that $f_1 = 1 - (f_0 + f_2)$.)

8. Find the extinction probability for $f_0 = 0.8, f_1 = 0.1, f_2 = 0.1$. (*Hint*: Look at the numbers before proceeding to calculations.)

9. Show that, if $f_5 > \frac{1}{5}$, the extinction probability is less than one.

10. In a country, practically all men are married, 10% of married couples do not have children, and 90% have three children. Each child is equally likely to be a girl or boy. Using software, estimate the probability that a married man's male line of descent will ever cease to exist.

11. Let $X_0 = 1$, and $S = \sum_{t=0}^{\infty} X_t$, the total number of all "particles" in all generations. Show that $E\{S\} = \infty$, if $\mu \geq 1$; and $E\{S\} = 1/(1-\mu)$, if $\mu < 1$. If $f_0 = \frac{3}{4}$, and $f_2 = \frac{1}{4}$, what is the total mean number of the particles?

12. Find $E\{X_t\}$ and $Var\{X_t\}$ in the case where ξ's are Poisson with parameter λ. Do the same for the case where X_0 is a Poisson r.v. with the same parameter. (*Hint*: The second question does not require calculations.)

13. (a) Explain that if X_0 is not random and greater than 1, then $E\{X_t\}$ and $Var\{X_t\}$ are given by the expressions (3.2) and (3.3), respectively, multiplied by X_0.

    (b) Consider the case of a random X_0. Write general formulas and consider the case where X_0 is (i) Poisson with parameter λ and (ii) geometric with parameter p. (*Advice*: Use the results of Section 6.3.)

14. Find the extinction probability for the case where ξ's are distributed as in Exercise 7 and X_0 is Poisson with parameter λ.

15. In the model of Section 4.1, set $p = 3/4$ and find the probability that the random walk will revisit zero at most 3 times.

16. We have proved in Section 4.1 that $E\{\tau\} = \infty$ for $p = 1/2$. Show that the same follows from (4.1.9).

17. Suppose we are tossing a regular coin 10,000 times. Guess how often, on the average, the number of heads will be equal to the number of tails. Check your guess.

18. In the model of Section 4.2.2, consider the probability that one of the players will be leading 9900 times out of 10,000. Do you expect this probability to be small? Estimate it and comment on the result.

Chapter 5

Markov Chains

1,2

In Probability Theory, any collection of r.v.'s $\{X_t\}$, where t is a running parameter, is called a *stochastic* or *random process*. Usually, though not always, t is a time parameter or is viewed as such, and X_t describes the evolution of a system in time. If $t = 0, 1, 2, \ldots$, we talk about a random sequence or a process in discrete time. In this chapter, we consider a particular but important type of such processes, namely, Markov chains.

1 DEFINITIONS AND EXAMPLES. THE PROBABILITY DISTRIBUTIONS OF MARKOV CHAINS

1.1 A preliminary example

In a country, people reside in either the urban or rural area. For brevity, let us call them area 0 and area 1. Each year, each citizen either keeps living in the same area where she/he is living now, or moves to the other area. For a citizen chosen at random, one of possible realizations of the migration process *may* look as in Fig. 1, where X_t denotes the number of the area to which the citizen moved in year t. For the particular realization in Fig. 1, at the initial moment $t = 0$, the citizen lived in area 0, then moved to area 1, stayed there one year more, moved to area 0, and came back to area 1.

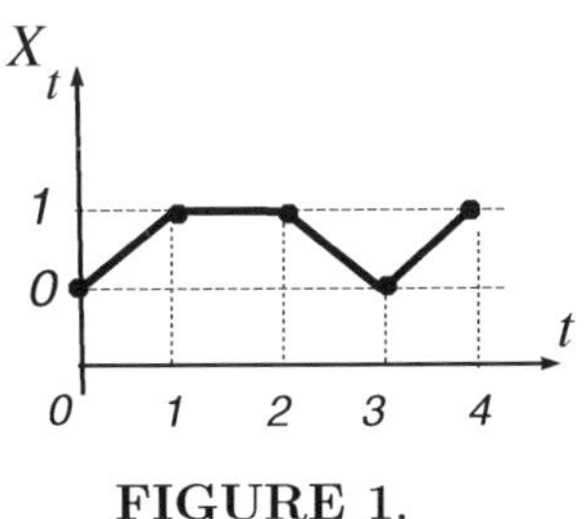

FIGURE 1.

For a citizen chosen at random, denote by p_{ij} the probability of moving from area i to area j; $i, j = 0, 1$. More precisely, this is the probability of moving to area j, *given* that the citizen resides in area i now. We assume this probability to be the same for all years.

Suppose also that *given* where a citizen lives now, the information about previous migration has no impact on the likelihood of the event that the citizen will move to the other area.

It is convenient to present the probabilities p_{ij} as a matrix

$$\mathcal{P} = \begin{Vmatrix} p_{00} & p_{01} \\ p_{10} & p_{11} \end{Vmatrix}. \tag{1.1.1}$$

If, for example,

$$\mathcal{P} = \begin{Vmatrix} 0.8 & 0.2 \\ 0.3 & 0.7 \end{Vmatrix}, \tag{1.1.2}$$

then, for instance, the probability of moving from area 0 to area 1 is 0.2, and the probability that a citizen will keep staying in area 0 is 0.8.

Clearly, the sum of the probabilities in each row in (1.1.1) should be equal to one.

Given in which area a citizen lived at the initial time, we can compute the probability of any realization (or path) with use of the multiplication rule. Consider, for example, the probability of moving from area 0 to area 1 and then staying there two years; denote this as the path $0 \to 1 \to 1$. By the multiplication rule, this probability equals $p_{01}p_{11}$, and if (1.1.2) holds, it is $0.2 \cdot 0.7 = 0.14$. The probability of the path in Fig. 1 (given that at the initial time, a citizen lived in area 0) is $p_{01}p_{11}p_{10}p_{01} = 0.2 \cdot 0.7 \cdot 0.3 \cdot 0.2 = 0.0084$.

However, a randomly chosen citizen may originally live in any of the two areas. Set $\pi_{0i} = P(X_0 = i)$, the probability that the citizen originally lived in area i, and the vector $\boldsymbol{\pi}_0 = (\pi_{00}, \pi_{01},)$. For example, if at time $t = 0$, forty percent of citizens lived in area 0, and sixty percent in area 1, then $\boldsymbol{\pi}_0 = (0.4, 0.6)$.

So, for the probability of the realization in Fig. 1, we write $\pi_{00}p_{01}p_{11}p_{10}p_{01} = 0.4 \cdot 0.2 \cdot 0.7 \cdot 0.3 \cdot 0.2 = 0.00168$.

1.2 A general framework

Consider the evolution of a system in discrete moments of time $t = 0, 1, 2, \ldots$. At each moment, the system may be in one of the *states* that usually, though not always, we will label $0, 1, 2, \ldots$. Denote by X_t the number of the state in which the system is at time t.

We suppose that once the *present* state of the system is known, any additional information about the behavior of the system in the *past* is redundant for predicting the *future* behavior of the system. Formally, this means that for any $t = 1, 2, \ldots$, and for any states $i_0, i_1, \ldots, i_{t-1}, i_t$ and j,

$$P(X_{t+1} = j \mid X_t = i_t, X_{t-1} = i_{t-1}, \ldots, X_1 = i_1, X_0 = i_0) = P(X_{t+1} = j \mid X_t = i_t). \tag{1.2.1}$$

To clarify this formula, suppose that t is the *present* time, and at this time, the value of the process (that is, the state at which the process has arrived) is known: it is i_t. Suppose also that the *past values* of the process at time moments $t = 0, 1, \ldots, t-1$, *are also known* and equal to $i_0, i_1, \ldots, i_{t-1}$, respectively. The l.-h.s. of (1.2.1) is the probability that at the *future* moment $t + 1$ the system will be in state j, *given the whole history* of the process by time t. Property (1.2.1) implies that this probability, as a matter of fact, depends only on the last (and present for us) value of X_t. This is indicated in the r.-h.s. of (1.2.1).

One may say that "*given the present, the future does not depend on the past.*"

The property (1.2.1) is called the *Markov property*, and the process we have defined—a *Markov chain* (after A.A. Markov who first considered such processes in the beginning of the twentieth century).

As we will see, by virtue of the Markov property, to determine the probabilities of possible realizations of the chain, it suffices to specify the following characteristics.

A. For all possible i, j, we define the *transition probability* $p_{ij} = P(X_{t+1} = j \mid X_t = i)$, the probability that the process, being in state i at time t, will make a transition to state j in the next step. In particular, p_{ii} is the probability of staying in the same state

i at the next time moment $t+1$. In general, the probabilities p_{ij} may depend also on time t; we restrict ourselves to what is called the *homogeneous* case where these probabilities are the same for all moments of time.

The matrix

$$\mathcal{P} = \|p_{ij}\| = \left\| \begin{matrix} p_{00} & p_{01} & p_{02} & \cdots \\ p_{10} & p_{11} & p_{12} & \cdots \\ p_{20} & p_{21} & p_{22} & \cdots \\ \vdots & \vdots & \vdots & \vdots \end{matrix} \right\|$$

is called a *transition probability matrix,* or briefly, a *transition matrix.* Since the process is always in some state, the sum of all entries in *each row* of the matrix $\mathcal{P}$ is equal to one. A matrix with non-negative elements and with such a property is also called *stochastic*.

B. Next, for all possible i, we define the *"initial probability"* $\pi_{0i} = P(X_0 = i)$, the probability of being in state i at the initial moment of time.

   We call the vector of probabilities $\boldsymbol{\pi}_0 = (\pi_{00}, \pi_{01}, \pi_{02}, ...)$ an *initial probability distribution.* If the system starts from a fixed state i_0, then $\boldsymbol{\pi}_0 = (...,0,1,0,...)$, where 1 is in the position i_0.

   Clearly, the sum of all entries in the vector π_0 is equal to one.

There are myriads of examples of Markov chains. We have considered one in Section 1.1 and consider three more below.

EXAMPLE 1. A police department provides daily reports on traffic accidents in a particular area. The intensity of daily accidents depends on weather conditions. The department distinguishes three types: normal, rainy, and icy-road conditions, which corresponds to three states we will label $0,1,2$. If the time period under consideration is not long, we may assume that the transition probabilities do not depend on time. For instance, for a soft winter, the transition matrix may look like

$$\mathcal{P} = \left\| \begin{matrix} 0.6 & 0.3 & 0.1 \\ 0.4 & 0.5 & 0.1 \\ 0.25 & 0.7 & 0.05 \end{matrix} \right\|. \tag{1.2.2}$$

Since among p_{ii}'s (the probabilities of staying in the same state as on the previous day), the probability $p_{00} = 0.6$ is the largest, the normal conditions appear to be the most stable.

The distribution $\boldsymbol{\pi}_0$ in this case characterizes the condition at the initial time. For example, if in the beginning of the season, the weather conditions are normal, $\boldsymbol{\pi}_0 = (1,0,0)$.

It makes sense to note that such a Markov model implicitly presupposes that *given* the current conditions, the information on the weather conditions on previous days is not needed for the weather forecast. This is certainly a simplification. However, the model is convenient for study purposes, and may also serve as a good first approximation in real situations.

EXAMPLE 2 (*Rearrangement* or *shuffling*). A stack of k books lies on a desk. You take a book, read what you need, and put the book on the top of the stack. The states of the system may be identified with the orders in which the books can be arranged, that is, with $k!$ permutations of the numbers $1,2,...,k$. The reader may replace the word "book" by "card", and "stack" by "deck." In this case, we are talking about one of the simplest ways of shuffling.

The process described cannot move from any state to any state in one step. For example, if $k = 3$, and we number books as 1, 2, 3, each state may be identified with an arrangement of numbers 1, 2, 3. Say, for the arrangement $(2, 1, 3)$, the second book lies on the top, and the third one in the bottom. The process can move from state (2, 1, 3) *only* to the same state (you took the book from the top and put it back on the top) and to states (1, 2, 3)—you took book 1 and put it on the top, and (3, 2, 1)—you took book 3 and put it on the top. If all books are equally likely to be chosen, the process can move from state (2, 1, 3) to each of the mentioned states with probability $1/3$, and with zero probability—to the other three states: (1, 3, 2); (2, 3, 1); (3, 1, 2). Applying the same argument to other states, we conclude that the transition matrix of our Markov chain is

$$\begin{matrix} & 123\downarrow & 132\downarrow & 213\downarrow & 231\downarrow & 312\downarrow & 321\downarrow & \\ \mathcal{P} = & \left\| \begin{matrix} 1/3 & 0 & 1/3 & 0 & 1/3 & 0 \\ 0 & 1/3 & 1/3 & 0 & 1/3 & 0 \\ 1/3 & 0 & 1/3 & 0 & 0 & 1/3 \\ 1/3 & 0 & 0 & 1/3 & 0 & 1/3 \\ 0 & 1/3 & 0 & 1/3 & 1/3 & 0 \\ 0 & 1/3 & 0 & 1/3 & 0 & 1/3 \end{matrix} \right\| & \begin{matrix} \leftarrow 123 \\ \leftarrow 132 \\ \leftarrow 213 \\ \leftarrow 231 \\ \leftarrow 312 \\ \leftarrow 321 \end{matrix} \end{matrix} \tag{1.2.3}$$

(the arrows $\leftarrow$and $\downarrow$ show to which state (permutation) a row and a column correspond).

If in the beginning all arrangements are equally likely, then $\boldsymbol{\pi}_0 = (1/6, ..., 1/6)$.

EXAMPLE 3 (*success runs*). Assume that the process may move from state $i = 0,1,2,...$ either to state $i+1$ with probability p, or to state 0 with probability $q = 1-p$. In other words, the process either "moves up" by one step, or "falls to the bottom" and starts from the state 0. Assume, for example, that we are dealing with repeated independent trials, and p is the probability of success at each trial. We are interested in the number of consecutive successes, or in other words, we consider the length of the success run. At each step, the length mentioned may either increase by one or drop to zero.

(Suppose, for instance, for an investor, each day is either lucky or not with probabilities p and q, respectively, and the investor is interested in the length of the period of the capital growth.)

The transition matrix in this case is

$$\mathcal{P} = \left\| \begin{matrix} q & p & 0 & 0 & \cdots \\ q & 0 & p & 0 & \cdots \\ q & 0 & 0 & p & \cdots \\ \vdots & \vdots & \vdots & \vdots & \vdots \end{matrix} \right\|. \tag{1.2.4}$$

EXAMPLE 4 concerns the simulation of Markov chains. A general simulation theory is presented in Section **7.4**, but we are able to consider a particular example right away.

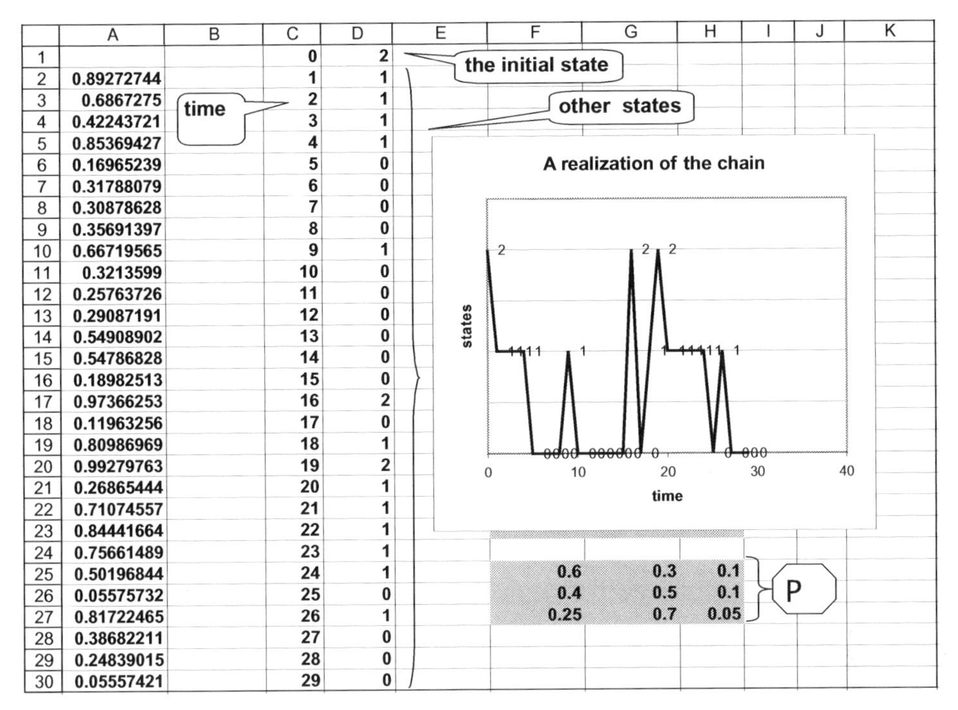

	A	B	C	D
1			0	2
2	0.89272744		1	1
3	0.6867275		2	1
4	0.42243721		3	1
5	0.85369427		4	1
6	0.16965239		5	0
7	0.31788079		6	0
8	0.30878628		7	0
9	0.35691397		8	0
10	0.66719565		9	1
11	0.3213599		10	0
12	0.25763726		11	0
13	0.29087191		12	0
14	0.54908902		13	0
15	0.54786828		14	0
16	0.18982513		15	0
17	0.97366253		16	2
18	0.11963256		17	0
19	0.80986969		18	1
20	0.99279763		19	2
21	0.26865444		20	1
22	0.71074557		21	1
23	0.84441664		22	1
24	0.75661489		23	1
25	0.50196844		24	1
26	0.05575732		25	0
27	0.81722465		26	1
28	0.38682211		27	0
29	0.24839015		28	0
30	0.05557421		29	0

	F	G	H
25	0.6	0.3	0.1
26	0.4	0.5	0.1
27	0.25	0.7	0.05

FIGURE 2. Simulation of the Markov chain from Examples 1,7 in Section 4.1

The main point is to arrange a random mechanism that allows us to choose states in accordance with transition probabilities. To this end, whatever software we use, we need a generator of random numbers; namely, a program or device that selects at random numbers from the interval $[0,1]$. For such a generator, the probability that the number will be selected from a subinterval $\Delta \subseteq [0,1]$ is equal to the length of Δ. (See Section **1**.2.3.) In Excel, to generate a random number Z, we go to Data $\Rightarrow$ Data Analysis $\Rightarrow$ Random Number Generation, and choose in *Distributions* the position "Uniform" with parameters 0 and 1.

Let us revisit Example 1. A simulation procedure is illustrated in the Excel worksheet in Fig. 2. The initial state is a free input entry in Cell D1. We choose the value 2. The matrix $\mathcal{P}$ is in the array F25:H27. Random numbers generated are placed in Column A, time moments t in Column C, the values of X_t in Column D.

Let us start the simulation. Since the initial state is 2, in the first step, we should consider the probabilities in the third row of the matrix $\mathcal{P}$, that is, 0.25, 0.7, 0.05. To simulate the motion of the chain in the first step, we generate a random number Z.

If Z is less than 0.25, (which occurs with probability 0.25), we choose a realization for which the process moves to state 0.

If $0.25 \leq Z < 0.95$ (which occurs with probability 0.7), we choose a realization for which the process moves to state 1. Otherwise, the process stays in state 2.

In the particular realization in Fig. 2, it turns out that the r.v. Z took on the value ≈ 0.893; see Cell A2. So $X_1 = 1$, and it is reflected in cell D2.

For the reader who is familiar with Excel, the command for Cell D2 is

```
=IF(D1=0,IF(A2<$F$25,0,IF(A2<$F$25+$G$25,1,2)),IF(D1=1, IF(A2<=$F$26,0,
IF(A2<=$F$26+$G$26,1,2)),IF(A2<=$F$27,0,IF(A2<=$F$27+$G$27,1,2))))
```

After the first step, we proceed similarly, depending on the current state. For example, since X_1 happened to be 1, to simulate X_2, we should consider the second row in the matrix $\mathcal{P}$ and generate another number from $[0,1]$ which is in Cell A3. This number turned out to be ≈ 0.687. This number is between 0.4 and 0.9, and the command in D3 similar to that in D2, leads to $X_2 = 1$. We continue in the same fashion.

The realization we got in Column D is graphed in the chart. Picking other numbers for Column A, we will get another realization.

Readers who are familiar with Excel, realize that, as a matter of fact, all numbers in column A have been generated by one command, and the commands in Column D were arranged by one-moment copying of the command in D2. □

The following important fact gives *more insight into the nature of Markov chains.* Let $\xi_0, \xi_1, \xi_2, ...$ be independent identically distributed r.v.'s assuming values $0, 1, 2, ...$ Let $g(m,n)$ be a function assuming integer values for integer m, n. Set

$$X_0 = \xi_0, \text{ and } X_t = g(X_{t-1}, \xi_t) \text{ for } t = 1, 2, ... \tag{1.2.5}$$

Since given X_{t-1}, the r.v. X_t does not depend on the values of $X_{t-2}, X_{t-3}, ...$, the sequence $\{X_t\}$ constitutes a Markov chain.

Moreover, it may be proved that, as a matter of fact, *any Markov chain admits the representation (1.2.5).* We will show this when considering the general case in Section **12**.5.

EXAMPLE 5. Let $X_t = \max\{\xi_0, \xi_1, ..., \xi_t\}$. Clearly,

$$X_t = \max\{\max\{\xi_0, \xi_1, ..., \xi_{t-1}\}, \xi_t\} = \max\{X_{t-1}, \xi_t\},$$

and (1.2.5) holds for $g(m,n) = \max\{m,n\}$. In Exercise 5, it is suggested to the reader to find the transition matrix for such a chain. □

In Exercises 6 and 7, the reader is invited to show that *the random walk and the branching process* (Sections **2**.2.4 and **4**.1, respectively) *are Markov chains.*

1.3 Transition in many steps

In Section 1.1, we have already considered an example of computing probabilities for particular realizations or paths. Other simple examples are given in Exercises 1 and 8. Now, let us consider transition probabilities in many steps.

Denote by $p_{ij}^{(2)}$ the probability of moving from state i to state j in two steps *whatever an intermediate state is.* Let the matrix $\mathcal{P}^{(2)} = \left\|p_{ij}^{(2)}\right\|$.

For fixed i and j, consider all possible two-step paths $i \to k \to j$; that is, all two-step paths starting from i, ending at j, and with an arbitrary intermediate state k. The probability of each such a path is $p_{ik}p_{kj}$. To compute $p_{ij}^{(2)}$, we should add up all these probabilities over all possible k's. Thus,

$$p_{ij}^{(2)} = \sum_k p_{ik}p_{kj}.$$

The r.-h.s. of the last formula is the (i, j)-element of the matrix product $\mathcal{P}\mathcal{P} = \mathcal{P}^2$. Thus,

$$\mathcal{P}^{(2)} = \mathcal{P}^2.$$

Calculations in the general case of t steps are similar, and may be provided by induction. So, we state

Proposition 1 *Let $p_{ij}^{(t)}$ be the probability of moving from state i to state j in t steps, and let $\mathcal{P}^{(t)} = \left\| p_{ij}^{(t)} \right\|$, the corresponding transition matrix. Then*

$$\mathcal{P}^{(t)} = \mathcal{P}^t, \tag{1.3.1}$$

where $\mathcal{P}^t$ is the t-th power of the matrix $\mathcal{P}$.

EXAMPLE 1. It is easy to calculate that for the matrix $\mathcal{P}$ in (1.1.2) in Example of Section 1.1,

$$\mathcal{P}^3 = \begin{Vmatrix} 0.65 & 0.35 \\ 0.525 & 0.475 \end{Vmatrix}.$$

(Certainly, it is convenient to use software for such calculations. In Excel, the command for matrix multiplication $\mathcal{P}_1\mathcal{P}_2$ is $=\left\{MMULT\left(\mathcal{P}_1, \mathcal{P}_2\right)\right\}$.)

So, for example, the probability that a person randomly selected from those who live in area 0 will live in the same area at the end of the third year is 0.65. This is, certainly, not the same as staying there all three years.

EXAMPLE 2. Consider the rearrangements (shuffling) of three books from Example 1.2-2. As is easy to verify, for $\mathcal{P}$ from (1.2.3),

$$\begin{matrix} & 123\downarrow & 132\downarrow & 213\downarrow & 231\downarrow & 312\downarrow & 321\downarrow & \end{matrix}$$

$$\mathcal{P}^2 = \begin{Vmatrix} 2/9 & 1/9 & 2/9 & 1/9 & 2/9 & 1/9 \\ 1/9 & 2/9 & 2/9 & 1/9 & 2/9 & 1/9 \\ 2/9 & 1/9 & 2/9 & 1/9 & 1/9 & 2/9 \\ 2/9 & 1/9 & 1/9 & 2/9 & 1/9 & 2/9 \\ 1/9 & 2/9 & 1/9 & 2/9 & 2/9 & 1/9 \\ 1/9 & 2/9 & 1/9 & 2/9 & 1/9 & 2/9 \end{Vmatrix} \begin{matrix} \leftarrow 123 \\ \leftarrow 132 \\ \leftarrow 213 \\ \leftarrow 231 \\ \leftarrow 312 \\ \leftarrow 321 \end{matrix}$$

Thus, all two-step transitions have positive probabilities. □

Now, let $\pi_{ti} = P(X_t = i)$, and $\boldsymbol{\pi}_t = (\pi_{t0}, \pi_{t1}, \pi_{t2}, ...)$, the probability distribution of the r.v. X_t, the state of the chain at time t. By the formula for total probability, for $t = 1$,

$$\pi_{1i} = P(X_1 = i) = \sum_k P(X_1 = i \mid X_0 = k)P(X_0 = k) = \sum_k p_{ki}\pi_{0k}. \tag{1.3.2}$$

This corresponds to the multiplication of the vector $\boldsymbol{\pi}_0$ by the matrix $\mathcal{P}$, and may be rewritten as

$$\boldsymbol{\pi}_1 = \boldsymbol{\pi}_0 \mathcal{P}. \tag{1.3.3}$$

It makes sense to emphasize that in (1.3.3) a row vector is multiplied by a square matrix on the right; so it looks like

$$\left\| \pi_{00}\ \pi_{01}\ \pi_{02}\ \dots \right\| \times \left\| \begin{matrix} p_{00} & p_{01} & p_{02} & \cdots \\ p_{10} & p_{11} & p_{12} & \cdots \\ p_{20} & p_{21} & p_{22} & \cdots \\ \vdots & \vdots & \vdots & \vdots \end{matrix} \right\|.$$

Replacing time moment 1 in (1.3.2) by time t, we have

$$\pi_{ti} = P(X_t = i) = \sum_k P(X_t = i \,|\, X_0 = k)P(X_0 = k) = \sum_k p_{ki}^{(t)} \pi_{0k}.$$

In matrix terms, this may be written as

$$\boldsymbol{\pi}_t = \boldsymbol{\pi}_0 \mathcal{P}^{(t)}.$$

From this and (1.3.1), we come to

Proposition 2 *For any* $t = 1, 2, \dots$,

$$\boldsymbol{\pi}_t = \boldsymbol{\pi}_0 \mathcal{P}^t. \tag{1.3.4}$$

EXAMPLE 3. We revisit Example in Section 1.1. Doing simple calculations or using software, we get

$$\boldsymbol{\pi}_3 = \boldsymbol{\pi}_0 \mathcal{P}^3 = \left\| 0.4\ \ 0.6 \right\| \times \left\| \begin{matrix} 0.8 & 0.2 \\ 0.3 & 0.7 \end{matrix} \right\|^3 = \left\| 0.4\ \ 0.6 \right\| \times \left\| \begin{matrix} 0.65 & 0.35 \\ 0.525 & 0.475 \end{matrix} \right\|$$
$$= \left\| 0.575\ \ 0.425 \right\|.$$

Thus, the probability that a randomly chosen citizen will live in area 0 at time $t = 3$ is 0.575. However, it would be a mistake to think that this means that in the third year 57.5% of people will live in area 0. The process is random, and the percentage mentioned is also random. The figure 57.5% is the *expected value* of the percentage of citizens of area 0 in the year mentioned. Let us show it rigorously.

Denote by n the total number of citizens, and let the r.v. $I_k = 1$ if the kth citizen will be in area 0 in three transitions, and $I_k = 0$ otherwise. Suppose that the migration of every citizen fits the above model. Then $E\{I_k\} = P(I_k = 1) = \pi_{30} = 0.575$.

The total number of people who live in area 0 at $t = 3$ is $\sum_{k=1}^{n} I_k$, and

$$E\left\{ \sum_{k=1}^{n} I_k \right\} = \sum_{k=1}^{n} E\{I_k\} = n \cdot 0.575.$$

So, the mean percentage is 57.5%. The reader is invited to look also at Exercise 10. □

Route 1 ⇒ page 147

2 FIRST STEP ANALYSIS. PASSAGE TIMES

2.1 An example

Consider a Markov chain with states $\{0,1,2\}$ and the transition matrix

$$\mathcal{P} = \begin{Vmatrix} 0.5 & 0.2 & 0.3 \\ 0.25 & 0.7 & 0.05 \\ 0.3 & 0.2 & 0.5 \end{Vmatrix}. \tag{2.1.1}$$

Assume that the chain starts from state 0, and denote by T_{02} the *transition time* it takes for the chain to go from state 0 to state 2. Our goal is to find the mean transition time $\mu_{02} = E\{T_{02}\}$. We will see that to this end, we should also consider the r.v. T_{12}, the time it takes to go from state 1 to state 2, and $\mu_{12} = E\{T_{12}\}$.

Given that the chain starts from state 0 and conditioning on X_1, by the formula for total expectation (**3**.6.2.2), we have

$$\begin{aligned} \mu_{02} = E\{T_{02}\} &= E\{T_{02}\,|\,X_1=0\}P(X_1=0\,|\,X_0=0) \\ &+ E\{T_{02}\,|\,X_1=1\}P(X_1=1\,|\,X_0=0) \\ &+ E\{T_{02}\,|\,X_1=2\}P(X_1=2\,|\,X_0=0). \end{aligned} \tag{2.1.2}$$

If at the first step, the chain remained in state 0, the process "starts over" from the same state, and consequently, the mean number of steps it takes to reach state 2, is $1+\mu_{02}$. So, $E\{T_{02}\,|\,X_1=0\} = 1+\mu_{02}$.

If the process moves to state 1, it starts over from this state; so, $E\{T_{02}\,|\,X_1=1\} = 1+\mu_{12}$.

If the process goes to state 2, there is no need in additional steps; so, $E\{T_{02}\,|\,X_1=2\} = 1$.

Thus, (2.1.2) implies

$$\begin{aligned} \mu_{02} &= (1+\mu_{02})P(X_1=0\,|\,X_0=0) + (1+\mu_{12})P(X_1=1\,|\,X_0=0) + P(X_1=2\,|\,X_0=0) \\ &= \mu_{02}P(X_1=0\,|\,X_0=0) + \mu_{12}P(X_1=1\,|\,X_0=0) + 1, \end{aligned} \tag{2.1.3}$$

because $P(X_1=0\,|\,X_0=0)+P(X_1=1\,|\,X_0=0)+P(X_1=2\,|\,X_0=0) = 1$.

In accordance with (2.1.1), the probabilities $P(X_1=0\,|\,X_0=0) = 0.5$, and $P(X_1=1\,|\,X_0=0) = 0.2$. So,

$$\mu_{02} = 0.5\mu_{02} + 0.2\mu_{12} + 1,$$

or

$$0.5\mu_{02} - 0.2\mu_{12} = 1. \tag{2.1.4}$$

Similarly, *assuming now that the chain starts from state* 1, we write

$$\begin{aligned} \mu_{12} &= E\{T_{12}\,|\,X_1=0\}P(X_1=0\,|\,X_0=1) \\ &+ E\{T_{12}\,|\,X_1=1\}P(X_1=1\,|\,X_0=1) \\ &+ E\{T_{12}\,|\,X_1=2\}P(X_1=2\,|\,X_0=1) \\ &= (1+\mu_{02})P(X_1=0\,|\,X_0=1) + (1+\mu_{12})P(X_1=1\,|\,X_0=1) + P(X_1=2\,|\,X_0=1) \\ &= \mu_{02}P(X_1=0\,|\,X_0=1) + \mu_{12}P(X_1=1\,|\,X_0=1) + 1 = 0.25\mu_{02} + 0.7\mu_{12} + 1, \end{aligned}$$

and hence,

$$-0.25\mu_{02} + 0.3\mu_{12} = 1. \tag{2.1.5}$$

Solving the system (2.1.4)–(2.1.5), we get $\mu_{02} = 5$, $\mu_{12} = 7.5$.

The method we applied is called the *first step analysis*. Its idea consists in conditioning with respect to what can happen at the first step. Before providing a general method of computing mean passage times, consider first

2.2 The case of absorbing states

A state i is called *absorbing* if $p_{ii} = 1$. One may say that the process gets stuck at state i.

EXAMPLE 1. Consider a medical insurance model with four states for an insured: healthy, sick, ceased paying, deceased. Suppose that for a particular group of customers, the matrix

$$\mathcal{P} = \begin{Vmatrix} 0.9 & 0.05 & 0.01 & 0.04 \\ 0.1 & 0.8 & 0.01 & 0.09 \\ 0 & 0 & 1 & 0 \\ 0 & 0 & 0 & 1 \end{Vmatrix} \tag{2.2.1}$$

gives transition probabilities corresponding to annual transition periods. For example, 0.1 is the probability that an insured being sick at the beginning of a year will recover by the beginning of the next year. State 3 ("deceased") is naturally absorbing. So is state 2 ("ceased paying") if we assume that in this case, the contract is terminated. □

Consider a chain for which states $i = 0, 1, ..., k$, are non-absorbing, and the last r states are absorbing, that is, $p_{ii} = 1$ for $i = k+1, ..., k+r$. (If the order of states is different, we can always renumerate them presenting non-absorbing states first.) Then, the transition matrix

$$\mathcal{P} = \begin{Vmatrix} \mathcal{A} & \mathcal{B} \\ \mathcal{O} & \mathcal{I}_r \end{Vmatrix}, \tag{2.2.2}$$

where $\mathcal{A}$ is a $(k+1) \times (k+1)$-matrix, $\mathcal{B}$ is a $(k+1)\times r$-matrix, $\mathcal{O}$ is a $r \times (k+1)$-matrix with zero elements, and $\mathcal{I}_r$ is a $r\times r$-identity matrix.

For instance, in Example 1,

$$\mathcal{A} = \begin{Vmatrix} 0.9 & 0.05 \\ 0.1 & 0.8 \end{Vmatrix}, \quad \mathcal{B} = \begin{Vmatrix} 0.01 & 0.04 \\ 0.01 & 0.09 \end{Vmatrix}. \tag{2.2.3}$$

Let $R = \{k+1, ..., k+r\}$, the set of all absorbing states. Assume that for each non-absorbing state, the probability of moving to R is positive: for each $i = 0, 1, ..., k$, there is a state j from R such that $p_{ij} > 0$. This implies, in particular, that the sum of all probabilities in each row of the matrix $\mathcal{A}$ is less than one.

Since in each step, the process can move to an absorbing state with a positive probability, at some random but finite time T, the process will arrive at an absorbing state and will never leave it (will get stuck in it). Intuitively it is clear; later, we will show it rigorously.

Denote by T_i the number of transitions it takes to go from a non-absorbing state i to set R; that is, T_i is the absorbtion time given $X_0 = i$. It is easy to write the distribution function of T_i.

Proposition 3 *For any $m = 1,2,...,$ and all $i = 0,...,k$,*

$$P(T_i \le m) = \sum_{j \in R} p_{ij}^{(m)}, \tag{2.2.4}$$

where, as was defined, $p_{ij}^{(m)}$ is the probability to move from state i to state j in m steps.

Proof. Once the chain arrives at the set R, it stays in it forever. The r.-h.s. of (2.2.4) is the probability that, given $X_0 = i$, the chain is in R at time $t = m$. But this event occurs if and only if the time it takes to arrive at R is less than or equal to m. ■

Our next goal is to find $\mu_i = E\{T_i\}$ for non-absorbing states i. (We may also write $\mu_i = E\{T \,|\, X_0 = i\}$.)

Let us apply the first step analysis. Conditioning on X_1, remembering that the chain starts from a non-absorbing state, and considering absorbing and non-absorbing states separately, for $i = 0,...,k$, we have

$$\mu_i = E\{T_i\} = \sum_{j=0}^{k} E\{T_i|X_1 = j\}P(X_1 = j|X_0 = i)$$

$$+ \sum_{j=k+1}^{k+r} E\{T_i|X_1 = j\}P(X_1 = j|X_0 = i). \tag{2.2.5}$$

We follow the same logic as in Section 2.1. If $j > k$ (i.e., j is an absorbing state), then $E\{T_i|X_1 = j\} = 1$. If $j \le k$ (i.e., j is a non-absorbing state), then $E\{T_i|X_1 = j\} = 1 + \mu_j$. Recall also that $P(X_1 = j|X_0 = i) = p_{ij}$. Thus, for $i = 0,...,k$,

$$\mu_i = \sum_{j=0}^{k}(1+\mu_j)p_{ij} + \sum_{j=k+1}^{k+r} p_{ij} = \sum_{j=0}^{k}\mu_j p_{ij} + \sum_{j=0}^{k+r} p_{ij} = \sum_{j=0}^{k}\mu_j p_{ij} + 1. \tag{2.2.6}$$

This is a system of equations for μ_i's. We may rewrite it in the following compact form.

Observe that not only i runs from 0 to k, but the summation in the very r.-h.s. of (2.2.6) is also over $j = 0,...,k$. This means that all probabilities p_{ij} involved in (2.2.6) are from the matrix $\mathcal{A}$ from (2.2.2).

Let vector $\boldsymbol{\mu} = (\mu_0, \mu_1, ... \mu_k)$, and $\boldsymbol{e} = (1,...,1)$, the vector consisting of ones. If the symbol ^{T} stands for the transpose operation, $\boldsymbol{\mu}^T$ is the same vector $\boldsymbol{\mu}$ viewed as a column vector.[1] We see that the r.-h.s. of (2.2.6) is the ith coordinate of the vector $\mathcal{A}\boldsymbol{\mu}^T + \mathbf{e}^T$. Then the system of equations (2.2.6) may be written in the compact form

$$\boldsymbol{\mu}^T = \mathcal{A}\boldsymbol{\mu}^T + \mathbf{e}^T. \tag{2.2.7}$$

[1]The notation ^{T} for matrix transposition is traditional, and so is the notation T for a transition time. The exposition below is designed in a way that, hopefully, this cannot cause inconvenience.

We can go further. Let the symbol $\mathcal{I}$ denote the $(k+1)\times(k+1)$ identity matrix, that is, the matrix of the same size as $\mathcal{A}$. Then the system of equations (2.2.7) may be rewritten as

$$(\mathcal{I}-\mathcal{A})\boldsymbol{\mu}^T = \mathbf{e}^T. \tag{2.2.8}$$

As was noted, the sum of the probabilities in each row of $\mathcal{A}$ is less than one. It is a known fact from Linear Algebra that for such a matrix $\mathcal{A}$, all eigenvalues are less than one, $\det(\mathcal{I}-\mathcal{A}) \neq 0$, and hence, the inverse matrix $(\mathcal{I}-\mathcal{A})^{-1}$ exists.

► (This fact may be derived, for example, from the following theorem. Consider the circles whose centers are the diagonal elements p_{ii}, and the radiuses are the sums of the non-diagonal elements that is, $\sum_{j:j\neq i} p_{ij}$, in the corresponding rows. Then all eigenvalues belong to the union of these circles; see, e.g., [22]. For our stochastic matrix, all points in these circles are less than one in the absolute value. Another proof is connected with the Perron–Frobenius theorem; see, e.g., [33, p.11, 141].) ◄

So, (2.2.8) implies

$$\boldsymbol{\mu}^T = (\mathcal{I}-\mathcal{A})^{-1}\mathbf{e}^T. \tag{2.2.9}$$

We see, in particular, that all $\mu_i = E\{T_i\}$ are finite, so we have proved along the way that the T_i's are finite with probability one.

EXAMPLE 1 revisited. We calculate $(\mathcal{I}-\mathcal{A})^{-1} = \begin{Vmatrix} 40/3 & 10/3 \\ 20/3 & 20/3 \end{Vmatrix}$. Then

$$\boldsymbol{\mu}^T = (\mathcal{I}-\mathcal{A})^{-1}\mathbf{e}^T = \begin{Vmatrix} 40/3 & 10/3 \\ 20/3 & 20/3 \end{Vmatrix} \cdot \begin{Vmatrix} 1 \\ 1 \end{Vmatrix} = \begin{Vmatrix} 50/3 \\ 40/3 \end{Vmatrix}.$$

We see that, for a sick person, the expected time of being a client is less than the same time for a healthy person. This is understandable. Let us look at the matrix (2.2.1). The probabilities of moving to the state "ceased paying" are the same for a sick or healthy person, while the probability of moving to the state "deceased" is larger for a sick person. So, a sick person "is likely to leave the system sooner." □

Now, let us recall that X_0 is also random, and its distribution is given by the vector $\boldsymbol{\pi}_0$. It is reasonable to assume that the chain does not start from absorbing states, and set $\pi_{0i} = 0$ for $i > k$. Denote by $\tilde{\boldsymbol{\pi}}_0 = (\tilde{\pi}_{00}, ..., \tilde{\pi}_{0k})$ the vector of the initial probabilities for the non-absorbing states. Then, for the absorbtion time T,

$$E\{T\} = \sum_{i=0}^{k} E\{T\,|\,X_0 = i\}P(X_0 = i) = \sum_{i=0}^{k} \mu_i\tilde{\pi}_{0i} = \langle \boldsymbol{\mu}, \tilde{\boldsymbol{\pi}}_0 \rangle, \tag{2.2.10}$$

where $\langle .,. \rangle$ stands for dot (or scalar) product.

EXAMPLE 1 revisited. Suppose that at the initial time, 94% of the clients are healthy. Then $\tilde{\boldsymbol{\pi}}_0 = (0.94, 0.06)$, and for a randomly chosen client, $E\{T\} = \langle \tilde{\boldsymbol{\pi}}_0, \mu \rangle = 0.94 \cdot \frac{50}{3} + 0.06 \cdot \frac{40}{3} \approx 16.5$. □

In conclusion, note that if absorbing states are not the last ones, to find the matrix $\mathcal{A}$ one can either renumerate states or simply eliminate from $\mathcal{P}$ all rows and columns corresponding to the absorbing states.

2.3 How to find the mean passage time

Consider a chain with a transition matrix $\mathcal{P}$ without absorbing states. We define the first passage time T_{ik} as the number of steps it takes for the chain to go from state i to state k. Set $\mu_{ik} = E\{T_{ik}\}$. The idea is to consider another chain with the transition matrix $\overline{\mathcal{P}}$ whose rows are the same as in $\mathcal{P}$ with one exception: the kth row is changed in a way that makes state k absorbing.

For instance, for the problem from Section 2.1 and $k = 2$, while

$$\mathcal{P} = \begin{Vmatrix} 0.5 & 0.2 & 0.3 \\ 0.25 & 0.7 & 0.05 \\ 0.3 & 0.2 & 0.5 \end{Vmatrix}, \quad \text{the matrix} \quad \overline{\mathcal{P}} = \begin{Vmatrix} 0.5 & 0.2 & 0.3 \\ 0.25 & 0.7 & 0.05 \\ 0 & 0 & 1 \end{Vmatrix}. \tag{2.3.1}$$

Let us return to the general case. In the new chain with the transition matrix $\overline{\mathcal{P}}$, the only absorbing state is state k, and hence the absorbtion time for the new chain is the passage time for the old. Consequently, we can use the results of the previous Section 2.2. The only thing we should require is that for the new chain, the conditions of Section 2.2 hold: for each state, the probability of moving to an absorbing state is positive. Hence, in the original chain, we require the probability of moving to state k to be positive.

For the chain in (2.3.1), it is true, and in the notation of the previous section, the matrix $\mathcal{A} = \begin{Vmatrix} 0.5 & 0.2 \\ 0.25 & 0.7 \end{Vmatrix}$. As is easy to calculate, $(\mathcal{I} - \mathcal{A})^{-1} = \begin{Vmatrix} 3 & 2 \\ 2.5 & 5 \end{Vmatrix}$, and proceeding from (2.2.9), we have

$$\boldsymbol{\mu}^T = (\mathcal{I} - \mathcal{A})^{-1}\mathbf{e}^T = \begin{Vmatrix} 3 & 2 \\ 2.5 & 5 \end{Vmatrix} \cdot \begin{Vmatrix} 1 \\ 1 \end{Vmatrix} = \begin{Vmatrix} 5 \\ 7.5 \end{Vmatrix}.$$

Naturally, this is the same answer as in Section 2.1. Looking over what we had done in that section, we see that now we did the same but in a compact form.

Let us come back to the general model. As was mentioned, if the state k of interest is not the last one, to determine $\mathcal{A}$ we should just eliminate the row and column corresponding the state k which we will "make" absorbing.

For instance, for the weather-condition chain from Example 1.2-1 and $k = 0$,

$$\mathcal{P} = \begin{Vmatrix} 0.6 & 0.3 & 0.1 \\ 0.4 & 0.5 & 0.1 \\ 0.25 & 0.7 & 0.05 \end{Vmatrix}, \quad \text{while} \quad \overline{\mathcal{P}} = \begin{Vmatrix} 1 & 0 & 0 \\ 0.4 & 0.5 & 0.1 \\ 0.25 & 0.7 & 0.05 \end{Vmatrix}. \tag{2.3.2}$$

In this case, the matrix $\mathcal{A} = \begin{Vmatrix} 0.5 & 0.1 \\ 0.7 & 0.05 \end{Vmatrix}$.

As is easy to calculate, in accordance with (2.2.9),

$$\boldsymbol{\mu}^T = (\mathcal{I} - \mathcal{A})^{-1}\mathbf{e}^T \approx \begin{Vmatrix} 2.35 & 0.25 \\ 1.73 & 1.23 \end{Vmatrix} \cdot \begin{Vmatrix} 1 \\ 1 \end{Vmatrix} = \begin{Vmatrix} 2.60 \\ 2.96 \end{Vmatrix}.$$

Thus, on the average, it takes ≈ 2.6 days to move from the rainy conditions to the normal, and ≈ 3 days from the icy-road conditions to the normal.

3 VARIABLES DEFINED ON A MARKOV CHAIN

3.1 A preliminary example

Consider the Markov chain from Example 1.2-1 concerning traffic accidents and weather conditions. The r.v. X_t shows the type of the weather conditions on day t. Denote by Y_{ti} the number of accidents on day t if the weather conditions correspond to state i.

Which particular r.v. Y_{ti} will "appear" at time t depends on the state at which the Markov chain X_t will arrive at this time. If X_t assumes a value of 2, the number of accidents will be equal to Y_{t2}, while if $X_t = 0$, we will deal with the r.v. Y_{t0}. So, as a matter of fact, the index i in Y_{ti} is random, and when modeling the evolution of the system, we should replace the index i by X_t. Thus, the number of accidents on day t is the r.v.

$$Z_t = Y_{tX_t}. \tag{3.1.1}$$

The r.v.'s $\{Z_0, Z_1, ...\}$ are said to be *defined on a Markov chain*.

Let $F_{ti}(x) = P(Y_{ti} \le x)$, the distribution function of Y_{ti}. Then

$$P(Z_t \le x) = \sum_i P(Z_t \le x \,|\, X_t = i)P(X_t = i) = \sum_i P(Y_{ti} \le x)\pi_{ti} = \sum_i F_{ti}(x)\pi_{ti}. \tag{3.1.2}$$

So, the distribution of Z_t is the weighted sum of the distributions F_{ti} with respect to the distribution $\boldsymbol{\pi}_t$. We did not specify the limits in the sum $\sum_i$ in (3.1.2) on purpose, in order to use this formula in the general model.

Now, assume that Y_{ti} is a Poisson r.v. with parameter λ_i, and $\lambda_0 = 2$, $\lambda_1 = 4$, $\lambda_2 = 8$. What is the probability that on the second day, there will be at most three accidents? We should consider

$$P(Z_2 \le 3) = \sum_{i=0}^{2} P(Y_{2i} \le 3)\pi_{2i}.$$

By (1.3.4), $\boldsymbol{\pi}_2 = \boldsymbol{\pi}_0 \mathcal{P}^2$, where $\mathcal{P}$ is given in (1.2.2).

If we assume, as in Example 1.2-1, the initial distribution $\boldsymbol{\pi}_0 = (1,0,0)$, we can easily compute, using any software we like, that $\boldsymbol{\pi}_2 = (0.505, 0.4, 0.095)$.

To compute the Poisson probabilities $P(Y_{2i} \le 3)$, one can also use software; in particular, the Excel commands for the Poisson distribution are given in Section **3**.4.5. So, we have

$$P(Z_2 \le 3) = \sum_{i=0}^{2} P(Y_{2i} \le 3)\pi_{2i} \approx 0.857 \cdot 0.505 + 0.433 \cdot 0.4 + 0.042 \cdot 0.095 \approx 0.610.$$

Our next object of study is

$$S_n = \sum_{t=0}^{n} Z_t = \sum_{t=0}^{n} Y_{tX_t}, \tag{3.1.3}$$

the total number of accidents during $n+1$ time periods (we count the initial period). We restrict ourselves to computing $E\{S_n\}$.

By the formula for total expectation (see (**3**.6.2.2)),

$$E\{Z_t\} = \sum_{i=0}^{2} E\{Y_{ti}\}P(X_t = i) = \sum_{i=0}^{2} \lambda_i \pi_{ti}.$$

The last sum above may be written as $\langle \boldsymbol{\lambda}, \boldsymbol{\pi}_t \rangle$, where the vector $\boldsymbol{\lambda} = (\lambda_0, \lambda_1, \lambda_2)$ and $\langle \cdot, \cdot \rangle$ stands for dot (or scalar) product. Then, since $\boldsymbol{\pi}_t = \boldsymbol{\pi}_0 \mathcal{P}^t$,

$$E\{S_n\} = \sum_{t=0}^{n} \langle \boldsymbol{\lambda}, \boldsymbol{\pi}_t \rangle = \langle \boldsymbol{\lambda}, \sum_{t=0}^{n} \boldsymbol{\pi}_t \rangle = \langle \boldsymbol{\lambda}, \sum_{t=0}^{n} \boldsymbol{\pi}_0 \mathcal{P}^t \rangle = \langle \boldsymbol{\lambda}, \boldsymbol{\pi}_0 \sum_{t=0}^{n} \mathcal{P}^t \rangle, \tag{3.1.4}$$

where, by convention, $\mathcal{P}^0 = \mathcal{I}$. Let us return to particular numbers and find $E\{S_2\}$, the expected number of accidents during three periods. By (3.1.4),

$$E\{S_2\} = \langle \boldsymbol{\lambda}, \boldsymbol{\pi}_0(\mathcal{I} + \mathcal{P} + \mathcal{P}^2) \rangle,$$

where $\boldsymbol{\lambda} = (2, 4, 8)$, $\boldsymbol{\pi}_0 = (1, 0, 0)$, and $\mathcal{P}$ is given in (1.2.2). Calculations are easy, especially if one uses software, and lead to $E\{S_2\} = 8.57$.

Calculations for larger n's are similar and tractable with a good computer program.

3.2 A general model

Certainly, the same scheme as above, or slightly modified, may be used in various applications. For example, X_t may indicate the health condition of a person and Y_{ti} the health care cost; or X_t indicates weather conditions, and Y_{ti} measures the energy consumption.

So, let X_t be a Markov chain, and Y_{ti} are some r.v.'s. We suppose that all Y's are independent of X's.

Let $c_{ti} = E\{Y_{ti}\}$, and vector $\mathbf{c}_t = (c_{t1}, c_{t2}, ...)$. As in the previous section, we set $Z_t = Y_{tX_t}$ and $S_n = \sum_{t=0}^{n} Z_t$. Similar to what we did above,

$$E\{Z_t\} = \sum_i E\{Y_{ti}\} P(X_t = i) = \sum_i c_{ti} \pi_{ti} = \langle \mathbf{c}_t, \boldsymbol{\pi}_t \rangle.$$

In view of (1.3.4), $E\{Z_t\} = \langle \mathbf{c}_t, \boldsymbol{\pi}_0 \mathcal{P}^t \rangle$, and in accordance with (3.1.3),

$$E\{S_n\} = \sum_{t=0}^{n} \langle \mathbf{c}_t, \boldsymbol{\pi}_0 \mathcal{P}^t \rangle. \tag{3.2.1}$$

If we cannot make an additional and simplifying assumption on $\mathbf{c}_t$, we should deal with (3.2.1) as is.

Below, we consider a particular but important case where

$$\mathbf{c_t} = v^t \mathbf{c}, \tag{3.2.2}$$

where a constant vector $\mathbf{c} = (c_1, c_2, ...)$, and a number v is such that $0 \leq v \leq 1$.

Assume, for example, that we are dealing with payments, and the mean payment at time t is equal to a constant value c_i if $X_t = i$. However, suppose we are evaluating the present value of this future payment from the standpoint of the initial time. The present value mentioned is $v^t c_i$, where v is a discount factor corresponding to a unit time interval. We discuss this issue in detail in the Appendix, Section 3. Eventually, we arrive at (3.2.2). Note, however, that this is one of possible interpretations; below, we simply adopt (3.2.2).

Substituting (3.2.2) into (3.2.1), we have

$$E\{S_n\} = \sum_{t=0}^{n} \langle v^t \mathbf{c}, \boldsymbol{\pi}_0 \mathcal{P}^t \rangle = \sum_{t=0}^{n} \langle \mathbf{c}, v^t \boldsymbol{\pi}_0 \mathcal{P}^t \rangle \tag{3.2.3}$$

$$= \sum_{t=0}^{n} \langle \mathbf{c}, \boldsymbol{\pi}_0 (v\mathcal{P})^t \rangle = \langle \mathbf{c}, \boldsymbol{\pi}_0 \sum_{t=0}^{n} (v\mathcal{P})^t \rangle. \tag{3.2.4}$$

The last sum $\sum_{t=0}^{n}(v\mathcal{P})^t$ is a *geometric series of matrices*, so it is tempting to use the formula for the sum of such a series.

For any *number* $r \neq 1$, we have $1+r+...+r^n = \dfrac{1-r^{n+1}}{1-r} \to \dfrac{1}{1-r}$, as $n \to \infty$, if $|r| < 1$; see the Appendix, (2.2.11).

For a matrix $\mathcal{C}$ and the identity matrix $\mathcal{I}$ of the same order as $\mathcal{C}$, we do have

$$\mathcal{I}+\mathcal{C}+...+\mathcal{C}^n = (\mathcal{I}-\mathcal{C})^{-1}(\mathcal{I}-\mathcal{C}^{n+1}) \to (\mathcal{I}-\mathcal{C})^{-1}, \tag{3.2.5}$$

if the inverse matrix $(\mathcal{I}-\mathcal{C})^{-1}$ exists. The limiting relation is true provided that the absolute values of all eigenvalues of $\mathcal{C}$ are less than one (see, e.g., [22], [27]).

However, we cannot use (3.2.5) in our problem if $v = 1$ because $\det(\mathcal{I}-\mathcal{P}) = 0$, and hence $(\mathcal{I}-\mathcal{P})^{-1}$ does not exist.

Indeed, let $\mathbf{e} = (1,1,...,1)$, and the transpose $\mathbf{e}^T$ is the corresponding column vector. Since the sum of the probabilities in each row of $\mathcal{P}$ is one, $\mathcal{P}\mathbf{e}^T = \mathbf{e}^T$; that is, one is an eigenvalue of $\mathcal{P}$. This means that $\det(\mathcal{I}-\mathcal{P}) = 0$.

So, for $v = 1$, we leave (3.2.4) as is.

For $0 \leq v < 1$, the sum of the entries in each row of the matrix $v\mathcal{P}$ is less than one. Hence, $\det(\mathcal{I}-v\mathcal{P}) \neq 0$ due to the fact from Linear Algebra which we already discussed in p.142.

Thus, for $0 \leq v < 1$, using formula (3.2.5) with $\mathcal{C} = v\mathcal{P}$, we get from (3.2.4) that

$$E\{S_n\} = \langle \mathbf{c}, \boldsymbol{\pi}_0(\mathcal{I}-v\mathcal{P})^{-1}(\mathcal{I}-(v\mathcal{P})^{n+1})\rangle. \tag{3.2.6}$$

If n is not large, we should stop at this stage and use the above formula. If we consider the process in the long run, for large n, we may use the approximation

$$\sum_{t=0}^{n}(v\mathcal{P})^t \to (\mathcal{I}-v\mathcal{P})^{-1} \text{ as } n \to \infty,$$

which implies that in this case,

$$E\{S_n\} \to \langle \mathbf{c}, \boldsymbol{\pi}_0(\mathcal{I}-v\mathcal{P})^{-1}\rangle \text{ as } n \to \infty. \tag{3.2.7}$$

EXAMPLE 1. An investor distinguishes five types of years: very bad, bad, moderate, good, and very good. The mean profit corresponding to the states mentioned is equal, in some units, to $-3, -1, 1, 2, 4$, respectively. The discount factor is $v = 0.97$.

Suppose that the change of investment conditions from year to year is well approximated by the Markov model with the transition matrix

$$\mathcal{P} = \begin{Vmatrix} 0.1 & 0.7 & 0.2 & 0 & 0 \\ 0.2 & 0.2 & 0.6 & 0 & 0 \\ 0.05 & 0.1 & 0.7 & 0.1 & 0.05 \\ 0 & 0.1 & 0.1 & 0.6 & 0.2 \\ 0 & 0 & 0.2 & 0.5 & 0.3 \end{Vmatrix}. \tag{3.2.8}$$

The present condition of the market is moderate. Estimate the mean present value of the profit in the long run.

Thus, $\boldsymbol{\pi}_0 = (0,0,1,0,0)$, $\mathbf{c} = (-3,-1,1,2,4)$, and the estimate of $E\{S_n\}$ for large n is $\langle \mathbf{c}, \boldsymbol{\pi}_0(\mathcal{I}-v\mathcal{P})^{-1}\rangle \approx 36.008$, which may be easily computed using, for example, the matrix-commands MMULT and MINVERSE in Excel.

1,2

4 ERGODICITY AND STATIONARY DISTRIBUTIONS

4.1 Ergodicity property

Let us come back to the two-area-migration problem from Section 1.1, and consider the consecutive powers of the transition matrix:

$$\mathcal{P} = \begin{Vmatrix} 0.8 & 0.2 \\ 0.3 & 0.7 \end{Vmatrix}, \quad \mathcal{P}^2 = \begin{Vmatrix} 0.7 & 0.3 \\ 0.45 & 0.55 \end{Vmatrix}, \quad \mathcal{P}^3 = \begin{Vmatrix} 0.65 & 0.35 \\ 0.525 & 0.475 \end{Vmatrix}, \ldots,$$

$$\mathcal{P}^6 = \begin{Vmatrix} 0.60625 & 0.39375 \\ 0.590625 & 0.409375 \end{Vmatrix}, \ldots, \mathcal{P}^{10} \approx \begin{Vmatrix} 0.6004 & 0.3996 \\ 0.5994 & 0.4006 \end{Vmatrix}, \ldots. \tag{4.1.1}$$

We see an explicit convergency pattern, and what is important is that the two rows are getting closer to each other. This is not accidental. Consider an arbitrary two-dimensional transition matrix which may be written as

$$\mathcal{P} = \begin{Vmatrix} 1-\alpha & \alpha \\ \beta & 1-\beta \end{Vmatrix},$$

where α, β are non-negative and are not greater than one.

Let $\alpha+\beta > 0$; that is, either α, or β, or both are positive. It is known that in this case,

$$\mathcal{P}^t = \frac{1}{\alpha+\beta}\begin{Vmatrix} \beta & \alpha \\ \beta & \alpha \end{Vmatrix} + \frac{(1-\alpha-\beta)^t}{\alpha+\beta}\begin{Vmatrix} \alpha & -\alpha \\ -\beta & \beta \end{Vmatrix}. \tag{4.1.2}$$

The reader may prove it by induction on her/his own or look at a proof, e.g., in [22] or [27].

Assume, in addition, $\alpha+\beta < 2$; that is, at least one number is not 1. Thus, $0 < \alpha+\beta < 2$, and hence $-1 < \alpha+\beta-1 < 1$. So, $|1-\alpha-\beta| < 1$, and the second term in (4.1.2) converges to zero as $t \to \infty$. Thus,

$$\mathcal{P}^t \to \frac{1}{\alpha+\beta}\begin{Vmatrix} \beta & \alpha \\ \beta & \alpha \end{Vmatrix} = \begin{Vmatrix} \pi_0 & \pi_1 \\ \pi_0 & \pi_1 \end{Vmatrix} \quad \text{as } t \to \infty, \tag{4.1.3}$$

where

$$\pi_0 = \frac{\beta}{\alpha+\beta}, \quad \pi_1 = \frac{\alpha}{\alpha+\beta}.$$

Note that the convergence is fast: $(1-\alpha-\beta)^t \to 0$ exponentially.

In the example above, $\alpha = 0.2$, $\beta = 0.3$, and $\pi_0 = \frac{0.3}{0.2+0.3} = 0.6$, $\pi_1 = \frac{0.2}{0.2+0.3} = 0.4$, which is consistent with (4.1.1).

What is remarkable in (4.1.3) is that the rows in the limiting matrix are identical. This means that, in the long run, asymptotically, the probability that the process will be in a particular state does not depend from which state the process has started at the initial time. The process is, so to say, "gradually forgetting" the past; look again at (4.1.1).

Such a property (with some variations in definitions) is called *ergodicity*, and the chain itself—*ergodic*.

We have established it for $0 < \alpha+\beta < 2$. If this is not true, ergodicity does not take place. Let $\alpha+\beta = 0$. Then, since α and β are non-negative, both numbers are zeros, and

$\mathcal{P} = \mathcal{I} = \begin{Vmatrix} 1 & 0 \\ 0 & 1 \end{Vmatrix}$, the identity matrix. In this case, $\mathcal{P}^t = \mathcal{I}$ for all t, and the process will never leave the initial state.

If $\alpha + \beta = 2$, which is possible only if both numbers, α and β, equal one, then $\mathcal{P} = \begin{Vmatrix} 0 & 1 \\ 1 & 0 \end{Vmatrix}$, and the process alternates between two states. (Starting from state 0, the chain moves to state 1, and then comes back to state 0, and keeps moving in the same fashion.) Excepting the cases

$$\mathcal{P} = \begin{Vmatrix} 1 & 0 \\ 0 & 1 \end{Vmatrix} \quad \text{or} \quad \begin{Vmatrix} 0 & 1 \\ 1 & 0 \end{Vmatrix}, \tag{4.1.4}$$

the chain is ergodic.

Certainly, the fact that we considered only two states is not essential.

EXAMPLE 1. Consider again the weather-condition chain from Example 1.2-1. Using even a calculator, it is easy to compute that

$$\text{while } \mathcal{P} = \begin{Vmatrix} 0.6 & 0.3 & 0.1 \\ 0.4 & 0.5 & 0.1 \\ 0.25 & 0.7 & 0.05 \end{Vmatrix}, \text{ the matrix } \mathcal{P}^6 \approx \begin{Vmatrix} 0.48218 & 0.42258 & 0.09524 \\ 0.48211 & 0.42265 & 0.09524 \\ 0.48208 & 0.42268 & 0.09524 \end{Vmatrix}.$$

We see that after just six steps, the rows differ from each other only in the fourth digit.

If in a problem, we want merely to estimate limiting probabilities, raising the transition matrix to a sufficiently large power may be enough. In the above example, the limiting probabilities are close to (0.482, 0.423, 0.095), and in the context of the problem (weather conditions) it is more than enough. □

However, such a heuristic approach does not lead to precise limiting distributions, and we proceed to rigorous constructions. In general, ergodicity takes place under some conditions, though as we will see, they are rather mild. Consider, first, finite chains.

Theorem 4 *Suppose that the number of states—denote it by d—is finite, and for a state j_0 and a positive integer v,*

$$p_{ij_0}^{(v)} > 0 \quad \text{for all } i. \tag{4.1.5}$$

Then there exists a row-vector $\boldsymbol{\pi} = (\pi_1, ..., \pi_d)$ such that $\pi_i \geq 0$ for all i, $\pi_1 + ... + \pi_d = 1$, and

$$\mathcal{P}^t \to \begin{Vmatrix} \boldsymbol{\pi} \\ \vdots \\ \boldsymbol{\pi} \end{Vmatrix} \quad \text{as } t \to \infty; \tag{4.1.6}$$

that is, in the limiting matrix, each row is equal to the vector $\boldsymbol{\pi}$.

We will prove it in Section 4.4.

Condition (4.1.5) is referred to as *Doeblin's condition.* It presupposes the existence of a state to which the process can move with a positive probability in a fixed number of steps from *any* state.

Algebraically, (4.1.5) means that for some v, the matrix $\mathcal{P}^v$ has at least one strictly positive column. Clearly, if all elements of $\mathcal{P}$ are positive, (4.1.5) holds automatically for $v = 1$. Such chains and corresponding transition matrices are called *regular*. For instance, this is the case for the matrix (1.2.2) in the weather-condition problem.

In the rearrangement problem of Examples 1.2-2 and 1.3-2, there is no strictly positive column in $\mathcal{P}$ itself. However, all elements of $\mathcal{P}^2$ are positive, so Doeblin's condition holds for $v = 2$.

Consider now $\boldsymbol{\pi}_t$, the distribution of the process at time t.

Corollary 5 *Under the conditions of Theorem 4, for any initial distribution* $\boldsymbol{\pi}_0$,

$$\boldsymbol{\pi}_t \to \boldsymbol{\pi} \text{ as } t \to \infty, \tag{4.1.7}$$

where $\boldsymbol{\pi}$ *is the same as in (4.1.6).*

***Proof*.** We have

$$\boldsymbol{\pi}_t = \boldsymbol{\pi}_0 \mathcal{P}^t \to \boldsymbol{\pi}_0 \cdot \begin{Vmatrix} \boldsymbol{\pi} \\ \vdots \\ \boldsymbol{\pi} \end{Vmatrix}.$$

As a matter of fact, the r.-h.s. does not depend on $\boldsymbol{\pi}_0$. Indeed, all elements in the ith column of the last matrix equal π_i. Hence, the ith element of the product is $\pi_{00}\pi_i + \pi_{01}\pi_i + ... = \pi_i(\pi_{00} + \pi_{01} + ...) = \pi_i \cdot 1 = \pi_i$. ■

Thus, whatever the initial distribution $\boldsymbol{\pi}_0$ is, asymptotically, as $t \to \infty$, the distribution of the process at time t converges to the same distribution $\boldsymbol{\pi}$. In particular, this implies that

> In the long run, the mean proportion of time when the process is in state j is approaching π_j, the jth coordinate of the vector $\boldsymbol{\pi}$.

Let us make it clearer and rigorous. Denote by S_{nj} the number of visits to state j during the time period from $t = 0$ to $t = n$. Set $\nu_{nj} = S_{nj}/n$, the proportion of visits to state j.

Corollary 6 *Under the conditions of Theorem 4, for any initial distribution* $\boldsymbol{\pi}_0$, *and any state* j,

$$E\{\nu_{nj}\} \to \pi_j \text{ as } n \to \infty, \tag{4.1.8}$$

where π_j *is the* j*th coordinate of the vector* $\boldsymbol{\pi}$.

Note that, as a matter of fact, not only $E\{\nu_{nj}\}$ but the r.v. ν_{nj} itself converges to π_j with probability one, but this would lead us too far astray.

***Proof of Corollary 6*.** Let $I_{tj} = 1$ if $X_t = j$, and 0 otherwise. (I_{tj}'s count visits.) Then $E\{I_{tj}\} = P(X_t = j) = \pi_{tj}$. On the other hand, $S_{nj} = \sum_{t=0}^{n} I_{tj}$, and

$$E\{\nu_{nj}\} = \frac{1}{n}\sum_{t=0}^{n} E\{I_{tj}\} = \frac{1}{n}\sum_{t=0}^{n} \pi_{tj} \to \pi_j$$

as $t \to \infty$. The last implication is based on the following fact from Calculus: if a sequence $a_t \to a$, then the average $\frac{1}{n}\sum_{t=0}^{n} a_t \to a$. ■

One may say, that if in the long run, the distribution $\boldsymbol{\pi}_t$ is approaching a distribution $\boldsymbol{\pi}$ that does not depend on $\boldsymbol{\pi}_0$, then the process approaches a *stationary regime*. We consider examples in the next section.

4.2 Stationary distribution

The next question seeks to find the limiting distribution $\boldsymbol{\pi}$. From (1.3.3), it follows that

$$\boldsymbol{\pi}_t = \boldsymbol{\pi}_{t-1}\mathcal{P} \tag{4.2.1}$$

(consider one step, viewing $t-1$ as the initial time). Since $\boldsymbol{\pi}_t \to \boldsymbol{\pi}$, and $\boldsymbol{\pi}_{t-1} \to \boldsymbol{\pi}$ as well, letting $t \to \infty$ in (4.2.1), we arrive at the fundamental equation for $\boldsymbol{\pi}$:

$$\boldsymbol{\pi} = \boldsymbol{\pi}\mathcal{P}. \tag{4.2.2}$$

The probability distribution $\boldsymbol{\pi}$ satisfying (4.2.2) is called a *stationary distribution*. The choice of the term is connected with the following fact.

Let $\boldsymbol{\pi}_0 = \boldsymbol{\pi}$, that is, the process starts from the distribution $\boldsymbol{\pi}$ from the very beginning. Then $\boldsymbol{\pi}_1 = \boldsymbol{\pi}_0\mathcal{P} = \boldsymbol{\pi}\mathcal{P} = \boldsymbol{\pi}$, $\boldsymbol{\pi}_2 = \boldsymbol{\pi}_1\mathcal{P} = \boldsymbol{\pi}\mathcal{P} = \boldsymbol{\pi}$, and so on: $\boldsymbol{\pi}_t = \boldsymbol{\pi}$ for *all* t.

Thus, for an arbitrary initial distribution $\boldsymbol{\pi}_0$, the distribution $\boldsymbol{\pi}_t$ is approaching the stationary distribution $\boldsymbol{\pi}$, while if the initial distribution is $\boldsymbol{\pi}$ itself, the distribution of the process is invariant through time, and the process runs in the *stationary regime* from the very beginning.

It is also worth noting that while any limiting distribution is stationary, a stationary distribution may exist when a limiting distribution does not. Consider two simple examples.

If $\mathcal{P} = \mathcal{I}$, the identity matrix, then $\boldsymbol{\pi}\mathcal{P} = \boldsymbol{\pi}$ for *any* π, though as we have seen, the chain is not ergodic.

Let $\mathcal{P} = \begin{Vmatrix} 0 & 1 \\ 1 & 0 \end{Vmatrix}$. Then for $\boldsymbol{\pi} = (\pi_1, \pi_2)$, we have $(\pi_1, \pi_2) \times \mathcal{P} = (\pi_2, \pi_1)$. Thus, the distribution $\boldsymbol{\pi} = (0.5, 0.5)$ is stationary, and any other distribution is not. Certainly, the chain itself is not ergodic.

More interesting examples may be found in Exercises 25 and 28; however, it is noteworthy that such situations are rare.

On the other hand, if a limiting distribution $\boldsymbol{\pi}$ exists, then it is the only stationary distribution. Indeed, suppose that there exists another stationary distribution $\widetilde{\boldsymbol{\pi}} \neq \boldsymbol{\pi}$. Then, for $\boldsymbol{\pi}_0 = \widetilde{\boldsymbol{\pi}}$, we have $\boldsymbol{\pi}_t = \widetilde{\boldsymbol{\pi}}$ for all t, and hence the limiting distribution is $\widetilde{\boldsymbol{\pi}}$ while it should be $\boldsymbol{\pi}$.

Let us come back to equation (4.2.2). Nowadays, solving it, at least numerically, is not a problem even for large dimension. We restrict ourselves to the following examples.

EXAMPLE 1. Consider the process of random rearrangements with $\mathcal{P}$ given in (1.2.3). It is straightforward to verify that for such a matrix, equation (4.2.2) is true for $\boldsymbol{\pi}=(\frac{1}{6},...,\frac{1}{6})$, meaning that, as $t \to \infty$, all arrangements are asymptotically becoming equally likely.

In Exercise 24, it is suggested to show that the same is true for any number of objects being shuffled. Let us call a way of shuffling perfect if it leads to equal probabilities of all possible permutations regardless of the initial arrangements of the objects. We see that a simple shuffling as in our example is close to a perfect one for a large number of iterations. □

EXAMPLE 2. Consider the weather-condition problem with matrix (1.2.2). Equation (4.2.2) may be written coordinatewise as follows:

$$0.6\pi_0 + 0.4\pi_1 + 0.25\pi_2 = \pi_0,$$
$$0.3\pi_0 + 0.5\pi_1 + 0.7\pi_2 = \pi_1,$$
$$0.1\pi_0 + 0.1\pi_1 + 0.05\pi_2 = \pi_2.$$

(We multiply $\mathcal{P}$ by $\boldsymbol{\pi}$ on the left!) Together with $\pi_0 + \pi_1 + \pi_2 = 1$, it yields

$$\pi_0 = \frac{81}{168}, \quad \pi_1 = \frac{71}{168}, \quad \pi_2 = \frac{16}{168}, \tag{4.2.3}$$

which is close to the estimates we obtained in Example 4.1-1. We can interpret this as in the long run, the proportion of days, for instance, with icy conditions, is $\pi_2 = 2/21$. □

EXAMPLE 3. Let a chain have one absorbing state, say,

$$\mathcal{P} = \begin{Vmatrix} * & * & * \\ * & * & * \\ 0 & 0 & 1 \end{Vmatrix}, \tag{4.2.4}$$

where the stars $*$ represent positive numbers. The reader is invited to verify that in this case the solution to (4.2.2) is $\boldsymbol{\pi} = (0, 0, 1)$, that is, the limiting distribution is concentrated at the last state. It is not surprising at all, since with probability one the process will arrive at the absorbing state. For us, it is worth noting that Theorem 4 covers such cases.

EXAMPLE 4. In a country, at the end of each year, the government arranges a poll where each interviewee evaluates her/his financial status as either "good," or "fair," or "bad". The actual income corresponding to these states may be different for different people, but we assume that for each citizen, the process of transition from one state to another constitutes a Markov chain with the same transition matrix

$$\mathcal{P} = \begin{Vmatrix} 0.6 & 0.4 & 0.0 \\ 0.2 & 0.6 & 0.2 \\ 0.0 & 0.6 & 0.4 \end{Vmatrix}.$$

What percentage of citizens will transit from "fair" to "good" in the stationary regime?

Similarly to what we did in Example 2, solving (4.2.2), one may easily find that the limiting probability π_1 of being in the state "fair " is $6/11 \approx 0.545$. Thus, in the stationary regime (or in the long run), each year, on the average, 54.5% of citizens consider their status "fair." Since $p_{10} = 0.2$, from these 54.5%, on the average, 20% will transit to the zero state ("good"). Thus, the solution is $\frac{6}{11} \cdot 0.2 = \frac{6}{55}$, or approximately 11%.

Route 1 ⇒ *page 161*

EXAMPLE 5. We include in Example 2 the information from the example of Section 3.1. Let Y_0, Y_1, Y_2 be independent Poisson r.v.'s with parameters $\lambda_0 = 2$, $\lambda_1 = 4$, $\lambda_2 = 8$, respectively. Since for large t, the probability of being in state j may be approximated by π_j, the r.v. Z_t, the number of accidents occurring on day t, may be approximated by a r.v.

$$Z = \begin{cases} Y_0 \text{ with probability } \pi_0, \\ Y_1 \text{ with probability } \pi_1, \\ Y_2 \text{ with probability } \pi_2. \end{cases}$$

In particular, as $t \to \infty$,

$$E\{Z_t\} \to E\{Z\} = \pi_0\lambda_0 + \pi_1\lambda_1 + \pi_2\lambda_2 \approx 3.42. \tag{4.2.5}$$

We can also estimate the expected total number of claims, S_n, for n days. Let $m = \pi_0\lambda_0 + \pi_1\lambda_1 + \pi_2\lambda_2$. Since $E\{S_n\} = \sum_{t=0}^{n} E\{Z_t\}$, in view of (4.2.5), $E\{S_n\} \sim mn = 3.42n$.

For example, for a season of $n = 60$ days, the estimate is $3.42 \cdot 60 = 205.2$. Omitting details, note that since $n = 60$ is "large," the last estimate is pretty good. This is connected with the fact that, as we will see in Section 4.4, the rate of convergence in (4.2.5) is exponential, which is fairly rapid. □

4.3 Limiting distributions for chains with an infinite number of states

The same results as above are true for infinite chains with a slight modification of Doeblin's condition.

Theorem 7 *(1) Suppose that for a state j_0, a positive integer v, and a number $\delta > 0$,*

$$p_{ij_0}^{(v)} > \delta \ \text{ for all } \ i. \tag{4.3.1}$$

Then there exists a row-vector $\boldsymbol{\pi} = (\pi_1, \pi_2, ...)$ such that $\pi_i \geq 0$ for all i, the sum $\pi_1 + \pi_2 + ... = 1$, and

$$\mathcal{P}^t \to \mathbf{\Pi} \ \text{ as } \ t \to \infty, \tag{4.3.2}$$

where $\mathbf{\Pi}$ is the matrix with all rows equal to $\boldsymbol{\pi}$.
(2) The vector $\boldsymbol{\pi}$ is a solution to the equation

$$\boldsymbol{\pi} = \boldsymbol{\pi}\mathcal{P}. \tag{4.3.3}$$

So, we need all probabilities $p_{ij_0}^{(v)}$ in column j_0 to be not only positive but separated from 0. If the number of states is equal to $d < \infty$, and all $p_{ij_0}^{(v)}$ are positive, this is true automatically: we can set $\delta = \min\{p_{ij_0}^{(v)};\ i = 0, ..., d\}$. If $d = \infty$, we cannot do that. Let, for example, column j_0 consist of numbers $\{1, \frac{1}{2}, \frac{1}{3}, ..., \frac{1}{n}, ...\}$. Then there is no $\delta > 0$ for which (4.3.1) would hold.

Equation (4.3.3) is derived exactly as in the case of finite chains. The proof of the first part is given in Section 4.4.

EXAMPLE 1. We revisit Example 1.2-3 concerning success runs with the transition matrix (1.2.4). It is natural to assume q in (1.2.4) to be positive, so Doeblin's condition holds (why?). In the system of equations $\boldsymbol{\pi} = \boldsymbol{\pi}\mathcal{P}$, the first equation may be written as

$$\pi_0 = q(\sum_{j=0}^{\infty} \pi_j), \tag{4.3.4}$$

and for the other equations,

$$\pi_{k+1} = \pi_k p \text{ for all } k = 0, 1, 2, \dots \,. \tag{4.3.5}$$

Since $\sum_{j=0}^{\infty} \pi_j = 1$, from (4.3.4) it follows that $\pi_0 = q$.

Then, by virtue of (4.3.5), $\pi_1 = \pi_0 p = qp$, $\pi_2 = \pi_1 p = qp^2$, and similarly,

$$\pi_i = qp^i, \;\; i = 0, 1, 2, \dots \,.$$

Thus, the limiting distribution is geometric (with switching p and q in the notation in contrast with the definition in Section **3**.4.3). In Exercise 34, the reader is invited to solve a problem where the probability of success depends on the current state. □

4.4 Proofs of Theorem 4 and 7

4.4.1 Proof of Theorem 4

Column j_0 of the matrix $\mathcal{P}^v$ is strictly positive, and the number of states d is finite. Let δ be the minimal value of $p_{ij_0}^{(v)}$. Then for all i,

$$p_{ij_0}^{(v)} \geq \delta > 0. \tag{4.4.1}$$

For any set of integers B, let $P_t(s, B) = P(X_t \in B \,|\, X_0 = s)$,

$$\overline{M}_t(B) = \max_s P_t(s, B), \;\text{ and }\; \underline{M}_t(B) = \min_s P_t(s, B).$$

Conditioning on the result of the first step, we may write $P_t(s,B) = \sum_m p_{sm} P_{t-1}(m,B)$. Then, keeping in mind that $\sum_m p_{sm} = 1$, we have

$$\overline{M}_t(B) = \max_s \sum_m p_{sm} P_{t-1}(m, B) \leq (\max_m P_{t-1}(m, B)) \max_s \sum_m p_{sm} = \overline{M}_{t-1}(B) \cdot 1 = \overline{M}_{t-1}(B).$$

Similarly,

$$\underline{M}_t(B) \geq \underline{M}_{t-1}(B).$$

Thus, for any B, the sequence $\overline{M}_t(B)$ is non-increasing, and the sequence $\underline{M}_t(B)$ is non-decreasing. Also,

$$\underline{M}_t(B) \leq P_t(s, B) \leq \overline{M}_t(B). \tag{4.4.2}$$

We are going to prove that

$$\overline{M}_t(B) - \underline{M}_t(B) \to 0 \quad \text{as} \quad t \to \infty. \tag{4.4.3}$$

Set $\Delta_{im}(t, B) = P_t(i, B) - P_t(m, B)$. Conditioning on the result of the first v steps, we write

$$\begin{aligned} \Delta_{im}(v+t, B) = P_{v+t}(i, B) - P_{v+t}(m, B) &= \sum_s p_{is}^{(v)} P_t(s, B) - \sum_s p_{ms}^{(v)} P_t(s, B) \\ &= \sum_s \left(p_{is}^{(v)} - p_{ms}^{(v)}\right) P_t(s, B). \end{aligned} \tag{4.4.4}$$

Let $l_{im}(s) = p_{is}^{(v)} - p_{ms}^{(v)}$. (In l_{im}, we suppress v in notation.) Denote by D^+ the set of integers s for which $l_{im}(s) \geq 0$, and by D^- the set of integers s for which $l_{im}(s) < 0$. (In D^+ and D^-, we suppress indices in notation.) Clearly, the union $D^+ \cup D^-$ constitutes the set of all states.

Since $\sum_s p_{is}^{(v)} = \sum_s p_{ms}^{(v)} = 1$,

$$0 = \sum_s p_{is}^{(v)} - \sum_s p_{ms}^{(v)} = \sum_s \left(p_{is}^{(v)} - p_{ms}^{(v)}\right) = \sum_s l_{im}(s) = \sum_{s \in D^+} l_{im}(s) + \sum_{s \in D^-} l_{im}(s),$$

and therefore

$$\sum_{s \in D^+} l_{im}(s) = -\sum_{s \in D^-} l_{im}(s). \tag{4.4.5}$$

Furthermore, because $\sum_{s \in D^+} p_{is}^{(v)} = 1 - \sum_{s \in D^-} p_{is}^{(v)}$,

$$\sum_{s \in D^+} \left(p_{is}^{(v)} - p_{ms}^{(v)}\right) = 1 - \sum_{s \in D^-} p_{is}^{(v)} - \sum_{s \in D^+} p_{ms}^{(v)}. \tag{4.4.6}$$

If $j_0 \in D^+$, we remove the first sum in the r.-h.s. of (4.4.6) and all terms in the second, save the term with $s = j_0$. This implies

$$\sum_{s \in D^+} \left(p_{is}^{(v)} - p_{ms}^{(v)}\right) \leq 1 - p_{mj_0}^{(v)} \leq 1 - \delta.$$

Similarly, if $j_0 \in D^-$, then we write

$$\sum_{s \in D^+} \left(p_{is}^{(v)} - p_{ms}^{(v)}\right) \leq 1 - p_{ij_0}^{(v)} \leq 1 - \delta.$$

So, in any case,

$$\sum_{s \in D^+} l_{im}(s) = \sum_{s \in D^+} \left(p_{is}^{(v)} - p_{ms}^{(v)}\right) \leq 1 - \delta. \tag{4.4.7}$$

From (4.4.4), (4.4.5), (4.4.7), and the fact that $l_{im}(s)$ is negative for $s \in D^-$, it follows that

$$\begin{aligned}
\Delta_{im}(t+v, B) &= \sum_{s \in D^+} l_{im}(s) P_t(s, B) + \sum_{s \in D^-} l_{im}(s) P_t(s, B) \\
&\leq \overline{M}_t(B) \sum_{s \in D^+} l_{im}(s) + \underline{M}_t(B) \sum_{s \in D^-} l_{im}(s) \\
&= \overline{M}_t(B) \sum_{s \in D^+} l_{im}(s) - \underline{M}_t(B) \sum_{s \in D^+} l_{im}(s) \\
&= (\overline{M}_t(B) - \underline{M}_t(B)) \sum_{k \in D^+} l_{im}(s) \leq (\overline{M}_t(B) - \underline{M}_t(B))(1 - \delta).
\end{aligned}$$

From this, we get

$$\begin{aligned}
\overline{M}_{t+v}(B) - \underline{M}_{t+v}(B) &= \max_i P_{t+v}(i, B) - \min_m P_{t+v}(m, B) = \max_{i,m} (P_{t+v}(i, B) - P_{t+v}(m, B)) \\
&= \max_{i,m} \Delta_{im}(t+v, B) \leq (\overline{M}_t(B) - \underline{M}_t(B))(1 - \delta).
\end{aligned}$$

Setting in the last inequality $t = (n-1)v$, we obtain that

$$\overline{M}_{nv}(B) - \underline{M}_{nv}(B) \le (\overline{M}_{(n-1)v}(B) - \underline{M}_{(n-1)v}(B))(1-\delta). \tag{4.4.8}$$

Similarly to what we did above, one can also show that

$$\overline{M}_v(B) - \underline{M}_v(B) \le 1-\delta.$$

From this and (4.4.8), we get by induction that

$$\overline{M}_{nv}(B) - \underline{M}_{nv}(B) \le (1-\delta)^n \to 0 \quad \text{as} \quad n \to \infty. \tag{4.4.9}$$

It remains to note that since the sequence $\overline{M}_t(B) - \underline{M}_t(B)$ is non-increasing, $\overline{M}_t(B) - \underline{M}_t(B) \to 0$ not only along the subsequence $t_n = nv$, but as $t \to \infty$ in an arbitrary way. The proof of (4.4.3) is completed.

Furthermore, the sequences $\overline{M}_t(B)$, $\underline{M}_t(B)$ converge as monotone sequences, and by virtue of (4.4.3), their limits coincide. Then, in view of (4.4.2), $P_t(s, B)$ also has the same limit, and it does not depend on s (the so called squeezing principle from Calculus). Thus, there exists a function $\boldsymbol{\pi}(B)$ such that for any s and any B,

$$P_t(s, B) \to \boldsymbol{\pi}(B) \quad \text{as} \quad t \to \infty. \tag{4.4.10}$$

Let $B = [j]$, the set consisting of one integer j. Then, (4.4.10) implies

$$p_{sj}^{(t)} \to \pi_j,$$

where $\pi_j = \boldsymbol{\pi}([j])$.

If the number of states k is finite, then

$$\sum_{j=0}^{k} \pi_j = \sum_{j=0}^{k} \lim_{t\to\infty} p_{sj}^{(t)} = \lim_{t\to\infty} \sum_{j=0}^{k} p_{sj}^{(t)} = \lim_{t\to\infty} 1 = 1,$$

which completes the proof of Theorem 4.

4.4.2 Proof of Theorem 7

We should just proceed directly from (4.4.1) and go along the proof above replacing the symbols max and min by sup and inf, respectively. This will lead us to (4.4.10).

It remains to prove that $\boldsymbol{\pi}(B)$ is a probability measure; that is, (a) $\boldsymbol{\pi}(B) \ge 0$; (b) $\boldsymbol{\pi}(B_0) = 1$, where B_0 is the set of all states; and (c) $\boldsymbol{\pi}(\cup B_i) = \sum_i \boldsymbol{\pi}(B_i)$ for all disjoint B_i.

Property (a) is obvious. Since $P_t(s, B_0) = 1$, to prove (b), it suffices to set in (4.4.10) $B = B_0$,

It is simple to show (c) for a finite number of B_i's. Consider, for instance, two disjoint events B_1 and B_2. We have

$$\begin{aligned}\boldsymbol{\pi}(B_1 \cup B_2) &= \lim_{t\to\infty} P_t(s, B_1 \cup B_2) = \lim_{t\to\infty} (P_t(s, B_1) + P_t(s, B_2)) \\ &= \lim_{t\to\infty} P_t(s, B_1) + \lim_{t\to\infty} P_t(s, B_2) = \boldsymbol{\pi}(B_1) + \boldsymbol{\pi}(B_2).\end{aligned}$$

If the number of sets B_i is not finite, then we should justify the possibility of interchanging limits and summation. This justification is beyond the scope of this book. The key point here is that the rate of convergence in (4.4.10) is the same for all B's; in other terms, the convergence is uniform.

4.4.3 An important remark

As a matter of fact, we have not only proved the above ergodic theorems, but established the rate of convergence in these theorems. Namely, from our proof it follows that there exist a constant $C \geq 0$ and a constant ρ such that $0 \leq \rho < 1$ and for any B and s,

$$|P_t(s, B) - \boldsymbol{\pi}(B)| \leq C\rho^t. \tag{4.4.11}$$

Indeed, first of all, (4.4.9) implies that $\overline{M}_{nv}(B) - \underline{M}_{nv}(B) \leq (1-\delta)^n = \left((1-\delta)^{1/v}\right)^{nv}$.

Secondly, we use the fact that $\overline{M}_t(B) - \underline{M}_t(B)$ is non-increasing in t. For a time t, let n be the largest integer for which $nv \leq t$. Then

$$\overline{M}_t(B) - \underline{M}_t(B) \leq \overline{M}_{nv}(B) - \underline{M}_{nv}(B) \leq \left((1-\delta)^{1/v}\right)^{nv} \leq \left((1-\delta)^{1/v}\right)^{t-v} = C\rho^t,$$

where $C = (1-\delta)^{-1}$, and $\rho = (1-\delta)^{1/v}$. Together with (4.4.2), this gives

$$P_t(s, B) - \boldsymbol{\pi}(B) \leq \overline{M}_t(B) - \boldsymbol{\pi}(B) \leq \overline{M}_t(B) - \underline{M}_t(B) \leq C\rho^t.$$

The lower bounded may be obtained similarly, and we have arrived at (4.4.11).

---

5 A CLASSIFICATION OF STATES AND ERGODICITY

This section is concerned with a more detailed analysis of states' properties. In particular, it will allow us to better understand the nature of ergodicity. As above, we consider homogeneous chains.

5.1 Classes of states

Consider a Markov chain X_t with a transition matrix $\mathcal{P} = \|p_{ij}\|$.

A state j is said to be *accessible* from a state i if $p_{ij}^{(m)} > 0$ for some m.

For instance, in Example 1.2-1, the chain can move from any state to any state with a positive probability in one step. In Example 1.2-2, the state (1,3,2) is not accessible from (1,2,3) in one step, but is accessible in two steps. So, we say that the former state is accessible from the latter (and, certainly, vice versa).

States i and j are said to *communicate* if they are accessible from each other. For instance, in the same Example 1.2-2, all states communicate.

It is not difficult to show (we skip formalities) that states of any chain can be partitioned into disjoint *classes* such that all states from the same class communicate, and any two states from different classes do not (since otherwise they would belong to the same class).

For example, a homogeneous chain with states labeled $0, 1, 2, 3$ and

$$\mathcal{P} = \begin{Vmatrix} 0.5 & 0.5 & 0 & 0 \\ 0.5 & 0.5 & 0 & 0 \\ 0.25 & 0.25 & 0.25 & 0.25 \\ 0 & 0 & 0 & 1 \end{Vmatrix} \tag{5.1.1}$$

has three classes: $\{0,1\}$, $\{2\}$, $\{3\}$.

The chain is said to be *irreducible* if it has only one class.

For instance, for the chains from Examples 1.2-1,2 all states communicate, and hence, the chains are irreducible. The same concerns the chain from Example 1.2-3 if $p > 0$ and $q > 0$. Indeed, in this case, any state is accessible from 0, and 0 is accessible from any state. Then any state i is accessible from any state j: with a positive probability the chain moves from j to 0, and then from 0 to i.

5.2 Recurrence property

Let f_i be the probability that, starting from state i, the chain will ever return to this state. State i is called *recurrent* if $f_i = 1$, and *transient* if $f_i < 1$.

For a recurrent state i, the process, starting from i, will return to i with probability one. Since the process is Markov, once it revisits i, the process will start over from i as from the beginning. After this, the process will again return to i with probability one, and so on, revisiting state i infinitely often with probability one. W. Feller, the author of one of best if not the best book on Probability Theory [13], called such a state *persistent* and regretted that in the first edition of his book, he had called it recurrent. However, the term recurrent is widely adopted.

Let M_i be the number of revisits to a state i given that the process has started from this state. We do not count the initial stay in i.

We saw that

If i is recurrent, then $M_i = \infty$ with probability one.

Let $f_i < 1$. Then the probability that the process will never return to i is $P(M_i = 0) = 1 - f_i$. The probability that the process will revisit i one time and then never come back is $P(M_i = 1) = f_i(1 - f_i)$, and similarly, $P(M_i = k) = f_i^k(1 - f_i)$. Thus, M_i has a geometric distribution and, in particular, $E\{M_i\} = f_i/(1 - f_i)$. (See, Section **3**.4.3.) Thus,

If i is transient, then $M_i < \infty$ with probability one. Moreover, $E\{M_i\} < \infty$. (5.2.1)

From (5.2.1), it almost immediately follows that

For any finite Markov chain, there exists at least one recurrent state. (5.2.2)

To show this, assume that all states are transient. Then, since all transient states may be visited only a finite number of moments of time, and the number of states is finite, after a finite number of steps no states will be visited, which is impossible.

Before turning to examples, consider two more facts.

Let us *fix* i for a while, and set $X_0 = i$ (the process starts from i). Let the indicator r.v. $I_n = 1$ or 0, depending on whether $X_n = i$ or not. Then $M_i = \sum_{n=1}^{\infty} I_n$.

On the other hand, $P(I_n = 1) = P(X_n = i) = p_{ii}^{(n)}$, and hence

$$E\{M_i\} = E\{\sum_{n=1}^{\infty} I_n\} = \sum_{n=1}^{\infty} E\{I_n\} = \sum_{n=1}^{\infty} p_{ii}^{(n)}. \qquad (5.2.3)$$

Thus,

$$\text{State } i \text{ is recurrent if and only if } \quad \sum_{n=1}^{\infty} p_{ii}^{(n)} = \infty. \qquad (5.2.4)$$

We will derive from this

Proposition 8 *If states i and j communicate, then i and j are either both recurrent or both transient.*

In other words,

The recurrence and transience properties are properties of classes. (5.2.5)

Proof of Proposition 8. If i and j communicate, by definition, there exist n_1 and n_2 such that $p_{ji}^{(n_1)} \geq \alpha$ and $p_{ij}^{(n_2)} \geq \beta$ for some $\alpha, \beta > 0$. One of possible paths that start from j and go back to j in $n_1 + n + n_2$ steps consists of moving in n_1 steps from j to i, returning to i in n steps, and moving from i to j in n_2 steps. Hence,

$$p_{jj}^{(n_1+n+n_2)} \geq p_{ji}^{(n_1)} p_{ii}^{(n)} p_{ij}^{(n_2)} = \alpha\beta p_{ii}^{(n)}. \qquad (5.2.6)$$

Now, because $\sum_{n=1}^{\infty} p_{jj}^{(n_1+n+n_2)} = \sum_{n=n_1+n_2+1}^{\infty} p_{jj}^{(n)}$, series $\sum_{n=1}^{\infty} p_{jj}^{(n_1+n+n_2)}$ and $\sum_{n=1}^{\infty} p_{jj}^{(n)}$ converge or diverge simultaneously. Consequently, from (5.2.6) it follows that if $\sum_{n=1}^{\infty} p_{jj}^{(n)}$ converges, then $\sum_{n=1}^{\infty} p_{ii}^{(n)}$ converges. Since i and j are arbitrary, we can switch i and j, and apply a similar argument. ■

From (5.2.2) and (5.2.5), it follows that

For any irreducible finite Markov chain, all states are recurrent.

For instance, the chains from Examples 1.2-1, 2 and 3 (for the last, if $0 < p < 1$) are irreducible, and hence all states are recurrent.

If a finite chain has several classes, for each class we should check whether the process may leave the class and never come back with a positive probability. For finite chains, it is usually easy.

For instance, for a chain with $\mathcal{P}$ from (5.1.1), classes $\{0,1\}$,and $\{3\}$ contain recurrent states, and class $\{2\}$ —transient. A chain with

$$\mathcal{P} = \left\| \begin{matrix} 0.5 & 0.5 & 0 & 0 \\ 0.5 & 0.5 & 0 & 0 \\ 0 & 0 & 0.75 & 0.25 \\ 0 & 0 & 0.3 & 0.7 \end{matrix} \right\|$$

clearly has two classes, and in each class, all states are recurrent.

If a chain is infinite, the question may be not so simple.

EXAMPLE 1. In Exercise 6, the reader is invited to make sure that a simple random walk is a Markov chain. We consider the simple case when the process does not stop at any barrier, and exclude the trivial case $p = 0$ or 1. Then all states communicate.

In Section **4**.4.1, we have already proved that the probability of revisiting zero is one if $p = \frac{1}{2}$, and it is less than one if $p \neq \frac{1}{2}$. Hence, state 0 is recurrent if and only if $p = \frac{1}{2}$, and in view of (5.2.5), this concerns all other states also. Let us look how the same follows from (5.2.4).

The random walk may revisit zero only in an even number of steps, so $p_{00}^{(2k+1)} = 0$. On the other hand, $p_{00}^{(2k)}$ is the probability that during $2k$ trials, there will be exactly k successes. Therefore, $p_{00}^{(2k)} = \binom{2k}{k} p^k (1-p)^k = \frac{(2k)!}{k!k!} p^k (1-p)^k$.

Applying Stirling's formula (see comments in p.49) $k! \sim \sqrt{2\pi k} k^k e^{-k}$, the reader can easily verify that

$$p_{00}^{(2k)} \sim \frac{1}{\sqrt{\pi k}} (4p(1-p))^k = \frac{1}{\sqrt{\pi k}} a_p^k,$$

where $a_p = 4p(1-p)$.

If $p = \frac{1}{2}$, then $a_p = 1$; while if $p \neq \frac{1}{2}$, then $a_p < 1$ (recall what is the maximum of $p(1-p)$). Then the series $\sum_{k=1}^{\infty} p_{00}^{(2k)} \sim \sum_{k=1}^{\infty} \frac{1}{\sqrt{\pi k}} a_p^k < \infty$ if and only if $p \neq \frac{1}{2}$. $\square$

The above classification and partitioning of the state space into classes is useful since, for many purposes, we can restrict our attention to one class. In particular, it is worth noting the following.

If a process starts from a transient state, it can leave the class to which this state belongs. For example, in the case (5.1.1), the process starting from state 2 can move to state 0, and then it will never leave the class $\{0, 1\}$.

However,

> If the process starts from a recurrent state, it (the process) will never leave the class from which it has started to evolve.

Indeed, let i be the initial state which is recurrent. The process can move to any state j which is accessible from i. If state j had not communicated with i, it would have meant that the probability to come back to i from j is zero. Then the probability of returning to i would not have been one, and i would not have been recurrent. Consequently, any state to which the process can move from i communicates with i. Such a state, by definition, belongs to the same class as i.

The same argument leads to the following fact:

> In a recurrent class, starting from any state, the process will arrive at any state with probability one. (5.2.7)

Indeed, if the probability to move from, say, state k to state i is less than one, then with a positive probability the process may move from i to k and not return to i, that is, the probability of returning to i, starting from i, would be less than one.

5.3 Recurrence and travel times

Let T_{ik} be the number of steps it will take for the process, starting from i, to reach state k. Then the r.v. $R_i = T_{ii}$ is a return (or recurrence) time. Set $m_{ik} = E\{T_{ik}\}$, and $m_i = m_{ii} = E\{R_i\}$.

If state i is transient, with the positive probability $1 - f_i$ defined above, the process will never come back to state i. So, with the mentioned positive probability, $R_i = \infty$ and, hence, $m_i = \infty$.

In general, if state i is recurrent, it also may happen that $m_i = \infty$. Such states are called *null recurrent*. Otherwise, the state is called *positive recurrent*.

A classical example of null recurrence concerns states in the symmetric random walk ($p = \frac{1}{2}$): in Section **4**.4.1, we proved that the mean time it takes to revisit zero is equal to infinity. Since for the random walk, any state may be viewed as initial, the same concerns all m_i.

The set of values that the recurrence times R_i can take on are also connected with the periodicity property. Assume that $p_{ii}^{(n)} = 0$ for all n not divisible by some number k. It means that, starting from i, the process can revisit i only at moments $k, 2k, \ldots$. We call such a state periodic and say that d is the *period of state* i if d is the largest number among all integers k with the property above. For example, if the chain may revisit the state i with positive probabilities only at moments $3, 6, 9, \ldots$, then $d = 3$. If $d = 1$, the state is called *aperiodic*. Clearly, if $p_{ii} > 0$, state i is aperiodic.

Obviously, both states of a chain with $\mathcal{P} = \begin{Vmatrix} 0 & 1 \\ 1 & 0 \end{Vmatrix}$ are periodic with a period of two. The same is true for the simple random walk.

It may be shown, by making use of (5.2.6), that periodicity is a class property, so for an irreducible chain all states are either aperiodic or periodic with the same period.

5.4 Recurrence and ergodicity

In conclusion, we state two ergodicity theorems connecting the classification of states with the ergodicity property.

Theorem 9 *For all states i, j of any irreducible recurrent aperiodic chain, there exists*

$$\lim_{t\to\infty} p_{ij}^{(t)} = \pi_j = \frac{1}{m_j}.$$

Analyzing this theorem, we can jump to the following conclusions.

- $\lim_{t\to\infty} p_{ij}^{(t)} = 0$ if and only if $m_j = \infty$, that is, only when the state j is null recurrent.

- Let $m_j = \infty$. Setting $i = j$, we get that $\lim_{t\to\infty} p_{jj}^{(t)} = 0$. On the other hand, all states in the chain under consideration communicate. Hence, in view of (5.2.6), if $p_{jj}^{(t)} \to 0$, then $p_{ii}^{(t)} \to 0$ for other states i, which means that $m_i = \infty$ for all states i. In other words, null recurrence is a class property, and for the chain under discussion (a) either all limits $\lim_{t\to\infty} p_{ij}^{(t)}$ are equal to zero, or all are positive; (b) either all expected recurrence (return) times m_j are infinite, or all are finite.

- Recall that for any chain $\sum_j p_{ij}^{(t)} = 1$. If the chain is finite, and d is the number of states, we can write $1 = \lim_{t\to\infty} \sum_{j=1}^{d} p_{ij}^{(t)} = \sum_{j=1}^{d} \lim_{t\to\infty} p_{ij}^{(t)} = \sum_{j=1}^{d} \pi_j$, having eventually $\sum_{j=1}^{d} \pi_j = 1$. (For an infinite chain we cannot bring $\lim_{t\to\infty}$ inside the sum without additional conditions.)

  But π_j's are either all zeros, or all positive. Since the sum equals one, the former is impossible. Thus,

  > For any finite irreducible recurrent chain, $\pi_j > 0$, and $m_j < \infty$ for all j.

Now we state an ergodicity theorem for positive recurrent states.

Theorem 10 *For any irreducible positive recurrent aperiodic chain and all states i, j*

$$\lim_{t\to\infty} p_{ij}^{(t)} = \pi_j = \frac{1}{m_j} > 0,$$
$$\sum_j \pi_j = 1, \tag{5.4.1}$$

and the limiting vector $\pi = (\pi_1, \pi_2, \ldots)$ *is a unique solution to the equation*

$$\boldsymbol{\pi} = \boldsymbol{\pi}\mathcal{P}, \tag{5.4.2}$$

satisfying (5.4.1).

(If a $\boldsymbol{\pi}$ satisfies (5.4.2), so does $c\boldsymbol{\pi}$ for any constant c. However, $\boldsymbol{\pi}$ satisfying (5.4.1) and (5.4.2) is unique.)

---

6 EXERCISES

The use of software for matrix operations below is strongly recommended.

Section 1

1. Jane is hiking in one of two areas; in both she may come across a rattlesnake. For the first area, the probability of seeing a rattle snake is 0.1, and for the second, it is 0.07. If Jane saw a rattle snake last time, she switches the areas; and if she did not see a snake, she keeps hiking

in the same area. (a) Write the transition matrix. (b) Given that Jane last time hiked in the first area, find the probability that during 4 hikes each time she will switch the areas.

2. *The Ehrenfest ball-box model.* Consider a fixed number n of balls distributed somehow in two boxes. One at a time, a ball is selected at random from the total of n balls and moved from its box to the other. Let X_t be the number of balls in, say, the first box at step t.

   Originally, this model was proposed as a model of heat exchange between two bodies isolated from the outside. At each step, an energy or temperature unit (a ball) is "transferred" from one body to the other. Another example may concern electrons independently occupying two orbits. Suppose that from time to time, one electron moves from its orbit to the other (due to receiving or emitting energy), and all electrons are equally likely to switch the orbits at the next step.

   Show that X_t is a Markov chain, and describe its transition matrix (a) for $n = 2, 3$ and 4; (b) for an arbitrary n. We return to this problem in Exercise 25 concerning the behavior of the process in the long run.

3. In a physical space, there is a site and particles are arriving at the site. At time moment $t = 0$, the site is vacant. At moments $t = 1, 2, ...,$ exactly one particle arrives with a probability of p_1. If the site is vacant, the particle occupies the site. If the site is occupied, arriving particles do not enter and are not taken into account. Once a particle occupied the site, at each future moment, it leaves the site with a probability of p_2 and remains there with probability $1 - p_2$, independently of the prehistory and of whether a new particle has arrived. If at a moment t, a particle left the site and a new particle has arrived, the latter occupies the site.

   In one of possible interpretations, the site is a server (say, an auto mechanic) and the particles are customers. In our model, if the server is busy, the customers leave the site. However, there are other possible interpretations. The particles may be electrons, and the site is an orbit (more precisely, a suborbit). The point is that not any electron may go to a suborbit that is already occupied; we skip details.

   Let X_t be the number of particles in the site (so, $X_t = 0$ or 1). Show that X_t is a Markov chain and write its transition matrix.

4. In the situation of Exercise 3, suppose that there are two sites, and the particles sitting in these sites exhibit independent behavior. Let X_t be the total number of particles in both sites. Show that X_t is a Markov chain and write its transition matrix.

5. Consider the chain in Example 1.2-5 and set $f_k = P(\xi_i = k)$, $k = 0, 1, ...$. (This probability is the same for all i). Find the transition matrix.

6. Show that the random walk process in Section **2**.2.4 is a Markov chain. Find the transition matrices for the following two cases.

   (a) The process does not stop upon reaching any level. In this case, X_t may assume all values $0, \pm 1, \pm 2, ...$.

   (b) The process comes to a stop when $X_t = 0$ or a, as it was assumed in Section **2**.2.4. In this case, X_t may assume values $0, 1, ..., a$.

7.* Show that the branching process from Section **4**.1 is a Markov chain. Describe how the transition matrix may look.

8. For Example 1.2-1, compute the probability of the path $0 \to 0 \to 1$, and the probability of the transition from state 0 to state 1 in two steps. Which probability is larger? Could we expect it without calculations?

9. We have mentioned that the Markov model in Example 1.2-1 may be too simple for describing the evolution of the weather conditions: in general, it is certainly useful to know the tendency of weather conditions in the past.

   However, we can keep using the Markov setup if we extend the set of possible states of the weather on a current day. Namely, we may include in the characterization of states information about the past. For example, if we want to take into account the information about the weather on two consecutive days, we may consider, *for a current day*, such states as "rainy today, and rainy yesterday," "rainy today, and normal yesterday," "icy today, and rainy yesterday," and so on. Such a model may turn out to be adequate enough.

   How many states, should we consider in such a modified model? What is the probability of moving from the state "rainy today, and rainy yesterday" to the state "rainy today, and normal yesterday?"

10. For the figure 57.5% at the end of Example 1.3-3 to be true, should we assume that the citizens are migrating independently?

11. (*A model of social mobility*). It concerns the transition of individuals between social classes defined on the basis of income or occupation. Sociologists distinguish inter-generational mobility from parents' class to child's class and intra-generational mobility over an individuals life course. The particular model below illustrates the former case. For the sake of simplicity, assume that in each generation, the members of any family belong to the same class, so we are talking about family transitions.

    Consider three classes: 0 (upper), 1 (middle), and 2 (lower). For a family chosen at random, let X_t be the family's social class in the t-th generation. Suppose that X_t is Markov chain with the transition matrix

$$\mathcal{P} = \begin{Vmatrix} 0.69 & 0.3 & 0.01 \\ 0.1 & 0.8 & 0.1 \\ 0.05 & 0.45 & 0.5 \end{Vmatrix}.$$

    (In the literature, one may find various figures, and the reader should not view the numbers above as real.) Suppose that in a particular year, the percentages of the families belonging to the classes mentioned is given by the vector (20%, 50%, 30%). Estimate the same vector for the great-grandchildren of the children born in this year.

12. Provide an Excel worksheet for Example 1.2-4 and play a bit, changing the transition matrix and watching what happens. Consider several realizations, generating different random numbers.

Sections 2 and 3**

13. For the situation of Exercise 6b, in the case of the symmetric walk, let T_u be the random time it takes for the process to reach either level 0 or a. Find $E\{T_u\}$ for $a = 4$ and $u = 0,1,2,3,4$. (*Hint*: Look at (2.2.9).)

14. Let us revisit Exercise 11 and consider a family from the middle class. In which generation on the average will the descendants of this family at the first time belong to the upper class? (*Advice*: Look over Section 2.3.)

15. Let us come back to the weather-condition example in Section 2.3. Suppose that either the today weather conditions are normal, or it is rainy. Without any calculations, show that the icy-road conditions will appear on the tenth day on the average. (*Hint*: $p_{02} = p_{12}$.)

16. For a health insurance plan, for every client, the company distinguishes three conditions at the beginning of each year: "healthy", "sick", and "deceased". The transition matrix is

$$\mathcal{P} = \begin{Vmatrix} 0.85 & 0.14 & 0.01 \\ 0.6 & 0.35 & 0.05 \\ 0 & 0 & 1 \end{Vmatrix}.$$

Assume that at the initial moment, 90% of the clients of the company are healthy.

(a) Find the expected lifetime of a client selected at random.

(b) Assuming that an insured was healthy at the beginning and not counting the initial state, find the expected length of the period when the insured remained healthy before switching to another state. (*Hint*: It is possible to avoid long calculations.)

(c) Suppose that the mean annual health care cost for the two first states—"healthy" and "sick"—equals 1 and 4 (units of money), respectively. Find the expected present value of the total cost (i.e., the expected discounted total cost) in the long run with a discount of 0.95. (*Advice*: Assign zero payment to the absorbing state "deceased.")

17. Let us come back to Section 1.1. Suppose that in area 0, the living expenses of a citizen chosen at random is a r.v. with a mean of 10 (units of money) and a standard deviation of 1. For area 1, the corresponding characteristics are 9 and 2, respectively. Find the mean and standard deviation of the living expenses of a randomly chosen citizen in the first year.

18. Mr. M. runs a business. Each year, with a probability of 0.1, Mr. M. quits this business. *Given* that this does not happen, Mr. M. faces either "bad" or "good" year with respective average incomes 1 or 2. The transition probabilities for these two states are specified by the matrix $\mathcal{P} = \begin{Vmatrix} 0.6 & 0.4 \\ 0.3 & 0.7 \end{Vmatrix}$, where zero state corresponds to a "bad" year. The initial year is "good." Evaluate the expected discounted total income in the long run for $v = 0.9$.

Section 4

19. In Exercise 1, using software or even a calculator, for a long period of time, *estimate* the proportion of the time when Jane hikes in the first area. Find the precise value.

20. Each weekend Joan is either jogging or swimming. If Joan swam last weekend, she switches to jogging with probability $2/3$. If she jogged, she jogs again with probability $1/2$. (a) Find the proportions of the time Joan swims and jogs in the long run. (b) For the stationary regime, find the probability that in a weekend, Jane switches from swimming to jogging.

21. Let the transition matrix of a chain is $\mathcal{P} = \begin{Vmatrix} p & 1-p \\ 1-p & p \end{Vmatrix}$, where $0 < p < 1$. Guess without calculations what is the limiting distribution? Check your guess in two ways: appealing to (4.1.2) and to (4.2.2). Comment the result from a common-sense point of view.

22. Assuming that in Exercise 3, $0 < p_1 < 1$, $0 < p_2 < 1$, can we say without calculations whether the ergodicity property holds? Find the limiting distribution. What is happening when p_1 gets smaller, very small? Explain the answer from a common-sense point of view.

23. Setting in Exercise 4 $p_1 = p_2 = 0.5$, guess which the stationary probability, π_0, π_1, or π_2, is the largest. Find these probabilities.

24. Consider Example 4.2-1 for k books. Show that in the long run, all $k!$ permutations are asymptotically equally likely.

25. Let us revisit Exercise 2.

(a) Explain why the limiting distribution does not exist. (*Hint*: Can X_t and X_{t+1} be, say, both even?)

(b) Nevertheless, we can figure out the explicit pattern of the behavior of X_t. Using Excel or another software, for $n = 4$, compute $\mathcal{P}^t$ for, say, $t = 2,...,8$, and comment on the results. ($\mathcal{P}$ is a 5×5-matrix.) Check that a stationary distribution does exist: it is $\boldsymbol{\pi} = (0.0625, 0.25, 0.375, 0.25, 0.0625)$. Looking at the calculations you have already done, make sure that the average of two neighbor matrices, $\frac{1}{2}\left(\mathcal{P}^{2k-1} + \mathcal{P}^{2k}\right)$, is approaching the matrix with all rows equal to $\boldsymbol{\pi}$ above.

(c) Check that a stationary distribution des exist and it is binomial: $\pi_i = \binom{n}{i}2^{-n}$ for (i) $n = 4$; (ii) for an arbitrary n.

26. Show that the transition matrices in (4.1.4) do not satisfy Doeblin's condition, while the matrix in Example 4.2-4 does.

27. Does Doeblin's condition of Theorem 4 hold for the transition matrix (4.2.4)? Connect it with what was said in Example 4.2-3.

28. Consider the transition matrix $\mathcal{P} = \begin{Vmatrix} 0.4 & 0.6 & 0 & 0 \\ 0.6 & 0.4 & 0 & 0 \\ 0 & 0 & 0.6 & 0.4 \\ 0 & 0 & 0.5 & 0.5 \end{Vmatrix}$.

(a) Argue why in this case we should not hope for ergodicity. How does the chain evolve in this case? Show that Doeblin's condition of Theorem 4 does not hold for this matrix.

(b) Let (π_0, π_1) and $(\tilde{\pi}_0, \tilde{\pi}_1)$ be the stationary distributions for the two-state chains with the transition matrices $\mathcal{R} = \begin{Vmatrix} 0.4 & 0.6 \\ 0.6 & 0.4 \end{Vmatrix}$, $\widetilde{\mathcal{R}} = \begin{Vmatrix} 0.6 & 0.4 \\ 0.5 & 0.5 \end{Vmatrix}$, respectively. Argue that these distributions are also the limiting distributions for the two-state chains mentioned.

(c) For $\alpha \in [0,1]$, let the vector $\boldsymbol{\pi}_{(\alpha)} = \alpha(\pi_0, \pi_1, 0, 0) + (1-\alpha)(0, 0, \tilde{\pi}_0, \tilde{\pi}_1)$.
Show that for any $\alpha \in [0,1]$, the distribution $\boldsymbol{\pi}_{(\alpha)}$ is stationary for the original chain.

(d) Let the initial distribution $\boldsymbol{\pi}_0 = (\pi_{00}, \pi_{01}, \pi_{02}, \pi_{03})$, and $\alpha = \pi_{00} + \pi_{01}$, the probability that the original chain started in one of the first two states. Show that $\boldsymbol{\pi}_t$, the distribution of the chain, converges to $\boldsymbol{\pi}_{(\alpha)}$.

(e) Explain all the facts above from a heuristic point of view. (*Hint*: The main point is that the sets of states $\{0,1\}$ and $\{2,3\}$ do not communicate; that is, the process starting in the set $\{0,1\}$ will never leave this set, and the same concerns $\{2,3\}$.)

29.* Consider the situation of the example on weather-condition in Section 2.3 in the stationary regime. Given that the today weather conditions are either rainy or icy-road ones, find the mean waiting time for the normal conditions. (*Advice*: Look also at Example 4.2-2.)

30.* Assume that in the situation of Example 4.2-5, in the stationary regime, a daily number of accidents occurred to be greater than six. Find the probability that the weather on this day corresponds to the icy road condition. Comment on that it is greater then π_2.

31. A stochastic matrix is said to be *doubly stochastic* if for all columns, the sum of the probabilities is also equal to one. Show that if a chain with such a matrix is finite and ergodic, in the long run, all states are equally likely.

32. Peter runs a small business classifying each day as good, moderate, or bad. The transition process is Markov with transition matrix (1.2.2). A daily income is a r.v. assuming values

1 or 2 on a bad day; 2 or 3 on a moderate day; and 3 or 4 on a good day; in all these three cases, with equal probabilities. Find (a) the expected daily income in the long run; (b) the probability that, in the stationary regime, the income on a particular day will exceed two units. (*Hint*: We have found a stationary distribution for (1.2.2) in Example 4.2-2.)

33. Argue that the Markov chain from Exercise 6b is not ergodic. Certainly, if this is true, Doeblin's condition must not hold. Show directly that this is indeed the case. (*Advice*: Try to avoid long calculations.)

34.* In the problem of Example 4.3-1, assume that from state i the process moves to state $i+1$ with probability $p_i = \frac{\lambda}{1+i}$, and to state 0 with probability $q_i = 1 - p_i$, where a number $\lambda \leq 1$, and $i = 0, 1, 2, \ldots$. (a) Proceeding from (4.3.3), find the limiting distribution $\boldsymbol{\pi}$. (b) Let T_0 be the return time to state 0 (not counting the initial stay at 0). Find the distribution of T_0 and its mean value.

Section 5*

35. Classify the states and classes of the chain in Example 2.2-1.

36. (a) Show that, if there are more than one state, and $p_{ii} = 1$ for some i, then the chain is reducible.

    (b) Give an example of a reducible chain for which *all* transition probabilities are less than one.

    (c) Show that in order to specify the classes of a chain, we do not need to know particular values of transition probabilities but only which of them are not equal to zero. Why does it not contradict the statement of Exercise 36a?

37. Prove that if the number of states is a finite k and state i is accessible from state j, then it is accessible in $k-1$ or a less number of steps.

38. Classify the states of $\mathcal{P} = \begin{Vmatrix} 0 & 0 & 0.5 & 0.5 \\ 1 & 0 & 0 & 0 \\ 0 & 0.75 & 0 & 0.25 \\ 0 & 1 & 0 & 0 \end{Vmatrix}$.

39. Figure out whether the matrices $\mathcal{P} = \begin{Vmatrix} 0 & 1/2 & 1/2 \\ 1/2 & 0 & 1/2 \\ 1/2 & 1/2 & 0 \end{Vmatrix}$ and $\widetilde{\mathcal{P}} = \begin{Vmatrix} 0 & 1/2 & 1/2 \\ 1 & 0 & 0 \\ 1/3 & 1/3 & 1/3 \end{Vmatrix}$ correspond to periodic chains. Find periods.

40. (*Multidimensional random walk*). Let $\mathbf{X}_t = \sum_{i=1}^{t} \boldsymbol{\xi}_i$, where independent d-dimensional r.vec.'s $\boldsymbol{\xi}_i$ take on 2^d vector values $(\pm 1, \ldots, \pm 1)$ with equal probabilities $1/2^d$. Another way to state the problem is to write $\boldsymbol{\xi}_i = (\xi_{i1}, \ldots, \xi_{id})$, where ξ_{ik}'s are independent r.v.'s assuming values ± 1 with equal probabilities $1/2$. Let $\mathbf{X}_0 = (0, \ldots, 0)$.

    For $d = 1$, this is a classical symmetric random walk we considered repeatedly.

    For $d = 2$, we may view it as the "concatenation" of two independent random walks: a particle moves independently in two perpendicular directions; one step "to right or left" in each direction with equal probabilities. The set of all possible values (or states) of the process $\mathbf{X}_t$ may be identified with vertices of a two-dimensional lattice with integer-valued coordinates. The interpretation for the dimension $d \geq 3$ is similar.

    Prove that all states are recurrent for $d = 1, 2$ and transient for $d \geq 3$.

41. Proceeding from Theorem 9, justify rigorously the ergodicity property for chains in Examples 1.2-1,2,3.

42. Show that the chain in Example 2.2-1 is not ergodic. (*Advice*: Use the first step analysis.)

Chapter 6

Continuous Random Variables

1,2

In Chapter 3, we classified r.v.'s as discrete and non-discrete and have considered the former case. In this chapter, we turn to non-discrete r.v.'s and explore the class of the so-called continuous r.v.'s. In Chapter 7, we will accomplish the consideration of possible types of r.v.'s.

The reader is recommended to watch, when reading, a straight analogy between the discrete and continuous cases and the main pattern: replacing summation by integration.

1 CONTINUOUS DISTRIBUTIONS

1.1 Probability density function

Below, we use the symbol B for sets in *the real line*. As a rule, it will be an interval, but we do not exclude more complex sets.

Consider a r.v. X, and set $F_X(B) = P(X \in B)$, the probability that the value of X will fall into a set B. The function of sets $F_X(B)$ is called the *probability distribution* (briefly, *distribution*) of X. If we deal with only one r.v. X, we will omit the index X in $F_X(B)$.

As has been already defined, a r.v. X and its distribution $F(B)$ are called *discrete* if the set of the values of X is either a finite or countably infinite set $\{x_1, x_2, ...\}$. In this case, we say that the distribution $F(B)$ is concentrated at points $x_1, x_2, ...$

A r.v. X and its distribution $F(B)$ are said to be *absolutely continuous* if there exists a non-negative function $f(x)$ such that for any B,

$$F(B) = \int_B f(x)dx. \tag{1.1.1}$$

In other words, $P(X \in B)$ is equal to the area above the set B and under $f(x)$ in the corresponding graph; see Fig. 1a. As a rule, we will skip the word "absolutely" and talk about continuous r.v.'s.

The function $f(x)$ above is called the *probability density function* of the r.v. X, or briefly, the *density* of X. Setting $B = (-\infty, \infty)$, we have

$$\int_{-\infty}^{\infty} f(x)dx = 1. \tag{1.1.2}$$

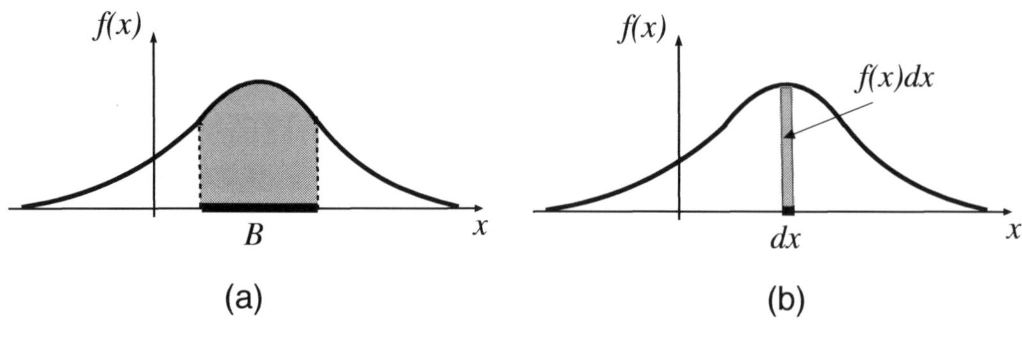

FIGURE 1.

We will also interpret the above definition as follows:

For an infinitesimally small interval $[x, x+dx]$, the probability $P(x \leq X \leq x+dx)$ may be represented as $f(x)dx$; (1.1.3)

see Fig. 1b. This is merely an interpretation but it will prove to be useful for further constructions.

Since a point is an interval of zero length, from (1.1.1) it follows that

For any continuous r.v. X and any number a,
$P(X = a) = 0.$ (1.1.4)

We already discussed this issue in Example 1.2.3-1. When we want to know the probability that tomorrow the highest air temperature T will equal $25C°$, we mean usually the probability that $25 \leq T < 26$, and this probability may be positive. However, the probability that T will be *exactly* equal to 25.000... equals zero, and this should not look strange.

So, r.v.'s for which (1.1.4) is not true, are not continuous. For instance, a Poisson r.v. is not continuous because it takes values $0, 1, \ldots$ with positive probabilities. Certainly, the same concerns any discrete r.v.

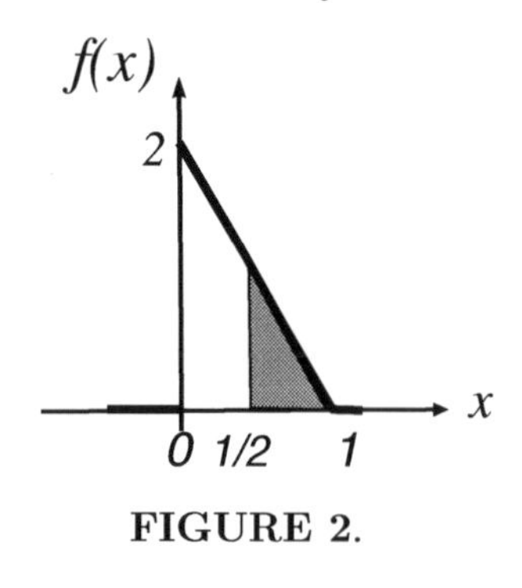

FIGURE 2.

EXAMPLE 1. Let the density function $f(x) = 2(1-x)$ for $0 \leq x \leq 1$, and $f(x) = 0$ otherwise. The graph is given in Fig. 2. It is easy to see that the area under $f(x)$ equals 1. Since $f(x) = 0$ for $x \notin [0, 1]$, the corresponding r.v. X assumes values from $[0, 1]$ with probability one. Since the density $f(x)$ is decreasing when x runs from 0 to 1, values from $[0, 0.5]$ must be more likely than values from $[0.5, 1]$. Indeed,

$$P\left(\frac{1}{2} \leq X \leq 1\right) = \int_{1/2}^{1} 2(1-x)dx = \frac{1}{4}$$

(see again Fig. 2), while $P(0 \leq X < \frac{1}{2}) = 1 - P(\frac{1}{2} \leq X \leq 1) = 1 - \frac{1}{4} = \frac{3}{4}$.

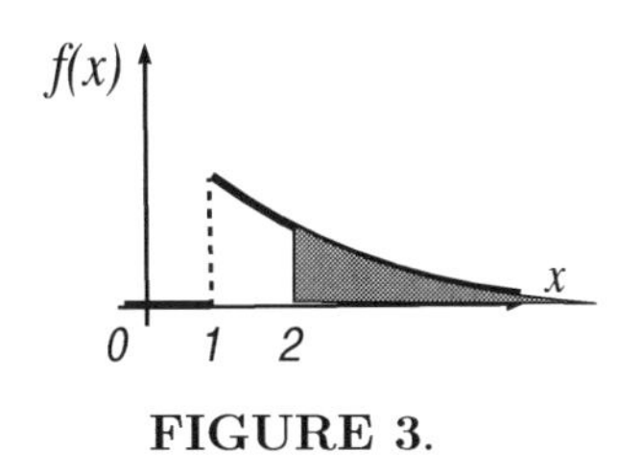

FIGURE 3.

EXAMPLE 2. Let $f(x) = 1/x^2$ for $x \geq 1$, and $f(x) = 0$ otherwise; see Fig. 3. The reader is invited to check that the area under the curve; that is, $\int_1^\infty f(x)dx = 1$. By straightforward integration,

$$P(X > 2) = \int_2^\infty f(x)dx = \int_2^\infty \frac{1}{x^2} f(x)dx = \frac{1}{2};$$

see again Fig. 3. So, the probability that a value of X will fall into $[1,2]$ is $1 - \frac{1}{2} = \frac{1}{2}$, and is the same as the probability that X will assume a value from $[2, \infty]$. It is also worth noting that since all probabilities are specified by integration, the value of $f(x)$ at the point $x = 1$ (which is a discontinuity point of $f(x)$; see Fig. 3) does not matter. The same concerns all other examples where the density is not everywhere continuous.

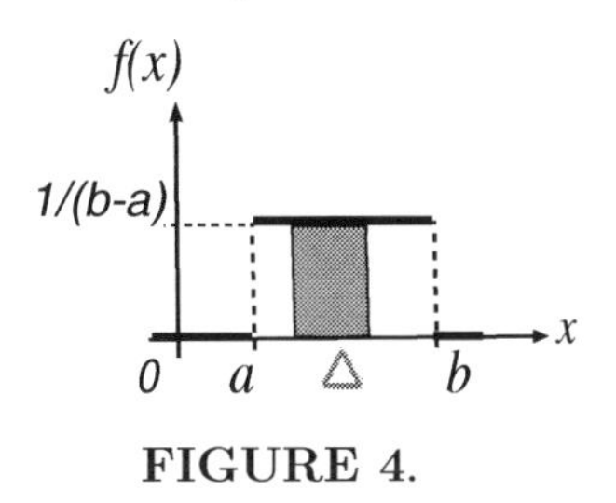

FIGURE 4.

EXAMPLE 3 (*the uniform distribution*). Suppose that a r.v. X takes on values only from an interval $[a, b]$, and all these values are equally likely to occur. Formally, we define such a r.v. as a r.v. having the density $f(x) = 0$ for $x \notin [a, b]$ and being constant on $[a, b]$; see Fig. 4. For the area under the graph to be equal to 1, we should set $f(x) = 1/(b-a)$ for $x \in [a, b]$; see Fig. 4.

Consider a subinterval $\Delta \subseteq [a, b]$. The area over this subinterval equals $\frac{|\Delta|}{b-a}$, where as usual, $|\Delta|$ is the length of Δ; see again Fig. 4. (Certainly, we could compute an integral, but in this simple situation, there is no need in it.) Thus, for any interval $\Delta \subseteq [a, b]$, the probability $P(X \in \Delta)$ is proportional to the length of Δ.

EXAMPLE 4. Let the density of a r.v. X be $f(x) = e^{-x}$ for $x \geq 0$. Find $f(x)$ for $x < 0$. The point is that $\int_0^\infty e^{-x}dx = 1$. Hence, $P(X \geq 0) = 1$, which implies $P(X < 0) = 0$. In this case, we set $f(x) = 0$ for $x < 0$.

EXAMPLE 5. Let a density $f(x) = c/(1+x^2)$ for a constant c. To find c, we proceed from (1.1.2). Recalling what the derivative of $\arctan x$ is, we have

$$\int_{-\infty}^{\infty} f(x)dx = c\int_{-\infty}^{\infty} \frac{1}{1+x^2}dx = c \cdot \arctan x\Big|_{-\infty}^{\infty} = c\left(\frac{\pi}{2} - \left(-\frac{\pi}{2}\right)\right) = c\pi.$$

Hence, in view of (1.1.2), $c = 1/\pi$, and

$$f(x) = \frac{1}{\pi(1+x^2)}. \tag{1.1.5}$$

The distribution with this density is called the *Cauchy distribution*. We will repeatedly come back to it.

EXAMPLE 6 (*The two envelopes problem*). You are given two closed envelopes containing different sums of money. You may choose one envelope and take the money it contains. You pick one envelope at random, open it and see that the envelope contains, say, c dollars. Then the host of the game offers you to switch to the other envelope. Is it beneficial to do it?

There are many examples of wrong reasoning driving to the conclusion that you should switch whatever c is. (One of such examples was discussed in Exercise **3**.42.) Certainly, this is not true because if it were true and the decision had not depended on c, then you would have had to switch even not opening the envelope you picked, which contradicts common sense.

However, surprisingly, there exists a strategy which takes into account the amount you found in the first envelope, and for which the probability to get a larger amount is greater than $\frac{1}{2}$. The only thing we should assume is that the amounts in the envelopes are different.

The main point is that this strategy is randomized: the decision making procedure involves a random mechanism.

More precisely, assume that the amounts in the envelopes are positive and different. Let us choose any non-negative continuous r.v. X whose density $f(x) > 0$ for all $x > 0$. For example, it may be $f(x) = e^{-x}$, but as we will see, this does not matter. Suppose that we can arrange a mechanism which allows us to get or simulate one particular (random!) value of X. In Section **7**.4, we discuss how to do that using a computer program, but now we just assume that we can get a random value of X.

The desirable strategy consists in the following. You pick an envelope and open it. Then you simulate (or get somehow by means of a random mechanism) a random number X. Suppose the envelope contains an amount of c.

If $c < X$, then you switch. If $c \geq X$, then you do not switch. Note also that because X is a continuous r.v., $P(X = c) = 0$. Let us compute the probability of choosing the "better" envelope for this strategy.

Let a and b be the amounts in the envelopes; $0 < a < b$. (You know that one of these values is c, but you do not know the other value, and whether it is larger or less than c.)

Let the events $B_1 = \{X \leq a\}$, $B_2 = \{a < X \leq b\}$, $B_3 = \{X > b\}$. Because the density $f(x) > 0$, all probabilities $P(B_i) > 0$. Let the event A consist in choosing the better envelope.

If B_1 occurs, then $c \geq X$, and you will not switch whatever envelope you selected. Hence, $P(A\,|\,B_1) = \frac{1}{2}$. If B_3 occurs, then $c < X$, and you will switch whatever envelope was picked, and hence, $P(A\,|\,B_3) = \frac{1}{2}$. However, if B_2 occurs, then you will switch only if you selected the envelope with the smaller amount a. Hence, $P(A\,|\,B_2) = 1$. Consequently,

$$P(A) = P(A\,|\,B_1)P(B_1) + P(A\,|\,B_2)P(B_2) + P(A\,|\,B_3)P(B_3) = \frac{1}{2}P(B_1) + P(B_2) + \frac{1}{2}P(B_3)$$
$$= \frac{1}{2}(P(B_1) + P(B_2) + P(B_3)) + \frac{1}{2}P(B_2) = \frac{1}{2} + \frac{1}{2}P(B_2).$$

Since $P(B_2) > 0$, we have $P(A) > \frac{1}{2}$.

Certainly, if we have an additional information about what a and b can be, then we can take it into account and choose X for which $P(B_2) = P(a < X \leq b)$ will be larger. So, in this case, the choice of $f(x)$ matters: it should reflect which information we possess. However, if we have no idea about what a and b are, we still can make $P(A)$ larger than one half; though perhaps not much larger. □

1.2 Cumulative distribution functions

As in Chapter 3, the cumulative distribution function (c.d.f.), or simply the distribution function (d.f.) of a r.v. X is the function $F_X(x) = P(X \leq x)$. If it cannot cause confusion, we will omit the index X.

When x is increasing, $P(X \le x)$ is getting larger or remains the same, so any distribution function (d.f.) $F(x)$ is non-decreasing. Also, setting $F(\pm\infty) = \lim_{x\to\pm\infty} F(x)$, we have $F(-\infty) = 0$, and $F(\infty) = 1$. So, a typical graph of a d.f. looks like in Fig. 5.

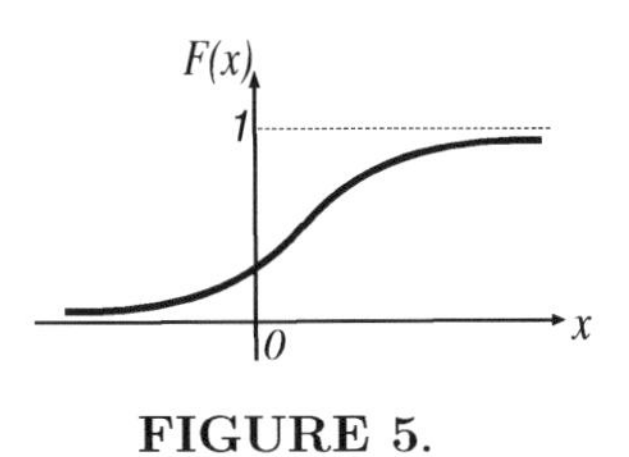

FIGURE 5.

For a r.v. X having a density $f(x)$, by the definition of a density,

$$F(x) = P(X \le x) = \int_{-\infty}^{x} f(u)du. \tag{1.2.1}$$

(Once we denote by x a limit of integration, we should use another letter inside the integral.)

From (1.2.1), it follows that the d.f. of a continuous r.v. is continuous (as an integral).

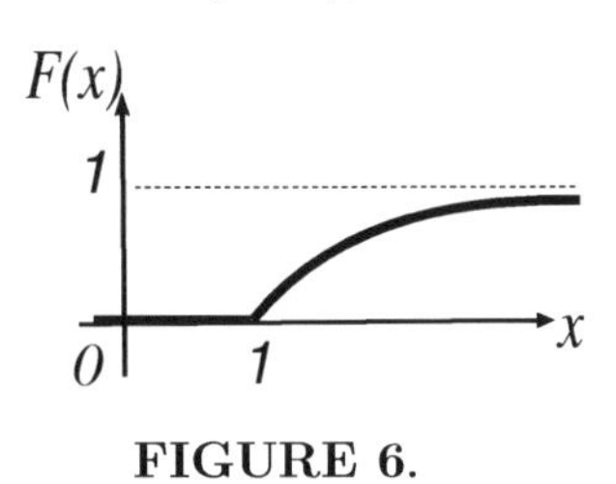

FIGURE 6.

EXAMPLE 1. Let us come back to the distribution in Example 1.1-2. Since $f(x) = 0$ for $x < 1$, the probability $P(X < 1) = 0$, and hence, the d.f. $F(x) = 0$ for $x < 1$ also. For $x \ge 1$, starting integrate from 1 (why?), we have

$$F(x) = \int_{1}^{x} \frac{1}{u^2} du = 1 - \frac{1}{x}.$$

The graph is given in Fig. 6. $\square$

From (1.2.1), it also follows that

$$f(x) = F'(x). \tag{1.2.2}$$

More precisely, (1.2.2) is true for all x's at which $f(x)$ is continuous since otherwise the derivative $F'(x)$ is not defined; see Example 2 below. However, as we will see in the same Example 2, it does not cause any difficulties.

EXAMPLE 2 (*the uniform distribution*). We come back to Example 1.1-3 and Fig. 4. Since $f(x) = 0$ for $x < a$, the d.f. $F(x) = 0$ for $x < a$. For $x \in [a, b]$, by (1.2.1),

$$F(x) = \int_{-\infty}^{x} f(u)du = \int_{a}^{x} \frac{1}{b-a} du = \frac{x-a}{b-a}.$$

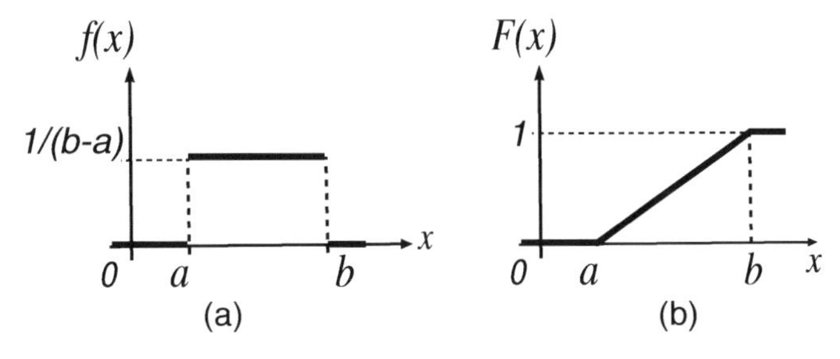

FIGURE 7. The density and the d.f. of a r.v. uniform on $[a, b]$.

For $x > b$, the density is again equal to zero, and hence, for $x \geq b$, the d.f. $F(x) = \int_a^x f(u)du = \int_{-\infty}^b f(u)du = 1$. The graph is depicted in Fig. 7b. In view of the importance of the uniform distribution, to have a complete picture, we added there the graph of the density in Fig. 7a.

We see in Fig. 7, that $F(x)$ is not differentiable at points a and b, and $f(x)$ is discontinuous at these points. However, both functions are simple, and regarding the points mentioned, *we can consider the right and left derivatives of $F(x)$ separately.*

EXAMPLE 3. We revisit John throwing a dart at a circular target with radius r in Examples **1**.1.1-7 and **1**.2.3-2. Let X be the distance the dart lands from the center. Then $P(X \leq x)$ is the probability that the dart will hit a point inside the disk with radius x and the center at the origin. As we defined in the examples mentioned, this probability is equal to the area of this disk divided by the total area of the target. This amounts to $\frac{\pi x^2}{\pi r^2} = \frac{x^2}{r^2}$. Thus, the d.f $F(x) = x^2/r^2$. Then, the density $f(x) = F'(x) = 2x/r^2$. □

Furthermore, for an interval [a, b], the probability $P(a \leq X \leq b) = P(X \leq b) - P(X < a)$. (We did not include the point a into the probability we *subtract*, because this point is included in $[a, b]$.) If the distribution is continuous, by virtue of (1.1.4), $P(X < a) = P(X \leq a)$. Therefore, for any r.v. X with a continuous d.f. $F(x)$, and any a and b,

$$P(a \leq X \leq b) = F(b) - F(a). \tag{1.2.3}$$

EXAMPLE 4. For a r.v. X having the Cauchy distribution from Example 1.1-5, find its d.f. and $P(-1 \leq X \leq 1)$. In view of (1.1.5), for all x's, the d.f.

$$F(x) = \int_{-\infty}^{x} \frac{du}{\pi(1+u^2)} = \frac{1}{\pi} \arctan |_{-\infty}^{x} = \frac{1}{\pi}\left(\arctan x - \left(-\frac{\pi}{2}\right)\right) = \frac{1}{\pi}\arctan x + \frac{1}{2}.$$

To find the desired probability, we could compute $\int_{-1}^{1} f(x)dx$, but once we have computed the d.f., we do not need to integrate:

$$P(-1 \leq X \leq 1) = F(1) - F(-1) = \frac{1}{\pi}(\arctan(1) - \arctan(-1)) = \frac{1}{\pi}\left(\frac{\pi}{4} - \left(-\frac{\pi}{4}\right)\right) = \frac{1}{2}. \ \square$$

1.3 Expected value

The notion of expected value and other characteristics we defined for discrete r.v.'s carry over to the continuous case. The transition is a matter of replacing the probabilities of the values of a discrete r.v. with the probability density of a continuous r.v., and summation with integration.

In Chapter 3, when defining the expected value of a discrete r.v., we multiplied a value x_i by its probability f_i, and considered the sum $\sum_i x_i f_i$. For a continuous r.v. X having a density $f(x)$, we view $f(x)dx$ as the probability that X will fall into an infinitesimally small interval $[x, x+dx]$. So, it is natural to multiply the value x by $f(x)dx$, and sum up over all x's. In the continuous case, summation amounts to integration, which leads to the definition of *expected value* as

$$E\{X\} = \int_{-\infty}^{\infty} xf(x)dx, \tag{1.3.1}$$

provided that this integral is finite. Sometimes, we will use also the term *"mean."*

EXAMPLE 1. We revisit Example 1.1-2; see also Fig. 2. Since $f(x) = 0$ for $x \notin [0,1]$,

$$E\{X\} = \int_0^1 x \cdot 2(1-x)dx = \frac{1}{3}.$$

The fact that the mean is less than the middle value $\frac{1}{2}$ is understandable: the values that belong to the first half of $[0,1]$ are more likely.

EXAMPLE 2. For the uniform distribution in Example 1.1-3, it is natural to expect that the mean value equals the middle value $(a+b)/2$. This is indeed true: $E\{X\} = \int_a^b x \cdot \frac{1}{b-a}dx = \frac{1}{b-a}\int_a^b x \cdot dx = \frac{1}{b-a}\frac{b^2-a^2}{2} = \frac{a+b}{2}$. □

The following formula may be helpful in computing the expectations of non-negative r.v.'s.

Proposition 1 *Let X be a non-negative r.v. with a d.f. $F(x)$. Then*

$$E\{X\} = \int_0^\infty (1-F(x))dx. \tag{1.3.2}$$

Proof. Remembering that X is non-negative, and changing the order of integration, we have

$$E\{X\} = \int_0^\infty uf(u)du = \int_0^\infty \left(\int_0^u dx\right) f(u)du = \int_0^\infty \left(\int_x^\infty f(u)du\right) dx$$
$$= \int_0^\infty \left(1 - \int_0^x f(u)du\right) dx = \int_0^\infty (1-F(x))dx. \blacksquare$$

In Chapter 7, we show that this formula is true for all (not necessarily continuous) non-negative r.v.'s. The reader who chose Route 2, may see that (1.3.2) is a continuous counterpart of formula (**3**.2.2.1) obtained for integer valued non-negative r.v.'s.

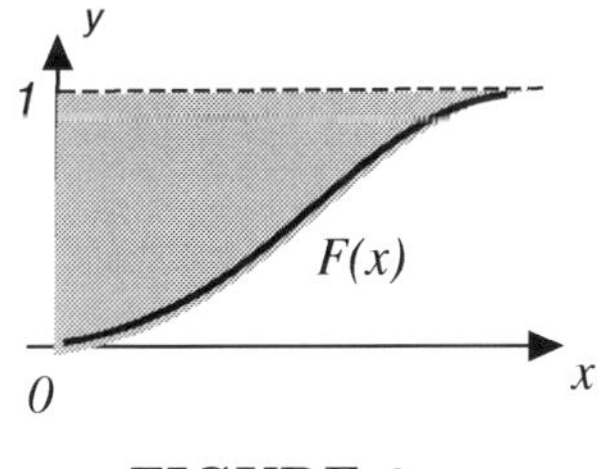

FIGURE 8.

Formula (1.3.2) has also a nice interpretation: it shows that the expected value equals the area between the graph of $y = F(x)$ and the line $y = 1$; see Fig. 8.

Another thing to note is that the integrand in (1.3.2), that is,

$$1 - F(x) = 1 - P(X \le x) = P(X > x). \tag{1.3.3}$$

In Section **3**.3.3, we called the function $\overline{F}(x) = 1 - F(x)$ the *tail of a distribution F.*

EXAMPLE 3 (the *Pareto distribution*). Suppose that for a positive r.v. X, its tail $\overline{F}(x) = 1/(1+x)^\alpha$ for $x \ge 0$ and a positive parameter α. This is a particular case of the Pareto distribution having many applications in various areas. We consider it in more detail in Exercise 9. To find the mean, we may compute the density $f(x) = F'(x) = -\overline{F}'(x)$, and then compute $\int_0^\infty xf(x)dx$. However, an easier way is to make use of (1.3.2), and write

$$E\{X\} = \int_0^\infty \frac{dx}{(1+x)^\alpha} = \frac{1}{\alpha - 1},$$

provided $\alpha > 1$. For $\alpha \le 1$, the above integral is infinite, and hence, so is $E\{X\}$. □

1.4 Transformations of r.v.'s

1.4.1 The distributions of transformations

Let X be a r.v. with a d.f. $F(x)$ and the density $f(x) = F'(x)$. How to find the same characteristics for a r.v. $Y = u(X)$, where $u(x)$ is a given function? We illustrate how it can be done in examples below.

EXAMPLE 1. Let X be uniform on $[0, \pi/2]$; say, we pick at random an angle between 0 and $\pi/2$. What is the distribution of $Y = \sin X$? To distinguish the d.f.'s of X and Y, we use the symbols $F_X(x)$ and $F_Y(y)$, respectively. From Example 1.2-2, it follows that $F_X(x) = \frac{2}{\pi} \cdot x$ for $x \in [0, \pi/2]$.

The function $\sin x$ is increasing on $[0, \pi/2]$, and Y assumes values from and only from $[0, 1]$. For $y \in [0, 1]$, the d.f. $F_Y(y) = P(Y \leq y) = P(\sin X \leq y) = P(X \leq \arcsin y) = F_X(\arcsin y) = \dfrac{2}{\pi} \cdot \arcsin y$. Also, $F_Y(y) = 0$ for $y < 0$ and $F_Y(y) = 1$ for $y > 1$.

The density $f_Y(y) = F'_Y(y) = \dfrac{2}{\pi\sqrt{1-y^2}}$ for $y \in [0, 1]$, and $f_Y(y) = 0$ otherwise. Thus,

$$F_Y(y) = \begin{cases} 0 & \text{if } y < 0, \\ \frac{2}{\pi} \arcsin y & \text{if } 0 \leq y \leq 1, \\ 1 & \text{if } y > 1. \end{cases} \quad \text{and} \quad f_Y(y) = \begin{cases} 0 & \text{if } y < 0, \\ \frac{2}{\pi\sqrt{1-y^2}} & \text{if } 0 \leq y \leq 1, \\ 0 & \text{if } y > 1. \end{cases}$$

The reader is invited to graph both functions. What is the $\lim_{y \to 1} f_Y(y)$?

EXAMPLE 2. Now, let X be uniform on $[-1, 1]$, and $Y = X^2$. Clearly, Y takes on values from $[0, 1]$. We have $F_Y(y) = P(X^2 \leq y) = P(-\sqrt{y} \leq X \leq \sqrt{y})$. We again may proceed from the d.f. F_X, but it is enough to observe that, since X is uniform, the last probability equals the ratio of the length of the interval $[-\sqrt{y}, \sqrt{y}]$ to the total length of $[-1, 1]$. Thus,

$$F_Y(y) = \frac{2\sqrt{y}}{2} = \sqrt{y} \quad \text{for} \quad y \in [0, 1].$$

Also, $F_Y(y) = 0$ for $y < 0$ and $F_Y(y) = 1$ for $y > 1$. The density $f_Y(y) = F'(y) = 1/(2\sqrt{y})$ for $y \in [0, 1]$; and $f_Y(y) = 0$, otherwise. Again, it is useful to sketch both graphs. □

1.4.2 Expectations

Next, we address the expected value of $Y = u(X)$. In Section **3**.2.1, we have shown that if X is discrete, and assumes values $x_1, x_2, ...$ with respective probabilities $f_1, f_2, ...$, then

$$E\{Y\} = E\{u(X)\} = \sum_i u(x_i) f_i. \tag{1.4.1}$$

So, it is natural to conjecture that if X has a density $f(x)$, then

$$E\{Y\} = E\{u(X)\} = \int_{-\infty}^{\infty} u(x) f(x) dx. \tag{1.4.2}$$

This is true, though it brings up questions. The main point is that we defined the notion of expectation for the discrete and continuous cases separately, but the r.v. $u(X)$ may turn out to be neither discrete nor continuous. (We discuss such r.v.'s later, in Chapter 7.) So, we do not know actually what $E\{u(X)\}$ is.

If $u(X)$ is a continuous or discrete r.v., then $E\{u(X)\}$ is defined, and (1.4.2) can be proved. In the former case, a proof may be based on (1.3.2); in the latter, it is straightforward; we discuss both cases below. However, we still do not know what $E\{u(X)\}$ is in the general case, so all of this looks somewhat messy.

In Chapter 7, we give a unified definition of expectation from which the particular definitions for the discrete and continuous cases will easily follow, and the same concerns (1.4.2). For now, we adopt (1.4.2) as a definition of $E\{u(X)\}$. From constructions of Chapter 7, it will follow that such a definition does not contradict the definitions of expectation we have already adopted.

► Nevertheless, to make (1.4.2) more plausible right now, let us derive it at least when Y is discrete, and when it is continuous and positive.

To see the point in the former case, it is enough to suppose that the function $y = u(x)$ takes on only two values: y_1 and y_2.

Let $D_1 = \{x : u(x) = y_1\}$, the set of all x's for which $u(x) = y_1$, and $D_2 = \{x : u(x) = y_2\}$. Then, by definition,

$$\begin{aligned} E\{Y\} &= y_1 P(Y = y_1) + y_2 P(Y = y_2) = y_1 P(X \in D_1) + y_2 P(X \in D_2) \\ &= y_1 \int_{D_1} f(x)dx + y_2 \int_{D_2} f(x)dx = \int_{D_1} u(x)f(x)dx + \int_{D_2} u(x)f(x)dx = \int_{-\infty}^{\infty} u(x)f(x)dx. \end{aligned}$$

Now, let $u(x) \geq 0$ for all x's, and the (non-negative) $Y = u(X)$ has a density $f_Y(y)$. Then, $E\{Y\}$ is well defined, and using Proposition 1 and changing the order of integration, we have

$$\begin{aligned} E\{Y\} &= \int_0^{\infty} (1 - F_Y(y))dy = \int_0^{\infty} P(Y > y)dy \\ &= \int_0^{\infty} \left(\int_{x:u(x)>y} f(x)dx \right) dy = \int_{-\infty}^{\infty} \left(\int_0^{u(x)} dy \right) f(x)dx = \int_{-\infty}^{\infty} u(x)f(x)dx. \end{aligned}$$ ◄

EXAMPLE 1. For Y in Example 1.4.1-1,

$$E\{Y\} = E\{\sin X\} = \int_{-\infty}^{\infty} (\sin x)f(x)dx = \int_0^{\pi/2} (\sin x)\frac{2}{\pi}dx = \frac{2}{\pi}\int_0^{\pi/2} \sin x dx = \frac{2}{\pi}.$$

EXAMPLE 2. In Example 1.4.1-2,

$$E\{Y\} = E\{X^2\} = \int_{-1}^{1} x^2 \frac{1}{2}dx = \frac{1}{3}. \ \square$$

EXAMPLE 3 (*An inventory problem*). A company should order in advance an amount of a product for the company's future needs. The amount that will be actually needed is not known, so we view it as a r.v. X. Suppose that proceeding from the past experience, we can determine the distribution of X. For simplicity, suppose that this distribution is continuous, and denote by $F(x)$ and $f(x)$ the d.f. and density, respectively.

Let c_1 be the expenses per unit of the product connected with purchasing and storing, and c_2 be the expenses per unit connected with the shortage of the product (if the company had ordered an amount that has turned out to be less than it is needed). The problem is meaningful if $c_2 > c_1$; otherwise the company should not purchase the product at all. What order is minimizing the mean cost?

Let s be the amount ordered. Then the total cost is

$$C = C(X) = c_1 s + \begin{cases} 0 & \text{if } X \le s, \\ c_2(X-s) & \text{if } X > s. \end{cases}$$

The expected cost is $E\{C\} = c_1 s + c_2 \int_s^\infty (x-s) f(x)dx$. Differentiating in s, we have

$$\frac{d}{ds}E\{C\} = c_1 + c_2\left(-0 \cdot f(s) - \int_s^\infty f(x)dx\right) = c_1 - c_2 P(X > s)$$
$$= c_1 - c_2(1 - F(s)) = c_1 - c_2 + c_2 F(s).$$

The derivative equals zero for s such that

$$F(s) = 1 - \frac{c_1}{c_2} = 1 - \delta, \qquad (1.4.3)$$

where $\delta = (c_1/c_2) < 1$ because we have assumed $c_2 > c_1$. In Exercise 10, it is suggested to consider two particular cases. □

Since the notion of expectation $E\{u(X)\}$ has been defined, the definitions of various characteristics of r.v.'s such as variance, standard deviation, moments of higher order, carry over from the discrete case to the continuous case in a straightforward fashion.

For example, exactly as in Section **3**.3, we define the variance of a r.v. X as

$$Var\{X\} = E\{(X - E\{X\})^2\} = E\{X^2\} - (E\{X\})^2.$$

EXAMPLE 4. Let X be uniform on $[0, 1]$, so the density $f(x) = 1$ for $x \in [0, 1]$. By (1.4.2),

$$E\{X\} = \int_0^1 x \cdot 1dx = \frac{1}{2}, \quad E\{X^2\} = \int_0^1 x^2 \cdot 1dx = \frac{1}{3},$$

and hence,

$$VarX = E\{X^2\} - (E\{X\})^2 = \frac{1}{3} - \left(\frac{1}{2}\right)^2 = \frac{1}{12}. \; \square \qquad (1.4.4)$$

Since the characteristics mentioned above have been defined for the continuous case, the general theorems of Chapter 3 pertaining to these characteristics and not dealing with discrete distributions specially, also carry over from the discrete to the continuous case. In particular, this concerns:

1. All properties of expectation and variance;
2. Inequalities for deviations including Markov's and Chebyshev's inequalities;
3. The definitions of convergence of r.v.'s; and
4. The law of large numbers.

The statements of these theorems and propositions are the same as in Chapter 3, and we do not repeat them.

The proofs of the theorems and propositions mentioned were specially designed in Chapter 3 in a way that they could be carried out for the continuous case by substituting integration for summation. On the other hand, the statements of these theorems in Chapter 3 did not involve references to the discrete nature of r.v.'s.

1.5 Linear transformation of r.v.'s. Normalization

Let X be a r.v. with a d.f. $F(x) = P(X \leq x)$. Denote by $f(x)$, m, and σ^2 the density, the expected value, and the variance of X, respectively. Let a r.v. $Y = a + bX$, where a and b are numbers, $b > 0$. Then the d.f. of Y is the function

$$F_Y(x) = P(Y \leq x) = P(a + bX \leq x) = P(X \leq (x-a)/b) = F((x-a)/b).$$

If the density $f(x)$ exists, then there exists the density of Y equal to $f_Y(x) = \frac{d}{dx}F_Y(x) = \frac{1}{b}f((x-a)/b)$. Using Property 4 in p.68, and Property (**3**.3.2.2) in p.78 for means and variances, we summarize all of this in the following table.

	the d.f.	the density	the mean	the variance
X	$F(x)$	$f(x)$	m	σ^2
$a+bX,\ b>0$	$F((x-a)/b)$	$\frac{1}{b}f((x-a)/b)$	$a+bm$	$b^2\sigma^2$

(1.5.1)

A memorizing rule is: if we multiply a r.v. by a number, then we divide the argument of the d.f. by this number; if we add a number to a r.v., then we subtract this number from the argument of the d.f.

Let $X' = X - m$. Then

$$E\{X'\} = E\{X\} - m = m - m = 0.$$

The r.v. X' is called *centered*, and the operation itself—*centering*.

Next, assuming $\sigma > 0$, consider the r.v.

$$X^* = \frac{X-m}{\sigma}. \tag{1.5.2}$$

Clearly, $E\{X^*\} = \frac{1}{\sigma}E\{X'\} = 0$, and $Var\{X^*\} = \frac{1}{\sigma^2}Var\{X\} = \frac{\sigma^2}{\sigma^2} = 1$. So,

$$E\{X^*\} = 0,\ Var\{X^*\} = 1. \tag{1.5.3}$$

We call such a r.v. *normalized*, and the operation (1.5.2)—standard normalization, or simply *normalization*. We will use this operation repeatedly. It is worth emphasizing that X^* may be viewed as the same r.v. X considered, so to speak, in a standard universal scale.

EXAMPLE 1. Bob ran 100 meters sprint in 12.1 seconds, and 5000 meters in 26.2 minutes. When did Bob do better? The question is meaningless unless we know the standards for a group to which Bob belongs. Let X and Y be the corresponding (random!) results for a person chosen at random from the group mentioned at a randomly chosen day. Suppose that for X (100m sprint), the mean and standard deviation are 12 secs and 1.2 secs, respectively, and for Y (the long distance run), these figures are 25.1 mins and 4.9 mins. If we proceed from these characteristics, we may compare Bob's results, normalizing them in accordance with (1.5.2). The corresponding normalized values are $x^* = \frac{12.1-12}{1.2} \approx 0.08$ and $y^* = \frac{26.2-25.1}{4.9} \approx 0.22$. So, in the sprint, Bob ran essentially better. □

---

2 SOME BASIC CONTINUOUS DISTRIBUTIONS

In this section, we list some distributions playing an important role in theory and applications. A summarizing table is given in the Appendix, Table 1.

2.1 The uniform distribution

We have been already considering the uniform on $[a, b]$ distribution in Examples 1.1-3, 1.2-1. Summarizing, the density and d.f are given by

$$f(x) = \begin{cases} \frac{1}{b-a} & \text{if } x \in [a,b], \\ 0 & \text{if } x \notin [a,b]; \end{cases} \quad F(x) = \begin{cases} 0 & \text{if } x < a, \\ \frac{x-a}{b-a} & \text{if } a \leq x \leq b, \\ 1 & \text{if } x > b; \end{cases} \tag{2.1.1}$$

see also Fig. 7.

Note also that a r.v. X uniformly distributed on $[a,b]$, may be represented as $X = a + (b-a)Z$, where a r.v. Z is uniformly distributed on $[0,1]$. Formally, it follows, for example, from (1.5.1), but it may be seen without calculations. If Z assumes values from $[0,1]$, then X takes on values from $[a,b]$. If values of Z from $[0,1]$ are equally likely, the same is true for values of X from $[a,b]$.

From the last fact, it follows that if we have a generator of random numbers from $[0,1]$, we do not need a special generator for simulating values of X. We would simulate values of Z and apply the linear transformation $a + (b-a)Z$.

In Example 1.3-2, we have shown that $E\{X\} = \frac{a+b}{2}$. In Example 1.4.2-4, we have computed that for the unit interval $[0,1]$, the variance equals $1/12$. Hence, in the general case,

$$E\{X\} = \frac{a+b}{2}, \quad Var\{X\} = (b-a)^2 Var\{Z\} = \frac{(b-a)^2}{12}. \tag{2.1.2}$$

It is worthwhile to warn the reader against the following widespread but incorrect reasoning.

Assume that we know that a r.v. ξ takes on values from, say, $[0,1]$, but we have no additional information about the distribution of ξ. Since we equally do not know anything about the likelihood of values from $[0,1]$, we may consider these values equally likely.

Hence, we can assume that ξ is uniform. Then, by (2.1.1), the d.f. of ξ is $F_\xi(x) = x$ for $x \in [0,1]$.

But the r.v. ξ^2 also takes on values from $[0,1]$, and we do not have any additional information about its distribution as well. Then, on the same grounds, we can set $F_{\xi^2}(x) = x$ and write $x = F_\xi(x) = P(\xi \leq x) = P(\xi^2 \leq x^2) = F_{\xi^2}(x^2) = x^2$. So, we have arrived at a false assertion that $x = x^2$.

The above reasoning was faulty because the absence of information does not allow us to jump to any particular conclusion about the distribution of ξ. Knowing nothing means that we cannot say anything about the distribution except that it is concentrated on $[0,1]$. Whereas the assertion on uniformity should be based on the rather concrete information that all values from $[0,1]$ are *equally* likely.

2.2 The exponential distribution

We call a continuous *non-negative* r.v. X_1 and its distribution *standard exponential* if the corresponding density $f_1(x) = e^{-x}$ for $x \geq 0$, and $f_1(x) = 0$ for $x < 0$. It is straightforward to compute that $E\{X_1\} = \int_0^\infty xe^{-x}dx = 1$, $E\{X_1^2\} = \int_0^\infty x^2e^{-x}dx = 2$. (The reader may calculate it by parts or just look at a table of integrals in her/his Calculus book.) Hence, $Var\{X_1\} = 2 - 1^2 = 1$.

Consider now the r.v. $X = X_a = X_1/a$ for a positive a. In accordance with (1.5.1), the density of X_a is

$$f_a(x) = \begin{cases} 0 & \text{if } x < 0, \\ ae^{-ax} & \text{if } x \geq 0, \end{cases} \tag{2.2.1}$$

and

$$E\{X_a\} = 1/a,\ Var\{X_a\} = 1/a^2. \tag{2.2.2}$$

Such a r.v. and its distribution are called *exponential* with a parameter a which, as we see, is a scale parameter: $X_a = X_1/a$.

Substituting the density $f_a(x)$ into (1.2.1), we readily get that the d.f.

$$F_a(x) = \begin{cases} 0 & \text{if } x < 0, \\ 1 - e^{-ax} & \text{if } x \geq 0. \end{cases} \tag{2.2.3}$$

Consequently,

$$P(X_a > x) = e^{-ax}, \tag{2.2.4}$$

from which the term "exponential" comes.

The exponential distribution has the unique

Lack-of-Memory (or Memoryless) Property: for any $x, y \geq 0$,

$$P(X > x + y \,|\, X > x) = P(X > y). \tag{2.2.5}$$

Indeed,

$$P(X > x + y \,|\, X > x) = \frac{P(X > x + y,\, X > x)}{P(X > x)} = \frac{P(X > x + y)}{P(X > x)}.$$

Then, in view of (2.2.4), $P(X > x+y\,|\,X > x) = \dfrac{e^{-a(x+y)}}{e^{-ax}} = e^{-ay} = P(X > y)$.

The property (2.2.5) is the continuous counterpart of a similar property for the geometric distribution in Section **3**.4.3. However, while in the discrete case, (2.2.5) was true for integer x and y, now it holds for all x, y.

The main point in (2.2.5) is that, while its l.-h.s. involves x, the r.-h.s. does not. Thus, given X has exceeded a certain level x, the overshoot over x has the same distribution as the r.v. X itself, and does not depend on the particular level x that has been exceeded.

Suppose, for example, that X is the duration of a job to be done. Assume that the job has already lasted x hours. Then the l.-h.s. of (2.2.5) is the conditional probability that the job will last y hours *more*. In the case of the memoryless property, this conditional probability does not depend on when the job began. One may view the situation as if at each moment the process starts over as from the beginning—so to say, the system "does not remember" what happened before.

The property under discussion is special, and it is important to keep in mind that

> The exponential distribution is the only distribution with the memoryless property.

We skip the proof; it may be found, e.g., in [13], [39], [42].

EXAMPLE 1. Ann is the first in line for a service. There are two service counters, and both are busy. Ann has noticed that one customer who was being served, was her friend Mary. Assume that the service times for both counters are independent exponential r.v.'s with the same mean. What is the probability that Ann will have been served before Mary?

The service times have the same distribution. (Since the mean is the same, the parameter a is also the same.) Due to the memoryless property, at each moment, "the process starts over as from the beginning." Then, regardless who started the service first, Mary and the customer at the other counter have equal chances to leave first. Hence, the probability, that Ann will start service when Mary is still being served is $1/2$.

If it happens, the process will again start over, and Mary and Ann will have equal chances to finish first. For Ann, the probability of this, is again $1/2$. So, the probability of interest is $\frac{1}{2} \cdot \frac{1}{2} = \frac{1}{4}$.

EXAMPLE 2. The duration X of a service is an exponential r.v., and the probability that the service will last more than 10 mins is 0.4. What is the probability that the service will last more that 20 mins? To answer this question, we do not have to compute the parameter a of the distribution: in accordance with (2.2.4), $P(X > 2x) = e^{-a2x} = (e^{-ax})^2 = (P(X > x))^2$, so the probability of interest is $(0.4)^2 = 0.16$. □

Route 1 ⇒ page 181

As was told, for a r.v. ξ different from an exponential, $P(\xi > x+y\,|\,\xi > x)$ depends on x. Let us consider it in more detail.

An important case is that where x is "large"; formally, we let $x \to \infty$ and set

$$Q(y) = \lim_{x\to\infty} P(\xi > x+y\,|\,\xi > x) = \lim_{x\to\infty} \frac{P(\xi > x+y,\, \xi > x)}{P(\xi > x)} = \lim_{x\to\infty} \frac{P(\xi > x+y)}{P(\xi > x)}.$$

We will consider three typical examples.

(a) $P(\xi > x) = (1+x)^{-\alpha}$ for an $\alpha > 0$. Dividing in the second step, the numerator and denominator by x, we have

$$Q(y) = \lim_{x\to\infty} \frac{(1+x)^\alpha}{(1+x+y)^\alpha} = \lim_{x\to\infty} \left(\frac{1+1/x}{1+(1+y)/x}\right)^\alpha = 1 \text{ for } any \ y.$$

So, for large x, the probability $P(\xi > x+y \,|\, \xi > x)$ is close to one for an *arbitrary large (!)* level y.

If, for example, X is the duration of a job, this means that, if the job has lasted for a long time, "with all probability" it will not be over soon (say, we may suspect that "something serious" has happened).

In this case, the *tail* $P(\xi > x)$ is classified as *heavy*.

(b) $P(\xi > x) = e^{-ax}$. This is an exponential case, and $Q(y) = e^{-ay}$; that is, the future duration of the job does not depend on how long it has already lasted.

(c) $P(\xi > x) = e^{-ax^\gamma}$, where $\gamma > 1$. In this case, the tail vanishes faster than any exponential function, and $Q(y) = 0$ for any y. For simplicity, let us show this for $\gamma = 2$. We have

$$Q(y) = \lim_{x\to\infty} \exp\{-a[(x+y)^2 - x^2]\} = \lim_{x\to\infty} \exp\{-a[2xy + y^2]\} = 0 \quad \text{for } any \quad y > 0.$$

So for large x, the probability $P(\xi > x+y \,|\, \xi > x)$ is close to zero for any, *arbitrary small (!)*, y. In the job example, this means that if the job has lasted for a long time, we may expect that it will end soon. (For the general case $\gamma > 1$, one may write $(x+y)^\gamma - x^\gamma = \left(\left(1+\frac{y}{x}\right)^\gamma - 1\right) \Big/ \left(\frac{1}{x}\right)^\gamma$ and apply L'Hôpital's rule.)

The cases (b) and (c) are classified as those of *light tails*.

2.3 The Γ(gamma)-distribution

1,2

First, for all $\nu > 0$, we define the function

$$\Gamma(\nu) = \int_0^\infty x^{\nu-1} e^{-x} dx,$$

which is called the *Γ(gamma)-function.*

The value $\Gamma(1) = \int_0^\infty e^{-x} dx = 1$. Using integration by parts, we have

$$\Gamma(\nu+1) = -\int_0^\infty x^\nu d(e^{-x}) = \int_0^\infty \nu x^{\nu-1} e^{-x} dx = \nu\Gamma(\nu), \tag{2.3.1}$$

because $\lim_{x\to\infty} x^\nu e^{-x} = 0$, and hence, $x^\nu e^{-x}|_0^\infty = 0$. Then for an integer $k \geq 1$, we have $\Gamma(k+1) = k\Gamma(k) = k(k-1)\Gamma(k-1) = ... = k(k-1)\cdot ... \cdot 1 \cdot \Gamma(1) = k!$. Thus, for $k = 0, 1, ...,$

$$\Gamma(k+1) = k!.$$

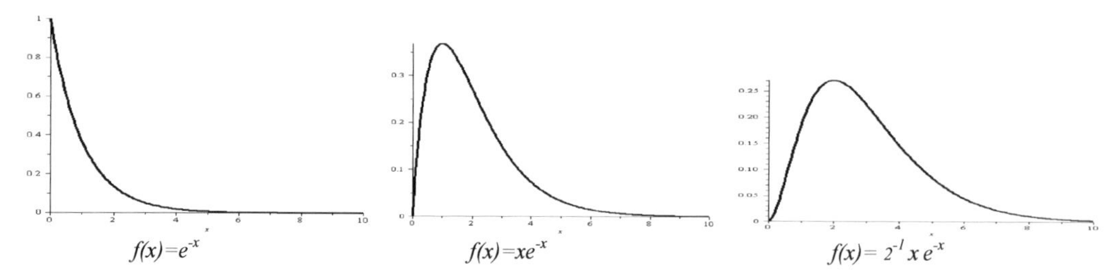

FIGURE 9. The Γ-densities for $\nu = 1,2,3$.

So, the Γ-*function may be viewed as a generalization of the notion of factorial.*

Consider a continuous r.v. $X_{1\nu}$ whose density is the function

$$f_{1\nu}(x) = \frac{1}{\Gamma(\nu)} x^{\nu-1} e^{-x} \text{ for } x \geq 0, \text{ and } = 0 \text{ otherwise.}$$

Due to $\Gamma(\nu)$ in the denominator, $\int_0^\infty f_{1\nu}(x)dx = \Gamma(\nu)/\Gamma(\nu)$, and is indeed equal to one.

For $\nu = 1$, the above density is standard exponential. In general, the parameter ν characterizes the type of the distribution. In Fig. 9, we demonstrate how this type depends on ν by sketching the graphs of $f_{1\nu}(x)$ for $\nu = 1,2,3$.

By virtue of (2.3.1), for a positive integer k, the kth moment

$$\begin{aligned} E\{X_{1\nu}^k\} &= \frac{1}{\Gamma(\nu)} \int_0^\infty x^k x^{\nu-1} e^{-x} dx = \frac{1}{\Gamma(\nu)} \int_0^\infty x^{k+\nu-1} e^{-x} dx \\ &= \frac{\Gamma(\nu+k)}{\Gamma(\nu)} = \frac{(\nu+k-1)\Gamma(\nu+k-1)}{\Gamma(\nu)} = \frac{(\nu+k-1)(\nu+k-2)\Gamma(\nu+k-2)}{\Gamma(\nu)} = \dots \\ &= \frac{(\nu+k-1)\cdot \ldots \cdot \nu\Gamma(\nu)}{\Gamma(\nu)} = (\nu+k-1)\cdot \ldots \cdot \nu, \end{aligned}$$

In particular,

$$E\{X_{1\nu}\} = \nu, \; E\{X_{1\nu}^2\} = (\nu+1)\nu, \text{ and hence, } Var\{X_{1\nu}\} = E\{X_{1\nu}^2\} - (E\{X_{1\nu}\})^2 = \nu. \tag{2.3.2}$$

Now, let $a > 0$, and the r.v. $X_{a\nu} = X_{1\nu}/a$. By (1.5.1), the density of $X_{a\nu}$ is

$$f_{a\nu}(x) = af_{a\nu}(ax) = \frac{a^\nu}{\Gamma(\nu)} x^{\nu-1} e^{-ax} \text{ for } x \geq 0, \text{ and } = 0 \text{ otherwise.} \tag{2.3.3}$$

The density $f_{a\nu}(x)$ is called the Γ-density with a *scale parameter a* and a parameter ν that may be called *essential* since it specifies the type of the distribution.

Note that *the exponential distribution is a particular case of the* Γ-*distribution* corresponding to $\nu = 1$.

From (2.3.2), it follows that

$$E\{X_{a\nu}\} = \frac{\nu}{a}, \; E\{X_{a\nu}^2\} = \frac{(\nu+1)\nu}{a^2}, \text{ and } Var\{X_{a\nu}\} = \frac{\nu}{a^2}. \tag{2.3.4}$$

The respective ***Excel commands*** for $P(X_{a\nu} \leq x)$ and the density $f_{a\nu}(x)$ are

=GAMMADIST(x, ν, $1/a$,TRUE) and =GAMMADIST(x, ν, $1/a$, FALSE).

We will see later how often the Γ-distribution proves to be useful in modeling phenomena of different nature.

2.4 The normal distribution

Even the name of this distribution points out an important role which it plays in theory and applications. It is also called Gaussian, and even the bell distribution (to emphasize the shape of its density curve), but the last term is not used in Probability Theory itself.

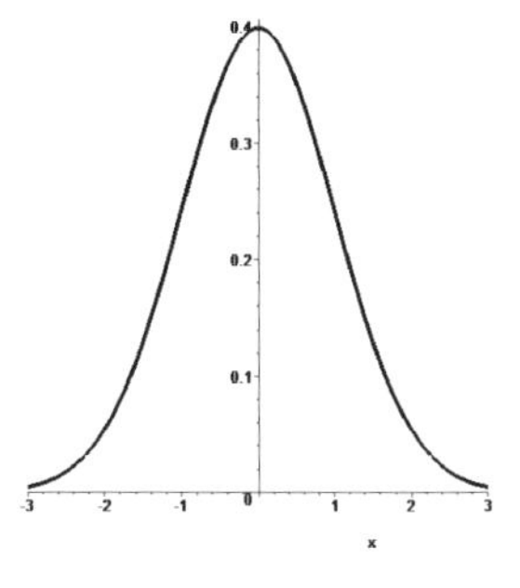

FIGURE 10.

A r.v. X and its distribution are called *standard normal* if the corresponding density is

$$\varphi(x) = \frac{1}{\sqrt{2\pi}} \exp\{-x^2/2\}. \tag{2.4.1}$$

The function $\varphi(x)$ is called the standard normal density; its graph for $-3 \leq x \leq 3$ is presented in Fig. 10.

▶ To show that $\int_{-\infty}^{\infty} \varphi(x)dx = 1$, we should prove that the integral $I = \int_{-\infty}^{\infty} e^{-x^2/2}dx = \sqrt{2\pi}$. Switching to the polar coordinates (r, θ) in the double integral below and setting $s = r^2/2$ at the very end, we have

$$I^2 = I \cdot I = \left(\int_{-\infty}^{\infty} e^{-x^2/2}dx\right)\left(\int_{-\infty}^{\infty} e^{-y^2/2}dy\right) = \int_{-\infty}^{\infty}\int_{-\infty}^{\infty} e^{-(x^2+y^2)/2}dxdy$$
$$= \int_0^{2\pi}\int_0^{\infty} e^{-r^2/2}rdrd\theta = \left(\int_0^{2\pi} d\theta\right)\int_0^{\infty} e^{-r^2/2}d(r^2/2) = 2\pi\int_0^{\infty} e^{-s}ds.$$

Since the last integral equals one, $I^2 = 2\pi$. ◀

Because the density $\varphi(x)$ is an even function, the integrand in the integral $\int_{-\infty}^{\infty} x\varphi(x)dx$ is an odd function, and hence this integral equals zero. This implies $E\{X\} = 0$.

For the variance, integrating by parts in an intermediate step, we write

$$Var\{X\} = E\{X^2\} = \int_{-\infty}^{\infty} x^2\varphi(x)dx = \frac{1}{\sqrt{2\pi}}\int_{-\infty}^{\infty} x^2 e^{-x^2/2}dx$$
$$= \frac{1}{\sqrt{2\pi}}\int_{-\infty}^{\infty} xd(-e^{-x^2/2}) = \frac{1}{\sqrt{2\pi}}\int_{-\infty}^{\infty} e^{-x^2/2}dx.$$

(The limits in the first part of the integration-by-parts operation equal zero.) We proved above that the last integral is equal to $\sqrt{2\pi}$. Hence, $E\{X^2\} = 1$. Thus,

$$E\{X\} = 0, \; Var\{X\} = E\{X^2\} = 1, \tag{2.4.2}$$

from which the term "standard" comes.

For an arbitrary m and $\sigma > 0$, the r.v. $Y = m + \sigma X$ and its distribution are called (m, σ^2)-*normal*. In view of (2.4.2) and (1.5.1), $E\{Y\} = m$, $Var\{Y\} = \sigma^2$, which justifies the choice of the notation m and σ^2. By the same rule (1.5.1), the density of Y is the function

$$\varphi_{m\sigma}(x) = \frac{1}{\sqrt{2\pi}\sigma} \exp\left\{-\frac{(x-m)^2}{2\sigma^2}\right\}. \tag{2.4.3}$$

The importance of the normal distribution is connected, first of all, with the so-called Central Limit Theorem (CLT) we consider in Chapter 9. Loosely put, the CLT says the following.

Consider the sum $X_1 + ... + X_n$ of independent r.v.'s X_i having the same (arbitrary) distribution with a finite variance. Then, for large n, the distribution of the sum is close to the normal distribution with the mean and variance equal to the mean and variance of the sum. We discuss this in detail in Chapter 9.

Let us come back to the normal distribution itself. In accordance with (1.2.1), the d.f. of a standard normal r.v. X is equal to

$$\Phi(x) = \frac{1}{\sqrt{2\pi}} \int_{-\infty}^{x} e^{-u^2/2} du.$$

The function $\Phi(x)$ is called the *standard normal d.f.* The integral above cannot be computed analytically, but any particular value of $\Phi(x)$ may be computed to a high degree of accuracy using the numerical integration technique.

In view of the importance of the normal distribution, practically all textbooks on Probability are provided with a table for the standard normal d.f. $\Phi(x)$. In this book, it is Table 3 in the Appendix. The first column there gives the integer value of x and the first digit after the point, and the other columns correspond to the second digit. For example, to find $\Phi(1.23)$, one should choose the row starting with 1.2 and go to the column with 0.03 at the top. The table gives $\Phi(1.23) = 0.8907$. Certainly, this is an approximate value.

The reader may notice that the table deals only with positive x's. It is enough because, by virtue of the symmetry of the standard normal distribution, $P(X > x) = P(X < -x)$, and hence,

$$\Phi(-x) = 1 - \Phi(x).$$

Say, $\Phi(-1.23) = 1 - \Phi(1.23) \approx 1 - 0.8907 = 0.1093$.

Table 4 in the Appendix gives the values of the inverse function $\Phi^{-1}(y)$ for some most frequently used values of y. The values $\Phi^{-1}(y)$ are also called y-quantiles (for the normal distribution); we discuss it in more detail in Chapters 7 and 18.

Using (1.5.1) one more time, we get that the d.f. of $Y = m + \sigma X$ is the function

$$\Phi_{m\sigma}(x) = \Phi\left(\frac{x-m}{\sigma}\right). \tag{2.4.4}$$

The respective ***Excel commands*** for the d.f. (2.4.4) and the density (2.4.3) are
=NORMDIST(x, m, σ,TRUE) and =NORMDIST(x, m, σ, FALSE).

Route 1 ⇒ page 185

As was mentioned, the standard normal d.f. $\Phi(x)$ cannot be presented in terms of elementary functions. Nevertheless, there are good approximations for the tail of this distribution. First, we present the following asymptotic approximation:

$$\Phi(-x) = 1 - \Phi(x) \sim \frac{1}{x}\varphi(x) = \frac{1}{\sqrt{2\pi}x} e^{-x^2/2} \text{ as } x \to \infty, \tag{2.4.5}$$

where $\varphi(x)$ is the standard normal density. (For functions $a(x)$, $b(x)$, the symbolism $a(x) \sim b(x)$ means that $\frac{a(x)}{b(x)} \to 1$.)

Relation (2.4.5) follows from a more precise

Proposition 2 *For any* $x > 0$,

$$\left(\frac{1}{x} - \frac{1}{x^3}\right)\varphi(x) \le 1 - \Phi(x) = \Phi(-x) \le \frac{1}{x}\varphi(x). \tag{2.4.6}$$

Proof. It is straightforward to verify that

$$\left(\frac{1}{x}\varphi(x)\right)' = -\left(1 + \frac{1}{x^2}\right)\varphi(x), \text{ and } \left(\left(\frac{1}{x} - \frac{1}{x^3}\right)\phi(x)\right)' = -\left(1 - \frac{3}{x^4}\right)\varphi(x).$$

Then, for $x > 0$,

$$1 - \Phi(x) = \int_x^\infty \varphi(y)dy \le \int_x^\infty \left(1 + \frac{1}{y^2}\right)\varphi(y)dy = -\int_x^\infty \left(\frac{1}{y}\varphi(y)\right)' dy = \frac{1}{x}\varphi(x).$$

Similarly,

$$1 - \Phi(x) = \int_x^\infty \varphi(y)dy \ge \int_x^\infty \left(1 - \frac{3}{y^4}\right)\varphi(y)dy = \left(\frac{1}{x} - \frac{1}{x^3}\right)\varphi(x). \blacksquare$$

---

3 CONTINUOUS MULTIVARIATE DISTRIBUTIONS

This section is concerned with random vectors (r.vec.'s). To make the exposition more explicit, we restrict ourselves to the two-dimensional case; the general case is treated similarly. The reader will see a straight analogy with the discrete case.

Let $\mathbf{X} = (X_1, X_2)$, where X_1 and X_2 are r.v.'s. The function of sets $F(B) = F_{\mathbf{X}}(B) = P(\mathbf{X} \in B)$, where now B is a set in the two-dimensional space $\mathbb{R}^2$, is called the distribution of the r.vec. $\mathbf{X}$. We will omit the index $\mathbf{X}$ in $F_{\mathbf{X}}(B)$ if this cannot cause confusion.

A r.vec. $\mathbf{X}$ and its distribution $F(B)$ are said to be *absolutely continuous* (briefly, continuous) if there exists a non-negative function $f(\boldsymbol{x})$, where $\boldsymbol{x} = (x_1, x_2)$, called a *joint density* and such that for any B from $\mathbb{R}^2$,

$$F(B) = \iint\limits_B f(\boldsymbol{x})d\boldsymbol{x} = \iint\limits_B f(x_1, x_2)dx_1dx_2. \tag{3.1}$$

The integral in (3.1) is a two-dimensional integral, and the differential $d\boldsymbol{x} = dx_1dx_2$. Setting $B = \mathbb{R}^2$ we have

$$\iint\limits_{\mathbb{R}^2} f(\boldsymbol{x})d\boldsymbol{x} = 1. \tag{3.2}$$

EXAMPLE 1. Let

$$f(\boldsymbol{x}) = 4(2 - x_1^2 - x_2^2)x_1x_2 \text{ for } 0 \le x_1 \le 1, 0 \le x_2 \le 1,$$

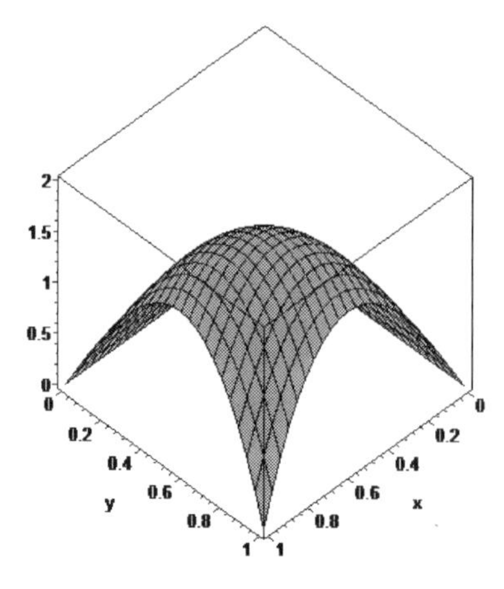

FIGURE 11.

and $f(\boldsymbol{x}) = 0$ otherwise. The graph of a density in the bivariate case is a surface in a three dimensional system of coordinates. The particular function above is graphed for $0 \le x_1 \le 1, 0 \le x_2 \le 1$ in Fig. 11.

The reader is invited to double check that (3.2) holds:

$$\iint_{\mathbb{R}^2} f(\boldsymbol{x})d\boldsymbol{x} = \int_0^1 \int_0^1 4(2 - x_1^2 - x_2^2)x_1 x_2 dx_1 dx_2 = 1.$$

Let us compute, for instance, $P(2X_2 \le X_1)$. Skipping elementary calculations, we have

$$P(2X_2 \le X_1) = \iint_{2x_2 \le x_1} f(\boldsymbol{x})d\boldsymbol{x} = \int_0^1 \left(\int_0^{x_1/2} 4(2 - x_1^2 - x_2^2)x_1 x_2 dx_2 \right) dx_1 = \frac{5}{32}. \ \square$$

Consider the *marginal distributions*; that is, the distributions of X_1 and X_2. If the vector $\mathbf{X}$ is continuous, its coordinates are continuous, and similar to the discrete case, the (marginal) densities of the r.v.'s X_1 and X_2 are given by

$$f_1(x_1) = \int_{-\infty}^{\infty} f(x_1, x_2)dx_2, \ \ f_2(x_2) = \int_{-\infty}^{\infty} f(x_1, x_2)dx_1;$$

compare with formulas $f_i^{(1)} = \sum_j f_{ij}, \ f_j^{(2)} = \sum_i f_{ij}$ in the discrete case (Section **3**.1.2).

Formally, it may be shown as follows. For any set D in the *real line*,

$$P(X_1 \in D) = \int_{-\infty}^{\infty} \int_D f(x_1, x_2)dx_1 dx_2 = \int_D \left(\int_{-\infty}^{\infty} f(x_1, x_2)dx_2 \right) dx_1.$$

So, if we denote by $f_1(x_1)$ the interior integral, we will have $P(X_1 \in D) = \int_D f_1(x_1)dx_1$ for any D. Hence, $f_1(x_1)$ is the density of X_1 by the definition of a density.

EXAMPLE 1 revisited. We have $f_1(x_1) = \int_0^1 4(2 - x_1^2 - x_2^2)x_1 x_2 dx_2 = 3x_1 - 2x_1^3$, and similarly, $f_2(x_2) = 3x_2 - 2x_2^3$. $\square$

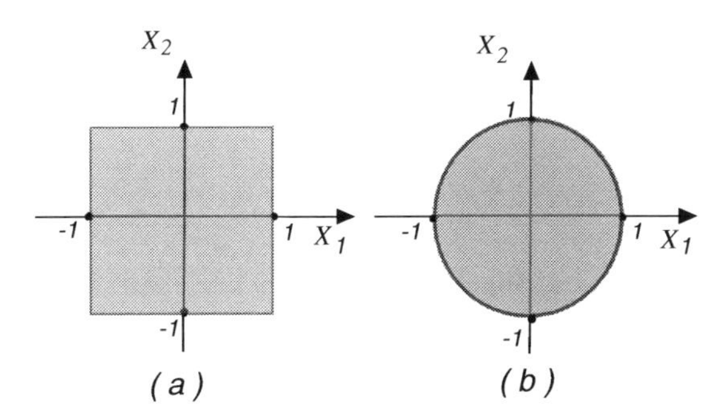

FIGURE 12.

The notion of independency of r.v.'s (and mutual independency, if we consider more than two r.v.'s) carry over to the continuous case literally, and there is no need to define them again.

Proposition 3 *An absolutely continuous r.vec.* (X_1, X_2) *have independent coordinates if and only if its density*

$$f(x_1, x_2) = f_1(x_1) f_2(x_2) \quad \textit{for any } x_1, x_2, \tag{3.3}$$

where $f_i(x)$ *is the marginal density of* X_i; $i = 1, 2$.

The proof repeats the proof of Proposition 2 in Chapter 3 with replacement summation by integration.

A similar result is true for mutual independency of more than two r.v.'s.

EXAMPLE 2. Let a r.vec. $\mathbf{X} = (X_1, X_2)$ take on values from the square in Fig. 12a, the joint density $f(x_1, x_2) = 1/4$ for all points $\boldsymbol{x} = (x_1, x_2)$ from this square, and $f(x_1, x_2) = 0$ otherwise.

(The area of the square is 4, and the total integral of $f(x_1, x_2)$ over the square should be one; see (3.2). Hence, if we set $f(x_1, x_2)$ equal to a constant, then this constant should be equal to one over the area of the square.)

Such a *multivariate* distribution is called *uniform* (in this particular case, uniform on a square), and we say that all points from the square are equiprobable.

For $|x_1| \leq 1$ and $|x_2| \leq 1$,

$$f_1(x_1) = \int_{-1}^{1} f(x_1, x_2) dx_2 = \int_{-1}^{1} \frac{1}{4} dx_2 = \frac{1}{2}, \quad f_2(x_2) = \int_{-1}^{1} f(x_1, x_2) dx_1 = \frac{1}{2}.$$

We see that the marginal distributions are uniform on $[-1, 1]$. Furthermore, $f(x_1, x_2) = \frac{1}{4} = \frac{1}{2} \cdot \frac{1}{2} = f_1(x_1) f_2(x_2)$. Hence, X_1, X_2 are independent, which could be predicted from the very beginning.

Now, let the distribution of $\mathbf{X}$ be uniform on the unit disk depicted in Fig. 12b. Formally, this means that the density is constant on the disk, and vanishes at all points not belonging to the disk.

In this case, X_1 and X_2 are dependent, since the value of X_1 determines the range within which X_2 can change. Say, if $X_1 = 0$, the r.v. X_2 can assume any value from $[-1, 1]$, while if $X_1 = 1$, the only possible value of X_2 is zero.

Let us find the marginal densities and show that in this case, (3.3) is indeed not true.

The disk in Fig. 12b consists of all points (x_1, x_2) for which $x_1^2 + x_2^2 \leq 1$. The area of the disk equals π. Hence, $f(x_1, x_2) = 1/\pi$ for all points in the disk, and $f(x_1, x_2) = 0$ otherwise. Consequently, for a fixed x_1, the density $f(x_1, x_2) = 0$ if $|x_2| > \sqrt{1 - x_1^2}$, and

$$f_1(x_1) = \int_{-\infty}^{\infty} f(x_1, x_2)dx_2 = \int_{-\sqrt{1-x_1^2}}^{\sqrt{1-x_1^2}} \frac{1}{\pi}dx_2 = \frac{2}{\pi}\sqrt{1 - x_1^2} \text{ if } |x_1| \leq 1, \text{ and } = 0 \text{ otherwise.}$$

Similarly, $f_2(x_2) = \frac{2}{\pi}\sqrt{1 - x_2^2}$ if $|x_2| \leq 1$, and $= 0$ otherwise.

So, $f_1(x_1)f_2(x_2) \neq \frac{1}{\pi} = f(x_1, x_2)$, which is consistent with Proposition 3. □

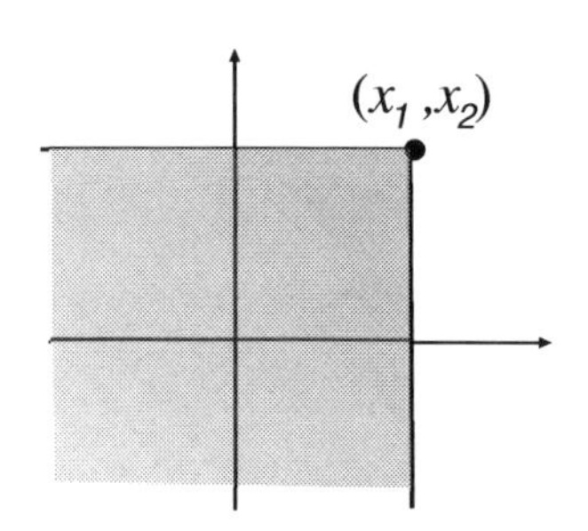

FIGURE 13.

In conclusion, we briefly address distribution functions in the multidimensional case. Similarly to the one-dimensional case, the d.f. of $\mathbf{X} = (X_1, X_2)$ is the function

$$F(\boldsymbol{x}) = F(x_1, x_2) = P(X_1 \leq x_1, X_2 \leq x_2).$$

This is the probability of a "corner" (see Fig. 13), and if $f(x_1, x_2)$ is the density, then

$$F(x_1, x_2) = \int_{-\infty}^{x_1} \int_{-\infty}^{x_2} f(u_1, u_2)du_1du_2. \tag{3.4}$$

Differentiating (3.4), we get the counterpart of the one-dimensional formula $f(x) = F'(x)$:

$$f(x_1, x_2) = \frac{\partial^2}{\partial x_1 \partial x_2} F(x_1, x_2). \tag{3.5}$$

The last formula turns out to be useful in many problems. However, the d.f.'s of r.vec.'s do not play as important a role as they do in the one-dimensional case. They represent the probabilities of "corners," and these sets in the multidimensional framework have no advantage over other sets, say, circles.

---

4 SUMS OF INDEPENDENT R.V.'S. CONVOLUTIONS

4.1 Convolution in the continuous case

Our object of study is the sum

$$S = S_n = X_1 + ... + X_n,$$

where X_i's are *independent* r.v.'s. Let $F_i(x)$ and $f_i(x)$ be the d.f. and density, respectively, of X_i.

First, consider the case $n=2$; so, $S = X_1 + X_2$.

Proposition 4 *If X_1 and X_2 are continuous, then S is also continuous, and its density*

$$f_S(x) = \int_{-\infty}^{\infty} f_1(x-y) f_2(y) dy. \tag{4.1.1}$$

Compare with the formula $\sum_{k=0}^{n} f_k^{(1)} f_{n-k}^{(2)}$ in the discrete case (Section **3**.1.2.3).

As in Section **3**.1.2.3, the operation itself is called *convolution (now, of densities)*. If we deal with a sum $S_3 = X_1 + X_2 + X_3$, we can write that $S_3 = S_2 + X_3$, and hence $f_{S_3} = f_{S_2} * f_3 = f_1 * f_2 * f_3$. In general, for an arbitrary integer n,

$$f_{S_n} = f_1 * \ldots * f_n.$$

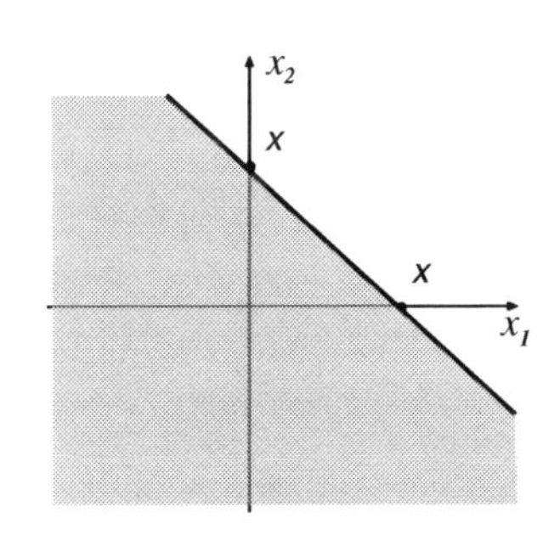

FIGURE 14.

Proof of Proposition 4. The idea is to consider the vector (X_1, X_2) whose values are points (x_1, x_2) in a plane. Then $P(X_1 + X_2 \le x)$ is the probability that the value of the vector (X_1, X_2) will be in the region $\{(x_1, x_2) : x_1 + x_2 \le x\}$; see Fig. 14. Since X_1, X_2 are independent, this probability may be written as the double integral

$$\iint_{x_1+x_2 \le x} f_1(x_1) f_2(x_2) dx_1 dx_2$$
$$= \int_{-\infty}^{\infty} \left(\int_{-\infty}^{x-x_2} f_1(x_1) dx_1 \right) f_2(x_2) dx_2$$
$$= \int_{-\infty}^{\infty} F_1(x - x_2) f_2(x_2) dx_2.$$

Replacing x_2 by y, we get that the d.f. of S is

$$F_S(x) = \int_{-\infty}^{\infty} F_1(x-y) f_2(y) dy.$$

Differentiating in x, for the density $f_S(x)$, we have

$$f_S(x) = \frac{dF_S(x)}{dx} = \frac{d}{dx} \int_{-\infty}^{\infty} F_1(x-y) f_2(y) dy$$
$$= \int_{-\infty}^{\infty} \frac{d}{dx} F_1(x-y) f_2(y) dy = \int_{-\infty}^{\infty} f_1(x-y) f_2(y) dy. \blacksquare$$

EXAMPLE 1. Let X_1 and X_2 be independent, X_1 be exponential with $E\{X_1\} = 1$, and X_2 be uniformly distributed on $[0,1]$. Since $f_2(y) = 1$ for $y \in [0,1]$ and $f_2(y) = 0$ otherwise, the density of the sum is

$$f_S(x) = \int_{-\infty}^{\infty} f_1(x-y) f_2(y) dy = \int_0^1 f_1(x-y) dy.$$

It remains to insert the concrete presentation for $f_1(x-y)$. The function $f_1(x) = e^{-x}$ for $x \geq 0$, and $f_1(x) = 0$ for $x < 0$. Consequently, the integrand $f_1(x-y) = e^{-(x-y)}$ if $y \leq x$, and $= 0$ otherwise. Hence, for $x \leq 1$, we may integrate only up to x, which implies that

$$f_S(x) = \int_0^x f_1(x-y)dy = \int_0^x e^{-(x-y)}dy = e^{-x}\int_0^x e^y dy = 1 - e^{-x}.$$

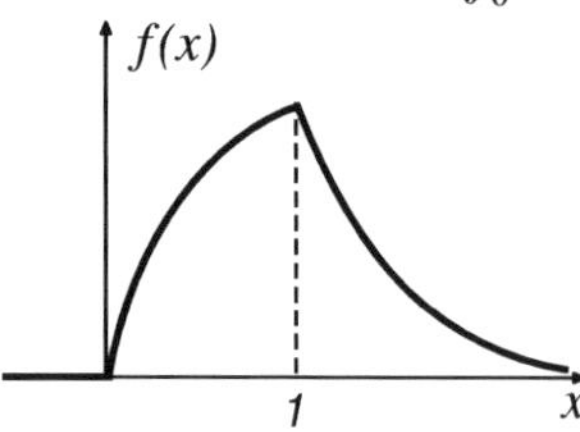

FIGURE 15.

For $x > 1$, we should consider the total $\int_0^1$, and

$$f_S(x) = \int_0^1 e^{-(x-y)}dy = e^{-x}(e-1).$$

The graph is given in Fig. 15.

EXAMPLE 2 is classical. Let X_1 and X_2 be independent and uniformly distributed on $[0,1]$. Obviously, $S_2 = X_1 + X_2$ takes on values from $[0,2]$, so it suffices to find the density $f_S(x)$ only on this interval. The densities $f_i(x) = 1$ for $x \in [0,1]$ and $= 0$ otherwise. Hence, by (4.1.1),

$$f_S(x) = \int_0^1 f_1(x-y)dy. \tag{4.1.2}$$

The integrand $f_1(x-y) = 1$ if $0 \leq x-y \leq 1$ which is equivalent to $x-1 \leq y \leq x$.

Let $0 \leq x \leq 1$. Then the inequality $x-1 \leq y$ holds automatically because $x-1 \leq 0$ while $y \geq 0$. So, $f_1(x-y) = 1$ if $y \leq x$, and $= 0$ otherwise. Hence, for $0 \leq x \leq 1$, we may integrate in (4.1.2) only over $[0,x]$, which implies that

$$f_{S_2}(x) = \int_0^x 1 \cdot dy = x.$$

On the other hand, in view of the symmetry of the distributions of the X's, the density $f_S(x)$ should be symmetric with respect to the center of $[0,2]$; that is, the point one (see Fig. 16a). So, because $f_S(x) = x$ for $0 \leq x \leq 1$, we should have $f_S(x) = 2-x$ for $1 \leq x \leq 2$; see again Fig. 16a. Eventually,

$$f_{S_2}(x) = \begin{cases} x & \text{if } 0 \leq x \leq 1, \\ 2-x & \text{if } 1 \leq x \leq 2, \\ 0 & \text{otherwise.} \end{cases} \tag{4.1.3}$$

This distribution is called *triangular*. We see that, while the values of the X's are equally likely, the values of the sum are not. In Exercise 60, the reader is invited to give a common-sense explanation of this fact.

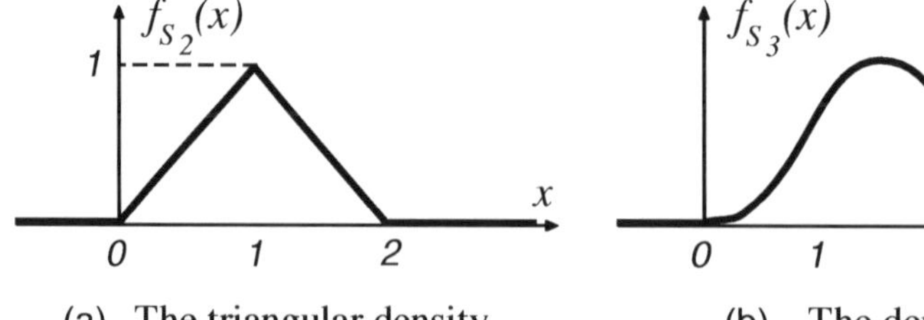

(a) The triangular density (b) The density of S_3

FIGURE 16.

Now, let X_3 be also uniformly distributed on $[0,1]$ and independent of X_1 and X_2. Obviously, the sum $S_3 = X_1 + X_2 + X_3$ assumes values from $[0,3]$. To find its density, we can again apply (4.1.1), replacing $f_2(y)$ by $f_3(y)$, and $f_1(x-y)$ by $f_{S_2}(x-y)$. Thus,

$$f_{S_3}(x) = \int_0^1 f_{S_2}(x-y)dy, \tag{4.1.4}$$

where f_{S_2} is given in (4.1.3). We relegate a bit tedious calculations to Exercise 60. The answer is

$$f_{S_3}(x) = \begin{cases} x^2/2 & \text{if } 0 \le x \le 1, \\ (-2x^2+6x-3)/2 & \text{if } 1 \le x \le 2, \\ (x-3)^2/2 & \text{if } 2 \le x \le 3, \\ 0 & \text{otherwise,} \end{cases}$$

The graph is sketched in Fig. 16b. □

4.2 Two more classical examples

In the examples above, we saw that the distribution of a sum may essentially differ from the distributions of the separate terms. Next, we consider cases where the convolution inherits properties of individual terms in the sum. The first such an example concerning the Poisson distribution has been explored in Section **3**.4.5.4. Now, we consider two continuous distributions.

I. Sums of normal r.v.'s.

Proposition 5 *Let X_1 and X_2 be independent and normal with expectations m_1 and m_2, and variances σ_1^2 and σ_2^2, respectively. Then the r.v. $S = X_1 + X_2$ is normal with expectation $m_1 + m_2$, and variance $\sigma_1^2 + \sigma_2^2$.*

In another notation, if φ_{m,σ^2} is the normal density with mean m and variance σ^2, then

$$\varphi_{m_1,\sigma_1^2} * \varphi_{m_2,\sigma_2^2} = \varphi_{m_1+m_2,\sigma_1^2+\sigma_2^2}.$$

Direct calculations with use of the convolution formula are somewhat tedious; we consider a particular case in Exercise 44 provided with a detailed advice. A short non-direct proof will be given in Section **8**.2.

II. Sums of Γ(Gamma)-distributed r.v.'s.

We call a random variable a Γ-r.v. if it has a Γ-distribution.

Proposition 6 *Let X_1 and X_2 be independent Γ-r.v.'s with parameters (a,ν_1) and (a,ν_2), respectively. (The scale parameter a is the same.) Then the r.v. $S = X_1 + X_2$ is a Γ-r.v. with parameters $(a,\nu_1+\nu_2)$.*

In other words, if $f_{a\nu}$ denotes the Γ-density with parameters (a,ν), then

$$f_{a\nu_1} * f_{a\nu_2} = f_{a,\nu_1+\nu_2}. \tag{4.2.1}$$

This is an important fact having a great many applications in diverse areas such as physics, economics, etc., and we will appeal to it repeatedly.

EXAMPLE 1. Rich is the nth in line for a taxi in an airport. Suppose that at each moment, the remaining waiting time for the next taxi does not depend on when the last and other taxis arrived. In other words, the memoryless property holds, and the periods between the consecutive arrivals of taxis—let us call them interarrival times—are independent exponential r.v.'s. Suppose that the mean interarrival time is, say, 2 minutes. What is the distribution of the waiting time for Rich?

Let us adopt 2 mins as a unit of time. Due to the memoryless property, we may think that when Rich took his place in line, the waiting process for the first person in line has started over as from the beginning. Then, if X_1 is the time of the first taxi arrival, and $X_2, X_3, ...$ are consecutive interarrival times, Rich's waiting time is the r.v. $S_n = X_1 + ... + X_n$, where the X's are exponentially distributed with the same parameter $a = 1$.

The exponential distribution is the Γ-distribution with parameter $\nu = 1$. By Proposition 6, the r.v. S_n has the Γ-distribution with parameters $a = 1$ and $\nu = \underbrace{1 + ... + 1}_{n \text{ times}} = n$. So,

$$f_{S_n}(x) = \frac{x^{n-1}}{\Gamma(n)} e^{-x} = \frac{x^{n-1}}{(n-1)!} e^{-x}.$$

For instance, if $n = 4$, then $f_{S_4}(x) = \frac{x^3}{3!} e^{-x} = \frac{x^3}{6} e^{-x}$. For example, the probability that being the fourth in line, Rich will wait more than 10 mins (i.e., 5 time units) is, as is easy to compute or estimate using software,

$$P(S_4 > 5) = 1 - \frac{1}{6} \int_0^5 x^3 e^{-x} dx \approx 0.27. \square$$

Proof of Proposition 6. Let f_S, f_1, and f_2 be the respective densities of S, X_1, and X_2. Without loss of generality, we can set the scale parameter $a = 1$. Taking into account that $f_1(x)$ and $f_2(x)$ equal zero for $x < 0$, and hence, $f(x-y) = 0$ for $y > x$, we have

$$f_S(x) = \int_{-\infty}^{\infty} f_1(x-y) f_2(y) dy = \int_0^{\infty} f_1(x-y) f_2(y) dy = \int_0^x f_1(x-y) f_2(y) dy$$
$$= \int_0^x \frac{1}{\Gamma(\nu_1)} (x-y)^{\nu_1 - 1} e^{-(x-y)} \cdot \frac{1}{\Gamma(\nu_2)} y^{\nu_2 - 1} e^{-y} dy = \frac{1}{\Gamma(\nu_1)\Gamma(\nu_2)} e^{-x} \int_0^x (x-y)^{\nu_1 - 1} y^{\nu_2 - 1} dy.$$

With the change of variables $y = xt$ in the last integral, we get

$$\begin{aligned} f_S(x) &= \frac{1}{\Gamma(\nu_1)\Gamma(\nu_2)} e^{-x} \int_0^1 (x - xt)^{\nu_1 - 1} (xt)^{\nu_2 - 1} x dt \\ &= \frac{1}{\Gamma(\nu_1)\Gamma(\nu_2)} x^{\nu_1 + \nu_2 - 1} e^{-x} \int_0^1 (1-t)^{\nu_1 - 1} t^{\nu_2 - 1} dt \\ &= \frac{1}{\Gamma(\nu_1 + \nu_2)} x^{\nu_1 + \nu_2 - 1} e^{-x} \frac{\Gamma(\nu_1 + \nu_2)}{\Gamma(\nu_1)\Gamma(\nu_2)} \int_0^1 (1-t)^{\nu_1 - 1} t^{\nu_2 - 1} dt \\ &= f_{1, \nu_1 + \nu_2}(x) \frac{\Gamma(\nu_1 + \nu_2)}{\Gamma(\nu_1)\Gamma(\nu_2)} B(\nu_1, \nu_2), \end{aligned} \tag{4.2.2}$$

where $f_{1,\nu_1+\nu_2}(x)$ is the Γ-density with the scale parameter 1 and the parameter $\nu = \nu_1 + \nu_2$, and

$$B(\nu_1, \nu_2) = \int_0^1 (1-t)^{\nu_1 - 1} t^{\nu_2 - 1} dt. \tag{4.2.3}$$

The last expression is referred to as the Beta-function, and it is known that

$$B(\nu_1, \nu_2) = \frac{\Gamma(\nu_1)\Gamma(\nu_2)}{\Gamma(\nu_1+\nu_2)}. \tag{4.2.4}$$

Substituting this into (4.2.2), we would accomplish the proof.

However, as a matter of fact, we may finish the proof *without knowing* (4.2.4), *and moreover, proving* (4.2.4) on the way.

Indeed, the very left side of (4.2.2) is a density, and so is $f_{1,\nu_1+\nu_2}(x)$. Hence, integrating the very left- and the very right-hand sides of (4.2.2) from 0 to ∞, we get that

$$1 = \frac{\Gamma(\nu_1+\nu_2)}{\Gamma(\nu_1)\Gamma(\nu_2)} B(\nu_1, \nu_2),$$

which leads to (4.2.4) and to the assertion of the proposition as well. ■

A shorter non-direct proof of Proposition 6 will be given in Section **8**.2.

In conclusion, we consider a nice example that we will use later in some applications.

EXAMPLE 2. Let $S = \sum_{i=1}^{N} X_i$, where $N, X_1, X_2, ...$ are independent r.v.'s, the X_i's have the exponential distribution with a parameter a, and N has the first version of the geometric distribution with a parameter p; that is, $P(N = k) = pq^{k-1}$, $k = 1, 2, ...$, and $q = 1 - p$.

For instance, a retired couple live in a resort until the first rainy day. Under the corresponding independency assumption, the number N of days spent in the resort (including the rainy day) has a geometric distribution. (If we call a rainy day success, the couple stays up to the first success.) If X_i is the expenses on the ith day, then S above is the total expenses.

Surprisingly, in this (very special) case, the distribution of S is simple: it is exponential with parameter ap. To show this, using the formula for total probability, we write

$$P(S \leq x) = \sum_{n=1}^{\infty} P(S \leq x \mid N = n) P(N = n) = \sum_{n=1}^{\infty} P(S_n \leq x) P(N = n). \tag{4.2.5}$$

By Proposition 6, S_n has the density $f_{an}(x)$. So, differentiating both parts of (4.2.5), we obtain that the density of S is

$$\begin{aligned} f_S(x) &= \sum_{n=1}^{\infty} f_{an}(x) P(N = n) = \sum_{n=1}^{\infty} \frac{a^n}{(n-1)!} x^{n-1} e^{-ax} pq^{n-1} \\ &= ape^{-ax} \sum_{n=1}^{\infty} \frac{(aqx)^{n-1}}{(n-1)!} = ape^{-ax} e^{aqx} = ape^{-ax(1-q)} = ape^{-apx}. \end{aligned}$$

The function we arrived at is the exponential density with parameter ap. □

Route 1 ⇒ page 195

4.3 An additional remark regarding convolutions: Stable distributions

The following notions help to come to a deeper understanding of Propositions 5-6 above.

We say that a class of distributions is *closed with respect to convolution* if for any two independent r.v.'s, X_1 and X_2, having distributions from this class, the distribution of their sum $X_1 + X_2$ also belongs to the same class. For example, as we saw, the class of all normal distributions is closed with respect to convolution, and the same is true for the class of all Poisson distributions, or the class of all Γ-distributions with a fixed scale parameter a, the same for all distributions in the class.

However, these classes have different structures, and to clarify this, we introduce one more notion.

Consider a r.v. X and the family of r.v.'s $Y = a + bX$ for *all* possible values of numbers a and $b > 0$. The family of the corresponding distributions is called a *type*. We refer to b as a scale factor, and a as a location constant. One may say that any two distributions from a type are the distributions of r.v.'s that may be linearly transformed to each other. In particular, this means that not only the r.v. X but any r.v. $Y = a + bX$ with $b > 0$ may serve as the "original" r.v. generating the type.

For example, since any (m, σ^2)-normal r.v. $Y = m + \sigma X$, where X is a standard normal r.v., normal distributions compose a type. The same is true for the family of all uniform distributions because the distribution uniform on $[s,t]$ may be represented as the distribution of a r.v. $Y = s + (t - s)X$, where X is uniform on $[0, 1]$.

On the other hand, Poisson distributions do *not* compose a type. Indeed, if X is a Poisson r.v., the r.v. $a + bX$ is a r.v. assuming values $a, a+b, a+2b, ...$ rather than $0, 1, 2, ...$, and hence, it is not Poisson. Therefore, the class of Poisson distributions is a class of another nature than a type.

The same is true for the Γ-distribution because two Γ-distributions with different values of the parameter ν cannot be reduced to each other by a change of scale. That is, applying (1.5.1) to a Γ-distribution we do not come to a Γ-distribution with another ν. See, for example, the graphs of Γ-densities for different ν's in Fig. 9; these functions are essentially different and cannot be reduced to each other by a linear transformation of the argument.

As we will see in Chapter 9, an important question is which *types* are closed with respect to convolutions. By Proposition 5, the normal type has such a property, while—as we saw in Section 4.1—the uniform type is not closed with respect to convolution.

It turns out that the closedness of a type with respect to convolution is a rare property and may be considered a characterization property of the normal distribution.

Proposition 7 *The normal type is the only type of distributions with finite variances, which is closed with respect to convolution.*

A proof will be given in Section **9**.1.3.

Thus, when the variances are finite, the sum has the same type as separate terms only in the case of normal distributions. Regarding the classes of Poisson or Γ-distributions that are closed with respect to convolution, we realize that these classes are not types, and the changes that convolution brings about are more essential than the change of scale.

Another matter is that there are distributions with infinite variances whose types have the property under discussion. These distributions are called *stable*. An example is the *Cauchy distribution* with density $f(x) = \frac{1}{\pi(1+x^2)}$ we considered in Sections 1.1 and 1.2. The theory of stable distributions may be found in many textbooks; see, e.g., [10], [13], [42], [47].

1,2

5 CONDITIONAL DISTRIBUTIONS AND EXPECTATIONS

We will see that the notions of conditional distribution and expectation carry over from the discrete to continuous case. For this reason, the exposition below is somewhat brief, and the reader is recommended to look over Section **3**.6 on conditioning in the discrete case.

5.1 Conditional density and expectation

Consider a random vector (r.vec.) (X,Y) having a joint density $f(x,y)$. (As in Section **3**.6, here we prefer the notation (X,Y) to (X_1,X_2).)

Denote by $f_X(x)$ the marginal density of X and consider the function

$$f_{Y|X}(y|x) = \frac{f(x,y)}{f_X(x)}, \tag{5.1.1}$$

provided $f_X(x) \neq 0$. If for some x, the density $f_X(x) = 0$, then we set $f_{Y|X}(y|x) = 0$ by definition.

We call $f_{Y|X}(y|x)$ *the conditional density of* Y given $X = x$. When it cannot cause confusion, we omit the index $Y|X$ in $f_{Y|X}(y|x)$.

To clarify the significance of definition (5.1.1), let us consider infinitesimally small intervals $[x, x+dx]$, $[y, y+dy]$ and assume $f_X(x) \neq 0$. Reasoning heuristically, we represent the probability $P(x \leq X \leq x+dx)$ as $f_X(x)dx$; see Section 1.1.

Similarly, we write $P(x \leq X \leq x+dx,\ y \leq Y \leq y+dy) = f(x,y)dxdy$. Then

$$P(y \leq Y \leq y+dy \,|\, x \leq X \leq x+dx) = \frac{P(y \leq Y \leq y+dy,\, x \leq X \leq x+dx)}{P(x \leq X \leq x+dx)}$$

$$= \frac{f(x,y)dxdy}{f_X(x)dx} = f(y|x)dy. \tag{5.1.2}$$

Since the interval $[x, x+dx]$ is infinitesimally small, we can view $P(y \leq Y \leq y+dy \,|\, x \leq X \leq x+dx)$ as $P(y \leq Y \leq y+dy \,|\, X = x)$, which leads to

$$P(y \leq Y \leq y+dy \,|\, X = x) = f(y|x)dy. \tag{5.1.3}$$

Thus, $f(y|x)$ indeed plays the role of the density of Y when X assumes a value x. Certainly, the informal relations (5.1.2) and (5.1.3) are just a clarification. Formally, we adopt (5.1.1) as a definition.

Now, again as a definition, we set the conditional expectation

$$m(x) = m_{Y|X}(x) = E\{Y \,|\, X = x\} = \int_{-\infty}^{\infty} y f(y|x)dy. \tag{5.1.4}$$

As in the discrete case, we define the expected value of Y given X as

$$E\{Y \,|\, X\} = m(X).$$

EXAMPLE 1. Let us revisit the second part of Example 3-2 and Fig. 12b, replacing the symbol X_1 with X, and X_2 with Y. Since the joint distribution is uniform, we can guess that the same is true for the conditional distribution of Y given X. Let us show it rigorously.

As was shown in Example 3-2, $f_X(x) = \frac{2}{\pi}\sqrt{1-x^2}$ for $|x| \leq 1$. Since $f(x,y) = \frac{1}{\pi}$ for $x^2+y^2 \leq 1$, by (5.1.1),

$$f(y\,|\,x) = f_{Y\,|\,X}(y\,|\,x) = \frac{f(x,y)}{f_X(x)} = \frac{1}{2\sqrt{1-x^2}} \quad \text{for} \quad -\sqrt{1-x^2} \leq y \leq \sqrt{1-x^2}.$$

If y is not in the interval $[-\sqrt{1-x^2}, \sqrt{1-x^2}]$, then $f(y\,|\,x) = 0$.

Thus, for a *fixed* x, the conditional density $f(y\,|\,x)$, *as a function of* y, is constant. Hence, the conditional distribution under consideration is indeed uniform on $[-\sqrt{1-x^2}, \sqrt{1-x^2}]$.

Let us consider conditional expectations $E\{Y\,|\,X\}$ and $E\{Y^2\,|\,X\}$. We can use the general formula (5.1.4), but it is not necessary.

For a r.v. ξ uniformly distributed on $[-a,a]$, the expected value $E\{\xi\} = 0$ and $E\{\xi^2\} = a^2/3$; see Section 2.1 or Table 1 in the Appendix. Hence, $E\{Y\,|\,X\} = 0$ (which also follows by the symmetry argument), and since in our case, $a = \sqrt{1-X^2}$,

$$E\{Y^2\,|\,X\} = (1-X^2)/3. \;\square$$

Let X_1, X_2 be independent r.v.'s with densities $f_1(x)$, $f_2(x)$, respectively, and let $S = X_1 + X_2$. In many problems, it is important to know the distribution of a term of the sum *given* the value of the sum. In the continuous case, the answer is given by the following formula for the conditional density:

$$f_{X_1\,|\,S}(x\,|\,s) = \frac{f_1(x)f_2(s-x)}{f_S(s)}. \tag{5.1.5}$$

Since the denominator is exactly what the definition (5.1.1) requires, to prove (5.1.5) it suffices to show that the joint density of the vector (X_1, S) is

$$f(x,s) = f_1(x)f_2(s-x). \tag{5.1.6}$$

Heuristically, it is almost obvious. For any r.v. X, its density $f(x)$ is connected with the probability that X will take on a value close to x. In our case, if $X_1 \approx x$ and $S \approx s$, then $X_2 \approx s-x$. Since X_1, X_2 are independent, $P(X_1 \approx x, S \approx s) = P(X_1 \approx x, X_2 \approx s-x) = P(X_1 \approx x)P(X_2 \approx s-x)$. This is reflected in (5.1.6).

We prove (5.1.6) rigorously at the end of this section.

EXAMPLE 2. Let us come back to Rich in Example 4.2-1 who was waiting for a taxi in an airport. Suppose that in the beginning, Rich was the second in line ($n = 2$), and his total waiting time turned out to be $s = 10$ minutes. Given this information, what is the distribution of the waiting time for being the first in line? Are the values between 0 and 10 minutes equally likely?

Let X_1 and X_2 be the same as in Example 4.2-1, and $S = X_1 + X_2$, Rich's waiting time. The object of interest is the distribution of X_1 given $S = s$. We apply (5.1.5). It suffices to consider $f_{X_1\,|\,S}(x\,|\,s)$ for $x \in [0,s]$. Indeed, X_1 and X_2 are positive, and given their sum equals s, both terms are not greater than s.

By Proposition 6 (see also Example 4.2-1), $f_S(s) = f_{a,2}(s)$, where $f_{a,2}$ is the Γ-density with parameters $(a,2)$. So, $f_S(s) = a^2 s e^{-as}$. By (5.1.5), for $x \le s$

$$f_{X_1 \mid S}(x \mid s) = \frac{ae^{-ax}ae^{-a(s-x)}}{a^2 s e^{-as}} = \frac{1}{s}.$$

Thus, the conditional distribution is uniform (!) on $[0,s]$. The result is nice: though the values of the exponential r.v. X_1 are not equally likely, given the information that the sum $X_1 + X_2$ is s, the r.v. X_1 may take on any value from $[0,s]$ with equal likelihood.

Once we know the conditional distribution, we can compute various expectations. In particular, $E\{X_1 \mid S\} = S/2$ (as the mean of the distribution uniform on $[0,S]$). If we define the conditional variance as the variance of the corresponding conditional distribution, we can write that $Var\{X_1 \mid S\} = S^2/12$. □

Proof of (5.1.6). Consider the distribution function of the vector (X_1, S), that is, $F_{X_1,S}(x,s) = P(X_1 \le x, S \le s) = P(X_1 \le x, X_1 + X_2 \le s)$. Thus,

$$F_{X_1,S}(x,s) = \iint\limits_{u_1 \le x,\, u_1+u_2 \le s} f_1(u_1) f_2(u_2) du_1 du_2 = \int_{-\infty}^{x} f_1(u_1) \left(\int_{-\infty}^{s-u_1} f_2(u_2) du_2 \right) du_1.$$

It remains to use (3.5). The reader is invited to provide the differentiation on her/his own, and get

$$f(x,s) = \frac{\partial^2}{\partial x \partial s} F_{X_1,S}(x,s) = f_1(x) f_2(s-x). \;\blacksquare$$

It is easy to verify—we skip technical details, that conditional expectation in the continuous case has the same properties that we stated for the discrete case in Section **3**.6.2. In particular, this concerns the *formula for total expectation*

$$E\{Y\} = E\{E\{Y \mid X\}\}. \tag{5.1.7}$$

5.2 Conditioning. Formulas for total probability and expectation

EXAMPLE 1. Let X and Y be independent *standard* normal r.v.'s, and a r.v. Z have an arbitrary distribution and do not depend on X and Y. Consider the r.v.

$$\xi = \frac{X + YZ}{\sqrt{1+Z^2}}.$$

Though it looks surprising at first glance, ξ has the standard normal distribution. For now, we give a slightly heuristic explanation. Given Z has assumed a value z, the r.v. $\xi = \dfrac{X + zY}{\sqrt{1+z^2}}$, and the values of X and Y do not depend on z. By Proposition 5, $X + zY$ is *normal* with a mean of zero, and a variance of $1 + z^2$. So, $X + zY$ has been divided by its standard deviation, and given $Z = z$, the r.v. ξ is a normalized r.v. Therefore, its mean is zero, its variance is equal to one, and hence ξ has the *standard normal distribution, whatever* z is. □

The procedure we used above is called conditioning. To make what we are doing more understandable and rigorous, we need some simple constructions.

First, we need to define the conditional expectation $m(x) = E\{Y \,|\, X = x\}$ in the case where X is continuous, while Y is discrete and takes on values $y_1, y_2, \ldots$. Let $f(j,x) = \frac{d}{dx}P(Y = y_j, X \leq x)$. With respect to X, it plays the role of a density as the derivative of $P(Y = y_j, X \leq x)$. With respect to Y, it concerns the probability of a particular value y_j.

We define the conditional probability of a value y_j as

$$f(j \,|\, x) = \frac{f(j,x)}{f_X(x)}$$

(compare with (5.1.1)), and set

$$m(x) = E\{Y \,|\, X = x\} = \sum_j y_j f(j \,|\, x).$$

It is straightforward to show that such a definition inherits all properties of conditional expectation we stated before, and in particular, the main property (5.1.7).

Let us return to the general scheme. Given $X = x$, the value of $E\{Y \,|\, X\}$ is $m(x) = E\{Y \,|\, X = x\}$, and consequently, if $f(x)$ is the density of X, then

$$E\{Y\} = \int_{-\infty}^{\infty} m(x) f(x) dx = \int_{-\infty}^{\infty} E\{Y \,|\, X = x\} f(x) dx. \tag{5.2.1}$$

This is one of the possible versions of the formula for total expectation in the continuous case.

The counterpart of the formula for total probability for the continuous case may look as follows: for any event A,

$$P(A) = \int_{-\infty}^{\infty} P(A \,|\, X = x) f(x) dx. \tag{5.2.2}$$

We see a similarity with the discrete formula $P(A) = \sum_i P(A \,|\, B_i) P(B_i)$. The role of i is played by x, the events B_i are connected with values of a r.v. X, and summation is replaced with integration involving $f(x)$.

To justify (5.2.2), first, we should rigorously define $P(A \,|\, X = x)$. To this end, as in Chapter 3, Section 2.1, we consider the *indicator* r.v.

$$I_A = \begin{cases} 1 & \text{if } A \text{ has occurred,} \\ 0 & \text{if } A \text{ has not occurred.} \end{cases} \tag{5.2.3}$$

Since A occurs with probability $P(A)$,

$$E\{I_A\} = P(A). \tag{5.2.4}$$

Then, it is natural to set, by definition,

$$P(A \,|\, X = x) = E\{I_A \,|\, X = x\}. \tag{5.2.5}$$

Substituting I_A for Y in (5.2.1), we get (5.2.2).

As above, replacing x by X, we may consider a conditional probability $P(A\,|\,X)$, the probability of A given X.

EXAMPLE 1 revisited. Let $A = \{\xi \le x\}$. Then, since X and Y do not depend on Z, we have $P(\xi \le x\,|\,Z=z) = P\left(\frac{X+zY}{\sqrt{1+z^2}} \le x\,|\,Z=z\right) = P\left(\frac{X+zY}{\sqrt{1+z^2}} \le x\right) = \Phi(x)$, where $\Phi(x)$ is the standard normal d.f. Assume, for example, that Z is continuous. Then, by (5.2.2),

$$P(\xi \le x) = \int_{-\infty}^{\infty} P(\xi \le x\,|\,Z=z) f_Z(z)dz = \int_{-\infty}^{\infty} \Phi(x) f_Z(z)dz = \Phi(x)\int_{-\infty}^{\infty} f_Z(z)dz = \Phi(x). \ \square$$

EXAMPLE 2. Suppose we choose at random a number θ from an interval $[0, 1]$ and run a sequence of n independent trials with the probability of success equal to the chosen number θ. (Usually, when the probability of success is certain, we denote it by p. Since now this probability is random, it makes sense to use another symbol.) In other words, the probability of success is the value of a r.v. Θ uniform on $[0, 1]$. Somewhat surprisingly, the values of the number of successes in this case are *equally* likely.

Indeed, let S be the number of successes. Then $P(S=k\,|\,\Theta=\theta) = \binom{n}{k}\theta^k(1-\theta)^{n-k}$. Hence, by (5.2.2),

$$\begin{aligned} P(S=k) &= \int_0^1 P(S=k\,|\,\Theta=\theta)\cdot 1\cdot d\theta = \int_0^1 \binom{n}{k}\theta^k(1-\theta)^{n-k}d\theta \\ &= \binom{n}{k} B(k+1, n-k+1), \end{aligned} \tag{5.2.6}$$

where the function $B(\nu_1, \nu_2)$ is the Beta-function defined in (4.2.4). By (4.2.4),

$$B(k+1, n-k+1) = \frac{\Gamma(k+1)\Gamma(n-k+1)}{\Gamma(n+2)} = \frac{k!(n-k)!}{(n+1)!}.$$

Substituting this into (5.2.6),we get

$$P(S=k) = \frac{1}{n+1} \quad \text{for all} \quad k = 0, ..., n. \ \square$$

Next, we consider the following general question. Let ξ_1, ξ_2 be independent continuous r.v.'s. How to find $P(\xi_2 > \xi_1)$? By virtue of independency and (5.2.2),

$$\begin{aligned} P(\xi_2 > \xi_1) &= \int_{-\infty}^{\infty} P(\xi_2 > \xi_1\,|\,\xi_1 = x) f_{\xi_1}(x)dx = \int_{-\infty}^{\infty} P(\xi_2 > x\,|\,\xi_1 = x) f_{\xi_1}(x)dx \\ &= \int_{-\infty}^{\infty} P(\xi_2 > x) f_{\xi_1}(x)dx. \end{aligned} \tag{5.2.7}$$

EXAMPLE 3. Suppose that ξ_1 and ξ_2 are exponential with respective parameters a_1 and a_2. Say, two customers are being served independently in an auto shop, and ξ_1 and ξ_2 are the respective service times. We assume that both r.v.'s are exponential, but the jobs are different, and so are the expected service times: $m_1 = 1/a_1$ and $m_2 = 1/a_2$.

What is the probability that the service with the larger expected time, in reality will take less time? From (5.2.7), we get

$$P(\xi_2 > \xi_1) = \int_0^{\infty} e^{-a_2x} a_1 e^{-a_1x} dx = a_1 \int_0^{\infty} e^{-(a_1+a_2)x} dx = \frac{a_1}{a_1+a_2} = \frac{m_2}{m_1+m_2} \tag{5.2.8}$$

because $m_i = 1/a_i$. Assume, for instance, that the mean service time of the second type is twice as large as of the first ($m_2 = 2m_1$). Then the probability that the second service will last longer than the first (i.e., $\xi_2 > \xi_1$) is 2/3, and $P(\xi_1 > \xi_2) = 1/3$. □

However, we should not expect that such a nice formula as (5.2.8) is true for other distributions—the exponential case is special.

EXAMPLE 4. Suppose ξ_1 and ξ_2 are uniform on $[0, c_1]$ and $[0, c_2]$, respectively, and $c_1 \leq c_2$. Then, for $0 \leq x \leq c_1$, the probability $P(\xi_2 > x) = 1 - \frac{x}{c_2}$, and (5.2.7) implies

$$P(\xi_2 > \xi_1) = \int_0^{c_1} \left(1 - \frac{x}{c_2}\right) \frac{1}{c_1} dx = 1 - \frac{c_1}{2c_2} = 1 - \frac{m_1}{2m_2}$$

because $E\{\xi_i\} = c_i/2$. and this is true for $m_1 \leq m_2$ since we have assumed $c_1 \leq c_2$. In particular, for $m_2 = 2m_1$, we get the probability $3/4$ instead of $2/3$ for the exponential case.

For $m_1 > m_2$, we switch the r.v.'s and write $P(\xi_1 > \xi_2) = 1 - m_2/(2m_1)$, and hence $P(\xi_2 > \xi_1) = 1 - P(\xi_1 \geq \xi_2) = m_2/(2m_1)$. □

Next, note that the conditioning procedure may concern densities. Namely, the following formula is true for the continuous r.vec. (X, Y):

$$f_Y(y) = \int_{-\infty}^{\infty} f_{Y|X}(y|x) f_X(x) dx. \tag{5.2.9}$$

To derive (5.2.9), it suffices to substitute into the r.-h.s. of (5.2.9) the definition of the conditional density, which leads to

$$\int_{-\infty}^{\infty} f_{Y|X}(y|x) f_X(x) dx = \int_{-\infty}^{\infty} \frac{f(x,y)}{f_X(x)} f_X(x) dx = \int_{-\infty}^{\infty} f(x,y) dx = f_Y(y).$$

EXAMPLE 5 (*The ratio of two normal r.v.'s*). Find the distribution of $Z = \xi_1/\xi_2$ for independent standard normal r.v.'s ξ_1 and ξ_2. Certainly, Z is well defined only if $\xi_2 \neq 0$, but since $P(\xi_2 = 0) = 0$, we can eliminate this case from consideration.

Consider the conditional density $f_{Z|\xi_2}(z|x)$. We will omit the index $Z|\xi_2$, writing just $f(z|x)$. Since ξ_1 and ξ_2 are independent, the density of Z given $\xi_2 = x \neq 0$ is the density of the r.v. ξ_1/x. If $x > 0$, then this is a normal r.v. with zero mean and variance $1/x^2$. A normal r.v. with zero mean is symmetric and multiplying it by -1, we do not change its distribution. So, if $x < 0$, then $(\xi_1/x) = -\xi_1/|x|$, and we again have a normal r.v. with zero mean and variance $1/x^2$. Thus, in both cases, $f(z|x) = (2\pi)^{-1/2}|x| \exp\{-z^2x^2/2\}$.

Furthermore, by (5.2.9),

$$f_Z(z) = \int_{-\infty}^{\infty} f_{Z|\xi_2}(z|x) f_{\xi_2}(x) dx = \int_{-\infty}^{\infty} (2\pi)^{-1/2}|x| \exp\{-z^2x^2/2\}(2\pi)^{-1/2} \exp\{-x^2/2\} dx$$
$$= (2\pi)^{-1} \int_{-\infty}^{\infty} |x| \exp\{-x^2(1+z^2)/2\} dx.$$

The integrand is an even function, so $\int_0^{\infty} = \int_{-\infty}^{0}$, and $\int_{-\infty}^{\infty} = 2\int_0^{\infty}$. By the variable change $y = x\sqrt{1+z^2}$, we have

$$f_Z(z) = \frac{1}{2\pi} \cdot \frac{1}{1+z^2} \cdot 2\int_0^{\infty} y \exp\{-y^2/2\} dy = \frac{1}{\pi(1+z^2)} \int_0^{\infty} d\left(-\exp\{-y^2/2\}\right)$$
$$= \frac{1}{\pi(1+z^2)} \left(-\exp\{-y^2/2\}\right)\Big|_0^{\infty} = \frac{1}{\pi(1+z^2)}.$$

This is the density of the Cauchy distribution. □

In conclusion, note that

> Conditional expectations $E\{Y\,|\,X\}$ and probabilities $P(A\,|\,X)$ in the case where X is a random vector rather than a r.v. are defined in a similar way. The same concerns the conditioning procedure. We skip formal technicalities.

(5.2.10)

EXAMPLE 6. This unexpected due to its result example is a generalization of Example 5. Consider the system of linear equations

$$\mathbf{AX} = \mathbf{b}, \tag{5.2.11}$$

where $\mathbf{A} = \{a_{ij}\}$ is a $n \times n$ matrix, $\mathbf{b} = (b_1, ..., b_n)$ is a n-dimensional vector, and the n-dimensional vector $\mathbf{X} = (X_1, ..., X_n)$ is the vector of unknowns; that is, $\mathbf{X}$ is a solution to (5.2.11). In (5.2.11), we view vectors $\mathbf{X}$ and $\mathbf{b}$ as column vectors.

Assume that all a_{ij} and b_i are independent standard normal r.v.'s. We prove that in this case, as in Example 5, each X_i has the Cauchy distribution.

In view of symmetry, it suffices to consider X_1. Let D_j be the cofactor of the element a_{1j}. By the well known formula for solutions of the system of linear equations (usually, it is called Cramer's formula), $X_1 = \dfrac{b_1D_1 + b_2D_2 + ... + b_nD_n}{a_{11}D_1 + a_{22}D_2 + ... + a_{1n}D_n}$.

Since all r.v.'s a_{1j} and $D_1, ..., D_n$ are continuous, the probability that the denominator will be equal to zero is zero, so we may exclude this case from consideration. For the same reason, we can divide the numerator and the denominator by $\sqrt{D_1^2 + ... + D_n^2}$ and write

$$X_1 = \frac{(b_1D_1 + b_2D_2 + ... + b_nD_n)/\sqrt{D_1^2 + ... + D_n^2}}{(a_{11}D_1 + a_{22}D_2 + ... + a_{1n}D_n)/\sqrt{D_1^2 + ... + D_n^2}}. \tag{5.2.12}$$

Consider the conditional distribution of X_1 given the vector $\mathbf{D} = (D_1, ..., D_n)$. Note that the b's and a_{1j}'s are independent of each other and do not depend on $\mathbf{D}$. Then, once $\mathbf{D}$ is given, the numerator and the denominator in (5.2.12) are independent.

Furthermore, given $\mathbf{D}$, the r.v. $b_1D_1 + b_2D_2 + ... + b_nD_n$ is the sum of independent normal r.v.'s. Hence, this sum is normal. Its mean is zero (since each a_{1j} has zero mean), and given $(D_1, ..., D_n)$, the conditional variance of this sum is $D_1^2 + ... + D_n^2$. Consequently, dividing by $\sqrt{D_1^2 + ... + D_n^2}$, we normalize the sum, making its variance equal to one (see also Example 1, where we used the same idea). Thus, given $\mathbf{D} = (D_1, ..., D_n)$, the numerator in (5.2.12) is a standard normal r.v.

The same argument applies to the denominator in (5.2.12). Thus, given $\mathbf{D}$, the conditional distribution of X_1 is the distribution of the ratio of two independent standard normal r.v.'s. By the result of Example 5, this is the Cauchy distribution.

So, the conditional distribution of X_1 given $\mathbf{D}$ does not depend on $\mathbf{D}$ at all and is equal to the Cauchy distribution. Then the unconditional distribution of X_1 must also be equal to the same Cauchy distribution. The formal proof is very similar to what we did when we revisited Example 1; we should only extend the formula for total probability to the case of conditioning with respect to a random vector. □

---

6 EXERCISES

Sections 1 and 2

1. Let a r.v. X have the density $f(x) = \begin{cases} cx^5 & \text{if } 0 < x < 2, \\ 0 & \text{otherwise.} \end{cases}$

   (a) Find c, and the probabilities $P(0.5 \leq X \leq 1)$, $P(X = 1)$, $P(X > 3)$, and $P(X < -1)$. (b) Find and graph the distribution function. (c) Find $E\{X\}$ and $E\{X^7\}$. (d) Find $Var\{X\}$.

2. Let a r.v. X have the density $f(x) = \frac{4}{\pi(1+x^2)}$ for $0 < x < 1$. Find $f(x)$ for all other x's.

3. Let a random variable X have the (cumulative) distribution function

$$F(x) = \frac{e^x - 1}{e - 1} \quad \text{if } 0 \leq x \leq 1.$$

   (a) Find $F(x)$ for $x < 0$, and for $x > 1$.

   (b) Find $P(0.5 \leq X \leq 1)$, $P(X = 1)$, $P(X = 0.25)$.

   (c) Find the density.

4. Let X be the r.v. from Exercise 1. Find the d.f. and density of the r.v. $Y = X^3$.

5. Let X be uniform on [-1,1]. Find the d.f. and density of the r.v. $Y = \arcsin X$. Do the same for the r.v. $Z = |Y|$. (*Hint*: Regarding Z, you can use what you got in the first part of the problem.)

6. Let the d.f. $F(x) = 1 - \cos x$ for $0 \leq x \leq \pi/2$. Check that this function has all properties of d.f.'s. What values would a corresponding r.v. assume? Figure out what $F(x)$ equals for $x \notin [0, \pi/2]$, and graph $F(x)$. Find $P(\frac{\pi}{6} \leq X \leq \frac{\pi}{3})$, $P(X = \pi/3)$. Find the density.

7. Let a r.v. X have the density $f(x) = \begin{cases} 3/4 & \text{if } 0 \leq x \leq 1, \\ 1/4 & \text{if } 1 < x \leq 2. \end{cases}$

   (a) Explain why $P(X \notin [0,2]) = 0$. Without any calculations, find $P(0 \leq X \leq 1)$ and $P(1 \leq X \leq 2)$.

   (b) Reasoning heuristically, explain why $E\{X\}$ should be equal to $\frac{1}{2} \cdot \frac{3}{4} + \frac{3}{2} \cdot \frac{1}{4} = \frac{3}{4}$. Calculate $E\{X\}$ rigorously.

   (c) State and answer similar questions for the case $f(x) = \begin{cases} 3/4 & \text{if } 0 \leq x \leq 1, \\ 1/4 & \text{if } 2 < x \leq 3. \end{cases}$

8. A random variable X has the density $f(x) = 0$ for $x < 0$, and for $x \geq 0$, the density is piecewise constant; namely, $f(x) = \frac{1}{2^{k+1}}$ for $x \in [k, k+1)$ and $k = 0, 1, ...$. Graph $f(x)$. What is $P(k \leq X \leq k+1)$ and $P(k \leq X \leq k + \frac{1}{2} | k \leq X \leq k+1)$?

   Show that the distribution of X coincides with the distribution of the r.v. $Z + K$, where the r.v.'s Z and K are independent, Z is uniform on $[0, 1]$, and K has the second version of the geometric distribution with parameter $p = \frac{1}{2}$. Namely, $P(K = k) = \frac{1}{2^{k+1}}$; see, for example, the Appendix, Table 1. Find $E\{X\}$ and $Var\{X\}$. (*Advice*: When considering the d.f. of X, use the formula for total probability.)

9. This exercise concerns the *Pareto distribution* that proved to be a good model for many real variables such as the sizes of towns, files of Internet traffic, meteorites, sand particles, etc.

   Consider a r.v. ξ_1 such that $P(\xi_1 > x) = x^{-\alpha}$ for $x > 1$ and a parameter $\alpha > 0$. We call a Pareto distribution the distribution of any linear transformations of ξ_1; more precisely, the distribution of any r.v. $\xi = b\xi_1 + d$ for arbitrary d and $b > 0$. The parameter b may be viewed as a scale parameter. Since ξ_1 assumes values from $[1, \infty)$ (*say why!*), the r.v. ξ takes on values from $[b+d, \infty)$, so $b+d$ may be called a location parameter.

   Often, the term "Pareto distribution" is applied to the distribution with the tail

$$\overline{F}(x) = P(X > x) = \left(\frac{\theta}{x+\theta}\right)^{\alpha} \quad \text{for } x \geq 0, \tag{6.1}$$

   where the parameter $\theta > 0$.

   (a) Show that (6.1) corresponds to the distribution of the r.v. $\theta(\xi_1 - 1)$.

   (b) Find the mean and variance of the distribution (6.1). When do they exist?

10. Consider the inventory problem from Example 1.4.2-3.

    (a) Explain why the solution to (1.4.3) is non-decreasing when δ is decreasing. Interpret this fact in the context of the problem.

    (b) Find the optimal size s of the order for two cases: (i) X is uniform on $[0, 2m]$; (ii) X is exponential with a mean of m. (So, the mean demand is the same for both cases.) Which optimal order is larger? Does the answer depends on δ?

11. Graph the density and the distribution function, and find the expectation and standard deviation for the distribution uniform on $[-3, 2]$. Calculate in mind the probability "to get into" the interval $[-2, -1]$.

12. Graph the density and the distribution function of the exponential distribution.

13. The continuous distribution with a density $f(x)$ is said to be *symmetric* if for a number s called a *center of symmetry*, and for any $x > 0$,

$$f(s+x) = f(s-x).$$

    (a) Which distributions considered in Sections 1.1 and 2 are symmetric? With respect to which centers of symmetry?

    (b) Give an example showing that the center of symmetry does not have to coincide with one of the values of the r.v.

    (c) Show that the above definition of symmetric distribution is equivalent to the requirement $P(X > s+x) = P(X < s-x)$ for any $x > 0$.

14. A number μ is called a *median* for a continuous r.v. X if $P(X < \mu) = P(X > \mu) = \frac{1}{2}$.

    (a) Show that for any symmetric distribution, its median, center of symmetry, and mean (if it exists) coincide. (*Advice*: First, switch to the r.v. $X - s$, where X is the r.v. in hand and s is its center of symmetry.)

    (b) Find a median for the exponential distribution, and for the distribution in Exercise 1.

15. A number q is called a *mode* of a continuous distribution with a density $f(x)$ if $f(x)$ attains its maximum at q. (So, q may be called the most probable value.) (a) Does a mode have to be unique? If not, give an example. (b) Find a mode for the normal, exponential, and Γ-distribution. (c) Find a mode for the uniform distribution.

16. Generalizing (1.3.2), prove that for any continuous r.v. with a d.f. $F(x)$ and a finite expected value,
$$E\{X\} = \int_0^\infty (1 - F(x))dx - \int_{-\infty}^0 F(x)dx. \tag{6.2}$$
What would we have for a non-positive r.v.? In Chapter 7, we show that (6.2) is true for any distribution.

17. Suppose that for r.v.'s X_1 and X_2 with respective d.f.'s $F_1(x)$ and $F_2(x)$, it is true that $F_1(x) \leq F_2(x)$ for *all* x's. Such a relation is referred to as the *first stochastic dominance* (FSD), and we say that X_1 (or its d.f. F_1) dominates X_2 (respectively, F_2) in the sense of the FSD.

    (a) Argue that if we follow the rule "the larger, the better" (for example, if the X's are r.v.'s of a future income), then the r.v. X_1 is preferable. (b) In particular, argue that one should expect that $E\{X_1\} \geq E\{X_2\}$. Prove it. (*Advice*: You may appeal to (6.2).) (c) Show that the FSD relation is true, for example, if X_1 and X_2 are uniform on $[0, a_1]$ and $[0, a_2]$, respectively, and $a_1 \geq a_2$; or if X_1 and X_2 are exponential with respective parameters a_1 and a_2, and $a_1 \leq a_2$. In both cases, provide the corresponding graphs.

18. Let X be an exponential r.v., and $P(X > 1) = 0.2$. Not computing the parameter a, find $P(X > 3)$ and $P(X > 1.5)$.

19. Ann is the first in line for a service. There are three service counters (which are busy), the service times are independent and have the same exponential distribution. Find the probabilities that Ann will be served after (a) all three customers who were being served when Ann took her place in line, (b) the customer at the first counter.

20. Compute the mean of the standard exponential distribution using formula (1.3.2) and compare your calculations with what we did in Section 2.2.

21. Let a r.v. X be exponentially distributed, and $E\{X\} = 2$. Write the density. Find the variance. Find $P(X \leq 4 \,|\, X > 1)$, $P(3 \leq X \leq 4 \,|\, X > 1)$.

22. For a random variable X having the-lack-of-memory property, let $P(X > 2) = 3/4$. Find the parameter a, $E\{X\}$, and the standard deviation of X.

23. The claim department of a company is receiving claims. The time between the arrivals of two consecutive claims (an interarrival time) is an exponential r.v., and the probability that this time is larger than 1 min. equals $1/e$. Say without any calculations what the mean interarrival time is. Suppose that it is 2 pm now, and we do know that the last call for a claim was 2 mins ago. What is the probability that the next claim will appear (a) not earlier than 2:02 pm; (b) before 2:02 pm; (c) at 2:02 pm?

24. Suppose you have found a median and mode of the Γ-distribution with the scale parameter $a = 1$. How to find the same characteristics for an arbitrary $a > 0$? Justify the answer.

25. For any distribution with mean $m \neq 0$ and standard deviation σ, the ratio σ/m is called a *coefficient of variation* (c.v.); see also Exercise **3**.53.

    (a) Will the c.v. change if we multiply the r.v. by a number $c \neq 0$?

    (b) Does the c.v. of a Γ-distribution depend on the scale parameter a?

(c) Compute the c.v. of the Γ-distribution. What can you say if it is equal to (i) one, (ii) one-fourth?

26. Let $X_{a,\nu}$ denote a r.v. having the Γ-distribution with parameters a and ν.

 (a) What is the difference between $P(X_{1,\nu} < x)$ and $P(aX_{a,\nu} < x)$?

 (b) For which ν does the Γ-density $f_{1,\nu}(x)$ converges to ∞ as $x \to 0$? For which ν is this limit equal to zero? Are there other possibilities? What about $\lim_{x\to\infty} f_{1,\nu}(x)$? Which of the answers above will be different for $f_{a,\nu}(x)$?

 (c) Show that $f_{1,\nu+1}(x) \leq f_{1,\nu}(x)$ for $x \leq \nu$, and $f_{1,\nu+1}(x) \geq f_{1,\nu}(x)$ for $x \geq \nu$. Sketch the graphs of both functions for $\nu = 2$ in the same picture. (*Hint*: By virtue of (2.3.1), $\frac{x^{\nu}}{\Gamma(\nu+1)} = \frac{x}{\nu} \cdot \frac{x^{\nu-1}}{\Gamma(\nu)}$.)

 (d) Using Excel or another software, find the probability that $X_{2,16}$ will fall within one standard deviation from its expected value.

27. When solving problems below, it makes sense to keep in mind (2.4.3).

 (a) Let X be standard normal. What is the distribution of $-X$? Generalize the problem considering a normal r.v. with zero mean. If X is normal with mean m and variance σ^2, then what is the distribution of $-X$?

 (b) In the same picture, sketch the graphs of the normal distributions with the unit variance and the means equal to 1 and -1, respectively.

 (c) In the same picture, sketch the graphs of the normal distributions with zero mean and the standard deviations equal to 1 and 2, respectively. (*Advice*: Compare the values of the two densities at $x = 0$ and for large x's.)

28. Let a r.v. Y be normal with an expectation of 3, and a variance of 16. Write a formula for $P(-2 \leq Y \leq 6)$. Estimate this probability using Table 3 from the Appendix.

29. Let X be a normal r.v. with mean 12 and variance 4. Write the normalized r.v. Estimate c such that $P(X < c) = 0.2$.

Sections 3 and 4

30. Let X_1 and X_2 be continuous r.v.'s with the joint density $f(x_1,x_2) = 1/2$ if $|x_1| + |x_2| \leq 1$, and $= 0$ otherwise. Explain why $f(x_1,x_2)$ is a density. Find $P(X_1^2 + X_2^2 \leq 1/2)$. (*Hint*: The answer will be almost immediate if we draw a picture.)

31. Let a three-dimensional r.vec. $\mathbf{X}$ be uniformly distributed in a unit ball with a center at the origin. Find the d.f., expectation and variance of $|\mathbf{X}|$, the length of $\mathbf{X}$.

32. Let the two-dimensional density of a r.vec. $\mathbf{X} = (X_1, X_2)$ be $f(\boldsymbol{x}) = \dfrac{1}{\pi(1+|\boldsymbol{x}|^2)^2}$, where $|\boldsymbol{x}| = \sqrt{x_1^2 + x_2^2}$, the length of the vector $\boldsymbol{x}$. Find $P(|\mathbf{X}| > 2)$, $E\{X_1\}$. (*Hint*: One of these quantities does not require calculations.)

33. Figure out whether the coordinates X_1, X_2 are independent in (a) Exercise 32, (b) Example 3-1.

34. A random vector (X,Y) has the joint density function

$$f(x,y) = \begin{cases} 6xy^2 & \text{if } 0 \leq x \leq 1,\ 0 \leq y \leq 1 \\ 0 & \text{otherwise.} \end{cases}$$

Show that $f(x,y)$ is a density. *Guess* whether X and Y are independent. Justify your answer. Write the marginal densities.

35. Suppose that a r.vec. $\mathbf{X} = (X_1, X_2)$ has the distribution uniform on the circle (not a disk!) $x_1^2 + x_1^2 = r^2$ for a fixed r. This means that the vector $\mathbf{X}$ takes values (x_1, x_2) only from the circle mentioned, and the probability that a value of $\mathbf{X}$ will fall into an arc of the circle is equal to the length of this arc divided by the length of the whole circle.

    Show that the distribution of the r.vec. $\mathbf{X}$ is not absolutely continuous; that is, the probability density function does not exist. Show that, nevertheless, the distribution of $\mathbf{X}$ may be identified with a one-dimensional uniform distribution. On which interval?

36. A random vector (X, Y) has the joint density function

$$f(x,y) = \begin{cases} c(x^2 + y^2) & \text{if } 0 \le x \le 1,\ 0 \le y \le 1 \\ 0 & \text{otherwise.} \end{cases}$$

    Find c. *Guess* whether X and Y are independent. Justify your answer. Write the marginal densities.

37. Let us revisit the record value problem of Section **3**.2.3. Make sure that all reasonings there remain true in the case where X's are continuous. Show that in this case $P(N > n) = \frac{1}{n+1}$, and that $P(N < \infty) = 1$; that is, N cannot take an infinite value. Show that $P(N = n) = \frac{1}{n(n+1)}$ for $n = 1, 2, ...$ (*Advice*: Show that, since X's are independent and continuous, with probability one, among the r.v.'s $X_1, ..., X_n$, only one r.v. attains the largest value.)

    We understand that we consider the r.v. N for the case when X_0 is random; that is, the value of X_0 is a future value for us, as well as values of other X's. What is the conditional distribution of N *given* that X_0 has assumed a particular value x_0?

38. (a) Give a common-sense explanation why values of S_2 in Example 4.1-2 are not equally likely.

    (b) Carry out calculations in (4.1.4).

39. (a) Can we switch f_1 and f_2 in (4.1.1)?

    (b) How should we change formula (4.1.1) to get the density of the r.v. $X_1 - X_2$?

40. Find and graph the probability density of the sum of independent r.v.'s X_1 and X_2 if

    (a) X_1 and X_2 are exponential with $E\{X_1\} = 1$ and $E\{X_2\} = 1/2$, respectively;

    (b) X_1 and X_2 are uniform on $[0,1]$ and $[0,2]$, respectively.

41. Let X_1, X_2 be independent standard exponential r.v.'s. Show that the density of the r.v. $X_1 - X_2$ is

$$f(x) = \frac{1}{2} e^{-|x|}.$$

    Graph it. The distribution with this density is called *two-sided exponential*. Show also that the same $f(x)$ is the density of the r.v. X equal to X_1 or $-X_2$ with equal probabilities.

42. A signal, when going through an electronic device, is distorted, which amounts to adding a random distortion error X. Suppose that X is normal with zero mean and a standard deviation of ε. The latter may be viewed as a measure of distortion. Assume that the signal is going through nine consecutive independently functioning devices identical to the device above. What is the distribution of the total distortion? Is the measure of total distortion (the standard deviation) nine times as large as this measure for one device? Find the probability that the absolute value of the total distortion error does not exceed 6ε.

43. Let X and Y be independent continuous random variables with the densities $e^{\pi(x-1)^2}$ and $\frac{1}{2\sqrt{\pi}}e^{-(x-2)^2/4}$, respectively. Using Table 3 from the Appendix, estimate $P(2 \leq S \leq 4)$, where $S = X + Y$.

44. Using the convolution formula, show that the sum of two independent standard normal r.v.'s is normal with mean 0 and variance 2 (which is consistent with Proposition 5). (*Advice*: First, show that $f_S(x) = \int_{-\infty}^{\infty} \varphi(x-y)\varphi(y)dy = (2\pi)^{-1}\int_{-\infty}^{\infty} \exp\left\{xy - y^2 - \frac{1}{2}x^2\right\} dy$.
Completing the square, write $xy - y^2 - \frac{1}{2}x^2 = -(y - x/2)^2 - \frac{1}{4}x^2$, and use the change of variables $y - x/2 = s/\sqrt{2}$.)

45. Let Z_n be a Γ-r.v. with a scale parameter of a and parameter ν equal to an integer n. Show that Z_n may be represented as $Y_1 + ... + Y_n$, where Y's are independent exponential r.v.'s with parameter a.

46. Let r.v.'s X_1 and X_2 be both Γ-distributed with a common scale parameter a and "essential" parameters ν_1 and ν_2; $\nu_1 > \nu_2$. Show that X_1 dominates X_2 in the sense of the FSD (see Exercise 17). Can we claim that $P(X_1 \geq X_2) = 1$? (*Advice*: To answer the second question, we may assume X_1 and X_2 to be independent.)

47. Consider Example 4.2-1. (a) What is the mean of the total waiting time $E\{S_4\}$? (b) What is the probability that S_4 will be equal to $E\{S_4\}$? (c) When computing $P(S_4 < E\{S_4\})$, can we set $a = 1$? Using software, compute $P(S_4 < E\{S_4\})$ and $P(S_4 > E\{S_4\})$, compare and connect the result of comparison with the fact that the Γ-distribution is skewed to the right (see, for example, Fig. 9).

48. Let $S_n = X_1 + ... + X_n$, where the X's are independent, and X_i has the Γ-distribution with parameters $(1, (\frac{1}{2})^i)$. To which distribution is the distribution of S_n close for large n? Write $\lim_{n\to\infty} f_{S_n}(x)$.

49. The densities of independent r.v.'s X_1 and X_2 are $f_1(x) = C_1x^5e^{-4x}$ and $f_2(x) = C_2x^8e^{-4x}$, respectively, where C_1 and C_2 are constants.

    (a) Do we need to calculate these constants in order to find $P(X_1 + X_2 > x)$?

    (b) Estimate $P(X_1 + X_2 > 3)$. (Use of software is recommended.)

    (c) Write C_1 and C_2.

Section 5

50. Using formula (5.1.5), prove that, if independent r.v.'s X_1 and X_2 are standard normal and $S = X_1 + X_2$, then the conditional distribution of X_1 given $S = s$, is normal with a mean of $s/2$ and a variance of $1/2$.

51. Let X_1 and X_2 have the Γ-distributions with the common scale parameter $a = 1$ and essential parameters ν_1 and ν_2, respectively. Let $S = X_1 + X_2$. Given $S = s$, the r.v. X_1 takes on values from $[0, s]$.

    (a) Using formula (5.1.5), prove that the conditional density

$$f_{X_1|S}(x|s) = \frac{1}{B(\nu_1,\nu_2)s^{\nu_1+\nu_2-1}}x^{\nu_1-1}(s-x)^{\nu_2-1} \text{ for } 0 \leq x \leq s, \tag{6.3}$$

    where the Beta-function $B(\nu_1, \nu_2)$ is defined in (4.2.3)-(4.2.4).
    Prove that the corresponding conditional density of the r.v. $Y = \frac{1}{s}X_1$ which takes on values from $[0, 1]$, is

$$f_{Y|S}(x|s) = \frac{1}{B(\nu_1,\nu_2)}x^{\nu_1-1}(1-x)^{\nu_2-1} \text{ for } 0 \leq x \leq 1, \tag{6.4}$$

and does not depend on s. The last distribution is referred to as the *Beta-distribution.*

(b) Which distribution are we dealing with in (6.4) when $\nu_1 = \nu_2 = 1$. Connect this fact with the result of Example 5.1-2.

(c) Proceeding from (6.4), find the conditional distribution of Y and $E\{Y\,|\,S\}$ for $\nu_1 = \nu$ and $\nu_2 = 1$. Does it grow when ν is increasing? To what does this distribution and $E\{Y\,|\,S\}$converge as $\nu \to \infty$. Give a common-sense interpretation.

52. Let the joint density of (X,Y) be $f(x,y) = x+y$ for $0 < x \leq 1, 0 \leq y \leq 1$, and $= 0$ otherwise. Find the conditional density $f(y\,|\,x)$, and $E\{Y\,|\,X\}$. At which x does the regression function $m(x) = E\{Y\,|\,X = x\}$ attain its maximum? Find $E\{X\,|\,Y\}$.

53. Let the joint density of (X,Y) be $f(x,y) = 2y\dfrac{e^{-x}}{x^2}$ for $0 \leq y \leq x$, and $= 0$ otherwise. Find the conditional density $f(y\,|\,x)$ and $E\{Y\,|\,X\}$.

54. A phone representative of a company is waiting for the next customer's call. Suppose that men and women call independently, the waiting time for the next man's call is exponentially distributed with a mean of m_1, while for women the corresponding r.v. is exponential with a mean of m_2. It is 2 pm now. Find the probability that the next call will be from a man.

55. Let us revisit Example 5-2. Suppose that Rich is the third in line. Given that Rich has been waiting t minutes, find the distribution of the waiting time for being the first in line. Find the conditional mean. (*Advice*: The reader who has solved Exercise 51 may just apply the result. The reader who skipped Exercise 51, may consider this particular case proceeding from the formula (5.1.5) where the role of X_1 should be played by S_2, the role of X_2 by X_3, and the total $S = S_3$. For the densities of S_2 and S_3, we may use Proposition 6.)

56. In Example 5.2-4, set $m_1 = m$, $m_2 = 1$ and graph $P(\xi_2 > \xi_1)$ as a function of m. (*Advice*: Pay attention to the behavior in a neighborhood of $m = 1$.)

57. R.v's ξ_1, ξ_2 are independent and exponential with a parameter a. Find the distribution (say, the density) of the r.v. $Z = \xi_1/\xi_2$. (*Advice*: First, realize that the distribution of Z should not depend on a.)

58. Let $\xi_1,...,\xi_n$ be i.i.d. r.v.'s with a continuous distribution. Explain why the r.v. $X_1 = \frac{\xi_1}{\xi_1+...+\xi_n}$ is well defined; that is, we should not worry about the case where the denominator is zero. Find $E\{X_1\}$ and $E\{X_1\,|\,\xi_1 + ... + \xi_n\}$, as well as the similar expectations of the other r.v.'s $X_i = \frac{\xi_i}{\xi_1+...+\xi_n}$. (*Advice*: Use the additivity property of conditional expectations.)

59. Let X_1 and X_2 be i.i.d. r.v.'s. Assume that we know the values of $X_{\min} = \min\{X_1, X_2\}$, and $X_{\max} = \max\{X_1, X_2\}$. What can we say about the expected value of, for instance, X_1 given this information?

60. Let Θ be a r.v. whose density $f(\theta) = 3\theta^2$ for $0 \leq \theta \leq 1$, and $f(\theta) = 0$ otherwise. Let N be the number of the first success in a sequence of independent trials with the probability of success equal to Θ. In other words, the probability of success is random being the same for all trials. Find $E\{N\}$ and $Var\{N\}$.

61. Let us come back to Example 1.1-6. Suppose that we know that the numbers a and b are chosen independently and at random from the interval $[0,1]$. Then it makes sense to choose X taking on values from $[0,1]$. Find $P(A)$ for X uniform on $[0,1]$. (*Advice*: Condition on a and b.)

Chapter 7

Distributions in the General Case. Simulation

1,2

In this chapter, we consider arbitrary r.v.'s, not necessarily purely discrete or continuous. As we will see, a typical case is that where a r.v. with a positive probability is discrete and with the complement probability—continuous.

EXAMPLE. Michael's trip to school takes from 20 to 25 minutes and all values are equally likely. However, with probability 0.05, Michael misses school. Then, with probability 0.05, the r.v. X, the time it takes to go to school on a randomly chosen day, equals a discrete r.v. $X_0 \equiv 0$, and with probability 0.95, X equals a continuous r.v. uniform on $[20, 25]$. □

1 DISTRIBUTION FUNCTIONS

1.1 Properties of d.f.'s

Let X be a r.v. and $F(x) = F_X(x) = P(X \leq x)$, its *(cumulative) distribution function.* As has been already noted in Chapters 3 and 6, each such a function is non-decreasing (when x is increasing, $P(X \leq x)$ is getting larger or remains the same), $F(-\infty) = \lim_{x \to -\infty} F(x) = 0$, and $F(\infty) = \lim_{x \to \infty} F(x) = 1$.

In the general case, $F(x)$ does not have to be continuous; for example, this is not true if X is discrete. However, $F(x)$ is *always right continuous.*

Indeed, let x converge to a number c from the right; that is, $x \to c$ and $x > c$. Then, for any such an x, the event $\{X \leq x\}$ contains the event $\{X \leq c\}$, and hence $F(x) = P(X \leq x) \to P(X \leq c) = F(c)$ as $x \to c$ and $x > c$; see also Fig. 1.[1]

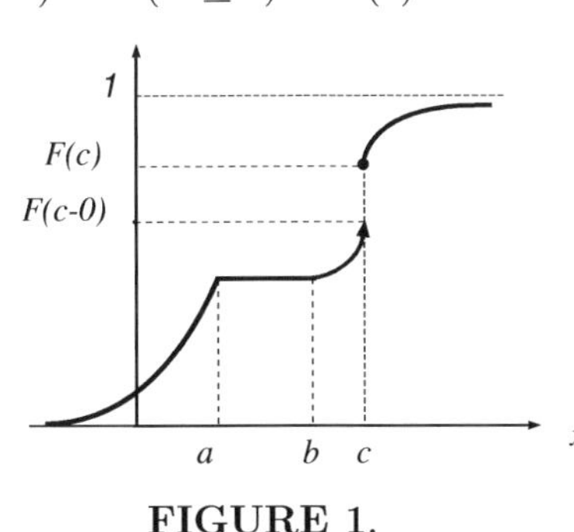

FIGURE 1.

If $x \to c$ from the left, i.e., if $x \to c$ and $x < c$, then the event $\{X \leq x\}$ does not include the event $\{X = c\}$, and hence, in this case, $F(x) = P(X \leq x) \to P(X < c)$.

Clearly, $P(X < c)$ is not equal to $P(X \leq c)$ if and only if X assumes the value c with a positive probability.

Set $F(c-0) = P(X < c)$ (see Fig. 1). Since $P(X = c) = P(X \leq c) - P(X < c)$, we have

$$P(X = c) = F(c) - F(c-0),$$

[1]For the reader who did not skip Section 1.2.6, note that we use the continuity of probability measures.

or (see also Fig. 1)

For any c, the probability $P(X=c)$ equals the jump of $F(x)$ at the point c. (1.1.1)

Consider the probabilities of intervals. For any interval $(a,b]$ (the point a is not included),

$$P(a<X\leq b)=P(X\leq b)-P(X\leq a)=F(b)-F(a). \tag{1.1.2}$$

(We *subtract* $P(X\leq a)$ rather then $P(X<a)$, because a is *not* included in $(a,b]$.)

If $F(x)$ is constant on $[a,b]$, then $F(a)=F(b)$, and we conclude that

If $F(x)$ is constant on an interval $[a,b]$, then $P(X\in(a,b])=0$. (1.1.3)

See again Fig. 1. For the other three types of intervals, we have

$$P(a\leq X\leq b)=P(X\leq b)-P(X<a)=F(b)-F(a-0), \tag{1.1.4}$$
$$P(a\leq X<b)=P(X<b)-P(X<a)=F(b-0)-F(a-0), \tag{1.1.5}$$
$$P(a<X<b)=P(X<b)-P(X\leq a)=F(b-0)-F(a). \tag{1.1.6}$$

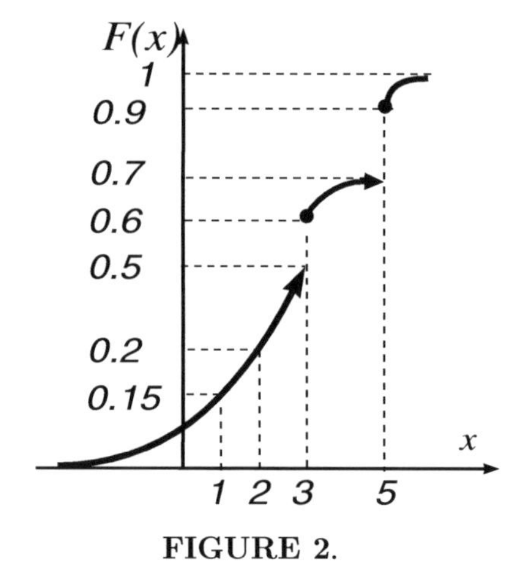

FIGURE 2.

EXAMPLE 1. Let a r.v. X have the d.f. graphed in Fig. 2. There are two values with positive probabilities: $P(X=3)=0.6-0.5=0.1$, and $P(X=5)=0.9-0.7=0.2$. Let us consider some intervals. At points $x=1$ and $x=2$, the function $F(x)$ is continuous, and hence, it does not matter whether we include the endpoints of the interval $[1,2]$ or not: $P(1<X\leq 2)=P(1<X<2)=P(1\leq X<2)=P(1\leq X\leq 2)=0.2-0.15=0.05$. For the interval $[3,5]$, using above formulas, we have

$$P(3<X\leq 5)=F(5)-F(3)=0.9-0.6=0.3;$$
$$P(3<X<5)=F(5-0)-F(3)=0.7-0.6=0.1;$$
$$P(3\leq X<5)=F(5-0)-F(3-0)=0.7-0.5=0.2;$$
$$P(3\leq X\leq 5)=F(5)-F(3-0)=0.9-0.5=0.4. \ \square$$

We conclude the description of properties of d.f.'s with

Proposition 1 *For any d.f. $F(x)$, the set of its discontinuity points (that is, points where the function jumps) is either finite or countably infinite.*

Proof. Let $D_n=\{x: F(x)-F(x-0)\geq 1/n\}$, the set of all points with jumps not less than $1/n$; $n=1,2,...$. The number of the points in D_n is not larger than n: otherwise, the sum of the jumps would be greater than one, and there would exist a point x at which $F(x)>1$. On the other hand, the set of all discontinuity points of $F(x)$ is the set $D=\cup_n D_n$. The union of finite sets is either finite or countably infinite, so D is countable. ■

1.2 Three basic propositions

The next question is whether we actually had a right to consider, for example, the function in Fig. 2 as a d.f. In other words, can *any* function of this type be the distribution function of some random variable? The positive answer is given by

Proposition 2 *For any non-decreasing right continuous function $F(x)$ defined on the real line and such that $F(-\infty) = 0$ and $F(\infty) = 1$, there exists a r.v. X for which $P(X \leq x) = F(x)$.*

The proof of Proposition 2 is constructive and turns out to be useful for many purposes.

To provide it, we consider the inverse function $F^{-1}(y)$, but because $F(x)$ may be discontinuous at some points and/or constant on some intervals, we should specify what we mean.

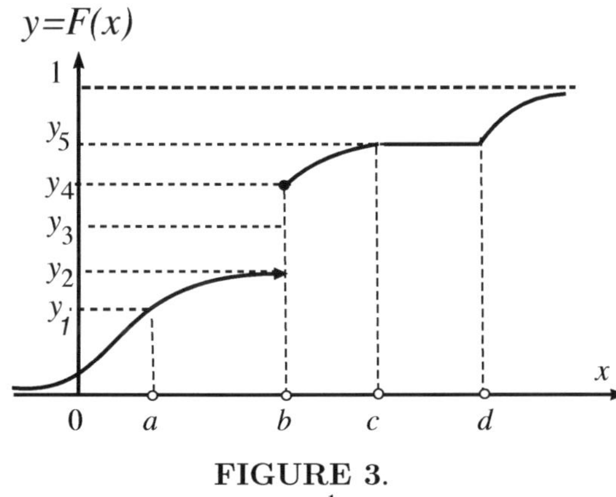

FIGURE 3.

If for a value y, there is only one x for which $y = F(x)$, we set $F^{-1}(y) = x$. For example, in Fig. 3, $F^{-1}(y_1) = a$.

If for a number y, there is no x for which $y = F(x)$, this means that $F(x)$ "jumps over" this value; see, for example, y_2 and y_3 in Fig. 3. In this case, we set $F^{-1}(y)$ equal to the point x at which $F(x)$ jumps over y. In Fig. 3, $F^{-1}(y_2) = F^{-1}(y_3) = F^{-1}(y_4) = b$.

If $F(x) = y$ for all x's from some interval, we choose as $F^{-1}(y)$ the right end point of the interval where $F(x) = y$. In Fig. 3, $F^{-1}(y_5)$ is *not* c but d, the right point of the interval $[c, d]$.

The value $F^{-1}(y)$ so defined is also called the *y-quantile* of the r.v. X and its distribution. If $F(x)$ is a continuous and increasing function, then for any y there is only one x for which $P(X \leq x) = y$, and such an x is the y-quantile of F. If $F(x)$ is discontinuous at some points, or it is constant on an interval, we should be more accurate.

► For the reader who is familiar with the notion of supremum, note that so defined the inverse $F^{-1}(y) = \sup\{x : F(x) \leq y\}$. ◄

Proposition 3 *Let $F(x)$ be a non-decreasing right continuous function such that $F(-\infty) = 0$ and $F(\infty) = 1$, and let Y be a r.v. uniform on $[0, 1]$. Then the distribution function of the r.v.*

$$X = F^{-1}(Y) \tag{1.2.1}$$

is equal to $F(x)$.

Clearly, Proposition 3 implies Proposition 2.

Above its theoretical significance, representation (1.2.1) suggests a way of simulating values of a r.v. with a (theoretically, arbitrary) d.f. $F(x)$. We will consider this in detail in Section 4.2.

Proof of Proposition 3. Formally, the proof is short. As we know, $P(Y \leq y) = y$ for $0 \leq y \leq 1$. Hence, for any x, we have $P(X \leq x) = P(F^{-1}(Y) \leq x) = P(Y \leq F(x)) = F(x)$.

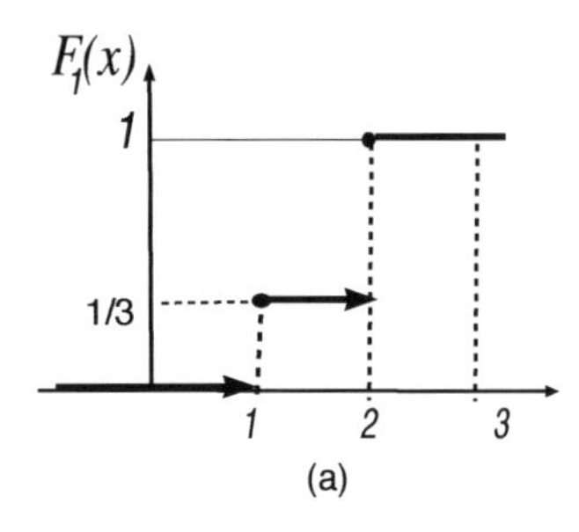

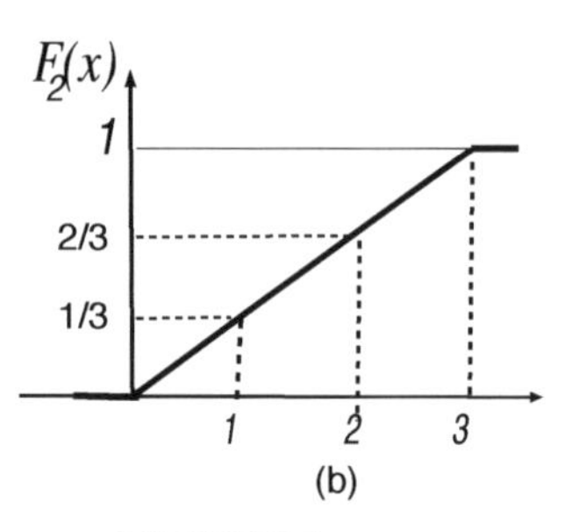

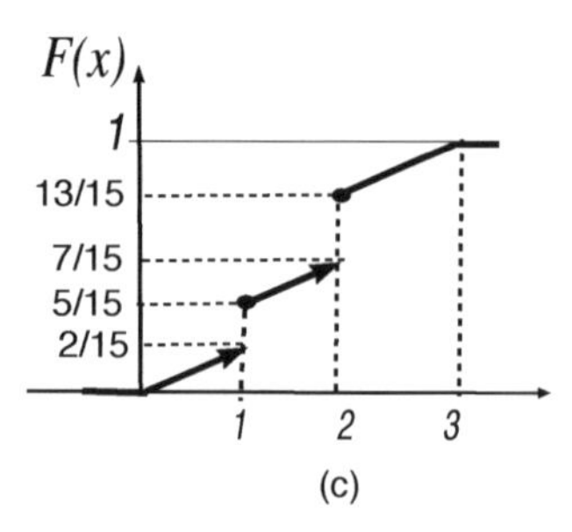

FIGURE 4.

However, it makes sense to clarify the chain of the equalities above, considering particular points in Fig. 3. The graph in this figure may be viewed as that of $F^{-1}(y)$ if we look at it "from the left." We can also think that the uniform r.v. Y assumes values from the interval $[0,1]$ in the y-axis. Then the r.v. X, as a function of Y, takes on values from the x-axis.

The event $\{X \le a\}$ corresponds to the event $\{Y \le y_1\}$. Since Y is uniform, the probability of the latter event is y_1, and $P(X \le a) = P(Y \le y_1) = y_1 = F(a)$, which is consistent with what we are proving.

The event $\{X \le b\}$ does not correspond to the events $\{Y \le y_2\}$ or $\{Y \le y_3\}$, but rather to the event $\{Y \le y_4\}$. The probability of this event is $y_4 = F(b)$.

The r.v. X does not assume values from inside $[c,d]$. So, $P(X \le c) = P(X \le d) = y_5$ and equals $P(Y \le y_5) = y_5$. Note also that $P(X = b) = P(y_2 \le Y \le y_4) = y_4 - y_2 = F(b) - F(b-0)$. ■

In conclusion, note the following. If we know the d.f. of a r.v. X, then proceeding from (1.1.2)–(1.1.6), we can find the probability $P(X \in B)$ for an arbitrary interval B. Then we can find the same probability for B being an arbitrary combination of arbitrary intervals. Eventually, this brings us to $P(X \in B)$ for any B for which this probability is well defined. So, we state

Proposition 4 *The d.f. $F(x)$ of a r.v. X completely determines probabilities $P(X \in B)$ for all sets B in the real line, for which these probabilities may be defined.*

It still needs a formal proof which we skip, but we will use this proposition repeatedly.

1.3 Decomposing into discrete and continuous components

EXAMPLE 1. Let a r.v. X_1 be discrete and take on values 1 and 2 with probabilities $\frac{1}{3}$ and $\frac{2}{3}$, respectively. The d.f. $F_1(x)$ of X_1 is graphed in Fig. 4a. Let X_2 be a (continuous) r.v. uniformly distributed on the interval $[0,3]$. The graph of its d.f., $F_2(x)$, is given in Fig. 4b. Consider a r.v.

$$X = \begin{cases} X_1 & \text{with probability } 3/5 \\ X_2 & \text{with probability } 2/5. \end{cases} \tag{1.3.1}$$

In other words, with probability $3/5$, we choose X_1 and deal with this r.v.; otherwise, we deal with X_2. (We skip formalities connected with defining these r.v.'s on the same sample space.) Then the d.f. of X is

$$\begin{aligned} F(x) &= P(X \le x) = P(X \le x \mid X_1 \text{ is chosen}) P(X_1 \text{ is chosen}) \\ &\quad + P(X \le x \mid X_2 \text{ is chosen}) P(X_2 \text{ is chosen}) \\ &= P(X_1 \le x) \cdot \frac{3}{5} + P(X_2 \le x) \cdot \frac{2}{5} = \frac{3}{5} F_1(x) + \frac{2}{5} F_2(x). \end{aligned} \tag{1.3.2}$$

The reader is invited to verify that substituting the functions from Fig. 4ab into (1.3.2), we come to the d.f. graphed in Fig. 4c. For example, $F(1) = \frac{3}{5}F_1(1) + \frac{2}{5}F_2(1) = \frac{3}{5} \cdot \frac{1}{3} + \frac{2}{5} \cdot \frac{1}{3} = \frac{5}{15} = \frac{1}{3}$, while $F(1-0) = \frac{3}{5}F_1(1-0) + \frac{2}{5}F_2(1-0) = \frac{3}{5} \cdot 0 + \frac{2}{5} \cdot \frac{1}{3} = \frac{2}{15}$. The value of the jump of $F(x)$ at point 2 is $\frac{3}{5} \cdot \frac{2}{3} = \frac{2}{5}$, which is also consistent with the numbers in Fig. 4c.

However, it is more interesting to look at it from the *opposite point of view*. Suppose we consider a r.v. X with a d.f. $F(x)$ (say, as in Fig. 4c), but we have no idea about the structure of X (as in (1.3.1)). Nevertheless, we can decompose $F(x)$ in a unique way into two components: discrete and continuous, and write

$$F(x) = \alpha F_d(x) + (1-\alpha)F_c(x), \tag{1.3.3}$$

where $F_d(x)$ is the d.f. of a discrete r.v. (for brevity, we say a discrete d.f.), $F_c(x)$ is a continuous d.f., and α may be called a weight corresponding to the discrete component.

Let us look at how it may be done in our particular example. The *total sum of the jumps* of $F(x)$ from Fig. 4c is $\frac{3}{15} + \frac{6}{15} = \frac{3}{5}$. This is α, the "probability mass" or "share" of the discrete component. Clearly, the jumps of the discrete component $F_d(x)$ should be at points 1 and 2. Since the second jump is twice as large as the first, the jumps of $F_d(x)$ should be $\frac{1}{3}$ and $\frac{2}{3}$, respectively.

So, we have come to the d.f. in Fig4a, and now we know $F(x)$, $F_d(x)$, and α. Then, we can find $F_c(x)$ proceeding from (1.3.3) and writing $F_c(x) = \frac{1}{1-\alpha}(F(x) - \alpha F_d(x))$. As is easy to verify, this will lead us to the continuous d.f. in Fig. 4b (equal to $x/3$), and eventually, to the decomposition in the very r.-h.s. of (1.3.2).

Once we have come to (1.3.2), we may *interpret* it in terms of (1.3.1). □

The similar result is true in the general case.

Theorem 5 *Let $F(x)$ be a non-decreasing right continuous function such that $F(-\infty) = 0$ and $F(\infty) = 1$. Then there exist a number $\alpha \in [0,1]$, a discrete d.f. $F_d(x)$ and a continuous d.f. $F_c(x)$ such that the decomposition (1.3.3) is true. This decomposition is unique.*

The distribution F in (1.3.3) is sometimes called a *mixture* of discrete and continuous distribution.

Proof is close to what we did in Example 1, so we give an outline. Let $x_1, x_2, \ldots$ be the discontinuity points of $F(x)$; in view of Proposition 1 the set of these points is finite or countable. We add up all jumps at these points which gives us α. If $\alpha = 1$, the original distribution is discrete, and we do not need decomposing. If $\alpha < 1$, we construct a discrete distribution F_d concentrated at points $x_1, x_2, \ldots$ with jumps proportional to the jumps of the original d.f. $F(x)$. The continuous component $F_c(x) = \frac{1}{1-\alpha}(F(x) - \alpha F_d(x))$. ■

The last question is whether the continuous component $F_c(x)$ always has a density. The answer is negative. We know that a density $f(x)$ is the derivative of the corresponding d.f., and we also know that the d.f. may be non-differentiable at some points: in these points the density is discontinuous. However, unfortunately (or fortunately, if we are glad to see the nature being not so simple), there are continuous d.f.'s that are nowhere differentiable.

Continuous distributions not having densities are called *singular*. They are rather special and very rarely appear in models of real phenomena. (See, e.g., [13], [24], [42], [47].)

> From now on, we consider only distributions satisfying representation (1.3.3) where the continuous component F_c is absolutely continuous, that is, has a density. (1.3.4)

Denote by $x_1, x_2, ...$ the points at which the discrete component $F_d(x)$ jumps, and by $f_1, f_2, ...$ the corresponding probabilities. Denote by $f(x)$ the density of the (absolutely) continuous component $F_c(x)$. Then, for a set B on the real line, for the corresponding r.v. X,

$$P(X \in B) = \alpha \sum_{i: x_i \in B} f_i + (1-\alpha) \int_B f(x)dx. \tag{1.3.5}$$

Indeed, let a r.v. X have the distribution given in (1.3.5). Substituting $B = (-\infty, x]$, we get that the d.f. of such an X satisfies (1.3.3). But, by Proposition 4, for all r.v.'s having the same d.f., the probabilities $P(X \in B)$ for arbitrary B's should be also the same.

---

2 EXPECTED VALUES

2.1 Definitions

Below, unless the contrary is stated, we assume that all series or integrals under consideration exist and are finite. We follow the notation of the previous section.

Let us come back to (1.3.5). In accordance with the definitions of expectation for the discrete and continuous cases, we define the *expected value*, or the *mean value*, as

$$E\{X\} = \alpha \sum_i x_i f_i + (1-\alpha) \int_{-\infty}^{\infty} x f(x)dx. \tag{2.1.1}$$

Note also that it does not matter whether we include or exclude from the integration the set of points $\{x_1, x_2, ...\}$: the value of the integral will be the same. In view of this, we can present (2.1.1) in the compact form

$$E\{X\} = \int_{-\infty}^{\infty} x dF(x), \tag{2.1.2}$$

where

> We view $dF(x)$ as $P(x \le X \le x + dx)$, the probability that X will assume a value from an infinitesimally small interval $[x, x+dx]$.

More precisely, we understand the "differential" $dF(x)$ as follows.

- At a discontinuity point x_i, the d.f. $F(x)$ has a jump of $\delta_i = F(x_i) - F(x_i - 0) = \alpha f_i$, and $P(X = x_i) = \delta_i$. So, we set $dF(x_i) = \delta_i$. Such a definition is natural for the following reason. The differential $dF(x)$ means the change of $F(x)$ in the infinitesimally small interval $[x, x + dx]$, but if $F(x)$ jumps at x_i, the change is not small and equals δ_i.

  So, the "part of the integral" in (2.1.2) at point x_i is $\alpha x_i f_i$.

- If $F(x)$ has a density $f(x)$, then $dF(x) = f(x)dx$. In the general case (1.3.5), if $F(x)$ is continuous at a point x (and hence, x does not coincide with any x_i), then we set $dF(x) = (1-\alpha)f(x)dx$.

  So, the "part of the integral" in (2.1.2) at point $x \neq x_i$ is $(1-\alpha)xf(x)dx$.

It also makes sense to emphasize that the r.-h.s. of (2.1.2) is *not a new expression or a new notion* but merely a compact (and convenient) form of presenting the r.-h.s. of definition (2.1.1). The integral in (2.1.2) is called a *Riemann–Stieltjes integral.*

Certainly, if $F(x)$ is either a purely discrete or purely absolutely continuous d.f. ($\alpha = 1$ or 0, respectively), definitions (2.1.1)–(2.1.2) coincide with the definitions of expectation in these cases.

It is straightforward to verify that the expectation we defined, has all the properties we established in the discrete and continuous cases.

EXAMPLE 1. Let us compute the expected value of the r.v. X from Example 1.3-1. In this case, the d.f.

$$F(x) = \frac{3}{5}F_d(x) + \frac{2}{5}F_c(x),$$

where $F_d(x)$ is the d.f. of a r.v. taking values 1 and 2 with probabilities $1/3$ and $2/3$, and $F_c(x)$ is the d.f. of the uniform distribution on $[0,3]$. Hence, in our example, $f(x) = \frac{1}{3}$ for $x \in [0,3]$, and $f(x) = 0$ otherwise. In accordance with (2.1.1),

$$E\{X\} = \frac{3}{5}\left(1 \cdot \frac{1}{3} + 2 \cdot \frac{2}{3}\right) + \frac{2}{5}\int_0^3 x \cdot \frac{1}{3}dx = \frac{3}{5} \cdot \frac{5}{3} + \frac{2}{5} \cdot \frac{3}{2} = 1.6. \ \square$$

Next, let $Y = u(X)$, where $u(x)$ is a function. The d.f. $F_Y(y)$ can also be decomposed into a discrete and continuous component, and, as we assumed, the latter has a density. Then $E\{Y\}$ is well defined, and we can state

Proposition 6 *For any $u(x)$ such that the expectations below are finite,*

$$E\{Y\} = E\{u(X)\} = \alpha \sum_i u(x_i) f_i + (1-\alpha)\int_{-\infty}^{\infty} u(x)xf(x)dx = \int_{-\infty}^{\infty} u(x)dF(x). \quad (2.1.3)$$

In the continuous and discrete cases, we considered similar representations in Sections **3**.2.1 and **6**.1.4.2, respectively. A formal proof of Proposition 6 is given in Section 2.3.

Next, consider a set B in the real line, and the indicator function

$$I_B(x) = \begin{cases} 1 \text{ if } x \in B \\ 0 \text{ if } x \notin B. \end{cases}$$

Clearly, $E\{I_B(X)\} = P(X \in B)$. Setting $u(x) = I_B(x)$ in (2.1.3), we have

$$P(X \in B) = E\{I_B(X)\} = \int_{-\infty}^{\infty} I_B(x) dF(x) = \int_B dF(x)$$

(since $I_B(x) = 0$ for $x \notin B$). We see that the d.f. $F(x)$ determines $P(X \in B)$ for any B, which confirms the claim of Proposition 4. This does not prove Proposition 4 though, because we use this proposition when proving Proposition 6.

Route 1 ⇒ *page 221*

2.2 A formula for the expectation of a positive r.v.

Next, for a non-negative r.v. X, we justify the formula

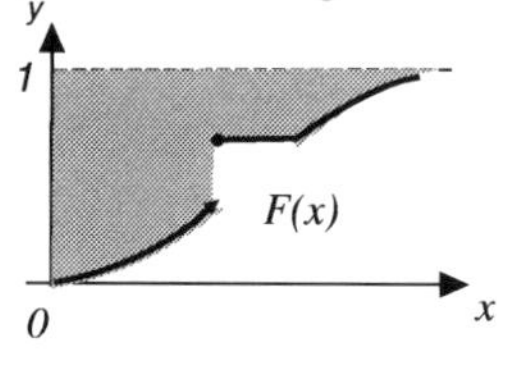

FIGURE 5.

$$E\{X\} = \int_0^{\infty} (1 - F(x)) dx. \tag{2.2.1}$$

In the continuous case, we proved it in Section **6**.1.3, and noted there that the r.-h.s. of (2.2.1) equals the area between the graph of $y = F(x)$ and the line $y = 1$. The same is true in the general case; see, in particular, Fig. 5, where on purpose, we chose, as $F(x)$, a mixture of continuous and discrete distributions.

First, note that we can come to (2.2.1) if in (2.1.2), we treat $dF(x)$ as a usual differential and integrate by parts as follows:

$$E\{X\} = \int_0^{\infty} x dF(x) = -\int_0^{\infty} x d[1 - F(x)]$$
$$= -\lim_{x \to \infty} x[1 - F(x)] + 0 \cdot [1 - F(0)] + \int_0^{\infty} (1 - F(x)) dx. \tag{2.2.2}$$

Since $1 - F(\infty) = 0$, in the expression $\lim_{x\to\infty} x[1 - F(x)]$, we have the indeterminate form $\infty \cdot 0$. As a matter of fact, we can set it equal to zero. To show this, first observe that, since we treat $dF(x)$ as a usual differential, we can write that $1 - F(x) = \int_x^{\infty} dF(y)$. Then

$$x[1 - F(x)] = x \int_x^{\infty} dF(y) \le \int_x^{\infty} y dF(y) \to 0 \text{ as } x \to \infty,$$

because we have assumed $E\{X\}$ to be finite, and hence $\int_0^{\infty} y dF(y)$ converges. So, the first two terms in (2.2.2) equal zero, and we come to (2.2.1).

However, such calculations are, though correct, somewhat heuristic because, formally speaking, dF is not a "usual" differential. Below, we provide a more formal

Proof of (2.2.1). As was told, we have proved (2.2.1) for the purely continuous case.

For integer valued r.v.'s, (2.2.1) is equivalent to formula $E\{X\} = \sum_{n=0}^{\infty} P(X > n)$ that we proved in Section **3**.2.2. Indeed, in this case, the d.f. $F(x)$ is constant on intervals $[n, n+1)$ for $n = 0, 1, \ldots$. Then, for any $x \in [n, n+1)$, we have $1 - F(x) = 1 - F(n) = P(X > n)$, and

$$\sum_{n=0}^{\infty} P(X > n) = \sum_{n=0}^{\infty} P(X > n) \int_n^{n+1} dx = \sum_{n=0}^{\infty} \int_n^{n+1} (1 - F(x)) dx = \int_0^{\infty} (1 - F(x)) dx.$$

The proof for an arbitrary positive discrete r.v. is similar to that in Section **3**.2.2. The only difference is that in this case, in the calculations in Section **3**.2.2, we should replace the formula $m = \sum_{k=1}^{m} 1$ by $x_m = \sum_{k=1}^{m}(x_i - x_{i-1})$, where $x_1, x_2, ...$ are the values of X, and $x_0 = 0$.

Once we proved (2.2.1) for the discrete and continuous cases separately, for the mixture (1.3.5), we can write

$$E\{X\} = \alpha\sum_i x_i f_i + (1-\alpha)\int_0^\infty xf(x)dx = \alpha\int_0^\infty (1-F_d(x))dx + (1-\alpha)\int_0^\infty (1-F_c(x))dx$$
$$= \int_0^\infty (1-(\alpha F_d + (1-\alpha)F_c(x)))dx = \int_0^\infty (1-F(x))dx. \blacksquare$$

2.3 Proof of Proposition 6

We use the representation

$$Y = Y^+ - Y^-, \tag{2.3.1}$$

where Y^+ and Y^- are the so-called positive and negative parts of Y; namely,

$$Y^+ = \begin{cases} Y & \text{if } Y \geq 0, \\ 0 & \text{if } Y < 0; \end{cases} \qquad Y^- = \begin{cases} 0 & \text{if } Y \geq 0, \\ |Y| & \text{if } Y < 0. \end{cases}$$

Note that both r.v.'s, Y^+ and Y^-, are positive. To compute $E\{Y^+\}$, we use (2.2.1):

$$E\{Y^+\} = \int_0^\infty (1-F_{Y^+}(y))dy = \int_0^\infty P(Y^+ > y)dy. \tag{2.3.2}$$

In accordance with (1.3.5) (which we proved using Proposition 4),

$$P(Y^+ > y) = \alpha \sum_{i:\, u(x_i)>y} f_i + (1-\alpha)\int_{x:\, u(x)>y} f(x)dx.$$

Inserting this into (2.3.2), we get

$$E\{Y^+\} = \alpha\int_0^\infty \left(\sum_{i:\, u(x_i)>y} f_i\right)dy + (1-\alpha)\int_0^\infty \left(\int_{x:\, u(x)>y} f(x)dx\right)dy$$
$$= \alpha \sum_{i:\, u(x_i)\geq 0} \left(\int_0^{u(x_i)} dy\right) f_i + (1-\alpha)\int_{x:\, u(x)\geq 0} \left(\int_0^{u(x)} dy\right) f(x)dx$$
$$= \alpha \sum_{i:\, u(x_i)\geq 0} u(x_i)f_i + (1-\alpha)\int_{x:\, u(x)\geq 0} u(x)f(x)dx = \int_{x:\, u(x)\geq 0} u(x)dF(x).$$

Because $Y^- = |Y|$ when $Y < 0$, similarly, $E\{Y^-\} = \int_{x:\, u(x)<0}(-u(x))dF(x)$, and

$$E\{Y\} = E\{Y^+\} - E\{Y^-\} = \int_{x:\, u(x)\geq 0} u(x)dF(x) - \int_{x:\, u(x)<0} (-u(x))dF(x)$$
$$= \int_{-\infty}^\infty u(x)dF(x). \blacksquare$$

2.4 A generalization: Infinite expectations

Here, we discuss the case when an expected value is infinite. We say that for a non-negative r.v. X, its expected value $E\{X\} = \infty$ if $\int_0^\infty x dF(x) = \infty$.

When dealing with r.v.'s taking positive and negative values, we should be more cautious, and define expectation in the following way.

Any r.v. X may be represented as $X = X^+ - X^-$, where X^+ and X^- are the positive and negative parts of X; see the comments on (2.3.1). Note again that both quantities, X^+ and X^-, are positive. We write $E\{X\} = E\{X^+\} - E\{X^-\}$, and understand it as follows.

- If $E\{X^+\} < \infty$, and $E\{X^-\} < \infty$, we define $E\{X\}$ in the usual way, and we say that $E\{X\}$ is finite. Certainly, in this case, $E\{X\} = E\{X^+\} - E\{X^-\}$.
- If $E\{X^+\} = \infty$, and $E\{X^-\} < \infty$, we say that $E\{X\} = \infty$.
- If $E\{X^+\} < \infty$, and $E\{X^-\} = \infty$, we say that $E\{X\} = -\infty$.
- If $E\{X^+\} = \infty$, and $E\{X^-\} = \infty$, we say that $E\{X\}$ is not defined.

2.5 On a general definition of expectation

The problems we encountered when defining expected values are connected with the fact that we did not proceed from the general definition of integral and expectation of Measure Theory. This definition is even simpler than what we were dealing with, but its rigorous justification requires some work and background in the theory of measure and integration. Below, we present the basic idea, which may make our constructions easier to grasp, and gives more insight into the nature of integration in the general case.

Consider *non-negative* r.v.'s $X = X(\omega)$ defined on a sample space Ω. In the general case, we follow the scheme of Section 2.4.

For a discrete r.v. X taking on different values $x_1, x_2, \ldots.$ with probabilities $f_1, f_2, \ldots$, respectively, we define the expected value as before: $E\{X\} = \sum_j x_j f_j$.

Note also that if the space Ω is discrete, that is, $\Omega = \{\omega_1, \omega_2, \ldots\}$, we can also define the expected value of a r.v. $X(\omega)$ as

$$E\{X\} = \sum_{\omega} X(\omega)p(\omega), \tag{2.5.1}$$

where $p(\omega)$ is the probability of the outcome ω. (Since the number of ω's is countable, we can use the summation operation and symbol $\sum_\omega$.)

To justify (2.5.1), consider events $A_i = \{\omega : X(\omega) = x_i\}$, and observe that $f_i = P(A_i) = \sum_{\omega \in A_i} p(\omega)$. Then

$$\sum_{\omega} X(\omega)p(\omega) = \sum_i \sum_{\omega \in A_i} X(\omega)p(\omega) = \sum_i \sum_{\omega \in A_i} x_i p(\omega) = \sum_i x_i \sum_{\omega \in A_i} p(\omega) = \sum_i x_i P(A_i) = \sum_i x_i f_i.$$

Let us consider an arbitrary non-negative r.v. The fact we use is that any r.v. $X = X(\omega)$ may be approximated to any degree of accuracy by a discrete r.v. Here is *one of the possible* techniques to accomplish this.

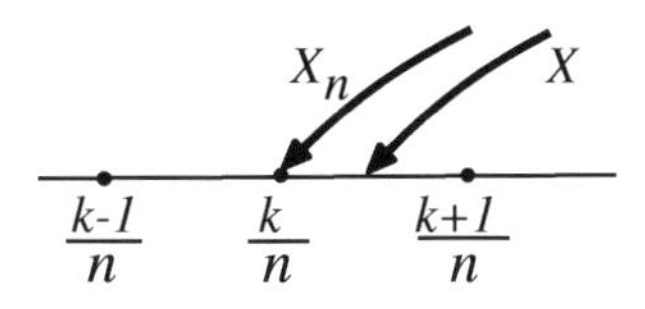

FIGURE 6.

For a fixed integer n, we divide the real line into intervals of the length $1/n$, say, into intervals $\Delta_k = [\frac{k}{n}, \frac{k+1}{n})$ where $k = 0, \pm 1, \pm 2, \dots$. Set $X_n(\omega) = \frac{k}{n}$, if $X(\omega) \in \Delta_k$; see Fig. 6. Then for all ω

$$|X_n(\omega) - X(\omega)| \leq 1/n,$$

so $X_n(\omega)$ approximates $X(\omega)$ with the accuracy $1/n$ uniformly in ω. If the expectations $E\{X_n\}$ exist, we set by definition

$$E\{X\} = \lim_{n\to\infty} E\{X_n\}. \tag{2.5.2}$$

We consider only non-negative r.v.'s and do not exclude the case where the above limit is infinite. It may be proved that

(a) the limit in (2.5.2) does not depend on the choice of approximating sequence X_n;

(b) the definition (2.5.2) leads to the representation $E\{X\} = \int_{-\infty}^{\infty} xdF(x)$ for all distributions considered in Section 2.1.

Proving all of this involves technical details which we omit here. Proofs may be found in many textbooks; see, e.g., [10], [13], [24], [42], [47].

The quantity (2.5.2) is called a *Lebesgue integral* (after the French mathematician Henri Lebesgue), and is often denoted by the following non-decomposable symbol:

$$\int_{\Omega} X(\omega)P(d\omega) \tag{2.5.3}$$

(compare with (2.5.1)). We can view (2.5.3) as an integral over Ω, where each value of $X(\omega)$ is multiplied by the probability $P(d\omega)$ of a "infinitesimally small neighborhood of ω." The integral may be viewed as the sum over all ω's.

However, it is worth emphasizing that this is merely an interpretation, and the expression (2.5.3) is just a symbol which denotes expectation in the sense of (2.5.2). This interpretation and the very symbolism (2.5.3) reflect our intuitive understanding of the notion of expectation.

---

3 ON CONVERGENCE OF DISTRIBUTIONS

3.1 Weak convergence

As we know, d.f.'s completely determine the corresponding probability distributions (that is, the probabilities of arbitrary sets in the real line (see Section 1.2)). Therefore, one of the natural ways to define convergence of distributions is to reduce it to convergence of distribution functions. However, there is a nuance: even in very simple situations, we cannot require the convergence of the distribution functions for all values of their argument.

EXAMPLE 1. Let $X_n \equiv \frac{1}{n}$ and $X \equiv 0$. These r.v.'s are actually non-random, but nothing prevents us from considering the convergence of their respective d.f.'s $F_n(x)$ and $F(x)$. Since $X_n \to X$ as $n \to \infty$ in any reasonable sense, any reasonable definition of convergence of d.f.'s should cover this case.

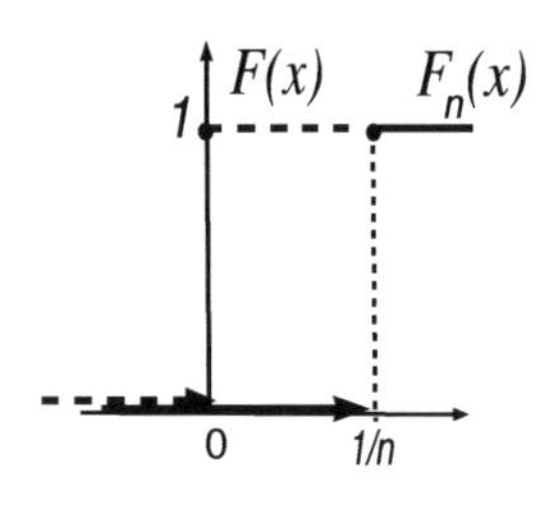

FIGURE 7.

However, let us look at the graphs of $F_n(x)$ and $F(x)$ in Fig. 7, where $F(x)$ is graphed by a dash line, and $F_n(x)$ by a solid line. (See also Fig. 3-2.)

Clearly, $F_n(x) = F(x) = 0$ for any $x < 0$.

Furthermore, $F(x) = 1$ for any $x \geq 0$, and $F_n(x) = 1$ for any $x \geq 1/n$. Hence, for any strictly positive x, for all n larger than some n_0, the d.f. $F_n(x) = 1$. Consequently, $F_n(x) \to F(x)$ for any $x > 0$.

However, $F(0) = 1$, while $F_n(0) = 0$ for all n. Hence, $F_n(0) \not\to F(0)$. So, we should exclude point 0 from consideration. It is noteworthy that 0 is the only discontinuity point of $F(x)$. □

In the general case, we say that distributions F_n *weakly* converge to a distribution F, as $n \to \infty$, if for the corresponding d.f.'s,

$$F_n(x) \to F(x) \quad \text{for any } \; x \text{ that is a continuity point of the function } \; F(x). \tag{3.1.1}$$

We say that r.v.'s X_n converge to a r.v. X *in distribution*, and we write it as $X_n \xrightarrow{d} X$, if (3.1.1) is true for the corresponding d.f.'s.

The next lemma is useful for future references and sheds additional light on the definition above.

Lemma 7 *Let $X_n \xrightarrow{d} X$. Then $P(X_n = a) \to 0$ for any point a at which the d.f. of X is continuous.*

Proof. Let $\varepsilon > 0$. If a is a continuity point of $F(x) = P(X \leq x)$, then there exists $\delta > 0$ such that $F(a) - F(a-\delta) < \varepsilon$. By Proposition 1, the set of discontinuity points of $F(x)$ is countable. Hence, there exists a continuity point b such that $a - \delta \leq b < a$. For such a point, $F(a) - F(b) \leq \varepsilon$ also. Since $b < a$,

$$\lim_{n\to\infty} P(X_n = a) \leq \lim_{n\to\infty} P(b < X_n \leq a) = \lim_{n\to\infty} [P(X_n \leq a) - P(X_n \leq b)] = F(a) - F(b) \leq \varepsilon.$$

Thus, $\lim_{n\to\infty} P(Y_n = a) \leq \varepsilon$ for an arbitrary small $\varepsilon > 0$. This is possible only if the limit under consideration equals zero. ■

3.2 Comparison with convergence in probability

As we defined in Section **3**.5.1, a sequence of r.v.'s X_n converges to a r.v. X *in probability* if $P(|X_n - X| < \varepsilon) \to 1$, as $n \to \infty$, for any arbitrary small $\varepsilon > 0$. We wrote this as $X_n \xrightarrow{P} X$.

Proposition 8 *Convergence* $X_n \xrightarrow{P} X$ *implies* $X_n \xrightarrow{d} X$.

The converse assertion is not true, for example, for the simple reason that two different r.v.'s may have the same distribution.

EXAMPLE 1. Let X_1 be uniform on $[-1,1]$, and $X_2 = -X_1$. The r.v. X_2 is also uniform on $[-1,1]$, but certainly, $X_2 \neq X_1$. Consider now the alternating sequence $X_1, X_2, X_1, X_2, \ldots$. The distribution of all terms is the same, so the distributions do converge to the distribution uniform on $[-1,1]$, but the r.v.'s themselves do not converge. □

However, if the distribution of X_n converges to the distribution of a degenerated (concentrated at one point) distribution, the convergence in probability does also take place.

Proposition 9 *For any number* c, *the relations* $X_n \xrightarrow{P} c$ *and* $X_n \xrightarrow{d} c$ *are equivalent.*

Proof of Proposition 8. Let $F_n(x)$ and $F(x)$ be the d.f.'s of X_n and X, respectively. For any $\varepsilon > 0$,

$$\begin{aligned} F_n(x) &= P(X_n \leq x) = P(X_n \leq x, |X_n - X| \leq \varepsilon) + P(X_n \leq x, |X_n - X| > \varepsilon) \\ &\leq P(X \leq x + \varepsilon) + P(|X_n - X| > \varepsilon) = F(x+\varepsilon) + P(|X_n - X| > \varepsilon). \end{aligned} \quad (3.2.1)$$

Since $P(|X_n - X| > \varepsilon) \to 0$ by assumption, from (3.2.1) it follows that $\lim_{n\to\infty} F_n(x) \leq F(x+\varepsilon)$ for any $\varepsilon > 0$.

Letting now $\varepsilon \to 0$, we obtain that $\lim_{n\to\infty} F_n(x) \leq F(x)$, provided that x is a continuity point of $F(x)$.

Switching X_n and X in inequality (3.2.1), we similarly get that $F(x) \leq F_n(x+\varepsilon) + P(|X_n - X| > \varepsilon)$, which implies that $\lim_{n\to\infty} F_n(x+\varepsilon) \geq F(x)$. We replace in the last inequality x with $x - \varepsilon$, which gives $\lim_{n\to\infty} F_n(x) \geq F(x-\varepsilon)$. If x is a continuity point, then letting $\varepsilon \to 0$, we get $\lim_{n\to\infty} F_n(x) \geq F(x))$.

Together with $\lim_{n\to\infty} F_n(x) \leq F(x)$, this gives $\lim_{n\to\infty} F_n(x) = F(x)$. ■

Proof of Proposition 9. It suffices to set $c = 0$, and prove that $X_n \xrightarrow{d} 0$ implies $X_n \xrightarrow{P} 0$.

So, $F(x)$ is the d.f. of $X \equiv 0$, and $F(x)$ is continuous at any point $x \neq 0$; see again Fig. 7.

Since we have assumed that $X_n \xrightarrow{d} 0$, we have $F_n(-\varepsilon) \to 0$ and $F_n(\varepsilon) \to 1$ for any $\varepsilon > 0$.

Note also that $F_n(\varepsilon - 0) \geq F_n(\varepsilon/2) \to 1$, so $F_n(\varepsilon - 0) \to 1$ also.

Then $P(-\varepsilon < X_n < \varepsilon) = F_n(\varepsilon - 0) - F_n(-\varepsilon) \to 1$. ■

---

4 SIMULATION. THE MONTE-CARLO METHOD

1,2

This section is concerned with generating (or simulating) values of r.v.'s with given distributions. We already slightly touched on this topic in Example **5**.1.2-4, when talking about simulating of trajectories of Markov chains. Here, we discuss this topic in more detail.

We will see that whatever simulation problem we consider, first, we need a generator of values of an original basic r.v. Usually, though not always, this is a r.v. uniform on $[0,1]$.

Since in this case all values are equally likely, the values generated are viewed as "purely random."

There are two principal approaches to simulating random numbers. The first is connected with physical phenomena that are expected to be random. It may be atomic or subatomic physical processes whose randomness is described by the laws of quantum mechanics, or exploding stars, "atmosphere noise," etc. Note that flipping a coin, rolling a die, or rotating a disk (say, a roulette) also correspond to this type of simulation. Say, if we number the cells in a roulette, it may be viewed as a generator of random integers.

The other widespread approach uses computational algorithms (programs) of different levels of complexity, which generate numbers that do not exhibit any visible pattern, look "chaotic," and may serve as random numbers not actually being random in the strict sense of this word. Such programs are usually called pseudorandom number generators. In spite of the fact that such numbers are only pseudorandom, the computational approach proved to be efficient enough in modeling various real phenomena.

In examples below, we illustrate simulation methods using Excel worksheets, and accordingly, the Excel (pseudo) random number generator. In Excel, to generate values of a r.v. uniformly distributed on $[0,1]$, we go to Data ⇒ Data Analysis ⇒ Random Number Generation, and choose in "*Distributions,*" the position "*Uniform*" with parameters 0 and 1. We can generate many (pseudo) independent numbers at once, providing an *array address* for the output.

4.1 Simulation of discrete r.v.'s

Suppose we want to simulate a value of a discrete r.v. X taking on values $x_1, x_2, x_3, ...$ with respective probabilities $f_1, f_2, f_3, ...$ To this end, at the first stage, using a random number generator, we simulate a (random) value Z_1 of a r.v. uniformly distributed on $[0,1]$.

If the value of Z_1 is less then f_1 (which occurs with probability f_1), we take, as a value of X, the value x_1. If $f_1 \leq Z_1 < f_1 + f_2$ (which occurs with probability f_2), we take, as a value of X, the value x_2; and we continue the procedure in the same fashion until for some k, we will have $f_1 + ... + f_{k-1} \leq Z_1 < f_1 + ... + f_{k-1} + f_k$. In this case, we choose x_k as a value of X.

So, the first value of X has been simulated. To simulate the second value, we select (simulate) the second (and independent of Z_1) value Z_2 and repeat the above procedure.

	A	B
1	0.366131	0
2	0.049928	0
3	0.744346	10
4	0.866604	10
5	0.054842	0
6	0.810999	10
7	0.12241	0
8	0.950499	15
9	0.208167	0
10	0.321146	0
11		
12	Values	Values
13	of Z	of X
14		

FIGURE 8.

To simulate n values, we should do the same n times, though, as a matter of fact, for practically any software including Excel, we can simulate all values simultaneously, just copying the command for the first simulation.

EXAMPLE 1. John sells T-shirts at the beach, priced \$10 or \$15. Each potential customer who stops at the stand does not buy a shirt with a probability of 0.7, buys a cheaper shirt with a probability of 0.2 and buys a shirt for \$15 with a probability of 0.1. Let us simulate a cash flow provided by the customers stopping at the stand. In other words, for some n, we should simulate a sequence $X_1, X_2, ..., X_n$ of independent and identically distributed r.v.'s assuming values $0, 10$ and 15 with the above probabilities. Let $n = 10$. We use Excel, and first simulate 10 uniformly distributed on $[0, 1]$ random numbers in Column A; see Fig. 8. In a cell of Column B, we "write" 0, if the corresponding random number in Column A is less than 0.7; we write 10, if this number is between 0.7 and 0.9; and we write 15, if this number is greater than 0.9. Note also that, since the uniform distribution is continuous, it does not matter in which subinterval we include the points 0.7 and 0.9. The command for, say, sell B1 is =IF(A1$<$ 0.7, 0, IF(A1$<$ 0.9, 10, 15)). □

Note that if a discrete r.v. X assumes an infinite number of values, the method may still work because, for a given value of Z, the search of k for which $f_1 + ... + f_{k-1} \leq Z_1 < f_1 + ... + f_{k-1} + f_k$ will take a finite (though random) time. However, it may turn out to be time consuming.

4.2 The cumulative distribution function method

Another term in use is the *inverse transformation method.* The method is universal, but usually is applied to continuous distributions. It is based on Proposition 3 that says that if $F(x)$ is a d.f., $F^{-1}(z)$ is its inverse, and a r.v. Z is uniform on $[0, 1]$, then the d.f. of the r.v. $X = F^{-1}(Z)$ is $F(x)$. So, if a r.v. X has a d.f. $F(x)$, and after the simulation of values of Z, we got numbers $z_1, z_2, ...$, then the numbers $x_1 = F^{-1}(z_1), x_2 = F^{-1}(z_2), ...$, may serve as values of X.

	A	B
1	0.119724	4.24513
2	0.283761	2.519245
3	0.752403	0.568966
4	0.133702	4.02429
5	0.119846	4.243092
6	0.796045	0.4562
7		
8	Values	Values
9	of Z	of X
10		

FIGURE 9.

EXAMPLE 1. You write a program for a game where monsters are jumping out of a cave. For the appearance of a monster to be unpredictable, the waiting time for the next monster should not depend on how long the player has been already waiting. That is, the lengths of the time periods between monsters' appearances (interarrival times) should be independent and exponential. We want to simulate a sequence of interarrival times.

Let X be exponential and $E\{X\} = m$. The d.f. $F(x) = 1 - e^{-x/m}$. Solving the equation $z = F(x)$ with respect to x, we have $F^{-1}(z) = -m\ln(1-z)$. So, we may proceed from the representation $X = -m\ln(1-Z)$, where Z is uniform on $[0, 1]$. Moreover, since $1 - Z$ is also uniformly distributed on $[0, 1]$, the r.v. $-m\ln(Z)$ is also exponential with a mean of m, so it suffices to use this expression. The simulation of six numbers for $m = 2$ is given in Fig. 9. Column A contains six numbers selected from $[0, 1]$. The command in Cell B1 is = − 2*LN(A1). □

4.3 Special methods for particular distributions

Though the two methods above are universal (the first—in the discrete case), their applications may cause difficulties. As was already mentioned, if a discrete r.v. X assumes an infinite number of values, the simulation process may be time consuming.

In the case of continuous r.v.'s, computing the inverse of a d.f. may turn out not to be easy; in particular, it may be impossible to present the inverse $F^{-1}(z)$ by an explicit formula. In this case, for each z, the search of x for which $z = F(x)$ has to be provided numerically.

Another matter is that for particular important distributions, say, the standard normal, there are detailed tables which may be incorporated into programs.

What is more important is that when dealing with particular distributions, the knowledge of their *special* features may lead to *special* and easier methods.

EXAMPLE 1. Let a r.v. X have the Γ-distribution with a scale parameter a and an *integer* parameter ν. The inverse of the corresponding d.f. is not a simple function and does not have an explicit representation. However, we may appeal to Proposition **6**.6 from which it follows that, if $\xi_1, \xi_2, ...$ are independent exponential r.v.'s with parameter a, then $\xi_1 + ... + \xi_\nu$ has the Γ-distribution mentioned. So, to get one value of X, it suffices to get ν independent values of ξ's and add them up.

	A	B	C
1	0.634846	0.908746	
2	0.732566	0.622403	
3	0.585925	1.069127	Values
4	0.607288	0.997505	of ξ
5	0.235786	2.889661	
6			
7	Values	6.487442	A value
8	of Z		of X
9			

FIGURE 10.

In Fig. 10, we provide one value of X for $\nu = 5$ and $a = 1/2$. Hence, the mean of ξ's is 2.

Column A contains five independent values of a uniformly distributed r.v. Cells B1-B5 contain values of ξ's computed as in Example 4.2-1. For instance, the command for B1 is = $-$ 2*LN(A1). Cell B7 contains the sum of the values in Cells B1-B5; the command for B7 is =SUM(B1:B5).

In Exercise 12, the reader is advised to provide an Excel file allowing to get several values of X simultaneously, but it makes sense to note that with a software more sophisticated than Excel, a similar program may be quicker and more convenient.

	A	B	C	D	E
1	0.784204	0.243087	0.243087		
2	0.194617	1.636724	1.879811	N_1=1	
3	0.795038	0.229366	2.109176	N_2=1	
4	0.71688	0.332847	2.442024		
5	0.62743	0.466123	2.908147		
6	0.832057	0.183855	3.092001	N_3=3,	
7	0.016266	4.118656	7.210657	N_4=0,N_5=0, N_6=0	
8	0.748375	0.289851	7.500508		
9	0.1601	1.831956	9.332464	N_7=2	

FIGURE 11.

EXAMPLE 2. Next, we simulate values of a Poisson r.v. with a parameter λ. Let $\xi_1, \xi_2, ...$ be exponential r.v.'s with parameter $a = \lambda$, and

$$S_k = \xi_1 + ... + \xi_k.$$

Since ξ's are positive, the sums S_k are increasing. Let N_1 be the index of the largest sum that still does not exceed 1; formally,

$$N_1 = \max\{k : S_k \leq 1\}.$$

In Section **13**.1.1, we will discuss in detail why N_1 has the Poisson distribution mentioned, and the significance of this fact. For now, we take it for granted and use it for simulation. Let us look at the Excel worksheet in Fig. 11.

For simplicity, we chose $\lambda = 1$. As above, Column A contains random numbers selected from $[0, 1]$, Column B—their natural logarithms multiplied by -1. For example, the command for Cell B1 is = - LN(A1). Column C contains consecutive sums of the numbers from Column B; for example C1=B1, C2=C1+B2, and C3=C2+B3. We see that the first sum that has exceeded 1 is S_2 in Cell C2, so the value of N_1 we get is 1.

To get the second value (independent of the first), we *may* start over and perform the same procedure again. However, as a matter of fact, we do not have to do that. By virtue of the memoryless property, the overshoot of S_{N_1+1} over 1, that is, the r.v. $S_{N_1+1}-1$ does not depend on S_{N_1} and has the same exponential distribution. So, we may just continue summation and watch when the consecutive sums exceed 2, 3, an so on. This issue will be also discussed in detail in Example **12**.2-1 and Section **13**.1.1; for now, let us consider the simulation procedure itself.

Let us come back to Fig. 11. Between 1 and 2, there is also only one sum; so the second value, $N_2 = 1$. However, between 2 and 3, there are three sums, and hence, the third simulated value is $N_3 = 3$. Between 3 and 7, there are no sums; so $N_4 = N_5 = N_6 = 0$. The reader is invited to realize that $N_7 = 2$. □

4.4 On the Monte Carlo Method

Simulation is directly connected with what is called Monte Carlo methods, a class of computational algorithms based on repeated random sampling. The methods are rather developed; here, we will consider only one problem to demonstrate the original idea.

Suppose we wish to estimate the value of an integral $\int_0^1 u(x)dx$ that cannot be calculated analytically. Let us use the fact that the integral may be viewed as the expectation $E\{u(Z)\}$, where Z is a r.v. uniformly distributed on $[0,1]$. Consider a sequence of independent r.v.'s $Z_1, Z_2, ...$ having the same distribution, and the r.v.

$$Y_n = \frac{1}{n}\left(u(Z_1) + ... + u(Z_n)\right). \tag{4.4.1}$$

By the law of large numbers, with probability one, $Y_n \to E\{u(Z)\} = \int_0^1 u(x)dx$, which gives a way of estimating the integral: we should simulate n random numbers and compute the average (4.4.1).

EXAMPLE 1. Let $u(x) = e^{-x^2}$. In Fig. 12, we see a part of an Excel worksheet. Column A contains 1000 values selected from $[0,1]$ (not all values are presented in the figure). Column B contains values of $u(Z)$.

	A	B	C
1	0.551103	0.738071	0.746932
2	0.160833	0.974465	
3	0.164708	0.973236	
4	0.594714	0.702096	
5	0.657247	0.649227	
6	0.88049	0.460583	
7	0.073519	0.99461	
8	0.120304	0.985631	
9	0.959044	0.398612	
10	0.435011	0.827592	

FIGURE 12.

The command for, say, B1 is =EXP((-1)*(A1)^2). In C1, we compute the average; the command is =AVERAGE(B1:B1000). So, the estimate for the integral is ≈ 0.7469. Maple (a popular software for mathematicians) gives ≈ 0.7468.

It makes sense to note that the result is still random (for another 1000 values, it may be slightly different), and the accuracy of estimation depends on the quality of the random number generator in hand. □

If we want to estimate $\int_a^b u(x)dx$, we may apply the variable change $y = \frac{x-a}{b-a}$, present the integral as $(b-a)\int_0^1 u(a+y(b-a))dy$, estimate $\int_0^1 u(a+y(b-a))dy$, and multiply the result by $b-a$.

5 HISTOGRAMS

In the conclusion of this chapter, we introduce the notion of a histogram, a graphical representation of the distribution of data. In the discrete distribution case, a histogram presents estimates of the probabilities of particular values. In the continuous case, it may be viewed as an estimate of the probability density. It suffices to consider examples. In real problems, when providing histograms, people work with real data. In the study examples below, we will deal with simulation results.

	A	B	C	D	E
1	0.5815	0		*Bin*	*Frequency*
2	0.174139	0	0	0	71
3	0.871609	10	10	10	20
4	0.553667	0	15	15	9
5	0.082156	0		More	0
6	0.578722	0			
7	0.924497	15			
8	0.636128	0			
9	0.988159	15			
10	0.769677	10			
11	0.517472	0			
12	0.682028	0			
13	0.409162	0			
14	0.730857	10			
15	0.259926	0			
16	0.680044	0			
17	0.42848	0			
18	0.564592	0			

FIGURE 13.

EXAMPLE 1. Consider the situation of Example 4.1-1. Let us simulate, say, $n = 100$, independent values $X_1, ..., X_{100}$ as we did in Example 4.1-1. They are in Column B of the worksheet in Fig. 13 (not all values are shown). For study purposes, let us view this as real data concerning the flow of customers.

We want to count how many values out of 100 equal 0, how many equal 10, and how many equal 15. The program did it for us, and presented in the array D2:E4. (In this case the term "bin" is used for values of the r.v.

At the end of the example, we will consider the program details; now let us look at results. In our particular realization, there were 71 zeros; 20 values were equal to 10, and 9 to 15. These numbers are called *frequencies*. Since the total size of the data is 100, the relative frequencies are $\frac{71}{100} = 0.71$, $\frac{20}{100} = 0.2$, and $\frac{9}{100} = 0.09$.

By the law of large numbers (see also Corollary **3**.17), for large n, these frequencies should serve as good estimates of the real probabilities; and we see that the results are not bad because the real probabilities are 0.7, 0.2, and 0.1.

A chart, as in Fig. 13, representing frequencies (or relative frequencies) is called a *histogram*.

In Excel, to obtain a histogram, first, we prepare the list of the values; see Column C. The program calls it bins.

Next, we go to Data ⇒ Data Analysis ⇒ Histogram, provide the input range (for us, this is B1:B100), and the bin range (for us, this is C2:C4). We should give an array address for the output (for us, this is D2:E4), and click OK. The program will give us the result as in Fig. 13. Then, we make a chart as in the same figure.

EXAMPLE 2. Consider the situation of Example 4.2-1. Now, we simulated 1000 values of the exponential r.v. X; see Column B in Fig. 14 (not all values are shown). It does not make sense to consider precise values of all these 1000 numbers. We restrict ourselves to a number of consecutive intervals and count the numbers of values that fell into these intervals.

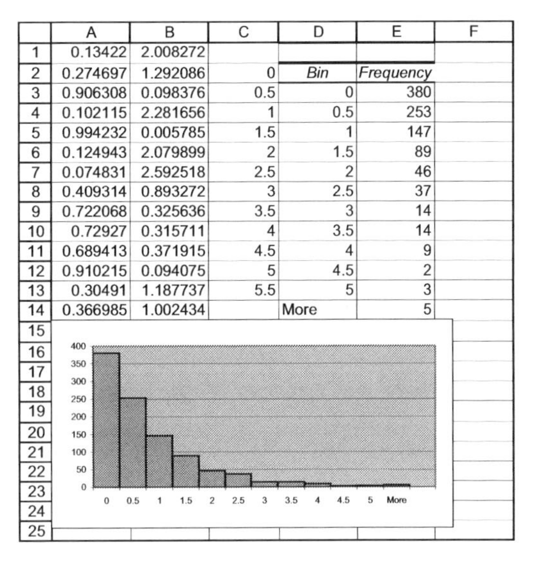

	A	B	C	D	E	F
1	0.13422	2.008272				
2	0.274697	1.292086	0	*Bin*	*Frequency*	
3	0.906308	0.098376	0.5	0	380	
4	0.102115	2.281656	1	0.5	253	
5	0.994232	0.005785	1.5	1	147	
6	0.124943	2.079899	2	1.5	89	
7	0.074831	2.592518	2.5	2	46	
8	0.409314	0.893272	3	2.5	37	
9	0.722068	0.325636	3.5	3	14	
10	0.72927	0.315711	4	3.5	14	
11	0.689413	0.371915	4.5	4	9	
12	0.910215	0.094075	5	4.5	2	
13	0.30491	1.187737	5.5	5	3	
14	0.366985	1.002434		More	5	
15						
16						
17						
18						
19						
20						
21						
22						
23						
24						
25						

FIGURE 14.

Suppose we decided to consider the intervals with endpoints $0, 0.5, 1, 1.5, ..., 5.5$, and the last interval is $[5.5, \infty)$. In Fig. 14, the end points of these intervals are provided in the cells C2–C13; this is the input for the bins the program needs. The results the program has calculated are in the array D2:E14. For example, 380 values out 1000 fell into the interval $[0, 0.5)$, and 253 into $[0.5, 1)$. The chart (histogram) is also given in Fig. 14. It is pretty consistent with the real density $f(x) = e^{-x}$. (In this case, the rectangles should touch each other to indicate that the r.v. is continuous.)

---

6 EXERCISES

1. Consider a r.v. X with the d.f. $F(x)$ graphed in Fig. 15. (a) Find $P(1 \le X \le 4)$, $P(1 < X \le 4)$, $P(1 \le X < 4)$, $P(1 < X < 4)$, $P(X = 1)$, $P(X = 0.5)$, $P(X = 2)$, $P(2 \le X \le 2.5)$, $P(3 \le X \le 5)$, $P(4.5 \le X \le 5)$, $P(4.5 < X < 5)$.

   (b) Compute in mind $P(X < 0.5 \,|\, X < 1)$, $P(X = 1 \,|\, 1 \le X < 3)$. Find $P(X = 1 \,|\, 1 \le X \le 3)$.

   (c) Specify the decomposition (1.3.3). Find $E\{X\}$.

2. Find a 0.2-quantile of a r.v. X taking values $0, 3, 7$ with probabilities $0.1, 0.3, 0.6$, respectively.

3. Find y-quantiles of a r.v. X that (a) uniform on $[a, b]$; (b) exponential with a mean of m.

4. Let q_y be a y-quantile of the (m, σ^2)-normal distribution. Show that $q_y = m + q_{ys}\sigma$, where q_{ys} is a y-quantile for the standard normal distribution. When $q_y < m$, and when $y > m$? Consider particular cases $y = 0.5, 0.3, 0.89$ with use of Table 3 from the Appendix.

5. Denote by $q_y(X)$ a y-quantile of a r.v. X. Let X have a continuous d.f. $F(x)$ strictly increasing at all x's such that $0 < F(x) < 1$. Show that $q_y(a - X) = a - q_{1-y}(X)$ for any a and $y \in (0, 1)$.

6. Show that if a r.v. X has a continuous d.f. $F(x)$, then the distribution of the r.v. $Y = F(X)$ coincides with the distribution uniform on $[0, 1]$. (The r.v. Y may not assume values 0 and 1, but for the uniform distribution, the probabilities of these values equal zero anyway.) Show that the continuity condition is necessary.

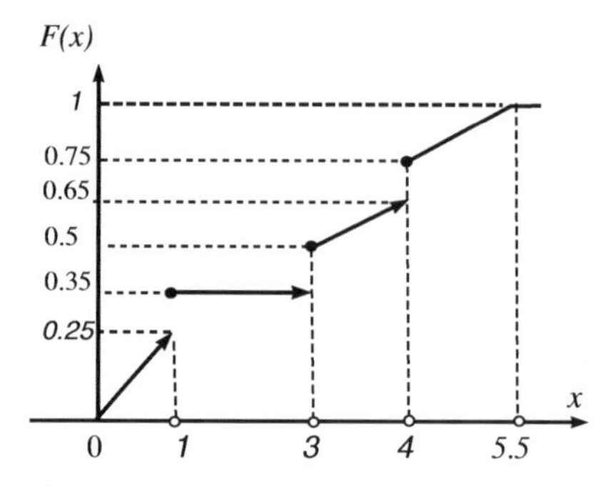

FIGURE 15.

7. Let two d.f.'s, $F_1(x)$ and $F_2(x)$, be such that $F_1(x) \leq F_2(x)$ for all x. This relation referred to as the *first stochastic dominance* (FSD) was already considered and interpreted in Exercise **6**-17. Show that we can *construct* a sample space and probability measure on it, and two r.v.'s, X_1 and X_2 defined on this space, having the d.f. $F_1(x)$ and $F_2(x)$, respectively, and such that $P(X_1 \geq X_2) = 1$. Is it true for *any* r.v.'s having the above d.f.'s?

8. For $F(x) = \begin{cases} \frac{1}{3} - \frac{1}{3}e^{-2x} & \text{for } 0 \leq x < 1, \\ \frac{7}{15} - \frac{1}{3}e^{-2x} & \text{for } 1 \leq x < 2, \\ 1 - \frac{1}{3}e^{-2x} & \text{for } 2 \leq x < \infty, \end{cases}$ specify the decomposition, the mean and variance.

9. For a d.f. $F(x)$, the jumps $F(k) - F(k-0) = \frac{1}{5}\left(\frac{2}{3}\right)^{k-1}$ for $k = 1, 2, \ldots$. At all other points, $F(x)$ is differentiable, and $F'(x) = a$ for some number a and $0 \leq x < 2$, and $F'(x) = 0$ for all other non-integer x's. Write the decomposition (1.3.3) and find the mean and variance.

10. Using Excel or another software, simulate the process of flipping a non-regular coin with a probability of heads p as an input parameter.

11. Using Excel or another software, simulate 100 values of a r.v. X assuming values -1, 0, and 1, with respective probabilities 0.3, 0.4, and 0.3. Compute $E\{X\}$ and compare it with the average of the numbers you simulated. Did you get what had to be expected? What does it say about the quality of the random number generator you used?

12. Using Excel or another software, simulate 5 values of (a) the Γ-r.v. from Example 4.3-1, (b) a r.v. X having the d.f. $F(x) = x^3/8$ for $x \in [0.2]$.

13. Provide an Excel sheet or write a program using another software to simulate a sequence of values of Poisson r.v., where λ is an input parameter.

14. Suggest simulation methods for (a) the Binomial distribution, and (b) the geometric distribution, proceeding from the independent-trials model.

15. Estimate by the Monte Carlo method $\int_0^2 x^3 dx$ and compare the result with the precise value of this integral.

16. Find a discrepancy between the histogram in Fig. 14 and the exponential density.

17. Simulate 1000 values of a r.v. with the d.f. $F(x) = \sqrt{x}$ on $[0, 1]$, build a histogram, and compare it with the real density.

Route 1 $\Rightarrow$ *page 249*

Chapter 8

Moment Generating Functions: The Theory and Applications

2

This chapter concerns one more analytical method: that of moment generating functions. The method may be viewed as a generalization of the generating functions method considered in Chapter 4. The approach we discuss proved to be very useful in many theoretical and applied problems.

1 DEFINITIONS AND PROPERTIES

1.1 Laplace transform

Below, we consider arbitrary distributions: continuous, discrete, or their combinations. For a r.v. X, its d.f. $F(x)$ and a function $u(x)$, we use the unified representation

$$E\{u(X)\} = \int_{-\infty}^{\infty} u(x)dF(x)$$

which was clarified in Section **7.2**

The *Laplace transform* of a r.v. X and its distribution F is the function

$$M_X(z) = E\{e^{zX}\} = \int_{-\infty}^{\infty} e^{zx}dF(x) \tag{1.1.1}$$

defined for all z for which the above expectation exists. When it cannot cause confusion, we will omit the index X in $M_X(z)$.

If the distribution is continuous with the density $f(x) = F'(x)$, then (1.1.1) may be rewritten as

$$M_X(z) = E\{e^{zX}\} = \int_{-\infty}^{\infty} e^{zx}f(x)dx. \tag{1.1.2}$$

If the distribution is discrete, then

$$M_X(z) = E\{e^{zX}\} = \sum_i e^{zx_i}f_i, \tag{1.1.3}$$

where x_i's are the values of X, and f_i are the corresponding probabilities.

Clearly, for $z = 0$,

$$M_X(0) = E\{e^{0 \cdot X}\} = E\{1\} = 1. \tag{1.1.4}$$

In general, the argument z is a complex number. If $z = it$, where t is real and the imaginary $i = \sqrt{-1}$, then the transform $M_X(it)$ is called the *characteristic function* of a r.v. X.

In this book, we consider real z's. In this case, the Laplace transform is usually called a *moment generating function* (m.g.f.). The term will be clarified after the following examples.

EXAMPLE 1. Let $X = a$ with probability one. Then $M_X(z) = E\{e^{zX}\} = E\{e^{za}\} = e^{za}$. If $a = 0$, then $M_X(z) \equiv 1$.

EXAMPLE 2. Let a r.v. $X = 1$ or 0 with respective probabilities p and $1-p$. Then

$$M_X(z) = e^{z \cdot 1} \cdot p + e^{z \cdot 0} \cdot (1-p) = e^z p + 1 - p = 1 + p(e^z - 1). \tag{1.1.5}$$

EXAMPLE 3. Let X be uniform on $[0,1]$. Then,

$$M_X(z) = \int_0^1 e^{zx} dx = \frac{e^z - 1}{z} \tag{1.1.6}$$

for $z \neq 0$. If $z = 0$, the above integral is equal to one, which is consistent with (1.1.4). Note also that the limit of the very r.-h.s. of (1.1.6), as $z \to 0$, equals one also (for example, one may apply L'Hôpital's rule). □

The term *moment* g.f. is related to the following fact. As above (e.g., in Section **3**.3), denote by m_k the kth moment $E\{X^k\}$. Making use of the Taylor expansion of the exponential function [see the Appendix, (2.2.4)], we can write

$$M_X(z) = E\{e^{zX}\} = E\left\{\sum_{k=0}^{\infty} \frac{(zX)^k}{k!}\right\} = \sum_{k=0}^{\infty} \frac{z^k E\{X^k\}}{k!} = \sum_{k=0}^{\infty} \frac{m_k}{k!} z^k. \tag{1.1.7}$$

(We omit the formal justification of passing the expectation operation inside the sum). This is the Taylor expansion in powers of z for $M_X(z)$, and it is given in terms of moments.

One more thing to note is that the integral in (1.1.1) may not exist for $z \neq 0$. Therefore, m.g.f.'s exist not for all r.v.'s.

EXAMPLE 4. For the Cauchy distribution with the density $f(x) = \dfrac{1}{\pi(1+x^2)}$, the integral $\int_{-\infty}^{\infty} e^{zx} \frac{1}{\pi(1+x^2)} dx = \infty$ for any $z \neq 0$. Indeed, if $z > 0$, then $\int_0^{\infty} e^{zx} f(x) dx = \infty$. For $z < 0$, this is true for $\int_{-\infty}^{0} e^{zx} f(x) dx = \int_{-\infty}^{0} e^{|zx|} f(x) dx$. □

Also, a m.g.f. $M(z)$ may exist for some but not all $z \neq 0$.

EXAMPLE 5. Consider the exponential distribution with the density $f(x) = ae^{-ax}$ for $x \geq 0$, and a parameter $a > 0$. Since $f(x) = 0$ for $x < 0$, the m.g.f.

$$M(z) = \int_0^{\infty} e^{zx} a e^{-ax} dx = a \int_0^{\infty} e^{-(a-z)x} dx = \frac{a}{a-z} = \frac{1}{1-z/a} \tag{1.1.8}$$

for $z < a$. For $z \geq a$ the integral above diverges, and hence the m.g.f. does not exist. □

If X assumes values $0, 1, \ldots$, and $f_k = P(X = k)$, it is convenient to consider real $z \leq 0$, and set $s = e^z$. Then $E\{e^{zX}\} = E\{s^X\}$, where $0 < s \leq 1$. Denoting the last function by $G_X(s)$, we can write

$$G_X(s) = E\{s^X\} = s^0 f_0 + s^1 f_1 + s^2 f_2 + \ldots = \sum_{k=0}^{\infty} f_k s^k.$$

This is a *probability generating function* (or simply a generating function) which we have considered in detail in Chapter 4.

The Laplace transform in its different versions has proven to be a powerful tool for solving a wide variety of problems. There are several reasons for this, but possibly the primary one is connected with the following property:

For any independent r.v.'s X_1 and X_2,
$$M_{X_1+X_2}(z) = M_{X_1}(z)M_{X_2}(z) \tag{1.1.9}$$
for all z for which the Laplace transforms above are finite.

Indeed, since X_1 and X_2 are independent,
$$M_{X_1+X_2}(z) = E\{e^{z(X_1+X_2)}\} = E\{e^{zX_1}e^{zX_2}\} = E\{e^{zX_1}\}E\{e^{zX_2}\} = M_{X_1}(z)M_{X_2}(z).$$

Before considering more examples, we establish two elementary and one non-elementary property.

A. *The Laplace transform of a linear transformation.*
$$\text{For } Y = a+bX, \quad \text{the Laplace transform } M_Y(z) = e^{az}M_X(bz). \tag{1.1.10}$$
Indeed, $M_Y(z) = E\{e^{z(a+bX)}\} = E\{e^{az}e^{bzX}\} = e^{az}E\{e^{(bz)X}\} = e^{az}M_X(bz)$.

B. *The Laplace transform of a linear combination of distributions is equal to the linear combination of the Laplace transforms.* More precisely, let $F_1(x)$, $F_2(x)$ be two d.f.'s, and $F(x) = k_1F_1(x) + k_2F_2(x)$, where k_1, k_2 are coefficients. (Unlike what we considered in Chapter 7, here we consider a mixture of arbitrary d.f.'s.) If k_1 and k_2 are positive, and $k_1 + k_2 = 1$, then $F(x)$ is also a d.f. For arbitrary k_1, k_2 it may be not true, but we need a general statement.

Let $M_1(z)$, $M_2(z)$, $M(z)$ be the Laplace transforms of F_1, F_2, F, respectively. Then
$$M(z) = k_1M_1(z) + k_2M_2(z). \tag{1.1.11}$$
This immediately follows from (1.1.1):
$$M(z) = \int_{-\infty}^{\infty} e^{zx}dF(x) = \int_{-\infty}^{\infty} e^{zx}d[k_1F_1(x) + k_2F_2(x)]$$
$$= k_1\int_{-\infty}^{\infty} e^{zx}dF_1(x) + k_2\int_{-\infty}^{\infty} e^{zx}dF_2(x).$$

The non-elementary property mentioned is the uniqueness property: r.v.'s with different distributions have different Laplace transforms.

Theorem 1 *If for two r.v.'s, X and Y, with the d.f.'s $F_X(x)$ and $F_Y(x)$, respectively, $M_X(z) = M_Y(z)$ for all z from a neighborhood of zero, then $F_X(x) = F_Y(x)$ for all x. The same is true when we consider only real z's, that is, moment generating functions, provided that the m.g.f.'s under consideration exist for all z's from a neighborhood of zero (i.e., from an interval $(-c, c)$ for a positive c).*

We skip the proof but will repeatedly use the theorem itself.

From now on, we consider m.g.f.'s.

1.2 The m.g.f.'s of some basic distributions

The results of this section are summarized in Table 2 in the Appendix. In all formulas below, z is real.

1.2.1 The binomial distribution

An easy way to find the m.g.f. of a binomial r.v. X is to use the representation

$$X = X_1 + ... + X_n,$$

where the r.v.'s X_i are independent and equal 1 or 0 with probabilities p and q, respectively. We have computed in (1.1.5) that the m.g.f. of each X_i is $1 + p(e^z - 1)$. Then, by property (1.1.9), the m.g.f. of X is the product of the m.g.f.'s of X_i's, so

$$M_X(z) = (1 + p(e^z - 1))^n.$$

1.2.2 The geometric and negative binomial distributions

In Section **3**.4.3, we introduced two versions of the geometric distribution. For a r.v. $K = 0, 1, ...$ with the corresponding probabilities $f_k = pq^k$,

$$M_K(z) = \sum_{k=0}^{\infty} e^{kz} pq^k = p \sum_{k=0}^{\infty} (qe^z)^k = \frac{p}{1 - qe^z}. \tag{1.2.1}$$

(In the last step, we used the formula for a geometric series; the Appendix, (2.2.11).)

To get the m.g.f. for the r.v. $N = K + 1$ (see Section **3**.4.3), we use (1.1.10), which leads to $M_N(z) = e^z M_K(z)$.

For the negative binomial distribution (**3**.4.4.3) of a r.v. K_ν from Section **3**.4.4, using expansion (2.2.9) from the Appendix, we have

$$\begin{aligned} M_{K_\nu}(z) &= \sum_{m=0}^{\infty} e^{mz} \binom{\nu + m - 1}{m} p^\nu q^m = p^\nu \sum_{m=0}^{\infty} \binom{\nu + m - 1}{m} (e^z q)^m \\ &= p^\nu (1 - qe^z)^{-\nu} = \left(\frac{p}{1 - qe^z} \right)^\nu. \end{aligned} \tag{1.2.2}$$

To get the result for $N_\nu = K_\nu + \nu$, we again use (1.1.10), which leads to $M_{N_\nu}(z) = e^{z\nu} M_{K_\nu}(z)$.

1.2.3 The Poisson distribution

For the Poisson distribution with parameter λ, the m.g.f.

$$M(z) = \sum_{k=0}^{\infty} e^{zk} e^{-\lambda} \frac{\lambda^k}{k!} = e^{-\lambda} \sum_{k=0}^{\infty} \frac{(\lambda e^z)^k}{k!} = e^{-\lambda} e^{\lambda e^z} = \exp\{\lambda(e^z - 1)\}. \tag{1.2.3}$$

Next, we consider continuous distributions.

1.2.4 The uniform distribution

In Example 1.1-3, we have shown that the m.g.f. of X uniformly distributed on $[0,1]$ is $M_X(z) = \frac{1}{z}(e^z - 1)$ for $z \neq 0$. Any r.v. Y uniformly distributed on $[a, b]$, may be presented as $Y = a + (b-a)X$. So, by (1.1.10), the m.g.f.

$$M_Y(z) = e^{az} M_X((b-a)z) = e^{az} \frac{1}{(b-a)z} (e^{(b-a)z} - 1) = \frac{e^{bz} - e^{az}}{z(b-a)} \tag{1.2.4}$$

if $z \neq 0$. For $z = 0$, in accordance with (1.1.4), we set $M(0) = 1$. Note that the limit of the r.-h.s. of (1.2.4), as $z \to 0$, equals one (for example, by L'Hôpital's rule).

1.2.5 The exponential and gamma distributions

In Example 1.1-5, we have already shown that for the exponential distribution with the density $f(x) = ae^{-ax}$ for $x \geq 0$, the m.g.f. $M(z) = \frac{1}{1-z/a}$ for $z < a$.

Recalling that the mean $m = 1/a$, we can rewrite it as

$$M(z) = \frac{1}{1-mz}. \tag{1.2.5}$$

In the general case, for the Γ-density $f_{a\nu}(x) = \dfrac{a^\nu x^{\nu-1}}{\Gamma(\nu)} e^{-ax}$ and $z < a$, making the change of variable $y = (a-z)x$, we have

$$\begin{aligned}
M(z) &= \int_0^\infty e^{zx} f_{a\nu}(x)dx = \frac{a^\nu}{\Gamma(\nu)} \int_0^\infty e^{zx} x^{\nu-1} e^{-ax} dx = \frac{a^\nu}{\Gamma(\nu)} \int_0^\infty x^{\nu-1} e^{-(a-z)x} dx \\
&= \frac{a^\nu}{\Gamma(\nu)} \frac{1}{(a-z)^\nu} \int_0^\infty y^{\nu-1} e^{-y} dy = \frac{a^\nu}{\Gamma(\nu)} \frac{1}{(a-z)^\nu} \Gamma(\nu) \\
&= \left(\frac{a}{a-z}\right)^\nu = \frac{1}{(1-z/a)^\nu}.
\end{aligned} \tag{1.2.6}$$

1.2.6 The normal distribution

For a standard normal r.v. X, completing the square in the third step below, we have

$$\begin{aligned}
M_X(z) &= \int_{-\infty}^\infty e^{zx} \frac{1}{\sqrt{2\pi}} e^{-x^2/2} dx = \frac{1}{\sqrt{2\pi}} \int_{-\infty}^\infty \exp\{zx - x^2/2\} dx \\
&= (2\pi)^{-1/2} \int_{-\infty}^\infty \exp\left\{\frac{1}{2}\left(z^2 - (x-z)^2\right)\right\} dx = (2\pi)^{-1/2} e^{z^2/2} \int_{-\infty}^\infty \exp\left\{-\frac{1}{2}(x-z)^2\right\} dx.
\end{aligned}$$

With the change of variable $y = x - z$, we get

$$M_X(z) = (2\pi)^{-1/2} e^{z^2/2} \int_{-\infty}^\infty \exp\{-y^2/2\} dy = e^{z^2/2} \int_{-\infty}^\infty (2\pi)^{-1/2} \exp\{-y^2/2\} dy.$$

The integrand in the last integral is the standard normal density. Hence, the integral itself equals one, and

$$M_X(z) = e^{z^2/2}.$$

For the (m, σ^2)-normal r.v. $Y = m + \sigma X$, by virtue of (1.1.10),

$$M_Y(z) = e^{mz} M_X(\sigma z) = \exp\{mz + \sigma^2 z^2/2\}. \tag{1.2.7}$$

1.3 Moment generating function and moments

Consider a r.v. X with a d.f. $F(x)$, and assume that for some $z_0 > 0$,

$$E\{e^{|zX|}\} < \infty \text{ for } |z| < z_0. \tag{1.3.1}$$

Since $E\{e^{zX}\} \leq E\{e^{|zX|}\}$, condition (1.3.1) ensures that the m.g.f. $M(z) = E\{e^{zX}\} = \int_{-\infty}^{\infty} e^{zx} dF(x)$ is well defined for all z from the interval $(-z_0, z_0)$.

It may be proved that under condition (1.3.1), we can differentiate $M(z)$ an arbitrary number of times, and we can do that by passing the operation of differentiation inside the expectation $E\{\cdot\}$. Differentiating $M(z)$ once, we get

$$M'(z) = \frac{d}{dz}E\{e^{zX}\} = E\left\{\frac{d}{dz}e^{zX}\right\} = E\{Xe^{zX}\} = \int_{-\infty}^{\infty} xe^{zx}dF(x). \tag{1.3.2}$$

In particular,

$$M'(0) = E\{Xe^{X\cdot 0}\} = E\{X\}. \tag{1.3.3}$$

The differentiation of (1.3.2) leads to

$$M''(z) = E\left\{\frac{d}{dz}Xe^{zX}\right\} = E\{X^2e^{zX}\} = \int_{-\infty}^{\infty} x^2e^{zx}dF(x), \tag{1.3.4}$$

and

$$M''(0) = E\{X^2\}. \tag{1.3.5}$$

Continuing in the same fashion, we get that the kth derivative $M^{(k)}(z) = E\{X^ke^{zX}\} = \int_{-\infty}^{\infty} x^ke^{zx}dF(x)$, and for all k,

$$M^{(k)}(0) = E\{X^k\} = m_k, \tag{1.3.6}$$

the kth moment of X.

(The same also follows from the Taylor expansion (1.1.7) if we compare it with the general Taylor formula (2.2.3) from the Appendix.)

From (1.3.4), it follows that $M''(z) \geq 0$ for all z, and hence,

Any moment generating function $M(z)$ is convex.	(1.3.7)

See typical graphs in Fig. 1.

EXAMPLE 1. What is the difference between r.v.'s having the m.g.f.'s as in Fig. 1? In view of (1.3.3), for the first, the expected value is positive, and for the second, it is negative.

EXAMPLE 2. We saw that $\exp\{z^2/2\}$ is the m.g.f. of the standard normal distribution. Can the function $g(z) = \exp\{z^4/2\}$ be the m.g.f. of a r.v. X? No, because $g''(0) = 0$, from which it would have followed that $E\{X^2\} = 0$. The last relation holds only if $X = 0$ with probability one, but then $M_X(z) = 1$ for all z's. The reader is invited to show that the same is true for $g(z) = \exp\{cz^4\}$ for any constant $c \neq 0$.

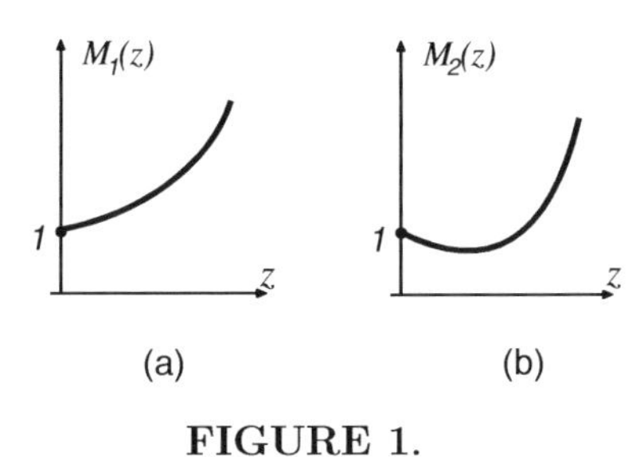

FIGURE 1.

EXAMPLE 3. In Example 1.1-2, we have shown that a r.v. X assuming values 1 and 0 with probabilities p and $1-p$, has the m.g.f. $M(z) = 1 + p(e^z - 1)$. *Any* derivative $M^{(k)}(z) = pe^z$; the same for all k. In particular, $M^{(k)}(0) = p$ for all k. It is consistent with the fact that, since $1^k = 1$ and $0^k = 0$, for our r.v., all moments $E\{X^k\} = p$. □

2 SOME EXAMPLES OF APPLICATIONS OF M.G.F.'S IN THEORY AND PARTICULAR MODELS

This section concerns various examples demonstrating the efficiency of m.g.f.'s in solving various problems. Many applications are based on the property that the m.g.f. of the sum of r.v.'s equals the product of the m.g.f's—property (1.1.9). We will also use this in Chapter 9 devoted to the central limit theorem. However, the usefulness of m.g.f's is, by no means, connected only with the property mentioned.

2.1 Sums of normal and gamma r.v.'s

We give new and shorter proofs of Propositions **6**.5 and **6**.6 concerning sums of normal and gamma r.v.'s.

Let X_1 and X_2 be independent normal r.v.'s with means m_1 and m_2, and variances σ_1^2 and σ_2^2, respectively. Let $S = X_1 + X_2$. Since the m.g.f. of a (m, σ^2)-normal r.v. is $\exp\{mz + \sigma^2 z^2/2\}$, by virtue of (1.1.9), the m.g.f.

$$M_S(z) = \exp\{m_1 z + \sigma_1^2 z^2/2\}\exp\{m_2 z + \sigma_2^2 z^2/2\} = \exp\{(m_1 + m_2)z + (\sigma_1^2 + \sigma_2^2)z^2/2\}.$$

This is the m.g.f. of the normal distribution with expectation $m_1 + m_2$ and variance $\sigma_1^2 + \sigma_2^2$. Then, by virtue of the uniqueness Theorem 1, the distribution of S is also normal with the mentioned mean and variance.

We see that, in comparison with the convolution method we used in Chapter 6, the method of m.g.f.'s gives a shorter and more explicit proof.

The same concerns the sum of Γ-distributed r.v.'s. Let X_1 and X_2 be Γ-r.v.'s with parameters (a, ν_1) and (a, ν_2), respectively. Since the m.g.f. of the Γ-r.v. with parameters (a, ν) is $(1 - z/a)^{-\nu}$, the m.g.f.

$$M_S(z) = (1 - z/a)^{-\nu_1}(1 - z/a)^{-\nu_2} = (1 - z/a)^{-(\nu_1 + \nu_2)}.$$

This is the m.g.f. of the Γ-distribution with parameters $(a, \nu_1 + \nu_2)$, which together with Theorem 1 proves Proposition **6**-6.

In Exercise 12, the reader is invited to do the same for the sum of Poisson r.v.'s.

2.2 An example where the m.g.f.'s are rational functions

Let X_1 and X_2 be exponential with $E\{X_1\} = 1$ and $E\{X_2\} = 2$, respectively; $S = X_1 + X_2$. In Exercise **6**.40a, the reader was invited to find the distribution of S computing the corresponding convolution. Now, we show how to find the same by the m.g.f.'s method. By virtue of (1.2.5), $M_{X_1}(z) = 1/(1-z)$ and $M_{X_2}(z) = 1/(1-2z)$. It is straightforward to verify that

$$M_S(z) = \frac{1}{1-z} \cdot \frac{1}{1-2z} = \frac{2}{1-2z} - \frac{1}{1-z}. \tag{2.2.1}$$

The idea is connected with the so-called method of partial fractions. If we had not known how to decompose $\frac{1}{(1-z)(1-2z)}$ in partial fractions, we could write $\frac{1}{(1-z)(1-2z)} = \frac{a}{1-2z} + \frac{b}{1-z}$, and look for a and b for which it is true. To find a and b, we put the r.-h.s. into a common denominator, and rewrite it as

$$\frac{1}{(1-z)(1-2z)} = \frac{a+b-(a+2b)z}{(1-z)(1-2z)}.$$

This implies $a+b=1$ and $a+2b=0$. Hence, $a=2$ and $b=-1$, which leads to (2.2.1).

Let us come back to (2.2.1). We see that $M_S(z)$ is a linear combination of two exponential m.g.f's: with a mean of two, and a mean of one. Hence, by virtue (1.1.11), the distribution of S is the linear combination with the same coefficients of the exponential distributions with the corresponding densities: $f_2(x) = \frac{1}{2}\exp\{-x/2\}$ and $f_1(x) = e^{-x}$, respectively, for $x \geq 0$. For $x < 0$, both densities are equal to zero.

Consequently, the density of S is $f_S(x) = 2f_2(x) - f_1(x) = e^{-x/2} - e^{-x}$ for $x \geq 0$. For $x < 0$, the density $f_S(x) = 0$. In Exercise 13, it is suggested to the reader to double-check that the function $f(x)$ above is indeed a density; that is, it is non-negative and the total integral of it is one.

2.3 The sum of a random number of r.v.'s

Let r.v.'s $X_1, X_2, ...$ and N be mutually independent, the r.v.'s $X_1, X_2, ...$ be identically distributed, and N be integer valued. Let

$$S = S_N = \sum_{j=1}^{N} X_j. \tag{2.3.1}$$

If N assumes zero value, we set $S = 0$. The following proposition is useful for exploring such sums.

Proposition 2 *For all z's for which the m.g.f.'s below are well defined, the m.g.f. of S is*

$$M_S(z) = M_N(\ln M_X(z)), \tag{2.3.2}$$

where $M_N(\cdot)$ is the m.g.f. of N, and $M_X(z)$ is the (common) m.g.f. of the r.v.'s X_i. In particular, if N is a Poisson r.v. with parameter λ,

$$M_S(z) = \exp\{\lambda(M_X(z) - 1)\}. \tag{2.3.3}$$

Proof. Conditioning on N, we have $M_S(z) = E\{e^{zS}\} = E\{E\{e^{zS}\,|\,N\}\}$.

In $E\{e^{zS}\,|\,N\}$, we may treat N as a given integer, and view the conditional expectation $E\{e^{zS}\,|\,N\}$ as the m.g.f. of a sum of a *fixed* number of independent r.v.'s. Hence, by the main property of m.g.f.'s (1.1.9), $E\{e^{zS}\,|\,N\} = (M_X(z))^N = e^{(\ln M_X(z))N}$, and

$$E\{e^{zS}\} = E\{e^{(\ln M_X(z))N}\}. \tag{2.3.4}$$

The r.-h.s. of (2.3.4) is the m.g.f. of N at the point $(\ln M_X(z))$, which implies (2.3.2).

If N is a Poisson r.v. with parameter λ, the m.g.f. $M_N(z) = \exp\{\lambda(e^z - 1)\}$. Replacing z by $\ln M_X(z)$, we come to (2.3.3). ■

EXAMPLE 1. Consider a register counting some particles. Assume the flow of particles to be Poisson. The particles may be of two types, and each particle registered belongs to the first type with a probability of p. Let N be the total number of particles during a time period, and $E\{N\} = \lambda$. For example, the particles are emitted from a nuclei as a result of nuclear instability, and there are two types of particles: alpha (helium nuclei) and beta (electrons or positrons).

Let N_1 and N_2 be the number of particles of the first and the second type, respectively; so $N = N_1 + N_2$. In Section **3**.4.5.5, we have proved that N_1 has the Poisson distribution with parameter $p\lambda$. Let us look at how it can be shown by the method of m.g.f.'s.

Set a r.v. $X_j = 1$ if the jth particle is of the first type, and $X_j = 0$ otherwise. Then S in the above framework is the number of the particles of the first type: $S = N_1$.

The m.g.f. $M_{X_j}(z) = M_X(z) = 1 + p(e^z - 1)$; see (1.1.5). Hence, in view of (2.3.3),

$$M_S(z) = \exp\{\lambda(1 + p(e^z - 1) - 1)\} = \exp\{p\lambda(e^z - 1)\}.$$

This is the m.g.f. of a Poisson r.v. with parameter $p\lambda$. (Compare this with a longer proof in Section **3**.4.5.5.)

EXAMPLE 2. This beautiful example is close to Example **6**.4.2-2. Now, we use m.g.f.'s. Let X_i be standard exponential, and N have the second version of the geometric distribution with parameter p; that is, $P(N = k) = pq^k$, where $k = 0, 1, ...$ and $q = 1 - p$. We have

$$M_X(z) = \frac{1}{1-z}, \quad M_N(z) = \frac{p}{1 - qe^z}.$$

Then, by (2.3.2),

$$\begin{aligned} M_S(z) &= \frac{p}{1 - qM_x(z)} = \frac{p}{1 - q(1-z)^{-1}} = \frac{p(1-z)}{1 - z - q} = \frac{p - pz}{p - z} = \frac{p^2 - pz}{p - z} + \frac{p - p^2}{p - z} \\ &= \frac{p(p-z)}{p-z} + \frac{p(1-p)}{p-z} = p + \frac{pq}{p-z} = p + q \cdot \frac{1}{1 - z/p}. \end{aligned} \tag{2.3.5}$$

The function $\frac{1}{1-z/p}$ is the m.g.f. of the exponential distribution with parameter $a = p$. Let a (non-random) variable $Y \equiv 0$. The m.g.f. $M_Y(z) \equiv 1$. We see that $M_S(z)$ is the linear combination (with the coefficients p and q) of two m.g.f.'s: $M_Y(z)$ and the exponential m.g.f. with parameter p.

Then $F_S(x)$, the d.f. of S, should be the linear combination, with the same coefficients, of the corresponding d.f.'s. Denote by $Q(x)$ the d.f. of the r.v. Y. That is, $Q(x) = 1$ for $x \geq 0$ and $= 0$ for $x < 0$; see also Fig. **3**-2.

Let $F_p(x)$ be the d.f. of the exponential distribution with parameter p. Then from (2.3.5), it follows that

$$F_S(x) = pQ(x) + qF_p(x). \tag{2.3.6}$$

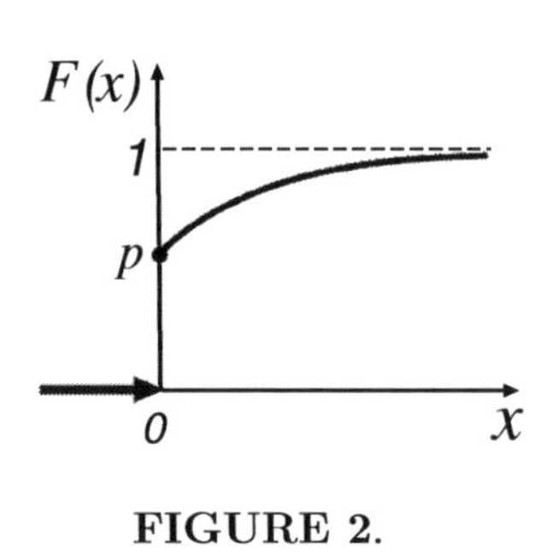

FIGURE 2.

Thus, $F_S(x) = 0$ for $x < 0$, and

$$F_S(x) = p + q(1 - e^{-px}) = 1 - qe^{-px} \quad \text{for} \quad x \geq 0. \tag{2.3.7}$$

See the graph in Fig. 2. In particular, the jump at point 0 equals p.

We may view (2.3.6) as if $S = \begin{cases} 0 & \text{with probability } p, \\ Z_p & \text{with probability } q, \end{cases}$

where Z_p is an exponential r.v. with parameter p.

Actually, it is amazing. Let $p = 0.01$. Then $E\{N\} = q/p = 99$. So, S is the sum of a large (random) number of exponential r.v.'s; on the average, in the sum S, there are 99 terms. Nevertheless, S has a simple structure. Namely, with probability 0.01, it is equal to zero, and with probability 0.99, it may be viewed as just one exponential r.v. with a mean of $\frac{1}{p} = 100$. □

2.4 M.g.f.'s in Finance

For certainty, we will talk about insurance but the reader will readily see that *the same or similar models may describe other financial operations.*

Consider an insurance which provides for payment of a single benefit (a sum insured) at some random time T in the future. Mostly, this concerns life insurance, although the same model may be applied, for example, for describing contracts offering warranties for machines. We take a year as a unit of time, and \$1 as a unit of money.

The main feature of any life insurance consists in the time lag between the moment of the benefit payment and the time of policy issue. The value of each \$1 to be paid in the future is less than \$1 at the moment when the insurance is purchased.

For the present value of \$1 to be paid at a future time t, we adopt the presentation $e^{-\delta t}$, where δ is an annual risk free interest rate. This issue is discussed in detail in the Appendix, Section 3, and the reader is recommended to look over it.

If the moment of payment is a r.v. T, then the present value of the future payment of a *unit of money* is also random and is equal to the r.v.

$$Z = e^{-\delta T}. \tag{2.4.1}$$

Then the expected present value of the future payment of the insurance company equals

$$A = E\{Z\} = E\{e^{-\delta T}\}. \tag{2.4.2}$$

The notation A is traditional. The quantity A is called the *actuarial present value* or the *net single premium* of the contract. The term "*single*" means that we are talking about a premium paid one time at the moment of policy issue. The term "*net*" is related to the fact that such a premium does not reflect the riskiness of the contract; it is equal to the mean present value of the obligation of the company. For the company to be able to function, the real premium should be at least slightly larger than the net premium. However, as a rule, when determining real premiums that will be charged, the companies proceed from net premiums, computing them and then adding to them an "additional payment for the risk incurred" called a security loading. Usually, such an addition constitutes 5–10% of the net premium. In Chapter 9, Example 1-5, we will consider one of approaches to computing security loadings.

An important—and nice from a mathematical point of view—fact is that, as a function of δ, the net premium A is the m.g.f. of the r.v. T. More precisely,

$$A = E\{e^{-\delta T}\} = M_T(-\delta), \tag{2.4.3}$$

where $M_T(s)$ is the m.g.f. of T.

Moreover, by (2.4.1), the kth moment

$$E\{Z^k\} = E\{e^{-k\delta T}\} = M_T(-k\delta).$$

So, if we know the m.g.f. $M_T(z)$, then we know all moments of Z. In particular, $E\{Z^2\} = M_T(-2\delta)$, and

$$Var\{Z\} = E\{Z^2\} - (E\{Z\})^2 = M_T(-2\delta) - (M_T(-\delta))^2. \tag{2.4.4}$$

EXAMPLE 1. Consider life insurance that provides for paying to a beneficiary a unit of money at the moment of the death of the insured. In this case, T is the *remaining* life time of the insured. Assume that the r.v. T is uniformly distributed on an interval $[0,d]$.

(For non-young people such an assumption is not as unrealistic as it might seem, and for old people it works pretty well. For example, the distribution uniform on $[0,20]$, surprisingly, proves to be a very good approximation for the remaining lifetime of a person taken at random from the population of people of age about 75. We will consider survival probabilities in more detail in Section **11**.3.)

In accordance with (1.2.4), the m.g.f. $M_T(z) = (e^{dz} - 1)/(dz)$, and

$$A = M_T(-\delta) = \frac{e^{-d\delta} - 1}{-d\delta} = \frac{1 - e^{-\delta d}}{\delta d}. \tag{2.4.5}$$

By (2.4.4) and (2.4.5),

$$Var\{Z\} = \frac{1 - e^{-2\delta d}}{2\delta d} - \left(\frac{1 - e^{-\delta d}}{\delta d}\right)^2. \tag{2.4.6}$$

Let $d = 40$, so we consider a group of customers for whom the *mean* remaining life time is 20 years. Let us adopt $\delta = 0.04$ (that is, the 4% interest rate), a realistic figure for the risk free interest rate for a not bad economy. Then, as is easy to compute, $A \approx 0.499$, $Var\{Z\} \approx 0.050$, and hence, the standard deviation $\sigma_Z \approx 0.225$. Thus, the present value of the obligation to pay \$1 at the moment of death is, on the average, about 50¢with a standard deviation of 22.5¢. Say, for a life insurance benefit of \$100,000, the net single premium is \$50,000.

EXAMPLE 2. A store offers a warranty for a product to be sold. Let T be the "life time" of the product until the first repair, and X be the cost of the repair. Assume that T is an exponential r.v. with a mean of $m = 5$ years, and X is uniform on $[1,3]$. (Say, we adopt \$100 as a unit of money.) Setting $\delta = 0.04$, let us estimate the cost of the warranty. The m.g.f. $M_T(z) = 1/(1-mz)$, and hence, the expected present value of a unit money to be paid at time T is

$$A = M_T(-\delta) = \frac{1}{1+m\delta}.$$

The cost of repair is X. If we assume X and T to be independent, and set $Y = Xe^{-\delta T}$, the present value of the future repair, we can write

$$E\{Y\} = E\{Xe^{-\delta T}\} = E\{X\}E\{e^{-\delta T}\} = E\{X\} \cdot \frac{1}{1+m\delta} = 2 \cdot \frac{1}{1+5\cdot(0.04)} = 1.66....$$

This is a net premium. If the company adds a security loading of, say, 5%, then the real cost of the warranty will be $(1.66...)\cdot 1.05 = 1.75$, or \$175 (since the unit of money is \$100). □

Next, we consider *annuities*. An annuity is a series of payments made at certain intervals during some period which is, as a rule, random. Suppose that an individual or a company—a general term is an *annuitant*—at each moment of time $t = 0, 1, ..., K$, is going to receive a *unit* of money. As is shown in the Appendix, Section 3, the present value of such a "cash flow" is equal to

$$Y = 1 + v + v^2 + ... + v^K = \frac{1 - v^{K+1}}{1-v},$$

where v is the discount factor corresponding to a unit time period. As was shown in the Appendix, Section 3, if a unit of time is a year, δ is an annual interest rate, and interest is compounded continuously, then $v = e^{-\delta}$. So, $v^{K+1} = e^{-\delta(K+1)}$, and

$$Y = \frac{1 - e^{-\delta(K+1)}}{1-v}.$$

If K is random, then so is Y, and the expected present value equals

$$a = E\{Y\} = \frac{1 - E\{e^{-\delta(K+1)}\}}{1-v} = \frac{1 - M_{K+1}(-\delta)}{1-v}. \tag{2.4.7}$$

The notation a is also traditional.

EXAMPLE 3. Let us come back to Example 1. Consider a more realistic situation when the insured pays not a total single premium at the moment of the purchase of the insurance, but provides regular payments for the insurance during her/his life. For simplicity, suppose that the insured pays an annual premium at the beginning of each year. In this case, the annuitant is the insurance company, and the moment of the last payment is the beginning of the year of the death of the insured. So, K is the integer part of T.

If T is uniform on $[0,d]$, and d is an integer, then K takes on values $0,1,...,d-1$ with equal probabilities $1/d$. Using the formula for a geometric series, we have

$$M_{K+1}(z) = \sum_{k=0}^{d-1} e^{z(k+1)}\frac{1}{d} = \frac{e^z}{d}\sum_{k=0}^{d-1} e^{zk} = \frac{e^z}{d}\cdot\frac{e^{dz}-1}{e^z-1} = \frac{e^{dz}-1}{d(1-e^{-z})}.$$

Then

$$M_{K+1}(-\delta) = \frac{1-e^{-\delta d}}{d(e^{\delta}-1)}.$$

Setting $d=40$ and $\delta = 0.04$, we get $M_{K+1}(-\delta) \approx 0.489$, and substituting this into (2.4.7), we obtain

$$a = \frac{1-M_{K+1}(-\delta)}{1-v} = \frac{1-M_{K+1}(-\delta)}{1-e^{-0.04}} \approx 13.03.$$

Now, let us recall that the quantity a corresponds to the payment of a *unit* of money at moments $t=0,1,...K$. We see that the expected present value of such a sequence of payments is equal to 13.03 units.

However, the insured pays not a unit of money but the premium the company charges. If we denote this premium by π, the present value of the insured's payments will be πY, and the expected present value is $E\{\pi Y\} = \pi a$. To find π, we proceed from what is called the *equivalence principle* based on the following scheme.

Let Z be the present value of what the company will pay. Then the present value of the company's loss is the r.v.

$$L = Z - \pi Y. \tag{2.4.8}$$

We call a premium π a *net premium* (another term is a *benefit premium*) if the expected loss $E\{L\} = 0$: on the average, the company and the insured "are even." As above, it is not a real premium the company charges. When determining the real premium, the company proceeds from the net premium adding to it a security loading, reflecting the value of the risk incurred.

From (2.4.8), it immediately follows that for $E\{L\} = 0$, we should set

$$\pi = \frac{E\{Z\}}{E\{Y\}} = \frac{A}{a}.$$

For the particular values we have chosen, $\pi = \dfrac{0.499}{13.02} \approx 0.038$.

It remains to recall that this concerns a unit benefit. If, as a matter of fact, the single benefit upon death equals, say, \$100,000, the annual *net* premium will be \$3,800. In reality, it would lead to a premium about \$4,000. The mean remaining lifetime is 20 years, so on the average, the insured will pay about \$80,000. The fact that this number is less than \$100,000 which the company will pay is not surprising at all: the insured pays earlier than the company. □

3 EXPONENTIAL (OR BERNSTEIN–CHERNOFF'S) BOUNDS

3.1 Main idea

This section concerns bounds for the tails of the distributions; that is, bounds for the probabilities $P(X > x)$ for large x. The main idea is to apply the inequality for deviations

$$P(X \geq x) \leq \frac{E\{u(X)\}}{u(x)} \tag{3.1.1}$$

we proved in Section **3**.3.3, and choose the function $u(x) = e^{zx}$, where a positive z is a free parameter. For such a function $u(x)$, (3.1.1) implies $P(X \geq x) \leq \frac{E\{e^{zX}\}}{e^{zx}} = e^{-zx}M_X(z)$. The l.-h.s. of the last inequality does not depend on z, so to optimize the bound, we choose a z minimizing the r.-h.s. This leads to the inequality

$$P(X \geq x) \leq \min_{z \geq 0} \left\{ e^{-zx} M_X(z) \right\}. \tag{3.1.2}$$

In particular problems, people often choose z that is not exactly a minimizer of $e^{-zx}M_X(z)$ but is close to it, and leads to a convenient form of the bound. In particular, this is what we will do in the next Section 3.2. However, first, consider an example with a precise minimizer.

EXAMPLE 1. Let X be a Poisson r.v. with a mean of λ. Since $M_X(z) = \exp\{\lambda(e^z - 1)\}$, for any $z > 0$,

$$P(X \geq x) \leq e^{-zx}M_X(z) = \exp\{\lambda(e^z - 1) - zx\}. \tag{3.1.3}$$

The minimization of the r.-h.s. is equivalent to minimizing $\lambda(e^z - 1) - zx$ in z. Simple calculus shows that the minimum is attained at $z = \ln(x/\lambda)$ provided $x > \lambda$. Substitution into (3.1.3) leads to $P(X \geq x) \leq e^{-\lambda}(e\lambda)^x/x^x$ for $x > \lambda$. For example, if the mean value $\lambda = 1$, then $P(X \geq x) \leq e^{x-1}/x^x$ for $x > 1$. Note that x^x converges to ∞ much faster than e^x.

Assume, for instance, that Mr. K. receives one e-mail message each half an hour on the average. Consider a particular half an hour, say, from 2 pm to 2:30 pm. Suppose that the random number X of messages during this period is a Poisson r.v. (with, as we have assumed, a mean of one). Then, the probability of getting more than, for example, five messages $P(X > 5) = P(X \geq 6) \leq (e^{6-1}/6^6) \approx 0.0032$. □

Bounds of such a type are called *exponential*, or *Bernstein's* or *Chernoff's* bounds. They are especially efficient when it concerns sums of r.v.'s. One of such bounds will be considered below.

3.2 A bound for large deviations for sums of r.v.'s

Consider a sequence of independent identically distributed r.v.'s $X_1, ..., X_n$ with $E\{X_j\} = 0$, $Var\{X_j\} = \sigma^2$, and suppose that for some constant $c > 0$, with probability one,

$$|X_j| \leq c. \tag{3.2.1}$$

Let $S_n = X_1 + ... + X_n$. Clearly, $E\{S_n\} = 0$ and $Var\{S_n\} = \sigma^2 n$. Consider the *normalized sum*

$$S_n^* = \frac{S_n - E\{S_n\}}{\sqrt{Var\{S_n\}}} = \frac{S_n}{\sigma\sqrt{n}}.$$

We know that $E\{S_n^*\} = 0$ and $Var\{S_n^*\} = 1$.

Theorem 3 *Let $k(n) = \sigma\sqrt{n}/c$. Then for $x \geq 0$,*

$$P(S_n^* \geq x) \leq \begin{cases} \exp\left\{-\frac{x^2}{2}\left(1 - \frac{x}{2k(n)}\right)\right\} & \text{if } x \leq k(n), \\ \exp\left\{-\frac{xk(n}{4}\right\} & \text{if } x \geq k(n). \end{cases} \quad (3.2.2)$$

We prove it in Section 3.3.

First of all, note that for $x = k(n)$, both bounds above coincide.

For a fixed x, the term $\left(1 - \frac{x}{2k(n)}\right) \to 1$ as $n \to \infty$, and $x \leq k(n)$ for sufficiently large n. So, for large n, we may use the *approximate* bound $\exp\{-x^2/2\}$.

Next, observe that in the first inequality in (3.2.2), $\left(1 - \frac{x}{2k(n)}\right) \geq \frac{1}{2}$ for $x \leq k(n)$. This leads to a less sharp but more convenient bound.

Corollary 4 *For $x > 0$,*

$$P(S_n^* \geq x) \leq \begin{cases} \exp\left\{-\frac{x^2}{4}\right\} & \text{if } x \leq k(n), \\ \exp\left\{-\frac{xk(n)}{4}\right\} & \text{if } x \geq k(n). \end{cases} \quad (3.2.3)$$

One more thing to note is that bounds (3.2.2)–(3.2.3) involve only the characteristics c and σ, and both characterize the distribution of $|X_j|$. Hence, the same bounds remain true for the "left tail" $P(S_n^* \leq -x)$, $x > 0$. Then for the two-sided probability $P(|S_n^*| \geq x)$ we may use the same bounds multiplying them by two.

EXAMPLE 1. Suppose that X_j's are uniform on $[-c, c]$. Then $\sigma^2 = \frac{4c^2}{12} = \frac{c^2}{3}$; see, e.g., the Appendix, Table 1, and we can use (3.2.2) with $k(n) = \sqrt{n/3}$.

Let us revisit, for instance, Exercise **3**-81 where Bob is checking whether the total charge in his credit card statement is plausible. To this end, Bob adds up all figures in the statement concerning the purchases made. To make it easier, Bob rounds each number to the nearest integer. In Exercise **3**-81 , we have assumed that the fraction part of the cost of each purchase is equally likely to be any number from $0, 1, ..., 99$ cents, and hence the rounding error for each figure takes on values $0, \pm 1, ..., \pm 50$ cents. To make calculations less tiresome, let us approximate such an error by a continuous r.v. uniform on $[-50, 50]$. The reader may make sure that the accuracy of our estimates practically does not change after such an approximation.

So, let n be the number of purchases, and $X_1, ..., X_n$ be the corresponding rounding errors. We assume X's to be independent and uniform on $[-50, 50]$. The cumulative error is $S_n = X_1 + ... + X_n$. The standard deviation of the sum is $\sigma_{S_n} = \sigma\sqrt{n}$, which may be viewed as the "order" of the total error. For example, for $n = 100$, we have $\sigma_{S_n} = \frac{50}{\sqrt{3}} \cdot \sqrt{100} \approx 289$ cents; that is, about \$3.

Let us compute

$$P(|S_n| \geq z) = P\left(\frac{|S_n|}{\sigma\sqrt{n}} \geq \frac{z}{\sigma\sqrt{n}}\right) = P(|S_n^*| \geq x(n)),$$

where $x(n) = \frac{z}{\sigma\sqrt{n}}$. We can use the first bound in (3.2.2) if $x(n) \leq \sqrt{n/3} \approx 5.77$ for $n = 100$. In this case,

$$P(|S_n| \geq z) \leq 2\exp\left\{-\frac{x^2(n)}{2}\left(1 - \frac{x(n)}{2\sqrt{n/3}}\right)\right\}.$$

The reader will readily double-check that for $z = 1000$ (i.e., \$10), $x(n) \approx 3.46$, and $P(|S_n| \geq z) \leq 0.03$. So, if Bob made 100 purchases (which may amount to a couple of thousands or more), the probability that the cumulative rounding error exceeds \$10 is not larger than 3%.

EXAMPLE 2 (*Estimating a probability*). This example concerns estimating an unknown probability of "success" in a sequence of independent trials. Suppose, for instance, we want to estimate the probability p of a particular side effect of a medical treatment. To this end, we consider n patients who received such a treatment, and compute the proportion of the patients who faced the side effect. Denote this proportion by $\hat{p}$, and set $\xi_j = 1$ if the jth patient experienced the side effect, and $\xi_j = 0$ otherwise. Then $\hat{p} = \frac{1}{n}\sum_{j=1}^{n}\xi_j$, $P(\xi_j = 1) = p$, $E\{\xi_j\} = p$, and $Var\{\xi_j\} = p(1-p)$.

Denote by δ a desired accuracy of the estimation. Then, for this accuracy to be achieved, we should have $|p - \hat{p}| \leq \delta$. We wish to evaluate the probability of this event, and set $r(\delta) = P(|p - \hat{p}| \leq \delta) = P(\hat{p} - \delta \leq p \leq \hat{p} + \delta)$. So, $r(\delta)$ is the probability that the unknown parameter p is within the interval $[\hat{p} - \delta, \hat{p} + \delta]$. We want this probability to be not less than a fixed number C which is called a *confidence level* of estimation. Say, if $C = 0.95$, we want the estimate to be within the desired accuracy with a probability not less than 95%.

The interval $[\hat{p} - \delta, \hat{p} + \delta]$ for which such a requirement is fulfilled is called a *confidence interval*.

Naturally, we want δ to be small, and C to be large. However, the smaller δ, the smaller the probability to get into the interval $[\hat{p} - \delta, \hat{p} + \delta]$, and $r(\delta)$ may turn out to be smaller than C. Consequently, the choice of C and δ (that is, the choice of the estimation strategy) is a tradeoff between the values of C and δ.

To estimate $r(\delta)$, observe that $\hat{p} - p = \frac{1}{n}\sum_{j=1}^{n}\xi_j - p = \frac{1}{n}\sum_{j=1}^{n}(\xi_j - p) = \frac{1}{n}S_n$, where $S_n = \sum_{j=1}^{n} X_j$ and $X_j = \xi_j - p$. We have $E\{X_j\} = 0$, $\sigma^2 = Var\{X_j\} = p(1-p)$.

Let $\bar{r}(\delta) = 1 - r(\delta) = P(|\hat{p} - p| > \delta)$. We have

$$\bar{r}(\delta) = P\left(\frac{1}{n}|S_n| > \delta\right) = P\left(\frac{|S_n|}{\sigma\sqrt{n}} > \frac{\delta\sqrt{n}}{\sigma}\right) = P\left(|S_n^*| > \frac{\delta\sqrt{n}}{\sqrt{p(1-p)}}\right).$$

Let us set

$$x(n) = \frac{\delta\sqrt{n}}{\sqrt{p(1-p)}}.$$

Assume for a while that $p \geq 0.5$.

Since $-p \leq X_j \leq 1-p$, in this case, $|X_j| \leq p$, and we may set the above parameter $c = p$. Consequently, in this case,

$$k(n) = \frac{\sqrt{p(1-p)n}}{p} = \sqrt{\frac{(1-p)n}{p}}. \tag{3.2.4}$$

Let us apply (3.2.3). It is easy to see that $x(n) \leq k(n)$ if and only if

$$\delta \leq 1-p. \tag{3.2.5}$$

Suppose that the last relation is true. Then, in accordance with (3.2.3),

$$\bar{r}(\delta) = P(|S_n^*| > x(n)) \leq 2\exp\left\{-\frac{x^2(n)}{4}\right\} = 2\exp\left\{-\frac{\delta^2 n}{4p(1-p)}\right\}.$$

The parameter p is unknown, so we cannot use the last bound as is. However, $p(1-p) \leq 1/4$ (why?), and if we consider the worst case, that is, if we replace $p(1-p)$ by $\frac{1}{4}$, we will get a bound that is true for all p.

Eventually, for all $p \geq 0.5$,

$$\bar{r}(\delta) \leq 2\exp\left\{-\delta^2 n\right\}. \tag{3.2.6}$$

If (3.2.5) is not true, then we apply the second inequality in (3.2.3). Moreover, in this case, $p+\delta > 1$, and hence , $P(\hat{p} > p+\delta) = 0$. So, we may apply the one-sided probability, and write

$$\bar{r}(\delta) \leq \exp\left\{-\frac{x(n)k(n)}{4}\right\} = \exp\left\{-\frac{\delta n}{4p}\right\}.$$

Since $p \leq 1$, this implies

$$\bar{r}(\delta) \leq \exp\left\{-\frac{\delta n}{4}\right\}. \tag{3.2.7}$$

The reader may double-check (it also suffices to imagine that we estimate $1-p$ rather than p) that the same bounds, (3.2.6) and (3.2.7), are true for $p < 0.5$. The only difference is that we should replace (3.2.5) by the inequality $\delta \leq p$.

If we have some additional information allowing us to figure out whether $p \geq 0.5$, and how p is related to δ, we can choose a correct bound from (3.2.6)-(3.2.7).

If we do not have any information, then we only know that one of bounds, (3.2.6) or (3.2.7), is true, and we should choose the worst.

It is easy to see that, if $\delta < 1/4$, then the r.-h.s. of (3.2.6) is larger than the r.-h.s. of (3.2.7) even if the latter multiplied by two. Since δ is an accuracy of the estimation of a probability, the requirement $\delta < 1/4$ may be adopted, and eventually we arrive at (3.2.6).

Consider how to apply such a bound. Suppose that we want to estimate p with an accuracy $\delta = 0.05$ and with a confidence level $C = 0.9$. To be sure that the estimation will meet these requirements, we should have $2e^{-\delta^2 n} \leq 1 - C = 0.1$. This is true if the number of observations (trials) $n \geq \dfrac{\ln(2/(1-C))}{\delta^2} = \dfrac{\ln 20}{(0.05)^2} \approx 1198.3$, or rounding, 1200.

If we do not have so many observations, say, if $n = 1000$, then for $2e^{-\delta^2 n} \leq 1 - C$, we should have $\delta > \sqrt{\frac{\ln(2/(1-C))}{n}} \approx 0.055$. So, we have to either reconcile with a worse accuracy, or accept a lower confidence level. □

Note also that in both above examples, we demonstrated how to use universal bounds. If we take into account particular features of the distributions under consideration, then the bounds may be better.

3.3 Proof of Theorem 3

Lemma 5 *Let X be a r.v. such that $|X| \leq c$, $E\{X\} = 0$, and $Var\{X\} = \sigma^2$. Then, for $0 < z \leq \frac{1}{c}$,*

$$M_X(z) \leq \exp\left\{\frac{z^2\sigma^2}{2}\left(1 + \frac{zc}{2}\right)\right\}. \tag{3.3.1}$$

Proof. By the Taylor expansion (the Appendix, (2.2.4)),

$$e^{zX} = 1 + zX + \frac{1}{2}(zX)^2 + \sum_{k=3}^{\infty} \frac{1}{k!}(zX)^k.$$

Since $|X| \leq c$ and $zc \leq 1$, for $k \geq 3$,

$$|(zX)^k| = X^2 z^k |X^{k-2}| \leq X^2 z^k c^{k-2} = X^2 z^3 c(zc)^{k-3} \leq X^2 z^3 c.$$

Now, since $E\{X\} = 0$, the second moment $E\{X^2\} = \sigma^2$, and by the above inequalities,

$$E\{e^{zX}\} \leq 1 + z \cdot 0 + \frac{1}{2}z^2\sigma^2 + \sum_{k=3}^{\infty} \frac{1}{k!}\sigma^2 z^3 c = 1 + \frac{1}{2}\sigma^2 z^2 + z^3 c\sigma^2 \sum_{k=3}^{\infty} \frac{1}{k!}.$$

Furthermore, $\sum_{k=3}^{\infty} \frac{1}{k!} = e - 1 - 1 - \frac{1}{2} \leq \frac{1}{4}$, and hence,

$$E\{e^{zX}\} \leq 1 + \frac{1}{2}\sigma^2 z^2 + \frac{1}{4}z^3 c\sigma^2 = 1 + \frac{\sigma^2 z^2}{2}\left(1 + \frac{zc}{2}\right) \leq \exp\left\{\frac{\sigma^2 z^2}{2}\left(1 + \frac{zc}{2}\right)\right\}$$

by virtue of the inequality $1 + x \leq e^x$. ■

Let us proceed to the theorem. Since $S_n^* = \sum_{j=1}^{n} \frac{X_j}{\sigma\sqrt{n}}$, by the properties of the m.g.f.'s,

$$M_{S_n^*}(z) = \prod_{j=1}^{n} M_{X_j/(\sigma\sqrt{n})}(z) = \prod_{j=1}^{n} M_{X_j}(z/(\sigma\sqrt{n})). \tag{3.3.2}$$

Let $\frac{z}{\sigma\sqrt{n}} \leq \frac{1}{c}$, which is equivalent to $z \leq k(n)$. Then, by Lemma 5,

$$M_{X_j}\left(\frac{z}{\sigma\sqrt{n}}\right) \leq \exp\left\{\frac{\sigma^2 z^2}{2\sigma^2 n}\left(1 + \frac{zc}{2\sigma\sqrt{n}}\right)\right\} = \exp\left\{\frac{z^2}{2n}\left(1 + \frac{z}{2k(n)}\right)\right\}$$

From this and (3.3.2), it follows that for $z \leq k(n)$,

$$M_{S_n^*}(z) \leq \exp\left\{\frac{z^2}{2}\left(1 + \frac{z}{2k(n)}\right)\right\}.$$

Then, by the basic inequality (3.1.2), for $z \leq k(n)$,

$$P(S_n^* \geq x) \leq e^{-zx} M_{S_n^*}(z) \leq \exp\left\{-zx + \frac{z^2}{2}\left(1 + \frac{z}{2k(n)}\right)\right\}. \tag{3.3.3}$$

We choose $z = x$ if $x \leq k(n)$, and we set $z = k(n)$ if $x > k(n)$. Certainly, the chosen $z \leq k(n)$, and we can substitute it into (3.3.3).

For $x \leq k(n)$ such a substitution leads exactly to the first bound in (3.2.2). If $x > k(n)$, then this substitution gives $-k(n)x + \frac{3}{4}k^2(n) \leq -k(n)x + \frac{3}{4}k(n)x = -\frac{xk(n)}{4}$, which leads to the second inequality in (3.2.2). ■

4 EXERCISES

1. Find the m.g.f. of a r.v. assuming values -1 and 1 with equal probabilities, and the m.g.f. of a r.v. uniform on $[-1, 1]$. Graph both m.g.f.'s in one picture and comment *in the terms of moments* why the derivatives at the point zero are the same, and why the first function is steeper.

2. Graph the m.g.f. of an exponential r.v. with a given mean. For which z's can we do it?

3. Show that for an integer ν, (1.2.2) immediately follows from (1.2.1). (*Advice*: Think about N_ν as the moment of the νth success.)

4. Let X_1, X_2 be independent standard exponential r.v.'s. In Exercise **6**.41, we proved that the r.v. $Y = X_1 - X_2$ has the two-sided exponential density $f(x) = \frac{1}{2}e^{-|x|}$. Find the m.g.f. and density of Y using the method of partial fractions we touched on in Section 2.2.

5. Which functions $g(z)$ in Fig. 3abc look as m.g.f.'s?

6. Show that if $P(X \geq 3) = 0.2$, then the corresponding m.g.f. $M(z) \geq 0.2e^{3z}$. What can we say about the distribution of X if $M(z) \sim 0.2e^{3z}$ as $z \to \infty$?

7. Let $M(z)$ be the m.g.f. of a r.v. X. When does there exist at least one number $z_0 > 0$ such that $M(z_0) < 1$? Justify your answer.

8. Let $M(z)$ be the m.g.f. of a r.v. X such that $E\{X\} = 1$. Is it true that $M(z) \geq 1$ for all $z \geq 0$? Answer the same question for the case $E\{X\} = -1$.

9. Let $M(z)$ be the m.g.f. of a r.v. X, $E\{X\} = m$, $Var\{X\} = \sigma^2$. Write $M'(0)$ and $M''(0)$.

10. Compare the means and *variances* of the r.v.'s whose m.g.f.'s are graphed in Fig. 3d.

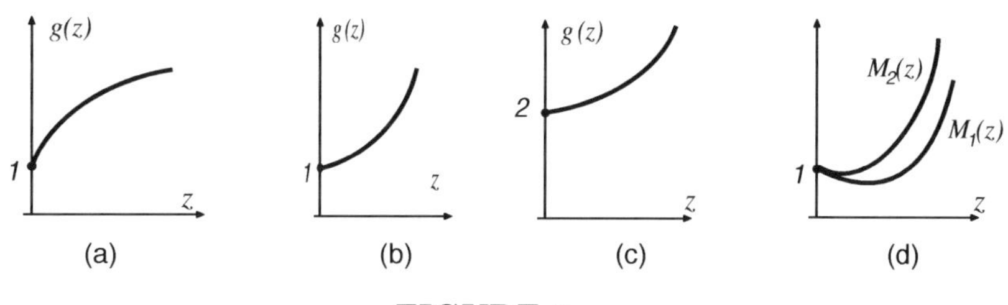

FIGURE 3.

11. (a) Can the function $1+z^4$ be the m.g.f. of a r.v.? (*Hint*: Think about $g''(0)$.)

    (b) In general, show that if $g(z) = 1+\varepsilon(z)z^2$, where $\varepsilon(z) \neq 0$ for $z \neq 0$, and $\varepsilon(z) \to 0$, as $z \to 0$, then $g(z)$ cannot be a m.g.f.

12. Prove that the sum of independent Poisson r.v.'s is Poisson by the method of m.g.f.'s.

13. We come back to Section 2.2. Show that the function $f_S(x) = e^{-x/2} - e^{-x}$ for $x \geq 0$ and $= 0$ otherwise is a probability density; that is, it is non-negative and the total integral is one.

14. Using the method of m.g.f.'s, find the density of the sum of two independent exponential r.v.'s with means 2 and 3, respectively.

15. If $S_n = X_1 + ... + X_n$, where the X's are i.i.d., then as we know, $M_{S_n}(z) = (M_X(z))^n$, where $M_X(z)$ is the m.g.f. of X_i. Show that Proposition 2 includes this case. (*Hint:* Write the m.g.f. of a (non-random) variable $N \equiv n$.)

16. The number of customers of a company for a randomly chosen day is a Poisson r.v. with parameter $\lambda = 50$. The amount spent by each customer is an exponential r.v. with a mean of $m = 100$. All the r.v.'s involved are independent. Write the m.g.f. of the total daily income.

17. Assume that, in an area, the number of traffic accidents on a randomly chosen day is a Poisson r.v. with $\lambda = 300$, and the probability that a separate accident causes serious injuries is $p = 0.07$. The outcomes of different accidents are independent. Find the probability that the number of accidents with serious injuries will exceed 15.

18. Assume that, in an area, the number of traffic accidents on a randomly chosen day is a Poisson r.v. with parameter λ_1, and the number of injuries a particular accident causes is a Poisson r.v. with parameter λ_2. Write the m.g.f. of the total number of injuries. The corresponding distribution is called *Poisson-Poisson*. Make up another example where such a distribution would serve as an appropriate model.

19. The number of customers of a company during a randomly chosen day is a r.v. K for which $P(K=k) = \frac{1}{3}(\frac{2}{3})^k$, $k = 0, 1, ...$ The amount spent by a particular customer is an exponential r.v. with a mean of \$1,000. All the r.v.'s involved are independent. Find the probability that the daily income exceeds \$3,000.

20. (a) Give an alternative proof (without m.g.f's) of the result of Example 2.3-2 using Example **6**.4.2-2 and Exercise **3**-65.

    (b) Generalize the result of Example 2.3-2 to the case where the X_i's are still exponential but the distribution of N is the second version of the negative binomial distribution with parameters (p, ν) (the distribution of K_ν in Section 1.2.2) where ν is a positive integer.

21. Solve the problems of Examples 2.4-1,3 (that is, find the quantities A, a, and π) for the case when the remaining lifetime T is exponential with parameter μ (we are using the symbol a for other purposes). For simplicity and to make formulas nicer, suppose, that the premium payment is provided continuously with a rate of π. That is, $Y = \int_0^T e^{-\delta t} dt = \frac{1}{\delta}(1 - e^{-\delta T})$, and $a = E\{Y\} = \frac{1}{\delta}(1 - M_T(-\delta))$.

22. Write a bound of the type (3.1.2) for the standard normal distribution and compare it with the bound (**6**.2.4.6). Which is better? Do you find the difference significant?

23. John and Mike play a game consisting in tossing a coin. If the coin comes up heads John pays to Mike \$1. Otherwise, it is Mike who pays \$1. Let S_n is Mike's gain (which may be negative) after n tosses. Using (3.2.3), write a bound for $P(S_n > x)$ and $x > 0$.

24. In a poll concerning future elections, 55% of 1,500 respondents said that they would vote for Candidate A. To what degree can we expect that A will win? (*Hint*: We want to know whether $p > 0.5$. Will you use a one- or two-sided bound?)

Chapter 9

The Central Limit Theorem for Independent Random Variables

1 THE CENTRAL LIMIT THEOREM FOR I.I.D. RANDOM VARIABLES

1.1 A theorem and examples

Loosely put, the Central Limit Theorem (CLT) says that the distribution of the sum of a large number of independent r.v.'s is close to a normal distribution. In the case of identically distributed r.v.'s, this may be stated as follows.

Consider a sequence of independent and identically distributed (i.i.d.) r.v.'s $X_1, X_2, \dots$ with finite variances and set $S_n = X_1 + \dots + X_n$. Let $m = E\{X_i\}$ and $\sigma^2 = Var\{X_i\}$. Then $E\{S_n\} = mn$, $Var\{S_n\} = \sigma^2 n$. We normalize S_n and consider the r.v.

$$S_n^* = \frac{S_n - E\{S_n\}}{\sqrt{Var\{S_n\}}} = \frac{S_n - mn}{\sigma\sqrt{n}}.$$

It is noteworthy that the normalized r.v. S_n^* may be viewed as the same sum S_n considered in an appropriate scale: after the normalization, $E\{S_n^*\} = 0$, and $Var\{S_n^*\} = 1$ (for detail, see Section **6**.1.5).

Theorem 1 *(The Central Limit Theorem). For any (finite or infinite) a and $b \geq a$,*

$$P(a \leq S_n^* \leq b) \to \Phi(b) - \Phi(a) = \frac{1}{\sqrt{2\pi}} \int_a^b e^{-x^2/2} dx \quad as \quad n \to \infty, \tag{1.1.1}$$

where $\Phi(x) = \dfrac{1}{\sqrt{2\pi}} \displaystyle\int_{-\infty}^{x} e^{-u^2/2} du$, the standard normal distribution function.
In particular, for any x,

$$P(S_n^* \leq x) \to \Phi(x) \quad as \quad n \to \infty. \tag{1.1.2}$$

A proof will be considered in Section 1.2.

In spite of its simple statement, the CLT deals with a deep and important fact. The theorem says that as the number of the terms in the sum S_n is getting larger, the influence of separate terms is diminishing and the distribution of S_n is getting close to a standard distribution (namely, normal) *regardless of which distribution the separate terms have.* It may be continuous or discrete, or neither of these, uniform, exponential, binomial, or anything else—provided that the variance of the terms is finite, the distribution of the sum S_n for large n may be well approximated by a normal distribution.

The theorem has an enormous number of applications because it allows one to estimate probabilities concerning sums of r.v.'s in situations where the distribution of the terms X's is not known; more precisely, when one knows or has estimated only the most "rough" characteristics: the mean and standard deviation.

The same theorem may be useful when formally the distribution of the separate terms is known, but it is difficult (if possible) to present a tractable representation for the distribution of the sum.

A natural question is how large should n be for the normal approximation to apply? Certainly, it depends on our requirements to the accuracy of approximation in each particular problem, but in any case, to choose an appropriate n, we should figure out somehow—theoretically or empirically—the rate of convergence in the CLT.

In Section 2.1, we present a universal bound for such a rate. However, in particular applications, it is important to keep in mind that while the limiting distribution does not depend on the distribution of the X's, the rate of convergence does. For symmetric and bounded X's, like uniform, the rate of convergence is high, and for some purposes, an appropriate n may turn out to be not large, even around 10. For non-symmetric and/or unbounded r.v.'s, like exponential, the rate is somewhat lower, but even $n = 20$ or 30 may be enough. In Exercises 3 and 14, the reader is invited to check the accuracy of the normal approximation numerically for two cases: uniform and exponential.

The fact that the limiting distribution turns out to be normal is not accidental; this is connected with the unique stability property of the normal distribution, which we discuss in Section 1.3.

The last thing to note is that the independence condition and that of the identity of the distributions may be essentially weakened, and, as a matter of fact, the CLT is true in a much more general situation than that of this section.

In Section 2.1, we consider independent but not identically distributed r.v's, and establish a sufficient condition that, loosely speaking, requires the sum not to include r.v.'s having the same "order" as the whole sum.

Sections **15**.4.3, **17**.3, and **17**.4.3 concern dependent r.v.'s. The theory of normal approximation in this case is now well developed and deals with a wide variety of types of dependency. Loosely put, the corresponding conditions of asymptotic normality require the dependence of random components to be "not too strong."

Now, let us consider examples concerning the situation of this section.

EXAMPLE 1. The time it takes for a mechanic in an auto shop to repair a randomly selected car is a r.v. with a mean of $m = 1.5$ hours and a standard deviation of $\sigma = 0.5$ hour. The service times for different cars are i.i.d r.v.'s. Given that a week consists of 40 working hours, estimate the probability that the mechanic will serve at least $n = 24$ customers during a week.

Let X_i be the ith service time, and $S_n = X_1 + ... + X_n$, the total time needed to serve n customers. Using the CLT, for the probability of interest, we have

$$P(S_n \leq 40) = P\left(\frac{S_n - mn}{\sigma\sqrt{n}} \leq \frac{40 - mn}{\sigma\sqrt{n}}\right) = P\left(S_n^* \leq \frac{40 - mn}{\sigma\sqrt{n}}\right)$$
$$= P\left(S_n^* \leq \frac{40 - 1.5 \cdot 24}{0.5 \cdot \sqrt{24}}\right) \approx P(S_n^* \leq 1.633) \approx \Phi(1.633) \approx 0.949. \ \square$$

In the above and next examples we restrict ourselves to a heuristic solution not estimating the error of the approximation and using the (not purely mathematical) sign $\approx$. In connection with this, it makes sense to repeat that the theory estimating the rate of convergence in the CLT is well developed, and as a matter of fact, in problems as above (and below), it is possible to get more precise results. One of results on the error of normal approximation is considered in Section 2.1.

EXAMPLE 2. How many times approximately should we toss a regular coin to get at least 100 heads with a probability that is not less than 0.9? If we toss a coin n times, the number of heads is $S_n = X_1 + ... + X_n$, where each $X_i = 1$ or 0 with probabilities $\frac{1}{2}$. We have $E\{S_n\} = \frac{n}{2}$, and $Var\{S_n\} = n \cdot \frac{1}{2} \cdot \frac{1}{2} = \frac{n}{4}$. Then

$$P(S_n \geq 100) = P\left(\frac{S_n - n/2}{\sqrt{n}/2} \geq \frac{100 - n/2}{\sqrt{n}/2}\right) = P\left(S_n^* \geq \frac{100 - n/2}{\sqrt{n}/2}\right)$$
$$\approx 1 - \Phi\left(\frac{100 - n/2}{\sqrt{n}/2}\right).$$

We wish that $P(S_n \geq 100) \geq 0.9$, and hence $\Phi\left(\frac{100-n/2}{\sqrt{n}/2}\right) < 0.1$. From the Appendix, Table 3, we get that the number q for which $\Phi(q) = 0.1$, is $q \approx -1.2816$. So, we take n for which

$$\frac{100 - n/2}{\sqrt{n}/2} < -1.28.$$

Setting $y = \sqrt{n}$, we will get a quadratic inequality for y. The reader is invited to double check that the corresponding quadratic equality has only one positive solution: ≈ 14.80. Since $(14.8)^2 = 219.04$, we should choose $n \geq 220$.

EXAMPLE 3. (*The normal approximation for the Poisson distribution.*) Let Z_λ be a Poisson r.v. with a parameter λ. Recall that its mean is λ and the standard deviation is $\sqrt{\lambda}$. By the CLT, for large λ, the distribution of Z_λ may be well approximated by the normal distribution with the same parameters: a mean of λ and a standard deviation of $\sqrt{\lambda}$. Indeed, first, suppose that λ is equal to an integer n. Then we may use the representation

$$Z_n = X_1 + ... + X_n,$$

where X_i's are independent r.v.'s having the standard Poisson distribution (the parameter equals 1). Then for the normalized r.v. $Z_n^* = \dfrac{Z_n - n}{\sqrt{n}}$, we may apply the CLT.

If λ is not an integer, we may write that $Z_\lambda = X_1 + ... + X_n + Y_{\lambda - n}$, where X_i's are the same as above, n is the integer part of λ, and $Y_{\lambda-n}$ is independent of X_i' s and has the distribution with parameter $\lambda - n$, the fractional part of λ. For large λ, the r.v. $Y_{\lambda-n}$ is negligible in comparison with the sum $X_1 + ... + X_n$; Exercise 12 contains detailed advice on how to show this rigorously.

In Exercise 11, the reader is invited to provide a worksheet illustrating the accuracy of the above approximation.

EXAMPLE 4. (*The normal approximation for the* Γ*-distribution.*) The same idea may be applied to a r.v. $Z_{a\nu}$ having the Γ-distribution with parameters (a, ν). Assuming ν to be equal to an integer n, we may write $Z_{an} = X_1 + ... + X_n$, where X_i's are independent exponential r.v.'s with the same parameter a; see Section **6**.4.2 for detail. Hence, for large n, we can approximate the distribution of Z_{an} by the normal distribution with the mean $\frac{n}{a}$ and the variance $\frac{n}{a^2}$.

In Exercise 13, the reader is asked to show the same for an arbitrary, not necessary integer, ν; in Exercise 14—to create a worksheet illustrating the accuracy of the approximation. Both exercises are provided with advice.

EXAMPLE 5 (*The CLT in insurance.*) Consider an insurance portfolio of n policies. Let X_i, $i = 1,...,n$, be the random value of the payment to the ith policyholder. Assume the X's to be independent and identically distributed. The latter may be interpreted as if the customers come from a "*homogeneous group.*" Let $m = E\{X_i\}$, $\sigma^2 = Var\{X_i\}$, and c be the premium for each policy (which should be the same since all X's are identically distributed).

The total premium the insurance company gets for this portfolio is nc, and the total payment is $S_n = X_1 + ... + X_n$.

Assume that the company specifies the lowest acceptable level β for the probability of not suffering a loss. For instance, the company wishes the mentioned probability to be not less than $\beta = 0.95$; in the worst case—to be equal to 0.95.

Then, if c is the least acceptable premium for the company, we may write that

$$\beta = P(nc - S_n \geq 0) = P(S_n - mn \leq nc - mn) = P\left(\frac{S_n - mn}{\sigma\sqrt{n}} \leq \frac{n(c-m)}{\sigma\sqrt{n}}\right)$$
$$= P\left(S_n^* \leq \frac{\sqrt{n}(c-m)}{\sigma}\right).$$

By the CLT, $P(S_n^* < x) \approx \Phi(x)$ for large n. Thus, $\beta \approx \Phi\left(\frac{\sqrt{n}(c-m)}{\sigma}\right)$. Let q_β be a number such that $\Phi(q_\beta) = \beta$; that is, q_β is the β-quantile of the standard normal distribution. Then, $\frac{\sqrt{n}(c-m)}{\sigma} \approx q_\beta$. From this, we get the following approximation for the least premium acceptable for the company:

$$c \approx m + \frac{q_\beta \sigma}{\sqrt{n}}. \qquad (1.1.3)$$

For $\beta = 0.95$, we have $q_\beta \approx 1.64$, and $c \approx m + \dfrac{1.64\sigma}{\sqrt{n}}$.

In particular, we see that the larger the number of customers n, the closer the premium can be to the mean payment.

Consider, for example, a special insurance that pays $b = \$150$ to passengers of an airline in the case of a serious flight delay. Assume that for each of $n = 10,000$ clients who bought such an insurance, the probability of a delay is $p = 0.1$. In this case,

$$X_i = \begin{cases} b & \text{with probability } p, \\ 0 & \text{with probability } 1-p, \end{cases}$$

$m = bp = 15, \sigma = b\sqrt{p(1-p)} = 45$. Then for $\beta = 0.95$, by (1.1.3), $c \approx 15 + \frac{1.64 \cdot 45}{100} \approx$ 15.74. So, a premium of \$16 would be enough for the company. □

The CLT also explains why in many applications, we deal with the so called *log-normal distribution*: the distribution of a r.v. Y whose logarithm $\ln Y$ is a normal r.v.

The point is that the log-normal distribution appears, in particular, when the r.v. of interest $Y = X_1 \cdot ... \cdot X_n$, where X_i's are i.i.d. positive r.v.'s and n is large. In this case, $\ln Y = \ln X_1 + ... + \ln X_n$, and for large n, is asymptotically normal by the CLT. Then, the distribution of Y itself is asymptotically log-normal.

EXAMPLE 6. (*Why and when the distribution of an investment return may be approximated by a log-normal distribution.*) In a simple investment model, we suppose that at each time moment—say, each day—the capital of the investor is multiplied by a r.v. R, a (daily) return per unit of money. This is what the investor would have if she/he invested one unit of money on the previous day. If $R > 1$, then the capital has increased; if $R < 1$, then the investor suffers a loss. (See also Example **3**.2.1-9 where X took on two values.)

Denote by C_0 the initial capital, and by R_i the return on the ith day after the initial investment. Then the capital on day n is the r.v. $C_n = C_0 R_1 R_2 \cdots R_n$, and $\ln C_n = \ln C_0 + \ln R_1 + \ln R_2 + ... + \ln R_n$. If the returns R_i's are i.i.d, which is relatively realistic assumption for a stable market, then $\ln C_n$ is asymptotically normal. □

Route 1 ⇒ page 262

1.2 On a proof of the CLT

2

The complete proof of the CLT is rather involved. It uses Laplace transforms and what is called the *continuity theorem*. This theorem claims that the convergence of Laplace transforms implies the convergence of the corresponding probability distributions; more precisely, the convergence of the d.f.'s for any continuity point of the limiting d.f. We considered this type of convergence, called *weak convergence*, in Section **7**.3 in detail.

In our case, the limiting d.f. is $\Phi(x)$ which is continuous at any point, so we consider the convergence $P(S_n^* \leq x) \to \Phi(x)$ for all x's.

We take the continuity theorem for granted; proofs may be found in many textbooks, e.g., [13], [24], [42], [47]. To consider the general case, one should use Laplace transforms with imaginary arguments, that is, characteristic functions which exist for all probability distributions (see Chapter 8).

To avoid complex numbers, we restrict ourselves to moment generating functions. Let $M(z)$ be the (common) m.g.f. of the i.i.d. X's. We assume that $M(z)$ exists for all z's from a neighborhood of zero. This means, in particular, that the X's have all moments, while for the theorem itself to be true, it suffices to assume only variances to be finite. Under this additional restriction, the proof is not lengthy.

Without loss of generality, we can also assume $m = E\{X_i\} = 0$; otherwise we would switch to centered r.v.'s. We have $M'(0) = 0,\ M''(0) = E\{X_i^2\} = Var\{X_i\} = \sigma^2$. Then, by the Taylor formula (see, e.g., the Appendix, (2.2.1)),

$$M(z) = 1 + \frac{1}{2}\sigma^2 z^2 + o(z^2), \tag{1.2.1}$$

where the symbol $o(z^2)$ stands for a function that converges to 0 faster than z^2; that is, $\frac{o(z^2)}{z^2} \to 0$ as $z \to 0$.

(See also the Appendix, Section 2.1. To be absolutely rigorous, for (1.2.1) to be true, $M''(z)$ should be continuous in a neighborhood of zero; see the Appendix, Section 2.2. Under conditions we imposed, this is indeed true. One may make it sure proceeding from representation (**8**.1.3.4).)

Since the X's are independent, $M_{S_n}(z) = (M(z))^n$. So, making use of (1.2.1) and properties of m.g.f.'s established in Chapter 8, we write

$$M_{S_n^*}(z) = M_{S_n}\left(\frac{z}{\sigma\sqrt{n}}\right) = \left(M\left(\frac{z}{\sigma\sqrt{n}}\right)\right)^n = \left(1+\frac{1}{2}\sigma^2\left(\frac{z}{\sigma\sqrt{n}}\right)^2 + o\left(\left(\frac{z}{\sigma\sqrt{n}}\right)^2\right)\right)^n$$

$$= \left(1+\frac{z^2}{2n}+o\left(\frac{z^2}{\sigma^2 n}\right)\right)^n \to e^{z^2/2} \text{ as } n\to\infty. \tag{1.2.2}$$

So, the m.g.f. of S_n^* converges to the m.g.f. of the standard normal distribution. By the continuity theorem mentioned above, this implies the convergence $P(S_n^* \le x) \to \Phi(x)$.

It remains to note how to get from this the convergence $P(a \le S_n^* \le b) \to \Phi(b) - \Phi(a)$. Since $P(a \le S_n^* \le b) = P(S_n^* \le b) - P(S_n^* < a)$, we should show that $P(S_n^* < a)$ and $P(S_n^* \le a)$ have the same limit. The reader who did not skip Section **7**.3 may use Lemma **7**.7. However, in this particular case, the proof is easier than that of the lemma mentioned.

Indeed, for any $\delta > 0$, we have $P(S_n^* = a) \le P(a-\delta < S_n^* \le a) = P(S_n^* \le a) - P(S_n^* \le a-\delta) \to \Phi(a) - \Phi(a-\delta)$ as $n\to\infty$. Hence, $\lim_{n\to\infty} P(S_n^* = a) \le \Phi(a) - \Phi(a-\delta)$. Since $\Phi(x)$ is continuous at any point, letting $\delta \to 0$, we get $\Phi(a) - \Phi(a-\delta) \to 0$. Thus, $\lim_{n\to\infty} P(S_n^* = a) \le 0$, which is possible only if this limit equals zero.

1.3 Why the limiting distribution is normal: The stability property

We already touched on this question in Section **6**.4.3 where, in particular, we have introduced the notion of a type: for a fixed distribution F, its type is the family of the distributions of *all* r.v.'s $a+b\xi$ where ξ has the distribution F, the number a is arbitrary, and b is an arbitrary positive number. In Section **6**.4.3, we also considered examples.

To simplify the exposition below, we write $\xi_1 \stackrel{d}{=} \xi_2$ if the r.v.'s ξ_1 and ξ_2 have the same distribution.

Let ξ_1 and ξ_2 be independent and have the same distribution F. The distribution F is said to be *stable* if for any positive $c_1,\ c_2$, there exist a number $b > 0$ and a number a such that

$$c_1\xi_1 + c_2\xi_2 \stackrel{d}{=} b\xi_1 + a. \tag{1.3.1}$$

In other words, the linear combination of independent r.v.'s with the distribution F have the same distribution F up to a linear transformation.

It is worth emphasizing that the stability property is that of a type: if ξ has a stable distribution, then so does $a+b\xi$. (Replacing ξ_1 and ξ_2 in (1.3.1) by $a+b\xi_1$ and $a+b\xi_2$ will merely change constants c_1 and c_2.) So, in other words, in the case of a stable distribution, summation preserves the type.

Clearly, this property may be extended to the case of many r.v.'s. Namely, if $\xi_1, \xi_2, \ldots$ are independent r.v.'s with the same stable distribution, and $Z_n = \xi_1 + \ldots + \xi_n$, then for any n, there exist numbers a_n and $b_n > 0$ such that

$$Z_n \stackrel{d}{=} b_n\xi_1 + a_n. \tag{1.3.2}$$

This may be derived from (1.3.1), for example, by induction.

We know that a linear combination of independent normal r.v.'s is normal, and hence the normal distribution is stable. Another example is the Cauchy distribution: it may be proved that it is stable. A complete description of stable distributions may be found, e.g., in [13], [42].

The two theorems below point out a unique property of the normal distribution and explain why the normal distribution appears as a limiting distribution in the CLT.

We call a distribution *non-degenerate* if it is not concentrated only at one point.

Theorem 2 *Let $X_1, X_2, \ldots$ be i.i.d. r.v.'s with an arbitrary (not necessary stable) distribution, and let $S_n = X_1 + \ldots + X_n$. Suppose that for some numerical sequences a_n and $b_n > 0$, and for some non-degenerated d.f. $F(x)$,*

$$P\left(\frac{S_n - a_n}{b_n} \leq x\right) \to F(x), \quad as \;\; n \to \infty, \tag{1.3.3}$$

for any x at which $F(x)$ is continuous. Then the distribution F is stable.

(Regarding the continuity-point issue, see Section **7**.3. As a matter of fact, this circumstance is not essential because it may be proved that all stable distributions are continuous anyway.) Thus, *any limiting distribution is stable.*

Theorem 3 *Any stable distribution with a finite variance is normal.*

We begin with

Proof of Theorem 3. Consider (1.3.2) in the case where ξ's have a stable distribution F with a finite variance. Without loss of generality, we can set $E\{\xi_i\} = 0$, and $Var\{\xi_i\} = 1$. If it is not so, we may switch to the normalized r.v.'s $\xi_i^* = \dfrac{\xi_i - E\{\xi_i\}}{\sqrt{Var\{\xi_i\}}}$, which are stable as well. (Recall also that the stability property is that of a type.)

Because the expectations and variances of both sides of (1.3.2) coincide, $0 = a_n$, and $1 \cdot n = b_n^2$. Then $a_n = 0,\; b_n = \sqrt{n}$, and $Z_n \stackrel{d}{=} \sqrt{n}\xi_1$. We rewrite it as

$$\frac{Z_n}{\sqrt{n}} \stackrel{d}{=} \xi_1.$$

The l.-h.s. of the last relation depends on n while the r.-h.s. does not. By the CLT, the distribution of the l.-h.s. converges to the standard normal distribution. So, letting $n \to \infty$, we jump to the conclusion that ξ_1 is standard normal. ■

Proof of Theorem 2. In the proof below, there will be one purely technical thing that we will take for granted. The rest of the proof is explicit and illustrative. First, for simplicity, we set $a_n = 0$. Since, when subtracting a_n, we merely change the origin, such an assumption does not restrict generality.

Set $S_{nm} = \sum_{i=n+1}^{n+m} X_i$. Then $S_{n+m} = S_n + S_{nm}$. Dividing by b_m, we have

$$\frac{S_{n+m}}{b_m} = \frac{S_n}{b_m} + \frac{S_{nm}}{b_m} = \frac{b_n}{b_m} \cdot \frac{S_n}{b_n} + \frac{S_{mn}}{b_m}.$$

We rewrite it as

$$\frac{b_{n+m}}{b_m} \cdot \frac{S_{n+m}}{b_{n+m}} = \frac{b_n}{b_m} \cdot \frac{S_n}{b_n} + \frac{S_{mn}}{b_m}. \tag{1.3.4}$$

It may be proved (see, e.g., [13], [42]) that under condition (1.3.3), for any $c > 0$, one may choose a sequence $m = m(n) \to \infty$ as $n \to \infty$ and such that

$$\frac{b_n}{b_m} = \frac{b_n}{b_{m(n)}} \to c \text{ as } n \to \infty. \tag{1.3.5}$$

The sum S_{nm} consists of m terms, and since X_j are identically distributed, S_{nm} has the same distribution as S_m. Hence, by (1.3.3), the d.f. of the r.v. S_{nm}/b_m converges to $F(x)$. The same is true for S_n/b_n. Also, the r.v.'s S_n and S_{nm} are independent because they involve non-overlapping sequences of X_i's. Thus, for the sequence $m(n)$ from (1.3.5), the distribution of the r.-h.s. of (1.3.4) converges to the distribution of a r.v. $c\xi_1 + \xi_2$, where ξ_1 and ξ_2 are independent and have the distribution F.

If the distribution of the r.-h.s. of (1.3.4) has a limit, so does the distribution of the l.-h.s. On the other hand, the distribution of S_{n+m}/b_{n+m} has the same limit F. Hence, the sequence b_{n+m}/b_m must have a limit too. Denoting it by r, we come to the conclusion that the limit of the distribution of the l.-h.s. coincides with the distribution of the r.v. $r\xi_1$.

Eventually, $r\xi_1 \stackrel{d}{=} c\xi_1 + \xi_2$. To arrive at property (1.3.1), it suffices to set $c = c_1/c_2$, and choose $b = rc_2$. ■

---

2 THE CLT FOR INDEPENDENT VARIABLES IN THE GENERAL CASE

2.1 The Lyapunov fraction and the Berry–Esseen theorem

This section concerns two issues: a version of the CLT for independent but not necessary identically distributed r.v.'s, and the rate of convergence in the CLT.

We still assume X_i's to be independent. To avoid somewhat cumbersome notations and without loss of generality, we suppose

$$E\{X_i\} = 0 \quad \text{for all} \quad i = 1, 2, ... \tag{2.1.1}$$

In particular problems where it is not so, we may switch to the centered r.v.'s $X_i' = X_i - E\{X_i\}$.

Set $\sigma_i^2 = Var\{X_i\}$ and $B_n^2 = \sigma_1^2 + ... + \sigma_n^2$ which is equal to the variance of $S_n = X_1 + ... + X_n$. We suppose that at least one σ_i is not zero, so $B_n > 0$. Then, in view of (2.1.1), the normalized r.v. $S_n^* = \dfrac{S_n}{B_n}$.

Our goal is to figure out when

$$P(S_n^* \le x) \to \Phi(x), \quad \text{as} \quad n \to \infty, \quad \text{for all } x, \tag{2.1.2}$$

and estimate the rate of convergence. First of all, note that (2.1.2) is not always true.

EXAMPLE 1. Let X_1 have a distribution different from the normal distribution, and $\sigma_1 \neq 0$, while all other $X_i = 0$, $i = 2,3,..,$ with probability one. Then $\sigma_i^2 = 0$ for $i = 2,3,...,$ and $B_n^2 = \sigma_1^2 + 0 + ... + 0 = \sigma_1^2$. Hence, $S_n^* = X_1/\sigma_1$ for all n, and consequently, (2.1.2) does not hold. □

The case of Example 1 is certainly extreme: one term constitutes the whole sum, and the contribution of the other terms vanishes. The next example is much more sophisticated.

EXAMPLE 2. The question we discuss here is whether the CLT is true in the case where the X's are non-identically distributed but have the same means and variances; say, if $E\{X_i\} = 0$, and $Var\{X_i\} = 1$ for all i. We will see that the identity of means and variances is not sufficient for normal convergence. Let $X_1, X_2, ...$ be independent, and

$$X_i = \begin{cases} i & \text{with probability } \frac{1}{2i^2} \\ 0 & \text{with probability } 1 - \frac{1}{i^2} \\ -i & \text{with probability } \frac{1}{2i^2} \end{cases}.$$

As is easy to calculate, $E\{X_i\} = 0$ and $Var\{X_i\} = 1$.

Let the event $A_i = \{X_i \neq 0\}$. Then $P(A_i) = 1/i^2$, and the series $\sum_{i=1}^{\infty} P(A_i) = \sum_{i=1}^{\infty} \frac{1}{i^2}$ converges. Hence, by the Borel-Cantelli Theorem **2**.1, with probability one, only a finite number of X's do not vanish, and consequently, the sum $S_\infty = \sum_{i=1}^{\infty} X_i$ is finite with probability one.

Moreover, since all X's are integer-valued, so is S_∞, and hence the distribution of S_∞ is not normal. Then, when considering the r.v.'s $S_n = X_1 + ... + X_n$, we do not need any normalization: the distribution of S_n converges to the distribution of S_∞.

Suppose that, nevertheless, we decided to normalize S_n. Since $E\{S_n\} = 0$ and $Var\{S_n\} = n$, the normalized sum $S_n^* = \dfrac{S_n}{\sqrt{n}} \to 0$ with probability one, because S_∞ is finite. So, the result of the normalization is trivial, and the limiting distribution is not standard normal as in the CLT. □

Next, we introduce a characteristic which, in a certain sense, measures to what extent the terms in the sum are different. To this end, we set $\mu_i = E\{|X_i|^3\}$, $i = 1,2,...,$ the third absolute moment (for short, the third moment) of the r.v. X_i, which we assume to be finite, and define the quantity

$$L_n = \frac{1}{B_n^3} \sum_{i=0}^{n} \mu_i. \tag{2.1.3}$$

The characteristic L_n, which is called *Lyapunov's fraction*, plays an essential role in the theory of normal approximation. Let us consider its properties and examples.

First, note that L_n is not sensitive to a change of scale: if we multiply all X's by a constant $c > 0$, then the denominator and the numerator in (2.1.3) will be multiplied by c^3, and L_n will not change. If we multiply X's by a negative c, the denominator and the numerator will be multiplied by $|c|^3$.

Accordingly, L_n does not have a dimension. For instance, if we measure the X's in inches, then the dimensions of the numerator and the denominator in (2.1.3) are the same: $(inch)^3$.

We will see that in the case of finite third moments, the Lyapunov fraction L_n specifies the rate of convergence to the normal distribution, and in particular, the normal convergence takes place if L_n converges to zero. First, let us come back to Examples 1-2, where the normal convergence does not hold, and make sure that in these cases $L_n \not\to 0$.

EXAMPLE 1 revisited. In this case, $L_n = \dfrac{1}{\sigma_1^3}\mu_1 \not\to 0$.

EXAMPLE 2 revisited. The reader is invited to calculate that $\mu_i = i$, and $B_n^2 = n$. Hence, using the formula $\sum_{i=1}^n i = \frac{n(n+1)}{2}$, we have $L_n = \dfrac{1}{n^{3/2}}\sum_{i=1}^n i = \dfrac{1}{n^{3/2}}\dfrac{n(n+1)}{2} \not\to 0$ also.

EXAMPLE 3. Consider the opposite, in a sense, case where X_i's are identically distributed. Then all μ_i equal some μ and all σ_i equal some σ. This implies that $B_n^2 = \sigma^2 n$, and hence, $B_n = \sigma\sqrt{n}$. So,

$$L_n = \frac{1}{\sigma^3 n^{3/2}}\mu n = \frac{1}{\sqrt{n}} \cdot \frac{\mu}{\sigma^3} \to 0 \text{ as } n \to \infty, \tag{2.1.4}$$

and the order of the convergence is $1/\sqrt{n}$. Note that, as we know, in this case, the CLT holds.

The same pattern of convergence takes place when X_i's are though non-identically distributed but have the "same order." More generally, suppose that all $\mu_i \leq \mu$ and $\sigma_i \geq \sigma$ for all i and some positive numbers μ and σ. Then $B_n^2 = \sum_{i=0}^n \sigma_i^2 \geq \sigma^2 n$ while $\sum_{i=0}^n \mu_i \leq \mu n$, and

$$L_n \leq \frac{1}{\sigma^3 n^{3/2}}\mu n = \frac{1}{\sqrt{n}} \cdot \frac{\mu}{\sigma^3} \to 0 \text{ as } n \to \infty.$$

All other cases are intermediate, and the order of L_n may be any between $1/\sqrt{n}$ and a constant (when $L_n \not\to 0$).

EXAMPLE 4. Let $Y_1, Y_2, \ldots$ be independent and *identically* distributed r.v.'s with zero means. Set $\mu = E\{|Y_i|^3\}$, and $\sigma^2 = Var\{Y_i\}$. Let $X_i = \dfrac{1}{i^a}Y_i$, and $a = 5/12$. We show that in this case, L_n has an order of $n^{-1/4}$, the intermediate, in a sense, in comparison with Examples 2 and 3. Indeed, $\sigma_i^2 = \dfrac{1}{i^{5/6}}\sigma^2$, and $\mu_i = \dfrac{1}{i^{5/4}}\mu$. Furthermore, for a constant C,

$$\sum_{i=1}^n \mu_i = \mu \sum_{i=1}^n \frac{1}{i^{5/4}} \leq C\mu,$$

since the above series converges. On the other hand, for another constant C_1,

$$B_n^2 = \sum_{i=1}^n \sigma_i^2 = \sigma^2 \sum_{i=1}^n \frac{1}{i^{5/6}} \sim \sigma^2 \int_1^n \frac{1}{x^{5/6}}dx \sim C_1\sigma^2 n^{1/6}.$$

Hence $L_n = \dfrac{1}{B_n^3}\sum_{i=1}^n \mu_i \sim C_2\dfrac{\mu}{\sigma^3} \cdot \dfrac{1}{n^{1/4}}$ for a constant C_2. □

Now, we turn to the estimation of the accuracy of normal approximation based on the following celebrated theorem.

Theorem 4 *(Berry–Esseen). Let $F_n^*(x) = P(S_n^* \leq x)$, the d.f. of the r.v. S_n^*. Then there exists an absolute constant C_0 such that for any x,*

$$|F_n^*(x) - \Phi(x)| \leq C_0 L_n. \tag{2.1.5}$$

A proof may be found, e.g., in [13].

It is worth noting that in the (most common) situation of Example 3, the rate of convergence given by (2.1.5) has an order of $1/\sqrt{n}$.

From Theorem 4, we immediately get *a sufficient condition for normal convergence for the general case of non-identically distributed r.v.'s.*

Corollary 5 *(Lyapunov's Theorem).* *If $L_n \to 0$ as $n \to \infty$, then for any x, the d.f. $F_n^*(x) \to \Phi(x)$.*

There has been a great deal of interest in calculating the constant C_0 in (2.1.5). It has been proved that it cannot be less than the number $\overline{C} = \frac{\sqrt{10}+3}{6\sqrt{2\pi}} \approx 0.40973$; and there is a quite plausible conjecture (still not confirmed) that (2.1.5) is true with $C_0 = \overline{C}$. The up-to-date results show that (2.1.5) holds with $C_0 = 0.5600$, and in the case of identically distributed r.v.'s—with $C_0 = 0.4784$. (See the history and details, e.g., in [26], [46], [52], [53].)

It is also worthwhile to emphasize that the bound (2.1.5) is universal: it is true for *all* X_i with finite third moments. Consequently, this bound points out the accuracy of normal approximation for the worst case; so to speak, with a leeway. As has been already noted, in particular situations, the accuracy of approximation may be better; especially for large x's. (Many results of this type may be found, e.g., in monographs [6], [35], [45], and survey [41].)

On the other hand, the order $1/\sqrt{n}$ in (2.1.5) cannot be improved in the sense that there exist X_i's for which the rate of convergence has the order mentioned. Moreover, this example concerns a simple classical scheme, perhaps, the simplest.

EXAMPLE 5. Let

$$X_i = \begin{cases} 1 & \text{with probability } 1/2 \\ -1 & \text{with probability } 1/2. \end{cases} \tag{2.1.6}$$

By virtue of symmetry, $P(S_n^* > 0) = P(S_n^* < 0)$. Then $P(S_n^* < 0) = \frac{1}{2}(P(S_n^* > 0) + P(S_n^* < 0)) = \frac{1}{2}(P(S_n^* > 0) + P(S_n^* < 0) + P(S_n^* = 0) - P(S_n^* = 0)) = \frac{1}{2}(1 - P(S_n^* = 0))$. This implies that

$$P(S_n^* \le 0) = P(S_n^* < 0) + P(S_n^* = 0) = \frac{1}{2}(1 - P(S_n^* = 0)) + P(S_n^* = 0) = \frac{1}{2}(1 + P(S_n^* = 0)). \tag{2.1.7}$$

To make the example more explicit, let us consider an even $n = 2m$. From (2.1.7), we have

$$P(S_{2m}^* \le 0) - \Phi(0) = P(S_{2m}^* \le 0) - \frac{1}{2} = \frac{1}{2}P(S_{2m}^* = 0). \tag{2.1.8}$$

To estimate the r.-h.s., note that $P(S_{2m}^* = 0) = P(S_{2m} = 0)$, and $S_{2m} = 0$ if and only if the number of "ones" in S_{2m} is equal to the number of "negative ones," and, hence, is equal to m. In (**2**.3.2), we have already obtained that $P(S_{2m} = 0) \sim \dfrac{1}{\sqrt{\pi m}}$. (One may write $P(S_{2m} = 0) = \frac{(2m)!}{m!m!}4^{-m}$ and apply Stirling's formula (**2**.3.1).) Consequently, for the particular scheme of this example, and for the particular value $x = 0$,

$$P(S_{2m}^* \le 0) - \Phi(0) = \frac{1}{2}P(S_{2m} = 0) \sim \frac{1}{2} \cdot \frac{1}{\sqrt{\pi m}} = \frac{1}{\sqrt{2\pi}} \frac{1}{\sqrt{2m}}, \tag{2.1.9}$$

which shows that, regarding the order of vanishing in n, the Berry–Esseen theorem gives a sharp bound for the rate of convergence. Moreover, it has been proved ([38], see also [35]) that, if we consider the behavior of $F_n^*(x) - \Phi(x)$ not for all n but merely for $n \to \infty$, then the optimal constant in the corresponding bound is $1/\sqrt{2\pi}$; that is, as in (2.1.9). Consequently, in the mentioned sense, the lowest rate in the CLT corresponds to the case (2.1.6). □

2.2 Necessary and sufficient conditions for normal convergence

This section is concerned with the most general conditions for normal approximation. We again assume $E\{X_i\} = 0$ and keep the notation of the previous section. Denote by $F_i(x)$ the d.f. of X_i, and for $\varepsilon > 0$, define the quantity

$$\Lambda_n(\varepsilon) = \frac{1}{B_n^2} \sum_{i=1}^{n} \int_{|x|>\varepsilon B_n} x^2 dF_i(x). \tag{2.2.1}$$

Theorem 6 *(Lindeberg). If for any* $\varepsilon > 0$,

$$\Lambda_n(\varepsilon) \to 0 \text{ as } n \to \infty, \tag{2.2.2}$$

then for any x, the d.f. $F_n^*(x) \to \Phi(x)$.

Condition (2.2.2), which is called *Lindeberg's condition*, does not involve any moments of orders higher than two, and is weaker (that is, better) than the conditions of normal convergence we considered before.

Indeed, let first the X_i's be i.i.d. Then all $F_i(x)$ equal some d.f. $F(x)$, and $B_n^2 = \sigma^2 n$, where $\sigma^2 = Var\{X_i\}$. In this case,

$$\Lambda_n(\varepsilon) = \frac{1}{\sigma^2 n} \cdot n \int_{|x|>\varepsilon\sigma\sqrt{n}} x^2 dF(x) = \frac{1}{\sigma^2} \int_{|x|>\varepsilon\sigma\sqrt{n}} x^2 dF(x) \to 0$$

because $\int_{-\infty}^{\infty} x^2 dF(x)$ is finite. So, *in the i.i.d. case, Lindeberg's condition holds automatically.* Next, suppose that the third moments μ_i are finite, and observe that for $\varepsilon > 0$,

$$\int_{|x|>\varepsilon B_n} x^2 dF_i(x) \le \int_{|x|>\varepsilon B_n} \frac{|x|}{\varepsilon B_n} \cdot x^2 dF_i(x) \le \frac{1}{\varepsilon B_n} \int_{-\infty}^{\infty} |x|^3 dF_i(x) = \frac{1}{\varepsilon B_n} \mu_i.$$

Substituting this into (2.2.1), we get that $\Lambda_n(\varepsilon) \le \frac{1}{B_n^2} \sum_{i=1}^{n} \frac{1}{\varepsilon B_n} \mu_i = \frac{1}{\varepsilon} L_n$.

Hence, if $L_n \to 0$, then so does $\Lambda_n(\varepsilon)$ for any $\varepsilon > 0$.

Next, we show that if Lindeberg's condition holds, then

$$\max_{i=1,\dots,n} \frac{\sigma_i}{B_n} = \frac{1}{B_n} \max_{i=1,\dots,n} \sigma_i \to 0. \tag{2.2.3}$$

Condition (2.2.3) is often referred to as that of *uniform asymptotic negligibility*. To clarify it, let us observe that the normalized sum $S_n^* = \sum_{i=1}^{n} \frac{X_i}{B_n}$, and the standard deviation of X_i/B_n

is σ_i/B_n. So, the last quantity may be viewed as a measure of the contribution of X_i to the total sum, and under condition (2.2.3), asymptotically, S_n^* is the sum of many "individually negligible components."

To prove that Lindeberg's condition implies (2.2.3), let us recall that $E\{X_i\} = 0$ and write

$$\sigma_i^2 = E\{X_i^2\} = \int_{-\infty}^{\infty} x^2 dF_i(x) = \int_{|x|\le\varepsilon B_n} x^2 dF_i(x) + \int_{|x|>\varepsilon B_n} x^2 dF_i(x)$$
$$\le \varepsilon^2 B_n^2 \int_{|x|\le\varepsilon B_n} dF_i(x) + \int_{|x|>\varepsilon B_n} x^2 dF_i(x) \le \varepsilon^2 B_n^2 + \int_{|x|>\varepsilon B_n} x^2 dF_i(x).$$

The second term in the very r.-h.s. is a "part" of $\Lambda_n(\varepsilon)$, and hence, $\dfrac{\sigma_i^2}{B_n^2} \le \varepsilon + \Lambda_n(\varepsilon)$.

The r.-h.s. of the last inequality does not depend on i. Letting $n \to \infty$, we get that under the Lindeberg condition, $\lim\limits_{n\to\infty} \max\limits_{i\le n} \dfrac{\sigma_i^2}{B_n^2} \le \varepsilon$ for an arbitrary small $\varepsilon > 0$. This means that this limit is equal to zero.

Now, let us realize that the Lindeberg condition is not necessary for normal convergence for the simple reason that

> If the r.v.'s X_i are normal, the distribution of the sum S_n is exactly normal for *any* n, whatever σ_i's are.

The next theorem clarifies the situation.

Theorem 7 *(Feller). Suppose (2.2.3) holds, and $F_n^*(x) \to \Phi(x)$ for all x. Then the Lindeberg condition is fulfilled.*

Theorems 6 and 7, when being considered together, are referred to as the Lindeberg–Feller theorem, which states that *under the asymptotic negligibility condition (2.2.3), the Lindeberg condition is necessary and sufficient for normal convergence.*

We see that the asymptotic normality of the sum of r.v.'s may take place for, at least, two reasons: either the sum is that of a large number of "small" with respect to the sum components, and we are watching the so called *accumulation effect*, or the components themselves are normal.

Regarding the latter issue, it is natural to conjecture that for asymptotic normality we need only the asymptotic proximity of the distribution of the components to a normal distribution. Theorems taking into account this factor are often called *non-classical*. Below, we state one of the versions of the final result on normal convergence.

First, we generalize the statement of the problem and assume that for each n, separate terms X_i may depend on n. So, we use the notation X_{in} and set

$$S_n = \sum_{i=1}^{n} X_{in}. \tag{2.2.4}$$

Note that once we have made such an assumption, we do not need to introduce the normalization operation in the general framework.

Indeed, in the previous framework, we dealt with the normalized sum $S_n^* = \frac{1}{B_n}\sum_{i=1}^n X_i = \sum_{i=1}^n \frac{X_i}{B_n}$. However, we may set $X_{in} = \frac{X_i}{B_n}$ and work with the sum (2.2.4).

We assume $E\{X_{in}\} = 0$, and set $\sigma_{in}^2 = Var\{X_{in}\}$. For example, for the previous framework, $\sigma_{in}^2 = \sigma_i^2/B_n^2$.

Since the normalization procedure has been already involved in our new setup (2.2.4), in the general framework, we assume from the very beginning that

$$\sum_{i=1}^{n} \sigma_{in}^2 = 1 \quad \text{for all } \; n. \tag{2.2.5}$$

Let $F_{in}(x)$ be the d.f. of X_{in}. Denote by $\Phi_{in}(x)$ the normal d.f. with zero mean and the variance σ_{in}^2, the same as for $F_{in}(x)$. Set

$$\Delta_n(\varepsilon) = \sum_{i=1}^{n} \int_{|x|>\varepsilon B_n} |x| \cdot |F_{in}(x) - \Phi_{in}(x)| dx.$$

First, let us realize that if X's are normal, then $F_{in}(x) = \Phi_{in}(x)$, and $\Delta_n(\varepsilon) \equiv 0$.

To compare $\Delta_n(\varepsilon)$ with the Lindeberg expression $\Lambda_n(\varepsilon)$, let us observe that for two d.f.'s, $F(x)$ and $G(x)$, the difference of the corresponding second moments may be written as

$$\int_{-\infty}^{\infty} x^2 dF(x) - \int_{-\infty}^{\infty} x^2 dG(x) = \int_{-\infty}^{\infty} x^2 d(F(x) - G(x)) = -2\int_{-\infty}^{\infty} x(F(x) - G(x))dx.$$

(We skip the proof that in integrating by parts the limits at $\pm\infty$ vanish.)

So, the expression $\Delta_n(\varepsilon)$ involves a sort of integration by parts, and instead of the d.f.'s $F_{in}(x)$, it deals with the differences $F_{in}(x) - \Phi_{in}(x)$.

Theorem 8 *(The non-classical CLT). The convergence $P(S_n \leq x) \to \Phi(x)$ for all x holds if and only if for any $\varepsilon > 0$,*

$$\Delta_n(\varepsilon) \to 0 \quad as\ n \to \infty. \tag{2.2.6}$$

Condition (2.2.6) takes into account both factors mentioned above: the accumulation effect and a possible proximity of the distributions of the components to the normal distribution. In particular, if X_i's are normal, then as was mentioned, $\Delta_n(\varepsilon) \equiv 0$.

More detailed comments and a derivation of the Lindeberg theorem from Theorem 8 may be found in [42]. For the general modern theory including references, see, e.g., [41], [42], [45], [59].

3 EXERCISES

1. On Sundays, Michael sells the Sunday issue of a local newspaper. The mean time between consecutive buyers is 3 min., and the corresponding standard deviation is 1.5 min. Estimate the time period that will be sufficient to sell 60 copies on 80% of Sundays on the average.
2. A surveyor is measuring the distance between two remote objects with use of geodesic instruments from 25 different positions. The errors of different measurements are independent r.v.'s with zero mean and a standard deviation of 0.1 length units. Estimate the probability that the average of the measurements lies within 0.05 of the true distance.

3. Let $S_n = X_1 + ... + X_n$, where X's are independent and uniform on $[0,1]$. Consider $S_n^* = (S_n - \frac{n}{2})/\sqrt{n/12}$. Explain why it makes sense to consider this r.v. Using software, for $n =$ 5, 10, 20, simulate a number of values (say, 1000) of S_n^*, and make histograms (see Section **7**.5). Compare these histograms and provide recommendations regarding the application of the CLT in this case. (*Advice*: If we use Excel, then, say for $n = 5$, we should arrange 5 consecutive *columns* with 1000 generated (from $[0,1]$) numbers in each. For each *row* (of five numbers), we compute the value of S_n^*; the results may be placed into a separate column.)

4. A regular die is rolled n times. Let S_n be the total sum of the numbers showed up. Say without calculations what S_n is equal to on the average. For $a > 0$, let $a_n^+ = 3.5n + a\sqrt{n}$, $a_n^- = 3.5n - a\sqrt{n}$, and $q_n(a) = P(a_n^- \leq S_n \leq a_n^+)$. (a) Show that $q_n(a) \to q(a)$ as $n \to \infty$, where $q(a)$ is a finite function of a. Find $q(a)$. (b) Estimate a for which $q(a) = 0.9$. (c) Proceeding from what you obtained, estimate z for which $P(-z \leq S_{100} - 350 \leq z) = 0.9$.

5. Let $X_1, X_2, ...$ be a sequence of independent r.v.'s uniformly distributed on the interval $[0,2]$, and let $S_n = X_1 + ... + X_n$. (a) Find $E\{S_n\}$ and $Var\{S_n\}$. (b) Using the Central Limit Theorem, estimate $P(n - \sqrt{n} \leq S_n \leq n + \sqrt{n})$. (c) Write down a formula for the limit of $P\left(\left|\frac{S_n}{n} - 1\right| \geq \frac{c}{\sqrt{3}\sqrt{n}}\right)$. Proceeding from the formula you have obtained, explain why the accuracy with which the average $\overline{X}_n = S_n/n$ is close to 1 has an order of $1/\sqrt{n}$.

6. Generalizing Exercise 5-c for r.v.'s X_i having a mean of m and a variance of σ^2, find the limit $P\left(\left|\frac{S_n}{n} - m\right| \geq \frac{c\sigma}{\sqrt{n}}\right)$. Comment on your result and connect it with the LLN.

7. Let $X_1, X_2, ...$ be a sequence of independent exponential r.v.'s with $E\{X_i\} = 2$, and let $S_n = X_1 + ... + X_n$. Find $\lim_{n\to\infty} P(S_n \leq 2n + \sqrt{n})$.

8. An energy company provides services for a town of $n = 10,000$ households. For a household, the size of monthly energy consumption is exponentially distributed with a mean of 800 kwh, and does not depend on the consumption of other households. The company has specified a consumption baseline d which is not exceeded by 70% of households *on the average*. (That is, the expected proportion of the households with consumption less than d, is 0.7.) The company charges 12¢ per one kwh the households with consumption less than d, and charges 15¢/kwh for a surplus over d. (a) Let M be the proportion of households with a consumption less than d, and let k be a number such that $M > k$ with probability 0.9. Do you expect k to be less or larger than 0.7? Estimate k using normal approximation. (b) The total payment of all households with probability 0.9 is larger than what amount? (*Hint*: We count households with the consumption less than d. What is the probability that a particular household will be counted? Use of software for (b) is recommended.)

9. Show that the negative binomial distribution with parameters (p, ν) may be well approximated by a normal distribution for large ν. State the CLT for this case.

10. Let Y be a standard normal r.v., and for a set B in the real line, let $\Phi(B) = P(Y \in B)$. From Theorem 1, we know that $P(S_n^* \in B) \to \Phi(B)$ as $n \to \infty$ if B is an interval. Is the same true for any B? (*Advice*: Consider $X_j = \pm 1$ with equal probabilities, and realize that in this case $S_n^* = \frac{1}{\sqrt{n}} S_n$. Let $\overline{B}$ be the set of *all* numbers that may be presented as $\frac{k}{\sqrt{n}}$, where $k = 0, \pm 1, \pm 2, ...$ and $n = 1, 2, ...$. What are $P(S_n^* \in \overline{B})$ and $\Phi(\overline{B})$ in this case?)

11. This exercise concerns Example 1.1-3. Using software, say Excel, provide a worksheet demonstrating the accuracy of the normal approximation of the Poisson distribution for large λ. (*Advice*: When comparing the distribution of the Poisson r.v. with parameter λ and the normal distribution with a mean of λ and a variance of λ, you may follow the scheme of Example **3**.4.5.1-4 and Fig. 7 there.)

12. Proceeding from Example 1.1-3, show that the distribution of Z_λ may be approximated by a normal distribution for any large, not necessary integer, λ. (*Advice*: First, $\lambda = [\lambda] + \{\lambda\}$, where $[\lambda]$ is the integer part of λ, and $\{\lambda\} = \lambda - [\lambda]$, the fractional part of λ. Accordingly, with $n = [\lambda]$, we may write $Z_\lambda = Z_n + Y_{\{\lambda\}}$, where Z_n and $Y_{\{\lambda\}}$, are independent Poisson r.v.'s with means n and $\{\lambda\}$ respectively. Clearly, $[\lambda] \to \infty$ as $\lambda \to \infty$. To apply the CLT, one can switch to the normalized r.v.'s, and write $Z_\lambda^* = \frac{Z_\lambda - \lambda}{\sqrt{\lambda}} = \sqrt{\frac{n}{\lambda}}\frac{Z_n - n}{\sqrt{n}} + \frac{Y_{\{\lambda\}} - \{\lambda\}}{\sqrt{\lambda}}$. It remains to show that (a) $(Z_n - n)/\sqrt{n}$ is asymptotically normal; (b) $\sqrt{[\lambda]/\lambda} \to 1$; and (c) $\frac{1}{\sqrt{\lambda}}(Y_{\{\lambda\}} - \{\lambda\}) \xrightarrow{P} 0$, (d) these three facts considered together imply what we are proving.)

13. Proceeding from Example 1.1-4, prove that the normalized r.v. $Z_{a\nu}^* = \left(Z_{a\nu} - \frac{\nu}{a}\right) \Big/ \left(\frac{\sqrt{\nu}}{a}\right)$ is asymptotically normal as $\nu \to \infty$. (*Advice*: We may use the scheme of Exercise 12.)

14. Consider a sequence of independent standard exponential r.v.'s $X_1, X_2, \ldots$, and the r.v. $S_n = X_1 + \ldots + X_n$. Let $\Gamma_{a\nu}(x)$ be the d.f. of a Γ-r.v. with parameters (a, ν), and as usual, $f_{a\nu}(x)$ be the corresponding density.

    (a) Show that $P(S_n \le x) = \Gamma_{1n}(x)$, and the d.f. of the r.v. S_n^* is $F_n^*(x) = \Gamma_{1n}(x\sqrt{n} + n)$.

    (b) Show that the density of the r.v. S_n^* is $f_n^*(x) = \sqrt{n} f_{1n}(x\sqrt{n} + n)$.

    (c) Using software, for example Excel, provide a worksheet including columns with the values of the standard normal density $\varphi(x)$ and d.f. $\Phi(x)$, and the density $f_n^*(x)$ and the d.f. $F_n^*(x)$. It suffices to consider $-3 \le x \le 3$, and n equal to, say, $10, 20, 100$. Consider the accuracy of the normal approximation; in particular, $\max_x |F_n^*(x) - \Phi(x)|$. (*Advice*: You may follow the scheme of Example **3**.4.5.1-4 and Fig. 7 there.)

15. Regarding Example 1.1-5, what would the probability that the company will not suffer a loss have been equal to if the premium had been equal to the mean payment?

16.* In the life insurance model in Example **8**.2.4-1, consider the probability that the company will not suffer a loss. Using the particular data from this example, for a portfolio of $n = 1000$ independent policies, find a single premium (that is, the premium paid at the moment of policy issue) for which this probability is not smaller than 0.9. (*Hint*: The evaluation should be carried out from the standpoint of the initial time. Use the calculations provided in the example mentioned.)

17.* Proceeding from Theorems 2 and 3, explain why the Cauchy distribution (which is stable) cannot be a limiting distribution for the normalized sum of i.i.d. r.v.'s with a finite variance.

18.* Under which conditions is (2.1.2) true if the X's are normal?

19.* Consider a generalization of Example 2.1-4 for an arbitrary parameter a. For which a does the CLT hold? Consider the order of the Lyapunov fraction L_n (and hence, the rate of convergence in the CLT) for different a's. (*Advice*: Consider the cases: $a > \frac{1}{2}$, $a = \frac{1}{2}$, $\frac{1}{3} < a < \frac{1}{2}$, $a = \frac{1}{3}$, $a < \frac{1}{3}$.)

20.* Show that if (2.2.3) holds, then all terms X_i/B_n are asymptotically negligible; more precisely, that $\max_{i=1,\ldots,n} P\left(\frac{|X_i|}{B_n} > \varepsilon\right) \to 0$ as $n \to \infty$, for any $\varepsilon > 0$.

21.* Lindeberg's condition (2.2.2) requires $\Lambda(\varepsilon) \to 0$ for any $\varepsilon > 0$. Can it be true for $\varepsilon = 0$? To what, as a matter of fact, is $\Lambda(0)$ equal?

Route 1 $\Rightarrow$ *page 295*

Chapter 10

Covariance Analysis. The Multivariate Normal Distribution. The Multivariate Central Limit Theorem

2

In this chapter, we assume all variances to be finite.

1 COVARIANCE AND CORRELATION

1.1 Covariance

Covariance is one of simplest measures of dependency between two r.v.'s. Let X_1, X_2 be r.v.'s, and $m_i = E\{X_i\},\ i = 1,2$. We call the covariance between X_1, X_2 the quantity

$$Cov\{X_1, X_2\} = E\{(X_1 - m_1)(X_2 - m_2)\}. \tag{1.1.1}$$

In particular,

$$\text{If } m_1 = m_2 = 0, \text{ then} \quad Cov\{X_1, X_2\} = E\{X_1 X_2\}. \tag{1.1.2}$$

Note also that, setting $X_1 = X_2 = X$, and $m = E\{X\}$, we have

$$Cov\{X, X\} = E\{(X-m)^2\} = Var\{X\}.$$

EXAMPLE 1. Let ξ_1, ξ_2, and η be mutually independent r.v.'s with zero means and unit variances. Let $X_1 = \xi_1 + \eta$ and $X_2 = \xi_2 + \eta$. Then $E\{X_1\} = E\{X_2\} = 0$, and

$$Cov\{X_1, X_2\} = E\{X_1 X_2\} = E\{(\xi_1 + \eta)(\xi_2 + \eta)\} = E\{\xi_1\xi_2\} + E\{\xi_1\eta\} + E\{\xi_2\eta\} + E\{\eta^2\}$$
$$= E\{\xi_1\}E\{\xi_2\} + E\{\xi_1\}E\{\eta\} + E\{\xi_2\}E\{\eta\} + E\{\eta^2\} = 0\cdot 0 + 0\cdot 0 + 0\cdot 0 + 1 = 1$$

because $E\{\eta^2\} = Var\{\eta\} = 1$.

EXAMPLE 2. Suppose a r.vec. (X_1, X_2) has the density $f(x_1, x_2) = \frac{3}{8}(x_1 - x_2)^2$ for $-1 \leq x_1 \leq 1$, and $-1 \leq x_2 \leq 1$, and $f(x_1, x_2) = 0$ otherwise. The reader may double-check that indeed

$$\int_{-\infty}^{\infty}\int_{-\infty}^{\infty} f(x_1, x_2)dx_1dx_2 = \int_{-1}^{1}\int_{-1}^{1} f(x_1, x_2)dx_1dx_2 = 1, \text{ and}$$
$$\int_{-\infty}^{\infty}\int_{-\infty}^{\infty} x_1 f(x_1, x_2)dx_1dx_2 = \int_{-1}^{1}\int_{-1}^{1} x_1 f(x_1, x_2)dx_1dx_2 = 0.$$

(In the latter case, one may use just the symmetry argument.)

Hence, $E\{X_1\}=0$, and similarly, $E\{X_2\}=0$. Then, as is easy to calculate,

$$Cov\{X_1,X_2\}=E\{X_1X_2\}=\int_{-1}^{1}\int_{-1}^{1}x_1x_2\frac{3}{8}(x_1-x_2)^2dx_1dx_2=-\frac{1}{3}. \ \square$$

Multiplying the variables in the parentheses in definition (1.1.1), we get

$$\begin{aligned}Cov\{X_1,X_2\}&=E\{X_1X_2-m_1X_2-m_2X_1+m_1m_2\}\\&=E\{X_1X_2\}-m_1E\{X_2\}-m_2E\{X_1\}+m_1m_2,\end{aligned}$$

from which it easily follows that

$$Cov\{X_1,X_2\}=E\{X_1X_2\}-m_1m_2. \quad (1.1.3)$$

In particular, if $X_1=X_2=X$, (1.1.3) becomes the known for us formula $Var\{X\}=E\{X^2\}-(E\{X\})^2$.

If X_1,X_2 are independent, then $Cov\{X_1,X_2\}=0$. Indeed, in this case, the r.-h.s. of (1.1.3) equals $E\{X_1\}E\{X_2\}-m_1m_2=m_1m_2-m_1m_2=0$.

Thus,

> If $Cov\{X_1,X_2\}\neq 0$, then the r.v.'s X_1,X_2 are dependent. (1.1.4)

EXAMPLE 1 revisited. Since the representations for X_1 and X_2 contain the common term η, one may conjecture that X_1 and X_2 are dependent. This is indeed true since $Cov\{X_1,X_2\}=1\neq 0$.

EXAMPLE 2 revisited. The density $f(x_1,x_2)=\frac{3}{8}(x_1-x_2)^2$ is getting larger when $|x_1-x_2|$ is increasing, so $|X_1-X_2|$ is likely to be large. Loosely put, this means that it is more likely that the values of X_1 are not close to those of X_2, and one may conjecture that X_1 and X_2 are dependent. This is confirmed by the fact that $Cov\{X_1,X_2\}=-\frac{1}{3}\neq 0$. The significance of the negative sign will be clarified a bit later. $\square$

It is important to emphasize that *the converse to assertion (1.1.4) is not true.*

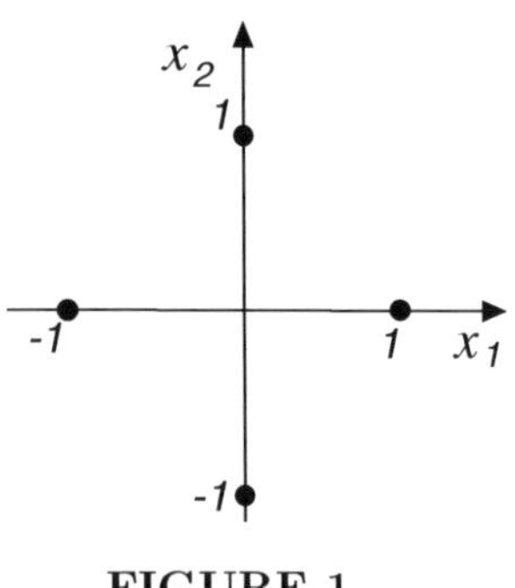

FIGURE 1.

EXAMPLE 3. Consider a r.vec. $\mathbf{X}=(X_1,X_2)$ that assumes four vector-values corresponding to four points in Fig. 1 with probabilities $1/4$. The reader is invited either to show that X_1,X_2 are dependent or to look it over in Example **3**.1.2.2-1. On the other hand, as one can compute or just proceed for the symmetry argument, $E\{X_1\}=E\{X_2\}=0$. Then, in view of (1.1.2), $Cov\{X_1,X_2\}=E\{X_1X_2\}$. In our example, for any point in Fig. 1, one of the coordinates, X_1 or X_2, equals zero, so $X_1X_2=0$. Hence, $Cov\{X_1,X_2\}=0$. $\square$

Thus, *independence implies zero covariance, but not vice versa.*

Next, we state two basic properties of covariance. For any constants c_1 and c_2,

$$Cov\{X_1+c_1,X_2+c_2\}=Cov\{X_1,X_2\}, \quad (1.1.5)$$

$$Cov\{c_1X_1,c_2X_2\}=c_1c_2Cov\{X_1,X_2\}. \quad (1.1.6)$$

Indeed, by definition, $Cov\{X_1+c_1,X_2+c_2\} = E\{(X_1+c_1-(m_1+c_1))(X_2+c_2-(m_2+c_2))\} = E\{(X_1-m_1)(X_2-m_2)\} = Cov\{X_1,X_2\}$. The reader is invited to prove (1.1.6) in a similar way.

From (1.1.5), it follows that if we consider the centered r.v.'s $X_1' = X_1 - m_1$ and $X_2' = X_2 - m_2$, then

$$Cov\{X_1',X_2'\} = Cov\{X_1,X_2\}. \tag{1.1.7}$$

On the other hand, $E\{X_1'\} = E\{X_2'\} = 0$, and hence

$$Cov\{X_1',X_2'\} = E\{X_1'X_2'\}. \tag{1.1.8}$$

Next, we prove that for *any* r.v.'s X_1, X_2 (with finite variances),

$$Var\{X_1+X_2\} = Var\{X_1\} + Var\{X_2\} + 2Cov\{X_1,X_2\}. \tag{1.1.9}$$

It suffices to prove this for the centered r.v.'s X_1', X_2' because they have the same variances and covariance as the original r.v.'s X_1, X_2. We have

$$\begin{aligned} Var\{X_1'+X_2'\} &= E\{(X_1'+X_2')^2\} = E\{(X_1')^2\} + E\{(X_2')^2\} + 2E\{X_1'X_2'\} \\ &= Var\{X_1'\} + Var\{X_2'\} + 2Cov\{X_1', X_2'\}. \end{aligned}$$

Thus, for the relation

$$Var\{X_1+X_2\} = Var\{X_1\} + Var\{X_2\} \tag{1.1.10}$$

to be true, we need only $Cov\{X_1,X_2\}$ to be zero, *which, as we saw, is a weaker property than independence.*

A generalization of (1.1.9) for the case of n r.v.'s $X_1,...,X_n$ may be written as follows:

$$Var\left\{\sum_{i=1}^{n} X_i\right\} = \sum_{i=1}^{n} Var\{X_i\} + 2\sum_{(i,j):i>j,i=1,...,n} Cov\{X_i,X_j\}. \tag{1.1.11}$$

In Exercise 13, the reader is asked to provide a formal proof which is very similar to that of (1.1.9).

To move further, we present a basic inequality.

1.2 The Cauchy–Schwarz inequality

Proposition 1 *For any r.v.'s* ξ *and* η *with finite second moments,*

$$(E\{\xi\eta\})^2 \le E\{\xi^2\}E\{\eta^2\}. \tag{1.2.1}$$

The strict equality holds if and only if with probability one

$$\xi = c\eta \quad \textit{for some number } c. \tag{1.2.2}$$

Proof. For any real t,

$$0 \leq E\{(\xi - t\eta)^2\} = E\{\xi^2\} - 2tE\{\xi\eta\} + t^2E\{\eta^2\}. \tag{1.2.3}$$

Denote by $Q(t)$ the r.-h.s. of (1.2.3). As a function of t, this is a quadratic function. Since $Q(t) \geq 0$ for all t, the discriminant of $Q(t)$ should be non-positive. The discriminant is equal to $4(E\{\xi\eta\})^2 - 4E\{\xi^2\}E\{\eta^2\} = 4[(E\{\xi\eta\})^2 - E\{\xi^2\}E\{\eta^2\}]$. We see that it is non-positive if and only if (1.2.1) is true.

It is straightforward to verify that if $\xi = c\eta$ for some number c, then (1.2.1) becomes a strict *equality*.

Conversely, let $(E\{\xi\eta\})^2 = E\{\xi^2\}E\{\eta^2\}$. Then the discriminant above equals zero, and the equation $Q(t) = 0$ has only one root. Denote it by c. By the definition of $Q(t)$, this means that $E\{(\xi - c\eta)^2\} = 0$. The r.v. $(\xi - c\eta)^2$ is non-negative. Its expectation may be equal to zero only if $\xi - c\eta = 0$ with probability one, which implies (1.2.2). ■

1.3 Correlation

Let $\sigma_i^2 = Var\{X_i\} > 0$, $i = 1,2$. The *correlation coefficient* of the r.v.'s X_1, X_2 (the preposition "between" is also in use), briefly the *correlation*, is the quantity

$$Corr\{X_1, X_2\} = \frac{Cov\{X_1, X_2\}}{\sigma_1 \sigma_2}. \tag{1.3.4}$$

Note that correlation is a dimensionless, or unitless, characteristic. Say, if we measure X_1, X_2 in inches, the dimension of $Cov\{X_1, X_2\}$ and $Var\{X_i\}$ is inch^2, while $Corr\{X_1, X_2\}$, as follows from (1.3.4), does not have a dimension. For this reason, correlation, which may be viewed as *normalized covariance*, is a more adequate measure of dependency.

Proposition 2 *The following properties of correlation are true.*
(A) For all X_1, X_2 (with finite variances),

$$-1 \leq Corr\{X_1, X_2\} \leq 1. \tag{1.3.5}$$

(B) If $Corr\{X_1, X_2\} = 1$, then $X_2 = b + cX_1$ for some b and a positive c.
(C) If $Corr\{X_1, X_2\} = -1$, then $X_2 = b + cX_1$ for some b and a negative c.

R.v.'s X_1, X_2 for which $Corr\{X_1, X_2\} = 0$ are called *non-correlated* or *uncorrelated.* R.v.'s for which the correlation is positive are called *positively correlated.* If $Corr\{X_1, X_2\} < 0$, then the r.v.'s X_1, X_2 are said to be *negatively correlated.* R.v.'s X_1, X_2 for which $Corr\{X_1, X_2\} = \pm 1$, are called *perfectly correlated.*

Set $\rho = Corr\{X_1, X_2\}$. We may interpret Proposition 2 as follows.

The larger $|\rho|$, to a greater extent the r.v.'s X_1, X_2 are dependent.

If $\rho > 0$, the association between X_1, X_2 is positive in the sense that the larger the value of X_1, the larger the value of X_2 should we expect on the average. If $\rho < 0$, the association is negative: if the value of X_1 increased, then we expect that, on the average, the value of X_2 will decrease.

If $\rho = \pm 1$, the r.v.'s X_1, X_2 are perfectly dependent in the sense that the value of X_1 is uniquely determined by the value of X_2, and vice versa. Moreover, the relation between X_1 and X_2 is linear, i.e., is determined by a *linear* equation.

EXAMPLE 1. In Example 1.1-1, $\sigma_1^2 = Var\{X_1\} = Var\{\xi_1\} + Var\{\eta\} = 1 + 1 = 2$. Similarly, $\sigma_2^2 = 2$ also. Then

$$\rho = \frac{Cov\{X_1, X_2\}}{\sigma_1 \sigma_2} = \frac{1}{\sqrt{2}\cdot\sqrt{2}} = \frac{1}{2}.$$

The correlation ρ is positive, which reflects the essence of the problem: since X_1, X_2 have the common term η, it is natural to expect a positive association.

Now, suppose that $X_1 = \xi_1 + \eta$ while $X_2 = \xi_2 - \eta$. Then it is reasonable to conjecture that the correlation is negative. This is, certainly, true. The variances σ_1^2, σ_2^2 are the same, but $Cov\{X_1, X_2\} = E\{(\xi_1 + \eta)(\xi_2 - \eta)\} = E\{\xi_1\}E\{\xi_2\} - E\{\xi_1\}E\{\eta\} + E\{\xi_2\}E\{\eta\} - E\{\eta^2\} = 0\cdot 0 - 0\cdot 0 + 0\cdot 0 - 1 = -1$. Then

$$\rho = \frac{-1}{\sqrt{2}\cdot\sqrt{2}} = -\frac{1}{2}.$$

EXAMPLE 2. In Example 1.1-2, by the symmetry argument, $Var\{X_2\} = Var\{X_1\} = E\{X_1^2\} = \int_{-1}^{1}\int_{-1}^{1} x^2 \frac{3}{8}(x_1 - x_2)^2 dx_1 dx_2 = \frac{7}{15}$, as is easy to compute. Hence, $\rho = \frac{-1/3}{7/15} = -\frac{5}{7}$, which reflects the dependency character in this case. □

Proof of Proposition 2. We switch to the centered r.v.'s $X_1' = X_1 - m_1$ and $X_2' = X_2 - m_2$ for which the covariance and the variances are the same as for X_1 and X_2. By the Cauchy-Schwarz inequality,

$$|E\{X_1'X_2'\}| \leq \sqrt{E\{(X_1')^2\}}\sqrt{E\{(X_2')^2\}}. \tag{1.3.6}$$

Furthermore,

$$\rho = \frac{Cov\{X_1, X_2\}}{\sqrt{Var\{X_1\}}\sqrt{Var\{X_2\}}} = \frac{Cov\{X_1', X_2'\}}{\sqrt{Var\{X_1'\}}\sqrt{Var\{X_2'\}}} = \frac{E\{X_1'X_2'\}}{\sqrt{E\{(X_1')^2\}}\sqrt{E\{(X_2')2\}}}. \tag{1.3.7}$$

From this and (1.3.6), it follows that $|\rho| \leq 1$.

By virtue of Proposition 1, the very r.-h.s. of (1.3.7) equals 1 or -1 if and only if $X_2' = cX_1'$ for a constant c. Since we have assumed $\sigma_i \neq 0$, the constant $c \neq 0$ because otherwise σ_2 would equal zero. From (1.3.7), it follows that in this case,

$$\rho = \frac{E\{cX_1'X_1'\}}{\sqrt{E\{(cX_1')^2\}}\sqrt{E\{(X_1')^2\}}} = \frac{cE\{(X_1')^2\}}{|c|(\sqrt{E\{(X_1')^2\}})^2} = \frac{c}{|c|}.$$

So, $\rho = 1$ if and only if $c > 0$. Then

$$X_2 = X_2' + m_2 = cX_1' + m_2 = c(X_1 - m_1) + m_2 = m_2 - cm_1 + cX_1,$$

which implies assertion B in Proposition 2. Assertion C is proved similarly. ■

1.4 A new perspective: Geometric interpretation

We establish an analogy between the above framework and that of the standard Euclidean geometry. Consider a non-random k-dimensional vector $\boldsymbol{x} = (x_1, ..., x_k)$. The reader can even restrict her/himself to the "usual" three-dimensional case $k = 3$; it does not lead to a great loss of generality.

Denote the length of $\boldsymbol{x}$ by $|\boldsymbol{x}|$; in the Euclidean space, $|\boldsymbol{x}| = \sqrt{x_1^2 + ... + x_k^2}$. The fact that for a number x, the symbol $|x|$ denotes the absolute value of x should not cause confusion: in the one-dimensional case, the two quantities coincide.

The distance between two vectors, $\boldsymbol{x} = (x_1, ..., x_n)$ and $\boldsymbol{y} = (y_1, ..., y_n)$ is the length of the vector $\boldsymbol{x} - \boldsymbol{y}$; that is, $|\boldsymbol{x} - \boldsymbol{y}| = \sqrt{(x_1 - y_1)^2 + ...(x_k - y_k)^2}$.

The scalar (or dot) product operation is defined as

$$\langle \boldsymbol{x}, \boldsymbol{y} \rangle = x_1 y_1 + ... + x_k y_k.$$

We know that

$$|\langle \boldsymbol{x}, \boldsymbol{y} \rangle| \leq |\boldsymbol{x}| \cdot |\boldsymbol{y}|, \tag{1.4.1}$$

and if $|\boldsymbol{x}| \neq 0, \ |\boldsymbol{y}| \neq 0$, then

$$\frac{\langle \boldsymbol{x}, \boldsymbol{y} \rangle}{|\boldsymbol{x}| \cdot |\boldsymbol{y}|} = \cos \varphi,$$

where φ is the angle between the vectors $\boldsymbol{x}$ and $\boldsymbol{y}$. (For $k > 3$, we still can talk about a "usual" angle between $\boldsymbol{x}$ and $\boldsymbol{y}$ because the (hyper) plane going through these *two* vectors is a two-dimensional plane.)

One of the main features of the modern Mathematics is that it carries over the notions of length, distance, and scalar product to objects of more complicated nature than finite-dimensional vectors; and the corresponding generalization inherits many properties of the Euclidean space. In particular, this concerns r.v.'s.

Let us consider the space (or collection) of all r.v.'s X for which $E\{X\} = 0$ and $E\{X^2\}$ is finite. The requirement $E\{X\} = 0$ does not restrict generality: if $E\{X\} \neq 0$, we can switch to the centered r.v. $X - E\{X\}$.

Let us view such r.v.'s as vectors and define the "length" of X as the quantity $\|X\| = \sqrt{E\{X^2\}}$ (compare with $|\boldsymbol{x}| = \sqrt{x_1^2 + ... + x_k^2}$). In this context, as a rule, the term "*norm*" is used instead of "length."

The difference between two r.v.'s, X and Y, is the usual difference $X - Y$, and the distance between these r.v.'s is defined (by analogy with the Euclidean geometry) as the "length" of the difference; that is,

$$\|X - Y\| = \sqrt{E\{(X - Y)^2\}}.$$

The scalar (or dot) product of two r.v.'s from our space is defined as the quantity $\langle X, Y \rangle = E\{XY\}$. By the Cauchy-Schwarz inequality,

$$|E\{XY\}| \leq \sqrt{E\{X^2\}} \sqrt{E\{Y^2\}},$$

which may be rewritten as

$$|\langle X, Y \rangle| \leq \|X\| \cdot \|Y\|. \tag{1.4.2}$$

So, we see a complete analogy with (1.4.1).

In the usual geometry framework, if the dot product $\langle \boldsymbol{x}, \boldsymbol{y} \rangle = 0$, the vectors $\boldsymbol{x}, \boldsymbol{y}$ are perpendicular (in another terminology, orthogonal). In our scheme, $\langle X, Y \rangle = E\{XY\} = Cov\{X, Y\}$

(since $E\{X\} = E\{Y\} = 0$), and hence uncorrelatedness may be interpreted as perpendicularity (or orthogonality) in terms of the dot product defined.

The correlation $\rho = \frac{\langle X,Y \rangle}{\|X\| \cdot \|Y\|}$ is a counterpart of the cosine of the angle between vectors, and in a certain sense, may be viewed as such.

1.5 The prediction problem: Linear regression

Consider the prediction problem that has already been touched on in Section **3**.6.4. Let X and Y be two r.v.'s. We view X as an observable r.v., and Y as a r.v. whose value we want to predict. For example, we may try to predict the future value of an economic index given the present value of this index; or knowing a high school student's Math score (say, SAT, if we are talking about USA), we want to predict her/his future University result (say, GPA).

A prediction is specified by a function $g(X)$, and the prediction is "good" if the difference $Y - g(X)$ is small in a certain sense. In our framework, it is natural to measure the accuracy of prediction by the distance $\|Y - g(X)\|$, or—which does not change the essence of the matter—by

$$\|Y - g(X)\|^2 = E\{(Y - g(X))^2\}. \tag{1.5.1}$$

In the above definition, we do *not* assume the r.v.'s under consideration to have zero means.

We have proved in Section **3**.6.4 that the function $g(x)$ that minimizes (1.5.1), is the *regression function* $g(x) = E\{Y \mid X = x\}$. In another terminology, $E\{Y \mid X = x\}$ is the *least square predictor*.

Now, let us consider only linear predictors, that is, functions $g(x) = \beta x + \alpha$, and find an optimal α and β. Let $E\{X\} = m_1$, $E\{Y\} = m_2$, $Var\{X\} = \sigma_1^2$, $Var\{Y\} = \sigma_2^2$. Suppose both variances to be positive, and set $Corr\{X, Y\} = \rho$.

Theorem 3 *The minimum of (1.5.1) over the class of linear functions is attained at the function*

$$g(x) = m_2 + \frac{\rho\sigma_2}{\sigma_1}(x - m_1). \tag{1.5.2}$$

The function (1.5.2) specifies the *linear regression* of Y on X. One may represent the function $y = g(x)$ in (1.5.2) in the following more illustrative form:

$$\frac{y - m_2}{\sigma_2} = \rho\frac{x - m_1}{\sigma_1}. \tag{1.5.3}$$

The graphs are given in Fig. 2. The line goes through the point (m_1, m_2). The slope equals $\frac{\rho\sigma_2}{\sigma_1}$. If $\rho = 0$, then the least square linear predictor gives just the mean value of Y whatever the value of X is: the variables are treated as independent. If $\rho \neq 0$, then the linear predictor takes into account the value of X.

EXAMPLE 1. Let X be the value of the SAT score evaluating the performance of a high school student living in a particular area and chosen at random. Let Y be the future university GPA score of the same person. Suppose statistical data give the following estimates: $m_1 \approx 1130, m_2 \approx 2.9, \sigma_1 \approx 150, \sigma_2 \approx 0.6$ and $\rho \approx 0.8$. Then the linear predictor

$$g(x) = 2.9 + 0.8 \cdot \frac{0.6}{150} \cdot (x - 1130) = -0.716 + 0.0032x.$$

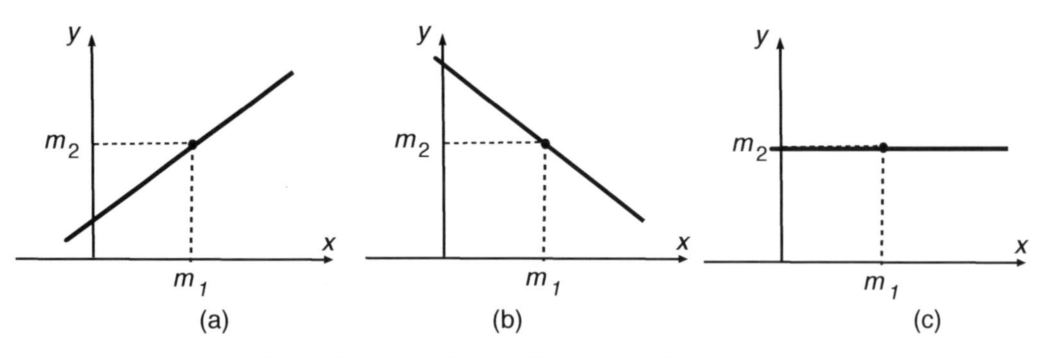

FIGURE 2. Graphs of linear predictors. The graphs (a)–(c) correspond to the cases $\rho > 0$, $\rho < 0$, and $\rho = 0$, respectively.

Assume that Matthew graduates from a high school, having a SAT score of 1145. Let us predict his future GPA. The predictor is $g(1145) = -0.716 + 0.0032 \cdot 1145 \approx 2.95$. □

Proof of Theorem 3. We have $E\{(Y - \beta X - \alpha)^2\} = E\{(Y - m_2 - \beta(X - m_1) + m_2 - \beta m_1 - \alpha)^2\} = E\{(Y - m_2 - \beta(X - m_1) + c)^2\}$, where $c = m_2 - \beta m_1 - \alpha$. Hence,

$$\begin{aligned} E\{(Y - \beta X - \alpha)^2\} &= E\{(Y - m_2)^2\} + \beta^2 E\{(X - m_1)^2\} + c^2 \\ &-2\beta E\{(X - m_1)(Y - m_2)\} + 2cE\{Y - m_2\} - 2c\beta E\{X - m_1\} \\ &= \sigma_2^2 + \beta^2\sigma_1^2 + c^2 - 2\beta Cov\{X, Y\} + 2c \cdot 0 - 2c\beta \cdot 0 \\ &= \sigma_2^2 + \beta^2\sigma_1^2 - 2\beta\rho\sigma_1\sigma_2 + (m_2 - \beta m_1 - \alpha)^2. \end{aligned}$$

We see that to minimize the last expression, we should set $\alpha = m_2 - \beta m_1$. The partial derivative in β equals $2\beta\sigma_1^2 - 2\rho\sigma_1\sigma_2 - 2m_1(m_2 - \beta m_1 - \alpha)$, and for the α chosen, it equals $2\beta\sigma_1^2 - 2\rho\sigma_1\sigma_2$. Setting it to be zero, we get that the optimal $\beta = \rho\sigma_2/\sigma_1$. It is easy to double-check that for the α and β found, the function $g(x) = \beta x + \alpha$ is the function in (1.5.2). ■

In conclusion, it makes sense to emphasize that the linear function (1.5.2) is an adequate characteristic when the regression function $E\{Y \mid X = x\}$ is linear or close to such. The same concerns correlation itself. It does not mean that we cannot use it in other cases. For example, if we figured out somehow that a correlation coefficient $\rho \neq 0$, then we can claim that the corresponding r.v.'s are dependent whatever the dependency structure is. However, in non-linear cases, the correlation analysis may lead to non-accurate estimates or even turn out to be misleading. Suppose, for instance, that $Y = u(X)$ where $u(x)$ is a non-linear function; say, $Y = X^3$. The dependence is perfect in the sense that a value of X uniquely determines the value of Y. However, in this case $\rho \neq \pm 1$ because otherwise, by Proposition 2, the regression would have been linear. We consider an example in Exercise 17.

2 COVARIANCE MATRICES AND SOME APPLICATIONS

2.1 Covariance matrices

We will not use the notion of norm $\|.\|$ anymore, and from now on, the symbol $\|.\|$ is used for presenting matrices.

For any matrix $\alpha = \|\alpha_{ij}\|$ with random entries α_{ij}, we define its expectation $E\{\alpha\}$ as the matrix $\|E\{\alpha_{ij}\}\|$. In particular, for a random vector (r.vec.) $\mathbf{X} = (X_1, ..., X_k)$, the expectation $E\{\mathbf{X}\} = (E\{X_1\}, ..., E\{X_k\})$.

For a r.vec. $\mathbf{X} = (X_1, ..., X_k)$, we define the covariance matrix $\mathcal{C} = \|c_{ij}\|$, where $c_{ij} = Cov\{X_i, X_j\}$; $i, j = 1, ..., k$. *One can view a covariance matrix as a counterpart of variance in the multivariate case.*

We know that $c_{ii} = Cov\{X_i, X_i\} = Var\{X_i\}$. If the dimension $k = 1$, then the covariance matrix consists of one element: the variance of X_1. In the general case, let us set $\sigma_i^2 = Var\{X_i\}$, and $\rho_{ij} = Corr\{X_i, X_j\}$, provided that all $\sigma_i \neq 0$. Then, since $c_{ij} = \rho_{ij}\sigma_i\sigma_j$,

$$\mathcal{C} = \|c_{ij}\| = \left\| \begin{array}{cccc} \sigma_1^2 & c_{12} & \dots & c_{1k} \\ c_{21} & \sigma_2^2 & \dots & c_{2k} \\ \vdots & \vdots & \vdots & \vdots \\ c_{k1} & c_{k2} & \dots & \sigma_k^2 \end{array} \right\|$$
$$= \left\| \begin{array}{cccc} \sigma_1^2 & \rho_{12}\sigma_1\sigma_2 & \dots & \rho_{1k}\sigma_1\sigma_k \\ \rho_{21}\sigma_2\sigma_1 & \sigma_2^2 & \dots & \rho_{2k}\sigma_2\sigma_k \\ \vdots & \vdots & \vdots & \vdots \\ \rho_{k1}\sigma_k\sigma_1 & \rho_{k2}\sigma_k\sigma_2 & \dots & \sigma_k^2 \end{array} \right\|.$$

It is worth emphasizing that the covariance matrix $\mathcal{C}$ is symmetric because $c_{ij} = c_{ji}$ (and hence $\rho_{ij} = \rho_{ji}$ also). In particular, in the case $k = 2$,

$$\mathcal{C} = \|c_{ij}\| = \left\| \begin{array}{cc} \sigma_1^2 & c_{12} \\ c_{12} & \sigma_2^2 \end{array} \right\| = \left\| \begin{array}{cc} \sigma_1^2 & \rho\sigma_1\sigma_2 \\ \rho\sigma_1\sigma_2 & \sigma_2^2 \end{array} \right\|. \tag{2.1.1}$$

EXAMPLE 1. In Example 1.1-1, $Var\{X_i\} = Var\{\xi_i\} + Var\{\eta\} = 1 + 1 = 2$, and

$$\mathcal{C} = \|c_{ij}\| = \left\| \begin{array}{cc} 2 & 1 \\ 1 & 2 \end{array} \right\|. \square$$

Set $\mathbf{m} = E\{\mathbf{X}\}$. To make the notation below more customary, let us view $\mathbf{X}$ and $\mathbf{m}$ as *column vectors.* Then the transpose $\mathbf{X}^T$ is a row vector. For further calculations, it is useful to keep in mind that

$$\mathcal{C} = \|c_{ij}\| = E\{(\mathbf{X} - \mathbf{m})(\mathbf{X} - \mathbf{m})^T\}, \tag{2.1.2}$$

and hence, if $\mathbf{m} = \mathbf{0} = (0, ..., 0)$, a zero vector, then

$$\mathcal{C} = \|c_{ij}\| = \|E\{X_iX_j\}\| = E\{\mathbf{X}\mathbf{X}^T\}. \tag{2.1.3}$$

To show this, it suffices to consider (2.1.3). The product

$$\mathbf{X}\mathbf{X}^T = \begin{Vmatrix} X_1 \\ X_2 \\ \vdots \\ X_k \end{Vmatrix} \times \begin{Vmatrix} X_1 & X_2 & \dots & X_k \end{Vmatrix}$$
$$= \begin{Vmatrix} X_1^2 & X_1X_2 & \dots & X_1X_k \\ X_2X_1 & X_2^2 & \dots & X_2X_k \\ \vdots & \vdots & \vdots & \vdots \\ X_kX_1 & X_kX_2 & \dots & X_k^2 \end{Vmatrix}.$$

In the case of $\mathbf{m} = \mathbf{0}$, computing the expectation of both sides leads to (2.1.3) .

To list some properties of covariance matrices, first let us recall that for any square matrix $\mathcal{A} = \|a_{ij}\|$ and a vector $\mathbf{t} = (t_1, ..., t_k)$, the quadratic form $\sum_i \sum_j a_{ij} t_i t_j$ may be represented in the compact form $\langle \mathcal{A}\mathbf{t}, \mathbf{t} \rangle$ if we view $\mathbf{t}$ as a column vector and $\langle . \rangle$ stands for dot product.

Properties of covariance matrices

A. Consider the r.v. $t_1X_1 + ... + t_kX_k$, the linear combination of r.v.'s $X_1, ..., X_k$, and observe that it may be written in the compact form $\langle \mathbf{t}, \mathbf{X} \rangle$, where $\mathbf{t} = (t_1, ..., t_k)$. For the variance, we have

$$Var\{\langle \mathbf{t}, \mathbf{X} \rangle\} = \langle \mathcal{C}\mathbf{t}, \mathbf{t} \rangle. \tag{2.1.4}$$

B. Let $\mathcal{B}$ be a $(k \times k)$-matrix, the r.vec. $\mathbf{Y} = \mathcal{B}\mathbf{X}$, and let $\mathcal{C}_{\mathbf{X}}$ and $\mathcal{C}_{\mathbf{Y}}$ be the covariance matrices of $\mathbf{X}$ and $\mathbf{Y}$, respectively. Then

$$\mathcal{C}_{\mathbf{Y}} = \mathcal{B}\mathcal{C}_{\mathbf{X}}\mathcal{B}^T. \tag{2.1.5}$$

Proofs of Properties A-B. It suffices to present a proof under condition $E\{\mathbf{X}\} = \mathbf{0}$; otherwise, we will switch to the centered r.v.'s. Then

$$Var\{\langle \mathbf{t}, \mathbf{X} \rangle\} = E\left\{\left(\sum_j t_j X_j\right)^2\right\} = \sum_i \sum_j t_i t_j E\{X_i X_j\} = \sum_i \sum_j t_i t_j c_{ij} = \langle \mathcal{C}\mathbf{t}, \mathbf{t} \rangle.$$

Next, using (2.1.3), we have

$$\mathcal{C}_{\mathbf{Y}} = E\{\mathbf{Y}\mathbf{Y}^T\} = E\{\mathcal{B}\mathbf{X}(\mathcal{B}\mathbf{X})^T\} = E\{\mathcal{B}\mathbf{X}\mathbf{X}^T\mathcal{B}^T\} = \mathcal{B}E\{\mathbf{X}\mathbf{X}^T\}\mathcal{B}^T = \mathcal{B}\mathcal{C}_{\mathbf{X}}\mathcal{B}^T. \blacksquare$$

Now, as *good examples* of Covariance Analysis, we consider two useful and well known models based on the above properties.

2.2 Principal components method

Since the variance is non-negative, Property A implies, in particular, that any covariance matrix $\mathcal{C}$ of a r.vec. $\mathbf{X}$ is non-negative definite. (That is, $\langle \mathcal{C}\mathbf{t}, \mathbf{t} \rangle \geq 0$ for all $\mathbf{t}$.) Along with the fact that $\mathcal{C}$ is symmetric, this implies that all its eigenvalues are real and non-negative; see the corresponding theory in any Linear Algebra text; e.g., in [27, Ch.7].

Without loss of generality, we may consider these eigenvalues in a descending order, so denote them by $\lambda_1^2 \geq \lambda_2^2, ..., \lambda_k^2 \geq 0$. From Property B, and in accordance with the general Linear Algebra rule of switching to a new basis, it also follows that there exists an orthogonal matrix $\mathcal{B}$ such that the covariance matrix of the vector $\mathbf{Y} = \mathcal{B}\mathbf{X}$ is diagonal:

$$\mathcal{C}_{\mathbf{Y}} = \begin{Vmatrix} \lambda_1^2 & 0 & \dots & 0 \\ 0 & \lambda_2^2 & \dots & 0 \\ \vdots & \vdots & \vdots & \vdots \\ 0 & 0 & \dots & \lambda_k^2 \end{Vmatrix},$$

see, e.g., again [27, Ch.7].

Hence, $Var\{Y_j\} = \lambda_j^2, \;\; j = 1, ..., k.$

We may view $\mathcal{B}$ as a transition matrix to a new basis, and view $\mathbf{Y}$ as the *same* vector $\mathbf{X}$ considered in the new basis corresponding to the transformation $\mathcal{B}$. In this basis, the coordinates are *uncorrelated*. Later, in Section 3, we will see that in the case of the normal distribution this is equivalent to independence.

(Actually, we can restrict ourselves to rotation matrices $\mathcal{B}$. As we know from Linear Algebra, the new basis in this case is an orthonormal basis composed by eigenvectors of $\mathcal{C}$, but we do not use this fact here.)

The r.v.'s $Y_1, ..., Y_k$ above are called *principal components*.

EXAMPLE 1. Consider Example 1.1-1 that has been already revisited in Example 1.3-1. Let

$$\mathcal{B} = \begin{Vmatrix} \frac{\sqrt{2}}{2} & \frac{\sqrt{2}}{2} \\ -\frac{\sqrt{2}}{2} & \frac{\sqrt{2}}{2} \end{Vmatrix},$$

which corresponds to the clockwise rotation by $\pi/4$.[1] Then

$$\mathbf{Y} = \mathcal{B}\mathbf{X} = \begin{Vmatrix} \frac{\sqrt{2}}{2} & \frac{\sqrt{2}}{2} \\ -\frac{\sqrt{2}}{2} & \frac{\sqrt{2}}{2} \end{Vmatrix} \cdot \begin{Vmatrix} \xi_1 + \eta \\ \xi_2 + \eta \end{Vmatrix} = \begin{Vmatrix} \frac{\sqrt{2}}{2}(\xi_1 + \xi_2 + 2\eta) \\ \frac{\sqrt{2}}{2}(\xi_2 - \xi_1) \end{Vmatrix}. \tag{2.2.1}$$

It is easy to compute that $\mathcal{C}_{\mathbf{Y}} = \begin{Vmatrix} 3 & 0 \\ 0 & 1 \end{Vmatrix}$.

Indeed, since $E\{Y_1\} = E\{Y_2\} = 0$, we have $Cov\{Y_1, Y_2\} = E\{(\frac{\sqrt{2}}{2}(\xi_1 + \xi_2 + 2\eta))(\frac{\sqrt{2}}{2}(\xi_2 - \xi_1))\} = \frac{1}{2}E\{\xi_2^2 - \xi_1^2 + 2\eta(\xi_2 - \xi_1)\} = \frac{1}{2}(1 - 1 + 2E\{\eta\}E\{\xi_2 - \xi_1\}) = \frac{1}{2}(2 \cdot 0 \cdot (0 - 0)) = 0.$ Now, $Var\{Y_1\} = (\frac{\sqrt{2}}{2})^2(1 + 1 + 2^2 \cdot 1) = 3$, and similarly, $Var\{Y_1\} = (\frac{\sqrt{2}}{2})^2(1 + 1) = 1$.

Another way to compute the same is to use (2.1.5). □

---

[1]The matrix of the counterclockwise rotation by an angle φ is the matrix $\begin{Vmatrix} \cos\varphi & -\sin\varphi \\ \sin\varphi & \cos\varphi \end{Vmatrix}$. Here, we set $\varphi = -\pi/4$.

Let us view the r.v.'s X_i's as characteristics of some phenomenon. For example, the X's may represent the concentration of different chemicals in air, characterizing the air pollution situation in a randomly chosen area; or the X's can be numerical characteristics of the performance of different stocks in a market on a randomly chosen day.

As a rule, such characteristics are dependent, and it is reasonable to switch to, at least, uncorrelated (and, as we will see, independent in the normal case) characteristics Y_j.

Moreover, suppose that among the k standard deviations $\lambda_1, ..., \lambda_k$, the last m are equal to zero or, at least, small. Then the variables $Y_{k-m+1}, ..., Y_k$ are non-random (or almost non-random), and may be excluded from analysis. In particular, if we switch to the centered r.v.'s $Y_j' = Y_j - E\{Y_j\}$, then for $j = k-m+1, ..., k$, the variables $Y_j' = 0$ (or ≈ 0), and it suffices to consider only the first $l = k - m$ principal components.

For example, a detailed description of an air pollution situation may involve more than 50 characteristics, and it is very difficult to compare the situations in, say, two different areas comparing two collections of 50 values in each. However, research shows that, as a rule, the whole collection of the principal components may be reduced to a much smaller number of components, say five, which leads to essential *dimension reduction.*

Now, let us recall that for any orthogonal matrix $\mathcal{B} = \|b_{ij}\|$, its inverse $\mathcal{B}^{-1} = \mathcal{B}^T$, the transpose, and hence $\mathbf{X} = \mathcal{B}^T\mathbf{Y}$. Suppose that $\lambda_i = 0$ for all $i = l+1, ..., k$. Then Y_i are constants for $i = l+1, ..., k$, and for all $i = 1, ..., k$,

$$X_i = b_{1i}Y_1 + ... + b_{ki}Y_k = b_{1i}Y_1 + ... + b_{li}Y_l + c_i, \tag{2.2.2}$$

where the constant (!) $c_i = b_{l+1,i}Y_{l+1} + ... + b_{ki}Y_k$. As was told, l may be essentially less than k, and all the k variables X_i may be completely determined by a few variables $Y_1, ..., Y_l$. In this context, the principal components $Y_1, ..., Y_l$ are called *factors.*

In other words, we present correlated characteristics as linear combinations of uncorrelated factors. The weight coefficients $b_{1i}, ..., b_{li}$ are referred to as *factor loadings* or *loading coefficients*: b_{si} measures the impact of the factor Y_s on the characteristic X_i. We continue the discussion on loadings in Exercise 26.

EXAMPLE 2. Consider statistical data on the performance of five financial assets. Namely, we deal with the *daily returns* per $1 in 252 working days in 2010 for Microsoft (MSFT), AT&T, American Airlines (AMR), Walmart (WMT), and the market index Standard & Poor's (S&P). For the notion of return in more detail, see p.70. Calculations were provided with use of the MINITAB software.

Regarding the numbers below, one should keep in mind that *daily* returns X_i per $1 are r.v.'s close to one, and the standard deviations for such r.v.'s are small. Say, we will see below that the variance of the daily return for Microsoft is $Var\{X_1\} = 0.000191$, and hence, the standard deviation is ≈ 0.014, i.e., ≈ 1.4 cents per dollar, which is not small if we deal with one day.

For the whole vector of the returns, the estimate of the covariance matrix is

	MSFT	*AT&T*	*AMR*	*WMT*	*S&P*
MSFT	0.000191				
AT&T	0.000005	0.0000190			
AMR	0.000171	0.000016	0.001003		
WMT	0.000052	0.000002	0.000065	0.000078	
S&P	0.000114	0.000004	0.000209	0.000040	0.000128

(Since a covariance matrix is symmetric, we do not have to present the values above the main diagonal.)

The MINITAB has calculated the following estimates for the eigenvalues and the proportion of each eigenvalue with respect to the total sum of the eigenvalues:

Eigenvalue	0.0010991	0.0002114	0.0000588	0.0000328	0.0000184
Proportion	0.774	0.149	0.041	0.023	0.013
Cumulative	0.774	0.923	0.964	0.987	1.000

(For instance, the ratio of λ_2^2 to the total sum of the eigenvalues is 0.149, while the cumulative figure 0.923 is the ratio of $\lambda_1^2 + \lambda_2^2$ to the total sum.) We see that the first two eigenvalues constitute about 92% of the total sum, and the first three 96%. So, we may restrict ourselves even to the first two factors.

The matrix $\mathcal{B}$ (which is the matrix of factor loadings) is estimated by MINITAB as shown below ("PC" stands for a principal component):

Variable	*PC*1	*PC*2	*PC*3	*PC*4	*PC*5
MSFT	0.212	−0.769	0.449	−0.399	−0.048
AT&T	0.016	−0.010	−0.008	−0.102	0.995
AMR	0.945	0.313	0.009	−0.087	−0.021
WMT	0.082	−0.318	−0.883	−0.331	−0.045
S&P	0.232	−0.457	−0.134	0.844	0.077

If we restrict ourselves to the first two principal components, then we will characterize the situation regarding *all the five* assets by the *two* factors Y_1 and Y_2, and we will present X's as

$$\begin{aligned} X_1 &= 0.212Y_1 - 0.769Y_2 + c_1, \ X_2 = 0.016Y_1 - 0.0.010Y_2 + c_2, \\ X_3 &= 0.945Y_1 + 0.313Y_2 + c_3, \ X_4 = 0.082Y_1 - 0.318Y_2 + c_4, \\ X_5 &= 0.232Y_1 - 0.457Y_2 + c_5, \end{aligned}$$

where c_i's are close to constants and may be also estimated proceeding from the same data. The uncorrelated factors Y_1 are Y_2 are combinations of X's:

$$\begin{aligned} Y_1 = & \quad 0212X_1 + 0.016X_2 + 0.945X_3 + 0.082X_4 + 0.232X_5, \\ Y_2 = & - \ 0.769X_1 - 0.010X_2 + 0.313X_3 - 0.318X_4 - 0.457X_5. \ \square \end{aligned}$$

2.3 A model of portfolio optimization

The seminal theory we touch on in this section was developed by H. Markowitz in 1952 when he was just a graduate student. We consider it here as *a very good example of an application of covariance analysis.*

Consider a financial market consisting of k assets with random returns (i.e., incomes per one unit of investment) $X_1, ..., X_k$. We also suppose that there exists a risk-free asset with a non-random return $X_0 = q$. Say, if $q = 1.04$, then the risk-free interest of the asset is 4%. This may concern, for example, term investments with a fixed guaranteed rate $r = q - 1$.

An investor distributes a *unit* of money between the assets. Let $t_j,\ j = 0, 1, ...k$, be the amount invested in the jth asset,

$$t_0 + t_1 + ... + t_k = 1. \tag{2.3.1}$$

In general, t_i's may be negative. For $t_0 < 0$, this means that the investor borrows money with an obligation to return q units for each unit borrowed. For other assets, $t_i < 0$ means the so called short selling, when the investor borrows somewhere (usually, from a broker) shares of security i (rather than money) and immediately sells them with the obligation of buying identical securities later and returning them to the lender.

We set $\mathbf{t} = (t_1, ..., t_k)$, $\bar{\mathbf{t}} = (t_0, t_1, ..., t_k) = (t_0, \mathbf{t})$, and view all vectors as column vectors. We call a vector $\bar{\mathbf{t}}$ an (investment) *portfolio*. For any portfolio $\bar{\mathbf{t}}$, the cumulative return

$$R_{\bar{\mathbf{t}}} = t_0 X_0 + t_1 X_1 + ... + t_k X_k = t_0 q + \langle \mathbf{t}, \mathbf{X} \rangle, \tag{2.3.2}$$

where $\mathbf{X} = (X_1, ..., X_n)$.

Let $\mathbf{m} = (m_1, ..., m_k) = E\{\mathbf{X}\}$ and $C = \|c_{ij}\|$ be the covariance matrix of $\mathbf{X}$. In accordance with properties we established in Section 2.1,

$$E\{R_{\bar{\mathbf{t}}}\} = t_0 q + t_1 m_1 + ... + t_k m_k = t_0 q + \langle \mathbf{t}, \mathbf{m} \rangle, \text{ and} \tag{2.3.3}$$

$$Var\{R_{\bar{\mathbf{t}}}\} = \langle C\mathbf{t}, \mathbf{t} \rangle. \tag{2.3.4}$$

The investor wants the mean return $E\{R_{\bar{\mathbf{t}}}\}$ to be as large as possible, and—adopting variance as a measure of riskiness—the variance $Var\{R_{\bar{\mathbf{t}}}\}$ to be as small as possible. However, portfolios with large mean returns may have large variances of returns; so the choice of an optimal investment is a matter of the tradeoff between the two characteristics mentioned. We adopt the simplest linear criterion

$$V_{\bar{\mathbf{t}}} = \tau E\{R_{\bar{\mathbf{t}}}\} - Var\{R_{\bar{\mathbf{t}}}\},$$

where the weight coefficient τ reflects the attitude of the investor to risk and is called a *risk tolerance*: the larger the τ, to the larger extent the investor takes into account the mean return (and to the less extent, she/he tries to diminish riskiness). The parameter τ is an *individual characteristic* of the investor.

Our goal is to find a portfolio maximizing $V_{\bar{\mathbf{t}}}$.

By (2.3.3)–(2.3.4),

$$V_{\bar{\mathbf{t}}} = \tau(t_0 q + \langle \mathbf{t}, \mathbf{m} \rangle) - \langle C\mathbf{t}, \mathbf{t} \rangle.$$

To find $\bar{\mathbf{t}}$ maximizing $V_{\bar{\mathbf{t}}}$ under condition (2.3.1), we use the method of Lagrange multipliers and maximize the function $V_{\bar{\mathbf{t}}} + \lambda(t_0 + t_1 + ... + t_k - 1)$, where λ is a Lagrange multiplier. The differentiation with respect to t_0 gives

$$\tau q + \lambda = 0. \tag{2.3.5}$$

Recalling that the quadratic form $\langle C\mathbf{t}, \mathbf{t} \rangle = \sum_i \sum_j c_{ij} t_i t_j$, and differentiating with respect to t_i, $i \neq 0$, we have

$$\tau m_i - 2 \sum_{j=0}^{k} c_{ij} t_j + \lambda = 0, \quad i = 1, ..., k. \tag{2.3.6}$$

(The multiplier "2" before the sum in the second term appears because we differentiate $c_{ij}t_it_j$ and $c_{ji}t_jt_i$ as well.) The k equations (2.3.6) may be rewritten in the vector form

$$\tau\mathbf{m} - 2C\mathbf{t} + \lambda\mathbf{e} = \mathbf{0}, \tag{2.3.7}$$

where $\mathbf{0}$ is a zero vector, and $\mathbf{e} = (1, ..., 1)$. (We view all vectors as column ones.)

From (2.3.5), it follows that $\lambda = -\tau q$. Substituting this into (2.3.7), it is easy to verify that, if C is invertible, a solution to this equation is the vector

$$\mathbf{t}^* = \frac{1}{2}\tau C^{-1}(\mathbf{m} - q\mathbf{e}) = \frac{1}{2}\tau C^{-1}\mathbf{p}, \tag{2.3.8}$$

where the vector $\mathbf{p} = (m_1 - q, ..., m_k - q)$ is called a risk-premium vector.

The last term may be clarified as follows. It is reasonable to assume that for any risky asset i, the mean return $m_i > q$, since otherwise it would not make any sense to invest in such an asset. Then the difference $m_i - q$ may be interpreted as a "premium" for risk incurred: *on the average*, the investor will gain more in comparison with the non-risky investment, but there is a risk to lose money.

We see that the larger the risk tolerance τ, the larger the investment portion that goes to the risky assets.

Certainly, τ has been introduced *merely for modeling purposes*: the investor "does not know her/his τ," and as a matter of fact, the optimal investment may run as follows.

In accordance with her/his attitude to risk, the investor chooses somehow t_0^*, the amount to be invested into the non-risky asset. The rest amount, $1 - t_0^*$, is distributed among the risky assets in accordance with the solution $\mathbf{t}^*$. A noteworthy feature is that, once t_0^* has been chosen, *the principle of the allocation of the remaining amount between the risky assets does not depend on τ at all, and is the same for all investors.*

Indeed, let $\mathbf{t}_{\text{opt}} = (t_{1,\text{opt}}, ..., t_{k,\text{opt}})$, where

$$t_{i,\text{opt}} = \frac{t_i^*}{t_1^* + ... + t_k^*}, \quad i = 1, ..., k, \tag{2.3.9}$$

the proportion of the total investment into the risky assets that goes to the asset i. Again, the vector $\mathbf{t}_{\text{opt}}$ concerns only the distribution of the amount that *has been already assigned to the risky assets*, and the term "optimal" concerns only the investment allocation among the risky assets.

If we set $\mathbf{z} = (z_1, ..., z_k) = C^{-1}\mathbf{p}$, then from (2.3.8) it will follow that $t_i^* = \frac{1}{2}\tau z_i$. After inserting this into (2.3.9), $\frac{1}{2}\tau$ cancels out, so

$$t_{i,\text{opt}} = \frac{z_i}{z_1 + ... + z_k} \tag{2.3.10}$$

and, hence, does not depend on τ.

EXAMPLE 1 serves merely for illustrative purposes. Suppose $k = 2$, $m_1 = 1.14$, $m_2 = 1.08$, and $q = 1.06$; that is, on the average, the risky assets give 14% and 8% of profit, respectively, while for the risk-free asset, the profit constitutes 6%. Suppose the standard deviations $\sigma_1 = 0.2$ and $\sigma_2 = 0.15$. So, the assets are pretty risky; say, the first gives, on the average, 14 ± 20 percent of profit. Let $\rho = Corr\{X_1, X_2\} = 0.5$.

We have $\mathbf{p} = (1.14 - 1.06,\, 1.08 - 1.06) = (0.08,\, 0.02)$, and the covariance matrix

$$\mathcal{C} = \left\| \begin{array}{cc} (0.2)^2 & 0.5 \cdot 0.2 \cdot 0.15 \\ 0.5 \cdot 0.2 \cdot 0.15 & (0.15)^2 \end{array} \right\| = \left\| \begin{array}{cc} 0.04 & 0.015 \\ 0.015 & 0.0225 \end{array} \right\|.$$

It is easy to calculate using any appropriate software, or even manually, that

$$\mathbf{z} = \mathcal{C}^{-1}\mathbf{p} \approx \left\| \begin{array}{c} 2.22 \\ -0.59 \end{array} \right\|, \text{ and } \mathbf{t}_{\text{opt}} = \frac{1}{z_1 + z_2} \left\| \begin{array}{c} z_1 \\ z_2 \end{array} \right\| \approx \left\| \begin{array}{c} 1.36 \\ -0.36 \end{array} \right\|.$$

Suppose that an investor is going to invest \$10,000 which we view as a unit of money. The investor has decided to deposit \$7,000 into the risk-free account with the 6% growth. Then, the allocation of the rest \$3,000 will be given by the vector

$$3000 \cdot \mathbf{t}_{\text{opt}} = (3000 \cdot 1.36, 3000 \cdot (-0.36)) = (4080, -1080).$$

This means the following. The investor borrows shares of the second security for \$1,080 and immediately sells them for this price. After that, she/he adds this money to \$3,000, and buys shares of the first security for \$4,080. □

Let us come back to the general model. Skipping calculations, we present here a summary of results.

Suppose for a while that we invested the whole unit of money only in the risky assets in accordance with a portfolio $\mathbf{t}$ for which now $\sum_{i=1}^{k} t_i = 1$. By (2.3.3)–(2.3.4), the mean and variance of the return in this case will be equal to $m_{\mathbf{t}} = \langle \mathbf{t}, \mathbf{m} \rangle$ and $\sigma_{\mathbf{t}}^2 = \langle \mathcal{C}\mathbf{t}, \mathbf{t} \rangle$, respectively. It may be shown, that in the diagram (σ, m), all points $(\sigma_{\mathbf{t}}, m_{\mathbf{t}})$ corresponding to *all possible* portfolios $\mathbf{t}$ constitute a region enclosed by a hyperbolic curve as it is depicted in Fig. 3 (the "curve with dots"). The upper part of this curve is called an *efficient frontier* and designated so in Fig. 3.

Portfolios corresponding to points in the efficient frontier (and points themselves) are called efficient; the term is connected with the following. The points representing *all* possible portfolios lie either in the efficient frontier or below it. For any point below the frontier, the point in the efficient frontier with the same σ is "better": for the latter point, σ is the same, while m is larger. The very left point in the efficient frontier corresponds to the minimal standard deviation; in Fig. 3, it is designated as "the minimal st.deviation portfolio."

Let the portfolio $\bar{\mathbf{t}}_{\text{opt}} = (0, \mathbf{t}_{\text{opt}})$ for which all the money go to risky assets, and are distributed in accordance with (2.3.10). Denote by R_* its (random) return, and set $m_* = E\{R_*\}$ and $\sigma_*^2 = Var\{R_*\}$. In Fig. 3, the corresponding point in the efficient frontier is designated as "the optimal portfolio."

Now, suppose that the investor invested t_0 into the risk-free asset and allocated the remaining amount $1 - t_0$ in accordance with the solution $\mathbf{t}_{\text{opt}}$. Then the return $R = t_0 q + (1 - t_0)R_*$, and for $m = E\{R\}$ and $\sigma^2 = Var\{R\}$, we have

$$m = t_0 q + (1 - t_0)m_*, \quad \sigma^2 = (1 - t_0)^2 \sigma_*^2. \tag{2.3.11}$$

Because $t_0 \leq 1$, from the last equation it follows that $\sigma = (1 - t_0)\sigma_*$, and hence, $t_0 = 1 - \sigma/\sigma_*$. Inserting it into the first equation in (2.3.11), we readily come to the equation of what is called a *capital market line* or *a trade-off line*:

$$m = q + \frac{m_* - q}{\sigma_*} \sigma. \tag{2.3.12}$$

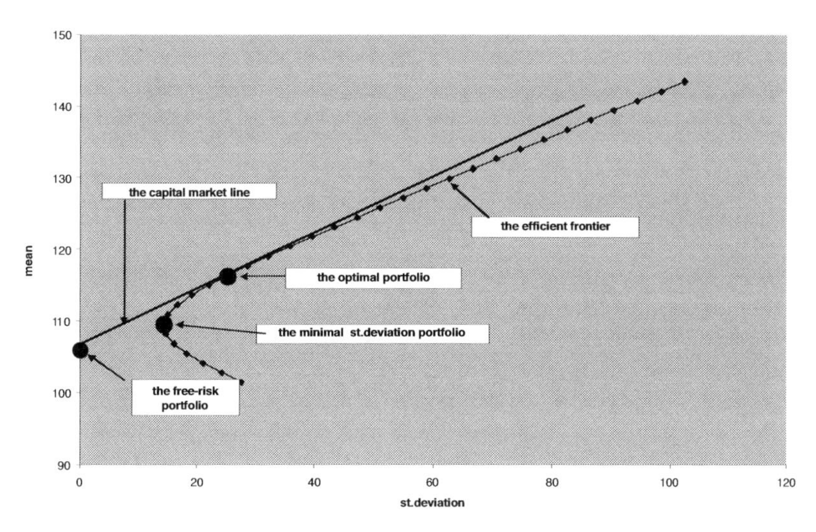

FIGURE 3. Particular curves and points in this figure correspond to the data from Example 1. For convenience, the values are given in percents.

It may be proved that this line is tangent to the efficient frontier at the point (σ_*, m_*).

In Fig. 3, this line starts from the point designated as "the risk-free portfolio" (it corresponds to the case $t_0 = 1$) and is tangent to the efficient frontier at the point corresponding to the "optimal portfolio."

Let us comment the qualitative picture using Fig. 3. If a risk-free asset had not existed, then the investor would have had to invest only in risky securities and to choose a portfolio corresponding to a point in the efficient frontier. A choice of this point would have depended on the investor's risk tolerance τ.

The presence of a risk-free asset dramatically changes the situation. Consider now points optimal with respect to the whole portfolio, including the investment in the risk-free asset. These points lie in the capital market line, and the choice of an investment strategy is a tradeoff between *only two fixed* portfolios:

(1) The non-risky portfolio corresponding to $t_0 = 1$ and the very left point in the line.

(2) The portfolio consisting of only risky assets ($t_0 = 0$) and corresponding to the unique vector $\mathbf{t}_{\text{opt}}$ and the point where the capital market line touches the frontier.

The intermediate points in the capital market line between the two mentioned points correspond to the case $0 < t_0 < 1$. Say, the middle point corresponds to $t_0 = 1/2$.

The points in the trade-off line above the tangency point correspond to the case $t_0 < 0$ where the investor borrows money and buys securities for the amount larger than she/he originally had.

3 THE MULTIVARIATE NORMAL DISTRIBUTION

3.1 Definitions and properties

As above, all vectors below are viewed as column vectors, though for the sake of convenience, we write down their coordinates as a row.

A r.vec. $\mathbf{X} = (X_1, ..., X_k)$ and its distribution are called *standard multivariate normal* if its coordinates are independent standard normal r.v.'s. Let $\varphi(x) = \frac{1}{\sqrt{2\pi}}\exp\{-x^2/2\}$, the standard univariate normal density, and as usual, let $\boldsymbol{x}$ stand for a vector $(x_1, ..., x_k)$. Then, in view of the independency of the coordinates, the joint density of $\mathbf{X}$ is

$$f_{\mathbf{X}}(\boldsymbol{x}) = \prod_{i=1}^{k}\varphi(x_i) = \prod_{i=1}^{k}\frac{1}{\sqrt{2\pi}}e^{x_i^2/2} = \frac{1}{(2\pi)^{k/2}}\exp\left\{-\frac{1}{2}\sum_{i=1}^{k}x_i^2\right\} = \frac{1}{(2\pi)^{k/2}}\exp\left\{-\frac{1}{2}|\boldsymbol{x}|^2\right\}, \tag{3.1.1}$$

where $|\boldsymbol{x}| = \sqrt{x_1^2 + ... + x_k^2}$ is the length of $\boldsymbol{x}$.

Since $E\{X_i\} = 0$, the mean $E\{\mathbf{X}\} = \mathbf{0}$. Because $Var\{X_i\} = 1$ for all $i = 1, ..., k$, and X_i's are independent, the covariance matrix $\mathcal{C}_{\mathbf{X}} = \mathcal{I}$, a $k \times k$-identity matrix.

Now, let the r.vec. $\mathbf{Y} = \mathbf{m} + \mathcal{B}\mathbf{X}$, where $\mathbf{m}$ is a non-random vector and $\mathcal{B}$ is an arbitrary (not necessary orthogonal) non-random invertible $k \times k$-matrix. We have

$$E\{\mathbf{Y}\} = \mathbf{m} + \mathcal{B}E\{\mathbf{X}\} = \mathbf{m} + \mathcal{B}\mathbf{0} = \mathbf{m}, \tag{3.1.2}$$

and the covariance matrix

$$\mathcal{C}_{\mathbf{Y}} = \mathcal{B}\mathcal{I}\mathcal{B}^T = \mathcal{B}\mathcal{B}^T = \mathcal{C} \tag{3.1.3}$$

if we denote $\mathcal{B}\mathcal{B}^T$ by $\mathcal{C}$.

We call the r.vec $\mathbf{Y}$ and its distribution normal with a mean of $\mathbf{m}$ and a covariance matrix of $\mathcal{C}$. Let us find the density of $\mathbf{Y}$.

For any set B in $\mathbb{R}^k$, we have $P(\mathbf{Y} \in B) = P(\mathbf{m} + \mathcal{B}\mathbf{X} \in B) = \int\cdots\int_{\boldsymbol{x}:\,\mathbf{m}+\mathcal{B}\boldsymbol{x}\in B} f_{\mathbf{X}}(\boldsymbol{x})d\boldsymbol{x}$. Using the variable change $\mathbf{y} = \mathbf{m} + \mathcal{B}\boldsymbol{x}$ and following the Calculus rules on how such a change should be carried out, we have $P(\mathbf{Y} \in B) = \frac{1}{|\det\mathcal{B}|}\int\cdots\int_B f_{\mathbf{X}}(\mathcal{B}^{-1}(\mathbf{y} - \mathbf{m}))d\mathbf{y}$.

Thus, for any set B, the probability $P(\mathbf{Y} \in B) = \int\cdots\int_B f_{\mathbf{Y}}(\mathbf{y})d\mathbf{y}$, where

$$f_{\mathbf{Y}}(\mathbf{y}) = \frac{1}{|\det\mathcal{B}|}f_{\mathbf{X}}(\mathcal{B}^{-1}(\mathbf{y} - \mathbf{m})). \tag{3.1.4}$$

Hence, by definition, the function (3.1.4) is the density of $\mathbf{Y}$.

The formula above is a general formula: it is valid regardless of which distribution $\mathbf{X}$ has. For the normal distribution, in accordance with (3.1.1),

$$f_{\mathbf{Y}}(\mathbf{y}) = \frac{1}{(2\pi)^{k/2}|\det\mathcal{B}|}\exp\left\{-\frac{1}{2}|\mathcal{B}^{-1}(\mathbf{y} - \mathbf{m})|^2\right\}. \tag{3.1.5}$$

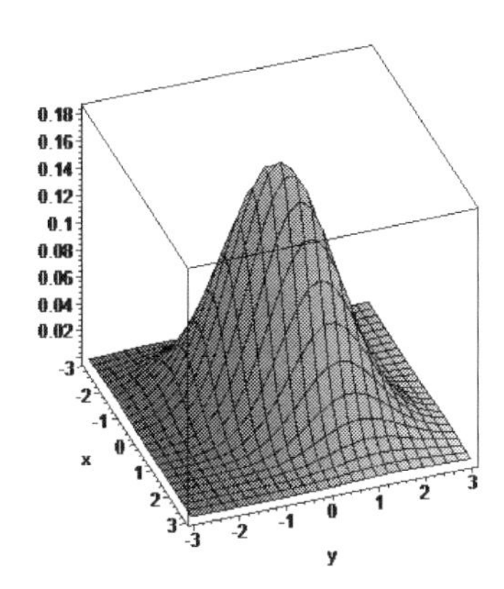

FIGURE 4.

Now, $\det \mathcal{C} = \det(\mathcal{B}\mathcal{B}^T) = (\det \mathcal{B})^2$, and hence, $|\det \mathcal{B}| = \sqrt{\det \mathcal{C}}$. The matrix $\mathcal{C}$ is nonnegative definite and even positive definite since we assumed $\mathcal{B}$, and hence $\mathcal{C}$, to be invertible. Consequently, $\det \mathcal{C} > 0$.

Furthermore, for any vector $\mathbf{z}$,

$$|\mathcal{B}^{-1}\mathbf{z}|^2 = \langle \mathcal{B}^{-1}\mathbf{z}, \mathcal{B}^{-1}\mathbf{z}\rangle = \langle (\mathcal{B}^{-1})^T \mathcal{B}^{-1}\mathbf{z}, \mathbf{z}\rangle = \langle (\mathcal{B}\mathcal{B}^T)^{-1}\mathbf{z}, \mathbf{z}\rangle = \langle \mathcal{C}^{-1}\mathbf{z}, \mathbf{z}\rangle.$$

Inserting all of this into (3.1.5), we have

$$f_{\mathbf{Y}}(\mathbf{y}) = \frac{1}{(2\pi)^{k/2}\sqrt{\det \mathcal{C}}} \exp\left\{-\frac{1}{2}\langle \mathcal{C}^{-1}(\mathbf{y}-\mathbf{m}), \mathbf{y}-\mathbf{m}\rangle\right\}. \tag{3.1.6}$$

Let us consider the particular case $k = 2$ and $\mathbf{m} = \mathbf{0}$. Keeping the notation of the previous section, we write $\mathcal{C} = \left\| \begin{matrix} \sigma_1^2 & \rho\sigma_1\sigma_2 \\ \rho\sigma_1\sigma_2 & \sigma_2^2 \end{matrix} \right\|$. The determinant $\det \mathcal{C} = \sigma_1^2\sigma_2^2(1-\rho^2) \neq 0$ if and only if $|\rho| < 1$. Under this condition, computing $\mathcal{C}^{-1}$, one will readily obtain

$$f_{\mathbf{Y}}(\mathbf{y}) = \frac{1}{2\pi\sigma_1\sigma_2\sqrt{1-\rho^2}} \exp\left\{-\frac{1}{2(1-\rho^2)}\left(\frac{y_1^2}{\sigma_1^2} - 2\rho\frac{y_1y_2}{\sigma_1\sigma_2} + \frac{y_2^2}{\sigma_2^2}\right)\right\}. \tag{3.1.7}$$

Fig. 4 presents a graph for the case $\rho = 0.5$, $\sigma_1 = \sigma_2 = 1$, and $y_1, y_2 \in [-3, 3]$.

Now, we come back to the general case of $\mathbf{Y} = (Y_1, ..., Y_k)$ having the density (3.1.6).

Theorem 4 *A. Any linear combination $\alpha_1 Y_1 + ... + \alpha_k Y_k$ is normal (and, hence, in particular, this is true for the marginal distributions of Y_i's).*

B. If the r.v.'s $Y_1, ..., Y_k$ are pairwise uncorrelated, then they are mutually independent.

C. Let $E\{\mathbf{Y}\} = \mathbf{0}$ and $Y_1, ..., Y_k$ be i.i.d. r.v.'s. Then the same is true in any orthonormal basis. In other words, the coordinates of the r.vec. $\mathbf{Z} = Q\mathbf{Y}$ are also i.i.d., provided that Q is orthogonal.

We prove it at the end of this section.

In particular, from Assertion B, it follows that

> In the normal case, the principal components method deals with independent factors.

Property C is a characteristic property: as a matter of fact, the normal distribution is the only distribution that preserves the independence of the coordinates with respect to orthogonal transformations, for example, rotations. This is stated in

Theorem 5 *Let the coordinates of a r.vec.* $\mathbf{Y} = (Y_1, ..., Y_k)$ *be independent r.v.'s,* $E\{\mathbf{Y}\} = \mathbf{0}$*, and suppose that for any orthogonal transformation of* $\mathbf{Y}$*, the coordinates of the resulting r.vec.* $\mathbf{Z} = (Z_1, ..., Z_k)$ *are also independent. Then* $Y_1, ..., Y_k$ *have the normal distribution with equal variances, and the same is true for* $Z_1, ..., Z_k$.

EXAMPLE 1 (*The velocity of a molecule in a gas*). Theorem 5 has a long history traced back to James Maxwell who, above other outstanding achievements in Physics, contributed to the *kinetic theory of gases.* A randomly chosen molecule of a gas is moving with a random velocity, and one of the Mechanics principles requires the coordinates of the velocity vector to be independent. If we do not take into account exogenous factors like gravitation or wind, then all directions in the space have "equal significance"; or in other words, no orthogonal basis has an advantage over others. Then, the independence of the coordinates should be true regardless of the basis chosen, which drives us to a remarkable conclusion: by virtue of Theorem 5, the distribution of the velocity vector is normal with independent coordinates having equal variances. Note also that in this case, the mean velocity vector $E\{\mathbf{Y}\} = \mathbf{0}$. Indeed, if we exclude exogenous factors, for any direction, the molecule equally likely moves in this direction or in the opposite. □

A proof of Theorem 5 may be found, e.g., in [13].

Proof of Theorem 4. Without a loss of generality, we can set $E\{\mathbf{Y}\} = \mathbf{0}$. Then $\mathbf{Y} = \mathcal{B}\mathbf{X}$, where $\mathbf{X}$ is a standard normal vector, and $\mathcal{B}$ is a matrix.

A. Let $\boldsymbol{\alpha} = (\alpha_1, ..., \alpha_k)$. Then $\alpha_1 Y_1 + ... + \alpha_k Y_k = \langle \boldsymbol{\alpha}, \mathbf{Y} \rangle = \langle \boldsymbol{\alpha}, \mathcal{B}\mathbf{X} \rangle = \langle \mathcal{B}^T \boldsymbol{\alpha}, \mathbf{X} \rangle$. Let $\beta_1, ..., \beta_k$ be the coordinates of the vector $\mathcal{B}^T \boldsymbol{\alpha}$. Then $\alpha_1 Y_1 + ... + \alpha_k Y_k = \beta_1 X_1 + ... + \beta_k X_k$ which has a normal distribution because X_i's are normal and independent.

B. Let $Y_1, ..., Y_k$ be uncorrelated, $\sigma_i^2 = Var\{Y_i\}$. Then

$$\mathcal{C} = \left\| \begin{array}{cccc} \sigma_1^2 & 0 & \dots & 0 \\ 0 & \sigma_2^2 & \dots & 0 \\ \vdots & \vdots & \vdots & \vdots \\ 0 & 0 & \dots & \sigma_k^2 \end{array} \right\|, \text{ and } \mathcal{C}^{-1} = \left\| \begin{array}{cccc} \sigma_1^{-2} & 0 & \dots & 0 \\ 0 & \sigma_2^{-2} & \dots & 0 \\ \vdots & \vdots & \vdots & \vdots \\ 0 & 0 & \dots & \sigma_k^{-2} \end{array} \right\|.$$

Then $\det\mathcal{C} = \sigma_1^2 \sigma_2^2 \cdot ... \cdot \sigma_k^2$, and the quadratic form

$$\langle \mathcal{C}^{-1}\mathbf{y}, \mathbf{y} \rangle = \frac{y_1^2}{\sigma_1^2} + \frac{y_2^2}{\sigma_2^2} + ... + \frac{y_k^2}{\sigma_k^2}.$$

Inserting this into (3.1.6), we get

$$\begin{aligned} f_{\mathbf{Y}}(\mathbf{y}) &= \frac{1}{(2\pi)^{k/2}\sigma_1\sigma_2\cdot\ldots\cdot\sigma_k}\exp\left\{-\frac{1}{2}\left(\frac{y_1^2}{\sigma_1^2}+\frac{y_2^2}{\sigma_2^2}+\ldots+\frac{y_k^2}{\sigma_k^2}\right)\right\} \\ &= \prod_{i=1}^{k}\frac{1}{\sqrt{2\pi}\sigma_i}\exp\left\{-\frac{y_i^2}{2\sigma_i^2}\right\} = \prod_{i=1}^{k} f_{Y_i}(y_i), \end{aligned}$$

which implies independence.

C. Suppose the coordinates $Y_1,...,Y_k$ are i.i.d., and $\sigma^2 = Var\{Y_i\}$; the same for all i's. Then the covariance matrix $\mathcal{C}$ is diagonal with the σ^2 in all positions of the diagonal. Then $\mathcal{C} = \sigma^2\mathcal{I}$, where $\mathcal{I}$ is the identity matrix. Then we may use the representation $\mathbf{Y} = \sigma\mathbf{X}$, where $\mathbf{X}$ is a standard normal vector.

Furthermore, let $\mathcal{Q}$ be an orthogonal matrix. Then the vector $\mathbf{Z} = \mathcal{Q}\mathbf{Y} = \mathcal{Q}(\sigma\mathbf{X}) = \sigma\mathcal{Q}\mathbf{X}$. Then the covariance matrix $\mathcal{C}_{\mathbf{Z}} = (\sigma\mathcal{Q})\mathcal{I}(\sigma\mathcal{Q})^T = \sigma^2\mathcal{Q}\mathcal{Q}^T = \sigma^2\mathcal{Q}\mathcal{Q}^{-1} = \sigma^2\mathcal{I} = \mathcal{C}_{\mathbf{Y}}$. Thus, $\mathbf{Z}$ and $\mathbf{Y}$ have the same distribution. ■

3.2 The χ^2-distribution

Let $\mathbf{X} = (X_1,...,X_k)$ be a standard normal r.vec. Our next object of interest is the square of the length of $\mathbf{X}$; more precisely, $|\mathbf{X}|^2 = X_1^2 + ... + X_k^2$. There is a special notation for this r.v., namely, χ_k^2, and such a r.v. is called a *chi-squared* r.v., or a χ^2-r.v., with *k degrees of freedom*. The same term is applied to the distribution of χ_k^2.

Note right away that since X_i's are standard normal, $E\{X_i^2\} = Var\{X_i\} = 1$, and hence,

$$E\{\chi_k^2\} = E\{|\mathbf{X}|^2\} = E\{X_1^2\} + ... + E\{X_k^2\} = k. \tag{3.2.1}$$

EXAMPLE 1 (*What is temperature?*). Let $\mathbf{Y} = (Y_1,Y_2,Y_3)$ be the velocity vector of a gas molecule chosen at random. In the previous subsection, we justified the model where all Y_i's are independent normal r.v.'s with the same variance, say, σ^2. By the symmetry argument (regarding the gas properties, no direction in the gas differs from the others), we assume $E\{Y_i\} = 0$. Then $\mathbf{Y} = \sigma\mathbf{X}$, where $\mathbf{X} = (X_1,X_2,X_3)$ is a standard normal vector.

The speed of the molecule is $|\mathbf{Y}| = \sigma|\mathbf{X}|$, and its kinetic energy is the r.v. $T = \dfrac{\mu|\mathbf{Y}|^2}{2} = \dfrac{\mu\sigma^2|\mathbf{X}|^2}{2}$, where μ is the mass of the molecule. Then $T = \dfrac{\mu\sigma^2}{2}\chi_3^2$; that is, up to a constant multiplier, T is a χ^2-variable with three degrees of freedom. In view of (3.2.1), the mean energy is $E\{T\} = \frac{\mu\sigma^2}{2}\cdot 3 = \frac{3\mu\sigma^2}{2}$. Up to units of measurement, this is exactly what is called *temperature* in Physics. □

Let us derive the χ_k^2-density.

Lemma 6 *Let a r.v. X be standard normal, and $Y = X^2$. Then the density of Y is $f_{\frac{1}{2},\frac{1}{2}}(y)$, the Γ-density with parameters $(\frac{1}{2},\frac{1}{2})$.*

Proof. Let, as usual, $\Phi(x)$ be the standard normal d.f. For $y \geq 0$, the d.f.

$$\begin{aligned} F_Y(y) &= P(Y \leq y) = P(X^2 \leq y) = P(-\sqrt{y} \leq X \leq \sqrt{y}) = \Phi(\sqrt{y}) - \Phi(-\sqrt{y}) \\ &= 2\Phi(\sqrt{y}) - 1 \end{aligned}$$

because $\Phi(-x) = 1 - \Phi(x)$. Then the density

$$f_Y(y) = \frac{d}{dy}F_Y(y) = 2\frac{1}{2\sqrt{y}}\Phi'(\sqrt{y}) = \frac{1}{\sqrt{y}}\frac{1}{\sqrt{2\pi}}e^{-y/2} = \frac{1}{\sqrt{\pi}}\left(\frac{1}{2}\right)^{1/2} y^{\frac{1}{2}-1}e^{-y/2}. \quad (3.2.2)$$

The r.-h.s. is a density, and the part depending on y is that of the Γ-density $f_{\frac{1}{2},\frac{1}{2}}(y)$. Then the r.-h.s. must be $f_{\frac{1}{2},\frac{1}{2}}(y) = \frac{1}{2^{1/2}\Gamma(1/2)}x^{\frac{1}{2}-1}e^{-x/2}$.

(So, on the way, we have proved that $\Gamma(1/2) = \sqrt{\pi}$.) ■

Thus, $\chi_k^2 = |\mathbf{X}|^2 = X_1^2 + ... + X_k^2$ is the sum of k independent r.v.'s with the density $f_{\frac{1}{2},\frac{1}{2}}$. As a convolution of Γ-densities (see Proposition **6**.6), the r.v. χ_k^2 has the Γ-density with parameters $(\frac{1}{2}, \frac{k}{2})$. More precisely,

$$f_{\chi_k^2}(x) = \frac{1}{2^{k/2}\Gamma(k/2)}x^{\frac{k}{2}-1}e^{-x/2}. \quad (3.2.3)$$

The EXCEL command for $P(\chi_k^2 > x)$ is =CHIDIST(x,k).

Now, consider $|\mathbf{X}|$. This r.v. and its distribution have the special notation χ_k, and its distribution is called the χ_k-distribution. Clearly, if we know the distribution of χ_k^2, we can find the distribution of χ_k, just writing

$$P(\chi_k \le x) = P(\chi_k^2 \le x^2).$$

In Exercises 35 and 36, the reader is invited to consider it in detail.

EXAMPLE 2 (*A random determinant*). Let an $n \times n$-matrix $\mathcal{A} = \|a_{ij}\|$, where a_{ij} are i.i.d. r.v.'s, and let $D = D_n = \det \mathcal{A}$. The problem of finding the distribution of the r.v. D_n is quite difficult even if we restrict ourselves to the asymptotic behavior for large n. For now, solutions are known only under some essential restrictions (see, e.g., [17]). However, in the case where a_{ij} are standard normal, the solution is explicit and elegant.

For a while and merely for simplicity, suppose that $n = 3$, and denote by $\boldsymbol{a}_1, \boldsymbol{a}_2, \boldsymbol{a}_3$ the three column-vectors of the matrix $\mathcal{A}$. We view them as random vectors in $\mathbb{R}^3$ emanating from the origin. As we know from Linear Algebra, the absolute value $|D|$ is the volume of the parallelepiped based on $\boldsymbol{a}_1, \boldsymbol{a}_2, \boldsymbol{a}_3$. Let us "compute" it.

The length of $\boldsymbol{a}_3$ has the χ_3-distribution. Let us keep χ_3 as the notation for this length.

Now, consider the (random) plane that goes through the origin and is perpendicular to $\boldsymbol{a}_3$, and consider the projection of $\boldsymbol{a}_2$ on this plane. The point is that a standard normal r.vec. has independent standard normal coordinates in any orthogonal basis (see Theorem 4-C). Therefore, the projection of such a vector on any plane has the two-dimensional standard normal distribution whatever the plane we have chosen. Hence, the projection of $\boldsymbol{a}_2$ on the (random) plane perpendicular to $\boldsymbol{a}_3$ has the two-dimensional standard normal distribution. The length of the projection mentioned does not depend on $\boldsymbol{a}_3$ and has the χ_2-distribution. We keep the same symbol for the length itself.

Now, we consider the line that goes through the origin and is perpendicular to both vectors $\boldsymbol{a}_2, \boldsymbol{a}_3$, and consider the projection of $\boldsymbol{a}_1$ on this line. Similarly to the above reasoning, we conclude that the length of this projection does not depend on $\boldsymbol{a}_2, \boldsymbol{a}_2$, and its length has the χ_1-distribution.

Thus, the volume of the parallelepiped may be presented as the product $\chi_1\chi_2\chi_3$, where the factors are *independent* r.v.'s.

Absolutely analogously, in the general n-dimensional case, we arrive at the presentation $|D_n| = \chi_1 \cdot ... \cdot \chi_n$ or, which is more convenient,

$$D_n^2 = \chi_1^2 \cdot ... \cdot \chi_n^2, \tag{3.2.4}$$

where $\chi_1^2, ..., \chi_n^2$ are independent r.v.'s with the corresponding distributions. Note that, since $E\{\chi_k^2\} = k$, from (3.2.4) and the independency of χ_k's, it follows that

$$E\{D_n^2\} = 1 \cdot ... \cdot n = n!.$$

► We may continue. Since $\ln(D_n^2) = \ln(\chi_1^2) + ... + \ln(\chi^2)$, the sum of independent r.v.'s, one may conjecture that $\ln(D_n^2)$ is asymptotically normal for large n by the CLT. This is indeed true but a direct verification (for example, with use of Lyapunov's theorem (Corollary **9**-5)) is cumbersome. A better non-direct way consists in writing

$$\ln(D_n^2) - \ln(n!) = \ln\left(\frac{D_n^2}{n!}\right) = \sum_{k=1}^{n} \ln\left(\frac{\chi_k^2}{k}\right) = \sum_{k=1}^{n} \ln\left(1 + \frac{\chi_k^2}{k} - 1\right) = \sum_{k=1}^{n} \ln(1 + \xi_k), \tag{3.2.5}$$

where $\xi_k = (\chi_k^2/k) - 1$.

Let us recall the representation $\frac{1}{k}\chi_k^2 = \frac{1}{k}(X_1^2 + ... + X_k^2)$, where X_i are independent and standard normal. Thus, by the LLN, $\frac{1}{k}\chi_k^2 \xrightarrow{P} 1$ as $k \to \infty$; and hence, $\xi_k \xrightarrow{P} 0$ as $k \to \infty$.

So, ξ_k is small for large k, and we can use Taylor's expansion for the logarithms (the Appendix, (2.2.8)). More precisely, proceeding from (3.2.5), we may write

$$\ln(D_n^2) - \ln(n!) = \sum_{k=1}^{n}\left(\xi_k - \frac{1}{2}\xi_k^2 + o(\xi_k^2)\right) = \sum_{k=1}^{n}\xi_k - \frac{1}{2}\sum_{k=1}^{n}\xi_k^2 + R_n, \tag{3.2.6}$$

where $o(\xi_k^2)$ is the remainder in the kth term, and R_n denotes the total remainder. The advantage of this approach is in the fact that it is much easier to calculate the variance and other moments of ξ_k than these characteristics for $\ln(1 + \xi_k)$. The rest consists in the following.

We should calculate the quantity $B_n^2 = \sum_{k=1}^{n} Var\{\xi_k\}$ (it may be shown that $B_n^2 \sim 2\ln n$), and taking into account that $E\{\xi_k\} = 0$, consider the normalized expression

$$\frac{1}{B_n}\left(\ln(D_n^2) - \ln(n!)\right) = \frac{1}{B_n}\sum_{k=1}^{n}\xi_k - \frac{1}{2B_n}\sum_{k=1}^{n}\xi_k^2 + \frac{R_n}{B_n}. \tag{3.2.7}$$

In Exercise 37, the reader is invited to show that the first sum is asymptotically standard normal, the second is close to a growing constant and the term $\frac{R_n}{B_n}$ converges to zero. ◄ □

3.3 The multivariate central limit theorem

Similarly to the one-dimensional results in Chapter 9, the distribution of the sum of independent random vectors (r.vec.'s) may be well approximated by a multivariate normal distribution if the number of the terms in the sum is sufficiently large. This may be stated as follows.

Let $\mathbf{X}_1, \mathbf{X}_2, ...$ be a sequence of independent k-dimensional identically distributed r.vec.'s. Denote by $\mathcal{C}$ their (common) covariance matrix, and set, for simplicity, $E\{\mathbf{X}_i\} = \mathbf{0}$, a zero vector. (If it is not so, we will consider the centered vectors $\mathbf{X}_i - E\{\mathbf{X}_i\}$.)

Let $\mathbf{S}_n = \sum_{i=1}^{n} \mathbf{X}_i$. In Chapter 9, in the case of the one-dimensional CLT, we normalized the sum dividing it by its standard deviation. We will see that in the multidimensional case, it is more convenient to divide the sum just by $\sqrt{n}$ and consider the r.vec. $\overline{\mathbf{S}}_n = \frac{1}{\sqrt{n}} \mathbf{S}_n$. Later, we will consider a way of normalization that is similar to what we did in the one-dimensional case.

Let $\mathbf{Y}$ be a normal vector with zero mean and the covariance matrix $\mathcal{C}$, and let $\Phi_{\mathcal{C}}(B) = P(\mathbf{Y} \in B)$, where B is a set in the k-dimensional Euclidean space $\mathbb{R}^k$. We call the function of sets $\Phi_{\mathcal{C}}(B)$ the normal distribution with zero mean and a covariance matrix $\mathcal{C}$.

In Chapter 9, when stating and proving the CLT, we considered the probabilities of intervals. In the multidimensional case, the role of intervals is played by convex sets.[2]

Theorem 7 *For any convex set $B \subseteq \mathbb{R}^k$,*

$$P(\overline{\mathbf{S}}_n \in B) \to \Phi_{\mathcal{C}}(B). \tag{3.3.1}$$

EXAMPLE 1. Consider the two-dimensional case of $k = 2$, and suppose all $\mathbf{X}_i$'s to be uniformly distributed on the unit disk with the center at the origin. Consider a separate term $\mathbf{X}$ skipping the index, and set $\mathbf{X} = (X_1, X_2)$, where X_1, X_2 are the coordinates of $\mathbf{X}$. The joint density $f(x_1, x_2) = 1/\pi$ if $x_1^2 + x_2^2 \leq 1$, and $= 0$ otherwise. By symmetry, $E\{\mathbf{X}\} = \mathbf{0}$, and by the same symmetry argument, $Corr\{X_1, X_2\} = E\{X_1 X_2\} = 0$. (Indeed,

$$E\{X_1 X_2\} = \iint_{x_1^2 + x_2^2 \leq 1} x_1 x_2 \frac{1}{\pi} dx_1 dx_2,$$

and the integrand above is positive in the first and third quadrant, and is negative in the second and fourth.)

The variance of the first coordinate is

$$Var\{X_1\} = E\{X_1^2\} = \frac{1}{\pi} \iint_{x_1^2 + x_2^2 \leq 1} x_1^2 dx_1 dx_2 = \frac{1}{\pi} \int_{-1}^{1} x_1^2 \cdot 2\sqrt{1 - x_1^2} dx_1 = \frac{1}{4}.$$

(The reader who wishes to compute this integral "manually," may use the change of variables $x_1 = \sin t$.)

[2] A set B in a k-dimensional Euclidean space $\mathbb{R}^k$ is convex if for any two points $\boldsymbol{x}_1$ and $\boldsymbol{x}_2$ from B, the segment connecting $\boldsymbol{x}_1$ and $\boldsymbol{x}_2$ lies in B.

Again by symmetry, $Var\{X_2\} = 1/4$ also, and hence,

$$\mathcal{C} = \left\| \begin{array}{cc} 1/4 & 0 \\ 0 & 1/4 \end{array} \right\|.$$

By (3.1.7), the normal density with this covariance matrix and zero mean is

$$f_{\mathbf{Y}}(\mathbf{y}) = \frac{2}{\pi} \exp\left\{-\frac{1}{2}\left(4y_1^2 + 4y_2^2\right)\right\} = \frac{2}{\pi} \exp\left\{-2\left(y_1^2 + y_2^2\right)\right\} = \frac{2}{\pi} \exp\left\{-2|\mathbf{y}|^2\right\},$$

where the vector $\mathbf{y} = (y_1, y_2)$.

Thus, by Theorem 7, for any convex set $B \subseteq \mathbb{R}^2$,

$$P(\overline{\mathbf{S}}_n \in B) \to \frac{2}{\pi} \iint_B e^{-2|\mathbf{y}|^2} d\mathbf{y}.$$

For example, for any $a \geq 0$,

$$P\left(|\mathbf{S}_n| > a\sqrt{n}\right) = P\left(|\overline{\mathbf{S}}_n| > a\right) \to \frac{2}{\pi} \iint_{|\mathbf{y}|>a} e^{-2|\mathbf{y}|^2} d\mathbf{y} = e^{-2a^2},$$

which is easy to calculate by switching to polar coordinates. Certainly, it would be difficult to get such an approximation without using the CLT. □

► To normalize $\mathbf{S}_n$ in a way that will lead to the standard normal distribution as the limiting one, we should consider a matrix $\mathcal{D}$ for which the r.vec.'s $\mathcal{D}\mathbf{X}_i$ has an identity covariance matrix. In accordance with (2.1.5), this means that $\mathcal{D}\mathcal{C}\mathcal{D}^T = \mathcal{I}$. Suppose that we have found such a $\mathcal{D}$. Then we may consider the normalized r.v.

$$\mathbf{S}_n^* = \frac{1}{\sqrt{n}} \mathcal{D} \sum_{i=1}^n \mathbf{X}_i = \frac{1}{\sqrt{n}} \sum_{i=1}^n \mathcal{D}\mathbf{X}_i. \tag{3.3.2}$$

Since the vectors $\mathcal{D}\mathbf{X_i}$ have zero mean and the identity covariance matrix, by virtue of (3.3.1), for any convex set $B \subseteq \mathbb{R}^k$,

$$P(\mathbf{S}_n^* \in B) \to \Phi(B),$$

where $\Phi(B)$ is the standard normal distribution; that is, $\Phi(B) = P(\mathbf{Y} \in B)$ in the case where $\mathbf{Y}$ is standard normal.

Finding the matrix $\mathcal{D}$ is a standard problem from Linear Algebra, and we skip the details. For $\mathcal{D}$ to exist, $\mathcal{C}$ should not be singular, and the problem is reduced to switching to a basis composed by eigenvectors of $\mathcal{C}$. On the other hand, this problem is time consuming (though nowadays one can use software), and this is why the statement of the CLT in the form of (3.3.1) may turn out to be more convenient. ◄

Before turning to a proof of Theorem 7, we define the notion of a moment generating function (m.g.f.) in the multidimensional case.

Let a non-random vector $\mathbf{z} = (z_1, ..., z_k)$. We define the m.g.f. of a r.vec. $\mathbf{X}$ as the function

$$M_{\mathbf{X}}(\mathbf{z}) = E\left\{e^{\langle \mathbf{z}, \mathbf{X} \rangle}\right\}, \tag{3.3.3}$$

where $\langle \cdot, \cdot \rangle$ stands for dot product.

Let the random variable $X_{\mathbf{z}} = \langle \mathbf{z}, \mathbf{X} \rangle$, and $M_{X_{\mathbf{z}}}(t) = E\{e^{tX_{\mathbf{z}}}\}$, the m.g.f. of the (one-dimensional) r.v. $X_{\mathbf{z}}$. (To distinguish the one- and multidimensional cases, for the former, we will use the letter t instead of z as in Chapters 8 and 9.) Then, from (3.3.3), it follows that the multidimensional m.g.f.

$$M_{\mathbf{X}}(\mathbf{z}) = M_{X_{\mathbf{z}}}(1), \tag{3.3.4}$$

and when working with multidimensional m.g.f.'s, we can switch to the one-dimensional case.

EXAMPLE 2. Let $\mathbf{Y}$ be a normal r.vec. with zero mean and a covariance matrix C, and the r.v. $Y_{\mathbf{z}} = \langle \mathbf{z}, \mathbf{Y} \rangle$. By virtue of Theorem 4, $Y_{\mathbf{z}}$ is normal. Furthermore, $E\{Y_{\mathbf{z}}\} = 0$, and in view of (2.1.4), $Var\{Y_{\mathbf{z}}\} = \langle C\mathbf{z}, \mathbf{z} \rangle$. Let $\sigma_{\mathbf{z}} = \sqrt{\langle C\mathbf{z}, \mathbf{z} \rangle}$, the standard deviation of $Y_{\mathbf{z}}$. Then,

$$M_{Y_{\mathbf{z}}}(t) = \exp\{\sigma_{\mathbf{z}}^2 t^2/2\} = \exp\{\langle C\mathbf{z}, \mathbf{z} \rangle t^2/2\},$$

and, by virtue of (3.3.4),

$$M_{\mathbf{Y}}(\mathbf{z}) = \exp\{\langle C\mathbf{z}, \mathbf{z} \rangle/2\}. \ \square \tag{3.3.5}$$

Proof of Theorem 7. It may be proved that the convergence of m.g.f.'s implies the convergence of the respective distributions for convex sets; we take this fact for granted. So, in order to prove (3.3.1), it suffices to prove that for all $\mathbf{z}$'s,

$$M_{\overline{\mathbf{S}}_n}(\mathbf{z}) \to \exp\{\langle C\mathbf{z}, \mathbf{z} \rangle/2\} \ \text{ as } n \to \infty. \tag{3.3.6}$$

On the other hand, in view of (3.3.4), $M_{\overline{\mathbf{S}}_n}(\mathbf{z})$ is the value at the point $t = 1$ of the m.g.f. of the r.v.

$$\langle \mathbf{z}, \overline{\mathbf{S}}_n \rangle = \frac{1}{\sqrt{n}} \langle \mathbf{z}, \sum_{i=1}^n \mathbf{X}_i \rangle = \frac{1}{\sqrt{n}} \sum_{i=1}^n \langle \mathbf{z}, \mathbf{X}_i \rangle = \frac{1}{\sqrt{n}} \sum_{i=1}^n X_{\mathbf{z}i},$$

where $X_{\mathbf{z}i} = \langle \mathbf{z}, \mathbf{X}_i \rangle$. Let us rewrite it as

$$\langle \mathbf{z}, \overline{\mathbf{S}}_n \rangle = \sigma_{\mathbf{z}} \cdot S_{\mathbf{z}n}^*, \tag{3.3.7}$$

where $\sigma_{\mathbf{z}} = \sqrt{Var\{X_{\mathbf{z}}\}} = \sqrt{\langle C\mathbf{z}, \mathbf{z} \rangle}$, and

$$S_{\mathbf{z}n}^* = \frac{1}{\sigma_{\mathbf{z}}\sqrt{n}} \sum_{i=1}^n X_{\mathbf{z}i}.$$

Clearly, $E\{X_{\mathbf{z}i}\} = 0$, and $Var\{X_{\mathbf{z}i}\} = \sigma_{\mathbf{z}}^2$, so $S_{\mathbf{z}n}^*$ is the normalized sum of i.i.d. r.v.'s. As we proved in Section **9**.1.2, the m.g.f. of such a sum converges to the standard normal m.g.f. $\exp\{t^2/2\}$. Then, in view of (3.3.7), the m.g.f. of $\langle \mathbf{z}, \overline{\mathbf{S}}_n \rangle$ converges to $\exp\{\sigma_{\mathbf{z}}^2 t^2/2\}$; see Property **A** of m.g.f.'s in Section **8**.1.1.

For $t = 1$, the last expression equals $\exp\{\sigma_{\mathbf{z}}^2/2\} = \exp\{\langle C\mathbf{z}, \mathbf{z} \rangle/2\}$, which implies (3.3.6). ■

In conclusion, note that one can find in the literature the generalization of Theorem 7 to the case of non-identically distributed r.vec.'s, as well as the multidimensional analog of the Berry-Esseen theorem (Theorem **9**.4).

4 EXERCISES

Section 1

1. Mark each statement below "true" or "false". Justify your answers.

   (a) $Cov\{X_1, X_2\} = Cov\{X_2, X_1\}$.

   (b) If two r.v.'s are independent, then they are uncorrelated.

   (c) If two r.v.'s are correlated, then they are dependent.

   (d) If two r.v.'s are uncorrelated, then they are independent.

   (e) A correlation coefficient may equal 2.56.

   (f) A covariance may equal 2.56.

   (g) A correlation coefficient may equal -0.56.

   (h) A covariance may equal -0.56.

   (i) A correlation coefficient is always less than 1.

2. (a) Show that for $Cov\{X_1, X_2\} = E\{X_1X_2\}$, it suffices that either X_1 or X_2 has zero mean.

   (b) Show that $Cov\{X+Y, Z\} = Cov\{X, Z\} + Cov\{Y, Z\}$.

3. (a) What is $Cov\{X, -X\}$, $Cov\{-X, -X\}$)?

   (b) Let $c = Cov\{X_1, X_2\}$. Write $Cov\{2X_1, 3X_2\}, Cov\{-2X_1, 3X_2\}, Cov\{-2X_1, -3X_2\}$.

   (c) Let $\rho = Corr\{X_1, X_2\}$. Write $Corr\{X_1, -X_2\}$, $Corr(\{X_1, 10 - X_2\}$, $Corr\{2X_1, 3X_2\}$, $Corr\{-2X_1, 3X_2\}$, $Corr\{-2X_1, -3X_2\}$. In general, how will $Corr\{X_1, X_2\}$ change if we multiply X_1 by a constant c_1, and X_2 by c_2?

4. Let a r.vec. (X, Y) be uniformly distributed on the disk $\{(x, y) : x^2 + y^2 \leq 4\}$. Find the covariance. Are X, Y independent? Do the same for the uniform distribution on $\{(x, y) : x^2 + x + y^2 - 3y \leq 20\}$. (*Hint*: The problem does not require calculations.)

5. Does it follow from the Cauchy-Schwartz inequality (Proposition 1) that for any r.v.'s ξ and η with finite second moments, $(E\{|\xi\eta|\})^2 \leq E\{\xi^2\}E\{\eta^2\}$? Which inequality is stronger: this or (1.2.1)?

6. Two dice are rolled. Let X and Y be the respective numbers on the first and the second die, and $Z = Y - X$. (a) Do you expect negative or positive correlation between X and Z? (b) Find $Corr\{X, Z\}$. (*Advice*: Use the particular information about the distribution of (X, Y) at the very end, *if needed*.) (c) What conditions on X and Y do we need as a matter of fact, for the answer to be the same as you obtained?

7. Let r.v.'s X and Y be independent, $Z_1 = X + Y$, $Z_2 = X - Y$. When are the r.v.'s Z_1 and Z_2 uncorrelated? Would it mean that they are independent? (*Hint*: Let, say, (X, Y) be uniformly distributed on the square $\{(x, y) : |x| \leq 1, |y| \leq 1\}$. If we know $X + Y$, does it give an additional information about $X - Y$?)

8. Let r.v.'s X and Y be independent, $Z = X + Y$. Find $Corr\{X, Z\}$ as a function of the ratio $k = Var\{Y\}/Var\{X\}$. Comment on the fact that this function is monotone.

9. Two balls, one at a time, are drawn without replacement from a box containing r red and b black balls.

(a) For $i = 1,2$, let $X_i = 1$ if the ith ball is red, and $X_i = 0$ otherwise. Do you expect negative or positive correlation between X_1 and X_2? Compute $Corr\{X_1,X_2\}$.

(b) Let Y_1 and Y_2 be the numbers of red and black balls in the sample, respectively. What is $Corr\{Y_1,Y_2\}$? (*Advice*: Try to avoid calculations.)

10. Suppose n balls are distributed at random into k boxes. Let $X_i = 1$ if box i is non-empty, and $X_i = 0$ otherwise. Argue that $Corr\{X_i,X_j\} = Corr\{X_1,X_2\}$ for all $i \neq j$ and find $Corr\{X_1,X_2\}$.

11. Let a r.vec. (X,Y) have the joint density $f(x,y) = \frac{3}{8}(x+y)^2$ for $|x| \leq 1$, $|y| \leq 1$, and $f(x,y) = 0$ otherwise. Do you expect the r.v.'s to be dependent? Find $Corr\{X,Y\}$. (Use of software is recommended.)

12. You are working with r.v.'s $X_1,X_2,X_3,\ldots$, and you are interested *only* in their variances and correlations. Can you switch to the centered r.v.'s and deal only with them?

13. Prove (1.1.11).

14. John invests one unit of money in two business projects; an amount of α goes to the first, and $1-\alpha$ goes to the second. For $i = 1,2$, let X_i be the (random) return (per \$1) of project i. It is known that the means of X_1 and X_2 are the same. The variances $\sigma_i^2 = Var\{X_i\}$ and $\rho = Corr\{X_1,X_2\} \neq \pm 1$ are known. Show that the mean return does not depend on α and find an α minimizing the variance of the return. (*Hint:* The return of the total investment is $\alpha X_1 + (1-\alpha)X_2$. In the case of investment into financial securities, α can be negative, which means the so called short selling or borrowing securities; see details in Section 2.3.)

15. Let X_1,X_2,X_3,X_4 be independent r.v.'s with variances $1,2,3,4$, respectively. Do you need to know the mean values of X_1,X_2,X_3,X_4 in order to compute the correlations of (a) X_1+X_2 and X_2+X_4; (b) X_1+X_2 and X_3+X_4? Compute the correlations mentioned.

16. Is it true that if $Corr(X_1,X_2) = 1$, then $X_1 = cX_2$ for a $c > 0$? Answer the same question under the additional condition $E\{X_1\} = E\{X_2\} = 0$. Let $m_i = E\{X_i\}$, $i = 1,2$. Is it true that if $Corr(X_1,X_2) = 1$, then $X_1 - m_1 = c(X_2 - m_2)$ for a $c > 0$?

17. (a) Let X be uniform on $[-1,1]$, and $Y = X^3$. Find $\rho = Corr\{X,Y\}$ and explain why $\rho \neq 1$ while the association (or dependence) is perfect and positive.

    (b) Solve the same problem for $Y = X^{2k+1}$, where k is a positive integer.

18. Represent the "beta of security" in Example **3**.6.2-2 in terms of the correlation coefficient $\rho = Corr\{X,Y\}$.

19. In a country, the mean weight of adult females is 110 lb, and the mean height is 155 cm. The respective standard deviations are 15 and 12. Ms. K weighs 114 lb, and her height is 159 cm. The same figures for Ms. S are 105 lb and 149 cm. Estimate the correlation proceeding from a linear regression relation.

Sections 2 and 3

20. For which ρ does there exist the inverse of $\mathcal{C}$ in (2.1.1)?

21. Given the covariance matrix $\mathcal{C}$ of a r.vec. $(X_1,\ldots,X_k)$, how to compute the variance of the sum $X_1+\ldots+X_k$? (*Hint*: The answer is simple and nice.)

22. Using a probability theory argument, show that the sum of all elements of any covariance matrix is non-negative.

23. Let $\mathbf{X}$ be a r.vec., $E\{\mathbf{X}\} = \mathbf{0}$, and $\mathbf{Y} = Q\mathbf{X}$, where Q is a non-random orthogonal matrix. Using a probability theory argument, prove that the covariance matrices $\mathcal{C}_{\mathbf{Y}}$ and $\mathcal{C}_{\mathbf{X}}$ have the same traces (the sums of the diagonal entries). (*Hint*: Think about $E\{|\mathbf{X}|^2\}$.)

24. Let $(X_1,...,X_k)$ be a r.vec. with a covariance matrix $\mathcal{C}$. Under which condition on $\mathcal{C}$ does there exist a linear combination $t_1X_1+...+t_kX_k$ with zero variance and with not all t's equal zero? (Say, you have estimated the covariance matrix of the returns of a collection of securities in a financial market, and you have a good software for matrix operations. What would you do to figure out whether it is possible to arrange a risk-free investment in these random securities?)

    Consider, in particular, the case $k=2$.

25. Let $\mathbf{X}=(X_1,...,X_k)$ be a random vector with a covariance matrix $\mathcal{C}$, and $\mathbf{t}=(t_1,...,t_k)$ and $\tilde{\mathbf{t}}=(\tilde{t}_1,...,\tilde{t}_k)$ be two non-random vectors. Consider two linear combinations: $\langle\mathbf{X},\mathbf{t}\rangle$ and $\langle\mathbf{X},\tilde{\mathbf{t}}\rangle$. Generalizing (2.1.4), prove that $Cov\{\langle\mathbf{X},\mathbf{t}\rangle,\langle\mathbf{X},\tilde{\mathbf{t}}\rangle\}=\langle\mathcal{C}\mathbf{t},\tilde{\mathbf{t}}\rangle$. Can we switch $\mathbf{t}$ and $\tilde{\mathbf{t}}$ in the r.-h.s. of this formula? Why?

26. Regarding relation (2.2.2), show that

$$Var\{X_i\}=b_{1i}^2Var\{Y_1\}+...+b_{ki}^2Var\{Y_k\}=b_{1i}^2Var\{Y_1\}+...+b_{li}^2Var\{Y_l\}.$$

    Explain why a squared loading coefficient may be viewed as the percentage of the factor variance "contributed" to the variance of the component. (*Hint*: The matrix $\mathcal{B}$ is orthogonal.)

27. This exercise concerns the theory of Section 2.3. Suppose that a financial market consists of two securities with expected returns of 6% and 8%, and standard deviations of 1% and 2%, respectively. Let the correlation $\rho=0.3$.

    (a) Consider two portfolios: $\mathbf{x}=(\frac{1}{2},\frac{1}{2})$, and $\mathbf{y}=(\frac{1}{3},\frac{2}{3})$. Let $R_\mathbf{x}$ and $R_\mathbf{y}$ be the returns corresponding to these portfolios. Find (a) $E\{R_\mathbf{x}\}$; (b) $Cov\{R_\mathbf{x},R_\mathbf{y}\}$. (The last question is relevant to Exercise 25.)

    (b) Now, we add an additional risk-free security with a risk-free interest of 3%. Find the optimal combination of the two risky securities. Derive and graph the capital market line. Find an optimal portfolio that invests 60% in the risk-free security. Where is the corresponding point on the trade-off line?

28. Let $f(x_1,x_2)$ be the density of a r.vec $\mathbf{X}=(X_1,X_2)$. We call a curve in a (x_1,x_2)-plane a *level curve* if it is determined by an equation $f(x_1,x_2)=c$, where c is a constant. Argue that all points in this curve are equally likely to be values of $\mathbf{X}$. Show that for a bivariate normal distribution, the level curves are ellipses. Graph typical level curves for the following covariance matrices:

$$\mathcal{C}_1=\mathcal{I}=\left\|\begin{array}{cc}1&0\\0&1\end{array}\right\|,\ \mathcal{C}_2=\left\|\begin{array}{cc}4&0\\0&1\end{array}\right\|,\ \mathcal{C}_3=\left\|\begin{array}{cc}1&0\\0&4\end{array}\right\|,\ \mathcal{C}_4=\left\|\begin{array}{cc}1&1/2\\1/2&1\end{array}\right\|.$$

    Describe verbally an analogous picture for the case of a larger dimension.

29. Let X and Y be independent standard normal r.v.'s, $Z_1=X+Y$, $Z_2=X-Y$. Are Z_1 and Z_2 independent? (*Hint*: The problem does not require long calculations.)

30. Let a r.vec (X_1,X_2) is normal with zero means and covariance matrix (2.1.1). Write the density of X_1+X_2.

31. Let $\mathbf{X}=(X_1,X_2)$ be a two-dimensional normal random vector; $E\{X_1\}=1$, $E\{X_1^2\}=5$, $E\{X_2\}=2$, $E\{X_1^2\}=13$, $E\{X_1X_2\}=-2$. (a) Write the density of the centered vector $\mathbf{Y}=\mathbf{X}-\mathbf{m}$, where the vector $\mathbf{m}=(1,2)$. Write the density of $\mathbf{X}$. (b) For $\mathbf{t}=(3,-4)$, (i) write the expectation, the variance and the density of $t_1X_1+t_2X_2$; (ii) write the expectation, the variance and the density of the projection of $\mathbf{X}$ on $\mathbf{t}$.

32. Let the covariance matrix of a normal r.vec.$\mathbf{Y} = (Y_1, Y_2, Y_3)$ be $C = \left\| \begin{array}{ccc} 1 & 1 & 0 \\ 1 & 4 & 0 \\ 0 & 0 & 9 \end{array} \right\|$.

    Explain without any calculations why Y_3 does not depend on the r.vec. (Y_1, Y_2).

    Assuming $E\{\mathbf{Y}\} = \mathbf{0}$, write the density of $\mathbf{Y}$.

33. Let us present a two-dimensional standard normal r.vec. $\mathbf{X}$ in the polar coordinates as $(\mathbf{R}, \mathbf{\Theta})$, where $\mathbf{R}$ is the length of $\mathbf{X}$ and $\mathbf{\Theta}$ is the angle between $\mathbf{X}$ and the first axis.

    (a) Are the r.v.'s $\mathbf{R}, \mathbf{\Theta}$ independent? What are the marginal distributions? (b) Would you expect the independence of $\mathbf{R}$ and $\mathbf{\Theta}$ in the case of a non-standard normal distribution?

34. Let $\mathcal{B}$ be a rotation matrix, $\mathbf{X}$ be a standard normal r.vec., and $\mathbf{Y} = \mathcal{B}\mathbf{X}$. What is the distribution of $|\mathbf{Y}|^2$?

35. In the framework of Section 3.2, find the density of the r.v. $\chi_k = |\mathbf{X}|$.

36. Show that $E\{\chi_k\} = \sqrt{2}\Gamma((k+1)/2)/\Gamma(k/2)$. (*Advice*: A simple way is to consider a r.v. U having the χ_k^2-distribution; that is, the Γ-distribution with parameters $(\frac{1}{2}, \frac{k}{2})$, and write the integral for $E\{U^{1/2}\} = \int_0^\infty x^{1/2} f_{\frac{1}{2}, \frac{k}{2}}(x) dx$.)

37. Continue the calculations in Example 3.2-2. (*Advice*: You may choose a slightly heuristic approach using the representation $\xi_k = \frac{1}{\sqrt{k}} Y_k$, where $Y_k = \frac{1}{\sqrt{k}} \sum_{i=1}^k (X_i^2 - 1)$, and X_i's are standard normal. For large k, by the CLT, Y_k is asymptotically normal, which gives a clue how to estimate the moments of ξ_k.)

38. John throws a dart at a circular target with a radius of r. Suppose that with respect to a system of coordinates with the origin at the center of the target, the point where the dart lands may be represented as a bivariate normal vector $\mathbf{Y} = \dfrac{r}{4}\mathbf{X}$, where $\mathbf{X}$ is a standard normal vector. (In particular, this means that John may not hit the target at all.) Let Z be the distance the dart lands from the center. Show that $E\{Z^2\} = \frac{r^2}{8}$. Find the probability that John will hit the target. Find $P(Z \le r/4)$. (The use of software is recommended).

39. In the case you solved Exercise 36, find the mean speed of a molecule in Example 3.2-1. (*Advice*: Use the fact that $\Gamma(t+1) = t\Gamma(t)$.)

40. Restate Theorem 7 for the one dimensional case. How does this restatement differ from Theorem 1 from Chapter 9? Show that both assertions are equivalent.

41. What is the matrix $\mathcal{D}$ in (3.3.2) if the matrix C in Section 3.3 is diagonal?

42. Suppose that the coordinates of a r.vec. $\mathbf{X} = (X_1, ..., X_k)$ are independent. Present the m.g.f $M_{\mathbf{X}}(\mathbf{z})$ in terms of the m.g.f.'s of the coordinates. Is $M_{\mathbf{X}}(\mathbf{z})$ uniquely determined by the m.g.f.'s of the coordinates in the general case? (*Advice*: As an example, we may consider the normal case, and formula (3.3.5).)

43. In Section 3.3, let $k = 2$, and separate terms $\mathbf{X}_i = (X_{i1}, X_{i2})$, where X_{i1}, X_{i2} are independent and have the standard exponential distribution. Let the vector $\mathbf{e} = (1, 1)$. For large n, estimate $P(|\mathbf{S}_n - n \cdot \mathbf{e}| \le \sqrt{n})$.

44. Is (3.3.1) true for the sets depicted in Fig. 5? (*Hint*: Theorem 7 is stated for convex sets, but it does not mean that (3.3.1) cannot be true for a non-convex set.)

FIGURE 5.

Chapter 11

Maxima and Minima of R.V.'s. Elements of Reliability Theory. Hazard Rate and Survival Probabilities

1 MAXIMA AND MINIMA OF INDEPENDENT R.V.'S. RELIABILITY CHARACTERISTICS

1.1 Main representations and examples

Let $X_1, X_2, ...$ be a sequence of independent r.v.'s with respective d.f.'s $F_1(x), F_2(x), ...$, and let

$$\widetilde{W}_n = \max\{X_1, ..., X_n\} \quad \text{and} \quad \underset{\sim}{W}_n = \min\{X_1, ..., X_n\}.$$

The distributions of $\widetilde{W}_n$ and $\underset{\sim}{W}_n$ have simple representations in terms of the d.f.'s F_i. Namely, since X_i's are independent,

$$P(\widetilde{W}_n \leq x) = P(X_1 \leq x, ..., X_n \leq x) = \prod_{i=1}^{n} P(X_i \leq x) = \prod_{i=1}^{n} F_i(x), \tag{1.1.1}$$

and

$$P(\underset{\sim}{W}_n > x) = P(X_1 > x, ..., X_n > x) = \prod_{i=1}^{n} P(X_i > x) = \prod_{i=1}^{n} (1 - F_i(x)). \tag{1.1.2}$$

For the d.f. of the minimum, we have

$$P(\underset{\sim}{W}_n \leq x) = 1 - P(\underset{\sim}{W}_n > x) = 1 - \prod_{i=1}^{n} (1 - F_i(x)). \tag{1.1.3}$$

If the r.v.'s X_i are identically distributed and $F_i(x) = F(x)$, then (1.1.1) implies

$$P(\widetilde{W}_n \leq x) = F^n(x), \tag{1.1.4}$$

and from (1.1.3), it follows that

$$P(\underset{\sim}{W}_n \leq x) = 1 - (1 - F(x))^n. \tag{1.1.5}$$

EXAMPLE 1 (*The last survivor annuity*). The term *annuity* means a sequence of payments. An elderly couple begins to get regular retirement annuity which terminates only when both spouses die. Suppose that X_1, X_2, the *remaining* lifetimes of the husband and

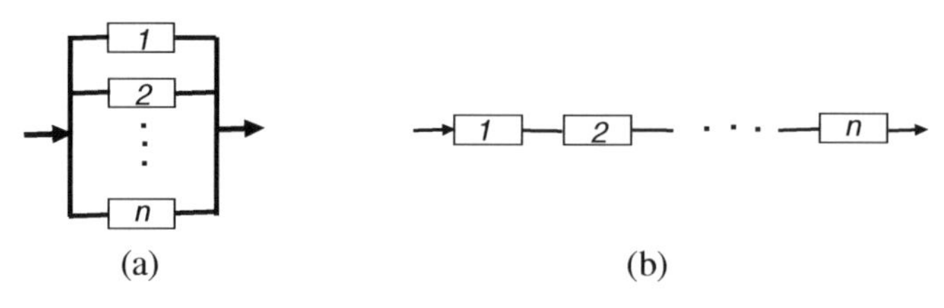

FIGURE 1.

wife, are independent r.v.'s uniformly distributed on $[0,15]$ and $[0,20]$, respectively, if measured in years. Clearly, the remaining duration of the annuity is the r.v. $\widetilde{W} = \max\{X_1, X_2\}$, and $P(\widetilde{W} \leq x) = F_1(x)F_2(x)$, where $F_1(x) = \frac{x}{15}$ for $x \in [0,15]$, and $F_2(x) = \frac{x}{20}$ for $x \in [0,20]$. For instance, the probability that the husband will live not longer than 10 years is $F_1(10) = 2/3$; the same probability for the wife is $F_2(10) = 1/2$; while the probability that the annuity will last not more than 10 years is $F_1(10)F_2(10) = 1/3$.

EXAMPLE 2. For a new car, let X_1 be the time to the first serious repair which is not connected with a traffic accident, and let X_2 be the time when an accident that causes a serious repair will occur (*if any*). Then the real time to the first repair is $\underset{\sim}{W} = \min\{X_1, X_2\}$. Suppose that the distribution of X_1 is well approximated by the distribution uniform on $[0,10]$, and the r.v. X_2 has the lack of memory property: the fact that an accident has not occurred yet has no impact on the chances of an accident in the future. Thus, X_2 is exponential. Suppose $E\{X_2\} = 15$. Then, assuming that X_1, X_2 are independent, for the probability that the first serious repair will not happen during the first 8 years, we have $P(\underset{\sim}{W} > 8) = (1 - F_1(8))(1 - F_2(8)) = (1 - \frac{8}{10}) \cdot e^{-8/15} \approx 0.12$.

EXAMPLE 3 concerns the reliability of a system consisting of components. In this example, we consider a parallel and series structure. A parallel structure may be illustrated as in Fig. 1a, and we may view it as electronic devices connected in parallel and receiving signals coming from the left. The signals reach the right end when at least one device functions.

This is only one of the possible interpretations. In general, a parallel system is defined as any system that works if and only if at least one of its components functions.

A series system stops functioning when just one component fails; it is illustrated in Fig. 1b. We again can think about signals, but now they go through only if no component has failed.

If we denote by X_i the time to failure of the ith component, then for the parallel system, the duration of the work of the whole system is $\widetilde{W}_n = \max\{X_1, ..., X_n\}$, while for the series system, the same duration is the r.v. $\underset{\sim}{W}_n = \min\{X_1, ..., X_n\}$.

In Reliability Theory, if T is the time to failure of a system, then the probability $P(T > t)$ is sometimes, though not often, called the *reliability function*. Let us find this function in our case, assuming that the X_i's are independent and uniformly distributed on $[0,1]$.

By (1.1.4) and (1.1.5), since $F(x) = x$ for $0 \leq x \leq 1$, for the same values of x,

$$
\begin{aligned}
P(\widetilde{W}_n > x) &= 1 - P(\widetilde{W}_n \leq x) = 1 - x^n, \\
P(\underset{\sim}{W}_n > x) &= 1 - P(\underset{\sim}{W}_n \leq x) = (1 - x)^n.
\end{aligned}
$$

To compute the expectations, we will use the formula (**6**.1.3.2) which, in our case, leads to

$$E\{\widetilde{W}_n\} = \int_0^\infty P(\widetilde{W}_n > x)dx = \int_0^1 (1-x^n)dx = 1 - \frac{1}{n+1} = \frac{n}{n+1}, \quad (1.1.6)$$

$$E\{\underset{\sim}{W}_n\} = \int_0^\infty P(\underset{\sim}{W}_n > x)dx = \int_0^1 (1-x)^n dx = \frac{1}{n+1}. \quad (1.1.7)$$

In particular, $E\{\widetilde{W}_n\} \to 1$ and $E\{\underset{\sim}{W}_n\} \to 0$ as $n \to \infty$. In Exercise 5 provided with a comment, the reader is invited to explain why the values of these limits do not look unexpected.

EXAMPLE 4. Let independent r.v.'s X_i, $i = 1,...,n$, have the exponential distributions with respective parameters a_i. Then, in accordance with (1.1.2),

$$P(\underset{\sim}{W}_n > x) = \prod_{i=1}^{n} \exp\{-a_i x\} = \exp\{-(a_1 + ... + a_n)x\}.$$

Thus, the minimum $\underset{\sim}{W}_n$ is also exponential r.v. with parameter $a = a_1 + ... + a_n$. One may say that the exponential distribution is stable with respect to minimization.

Consider, for instance, the series system in Fig. 1b, and assume all X_i's, the durations of the work of separate components, to be standard exponential ($a_i = 1$). Then, the duration of the work of the whole system has the exponential distribution with parameter n. It may also be presented as the distribution of the r.v. $\frac{1}{n}Z$, where Z is standard exponential. (Recall that the parameter of the exponential distribution is a scale one.)

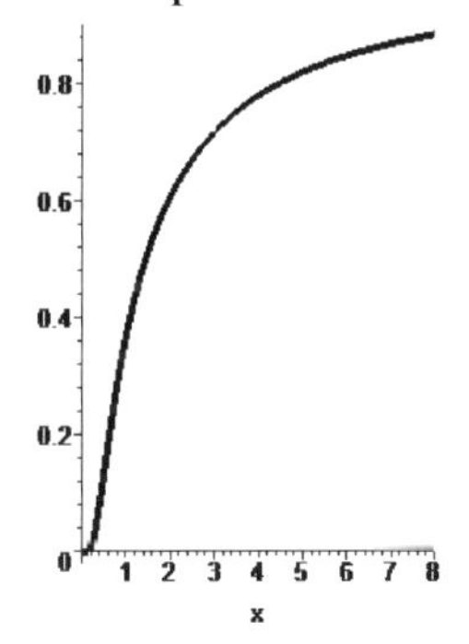

FIGURE 2.

EXAMPLE 5. Consider the d.f. $F_c(x) = e^{-c/x}$ for $x > 0$, where c is a positive parameter. Note that $F_c(x) \to 0$ as $x \to 0$ from the right ($x > 0$). Then, we should set $F_c(x) = 0$ for $x \leq 0$ also. The reader is invited to make sure that for $c = 1$ and $x \geq 0$, the graph of $F_c(x)$ looks as in Fig. 2.

Now, let r.v.'s X_i, $i = 1,...,n$, have respective d.f.'s $F_{c_i}(x)$, where c_i's are positive parameters. In accordance with (1.1.1),

$$P(\widetilde{W}_n \leq x) = \prod_{i=1}^{n} \exp\{-c_i/x\} = \exp\{-c/x\},$$

where $c = c_1 + ... + c_n$. So, we have come to a distribution of the same type. We may say that distributions of this type are stable with respect to maximization.

Consider the parallel system in Fig. 1a, and suppose all X_i's have the d.f. $F_1(x) = \exp\left\{-\frac{1}{x}\right\}$ (all $c_i = 1$). Then, $c = n$ and the duration of the work of the whole system has the d.f. $F_n(x) = \exp\{-n/x\}$. By the general rule for the linear transformations of r.v.'s (see (**3**.1.5.1)), this is the distribution of the r.v. nZ, where Z has the d.f. $F_1(x)$. □

1.2 More about reliability characteristics

Certainly, there are much more complicated configurations than those above. We restrict ourselves to particular examples.

First, consider the configuration in Fig. 3a, and denote by $X_{11},...,X_{1n}$ and $X_{21},...,X_{2m}$ the times to failure of the n parallel components in the first block and the m parallel components in the second block, respectively. We suppose all r.v.'s to be independent.

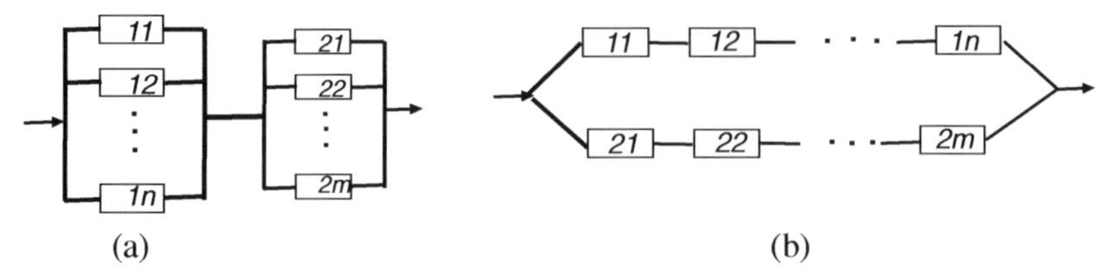

FIGURE 3.

Let $F_{ij}(x)$ be the corresponding d.f.'s. Denote by W_1 and W_2 the time to failure of the first and the second block respectively, and by W the time to failure for the whole system. Then

$$W = \min\{W_1, W_2\} = \min\{\max\{X_{11}, ..., X_{1n}\}, \max\{X_{21}, ..., X_{2m}\}\},$$

and, since all X's are independent,

$$P(W > x) = P(W_1 > x)P(W_2 > x) = \left(1 - \prod_{i=1}^{n} F_{1i}(x)\right)\left(1 - \prod_{i=1}^{m} F_{2i}(x)\right). \tag{1.2.1}$$

EXAMPLE 1. Assume that there are $n = 3$ parallel components in the first block and $m = 2$ parallel components in the second. Suppose all X's are exponential with an expected value of one unit of time in the first block and a half unit of time in the second. Say, the unit of time is ten thousand hours. From (1.2.1), to follows that

$$P(W > x) = \left(1 - (1 - e^{-x})^3\right)\left(1 - (1 - e^{-2x})^2\right).$$

Setting $x = 0.5$, we get that the probability that the system will work more than 5,000 hours equals ≈ 0.56. □

Now we consider the configuration in Fig. 3b keeping the notation X_{1i}, X_{2j}, $i = 1, ..., n$, $j = 1, ..., m$, for the components of the first and second series, respectively. Denote by V_1 and V_2 the time of work for the first and second series, and by V the time to failure for the whole system. Then

$$V = \max\{V_1, V_2\} = \max\{\min\{X_{11}, ..., X_{1n}\}, \min\{X_{21}, ..., X_{2m}\}\},$$

and

$$P(V \le x) = P(V_1 \le x)P(V_2 \le x) = \left(1 - \prod_{i=1}^{n}(1 - F_{1i}(x))\right)\left(1 - \prod_{i=1}^{m}(1 - F_{2i}(x))\right). \tag{1.2.2}$$

EXAMPLE 2. Again, let $n = 3$, $m = 2$, all X's are exponential with a mean of one in the first series, and 0.5 in the second. Then, from (1.2.2), it follows that

$$P(V \le x) = \left(1 - (e^{-x})^3\right)\left(1 - (e^{-2x})^2\right) = \left(1 - e^{-3x}\right)\left(1 - e^{-4x}\right).$$

For $x = 0.5$, we have $P(V \le 0.5) \approx 0.67$, and $P(V > 0.5) \approx 0.33$.

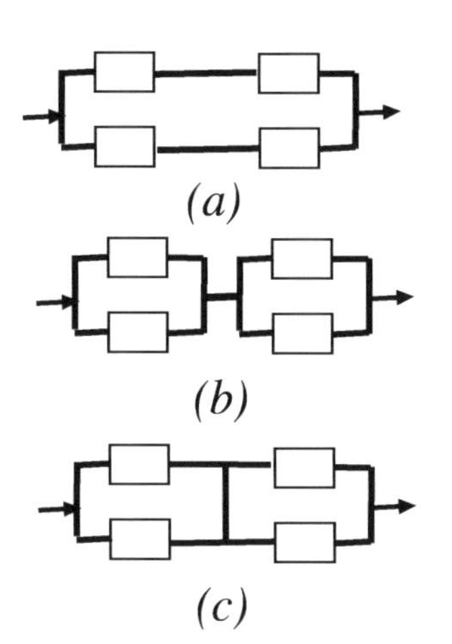

FIGURE 4.

EXAMPLE 3. Let us compare the configurations in Fig. 4a and 4b. Which configuration is more reliable? One may conjecture that it is the latter, and this becomes clear if we present the configuration 4b as it is done in Fig. 4c. For the last configuration, consider the events $A = \{$*at least one of possible paths works*$\}$ and $B = \{$*at least one of two possible paths (upper and lower) corresponding to configuration 4a works*$\}$. Clearly, $A \supseteq B$, and hence $P(A) \geq P(B)$.

Route 1 ⇒ page 305

2 LIMIT THEOREMS FOR MAXIMA AND MINIMA

In this section, we consider the behavior of the r.v.'s $\widetilde{W}_n$ and $\underset{\sim}{W}_n$ for large n. Assume X_i's to be independent and identically distributed. As above, we denote by $F(x)$ the d.f. of X_i, and set $\overline{F}(x) = 1 - F(x) = P(X_i > x)$, the tail of the distribution F.

2.1 Preliminary limit theorems: Convergence to bounds

First, suppose that X_i's are bounded. Let $\tilde{a}$ be an essential upper bound; that is, X_i is not larger than $\tilde{a}$ but may take on values arbitrarily close to $\tilde{a}$ or/and $\tilde{a}$ itself. More rigorously, $P(X_i \leq \tilde{a}) = 1$, and $P(X_i > \tilde{a} - \varepsilon) > 0$ for any $\varepsilon > 0$.

One may conjecture that for large n, at least one variable from $X_1, ..., X_n$ will take a value close to $\tilde{a}$, so $\widetilde{W}_n$ should be close to $\tilde{a}$. Formally, this would mean that

$$\widetilde{W}_n \xrightarrow{P} \tilde{a} \quad \text{as } n \to \infty. \tag{2.1.1}$$

This is indeed true. By (1.1.4), for any $\varepsilon > 0$,

$$P(\widetilde{W}_n \leq \tilde{a} - \varepsilon) = (F(\tilde{a} - \varepsilon))^n = (1 - \overline{F}(\tilde{a} - \varepsilon))^n \to 0$$

because $\overline{F}(\tilde{a} - \varepsilon) = P(X_i > \tilde{a} - \varepsilon) > 0$. Since, on the other hand, $P(\widetilde{W}_n \leq \tilde{a}) = 1$, for any $\varepsilon > 0$,

$$P(\tilde{a} - \varepsilon < \widetilde{W}_n \leq \tilde{a}) = P(\widetilde{W}_n \leq \tilde{a}) - P(\widetilde{W}_n \leq \tilde{a} - \varepsilon) \to 1 - 0 = 1,$$

which implies (2.1.1).

Similarly, let $\underset{\sim}{a}$ be an essential lower bound for X_i; that is, $P(X_i \geq \underset{\sim}{a}) = 1$, and $P(X_i < \underset{\sim}{a} + \varepsilon) > 0$ for any $\varepsilon > 0$. Then

$$\underset{\sim}{W}_n \xrightarrow{P} \underset{\sim}{a} \quad \text{as } n \to \infty. \tag{2.1.2}$$

The proof is similar.

EXAMPLE 1. Let us come back to Example 1.1-3. By virtue of what we have just proved, $\widetilde{W}_n \xrightarrow{P} 1$ and $\underset{\sim}{W}_n \xrightarrow{P} 0$ as $n \to \infty$, which is consistent with the formulas for the expectations in (1.1.6) and (1.1.7). □

Assume now that the r.v.'s X_i are not bounded from above; more precisely, that $P(X_i > a) > 0$ for any arbitrarily large a. (For example, the X's are exponential.) Then

$$\widetilde{W}_n \xrightarrow{P} \infty \quad \text{as } n \to \infty;$$

that is, $P(\widetilde{W}_n > a) \to 1$, as $n \to \infty$, for any (arbitrarily large) a. Indeed, by (1.1.4), $P(\widetilde{W}_n \leq a) = (F(a))^n \to 0$ because $F(a) = 1 - P(X_i > a) < 1$ for all a.

Analogously, if $P(X_i < -a) > 0$ for any arbitrarily large positive a, then

$$\underset{\sim}{W}_n \xrightarrow{P} -\infty \quad \text{as } n \to \infty.$$

2.2 A limit theorem for the minima of bounded variables

Next, we consider more sophisticated limit theorems showing *how* the r.v.'s $\underset{\sim}{W}_n$ and $\widetilde{W}_n$ converge to their (finite or infinite) limits. The most interesting thing here is that as in the CLT for sums, after a proper normalization, the limiting distributions for $\underset{\sim}{W}_n$ and $\widetilde{W}_n$ are standard and depend only on some particular characteristics of the r.v.'s X_i.

It is convenient to begin with minima and r.v.'s bounded from below. To simplify formulations and without loss of generality, we can set the bound $\underset{\sim}{a} = 0$. If it is not so, instead of X_i, we may consider the r.v.'s $X_i - \underset{\sim}{a}$.

Thus, with probability one, all $X_i \geq 0$, and $P(X_i \leq \varepsilon) > 0$ for any $\varepsilon > 0$. As we have already known, in this case, $\underset{\sim}{W}_n \xrightarrow{P} 0$ as $n \to \infty$.

We are interested in the order of this convergence and the limiting distribution after a proper normalization. It is natural to expect that the specific features of the convergence under consideration are determined by the probabilities of values of X_i close to zero. Assume that

$$F(x) \sim cx^{\alpha}, \quad \text{as } x \to 0, \tag{2.2.1}$$

for some positive c and α. (As usual, the symbol $a(x) \sim b(x)$ means that $\frac{a(x)}{b(x)} \to 1$ as x converges to a designated limit; in our case, as $x \to 0$.)

The most typical case is that of $\alpha = 1$. Suppose that X_i has a density $f(x)$ and $f(0) \neq 0$. Then the d.f. $F(x) \sim F'(0)x = f(0)x$ as $x \to 0$, and (2.2.1) holds with $c = f(0)$ and $\alpha = 1$.

We will see soon that the convergence of $\underset{\sim}{W}_n$ to zero has an order of $1/n^{1/\alpha}$. For this reason, we introduce the normalized r.v.

$$\underset{\sim}{W}_n^* = n^{1/\alpha} \underset{\sim}{W}_n. \tag{2.2.2}$$

It is worth emphasizing that the normalization we use, is different from what we applied in the CLT for sums. In the latter case, we subtracted means and divided by standard deviations; in the case of minima (and maxima below), we use characteristics of the behavior of $F(x)$ in a neighborhood of the bound; in our particular case, in a neighborhood of zero.

To move further, note also that if $a(x) \sim b(x)$, then $a(x) = b(x) + o(b(x))$, where the symbol $o(z)$ stands for a function negligibly small with respect to z for small z; rigorously,

$\frac{o(z)}{z} \to 0$ as $z \to 0$. (See also the Appendix, Section 2.1.) Indeed, $a(x) = b(x)\frac{a(x)}{b(x)} = b(x) + b(x)\left(\frac{a(x)}{b(x)} - 1\right)$ and the second term is small with respect to the first.

In particular, from (2.2.1) it follows that $F(x) = cx^{\alpha} + o(cx^{\alpha})$.

Making use of (1.1.5), for $x \geq 0$, we have

$$P(\underset{\sim}{W}_n^* > x) = P\left(\underset{\sim}{W}_n > \frac{x}{n^{1/\alpha}}\right) = \left(1 - F\left(\frac{x}{n^{1/\alpha}}\right)\right)^n$$

$$= \left(1 - c\left(\frac{x}{n^{1/\alpha}}\right)^{\alpha} - o\left(c\left(\frac{x}{n^{1/\alpha}}\right)^{\alpha}\right)\right)^n = \left(1 - \frac{cx^{\alpha}}{n} - o\left(\frac{cx^{\alpha}}{n}\right)\right)^n \to \exp\{-cx^{\alpha}\}.$$

Thus, $P(\underset{\sim}{W}_n^* \leq x) \to 1 - \exp\{-cx^{\alpha}\}$ for $x \geq 0$. Note also that, since X_i's are assumed to be non-negative, $P(\underset{\sim}{W}_n^* \leq x) = 0$ for $x < 0$. So, we have proved

Proposition 1 *Under condition (2.2.1), for all x's,*

$$P(\underset{\sim}{W}_n^* \leq x) \to Q_{c\alpha}(x), \tag{2.2.3}$$

where the distribution function $Q_{c\alpha}(x) = 0$ *for* $x < 0$, *and for* $x \geq 0$,

$$Q_{c\alpha}(x) = 1 - \exp\{-cx^{\alpha}\}. \tag{2.2.4}$$

From (2.2.2)–(2.2.3), it indeed follows that $\underset{\sim}{W}_n$ approaches zero at a rate of $1/n^{1/\alpha}$. More precisely, by virtue of (2.2.2),

$$\underset{\sim}{W}_n = \frac{1}{n^{1/\alpha}}\underset{\sim}{W}_n^*, \tag{2.2.5}$$

where the r.v. $\underset{\sim}{W}_n^*$ is *neither small nor large*, and for large n, its distribution is approaching a standard distribution (namely, $Q_{c\alpha}$) depending on the behavior of $F(x)$ *only* in a neighborhood of zero.

The distribution with the d.f. $Q_{c\alpha}(x) = 1 - \exp\{-cx^{\alpha}\}$, where c and α are positive parameters, is called *Weibull's distribution.*

When $\alpha = 1$, the limiting distribution Q_{c1} is exponential with parameter c, and as has been already noted, this is the most typical case.

EXAMPLE 1. Suppose that the X_i's are uniformly distributed on $[0,1]$. Then $f(x) \equiv 1$ for $x \in [0,1]$, and hence $c = 1$, $\alpha = 1$, and in accordance with (2.2.5),

$$\underset{\sim}{W}_n = \frac{1}{n}\underset{\sim}{W}_n^*,$$

where the distribution of $\underset{\sim}{W}_n^*$ is close to the standard exponential distribution for large n.

Say, in Example 1.1-3, if the parameter n, the number of the components, is large, then the duration of the work of the series system may be approximated by the r.v. $\frac{1}{n}Z$, where Z is standard exponential. The reader is invited to make sure that $\frac{1}{n}Z$ is an exponential r.v with parameter n, and hence, a mean of $1/n$. □

In conclusion, consider the case $\underset{\sim}{a} \neq 0$. As was already mentioned, we may set $Y_i = X_i - \tilde{a}$ and work with $\underset{\sim}{W}_{Yn}^* = \min\{Y_1, ..., Y_n\}$. Then, representation (2.2.5) may be replaced by

$$\underset{\sim}{W}_n = \min\{X_1, ..., X_n\} = \underset{\sim}{a} + \min\{Y_1, ..., Y_n\} = \underset{\sim}{a} + \frac{1}{n^{1/\alpha}}\underset{\sim}{W}_{Yn}^*, \tag{2.2.6}$$

where c and α are the respective characteristics of the r.v.'s Y_i, and the distribution of $\underset{\sim}{W}{}^*_{Yn}$ converges to $Q_{c\alpha}$.

2.3 A limit theorem for the maxima of bounded variables

Next, we consider maxima and r.v.'s bounded from above by an essential bound $\tilde{a}$.

Proposition 2 *Suppose that all $X_i \leq \tilde{a}$, and for $x > 0$,*

$$P(\tilde{a} - x \leq X_i \leq \tilde{a}) \sim cx^{\alpha}, \quad as \ \ x \to 0, \tag{2.3.1}$$

for some positive c and α. Let

$$\widetilde{W}_n^* = n^{1/\alpha}\left(\tilde{a} - \widetilde{W}_n\right). \tag{2.3.2}$$

Then, for all x's,

$$P(\widetilde{W}_n^* \leq x) \to Q_{c\alpha}(x). \tag{2.3.3}$$

We will prove this at the end of this section.

So, in accordance with (2.3.2),

$$\widetilde{W}_n = \tilde{a} - \frac{1}{n^{1/\alpha}}\widetilde{W}_n^*, \tag{2.3.4}$$

and the distribution of $\widetilde{W}_n^*$ is approaching the Weibull distribution with parameters c and α.

EXAMPLE 1. Let the X's be uniform on $[0,1]$. Then $\tilde{a} = 1$. The probability $P(1 - x \leq X_i \leq 1) = x$, so $c = 1$ and $\alpha = 1$ also. Consequently,

$$\widetilde{W}_n = 1 - \frac{1}{n}\widetilde{W}_n^*,$$

where the distribution of $\widetilde{W}_n^*$ converges to the standard exponential distribution.

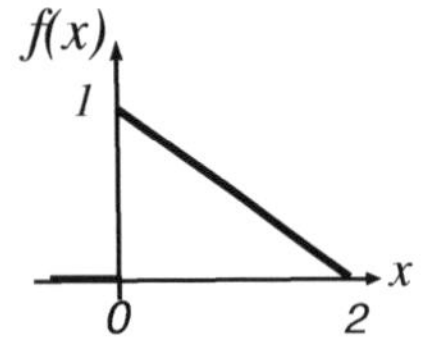

EXAMPLE 2. Let the X's have the density $f(x) = \frac{1}{2}(2-x)$ for $x \in [0,2]$; see Fig. 5. Since $\int_0^2 f(x)dx = 1$, the r.v.'s X_i assumes values from $[0,2]$, and the essential upper bound $\tilde{a} = 2$. It is straightforward to verify that $P(2-x \leq X_i \leq 2) = \int_{2-x}^{2} f(u)du = x^2/4$, so $c = 1/4$ and $\alpha = 2$. Hence,

FIGURE 5.

$$\widetilde{W}_n = 2 - \frac{1}{\sqrt{n}}\widetilde{W}_n^*,$$

and the d.f. of $\widetilde{W}_n^*$ is approaching the d.f. $Q_{\frac{1}{4},2}(x) = 1 - \exp\{-x^2/4\}$.

EXAMPLE 3 (*The estimation of a parameter of the uniform distribution*). This is a known and important example from Statistics. Assume X_i's to be uniformly distributed on an interval $[0,\theta]$, where θ is an unknown parameter. Suppose we have at our disposal n observations $X_1,...,X_n$, and want to estimate θ basing on these observations. There are, at least, two ways to do it.

First, we can compute the average observation $\overline{X}_n = \frac{1}{n}\sum_{i=1}^n X_i$. Since $E\{X_i\} = \theta/2$, by the law of large numbers, $\overline{X}_n \xrightarrow{P} \frac{\theta}{2}$. Hence, the estimate $Y_n = 2\overline{X}_n \xrightarrow{P} \theta$ and may serve as a good estimate for large n.

On the other hand, as we know, $\widetilde{W}_n$ also converges to θ as $n \to \infty$; so $\widetilde{W}_n$ is another candidate for an estimator. Which is better?

Let $\sigma^2 = Var\{X_i\} = \theta^2/12$, and $S_n = \sum_{i=1}^n X_i$. We have

$$Y_n = 2\overline{X}_n = \frac{2}{n}S_n = \theta + \frac{2}{n}\left(S_n - n\frac{\theta}{2}\right) = \theta + \frac{2\sigma}{\sqrt{n}} \cdot \frac{1}{\sigma\sqrt{n}}(S_n - n\theta/2) = \theta + \frac{2\sigma}{\sqrt{n}} \cdot S_n^*, \quad (2.3.5)$$

where the normalized sum $S_n^* = \frac{1}{\sigma\sqrt{n}}(S_n - n\theta/2)$. By the CLT, S_n^* is asymptotically standard normal.

Now, let us write the version of (2.3.4) for our case. Since X_i's are uniform, $P(\theta - x < X_i \leq \theta) = \frac{x}{\theta}$. So, when applying condition (2.3.1), we set $\tilde{a} = \theta$, $c = \frac{1}{\theta}$, and $\alpha = 1$. Then, (2.3.4) gives

$$\widetilde{W}_n = \theta - \frac{1}{n}\widetilde{W}_n^*, \quad (2.3.6)$$

where the distribution of $\widetilde{W}_n^*$ is close to the Weibull distribution with parameters$1/\theta$ and 1; that is, to the exponential distribution with a mean of θ.

Comparing (2.3.6) and (2.3.5), we see that $\widetilde{W}_n$ converges to the unknown parameter θ essentially faster: in (2.3.5), the order of convergence is $1/\sqrt{n}$, while in (2.3.6), it is $1/n$. See also Exercise 18 for a comment regarding the means of the estimates. □

Proof of Proposition 2. Set $Y_j = \tilde{a} - X_j$ and observe that

$$\widetilde{W}_n = \max\{X_1, ..., X_n\} = -\min\{-X_1, ..., -X_n\} = \tilde{a} - \min\{\tilde{a} - X_1, ..., \tilde{a} - X_n\} = \tilde{a} - \underset{\sim}{W}_{Yn},$$

where $\underset{\sim}{W}_{Yn} = \min\{Y_1, ..., Y_n\}$.

Clearly, zero is an essential lower bound for Y's, and $P(Y_j \leq x) = P(\tilde{a} - x \leq X_j \leq \tilde{a})$. So, condition (2.3.1) for the X's implies condition (2.2.1) for the Y's. By Proposition 1,

$$P\left(n^{1/\alpha}\underset{\sim}{W}_{Yn} \leq x\right) \to Q_{c\alpha}(x).$$

It remains to observe that $\widetilde{W}_n^* = n^{1/\alpha}\left(\tilde{a} - \widetilde{W}_n\right) = n^{1/\alpha}\underset{\sim}{W}_{Yn}$. ■

2.4 A limit theorem for unbounded variables

Next, we consider unbounded r.v.'s. Let the d.f. $\widetilde{G}_{c\alpha}(x) = 0$ for $x \leq 0$, and for $x > 0$,

$$\widetilde{G}_{c\alpha}(x) = \exp\{-c/x^{\alpha}\}, \quad (2.4.1)$$

where c and α are positive parameters. The graph for $c = 1$ and $\alpha = 2$ is given in Fig. 6a, the case $\alpha = 1$ was considered in Example 1.1-5.

Note that $\widetilde{G}_{c\alpha}(x) \to 0$ as $x \to 0$, which is reflected in the graph.

Along with this function, we consider the d.f. $\underset{\sim}{G}_{c\alpha}(x) = 1$ for $x \geq 0$, and

$$\underset{\sim}{G}_{c\alpha}(x) = 1 - \exp\{-c/|x|^{\alpha}\} \quad (2.4.2)$$

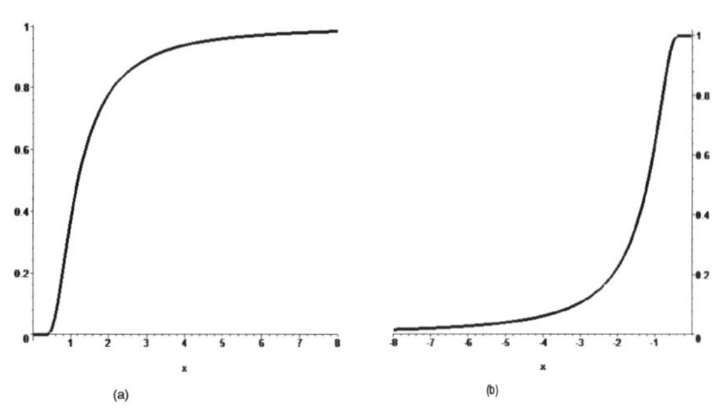

FIGURE 6.

for $x < 0$. The graph for $c = 1$ and $\alpha = 2$ is given in Fig. 6b. Note that $\underset{\sim}{G}_{c\alpha}(x) \to 1$ as $x \to 0$.

Proposition 3 *Suppose that*

$$\overline{F}(x) = 1 - F(x) \sim \frac{c}{x^{\alpha}}, \quad as \ x \to \infty, \tag{2.4.3}$$

for some positive c and α. *Let*

$$\widetilde{W}_n^* = \frac{1}{n^{1/\alpha}} \widetilde{W}_n. \tag{2.4.4}$$

Then, for all x's,

$$P(\widetilde{W}_n^* \leq x) \to \widetilde{G}_{c\alpha}(x). \tag{2.4.5}$$

Proposition 4 *Suppose that*

$$F(x) \sim \frac{c}{|x|^{\alpha}}, \quad as \ x \to -\infty, \tag{2.4.6}$$

for some positive c and α. *Let* $\underset{\sim}{W}_n^* = \frac{1}{n^{1/\alpha}} \underset{\sim}{W}_n$. *Then, for all x's,*

$$P(\underset{\sim}{W}_n^* \leq x) \to \underset{\sim}{G}_{c\alpha}(x). \tag{2.4.7}$$

So,

$$\widetilde{W}_n = n^{1/\alpha} \widetilde{W}_n^*, \tag{2.4.8}$$

where $\widetilde{W}_n^*$ is a r.v. having a distribution close to $\widetilde{G}_{c\alpha}$. A similar representation is true for $\underset{\sim}{W}_n$.

EXAMPLE 1. Let X_i's have the density $f(x) = \dfrac{3}{2(1+|x|)^4}$. Because $f(x) \sim \dfrac{3}{2x^4}$ as $x \to \infty$, the tail $\overline{F}(x) = \int_x^{\infty} f(u)du \sim \frac{3}{2}\int_x^{\infty} \frac{1}{u^4} du = \frac{1}{2x^3}$. Thus, $c = 1/2$, $\alpha = 3$, and

$$\widetilde{W}_n = n^{1/3} \widetilde{W}_n^*,$$

where $\widetilde{W}_n^*$ has the limiting distribution $\widetilde{G}_{\frac{1}{2},3}(x) = \exp\{-1/(2x^3)\}$ for $x > 0$.

The distribution of X_i is symmetric, so the constants c and α in (2.4.6) are the same as above, and

$$\underset{\sim}{W}_n = n^{1/3}\underset{\sim}{W}_n^*,$$

where $\underset{\sim}{W}_n^*$ has the limiting distribution $\underset{\sim}{G}_{\frac{1}{2},3}(x) = 1-\exp\{-1/(2|x|^3)\} = 1-\exp\{1/(2x^3)\}$ for $x < 0$. □

Proof of Proposition 3. In the case of positive X's (which actually does not restrict generality since $\widetilde{W}_n \to \infty$ anyway), we may again appeal to Proposition 1 writing $\widetilde{W}_n = 1/\min\{1/X_1,...,1/X_n\}$ and considering the r.v.'s $Y_i = 1/X_i$.

However, to avoid formal restrictions or additional reasoning, we present a direct proof. Similarly to the proof of Proposition 1, by (1.1.4) and condition (2.4.3), for $x > 0$

$$P(\widetilde{W}_n^* \le x) = P(\widetilde{W}_n \le xn^{1/\alpha}) = \left(F(xn^{1/\alpha})\right)^n = \left(1-\overline{F}(xn^{1/\alpha})\right)^n$$

$$= \left(1 - \frac{c}{(xn^{1/\alpha})^\alpha} - o\left(\frac{c}{(xn^{1/\alpha})^\alpha}\right)\right)^n = \left(1 - \frac{c}{x^\alpha n} - o\left(\frac{c}{x^\alpha n}\right)\right)^n \to \exp\{-c/x^\alpha\}.$$

Regarding $x \le 0$, we know that $\widetilde{W}_n \xrightarrow{P} \infty$, which in particular, implies that $P(\widetilde{W}_n^* \le 0) = P(\widetilde{W}_n \le 0) \to 0$. ■

The proof of Proposition 4 is similar. We also may write $\underset{\sim}{W}_n^* = -\max\{-X_1,...,X_n\}$ and apply Proposition 3.

In conclusion, note that the results above do not solve the problem in full because the behavior of the d.f. $F(x)$ as $x \to 0$, or of the tail $\overline{F}(x)$ as $x \to \infty$ may be determined by functions different from power functions. For example, it may turn out that $\overline{F}(x) = h(x)/x^\alpha$, where $h(x)$ is a slowly varying function; say, $\ln x$. It may be shown that in such cases, the proper normalization is slightly different, but the limiting distributions are the same.

The case when $\overline{F}(x) \to 0$, as $x \to \infty$, faster than a power function—say, as an exponential function—is also not covered above, but is touched on in Exercise 25.

Limit theorems for maxima and minima in a more general situation may be found, e.g., in [13], [42].

---

3 HAZARD RATE AND RELIABILITY. SURVIVAL PROBABILITIES

1,2

3.1 A definition and properties

Consider a positive r.v. X with a d.f. $F(x)$. In this section, we will interpret X as the lifetime of a system, or a product, or a human being. In the first two cases, we will also refer to X as "the time to failure." As was already mentioned, the function $s(x) = P(X > x) = 1 - F(x)$ is called sometimes a *reliability function*. In the case where X is the lifetime of a human being, $s(x)$ is called a *survival* function. Note that $s(x)$ is non-increasing, and $s(0) = 1$ because X is positive: once a system exists, its lifetime is positive.

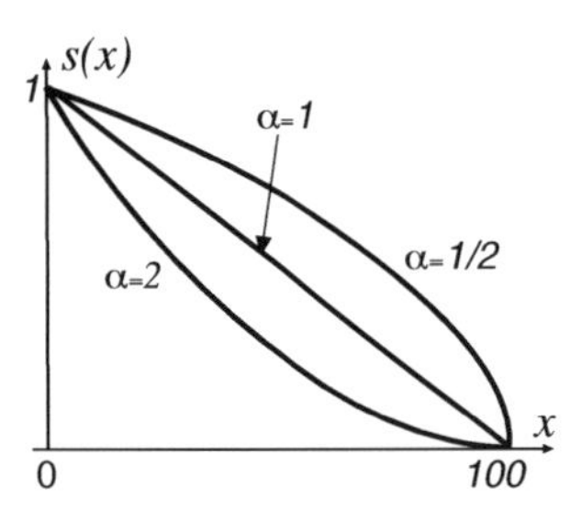

FIGURE 7.

EXAMPLE 1 serves merely for study purposes. In three different countries, typical survival functions $s(x) = [1 - x/100]^{\alpha}$ for $0 \leq x \leq 100$, where $\alpha = 0.5$, 1 and 2, respectively, and time x is measured in years. In which country do people live the longest?

Since $s(100) = 0$, people in these countries do not live more than 100 years, and $s(x) = 0$ for $x > 100$. The graphs are given in Fig. 7. For $\alpha = 1$, the distribution is uniform on $[0,100]$. (In this case, $P(X > x) = \frac{100-x}{100} = 1 - x/100$.)

For *any* x, the probability to survive x years is larger for $\alpha = 1/2$ than for other α's. For example, $s(90)$ equals ≈ 0.32, 0.1, and 0.01 for $\alpha = 0.5$, 1, and 2, respectively, so the first country is "much better" in terms of longevity. □

Henceforth, we assume that there exists the density $f(x) = F'(x)$. For an infinitesimal interval $(x, x+dx]$, as usual, we write that

$$P(x < X \leq x + dx) = f(x)dx. \tag{3.1.1}$$

Consider the conditional probability $P(x < X \leq x + dx \,|\, X > x)$, the probability that the system under consideration will fail within the interval $(x, x+dx]$, given that the system has survived time x. In view of (3.1.1),

$$P(x < X \leq x + dx \,|\, X > x) = \frac{P(x < X \leq x + dx,\, X > x)}{P(X > x)} = \frac{P(x < X \leq x + dx)}{P(X > x)} = \frac{f(x)dx}{s(x)}, \tag{3.1.2}$$

provided that $s(x) \neq 0$. Set

$$\mu(x) = \frac{f(x)}{s(x)}, \tag{3.1.3}$$

again assuming that $s(x) \neq 0$. If $s(x) = 0$, we set $\mu(x) = \infty$ by definition.

From (3.1.2), it follows that for $s(x) \neq 0$,

$$P(x < X \leq x + dx \,|\, X > x) = \mu(x)dx. \tag{3.1.4}$$

In the general probability theory, the function $\mu(x)$ is called a *hazard rate function* or shortly a *hazard rate*. In the case where the r.v. X is the lifetime of a human being, $\mu(x)$ is usually called a *force of mortality*.

The larger $\mu(x)$, the larger the probability that the system of age x will fail "soon"; i.e., within a small time interval.

Since $f(x) = F'(x)$ and $F(x) = 1 - s(x)$, we can also write that

$$\mu(x) = -\frac{s'(x)}{s(x)}. \tag{3.1.5}$$

Sometimes, it is convenient to present (3.1.5) in the form

$$\mu(x) = -\frac{d}{dx} \ln s(x). \tag{3.1.6}$$

Three examples below illustrate possible situations and are relevant to the classification of tails considered in Section **6**.2.2.

EXAMPLE 2. Let X be exponential with parameter a. Then $s(x) = e^{-ax}$, $f(x) = ae^{-ax}$ and by (3.1.3), $\mu(x) = a$. Thus,

> In the lack-of-memory case, the hazard rate is constant.

EXAMPLE 3. Now, consider the case when $s(x)$ is decreasing slower than any exponential function; for example, $s(x) = 1/(1+x)^{\alpha}$ for some $\alpha > 0$ and all $x \geq 0$. By (3.1.5),

$$\mu(x) = \frac{\alpha/(1+x)^{\alpha+1}}{1/(1+x)^{\alpha}} = \frac{\alpha}{1+x}.$$

For a time to failure, this is a non-realistic model: this would mean that the larger the time that has already lasted, the less the chances of failure.

EXAMPLE 4. Let now $s(x)$ be decreasing faster than any exponential function, for instance, $s(x) = e^{-x^2}$. Then, by (3.1.5), $\mu(x) = \dfrac{2xe^{-x^2}}{e^{-x^2}} = 2x \to \infty$ as $x \to \infty$, which is much more realistic for the lifetime of a system or an individual. $\square$

Now, assume that the hazard rate $\mu(x)$ is given, and we want to find the survival function $s(x)$. Certainly, $\mu(x)$ should be non-negative. Since $\mu(x)$ is given, (3.1.5) or (3.1.6) may be considered an equation for $s(x)$. Because $s(0) = 1$, from (3.1.6) it follows that $\ln s(x) = -\int_0^x \mu(z)dz$, and

$$s(x) = \exp\left\{-\int_0^x \mu(z)dz\right\}. \tag{3.1.7}$$

(The reader may also check that (3.1.7) is a solution to (3.1.5) by substitution.)

The representation (3.1.7) is nice and convenient. Above all, it may be viewed as a generalization of the representation for the exponential distribution. If the hazard rate $\mu(x)$ equals some $a > 0$, then (3.1.7) implies that $s(x) = \exp\{-\int_0^x adz\} = e^{-ax}$; that is, we are dealing with an exponential distribution. In general, the integral in the exponent in (3.1.7) is a non-linear and non-decreasing (since $\mu(x) \geq 0$) function of x.

EXAMPLE 5. A company is willing to offer a warranty on a particular product for k integer years, in a way that the chance of product failure during this period does not exceed 2%. Suppose that the hazard function for the time to failure is $\mu(t) = 0.004t$. Then the tail probability $s(x) = \exp\left\{-\int_0^x 0.004tdt\right\} = \exp\{-0.002x^2\}$. We should specify the largest integer x for which $s(x) \geq 0.98$. This inequality is true for $x \leq \left(\dfrac{-\ln(0.98)}{0.002}\right)^{1/2} \approx 3.18$. So, $s(3) > 0.98$, while $s(4) < 0.98$. The warranty should cover three years.

EXAMPLE 6 (*Analytical laws of mortality*). Over the years, there has been a great deal of interest in finding analytical representations for survival functions of "real" people. Such attempts were based on the belief that the duration of human life is subject to some

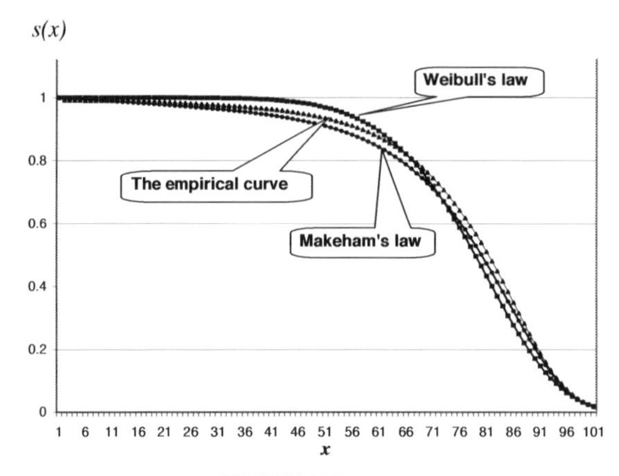

FIGURE 8.

universal laws. These laws, if any, have not been found yet, but some good approximations of real survival functions were suggested, and these approximations were given in terms of forces of mortality.

The first approximation is referred to as the *Gompertz–Makeham law* and consists in the presentation $\mu_{GM}(x) = Be^{\alpha x} + A$, where A, B, and α are parameters which may be estimated basing on statistical data.

Another, also well known, approximation deals with the already familiar for us *Weibull distribution* and uses the presentation $\mu_W(x) = Cx^{\beta}$, where C and β are also parameters to be estimated.

In Exercise 32, the reader is invited to double-check by using (3.1.7) that these representations lead to the respective survival functions

$$s_{\text{GM}}(x) = \exp\{-Ax - B(e^{\alpha x} - 1)/\alpha\}, \quad \text{and} \quad s_{\text{W}}(x) = \exp\{-Cx^{\beta+1}/(\beta+1)\}.$$

In Fig. 8, we present three curves. The first is the graph of the empirical survival function based on the real statistical data from the "United States Life Tables, 2007" [3]. The second curve is the graph of $s_{GM}(x)$ with values of the parameters A, B, and α, which fit the real data. The third curve is the graph of $s_W(x)$ with the parameters C and β adjusted to the real data. Namely, we took

$$s_{\text{GM}}(x) = \exp\{-0.001x - 0.0005(e^{0.09x} - 1)\}, \quad s_{\text{W}} = \exp\{-4(x/100)^7\}.$$

We see that visually both curves fit reasonably well. The parameters chosen are not optimal and serve merely for illustrative purposes. □

Now, consider representation (3.1.7) for $x = \infty$, defining $s(\infty)$ as $\lim_{x\to\infty} P(X > x)$. We interpret $s(\infty)$ as the probability of "existing forever." If X is the lifetime of an individual or a system, in order to be realistic, we should set $s(\infty) = 0$. It follows from (3.1.7) that for this to be true, we should have $\lim_{x\to\infty} \int_0^x \mu(z)dz = \infty$, or in another notation,

$$\int_0^{\infty} \mu(z)dz = \infty. \tag{3.1.8}$$

EXAMPLE 7. Can a force of mortality $\mu(x)$ equal $1/(1+x^2)$? Certainly not. If it were true, we would have had

$$s(\infty) = \exp\left\{-\int_0^\infty \frac{1}{1+z^2}dz\right\} = \exp\{-\arctan z|_0^\infty\} = e^{-\pi/2} \neq 0. \ \square$$

If X is a r.v. in a more general setting, condition (3.1.8) may be non-necessary. For example, in the case of an insurance policy covering accidental death, the company is interested only in the random time X when a lethal accident will occur. Since such an accident may not happen at all, there is nothing unnatural in the assumption $P(X < \infty) < 1$. (Such a r.v. and its distribution is called *defective* or *improper*).

Now, we consider n independent positive r.v.'s $X_1, ..., X_n$ with respective hazard rate functions $\mu_1(x), ..., \mu_n(x)$ and observe the following fact.

Proposition 5 *The hazard rate of the r.v.* $\underset{\sim}{W}_n = \min\{X_1, ..., X_n\}$ *is the function*

$$\mu(x) = \mu_1(x) + ... + \mu_n(x). \tag{3.1.9}$$

Proof is short: in accordance with (1.1.2), the survival function for $\underset{\sim}{W}_n$ is

$$\begin{aligned} P(\underset{\sim}{W}_n > x) &= \prod_{i=1}^n P(X_i > x) = \prod_{i=1}^n \exp\left\{-\int_0^x \mu_i(z)dz\right\} = \exp\left\{-\sum_{i=1}^n \int_0^x \mu_i(z)dz\right\} \\ &= \exp\left\{-\int_0^x \left(\sum_{i=1}^n \mu_i(z)\right) dz\right\}. \ \blacksquare \end{aligned}$$

EXAMPLE 8. Suppose that there are n possible causes of the failure of a system. Assume that the causes "act" independently, and denote by X_i the time of the failure under the assumption that all causes excepting i do not act. Then the real time to failure is $\underset{\sim}{W}_n = \min\{X_1, ..., X_n\}$. We see that the hazard rates for different independent causes are being added up.

EXAMPLE 9. Suppose $X_1, ..., X_n$ are exponential with respective parameters $a_1, ..., a_n$. Then $\mu_i(x) = a_i$ for all i, and the hazard rate for the minimum $\underset{\sim}{W}_n$ is $a_1 + ... + a_n$; that is, constant. Thus, the minimum of independent exponential r.v.'s is an exponential r.v., which we have already proved in Example 1.1-4. Proposition 5 generalizes the result of this example. $\square$

3.2 "Remaining lifetimes"

Let again X be the lifetime of a system that may be a product, or a device, or an individual if we deal with people. Suppose that the system has attained an age x, and consider the future (remaining) lifetime $T(x)$. The distribution of the r.v. $T(x)$ is the *conditional* distribution of $X - x$ given $X > x$. In particular, the survival probability

$$P(T(x) > t) = P(X > x + t \,|\, X > x).$$

This is the probability that a system of age x will "function" during a period of a length of t or longer. In some areas such as Demography or Actuarial Modeling, the traditional notation for this *conditional* survival function is ${}_tp_x$.

Given a survival function $s(x)$,

$$ {}_tp_x = P(X > x+t \mid X > x) = \frac{P(X > x+t)}{P(X > x)} = \frac{s(x+t)}{s(x)}, \tag{3.2.1} $$

provided $s(x) \neq 0$.

From (3.2.1) and (3.1.7), it follows that

$$ {}_tp_x = \frac{s(x+t)}{s(x)} = \frac{\exp\{-\int_0^{x+t}\mu(z)dz\}}{\exp\{-\int_0^{x}\mu(z)dz\}} = \exp\left\{-\int_x^{x+t}\mu(z)dz\right\}. \tag{3.2.2} $$

EXAMPLE 1. What will happen if the force of mortality is doubled? From (3.2.2), it follows that if the new force of mortality, say, $\mu^*(x) = 2\mu(x)$, then the new probability ${}_tp_x^* = ({}_tp_x)^2$. A traditional example concerns non-smokers and smokers. Assume that in a country, the force of mortality for non-smokers is half that of smokers for all x's. Assume that for 20-year-old non-smokers, the probability of attaining age 70, that is, ${}_{50}p_{20} = 0.95$. Then for smokers it is $(0.95)^2 = 0.9025$, so the difference is not dramatic. However, if for a 65-year-old non-smoker the probability to live at least 15 years more, i.e., ${}_{15}p_{65} = 0.4$, then for smokers, this probability will be much less: $0.4^2 = 0.16$. □

EXAMPLE 2. Usually, in mortality tables, the original information is given in terms of probabilities $p_x = {}_1p_x$ (the index 1 is, as a rule, omitted), and other probabilities are computed with use of the recurrence relation

$$ {}_tp_x = p_x \cdot {}_{t-1}p_{x+1}. \tag{3.2.3} $$

From a heuristic point of view, (3.2.3) is almost obvious. To attain age $x+t$, the person of age x should survive the first year (the probability of this is p_x) and *after that*, being $x+1$ years old, the person should live at least $t-1$ years. In Exercise 38, the reader is invited to carry out a formal proof using conditional probabilities.

Applying (3.2.3) consecutively, we obtain that

$$ {}_tp_x = p_x \cdot {}_{t-1}p_{x+1} = p_x \cdot p_{x+1} \cdot {}_{t-2}p_{x+2} = \dots = p_x \cdot p_{x+1} \cdot \dots \cdot p_{x+t-1}. \tag{3.2.4} $$

EXAMPLE 3. Suppose that we are dealing with an individual chosen at random from a population group. Find ${}_{10}p_{30}$ if the force of mortality $\mu(x) = 1/70$ for all x's. We saw that a constant force of mortality corresponded to the exponential distribution with the parameter equal to the (single) value of $\mu(x)$. Thus, X is exponential with parameter $a = 1/70$. In view of the lack-of-memory property, ${}_tp_x = P(X > x+t \mid X > x) = P(X > t) = s(t) = e^{-at}$. Thus, ${}_{10}p_{30} = \exp\{-\frac{1}{70} \cdot 10\} \approx 0.866$.

To what extent is this example realistic? Certainly, we cannot assume that the lack of memory property is true for the total lifetime. How long a person will live does depend on her/his age. Consider, however, a young person and the probability that she/he will not die within a fixed and relatively short period of time. Above, it was the probability for a 30-year-old person to live at least 10 years more. In this case, the assumption of the constancy of the mortality rate is not artificial since the causes of death in this case are weakly related to age. We continue this discussion in Exercise 40.

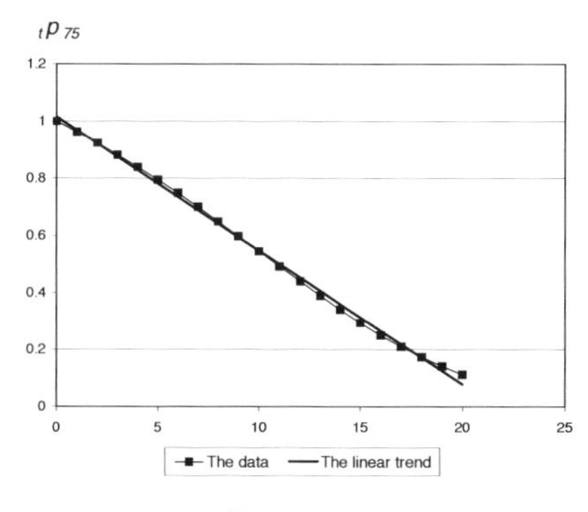

FIGURE 9.

EXAMPLE 4. The curve with dots in Fig. 9 presents an estimate of the survival function $_tp_x$ for $x = 75$, based on the data from the already mentioned "United States Life Tables, 2007" [3]. The straight line indicates the linear trend evaluated by Excel. We know (see, e.g., Example 3.1-1) that the linear survival function corresponds to the uniform distribution. So, we watch, actually, an amazing fact: the distribution of the remaining lifetime $T(75)$ of people aged 75 is very close to the distribution uniform on $[0,20]$. For example, it is almost equally likely that a chosen at random 75-year-old person will die at age 76 or at age 94. See also Exercise 44. □

---

4 EXERCISES

Section 1

1. For a husband and wife, the future lifetimes are independent continuous r.v.'s X_1 and X_2. Suppose the tail functions $\overline{F}_1(x) = P(X_1 > x)$ and $\overline{F}_2(x) = P(X_2 > x)$ are given. Find the probabilities that (a) both will survive n years; (b) exactly one will survive n years; (c) no one will survive n years; (d) at least one will survive n years; (e) both will die in the nth year.

2. (a) In a closed auction, the buyers submit sealed bids so that no bidder knows the bids of the others. The highest bid wins. Suppose that the bids are independent r.v.'s uniformly distributed on $[a, a+d]$. For the case of n bidders, find the distribution and expectation of the price for which the item will be sold. (*Advice*: It is convenient to consider "surplus" values uniformly distributed on $[0,d]$, and add a at the end.)

   (b) (i) Do the same if the number of bidders is a Poisson r.v. N with parameter λ. (*Advice*: Condition on N.) (ii) Assuming λ to be integer, compare the mean value with that of part (a) for $n = \lambda$. (iii) The reader who did Exercise **6**-17, may compare the result with that of part (a) in the sense of the first stochastic dominance.

3. Each day, John is jogging the same route. The times it takes are i.i.d. r.v.'s assuming values between 30 and 31 minutes. Find the distribution of the best result in a week and its expected value in the case where for each separate day (a) the distribution is uniform on $[30,31]$; (b) the distribution has the density $f(x) = 2(x-30)$ for $x \in [30,31]$. Compare the expectations and interpret the difference.

4. Let $X_1, X_2, X_3, ...$ be independent exponential r.v.'s, and let $E\{X_i\} = 2^i$. Find the distribution of $\min\{X_1, X_2, ...\}$ or, more precisely, the limiting distribution of $\underset{\sim}{W}_n$ for $n \to \infty$.

5. Comment on the behavior of the expectations in Example 1.1-3 for large n. If you pick (or simulate) 100 independent numbers from $[0,1]$, to what will the largest and the smallest number be close? The reader who chose Route 2, may also come back to this question after she/he reads Section 2.1.

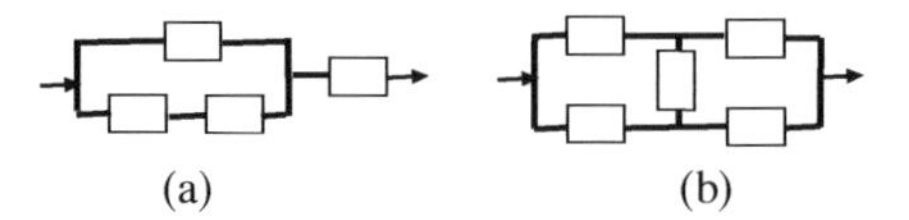

FIGURE 10.

6. The time to failure for each of particular standard electronic devices has the density $f(x) = 64/x^5$ for $x \geq 2$. Show that the corresponding r.v.'s are larger than 2 with probability one. Let T_n be the time to failure for a series of n such devices. Guess to what $E\{T_n\}$ should converge as $n \to \infty$. Find $E\{T_n\}$ and check your guess.

7. Consider the system with the configuration presented in Fig. 10a and suppose that all components work independently. Let the time to failure for a separate component have a d.f. $F(x)$. Find the d.f. of the time to failure for the whole system. Will your answer change if the signal moves in the opposite direction? Justify the answer rigorously. Compute the expected time for the whole system for the distribution F uniform on $[0,1]$.

8. A system consists of two components, and for the system to work, at least one component should work. The times to failure for the components are independent and exponential with a mean of one unit of time. For which time period is the probability that the system will not fail during this period is larger than 0.95?

9. Consider the configurations in Fig. 10b, suppose that the lifetimes of the separate components are independent, denote by $G(x)$ the d.f. of the lifetime of the "middle" component, and suppose that the other components have the same lifetime d.f. $F(x)$. Find the d.f. of the system lifetime. (*Advice*: Condition on the middle component.)

10. Consider i.i.d. r.v.'s $X_1, X_2, \ldots$ assuming a finite number of values $x_1 < x_2 < \ldots < x_m$. (a) Describe the behavior of the r.v.'s $\underset{\sim}{W}_n$ and $\widetilde{W}_n$ heuristically. (b) Provide a rigorous explanation. How "fast" do the distributions of $\underset{\sim}{W}_n$ and $\widetilde{W}_n$ approach their limits (if any)? (*Hint*: You will need the notations $f_1 = P(X_i = x_1)$ and $f_m = P(X_i = x_m)$.) (c) Describe the cases where we cannot expect a similar behavior of $\underset{\sim}{W}_n$ and $\widetilde{W}_n$.

*Section 2**

11. Prove (2.1.2).

12. What is α and the distribution of $\underset{\sim}{W}_n^*$ in the case (2.2.5) if the X's are exponential?

13. Write the limiting distribution and representation (2.2.5) for the X's having (a) the density $f(x) = 2x$ for $x \in [0,1]$; (b) the density $f(x) = \frac{1}{2}(2-x)$ for $x \in [0,2]$.

14. Write the limiting distribution and representation (2.2.6) for the X's having the density $f(x) = 2(x-1)$ for $x \in [1,2]$. Compare with Exercise 13a.

15. Write the limiting distribution and representation (2.3.4) for the following situations: (a) the density of the X's is $f(x) = 2(x-1)$ for $x \in [1,2]$; (b) the density of the X's is $f(x) = 3(2-x)^2$ for $x \in [1,2]$; (c) the r.v. $X_i = -\xi_i$, $i = 1,2,\ldots$, where ξ_i are independent standard exponential r.v.'s.

16. Show that if in the normalization procedure (2.2.2), we multiply $\underset{\sim}{W}_n$ by $(cn)^{1/\alpha}$, then the limiting distribution will be the Weibull distribution $Q_{1\alpha}(x) = 1 - \exp\{-x^{\alpha}\}$.

17. Regarding Propositions 1 and 2, does the increase of the parameter α lead to the faster or slower rate of convergence to the limit? Give a rigorous answer and a heuristic common-sense interpretation.

18. Revisit Example 2.3-3 and prove that the expected value of the estimate $\widetilde{W}_n$, that is, $E\{\widetilde{W}_n\} = \frac{n}{n+1}\theta$. Such an estimate is called biased: its expected value is not equal to the parameter we are estimating. By what should we multiply $\widetilde{W}_n$ for the new estimate, $\widetilde{V}_n$, to have the following two properties: first, $\widetilde{V}_n$ is unbiased, that is, $E\{\widetilde{V}_n\} = \theta$; secondly, the convergence $\widetilde{V}_n \xrightarrow{P} \theta$ still has an order of $1/n$?

19. Let $X_1,...,X_n$ have Weibull's distributions $Q_{c_1\alpha},...,Q_{c_n\alpha}$, respectively. Show that the $\min\{X_1,...,X_n\}$ has the distribution $Q_{c\alpha}$ with $c = c_1 + ... + c_n$. That is, the Weibull distribution is stable with respect to minimization. Connect this fact with the result of Proposition 5.

20. Make sure that you understand why the graphs of $\widetilde{G}_\alpha$ and $\underset{\sim}{G}_\alpha$ indeed look as they are depicted in Fig. 6.

21. Write the limiting distribution and representation (2.4.8) for the X's having the density $f(x) = 64/x^5$ for $x \geq 2$.

22. Write the limiting distribution, representation (2.4.8) and a similar representation for $\underset{\sim}{W}_n$ for X_i's having the Cauchy distribution (that has the density $f(x) = 1/(\pi(1+x^2))$.

23. Regarding Proposition 3, does the increase of the parameter α lead to the faster or slower rate of convergence to $\pm\infty$? Give a rigorous answer and a heuristic common-sense interpretation.

24. (a) Show that if in the normalization procedure (2.4.4), we divide $\widetilde{W}_n$ by $(cn)^{1/\alpha}$, then the limiting distribution will be the distribution $\widetilde{G}_{1\alpha}(x)$.

    (b) Let $X_1,...,X_n$ have the distributions $\widetilde{G}_{c_1\alpha},...,\widetilde{G}_{c_n\alpha}$, respectively. Show that then $\max\{X_1,...,X_n\}$ has the distribution $\widetilde{G}_{c\alpha}$ with $c = c_1 + ... + c_n$. One may say that $\widetilde{G}_{b\alpha}$ is stable with respect to maximization.

    (c) State a similar property for $\underset{\sim}{G}_{c\alpha}(x)$.

25. Let $X_1,...,X_n$ be independent standard exponential r.v.'s, and $\widetilde{W}_n = \max\{X_1,...,X_n\}$. Prove that $\widetilde{W}_n$ grows as $\ln n$; more precisely, $\widetilde{W}_n = \ln n + V_n$, where V_n is a r.v. whose d.f. $P(V_n \leq x) \to G(x) = \exp\{-e^{-x}\}$. This distribution is called *the double exponential distribution.* Graph the d.f. $G(x)$. Is it the d.f. of a positive r.v.? Compare the order of the growth and the type of normalization with these in Proposition 3. Give some heuristic comments on it. (*Hint*: $P(\widetilde{W}_n \leq x) = (1-e^{-x})^n$, and hence, $P(\widetilde{W}_n - \ln n \leq x) = (1-\exp\{-(x+\ln n)\})^n$.)

26. Do Propositions 1 and 2 cover the situation of Exercise 10? (*Hint:* To which distribution does the exponential distribution $Q_{c\alpha}$ converge as $c \to \infty$?)

Section 3

27. Suppose that in a country, for people who survived 30 years, the probability of dying before 40 years is negligible. How does the survival function look in this case?

28. Show by differentiation that (3.1.7) is a solution to (3.1.5).

29. Is it true that a hazard rate function $\mu(x)$ is constant if and only if the distribution is exponential?

30. Find the force of mortality function for the cases of Example 3.1-1. Write $s(120)$.

31. Find the hazard rate function $\mu(x)$ for the distribution uniform on $[0,a]$. What is $\lim_{x\to a}\mu(x)$? Give a common-sense interpretation.

32. Find the survival functions for the Gompertz–Makeham and Weibull laws in Example 3.1-6.

33. Let $\mu(x) = (1+x)^{-1}$. Find the survival function. What is noteworthy?

34. (a) How should the mortality force change in order that the percent of newborns who will survive the first year will increase twice? (b) Do the same for the case of the first k years.

35. In a country, for a typical person of age 50, the probability to survive 80 years equals 0.5. Find the same probability for a country where (a) the force of mortality for people of age fifty and older is three times higher; (b) the force of mortality for people of age fifty and older is 0.01 less.

36. Consider the r.v. $\underset{\sim}{W}_n$ in Proposition 5. For $P(\underset{\sim}{W}_n < \infty) = 1$, should $\int_0^\infty \mu_i(x)dx = \infty$ for all $i = 1, ..., n$? Give a rigorous proof and a common-sense explanation.

37. Prove that for X uniform on $[0, a]$, the r.v. $T(x)$ is uniformly distributed on $[0, a-x]$. Comment on this result.

38. Prove (3.2.3) rigorously. (*Advice*: You may start with ${}_tp_x = P(X > x+t \,|\, X > x) = \frac{P(X>x+t)}{P(X>x)} = \frac{P(X>x+1)}{P(X>x)} \cdot \frac{P(X>x+t)}{P(X>x+1)} = P(X > x+1 \,|\, X > x)P(X > x+t \,|\, X > x+1) = ...$)

39. In a country, the survival function for women is closely approximated by $s_f(x) = (1 - x/100)^{1/2}$, while for men, it is $s_m(x) = (1 - x/90)^{1/2}$. We assume that the probability of the birth of a boy is $1/2$. Let N_m and N_f be the number of men and women of age x. Estimate $E\{N_f\}/E\{N_m\}$ for $x = 0$ and $x = 50$. Give a heuristic explanation for why the ratio mentioned is greater for $x = 50$.

40. (a) For young people, causes of death are related mostly to accidents. Proceeding from this, explain why the assumption of the approximate constancy of $\mu(x)$ may look reasonable for x's between 30 and 40 years.

    (b) For what x's do we need to know values of $\mu(x)$ to compute the probability that a thirty year old person will survive 40 years? Which formula shows it explicitly? Relate this question to part (a).

41. Which is larger: the probability that a system will work 100 hours given that it has been working 50 hours or the same probability given that the system has been working 60 hours? State the question in general, using letters rather than numbers, and give a rigorous proof. (*Advice*: Show that ${}_tp_x = {}_sp_x \cdot {}_{t-s}p_{x+s}$ for $s \le t$.)

42. In a city, homeowners use two types of heaters. The heaters of the first type constitute 30%, and their lifetimes are exponentially distributed with a mean of 10 years. The heaters of the second type are cheaper, and their lifetime is also exponential but the mean time is 8 years. Thus, the distribution of the lifetime of a new heater that has been just bought is a mixture of exponential distributions. (a) Is the same true for a heater chosen at random from those whose "age" is five years? (b) Find the distribution of the remaining lifetime of such a heater. Interpret your results from a common-sense point of view. (Advice: Use (3.2.1).)

43. For a r.v. $|X|$ where X is standard normal, describe the behavior of the hazard rate $\mu(x)$ for large x's. (*Advice*: Use Proposition 2 from Section **6**.2.4.)

44. Proceeding from Fig. 8, argue that the phenomenon described in Example 3.2-4, does not take place for a lower age; say, for $x = 20$, $x = 50$, or even for $x = 60$.

Chapter 12

Stochastic Processes in a General Setup: Preliminaries and Some Examples

1,2

This chapter is concerned with a general framework in which we consider stochastic processes. We also discuss some general properties of processes. In Chapter 13, we turn to concrete schemes.

1 GENERAL DEFINITION

As we have already defined in Chapter 5, a *stochastic or random process* is a collection of r.v.'s $\{X_t\}$, where t is a running parameter. If we view t as time, the process X_t may present the evolution of a characteristic of a phenomenon in time. However, in general, t may be of any nature. For instance, t can represent the distance from the beginning of a trench dug by a gold miner to a particular place in the trench. Then X_t could represent the (random) concentration of gold at point t.

In Chapter 5, we considered a discrete parameter $t = 0, 1, 2, \ldots$, usually interpreting t as a discrete moment of time.

Below, unless the opposite is stated, we also view t as a time parameter, but as a rule suppose that t may take on all values from an interval. In this case, we say that X_t is a process in continuous time. For example, X_t may represent the change of the air temperature during a day.

Let us discuss the definition of a process in more detail. In Chapters 3 and 6, we defined a r.v. as a function $X(\omega)$ on a space $\Omega = \{\omega\}$ of elementary outcomes ω. As a rule, we omitted the argument ω and wrote X instead of $X(\omega)$.

As was already stated, a *stochastic or random process* is a collection of r.v.'s $X_t(\omega)$. So, $X_t(\omega)$ is a function of two arguments: ω and t. For a *fixed* t, the function $X_t(\omega)$, as a function of ω, is a random variable. For a *fixed* ω, we deal with a function of t.

Given ω, the function $X_t = X_t(\omega)$ is also called a (particular) *realization or a trajectory* of the random process.

Thus, when we are dealing with a r.v. $X = X(\omega)$, for each outcome ω, we deal with a *number*: a particular value of X in the case where ω occurred. When we consider a stochastic process $X_t = X_t(\omega)$, for each outcome ω, we deal with a function (or a curve): a particular trajectory in the case where ω occurred.

EXAMPLE 1. Suppose that the sample space Ω consists of only two elementary outcomes: ω_1 and ω_2. Say, we toss a coin. Assume that $X_t(\omega_1) = t$ while $X_t(\omega_2) = t^2$. So,

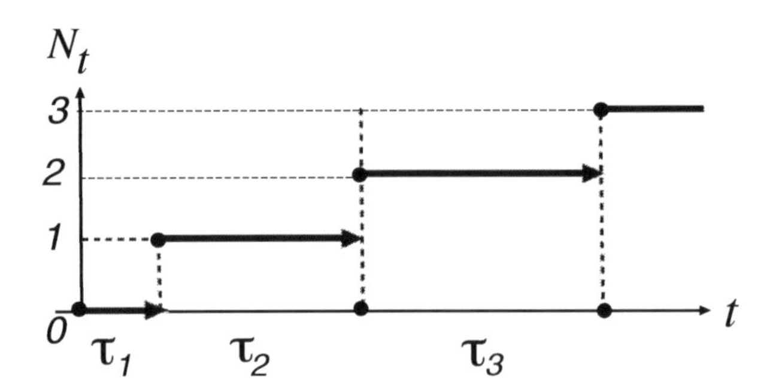

FIGURE 1. A typical realization of a counting process; $\tau_1, \tau_2, \ldots$ are interarrival times

depending on which ω occurs, we deal with either a linear function or a parabola.

In the case of a regular coin, it amounts to a random selection, with equal probabilities, of one out of two functions (rather than numbers): either t or t^2. □

As we did it when dealing with r.v.'s, we again omit the argument ω and write just X_t.

Consider a more realistic

EXAMPLE 2 (A *counting process*). One of simple processes in continuous time is a process $N_t, t \geq 0$, representing the total number of events of a certain type that have occurred prior to or at time t. For example, N_t may be the number of electrical breakdowns in an electric power system by time t, or the number of atomic nuclei that decayed by time t in a piece of radioactive substance.

For brevity, we call the occurrences of the event of interest *arrivals*, so N_t "counts" arrivals. Time-intervals between consecutive arrivals are called *interarrival times* or *sojourn times*.

A *typical realization* of a counting process is shown in Fig. 1. For any t, the value of N_t shows how many arrivals occurred by time t. The symbols τ_i there denote interarrival times. Note that the realization graphed in Fig. 1 is continuous from the right. This reflects the fact that when defining N_t as the number of arrivals during an interval $(0,t]$, we include the end point t. So, if an arrival occurs at the last moment t of the time interval $(0,t]$, we count this arrival. Therefore, for example, in Fig. 1, $N_{\tau_1} = 1$ rather than 0. □

Next, we consider two particular but important types of processes.

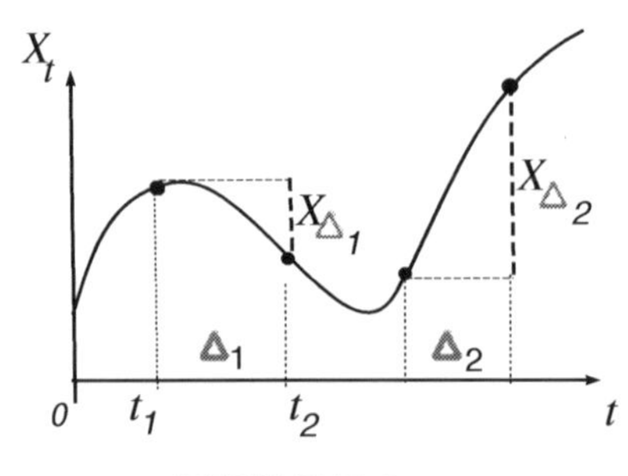

FIGURE 2.

2 PROCESSES WITH INDEPENDENT INCREMENTS

For a half-open interval $\Delta = (t_1, t_2]$, we define the increment of a process X_t as the r.v. $X_\Delta = X_{t_2} - X_{t_1}$. For instance, in Fig. 2 depicting a *particular* realization of a process in continuous time, we consider two intervals: Δ_1 and Δ_2. For each interval, we designate the corresponding increments. Note that for the realization in Fig. 2, $X_{\Delta_1} < 0$ and $X_{\Delta_2} > 0$.

Suppose that X_t is the air temperature at time t. Then for $\Delta = (t_1, t_2]$, the r.v. $X_\Delta = X_{(t_1,t_2]}$ presents the change of the temperature during the period from t_1 to t_2.

A process X_t, $t \geq 0$, is called a process *with independent increments* if for any collection of *disjoint* (or *non-overlapping*) intervals $\Delta_1, ..., \Delta_k$, the r.v.'s X_0, $X_{\Delta_1}, ..., X_{\Delta_k}$ are mutually independent. Say, if the process in Fig. 2 is that with independent increments, then the r.v.'s X_{Δ_1} and X_{Δ_2} are independent. (In Fig. 2, we see only particular values of these r.v.'s.)

Note that since we included X_0 in the definition, all increments of the process do not depend on the *initial value* X_0 of the process.

Suppose again that X_t is the air temperature. Consider two moments: t_1 and $t_2 > t_1$. Clearly, $X_{t_2} = X_{t_1} + X_{(t_1,t_2]}$. Certainly, X_{t_2}, the temperature at time t_2, does depend on X_{t_1}, the temperature at moment t_1, but if X_t is a process with independent increments, then $X_{(t_1,t_2]}$, the *change* of the temperature, does not depend on X_{t_1}. (Of course, we do not claim that this is always the case in reality.)

Consider now a concrete model.

EXAMPLE 1 (The *simplest counting process*). We call a counting process N_t the *simplest* if all interarrival times are independent identically distributed exponential r.v.'s. Let us realize that the process so defined is that with independent increments.

Suppose that we have been watching the process until time t. In particular, at time t, we know how much time has elapsed since the last arrival. However, by virtue of the lack of memory property of the exponential distribution (see Section **6**.2.2), the time we must wait for the next arrival after time t does not depend on how long we have already been waiting: the process starts over as from the very beginning. So, we know that the remaining time until the next arrival has an exponential distribution, and the parameter of this distribution is the same as for a separate interarrival time. This parameter is fixed since we have assumed the interarrival times to be identically distributed.

Furthermore, the next interarrival time also does not depend on what happened before, since the interarrival times are independent and identically distributed (that is, the distribution of an interarrival time does not depend on how many arrivals occurred before). Thus, for any interval $\Delta = (t, t']$, the increment N_Δ does not depend on the evolution of the process before t, and hence N_t is a process with independent increments.

We consider this basic process in detail in Section **13**.1.1 and in particular, show that it is strongly connected with the Poisson distribution. □

Route 1 ⇒ *page 321*

2

3 ONE MORE EXAMPLE: BROWNIAN MOTION

In the case of continuous time, we call a process *continuous* if with probability one its realizations are continuous functions. A counting process (see, for example, Fig. 1) is, clearly, not continuous. The next scheme, in a certain sense, is contrasting to that of counting processes.

Let $w_t, t \geq 0$, be a continuous random process with independent increments such that $w_0 = 0$, and for any interval Δ, the increment w_Δ is a normal r.v. with zero mean and variance $|\Delta|$, where $|\Delta|$ stands for the length of Δ. Such a process is called the *standard Wiener process* or *Brownian motion*.

It is noteworthy that we postulate rather than derive the independence of increments. In Example 1 below, we will discuss how to simulate trajectories of such a process.[1]

Since $w_0 = 0$, we can write $w_t = w_t - w_0 = w_{(0,t]}$, the increment over $(0,t]$. Hence, w_t is a normal r.v. with mean zero and variance t. In particular, w_1 is a standard normal r.v., and for this reason we use in the definition of w_t the term "standard."

The process w_t may be considered a counterpart of a standard normal r.v. when we deal with processes. Originally, the term Brownian motion was used for the motion of a particle totally immersed in a liquid. The name comes from the botanist R. Brown who first considered this phenomenon. It proved that processes of the type $a + bw_t$, where a, b were real numbers, were suitable for modeling such a physical motion. The mathematical theory was built mainly by A. Einstein and N. Wiener. Nowadays, Wiener processes, or Brownian motion, are widely used for modeling various phenomena of different nature, such as diffusion in Physics, certain processes in quantum mechanics, the evolution of stock prices, surplus processes in economics, etc.

EXAMPLE 1. To understand how realizations of w_t look, let us choose a real number $\delta > 0$ and consider points $t_k = k\delta$, where $k = 0, 1, \ldots$; that is, $t_0 = 0, t_1 = \delta, t_2 = 2\delta, \ldots$. Let an interval $\Delta_k = (t_{k-1}, t_k]$. Then

$$w_{t_k} = w_{\Delta_1} + \ldots + w_{\Delta_k}. \tag{3.1}$$

The r.v.'s w_{Δ_k} are independent normal r.v.'s with zero mean and variance δ, because the length of each interval is δ. Set

$$\xi_k = \frac{1}{\sqrt{\delta}} w_{\Delta_k}$$

for $k = 1, 2, \ldots$. Since we divided w_{Δ_k} by its standard deviation, the r.v.'s ξ_k are *standard* normal (see Section (**6**.1.5)) and independent because the intervals Δ_k do not overlap.

Since $w_{\Delta_k} = \sqrt{\delta}\xi_k$, from (3.1) it follows that

$$w_{t_k} = \sqrt{\delta}(\xi_1 + \ldots + \xi_k). \tag{3.2}$$

[1]If we want to be absolutely rigorous, we should prove that such a process exists; that is, may be properly defined as a function $w_t(\omega)$ on a probability space Ω endowed with a probability measure $P(A)$. This is beyond the scope of this book, and we take it for granted. A formal proof can be found in a number of advanced textbooks on random processes, e.g., in [24], [47]. However, the simulation procedure we discuss below makes the existence of the Wiener process more than just plausible.

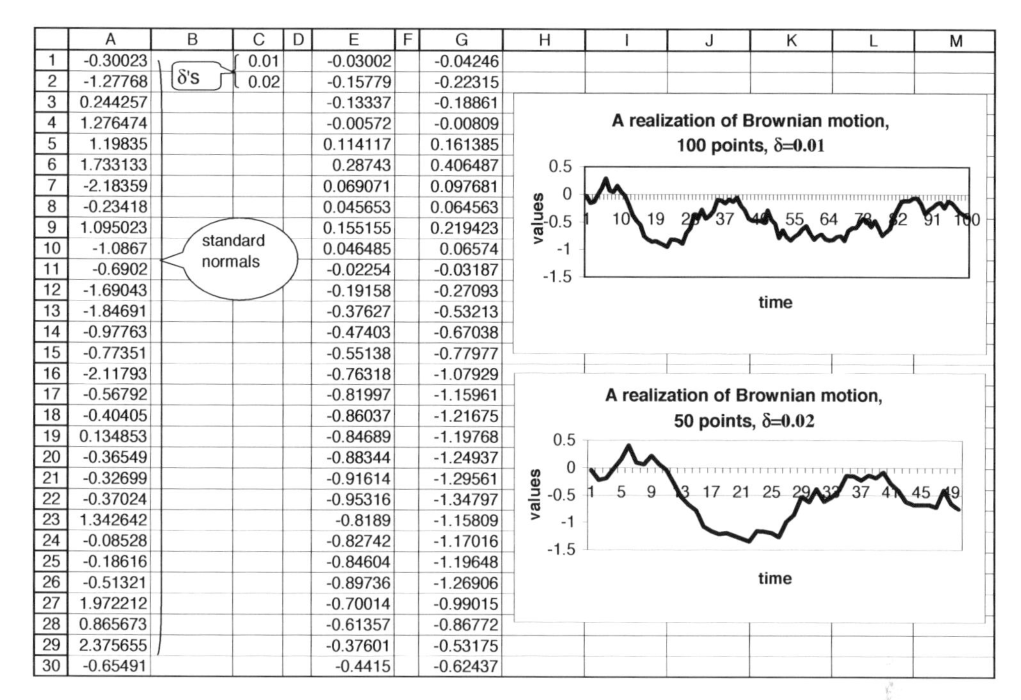

	A	B	C	D	E	F	G	H	I	J	K	L	M
1	-0.30023		0.01		-0.03002		-0.04246						
2	-1.27768		0.02		-0.15779		-0.22315						
3	0.244257				-0.13337		-0.18861						
4	1.276474				-0.00572		-0.00809						
5	1.19835				0.114117		0.161385						
6	1.733133				0.28743		0.406487						
7	-2.18359				0.069071		0.097681						
8	-0.23418				0.045653		0.064563						
9	1.095023				0.155155		0.219423						
10	-1.0867				0.046485		0.06574						
11	-0.6902				-0.02254		-0.03187						
12	-1.69043				-0.19158		-0.27093						
13	-1.84691				-0.37627		-0.53213						
14	-0.97763				-0.47403		-0.67038						
15	-0.77351				-0.55138		-0.77977						
16	-2.11793				-0.76318		-1.07929						
17	-0.56792				-0.81997		-1.15961						
18	-0.40405				-0.86037		-1.21675						
19	0.134853				-0.84689		-1.19768						
20	-0.36549				-0.88344		-1.24937						
21	-0.32699				-0.91614		-1.29561						
22	-0.37024				-0.95316		-1.34797						
23	1.342642				-0.8189		-1.15809						
24	-0.08528				-0.82742		-1.17016						
25	-0.18616				-0.84604		-1.19648						
26	-0.51321				-0.89736		-1.26906						
27	1.972212				-0.70014		-0.99015						
28	0.865673				-0.61357		-0.86772						
29	2.375655				-0.37601		-0.53175						
30	-0.65491				-0.4415		-0.62437						

FIGURE 3. Realizations of Brownian motion on [0,1] with the step $\delta = 0.01$ (100 points) and 0.02 (50 points). In the figure, Columns A, E, and G are truncated (not all numbers are shown).

The last representation gives a method of simulating realizations of Brownian motion.

Let, say, $\delta = 0.01$ and $k = 0, 1, 2, ..., 100$, which means that we divide the interval $[0, 1]$ into 100 equal subintervals, and consider 100 equally spaced points in $[0, 1]$. The first point $t_1 = 1 \cdot 0.01 = 0.01$, $t_2 = 2 \cdot 0.01 = 0.02$, and so on; the last point $t_{100} = 100 \cdot 0.01 = 1$.

In the Excel worksheet in Fig. 3, Column A contains 100 standard normal numbers generated by Excel (not all numbers are shown in the figure). They represent the ξ's.

(To get all numbers in Column A at once, we go to Data=>Data Analysis=>Random Number Generation, and choose "Normal" with parameters 0 and 1 in "Distribution." Then in "Output range", we type the array address A1:A100.)

Cell C1 contains the value of $\delta = 0.01$. The value in E1 is the value of $w_{t_1} = \sqrt{\delta}\xi_1$, and the corresponding command is =SQRT($C$1)*A1. (The parameter δ in C1 is an unchangeable fixed parameter, and in this case, Excel requires to write $ before the row and column numbers.)

The value in E2 =E1+SQRT($C$1)*A2, which comes from $w_{t_2} = w_{t_1} + \sqrt{\delta}\xi_2$. In general, the cell E$k$ equals E($k-1$)+SQRT($C$1)*Ak, which reflects the formula $w_{t_k} = w_{t_{k-1}} + \sqrt{\delta}\xi_k$. The graph of the values in Column E is given in the first chart.

This is the graph of a realization of the Wiener process.

Cell C2 and Column G correspond to the similar simulation for $\delta = 0.02$ and, accordingly, 50 points in the same interval. It is worth emphasizing that this not a part of the previous realization: we took the first 50 values in Column A, but assigned them to 50 equally spaced points in $[0, 1]$. For example, while in the first case the number in Cell

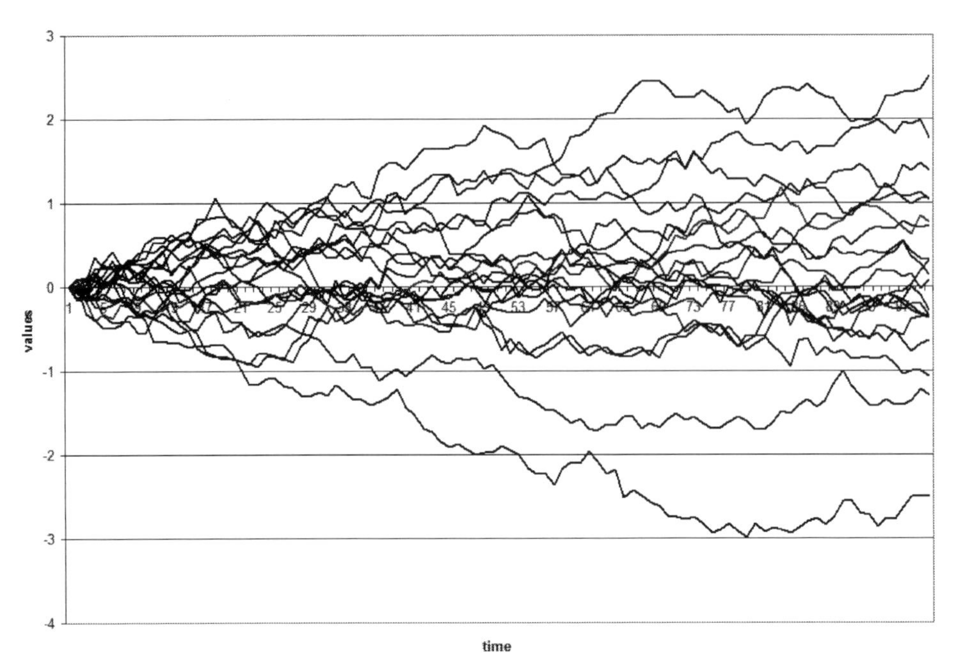

FIGURE 4. Twenty independent realizations of Brownian motion on [0,1] with the same step $\delta = 0.01$ (100 points)

E1, -0.03002, is the value of the process at the point $t = 0.01$, in the second case, the number in Cell G1, -0.04246, is the value of the process at $t = 0.02$ (the command is =A1*SQRT($C$2). Moreover, since in the second realization, we skipped the point $t = 0.01$, the values in Cells G1 and E2 (the value at $t = 0.2$ for the first realization) do not coincide too.

Thus, the second graph may be viewed as a new realization. We see that it looks smoother, which is understandable: taking only 50 points in the same interval, we eliminate 50 independent fluctuations between the chosen points.

The graphs in Fig. 3 represent just two possible realizations of the process. If we regenerate in Column A other normal numbers, the realization will be different. Fig. 4 represents twenty independent realizations with the step $\delta = 0.01$.

The envelopes of the twenty curves in Fig. 4 are consistent with the theory. The distribution of w_t is normal with zero mean and variance t, and hence the distribution of w_t coincides with the distribution of $\sqrt{t}Z$ where Z is standard normal. One may say that w_t has the order $\pm\sqrt{t}$, which is reflected in Fig. 4. See also Exercise 4. □

It may be shown that in the continuous time case, *any process* with independent increments may be represented as a certain combination of the Wiener process and processes of the type bN_t, where b is a number, and N_t is the simplest (later, we will call it Poisson) process from the previous section. Such combinations are referred to as *Lévy's processes*; see, e.g., [13], [24].

We consider the Wiener process in much more detail in Chapter 16.

4 MARKOV PROCESSES

We have already considered the *Markov property* in the case of discrete time, and it is recommended to the reader to look over the definition of this property in Section **5**.1.2. Now, we generalize the notion of the Markov property.

Consider a process $X_t, t \geq 0$, where, just for convenience, t is interpreted as time, and may be discrete or continuous as well. Let us consider time moments $t_1 < ... < t_{n-1} < t_n$. We view t_n as the present time, and $t_1, ..., t_{n-1}$ as time moments in the past. Let $x_1, ..., x_{n-1}, x_n$ be the values of the process at the moments $t_1, ..., t_{n-1}, t_n$, respectively, and let B be a set of real numbers; i.e., a set in the real line.

The process X_t is called *Markov* if, for any $t_1, ..., t_n$, any $x_1, ..., x_n$, any B, and any $s > 0$,

$$P(X_{t_n+s} \in B \,|\, X_{t_n} = x_n, X_{t_{n-1}} = x_{n-1}, ..., X_{t_1} = x_1) = P(X_{t_n+s} \in B \,|\, X_{t_n} = x_n). \tag{4.1}$$

Since we view t_n as the present time, the moment $t_n + s$ belongs to the future, and we see that in the Markov case, *given the value of the process at the present time*, the information of the values of the process in the past is redundant for the prediction of the future values of the process. "Given the present, the future does not depend on the past".

In Chapter 13 and others, we consider many particular examples of Markov processes, but first of all, note that

> Any process with independent increments is a Markov process.

Indeed, consider two moments of time: the "present time" t and the future time moment $t + s$, and observe that $X_{t+s} = X_t + X_{(t,t+s]}$. Hence, X_{t+s} depends only on X_t and the increment $X_{(t,t+s]}$ which, if the process is that with independent increments, does not depend on the values X_u for $u \leq t$ at all.

The converse assertion is not true.

EXAMPLE 1. Let X_t be a process with independent increments, and let $Y_t = e^{X_t}$. In Chapter 16, we use such a process for modeling the evolution of stock prices. We have

$$Y_{t+s} = \exp\{X_{t+s}\} = \exp\{X_t\}\exp\{X_{t+s} - X_t\} = Y_t \exp\{X_{(t,t+s]}\}. \tag{4.2}$$

Since X_t is a process with independent increments, $X_{(t,t+s]}$ does not depend on the values X_u for $u \leq t$, and hence, on the values Y_u for $u \leq t$. Thus, for any fixed t and s, given the value of the r.v. Y_t, the r.v. Y_{t+s} does not depend on Y_u for $u < t$. Consequently, the process Y_t is Markov.

Now, we show that increments of this process may be dependent. Consider intervals $(0,1]$ and $(1,2]$. By definition, $Y_{(0,1]} = Y_1 - Y_0$, and in view of (4.2),

$$Y_{(1,2]} = Y_2 - Y_1 = Y_1 \left(\exp\{X_{(1,2]}\} - 1\right).$$

The r.v. $\exp\{X_{(1,2]}\}$ does not depend on X_1 (and hence on Y_1). On the other hand, both r.v., $Y_{(0,1]}$ and $Y_{(1,2]}$, involve Y_1, and hence may be dependent. □

Route 1 ⇒ *page 324*

2

5 REPRESENTATION AND SIMULATION OF MARKOV PROCESSES IN DISCRETE TIME

To give more insight into the nature of Markov processes, let us consider the discrete time case $t = 0, 1, ...$. In this case, property (4.1) implies that in order to completely determine the joint distribution of all r.v.'s $X_0, X_1, X_2, ...$, it suffices to know the distribution of the "initial" r.v. X_0, and the conditional distribution of X_{t+1} given X_t for all t; say, the conditional distribution function

$$P(X_{t+1} \leq z \,|\, X_t = x). \tag{5.1}$$

(Certainly, once we used the letter x for a value of X_t, we should use another letter when considering values of X_{t+1}.)

We suppose that the conditional d.f. (5.1) does not depend on t. We interpret it as the dependence of the value of the process on its previous value has the same structure for all t's. In this case, we call the process *homogeneous*.

Now, consider the following construction. Let $\xi_1, \xi_2, ...$ be independent r.v.'s , and let $h(x,z)$ be a function of two variables. Let us define a process Y_t, $t = 0, 1, ...$ as follows. The Y_0 is a given r.v.; as above, we call it an *initial value*. For $t = 1, 2, ...$, we define Y_t's by the recurrence relation

$$Y_{t+1} = h(Y_t\,, \xi_{t+1}). \tag{5.2}$$

For example, let Y_t be the capital of a person at time t, and the value of the capital at the next time, i.e., $Y_{t+1} = (1 + \xi_{t+1})Y_t$, where ξ_{t+1} is a random interest over the period $(t, t+1]$.

If we assume ξ's to be independent of the previous history of the process, we will come to the model (5.2) with $h(x,z) = (1+z)x$.

In general, because ξ's are independent, Y_{t+1} depends only on the previous value of the process Y_t and the r.v. ξ_{t+1} which does not depend on the present or the past values of the process. Thus, the process Y_t so defined is Markov.

We have already touched on a representation similar to (5.2) in the case of Markov chains in Section **5**.1.2; now we consider the case when the process Y_t may assume any values.

It is important that, as a matter of fact,

> Any homogeneous discrete-time Markov process admits representation (5.2).

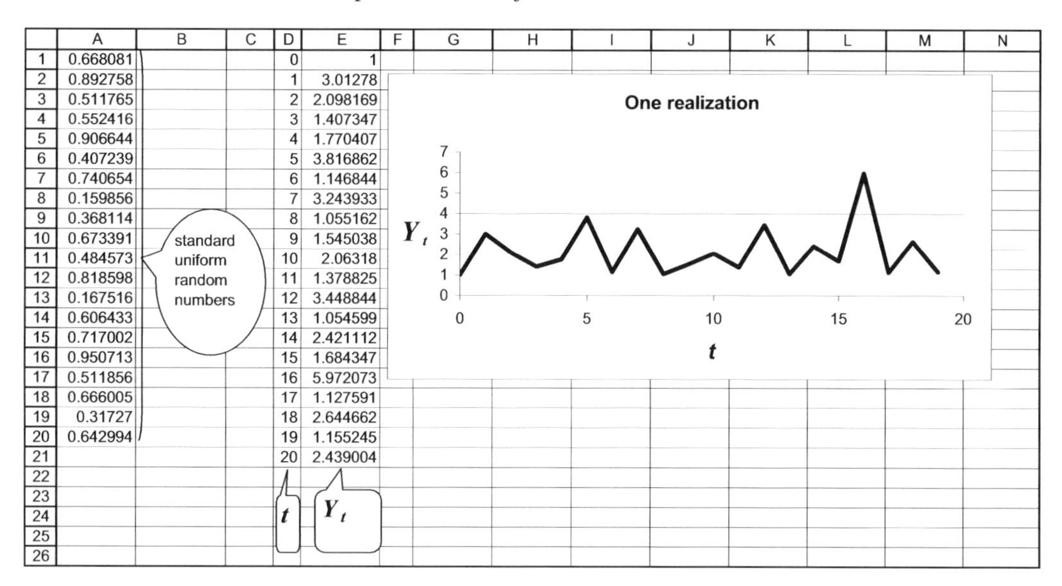

	A	B	C	D	E
1	0.668081			0	1
2	0.892758			1	3.01278
3	0.511765			2	2.098169
4	0.552416			3	1.407347
5	0.906644			4	1.770407
6	0.407239			5	3.816862
7	0.740654			6	1.146844
8	0.159856			7	3.243933
9	0.368114			8	1.055162
10	0.673391			9	1.545038
11	0.484573			10	2.06318
12	0.818598			11	1.378825
13	0.167516			12	3.448844
14	0.606433			13	1.054599
15	0.717002			14	2.421112
16	0.950713			15	1.684347
17	0.511856			16	5.972073
18	0.666005			17	1.127591
19	0.31727			18	2.644662
20	0.642994			19	1.155245
21				20	2.439004

FIGURE 5. Simulation of the Markov process from Example 1. The worksheet and the graph of one realization

So, (5.2) is not an example but, in a certain sense, another definition. The same is true for a non-homogeneous process, but in this case, the function $h(x,z)$ in (5.2) should depend on time t.

The proof of this fact is constructive, and gives a way to simulate Markov processes.

Let us come back to homogeneous Markov process X_t we have started with at the beginning of the section. Denote by $F(z\,|\,x)$ the conditional d.f. (5.1). (Since, for simplicity, we restrict ourselves to a homogeneous process, this d.f. does not depend on t.)

In accordance with Proposition 3 in Section **7**.1.2, if $F(z)$ is a distribution function, $F^{-1}(z)$ is its inverse, and ξ is uniformly distributed on $[0,1]$, then the r.v. $F^{-1}(\xi)$ has the distribution function $F(z)$.

Let r.v.'s $\xi_1, \xi_2, ...$ be independent and uniformly distributed on $[0,1]$, and let the function $h(x,z)$ from (5.2) be defined by

$$h(x,z) = F^{-1}(z\,|\,x),$$

where $F^{-1}(z\,|\,x)$ is the inverse of $F(z\,|\,x)$ with respect to z for a fixed x. Then, by virtue of the proposition mentioned, the r.v. $h(x, \xi_{t+1})$ has the distribution function $F(z\,|\,x)$.

Now, consider a process Y_t such that Y_0 has the same distribution as X_0, and all the other Y_t's are subject to the recurrence relation (5.2) with the function $h(x,z)$ chosen above. Formally, the process Y_t is not the original process X_t, but both processes have identical probability distributions. As was told, the same may be done for a non-homogeneous process, but in this case, $F(z\,|\,x)$ (and hence $h(x,z)$) will depend on t.

The described construction gives a way for simulating Markov processes in discrete time.

EXAMPLE 1. Consider a Markov process X_t for which $X_0 = 1$, and given $X_t = x$, the d.f. of X_{t+1} is

$$F(z\,|\,x) = 1 - z^{-x} \quad \text{for } z \geq 1.$$

This is the *Pareto distribution* with a parameter x; we have already considered it in Example **6**.1.3 and Exercise **6**-9. This distribution is used in a great many of applications. Here, we consider it just as an example.

First of all, note that since $P(X_{t+1} \leq 1 \,|\, X_t = x) = F(1\,|\,x) = 0$, all r.v.'s $X_t \geq 1$.

Next, in the case under consideration,

$$h(x,z) = F^{-1}(z\,|\,x) = (1-z)^{-1/x}.$$

(Indeed, let us write $y = 1 - z^{-x}$. From this, we get $z = (1-y)^{-1/x}$. Hence, $F^{-1}(y\,|\,x) = (1-y)^{-1/x}$. It remains to replace in the last expression, y by z.)

So, we set $Y_0 = 1$, and

$$Y_{t+1} = (1 - \xi_{t+1})^{-1/Y_t}. \tag{5.3}$$

A corresponding Excel worksheet is presented in Fig. 5. Column A contains twenty values of uniform ξ's; time moments t are in Column D; and twenty values of the process are in Column E. For example, the command for Cell E3, in accordance with the recurrence formula (5.3), is =(1-A2)^(-1/E2). The chart in Fig. 5 gives the graph of the realization obtained.

Note also that, because the r.v.'s $1 - \xi_t$ are also uniformly distributed on $[0,1]$, in the above construction we could consider just the relation $Y_{t+1} = (\xi_{t+1})^{-1/Y_t}$. $\square$

---

6 EXERCISES

1. Consider a counting process N_t for which interarrival times are independent and uniform on $[0,1]$. Show that in this case (a) increments of the process are dependent; (b) the process is not Markov. Does your answer to the second question answer the first? Give an answer to the first question proceeding directly from the distribution of increments. Give a common-sense explanation of your results. (*Advice*: For intervals $\Delta_1 = [0, 0.5]$ and $\Delta_2 = (0.5, 1]$, consider $P(N_{\Delta_2} = 0 \,|\, N_{\Delta_1} = 0)$ and $P(N_{\Delta_2} = 0 \,|\, N_{\Delta_1} = 1)$. Also, compare $P(N_2 = 2 \,|\, N_{1.5} = 2, N_1 = 2)$ and $P(N_2 = 2 \,|\, N_{1.5} = 2)$.)

2. Let the independent exponential interarrival times in Example 2-1 be not identically distributed. Give an heuristic argument on whether the counting process N_t still have independent increments. Compare it with the case of identically distributed interarrival times. (*Advice*: Assume, for example, that the expected value of the kth interarrival time equals k, and take into account that at any time you know how many arrivals have already occurred. Does this information matter for understanding how long we will wait for the next arrival? Say, for $\Delta_1 = [0,1]$ and $\Delta_2 = (1,2]$, compare $P(N_{\Delta_2} = 0 \,|\, N_{\Delta_1} = 1)$ and $P(N_{\Delta_2} = 0 \,|\, N_{\Delta_1} = 1000)$.)

3. * Provide an Excel worksheet with realizations of Brownian motion. Play a bit with it, considering—first of all—different normal values generated and different δ's.

4. * Show that $\frac{1}{t} w_t \xrightarrow{P} 0$ as $t \to \infty$. In other words, w_t is growing slower than t. For which functions $g(t)$ different from t it is also true; that is, when $\frac{1}{g(t)} w_t \xrightarrow{P} 0$?

5. * Using Excel, as we did in Example 5-1, or another software, simulate a realization of a Markov process X_t, $t = 0, 1, ...$, such that $X_0 = 1$, and given $X_t = x$, the r.v. X_{t+1} is uniform on $[\frac{1}{1+x}, 1]$.

Chapter 13

Counting and Queueing Processes. Birth and Death Processes: A General Scheme

1 POISSON PROCESSES

1.1 The homogeneous Poisson process

Consider a counting process N_t, and denote by $\tau_1, \tau_2, ...$ its consecutive interarrival times. A typical realization is shown in Fig. **12**-1 on p.316 where a counting process was defined.

In this section, we suppose that τ_i's are independent and exponentially distributed with the same parameter. In Example **12**.2-1, we have shown that in this case, N_t is a process with independent increments.

Denote the common parameter above by λ, and set $T_n = \tau_1 + ... + \tau_n$, the time of the nth arrival.

Clearly, N_t, the number of arrivals by time t, is not less than n if and only if the nth arrival occurred prior to or at time t. Consequently,

$$P(N_t \geq n) = P(T_n \leq t), \tag{1.1.1}$$

and to find the distribution of N_t, it suffices to find the distribution of T_n.

Each exponentially distributed τ has the Γ-density $f_{\lambda 1}$ in the notation of (**6**.2.3.3). By Proposition **6**.6, the density of T_n is the Γ-density with parameters (λ, n). Again using (**6**.2.3.3), we arrive at the density of T_n:

$$f_{T_n}(x) = \frac{\lambda^n}{(n-1)!} x^{n-1} e^{-\lambda x} \text{ for } x \geq 0. \tag{1.1.2}$$

Hence,

$$\begin{aligned} P(T_n \leq t) &= \int_0^t f_{T_n}(x)dx = \frac{\lambda^n}{(n-1)!}\int_0^t x^{n-1}e^{-\lambda x}dx \\ &= 1 - e^{-\lambda t}\left(1 + \frac{\lambda t}{1!} + ... + \frac{(\lambda t)^{n-1}}{(n-1)!}\right). \end{aligned} \tag{1.1.3}$$

The last step can be justified in two ways. First, the integral $\int_0^t x^{n-1}e^{-\lambda x}dx$ is standard, may be computed by integration by parts, and the result may be found in practically any Calculus textbook. Inserting the resulting formula for this integral, we will come to the very right-hand side of (1.1.3).

Another way to check (1.1.3) is to compute the derivative of its very r.-h.s. and make sure that it equals $f_{T_n}(t)$, that is, the derivative of the l.-h.s. The reader is invited to consider this in detail in Exercise 1. When doing it, she/he will see that after the differentiation, all terms cancel out except one.

So, both sides have the same derivatives. On the other hand, at $t = 0$, the r.-h.s. and the l.-h.s. themselves are equal to zero, and hence, are equal to each other. Together with the identity of the derivatives, this implies the identity of the r.-h.s. and the l.-h.s. for all t.

Now, let us observe that if we write representation (1.1.3) for $P(T_{n+1} \le t)$, it will differ from the representation for $P(T_n \le t)$ only by the last term in the parentheses: for the former representation, we should add the term $(\lambda t)^n/n!$.

Making use of this fact, and combining (1.1.1) and (1.1.3), we can write

$$P(N_t = n) = P(N_t \ge n) - P(N_t \ge n+1) = P(T_n \le t) - P(T_{n+1} \le t) = e^{-\lambda t}\frac{(\lambda t)^n}{n!}. \quad (1.1.4)$$

Thus, N_t has the Poisson distribution with parameter λt. In particular,

$$E\{N_t\} = \lambda t. \quad (1.1.5)$$

The last formula looks very natural: the mean number of arrivals during time t is proportional to t. The parameter λ has turned out to be the mean number of arrivals during a unit of time. On the other hand, since τ_i's are exponential with parameter λ, the mean interarrival time is $E\{\tau_i\} = 1/\lambda$. Hence,

$$E\{N_1\} = 1/E\{\tau_i\}. \quad (1.1.6)$$

For example, if the mean interarrival time $E\{\tau_i\} = \frac{1}{4}$ hour, then the mean number of arrivals during an hour is 4, which again sounds quite natural.

However, the reader should not be misled: such a simple formula is true for the basic process we are considering, but it is not true in the general case.

Certainly, if the interarrival times are not random and are equal to the same certain number τ, the number of arrivals during a unit of time is $1/\tau$. (If we accept a non-integer answer; otherwise we should consider the integer part of this.)

But if the process is random, we are dealing with different and random interarrival times $\tau_1, \tau_2, ...$, so the situation is more complicated. Moreover, if even all interarrival times had been equal to the same r.v. τ, the number of arrivals could be written as $\frac{1}{\tau}$, but $E\left\{\frac{1}{\tau}\right\} \ne \frac{1}{E\{\tau\}}$.

Thus, (1.1.6) is a special property of the process under consideration. If the τ's are not exponential, one can hope only for the asymptotic relation

$$\frac{1}{t}E\{N_t\} \to 1/E\{\tau_i\} \text{ as } t \to \infty.$$

We discuss this in detail in Section **14**.2.

Let us return to our process, consider an interval $\Delta = (t, t+s]$ and the increment N_Δ, the number of arrivals during Δ. In view of the memoryless property, we do not have to calculate the distribution of N_Δ: at any moment t, the process starts over, and the evolution of the process does not depend on what happened prior to t. Consequently, for $P(N_\Delta = n)$, we may use the same formula (1.1.4) just replacing t by $|\Delta|$, the length of the interval Δ. This gives

$$P(N_\Delta = n) = e^{-\lambda|\Delta|}\frac{(\lambda|\Delta|)^n}{n!}. \tag{1.1.7}$$

The process N_t we described is called a *homogeneous Poisson process*. The term homogeneous reflects the fact that the mean number of arrivals during a unit of time (that is, λ) does not vary in time.

We have established the following properties of N_t:

P1. $N_0 = 0$ with probability one (since the time of the first arrival, τ_1, is positive with probability one, and at the initial time, we do not have an arrival).

P2. N_t is a counting process with independent increments.

P3. For any interval Δ, the r.v. N_Δ has the Poisson distribution with parameter $\lambda|\Delta|$.

These properties may be considered a new definition of the process N_t since they are equivalent to the original definition in terms of the interarrival times τ_i. Indeed, from Property P3 it follows that for the first arrival

$$P(\tau_1 > t) = P(N_t = 0) = e^{-\lambda t}, \tag{1.1.8}$$

that is, τ_1 is exponentially distributed. For τ_2, using Property P2 and (1.1.7), we have

$$P(\tau_2 > t \,|\, \tau_1 = s) = P(\text{no arrivals in } (s, s+t] \,|\, \tau_1 = s) = P(\text{no arrivals in } (s, s+t]) = e^{-\lambda t}.$$

Consequently, τ_2 is also exponential with the same parameter λ and independent of τ_1. Other τ's are considered similarly.

When dealing with a Poisson process, we will call the corresponding *flow of arrivals* a *Poisson flow*. The parameter λ is called the *rate* of the Poisson process or the *intensity* of the corresponding flow of arrivals. When talking about a Poisson flow, it is also customary to say that arrivals are occurring at a *Poisson rate* of λ.

EXAMPLE 1. Vehicles are arriving at a service facility. The process of the arrivals is Poisson, and the mean time between two consecutive arrivals is half an hour.

(a) Find the expected value and variance of the number of vehicles arrived during the period between 2pm and 6pm. Since $E\{\tau_i\} = (1/\lambda) = 1/2$, we have $\lambda = 2$, and $E\{N_{(2,6]}\} = 4\lambda = 8$. Since we deal with a Poisson distribution, the variance $Var\{N_{(2,6]}\} = E\{N_{(2,6]}\} = 8$ too.

(b) For the same period, find the probability that there will be exactly 10 vehicles, and the probability that there will be at most 10 vehicles. By (1.1.7), $P(N_{(2,6]} = 10) = e^{-2\cdot 4}(2\cdot 4)^{10}/10! \approx 0.099$. The probability $P(N_{(2,6]} \leq 10) = PoissonDist(10; 8)$, where the symbol $PoissonDist(x; \lambda)$ stands for the Poisson distribution function with a parameter λ. We will use this type of symbols (though they do not look "mathematical") when direct calculations are lengthy, and require software. Such symbols coincide with or are close to the corresponding commands in popular programs like Excel.

In our case, *PoissonDist*$(10;8) = \sum_{k=0}^{10} e^{-8}(8)^k/k! \approx 0.816$, which may be obtained by any appropriate software. (For Excel commands, see Section **3**.4.5.1.)

(c) Find the probability that if we start to count at 2pm, the seventh vehicle will arrive after 5pm. Let us consider 2pm an initial time $t = 0$. Then, in view of (1.1.1), the probability under consideration is $P(T_7 > 3) = P(N_3 < 7)$. We can compute either side of this equality. If we choose the left, we can use (1.1.2) with $n = 7$, $\lambda = 2$, and write $P(T_7 > 3) = 1 - \int_0^3 f_{T_7}(x)dx = 1 - \int_0^3 \frac{2^7}{6!}x^6 e^{-2x}dx = 1 - GammaDist(3;2,7) \approx 0.606$, where $GammaDist(x;a,\nu)$ stands for the Γ-distribution function with argument x and parameters a,ν. (For Excel commands see Section **6**.2.3.)

If we prefer to compute $P(N_3 < 7)$, we write $P(N_3 < 7) = P(N_3 \leq 6) = PoissonDist(6;6)$, since $E\{N_3\} = 2 \cdot 3$. The answer is, naturally, the same. □

EXAMPLE 2. Assume now that for the same problem, the information given concerns the number of vehicles rather than interarrival times. For instance, suppose it is given that on the average, the facility receives 5 vehicles each 6 hours. This means that $E\{N_6\} = 5 = \lambda \cdot 6$, so $\lambda = 5/6$, and we may repeat all calculations with the new $\lambda = 5/6$. □

EXAMPLE 3. Consider the same type of a process as in the above examples. Suppose it was noticed that, on the average, in half of the cases, half an hour or more elapses before the next vehicle arrives. Translating into notation, it says that $P(\tau_i > 1/2) = 1/2$. Hence, $e^{-\lambda/2} = 1/2$, and $\lambda = 2\ln 2 \approx 1.38$. After that, we proceed as above. □

Route 1 ⇒ *page 333*

2

1.2 The non-homogeneous Poisson process

The distribution in (1.1.7) depends on the length of the interval but not on its location, which indicates that the intensity of arrivals does not change in time. In many applications, this is not the case, especially when we deal with a long period of time. For example, the intensity of the flow of customers entering a store, or the traffic intensity on a highway may be different at different moments of a day, on days of a week, and in seasons of a year.

To model this, we introduce a function $\lambda(t)$ which is interpreted as the instantaneous intensity of the arrival flow at time t. (The significance of the word " instantaneous" is the same as for the speed at a particular moment of a vehicle moving with a varying speed). We call the function $\lambda(t)$ an *intensity function* of the process of arrivals; or simply *intensity*.

Let

$$\chi(t) = \int_0^t \lambda(s)ds. \tag{1.2.1}$$

If $\lambda(s)$ is equal to a constant λ for all s, then $\chi(t) = \lambda t$.

For an arbitrary interval $\Delta = (t, t+u]$, set

$$\chi_\Delta = \int_\Delta \lambda(s)ds = \int_t^{t+u} \lambda(s)ds = \int_0^{t+u} \lambda(s)ds - \int_0^t \lambda(s)ds = \chi(t+u) - \chi(t). \tag{1.2.2}$$

When $\lambda(s) = \lambda$, we have $\chi_\Delta = \lambda u = \lambda|\Delta|$, where $|\Delta|$ stands for the length of Δ.

By analogy with Properties P1–P3 from Section 1.1, we call a process N_t, $t \geq 0$, a *non-homogeneous Poisson process* if it has the following properties.

PN1. $N_0 = 0$ with probability one;

PN2. N_t is a counting process with independent increments;

PN3. There exists a non-negative function $\lambda(t)$ such that for any interval Δ, the r.v. N_Δ has the Poisson distribution with parameter χ_Δ defined in (1.2.2).

Thus, the number of arrivals during any time interval is a Poisson variable whose mean value is determined by the intensity of arrivals in this period. In particular, in accordance with (1.2.2),

$$E\{N_\Delta\} = \chi_\Delta.$$

Considering an interval $\Delta = [0,t]$, we have

$$P(N_t = n) = \exp\{-\chi(t)\}\frac{\chi^n(t)}{n!}. \tag{1.2.3}$$

The homogeneous case corresponds to a constant intensity $\lambda(t)=\lambda$: in this case $\chi(t)=\lambda t$, and we return to (1.1.4).

EXAMPLE 1. Consider again vehicles arriving at a service facility, and suppose now that the process of arrivals is a non-homogeneous Poisson process. Assume that in the period between 9am and 11am, the rate equals 3 customers/hour, and after 11am, the rate is linearly decreasing from 3 at 11am to zero at 5pm. Find the probability that there will be not more than 15 vehicles between 10am and 4pm.

Taking 9am as an initial time, we have $\lambda(t) = 3$ for $t \in [0,2]$, and $\lambda(t) = (8-t)/2$ for $t \in [2,8]$. For the interval $\Delta = [1,7]$, the expected number of customers $\chi_\Delta = \int_1^7 \lambda(s)ds$. As is easy to calculate, inserting the $\lambda(s)$ defined above, we get $\chi_\Delta \approx 11.75$. With use, for example, of Excel, it leads to $P(N_\Delta \leq 15) = PoissonDist(15; 11.75) \approx 0.862$. $\square$

In the general case, when the rate can change in time, the number of arrivals may be small during a long period, and may be large during a short period.

EXAMPLE 2. (*Can the number of arrivals be finite during an infinite period of time?*) The question concerns the value of

$$E\{N_{[0,\infty)}\} = \chi_{[0,\infty)} = \int_0^\infty \lambda(s)ds.$$

If the integral above converges, the answer is positive. Suppose, for instance, that $\lambda(s) = e^{-s}$. Then the mean $E\{N_{[0,\infty)}\} = \int_0^\infty e^{-s}ds = 1$, so during the *infinite* period, there will be on the average *only one* arrival.

One may consider any example for which the integral above converges; say, $\lambda(s) = 1/(1+s^2)$. Then the r.v. $N_{[0,\infty)}$ is again a Poisson r.v. with the finite mean value equal to $\int_0^\infty \frac{1}{1+s^2}ds = \arctan s|_0^\infty = \pi/2$.

EXAMPLE 3 (*An explosion*). Suppose that the rate of arrivals begins to grow very fast. This may be modeled by letting $\lambda(t)$ converge to infinity during a finite period. For instance, let $\lambda(t) = 1/(1-t)$ for $t < 1$. Then, for $t < 1$, the mean number of arrivals $\chi(t) = \int_0^t (1-s)^{-1} ds = -\ln(1-t) \to \infty$ as $t \to 1$. The number of arrivals itself also goes to infinity. More precisely, $P(N_t \leq m) \to 0$ as $t \to 1$ for an *arbitrary large m*. Indeed, $P(N_t \leq m) = e^{-\chi(t)} \sum_{k=0}^{m} \frac{\chi^k(t)}{k!} \to 0$ because the exponential function converges to infinity faster than any polynomial function. □

Consider now how interarrival times look in the non-homogeneous case. We will see that, in general, they are no longer exponential or independent.

Indeed, for the first arrival,

$$P(\tau_1 > t) = P(N_t = 0) = \exp\{-\chi(t)\}. \tag{1.2.4}$$

If $\lambda(t)=\lambda$, we come to the exponential distribution (1.1.8), however if, for instance, $\lambda(s)=s$, then $P(\tau_1 > t) = \exp\left\{-\int_0^t s ds\right\} = \exp\{-t^2/2\}$. In this case, τ_1 has a distribution different from the exponential distribution. (Namely, it is Weibull's distribution with parameters $c = \frac{1}{2}$ and $\alpha = 2$; see p.301.)

Furthermore, for the nth interarrival time τ_n given the time T_{n-1} of the previous arrival, the conditional probability

$$P(\tau_n > s \,|\, T_{n-1} = t) = P(N_{(t,t+s]} = 0 \,|\, T_{n-1} = t) = P(N_{(t,t+s]} = 0) = \exp\{-\int_t^{t+s} \lambda(u) du\}. \tag{1.2.5}$$

We see that the above probability may depend on t, the time of the previous arrival. (If $\lambda(u) = \lambda$, it does not.)

However, it should not look scary because the distribution of the nth arrival is still tractable: the formula (1.1.1) is true in the general case, and

$$P(T_n \leq t) = P(N_t \geq n) = 1 - P(N_t < n), \tag{1.2.6}$$

which may be computed as a Poisson probability, say, with use of software.

EXAMPLE 4. Suppose that the intensity $\lambda(t) = 9(8-t)^2/64$ for $0 \leq t \leq 8$, that is, starting from 9, the intensity decreases as a parabola and equals zero at time 8. The expected number of customers during the whole period is $\chi(8) = \frac{9}{64}\int_0^8 (8-t)^2 dt = 24$.

So, on the average, we have 24 customers. What is the probability that at least 20 customers will come within the first half of the period? The intensity is rapidly decreasing, so the question is reasonable.

We have $E\{N_4\} = \chi(4) = \frac{9}{64}\int_0^4 (8-t)^2 dt = 21$, so the probability should be more than 0.5. More precisely, by (1.2.6), $P(T_{20} \leq 4) = P(N_4 \geq 20) = 1 - P(N_4 \leq 19) = 1 - \mathit{PoissonDist}(19;21) \approx 0.615$. □

1.3 Another perspective: Infinitesimal approach

Conditions P3 and PN3 in the above definitions of homogenous and non-homogeneous processes, respectively, are clear; but its verification in particular cases may cause difficulties. The equivalent definition below proceeds from the behavior of the process merely in small time intervals. It is useful and gives an additional insight into the nature of Poisson processes.

Below, the symbol $o(\delta)$ stands for a function which is "much smaller" than δ for small δ; formally, $\frac{1}{\delta}o(\delta) \to 0$ as $\delta \to 0$. It is important to emphasize that in different formulas $o(\delta)$ may denote different functions but since we do not need to specify them, we will use the same symbol in different formulas. The reader may find more comments regarding this notation in the Appendix, Section 2.1.

We define N_t from scratch. Suppose that it is a *counting process* N_t with *independent increments* such that $N_0 = 0$, and for a function $\lambda(t) \geq 0$ and any positive t and δ,

$$P(N_{(t,t+\delta]} = 1) = \lambda(t)\delta + o(\delta) \text{ as } \delta \to 0, \tag{1.3.1}$$

$$P(N_{(t,t+\delta]} > 1) = o(\delta) \text{ as } \delta \to 0. \tag{1.3.2}$$

Again, $o(\delta)$ in (1.3.1) may differ from $o(\delta)$ in (1.3.2). In (1.3.1), the term $o(\delta)$ is negligible with respect to $\lambda(t)\delta$; relation (1.3.2) means that for small δ (i.e., for small intervals), $P(N_{(t,t+\delta]} > 1)$ is negligibly small.

This may be understood as follows. For a small time interval, the probability of an arrival is proportional to the length of the interval (up to a negligible remainder), and the probability that more than one arrival will occur is negligible. The coefficient of proportionality, $\lambda(t)$, depends on time and, as in Section 1.2, is interpreted as the mean number of arrivals per unit of time in a neighborhood of t. It is called the *rate* or the *intensity* at time t.

We define $\chi(t)$ and χ_Δ as in (1.2.1) and (1.2.2), respectively.

Proposition 1 *For the process N_t defined above and any interval Δ,*

$$P(N_\Delta = n) = e^{-\chi_\Delta}\frac{\chi_\Delta^n}{n!}. \tag{1.3.3}$$

A proof will be given at the end of this section.

The infinitesimal approach allows us to easily solve many problems involving varying intensities.

Assume, for instance, that we count only a particular type of arrival. (For example, when counting customers entering a store, we count men and women, separately.) Suppose that each particular arrival is counted only with a probability p, perhaps depending on time: $p = p(t)$. In this case, the only change needed is to replace the intensity $\lambda(t)$ by $p(t)\lambda(t)$ in condition (1.3.1).

Indeed, for an arrival to be counted in a period $[t, t+\delta]$, first, an arrival must occur—the probability of this is $\lambda(t)\delta + o(\delta)$. Secondly, once an arrival has occurred, the probability that it belongs to the first type is $p(t)$. If we count only this type of arrival, we should multiply $\lambda(t)\delta + o(\delta)$ by $p(t)$. This gives $(\lambda(t)\delta + o(\delta))p(t) = p(t)\lambda(t) + p(t)o(t)$. The last term is negligible with respect to the second, and we may rewrite the whole expression as $p(t)\lambda(t) + o(t)$.

This is a particular case of the *marked Poisson process*. We considered a similar problem in a static framework in Sections **3**.4.5.5 and **8**.2.3.

EXAMPLE 1. After some moment which we view as initial, the flow of customers entering a service facility started to grow with intensity $\lambda(t) = 3(1+t)$. In order to keep the rate equal to the initial rate 3, the management decides to refuse some customers service depending on the character of the job to be done. Since the type of the next customer is not known in advance, the process of refusals is random. With what probability should the facility accept the claim arrived at time t? With probability $p(t) = 1/(1+t)$, since then $p(t)\lambda(t) = 3$.

EXAMPLE 2 is a dynamic counterpart of Example **3**.4.5.5-1. A company offers two products, for \$100 and \$150. Each day, the company deals with a Poisson flow of customers with intensity λ, and the probability that a current customer is from those who buy the first product is a given p. Denote by N_t the total number of customers arrived by time t, and by N_{t1}, N_{t2} the number of customers who bought the first and second product respectively, before or at time t.

The processes N_{t1}, N_{t2} are Poisson processes with the respective intensities $p\lambda$ and $(1-p)\lambda$, and as was shown in Example **3**.4.5.5-1, the r.v.'s N_{t1}, N_{t2} are independent. The total income of the company by time t is the r.v. $100 \cdot N_{t1} + 150 \cdot N_{t2}$. $\square$

Proof of Proposition 1. Set $p_n(t) = P(N_t = n)$, and for a $\delta > 0$, consider two moments of time: t and $t+\delta$. Note that if $N_{t+\delta} = n$, and $N_{(t,t+\delta]}$ equals some $k \leq n$, then N_t should be equal to $n-k$. Hence, for $n = 0, 1, 2, \ldots$, in view of the independence of increments,

$$\begin{aligned}
p_n(t+\delta) &= P(N_{t+\delta} = n) = \sum_{k=0}^{n} P(N_t = n-k, N_{(t,t+\delta]} = k) \\
&= \sum_{k=0}^{n} P(N_t = n-k)P(N_{(t,t+\delta]} = k) = P(N_t = n)P(N_{(t,t+\delta]} = 0) \\
&+ P(N_t = n-1)P(N_{(t,t+\delta]} = 1) + \sum_{k=2}^{n} P(N_t = n-k)P(N_{(t,t+\delta]} = k).
\end{aligned}$$

We are going to use (1.3.1)–(1.3.2). As was noted, the symbol $o(\delta)$ in these formulas may denote different functions. Nevertheless, we can combine them, and if for example, we consider the sum of remainders, then we arrive at another remainder which may be again denoted by $o(\delta)$. (See also the Appendix, Section 2.1 for comments.) First,

$$\begin{aligned}
P(N_{(t,t+\delta]} = 0) &= 1 - P(N_{(t,t+\delta]} = 1) - P(N_{(t,t+\delta]} > 1) = 1 - (\lambda(t)\delta + o(\delta)) - o(\delta) \\
&= 1 - \lambda(t)\delta + o(\delta),
\end{aligned}$$

and, secondly,

$$\sum_{k=2}^{n} P(N_t = n-k)P(N_{(t,t+\delta]} = k) \leq \sum_{k=2}^{n} P(N_{(t,t+\delta]} = k) = P(N_{(t,t+\delta]} > 1) = o(\delta).$$

Hence, $p_n(t+\delta) = p_n(t)(1-\lambda(t)\delta + o(\delta)) + p_{n-1}(t)(\lambda(t)\delta + o(\delta)) + o(\delta) = p_n(t)(1-\lambda(t)\delta) + p_{n-1}(t)\lambda(t)\delta + o(\delta)$. We rewrite it as

$$\frac{1}{\delta}[p_n(t+\delta) - p_n(t)] = -\lambda(t)p_n(t) + \lambda(t)p_{n-1}(t) + \frac{1}{\delta}o(\delta).$$

Letting $\delta \to 0$, and recalling that $\frac{1}{\delta}o(\delta) \to 0$, we come to the differential equation

$$p_n'(t) = -\lambda(t)p_n(t) + \lambda(t)p_{n-1}(t). \tag{1.3.4}$$

For $n = 0$, since $p_{-1}(t) = P(N_t = -1) = 0$,

$$p_0'(t) = -\lambda(t)p_0(t). \tag{1.3.5}$$

A solution to the last equation is

$$p_0(t) = \exp\{-\chi(t)\}, \tag{1.3.6}$$

where $\chi(t)$ is defined in (1.2.1). The reader who does not remember how to solve equations of this type can easily verify (1.3.6) by substitution. The reader who remembers this should take into account that $p_0(0) = 1$ (we start with no claims), and together with this initial condition, solution (1.3.6) with $\chi(t)$ given in (1.2.1) is the unique solution to the ordinary differential equation (1.3.5).

Once we know $p_0(t)$, we get from (1.3.4) that

$$p_1'(t) = -\lambda(t)p_1(t) + \lambda(t)\exp\{-\chi(t)\}, \tag{1.3.7}$$

which leads to

$$p_1(t) = \chi(t)\exp\{-\chi(t)\}.$$

The reader can again check it by substitution. Continuing the same procedure by induction, we come to solutions for any n; namely, to (1.2.3).

Consider an arbitrary interval $\Delta = [t, t+u]$. In this case, proceeding as in the previous section from the independence of increments, we can view the point t as an initial point and use the same formula (1.2.3) with $\chi(t)$ replaced by χ_Δ. ■

---

2 BIRTH AND DEATH PROCESSES

We begin with a particular but important scheme; after that we will turn to more general constructions.

2.1 A Markov queuing model with one server (M/M/1)

1,2

Let us consider a homogenous Poisson process with a parameter λ and view arrivals as those of "particles" entering a system. We suppose that in mean time the particles leave the system, and the process of departures is also a homogeneous Poisson process with a parameter μ. Certainly, the latter process runs unless there are particles in the system.

In the most common example, the particles are viewed as "customers" that receive a service and then leave the system. The term "customers" is understood in a broad sense: it may indeed refer to the customers of a company or, say, it may mean batches of data being received and analyzed by a computer.

Suppose that there is only one server, and the service times are independent exponential r.v.'s with parameter μ. Then, when the server is not idle, the times between consecutive departures are i.i.d. exponential r.v.'s, and hence, the process of departures is Poisson. If the server is busy, arriving customers form a queue.

Such a *queueing system* is referred to as M/M/1, where the M's come after the Markov property, for the arrivals and departures, and "1" indicates the number of servers. In Section 2.3, we will consider a more general arrival-departure scheme and, in particular, more complicated queueing systems.

Let us come back to the model of this section and denote by X_t the number of particles in the system at time t. In general, $P(X_t = k)$ may depend on t. For example, if the process starts at time $t = 0$ with no particles ($X_0 = 0$), then for small t, the probability that there will be many particles in the system is small. However, it may be shown that in the long run, the situation is stabilizing, and the probabilities $P(X_t = k)$ approaching some stable values. More precisely, there exist the limiting probabilities

$$\pi_k = \lim_{t\to\infty} P(X_t = k). \tag{2.1.1}$$

One may say that in the long run, the system is entering into a stationary or steady state regime. Sometimes, the "long-run" limiting probabilities π_k are called *equilibrium* or *steady-state probabilities.*

It is worth noting that this phenomenon has a straight analogy with the ergodicity property of Markov chains we considered in Chapter 5. However, Chapter 5 dealt with discrete time, while we consider a process in continuous time. To make the exposition less complicated, we take the validity of the ergodicity property (2.1.1) for granted.

Moreover, to make the exposition more explicit (though it will lead to a slightly heuristic way of reasoning), we suppose that the system is already in the steady state condition:

$$P(X_t = k) = \pi_k \text{ and, hence, does not depend on } t. \tag{2.1.2}$$

Our goal is to find π_k's. To this end, we introduce the quantity

$$\rho = \lambda/\mu$$

which sometimes is referred to as an *arrival/service ratio* or a *traffic intensity.*

We will see that if $\lambda \geq \mu$, then X_t is unlimitedly growing; that is, for large t, the number of the particles, X_t, becomes larger than an arbitrary large k. This means that for any particular k, the limiting probabilities $\pi_k = 0$. This is not surprising, at least, in the case $\lambda > \mu$, since if the intensity of arrivals is larger than that of departures, one cannot hope that the number of particles in the system will not grow.

The case $\lambda = \mu$ is less trivial, so the result mentioned means that if the number of arrivals equals that of the departures *on the average*, it is not enough for the number of particles not to grow.

If $\rho < 1$, then the intensity of arrivals is less than the intensity of departures, and we can hope for a stationary behavior of the system in the long run.

Proposition 2 *If* $\rho < 1$, *then for* $k = 0, 1, \ldots$,

$$\pi_k = (1-\rho)\rho^k. \tag{2.1.3}$$

A proof is given in Section 2.2.

Thus, in the stationary regime, X_t has the geometric distribution with parameter $1-\rho$. In particular, in the stationary regime, the mean number of particles in the system is

$$E\{X_t\} = \frac{\rho}{1-\rho}. \tag{2.1.4}$$

(see (**3**.4.3.7)). Let us view the system as a queueing one. Then, the probability that at a particular time, there will be no queue is $\pi_0 + \pi_1 = 1-\rho+(1-\rho)\rho = 1-\rho^2$, and the mean queue length is

$$0\cdot\pi_0 + \sum_{k=1}^{\infty}(k-1)P(X_t=k) = \sum_{k=1}^{\infty} kP(X_t=k) - \sum_{k=1}^{\infty}\pi_k$$

$$= E\{X_t\} - (1-\pi_0) = \frac{\rho}{1-\rho} - \rho = \frac{\rho^2}{1-\rho}. \tag{2.1.5}$$

EXAMPLE 1. Customers arrive at a service facility at an average rate of one customer each half an hour. A service time of the single server is exponential with a mean of 20 minutes. For the stationary regime, what is the probability that at a particular time, there will be no queue, what is the mean number of the customers in the system, and what is the mean length of the queue?

The mean interarrival time is 0.5 hour, so $\lambda = 2$, while the mean service time is $1/3$ hour, so $\mu = 3$. Hence, $\rho = 2/3$, and the probability that at a particular time, there will be no queue is $\pi_0 + \pi_1 = 1-\rho^2 = 5/9$.

By (2.1.4), in the stationary regime, $E\{X_t\} = 2$. (If the process is not in a stationary regime but is just approaching it, the last result gives the limiting number: $\lim_{t\to\infty} E\{X_t\} = 2$.)

By (2.1.5), the mean length of the queue is $\frac{\rho^2}{1-\rho} = 4/3$. □

Similarly to what we had in the case of limiting distributions for Markov chains (Section **5**.4.1),

> The probability π_k coincides with the expected proportion (in the long run) of the time during which there are exactly k particles in the system. (2.1.6)

Indeed, let the indicator process $I_{tk} = 1$ if $X_t = k$, and $= 0$ otherwise. Then, the r.v. $Y_{Tk} = \frac{1}{T}\int_{t=0}^{T} I_{tk}dt$ is the (random) proportion of the time, during a period $[0,T]$, when there are k particles in the system. Since $E\{I_{tk}\} = P(X_t = k)$,

$$E\{Y_{Tk}\} = \frac{1}{T}\int_0^T E\{I_{tk}\}dt = \frac{1}{T}\int_0^T P(X_t=k)dt = \frac{1}{T}\int_0^T \pi_k dt = \pi_k$$

by virtue of (2.1.1). It is worth noting that if we do not adopt (2.1.1) but just assume that $P(X_t = k) \to \pi_k$, then we will get that

$$E\{Y_{Tk}\} = \frac{1}{T}\int_0^T P(X_t=k)dt \to \pi_k.$$

(We use the fact from Calculus that if $f(t)$ is integrable on any finite interval and $f(t) \to c$ as $t \to \infty$, then $\frac{1}{T}\int_0^T f(t)dt \to c$ as $T \to \infty$).

EXAMPLE 1 revisited. In the long run, the proportion of time when there is no queue is $5/9$, or $\approx 55.5\%$. □

The proof of Proposition 2 to which we proceed, is not purely technical and reflects the essence of the matter.

2.2 Proof of Proposition 2

To make it more explicit, we use the customer-server interpretation. In the first reading, the reader, when going through the calculations below, can even replace all terms negligible with respect to δ just by 0.

Let an interval $\Delta = [t, t+\delta]$, and as usual, N_Δ be the number of arrivals during Δ. By Taylor's formula (see, e.g., the Appendix, (2.2.5)),

$$P(N_\Delta = 1) = \lambda\delta e^{-\lambda\delta} = \lambda\delta + o(\delta), \tag{2.2.1}$$

where $o(\delta)$ is a remainder negligible as compared with δ for small δ. Formally, $\frac{1}{\delta}o(\delta) \to 0$ as $\delta \to 0$; see the Appendix, Section 2.1 for detail. As was mentioned, without an essential loss in understanding, the reader may even set $o(\delta) = 0$ viewing $o(\delta)$ as "very small."

Furthermore,

$$P(N_\Delta > 1) = \sum_{k=2}^{\infty} e^{-\lambda\delta}\frac{(\lambda\delta)^k}{k!} = (\lambda\delta)^2 e^{-\lambda\delta}\sum_{k=2}^{\infty}\frac{(\lambda\delta)^{k-2}}{k!} = o(\delta) \tag{2.2.2}$$

because we have δ^2 in the r.-h.s. Another way to prove (2.2.2) is to write that $P(N_\Delta > 1) = 1 - e^{-\lambda\delta} - \lambda\delta e^{-\lambda\delta}$ and use again the same Taylor formula. Note also that the reader who did not skip Section 1.3, may just appeal to (1.3.1)–(1.3.2).

The departure process may be considered similarly, and if the server is not idle,

$$P(\text{one departure during } \Delta) = \mu\delta + o(\delta), \tag{2.2.3}$$

and

$$P(\text{more than one departure during } \Delta) = o(\delta). \tag{2.2.4}$$

Also, the probability that during the time interval Δ, there will be simultaneously at least one arrival and one departure is $(\lambda\delta + o(\delta))(\mu\delta + o(\delta)) = \lambda\mu\delta^2 + (\lambda+\mu)\delta o(\delta) + o^2(\delta)$, which is negligible with respect to δ.

Thus, for $k \geq 1$, there will be k customers in the system at time $t+\delta$ (that is, $X_{t+\delta} = k$) if

- either $X_t = k$, and there will be neither arrivals nor departures during Δ;
- or $X_t = k-1$, and there will be one arrival and no departures during Δ;
- or $X_t = k+1$, and there will be one departure and no arrivals during Δ;

and the probabilities of all other events leading to $X_{t+\delta} = k$ are negligible.
Accordingly,

$$\begin{aligned} P(X_{t+\delta} = k) &= P(X_t = k) \cdot (1 - \lambda\delta + o(\delta))(1 - \mu\delta + o(\delta)) \\ &+ P(X_t = k-1) \cdot (\lambda\delta + o(\delta))(1 - \mu\delta + o(\delta)) \\ &+ P(X_t = k+1) \cdot (1 - \lambda\delta + o(\delta))(\mu\delta + o(\delta)) + o(\delta). \end{aligned} \tag{2.2.5}$$

If, proceeding from (2.1.2), we write for $P(X_{t+\delta} = k)$ and $P(X_t = k)$ the same symbol π_k, and if we write the negligible term $o(\delta)$ just one time at the end, then we can simplify (2.2.5) as follows:

$$\pi_k = (1 - \lambda\delta)(1 - \mu\delta)\pi_k + \lambda\delta(1 - \mu\delta)\pi_{k-1} + (1 - \lambda\delta)\mu\delta\pi_{k+1} + o(\delta).$$

Canceling one time π_k's in the l.-h.s and r.-h.s., and dividing everything by δ, we get

$$0 = -(\lambda + \mu)\pi_k + \lambda\pi_{k-1} + \mu\pi_{k+1} + \lambda\mu\delta\pi_k - \lambda\mu\delta(\pi_{k-1} + \pi_{k+1}) + \frac{1}{\delta}o(\delta).$$

Eventually, letting $\delta \to 0$, for $k \geq 1$, we obtain

$$\lambda\pi_{k-1} + \mu\pi_{k+1} - (\lambda + \mu)\pi_k = 0. \tag{2.2.6}$$

Considering $k = 0$ separately, observe that if $X_t = 0$, then the probability that a departure occurs during the interval Δ is negligible (since for this to happen, there should at least one arrival and a departure after that). The reader is invited to check this, and proceeding in the same fashion, to get that $-\lambda\pi_0 + \mu\pi_1 = 0$, or

$$\pi_1 = \frac{\lambda}{\mu}\pi_0 = \rho\,\pi_0. \tag{2.2.7}$$

Next, we prove by induction that the same pattern takes place for all $k = 0, 1, ...$; namely,

$$\pi_{k+1} = \rho\,\pi_k. \tag{2.2.8}$$

For $k = 0$, relation (2.2.8) becomes (2.2.7) which has been established. Suppose that (2.2.8) is true for some k. Then, replacing k by $k+1$ in (2.2.6), solving it with respect to π_{k+2}, and using the induction assumption, we get

$$\pi_{k+2} = \frac{1}{\mu}[(\lambda + \mu)\pi_{k+1} - \lambda\pi_k] = (\rho + 1)\pi_{k+1} - \rho\pi_k = (\rho + 1)\pi_{k+1} - \rho\left(\frac{1}{\rho}\pi_{k+1}\right) = \rho\pi_{k+1}.$$

So, we come to (2.2.8) with replacement k by $k+1$. Thus, (2.2.8) is true, and

$$\pi_k = \rho\pi_{k-1} = \rho\rho\pi_{k-2} = ... = \rho^k\pi_0.$$

On the other hand, the sum $\sum_{k=0}^{\infty}\pi_k = 1$, which together with (2.2.8), for $\rho < 1$, implies

$$1 = \sum_{k=0}^{\infty}\pi_k = \pi_0\,(1 + \rho + \rho^2 + ...) = \pi_0\frac{1}{1-\rho}.$$

Thus, $\pi_0 = 1-\rho$, and $\pi_k = \rho^k(1-\rho)$. ■

Note that if $\rho \geq 1$, then $\sum_{k=0}^{n} \pi_k = \pi_0 \sum_{k=0}^{n} \rho^k \to \infty$ if $\pi_0 > 0$. Hence, π_0 must equal zero, and then, by (2.2.7)–(2.2.8), so do all π_k. Then, for an arbitrarily large n, the probability $P(X_t \leq n) \to \sum_{k=0}^{n} 0 = 0$, which means that X_t is unlimitedly growing.

Route 1 ⇒ *page 344*

2

2.3 Birth and death processes: A general scheme

As in Section 2.1, we consider a system characterized by two processes: that of arrivals and that of departures. However, now we assume that the intensities of these processes depend on the current number of particles in the system. More precisely, we will work with the following scheme.

First, assume that the process X_t, the number of particles at time t, is Markov. Consider a fixed time moment t, and suppose that $X_t = k$. In this case, we will also say that the system is in state k. Let τ be the (random) time it will take for the system to switch either to state $k+1$ (an arrival occurs first) or to state $k-1$ if $k \neq 0$ (a departure occurs first).

Due to the Markov property, given $X_t = k$, the evolution of the process before t does not impact the distribution of τ. Moreover, due to the same property, if at a time $t+s$, it turns out that the process is still in the state k, then "everything will start over as from the beginning." In other words, τ has the memoryless property, and hence τ is an exponential r.v.

Next, we introduce two parameters, λ_k and μ_k, call them the intensities or rates of arrivals and departures, respectively, and set $\mu_0 = 0$. Suppose that

A. The parameter of the (exponential) distribution of τ equals $\lambda_k + \mu_k$;

B. The system will switch to the state $k+1$ (an arrival comes first) with the probability $\lambda_k/(\lambda_k+\mu_k)$, and to the state $k-1$ with the probability $\mu_k/(\lambda_k+\mu_k)$.

Formally, the above conditions are postulated. However, they may be clarified in the following way.

Consider independent exponential r.v.'s τ_A and τ_D with respective parameters λ_k and μ_k. We interpret τ_A as the time it will take for the first arrival to occur if there had been no departures, and τ_D as the similar time for the first departure. Then the time it takes for the system to switch to either the state $k+1$ or the state $k-1$ is $\tau = \min\{\tau_A, \tau_D\}$. We use the general fact that has been already proved in Example **11**.1.1-4, and which, for further references, we state here as

Lemma 3 *Let $\xi_1,...,\xi_n$ be independent exponential r.v.'s with respective parameters $\lambda_1,...,\lambda_n$. Then the r.v.* $\min\{\xi_1,...,\xi_n\}$ *is exponential with the parameter $\lambda_1+...+\lambda_n$.*

By virtue of Lemma 3, the r.v. τ above has the exponential distribution with the parameter $\lambda_k+\mu_k$. Furthermore, as it follows from Example **6**.5.2-3, the probability that an arrival will come first is $P(\tau_A < \tau_D) = \lambda_k/(\lambda_k+\mu_k)$. (Because τ_A, τ_D are continuous, $P(\tau_A = \tau_D) = 0$.)

So, the process X_t has been defined and clarified. It only makes sense to emphasize that the above construction with the r.v.'s τ_A, τ_B was presented merely to clarify the significance of Conditions A–B.

EXAMPLE 1. In the case $\lambda_k = \lambda$, and $\mu_k = 0$ for all k (only arrivals, no departures), we deal with a homogeneous Poisson process.

EXAMPLE 2 (*A pure birth process (or Yule's process)*). Consider a population whose members never die, and each member independently of the others produces offsprings, one at a time, in a way that the time between the consecutive births of the offsprings of a particular member are independent exponential r.v.'s with a parameter λ.

Suppose that, at a time t, the size of the population $X_t = k$. Due to the memoryless property, at each time moment, for each member, the process of waiting for the next offspring starts over. Denote by τ_i, $i = 1, ..., k$, the time it takes (after time t) for the ith member to produce the next offspring. The τ_i's are exponential with parameter λ, and the time it will take for the first next member of the population to appear is $\min\{\tau_1, ..., \tau_k\}$. By Lemma 3, it has the exponential distribution with the parameter $\lambda_k = \lambda k$. All $\mu_k = 0$.

EXAMPLE 3. In the case $\lambda_k = \lambda$ for all k, $\mu_0 = 0$, and $\mu_k = \mu$ for all $k \geq 1$, we come to the M/M/1 scheme of Section 2.1.

EXAMPLE 4. (*The M/M/1 queueing system with finite capacity*). Consider the same queueing system with one server as in Example 3, but suppose that the system cannot receive more than a particles: if $X_t = a$, then new particles do not enter. In this case, $\lambda_k = \lambda$ for $k = 0, ..., a-1$, and $\lambda_k = 0$ for $k \geq a$. Clearly, in this case, $P(X_t > a) = 0$, $\mu_0 = 0$, and $\mu_k = \mu$ for $k = 1, ..., a$.

EXAMPLE 5. (*The M/M/s queueing scheme*). As in Section 2.1, let us view particles as customers, assume the process of arrivals to be a homogeneous Poisson process with a rate of λ, and suppose that there are s independent servers for which service times are exponential r.v.'s with a parameter of μ. We do not exclude the case $s = \infty$.

Since the service times are i.i.d., it does not matter in which order the servers are chosen to serve the customers. Suppose that at a time moment, there are $k \geq 1$ customers in the system, and $k \leq s$. Then, k servers are busy, and similarly to what we did in Example 2, using Lemma 3, we get that the time it takes for the first customer to leave the system is exponential with the parameter $\mu_k = \mu k$. If $k > s$, then $k - s$ customers are in line, and $\mu_k = \mu s$. Also, $\mu_0 = 0$, and $\lambda_k = \lambda$ for all k. □

2.4 Limiting probabilities

Our next goal is to find the limiting (or steady-state) probabilities $\pi_k = \lim_{t \to \infty} P(X_t = k)$.

As in Section 2.1, we simplify the problem, and suppose that the system is already in the steady state condition. So,

$$P(X_t = k) = \pi_k \text{ and, hence, does not depend on } t. \tag{2.4.1}$$

As we saw, in the simple M/M/1 model, the steady-state probabilities π_k are proportional to the powers ρ^k where the arrival/service ratio $\rho = \lambda/\mu$. We will show that in the general case, the counterparts of the powers ρ^k are the quantities

$$\theta_0 = 1 \quad \text{and} \quad \theta_k = \frac{\lambda_0\lambda_1\cdots\lambda_{k-1}}{\mu_1\mu_2\cdots\mu_k} \quad \text{for } k \geq 1, \tag{2.4.2}$$

provided $\mu_k \neq 0$ for $k \geq 1$.

Proposition 4 *Let $\mu_k \neq 0$ for $k \geq 1$. If*

$$\sum_{k=0}^{\infty} \theta_k < \infty, \tag{2.4.3}$$

then for $k = 0, 1, ...$

$$\pi_k = \frac{\theta_k}{\sum_{k=0}^{\infty} \theta_k}. \tag{2.4.4}$$

If (2.4.3) does not hold, then $\pi_k = 0$ for all k.

A proof will be given in Section 2.5.

EXAMPLE 1. For the M/M/1 model with $\lambda_k = \lambda$, and $\mu_k = \mu$ for $k \geq 1$, we have $\theta_k = \dfrac{\lambda^k}{\mu^k} = \rho^k$ for $\rho = \lambda/\mu$. Hence,

$$\sum_{k=0}^{\infty} \theta_k = \sum_{k=0}^{\infty} \rho^k = \frac{1}{1-\rho} \quad \text{for } \rho < 1,$$

and we come to the limiting geometric distribution as in Proposition 2.

If $\rho \geq 1$, then

$$\sum_{k=0}^{\infty} \theta_k = \infty,$$

and $\pi_k = 0$, which is consistent with what we obtained in Sections 2.1–2.2.

EXAMPLE 2. Suppose that $\mu_k = \mu$ for all $k \geq 1$, and for some fixed $\tilde{k}$,

$$\lambda_k \geq \mu \quad \text{once} \quad k \geq \tilde{k}. \tag{2.4.5}$$

For instance, assume that $\lambda_k = \lambda k$ as in Example 2.3-3. Then, starting from some $\tilde{k}$, (2.4.5) will be true.

Our intuition tells us that in the case (2.4.5), the number of particles in the system will unlimitedly grow. Let us make sure that it follows from Proposition 4.

To use characteristics θ_k, assume, first, $\mu > 0$. Relation (2.4.5) implies that for $k > \tilde{k}$,

$$\theta_k = \frac{\lambda_0\lambda_1\cdots\lambda_{\tilde{k}-1}\lambda_{\tilde{k}}\cdots\lambda_{k-1}}{\mu^k} \geq \frac{\lambda_0\lambda_1\cdots\lambda_{\tilde{k}-1}\mu\cdots\mu}{\mu^k} = \frac{\lambda_0\lambda_1\cdots\lambda_{\tilde{k}-1}}{(\mu)^{\tilde{k}}}.$$

So, if all λ_i's are positive, then θ_k's are greater than a positive constant, and hence, again $\sum_{k=0}^{\infty} \theta_k = \infty$. Consequently, all $\pi_k = 0$.

If the departure rate $\mu = 0$ (which, for example, is true for the pure birth process from Example 2.3-3), then formally we cannot apply Proposition 4. However, the less the departure intensity μ, the larger the number of the particles in the system; we skip a formal statement. So, if X_t unlimitedly grows for a positive μ, the same is true for $\mu = 0$.

EXAMPLE 3 (*The M/M/1 queueing system with finite capacity*). For the values of λ_k and μ_k from Example 2.3-4, by (2.4.2), we have $\theta_k = 0$ for $k > a$, and for $k = 0, ..., a$,

$$\theta_k = \frac{\lambda^k}{\mu^k} = \rho^k,$$

where again $\rho = \lambda/\mu$. For $\rho \neq 1$,

$$\sum_{k=0}^{\infty} \theta^k = \sum_{k=0}^{a} \rho^k = \frac{1-\rho^{a+1}}{1-\rho}.$$

Note that the numerator and denominator above are either both positive or both negative. By (2.4.4), $\pi_k = 0$ for $k > a$, and for $k = 0, ..., a$,

$$\pi_k = \frac{\theta_k}{\sum_{k=0}^{\infty} \theta_k} = \frac{\rho^k(1-\rho)}{1-\rho^{a+1}}. \tag{2.4.6}$$

If $\rho = 1$, then $\theta_k = 1$ for $k \leq a$, the sum

$$\sum_{k=0}^{\infty} \theta^k = \sum_{k=0}^{a} 1 = a+1,$$

and

$$\pi_k = \frac{1}{a+1}$$

for $k = 0, ..., a$. So, in this case, all the states $0, ..., a$ are equally likely.

Note also that we did not have to impose the condition $\rho < 1$, which is natural: the capacity is bounded, and hence there cannot be an unlimited queue. □

Next, we consider the M/M/s queueing scheme, and first, the case of infinitely many servers.

EXAMPLE 4 (*The M/M/∞ scheme*). Let the arrival process be homogeneous Poisson with parameter λ, and suppose there are infinitely many servers whose service times are exponential with parameter μ. As was shown in Example 2.3-5, in this case, $\lambda_k = \lambda$, and $\mu_k = \mu k$ for *all* $k \geq 1$. Then, in accordance with (2.4.2), for $k \geq 1$,

$$\theta_k = \frac{\lambda^k}{\mu \cdot (\mu 2) \cdots (\mu(k-1)) \cdot (\mu k)} = \frac{\lambda^k}{\mu^k k!} = \frac{\rho^k}{k!}, \tag{2.4.7}$$

where again $\rho = \lambda/\mu$. Since $\theta_0 = 1$, we have

$$\sum_{k=0}^{\infty} \theta_k = \sum_{k=0}^{\infty} \frac{\rho^k}{k!} = e^{\rho}.$$

Substituting into (2.4.4), we have

$$\pi_k = e^{-\rho} \frac{\rho^k}{k!}. \tag{2.4.8}$$

Thus, we have come to the Poisson distribution with parameter ρ. The result is true for any ρ, which is also not surprising: since there are an infinite number of servers, there is no queue.

EXAMPLE 5 (*The M/M/s scheme*). The case when s, the number of servers, is finite is a bit more complicated. As we saw in Example 2.3-5, in this case, $\mu_k = \mu k$ for $k = 1,...,s$, and $\mu_k = \mu s$ for $k \geq s$. Hence, for $k = 1,...,s$, we still have (2.4.7), while for $k > s$,

$$\theta_k = \frac{\lambda^k}{\mu \cdot (\mu 2) \cdot \dots \cdot (\mu s)(\mu s)^{k-s}} = \frac{1}{s!}\rho^s (\rho/s)^{k-s}, \tag{2.4.9}$$

where, as above, $\rho = \lambda/\mu$. Set

$$A_{s\rho} = \sum_{k=0}^{s-1} \frac{\rho^k}{k!}.$$

Then, since $\theta_0 = 1$,

$$\sum_{k=0}^{\infty} \theta^k = A_{s\rho} + \sum_{k=s}^{\infty} \frac{1}{s!}\rho^s (\rho/s)^{k-s} = A_{s\rho} + \frac{1}{s!}\rho^s \sum_{k=s}^{\infty} (\rho/s)^{k-s} = A_{s\rho} + \frac{1}{s!}\rho^s \sum_{m=0}^{\infty} (\rho/s)^m.$$

(We used the change of variable $m = s - k$). The last sum is that of a geometric series and converges if

$$\rho < s. \tag{2.4.10}$$

So, eventually, if $\rho < s$, then

$$\sum_{k=0}^{\infty} \theta^k = A_{s\rho} + \frac{1}{s!}\rho^s \frac{1}{1-\rho/s}. \tag{2.4.11}$$

Furthermore,

$$\pi_0 = \frac{1}{\sum_{k=0}^{\infty} \theta^k} = \left(A_{s\rho} + \frac{1}{s!}\rho^s \frac{1}{1-\rho/s}\right)^{-1}. \tag{2.4.12}$$

To get π_k for $k > 0$, we should combine (2.4.4) and (2.4.7) for $k \leq s$, and combine (2.4.4) and (2.4.9) for $k > s$.

Eventually,

$$\pi_k = \begin{cases} \frac{1}{k!}\rho^k \pi_0 & \text{if } k \leq s, \\ \frac{1}{s!}\rho^s (\rho/s)^{k-s} \pi_0 & \text{if } k > s, \end{cases}$$

where π_0 is given in (2.4.12).

We see that in the case of s servers, the stable distribution exists under the condition

$$\rho < s.$$

For $s = 1$, we come to $\rho < 1$, which is consistent with the corresponding condition for the M/M/1 model. □

In conclusion, note that as for the M/M/1 scheme, in the general case,

> The probability π_k concides with the expected proportion (in the long run) of the time during which there are exactly k particles in the system.

The proof is absolutely similar to that of (2.1.6).

2.5 Proof of Proposition 4

The reasoning below is very similar to what we did in Section 2.2 with one exception: we should reflect the fact that the intensities of the arrival and departure processes depend on the current state of the system.

Consider a time moment t and a "small" interval $\Delta = [t, t+\delta]$. Instead of (2.2.1), now we write

$$P(N_\Delta = 1 \,|\, N_t = k) = \lambda_k \delta + o(\delta), \tag{2.5.1}$$

and instead of (2.2.2),

$$P(N_\Delta > 1 \,|\, N_t = k) = o(\delta). \tag{2.5.2}$$

(As in Section 2.2, $o(\delta)$ is a remainder negligible as compared with δ for small δ. Formally, $\frac{1}{\delta} o(\delta) \to 0$ as $\delta \to 0$.)

For the departure process and $k \geq 1$, instead of (2.2.3)-(2.2.4), we have respectively

$$P(\text{one departure during } \Delta \,|\, N_t = k) = \mu_k \delta + o(\delta), \tag{2.5.3}$$

and

$$P(\text{more than one departure during } \Delta \,|\, N_t = k) = o(\delta). \tag{2.5.4}$$

Also, similarly to what was shown in Section 2.2, the probability that during the time interval Δ, there will be simultaneously at least one arrival and one departure has an order of δ^2 and is negligible with respect to δ.

Thus, for $k \geq 1$, a counterpart of (2.2.5) may be written as follows:

$$\begin{aligned} P(X_{t+\delta} = k) &= P(X_t = k) \cdot (1 - \lambda_k \delta + o(\delta))(1 - \mu_k \delta + o(\delta)) \\ &+ P(X_t = k-1) \cdot (\lambda_{k-1} \delta + o(\delta))(1 - \mu_{k-1} \delta + o(\delta)) \\ &+ P(X_t = k+1) \cdot (1 - \lambda_{k+1} \delta + o(\delta))(\mu_{k+1} \delta + o(\delta)) + o(\delta). \end{aligned}$$

Then, similarly to how we have arrived at (2.2.6), for $k \geq 1$, we obtain the equation

$$\lambda_{k-1} \pi_{k-1} + \mu_{k+1} \pi_{k+1} - (\lambda_k + \mu_k) \pi_k = 0. \tag{2.5.5}$$

For $k = 0$, in the same fashion we get that

$$0 = -\lambda_0 \pi_0 + \mu_1 \pi_1, \tag{2.5.6}$$

which may be also obtained from (2.5.5) if we set there $k = 0$, $\pi_{-1} = 0$, and $\mu_0 = 0$.

Again similar to what we did in Section 2.2, by induction, we come to the recurrence relation

$$\pi_{k+1} = \frac{\lambda_k}{\mu_{k+1}} \pi_k \;\text{ for all }\; k = 0, 1, \ldots. \tag{2.5.7}$$

Hence,

$$\pi_k = \frac{\lambda_{k-1}}{\mu_k} \pi_{k-1} = \frac{\lambda_{k-1}}{\mu_k} \cdot \frac{\lambda_{k-2}}{\mu_{k-1}} \pi_{k-2} = \ldots = \frac{\lambda_{k-1}}{\mu_k} \cdots \frac{\lambda_0}{\mu_1} \pi_0 = \theta_k \pi_0. \tag{2.5.8}$$

Then $\sum_{k=0}^{\infty} \pi_k = \pi_0 \sum_{k=0}^{\infty} \theta_k$. Thus, if $\sum_{k=0}^{\infty} \theta_k$ converges, then

$$\pi_0 = \left(\sum_{k=0}^{\infty} \theta_k \right)^{-1}.$$

Together with (2.5.8), this leads to the first part of the proposition.

If $\sum_{k=0}^{\infty} \theta_k = \infty$, then we must set $\pi_0 = 0$, which along with (2.5.8) implies $\pi_k = 0$ for all k. ■

1,2

3 EXERCISES

Use of software in calculations is strongly recommended.

Section 1

1. Verify (1.1.3).
2. Customers arrive at a service facility according to a Poisson process with an average rate of 5 per hour. Find
   (a) the probabilities that (i) during 6 hours no customers will arrive, (ii) at most, twenty-five customers will arrive;
   (b) the probabilities that the waiting time between the third and the fourth customers will be (i) greater than 30 min., (ii) equal to 30 min., (iii) greater than or equal to 30 min.;
   (c) the probability that after the first customer has arrived, the waiting time for the fifth customer will be greater than an hour;
   (d) for the same waiting time, its expected value and standard deviation.
3. How would you answer questions similar to those in Exercise 2 for the case where the mean interarrival time is 30 min.?
4. During an eight-hour work day, starting from 9am, customers of a company arrive at a Poisson rate of two per hour.
   (a) Assume that during the period from 10 am to 11 am, no customers arrived. Find the probability that during the next half an hour there will be no customers either.
   (b) Assume that the fifteenth customer has arrived at 4 pm. (i) What is the probability that the seventeenth customer will come in less than one hour? (ii) Suppose that customers may arrive after the office is closed. The clerk of the company decides to wait for the seventeenth customer after the work day is over. What is the expected waiting time? How long will the clerk wait for the seventeenth customer on the average if she/he does not work after hours and leaves at 5 pm?
5. Let N_t be a Poisson process with a rate of λ, and let T_i be the time of the ith arrival. Write
   (a) $E\{N_t N_{t+s}\}$;
   (b) $E\{T_{n+m} | N_t = n\}$, $Var\{T_{n+m} | N_t = n\}$;

(c) $Corr\{N_2, N_4 - N_2\}$; $Corr\{N_t, N_{t+s}\}$.
(*Hint*: $N_{t+s} = N_t + N_{(t,t+s]}$; to avoid long calculations, use the memoryless property.)

6. Consider a homogeneous Poisson process N_t with a rate of λ. The time unit is an hour. Given that there were three arrivals during the first three hours, find the probability that (a) all three occurred during the first hour; (b) at least one arrival occurred during the first hour.

   Consider the general case and find the distribution of N_t given $N_{t+s} = n$. Does the answer depend on λ? (*Advice*: Look over Section **3**.4.5.5.)

7. John is waiting for his date Mary at a street corner. People are passing by at a Poisson rate of λ per minute. Suppose that John's waiting time is uniformly distributed on the interval from a to $a+b$ minutes (say, from 5 to 35 minutes). Find the expected value and variance of the number of people who will pass by during the time John will be waiting.

8. Ann is receiving telephone calls from customers of a company. Calls come at a Poisson rate of one each 15 min. Consider a time interval and the probability that there will be no more than one call during this interval. What length should the interval have for this probability to be greater than 0.8?

9.* Suppose that the model of Section 1.2 is used for describing the flow of customers entering a store. We consider all $t \in [0, \infty)$. Which function $\lambda(t)$ should we choose if the store is open from 10am to 6pm, the intensity of arrivals equals $c_1 = 1$ from 10 am to 2 pm, is linearly growing with a slope of $c_2 = 2$ from 2 pm to 5 pm, and remains constant during the last hour of the work day. Graph $E\{N_t\}$ as a function of t for $t \in [0, 24]$. How will the graph look for $t \in [0, \infty)$?

10.* For a non-homogeneous Poisson flow of customers, the intensity $\lambda(t)$ during the first 8 hours is increasing as $10(t/8)^2$ [ending up with 10 customers/hour at the end of the period]. Find the expected value and the standard deviation of the number of customers during the whole period. Given that the fifth customer arrived at $t = 5$ h, find the probability that the sixth customer will arrive after 5h 6min.

11.* A flow of arrivals N_t is a non-homogeneous Poisson process with the periodical intensity $\lambda(t) = |\sin \pi t|$. The unit of time is a day.

    (a) What is the intensity of arrivals at the end and at the beginning of each day? When is the intensity the largest?

    (b) What are the mean and variance of the number of arrivals during a year?

    (c) *Estimate* without long calculations $P\left(|N_t - E\{N_t\}| > \sqrt{Var\{N_t\}}\right)$ for $t = 1$ year.

12.* Give an example of an intensity $\lambda(t)$ for which the probability that no arrival will ever occur is $1/e$.

13.* Let the intensity of a non-homogeneous Poisson process $\lambda(t) = 1$ for $t \in [0, 1]$, and $\lambda(t) = 100$ for $t \in [1, 2]$. Explain heuristically and rigorously that the interarrival times τ_1 and τ_2 are dependent. (*Advice*: Consider the conditional distribution of τ_2 given $\tau_1 = a$ for, say, $a = 1/2$ and $a = 1$.)

14.* In an area, the process of the occurrences of traffic accidents is a Poisson process with a rate of $\lambda = 30$ per day. The probability that a separate accident causes serious injuries is $p = 0.1$. The outcomes of different accidents are independent. Estimate the probability that during a month, the number of accidents with serious injuries will exceed 100.

Section 2

15. Argue that the M/M/1 model is not appropriate for when the service times, though being random, are close to a positive and most probable number; say, as in a ticket booth.

16. Customers arrive at a service facility with one server according to a Poisson process with a rate of 5 per hour. The service times are i.i.d. exponential r.v.'s, and on the average, the server can serve 7 customers per hour. Suppose that the system is in the stationary regime.

    (a) What is the probability that at a particular time moment, there will be no queue?

    (b) What is the probability that at a particular time moment, there will be more than three customers waiting for service?

    (c) What is the mean proportion of time when the server is idle?

17. In the long run, the mean proportion of time when the server in a M/M/1 system is idle, is 0.9. Find the arrival/service ratio. Find the same if 0.9 is the proportion of time when there is no queue.

18. For the M/M/1 scheme, find the variance of the number of particles in the system.

19. (a) For the scheme of Section 2.1, considering the stationary regime, graph the mean of X_t as a function of ρ. (b) Do the same for the standard deviation.

20. For the scheme M/M/1, show that the mean length of the queue, given that the server is busy, is equal to the mean number of particles in the system. Argue that the answer is not so surprising as it looks at first glance, showing that this is a special property of the process we consider. (*Advice*: Regarding, the last question, assume that the conditional expectation mentioned equals the mean number of particles in the system, and show that in this case, both are equal to $(1-\pi_0)/\pi_0$, which is consistent with (2.1.4).)

21. Consider the scheme M/M/1.

    (a) Does the mean number of customers in the queue increase or decrease when ρ is increasing?

    (b) Find $P(X_t > k)$ in the stationary regime. Does this probability increase or decrease when ρ is increasing?

    (c) Argue that both your answers are consistent with the significance of the characteristic ρ. So, for a larger ρ, does the system perform worse or better, and in what sense?

22. Consider the M/M/1 model, and suppose that the server started to serve a new customer.

    (a) Find the probability that during the service time of a customer, no new customer will arrive. (*Advice*: Use conditioning.)

    (b) Do you expect the mean number of new customers who will arrive during the service period to be less than one? Find the mean mentioned.

23. Consider the M/M/1 queueing model, and denote by T the amount of time a randomly chosen customer spends in the system. We consider the stationary regime and take for granted that the r.v. K, the number of customers that are ahead of a newly arriving customer, has the same distribution as X_t in the stationary regime.

    (a) Give a heuristic argument that in the stationary regime,

$$E\{T\} = \frac{1}{\mu - \lambda}, \tag{3.1}$$

provided $\mu > \lambda$. Comment on the fact that $E\{T\} \to \infty$ as λ is approaching μ. (*Hint*: A new customer should wait for $K+1$ services to be completed: K for the customers ahead, and one more for her/himself. The mean time of each service is $1/\mu$.

(b) Find the expecting waiting time $E\{T\}$ in the situation of Exercise 16.

(c) When writing that $E\{T\} = \frac{1}{\mu}E\{K+1\}$, we had to justify this step because, as a matter of fact, T is the sum of a *random* number of r.v.'s; namely, $T = \sum_{m=1}^{K+1} Y_i$, where Y_i are exponential r.v.'s. Realize that to make the proof of (3.1) rigorous, we may use the result of Example **3**.6.2-1.

24. It may be proved that for $\mu > \lambda$, the waiting time T defined in Exercise 23 has the exponential distribution with the parameter $\mu - \lambda$. Argue that this is consistent with (3.1). For the situation of Exercise 16, compute the probability that the waiting time will exceed one hour. Suppose that in the same situation, a customer has been already waiting half an hour in line. What is the probability that she/he will be waiting at least an hour more?

25. Consider the M/M/1 queueing model, and as in Exercise 23, take for granted that the r.v. K, the number of customers that are ahead of a newly arriving customer, has the same distribution as X_t in the stationary regime. Prove the fact that we use in Exercise 24. (*Advice*: Use the result of Example **6**.4.2-2.) and a remark in Exercise 23.)

26.* Describe the behavior of the process X_t in Section 2.3 for (a) $\lambda_0 = \lambda_1 = \lambda > 0$, $\lambda_k = 0$ for all $k \geq 2$, and $\mu_k = 0$ for all k; (b) $\lambda_0 = \lambda_1 = \lambda > 0$, $\lambda_k = 0$ for all $k \geq 2$, and $\mu_0 = 0, \mu_1 = \mu > 0$, $\mu_2 = 2\mu$.

27.* Consider the M/M/1 scheme with a capacity of a.

(a) Make sure that the limiting probabilities in (2.4.6) are positive for all $\rho \neq 1$.

(b) Figure out for which ρ the probability π_k, as a function of k, is increasing (certainly, for $k = 0, ..., a$); for which ρ it is decreasing; and for which ρ the probability π_k does not depend on k for $k = 0, ..., a$. Interpret the result proceeding from the significance of the arrival/service ratio ρ.

(c) How do the probabilities $\pi_0, ..., \pi_a$ look for large ρ; formally, for $\rho \to \infty$? Interpret the answer.

28.* (a) For a M/M/1 service facility with a capacity of two, find π_0, π_1, π_2 in two cases: (i) $\lambda = 3$, $\mu = 6$ and (ii) $\lambda = 6$, $\mu = 3$. Compare the results.

(b) Establish a general pattern; namely, show that if π_k^* is the limiting probability corresponding to the replacement of ρ with $1/\rho$, then $\pi_k = \pi_{a-k}^*$.

29.* Find the coefficient of variation for the number of particles in the system in the stationary regime for the M/M/∞ scheme with an arrival/service ratio of ρ.

30.* In a large store, there are many cashier counters. Customers arrive at the checkout area at a Poisson rate of five per minute. The service times are independent and exponential with a mean of one minute. Let k be a number such that the probability that the number of busy counters exceeds k is not larger than 0.05. Say without calculations whether k can be equal to 5 or to 7. Calculate k.

31.* Customers arrive at a single-server facility at a Poisson rate of λ; the service time is exponential with parameter μ. However, if a customer arrives when the server is busy and there are k people in line ($k = 0, 1, ...$), then the customer decides to wait with a probability of $1/(k+1)$. Find limiting probabilities.

32.* Consider two queueing systems. The first has two servers; the service time for *each* is exponential with parameter μ. The second system has a single server whose service time is exponential with parameter 2μ; that is, the single server in the second system works twice as fast as each server in the first system. The flows of arrivals for both systems are Poisson with the same parameter $\lambda < 2\mu$. Denote by X_t and $\widetilde{X}_t$ the respective numbers of customers at time t in the above systems with two and one servers. Regarding characteristics of these processes, like θ_k and pi_k, those that correspond to the second process, will also be marked by tilde $\tilde{}$.

(a) Reasoning heuristically, conjecture in which system we should expect more customers. Where do you expect a larger queue? (*Advice*: Compare the arrival and departure rates for both systems for all possible states: $0, 1, 2, \ldots$.)

(b) Not computing precisely the stationary distributions, show that in the stationary regimes, for any k,

$$P(X_t \leq k) \leq P(\widetilde{X}_t \leq k). \tag{3.2}$$

Comment on this result. (The reader may recollect the notion of the first stochastic dominance (FSD); see Exercise **6**-17.) Show that, in particular, from (3.2) it follows that $E\{X_t\} \geq E\{\widetilde{X}_t\}$. (Advice: First, realize that $\theta_k = 2\tilde{\theta}_k$ for all $k \geq 1$. Show that this implies that $\pi_0 < \tilde{\pi}_0$, and $\pi_k > \tilde{\pi}_k$ for all $k \geq 1$. Keep also in mind that $P(X_t \leq k) = 1 - \sum_{m=k+1}^{\infty} \pi_m \leq 1 - \sum_{m=k+1}^{\infty} \widetilde{\pi}_m$.)

(c) Find the stationary distributions for both systems.

(d) Let Y_t and $\widetilde{Y}_t$ be the respective sizes of the queues. Prove that in the stationary regimes, for any k,

$$P(Y_t \leq k) \geq P(\widetilde{Y}_t \leq k). \tag{3.3}$$

Comment on this result. Show that $E\{Y_t\} \leq E\{\widetilde{Y}_t\}$. (*Advice*: Show that for *all* $k \geq 1$, the probability $P(Y_t = k) = \pi_{k+2} \geq \widetilde{\pi}_{k+1} = P(\widetilde{Y}_t = k\}$.)

(e) Compute and compare the expected times newly arriving customers spend in each system. (*Hint*: If both systems are in the same state $k \geq 2$, the time a new customer waits for the beginning of service has the same distribution for both systems. However, in the first system, the service itself lasts twice longer on the average.)

33.* Make sure that the general formula for the steady-state distribution for the M/M/s system leads to the corresponding distributions for the particular cases $s = 1$ and $s = \infty$.

34.* A receiver of "particles" has two cells each of which may contain only one particle. The cells are empty at the initial time. Particles arrive at a Poisson rate of λ. If one cell is occupied, no departures are possible. Once both cells are occupied, particles from outside do not enter the receiver; while each particle in the system, being considered separately and independently of the other, leaves its cell in the random time exponentially distributed with a parameter of μ. However, once one particle leaves the receiver (whichever comes first), the departure for the other particle is terminated, and the system comes back to the state with one particle. Describe how the process X_t will run, and find the limiting probabilities. (*Advice*: You may use Proposition 4 assuming $\mu_1 = \varepsilon > 0$ and letting $\varepsilon \to 0$ at the very end.)

Route 1 $\Rightarrow$ *page 357*

Chapter 14

Elements of Renewal Theory

2

1 PRELIMINARIES

As in Section **13**.1.1, we consider a counting process N_t, with $N_0 = 0$, and denote by $\tau_1, \tau_2, ...$ the consecutive interarrival times. We again suppose that τ_i's are independent and identically distributed positive r.v's, but now, we do *not* assume the distribution of τ's to be exponential. In this general case, a counting process is usually called *renewal*, and the term comes from the following typical interpretation.

Suppose that from time to time, you replace a broken device by a new identical one; for example, a washer in your kitchen faucet, or a bulb (a popular example), say, in the lamp on your desk. In this context, the "arrivals" being counted are called renewals.

As we saw in Chapter 12 (see, e.g., Exercise **12**-1), when the distribution of interarrival times is arbitrary, the process N_t does not have to be that with independent increments.

Let $T_n = \tau_1 + ... + \tau_n$, the time of the nth arrival (or renewal). Clearly, the main relation

$$P(N_t \geq n) = P(T_n \leq t), \tag{1.1}$$

remains to be true. The reader may look over the explanation in Section **13**.1.1.

In Section **13**.1.1, we have shown that if the distribution of τ's is exponential, the relation above leads to a precise formula for the distribution of N_t.

In general, such cases are rare. Another matter is that, if t is large, then the r.v. N_t takes on large values, and hence, it is reasonable to consider the probabilities in (1.1) for large n. In this case, we can appeal to the CLT and use the normal approximation.

We discuss this in the next section, and for now consider the mean value $m(t) = E\{N_t\}$ which is referred to as a *renewal function*. Using (1.1) and the general formula $E\{N\} = \sum_{n=0}^{\infty} P(N > n)$ (see (**3**.2.2.1)), we have

$$m(t) = \sum_{n=0}^{\infty} P(N_t > n) = \sum_{n=0}^{\infty} P(N_t \geq n+1) = \sum_{n=1}^{\infty} P(N_t \geq n) = \sum_{n=1}^{\infty} P(T_n \leq t). \tag{1.2}$$

Let $F(t) = P(\tau_i \leq t)$, the (common) distribution function of the r.v.'s τ_i. Then $P(T_n \leq t) = F^{*n}(t)$, where—as in Sections **3**.1.2.3 and **6**.4.1—the symbol $*$ stands for the convolution operation, and $F^{*n} = \underbrace{F * ... * F}_{n \text{ times}}$. So, we can rewrite (1.2) as

$$m(t) = \sum_{n=1}^{\infty} F^{*n}(t). \tag{1.3}$$

We see that there is a correspondence between the functions $m(t)$ and $F(t)$. It may be proved that this correspondence is one-to-one; that is, given $m(t)$, there is only one $F(t)$ satisfying (1.3), and hence $m(t)$ uniquely determines the renewal process.

EXAMPLE 1. Suppose that $m(t)$ is a linear function. Since $m(0) = E\{N_0\} = 0$, we have $m(t) = \lambda t$ for a non-negative coefficient λ. We know that this is the renewal function for a homogeneous Poisson process with parameter λ. Hence, our process is exactly such a process, and the interarrival times are exponential with parameter λ. □

---

2 LIMIT THEOREMS

In Section **13**.1.1, we have shown that if N_t is a homogeneous Poisson process, then $E\{N_1\} = 1/\mu$, where $\mu = E\{\tau_i\}$. That is, the expected number of arrivals during a unit of time is the reciprocal of the mean interarrival time. We have also noted there that this was not true in the general case, and the relation mentioned held only asymptotically in the long run.

Before considering the corresponding theorem below, let us realize that the convergence of a sequence of r.v.'s X_n to a r.v. X does not automatically imply the convergence of the corresponding expectations.

EXAMPLE 1. Let Z be uniform on $[0,1]$, and $X_n = 0$ if $0 \leq Z < 1 - 1/n$, and $X_n = n$ otherwise. If $0 \leq Z < 1$ (which occurs with probability one), then there exists an m such that $Z < 1 - 1/m$. (As m, we may take the smallest integer $m > 1/(1-Z)$.) Hence, for all $n \geq m$, the r.v. $X_n = 0$. This implies that $X_n \to 0$ once $0 \leq Z < 1$, and as was told, this event occurs with probability one.

On the other hand, $E\{X_n\} = 0 \cdot (1 - \frac{1}{n}) + n \cdot \frac{1}{n} = 1$ for all n. □

We return to the renewal process N_t.

Theorem 1 *Let $\mu = E\{\tau_i\} > 0$. Then the following is true.*
(a) (The LLN for renewal processes). With probability one, as $t \to \infty$,

$$\frac{N_t}{t} \to \frac{1}{\mu}. \tag{2.1}$$

(b) (The elementary renewal theorem). As $t \to \infty$,

$$\frac{m(t)}{t} = \frac{E\{N_t\}}{t} \to \frac{1}{\mu}. \tag{2.2}$$

First of all, we understand now that (2.1) does not imply (2.2) automatically, and this is why we stated (2.2) separately.

Moreover, though by tradition, the assertion (2.2) is called the *elementary renewal theorem*, its proof is more complicated than that of (2.2) and requires some general facts from Mathematical Analysis which are beyond the scope of this book. So, we take (2.2) for granted. The proof of (2.1) will be considered in Section 3.

EXAMPLE 2. For a commercial car of a company, the times between consecutive oil changes are uniformly distributed between two and three months. (For simplicity, we view months as time periods with identical lengths.) Let N_t be the number of oil changes by time t. By Theorem 1, for large t, with probability close to one, $\frac{1}{t}N_t \approx \frac{1}{2.5} = 0.4$. Consider, for example, five years. Then, we can multiply 0.4 by 60, getting 24 oil changes for the whole period, but we should also remember that when considering the ratio N_t/t, we are figuring out merely the average number of arrivals per unit of time. So, the number 24 should be viewed merely as an approximation. □

To estimate the probability distribution of N_t (rather than only its expectation), we should consider a more sophisticated theorem, and to this end, we use the normal approximation. If we had known the precise expressions for $E\{N_t\}$ and $Var\{N_t\}$, we would have considered the normalized r.v.

$$N_t^* = (N_t - E\{N_t\})/\sqrt{Var\{N_t\}},$$

and have tried to prove that the distribution of N_t^* converges to the standard normal. However, $E\{N_t\}$ and $Var\{N_t\}$ depend on the distribution of the interarrival times τ's in a complicated way, and we can get only asymptotic presentations for $E\{N_t\}$ and $Var\{N_t\}$.

By Theorem 1, the quantity t/μ may serve as a good approximation for $E\{N_t\}$. To find an asymptotic expression for $Var\{N_t\}$, we may reason as follows. Let us look at the r.-h.s. of (1.1). First, $E\{T_n\} = \mu n$ and $Var\{T_n\} = \sigma^2 n$, where $\mu = E\{\tau_i\}$ and $\sigma^2 = Var\{\tau_i\}$. Set

$$t = \mu n + x\sigma\sqrt{n}, \tag{2.3}$$

where x is an arbitrary number. Then

$$P(T_n \leq t) = P\left(T_n \leq \mu n + x\sigma\sqrt{n}\right) = P\left(\frac{T_n - \mu n}{\sigma\sqrt{n}} \leq x\right) \to \Phi(x), \text{ as } n \to \infty,$$

where $\Phi(x)$ is the standard normal d.f. Consequently, the l.-h.s. of (1.1) should converge to the same limit if (2.3) is true.

Setting $z = \sqrt{n}$, we rewrite (2.3) as the quadratic equation $t = \mu z^2 + x\sigma z$. This equation has only one positive root, and choosing it, we get

$$n = \left(\frac{-\sigma x + \sqrt{\sigma^2 x^2 + 4\mu t}}{2\mu}\right)^2.$$

In view of Theorem 1, we believe that the main term for the asymptotic expansion for n should be t/μ. Therefore, we consider the expression $n - t/\mu$ which may be written as

$$n - \frac{t}{\mu} = \frac{\sigma^2 x^2 + \sigma^2 x^2 + 4\mu t - 2\sigma x\sqrt{\sigma^2 x^2 + 4\mu t}}{4\mu^2} - \frac{t}{\mu} = \frac{\sigma^2 x^2 - \sigma x\sqrt{\sigma^2 x^2 + 4\mu t}}{2\mu^2}.$$

As $t \to \infty$, the main term in this expression is $-\frac{\sigma x}{2\mu^2}\sqrt{4\mu t} = -x\sqrt{\frac{t\sigma^2}{\mu^3}}$. Hence, as $t \to \infty$,

$$n - \frac{t}{\mu} \sim -x\sqrt{\frac{t\sigma^2}{\mu^3}}. \tag{2.4}$$

Proceeding from (2.4), we conjecture that, once we consider the asymptotic behavior of N_t for large t, we may choose as a normalized r.v., the r.v.

$$N_t^* = \frac{N_t - t/\mu}{\sqrt{t\sigma^2/\mu^3}}. \tag{2.5}$$

In particular, from (2.5) we get that

$$Var\{N_t^*\} = \frac{Var\{N_t\}}{t\sigma^2/\mu^3}.$$

Therefore, if the distribution of N_t^* converges to the standard normal distribution (and we will see that this is the case), then one may conjecture that $Var\{N_t^*\} \to 1$, and hence,

$$\frac{Var\{N_t\}}{t} \to \frac{\sigma^2}{\mu^3}. \tag{2.6}$$

This is indeed true. We skip the formal proof of (2.6) but will show the asymptotic normality of N_t; more precisely, that for large t, the r.v. N_t is asymptotically normal with a mean of t/μ and a variance of $t\sigma^2/\mu^3$.

Theorem 2 *(The CLT for renewal processes). As $t \to \infty$, for any x,*

$$P\left(\frac{N_t - t/\mu}{\sqrt{t\sigma^2/\mu^3}} \le x\right) \to \Phi(x), \tag{2.7}$$

where $\Phi(x)$ is the standard normal d.f.

We sketch a proof of (2.7) in Section 3.

EXAMPLE 3. Let us come back to Example 2. The mean and standard deviation of the number of oil changes during $t = 60$ months are approximately equal to $\frac{t}{\mu} = \frac{60}{2.5} = 24$ and $\sqrt{t\frac{\sigma^2}{\mu^3}} = \sqrt{\frac{60\cdot(1/12)}{2.5^3}} \approx 0.57$. Hence, for $t = 60$,

$$P\left(N_t \le \frac{t}{\mu} + x\sqrt{\frac{t\sigma^2}{\mu^3}}\right) \approx P(N_{60} \le 24 + 0.57\cdot x) \approx \Phi(x). \tag{2.8}$$

Let us choose $x = 1.76$. Since $0.57 \cdot 1.76 = 1.0032$, and N_t is an integer-valued r.v., using Table 3 from the Appendix, we can write

$$P(N_{60} \le 25) = P(N_{60} \le 25.0032) = P(N_{60} \le 24 + 0.57\cdot 1.76) \approx \Phi(1.76) \approx 0.9608.$$

On the other hand, taking $x = 3.5$, because $0.57 \cdot 3.5 = 1.995$, we may write

$$P(N_{60} \le 25) = P(N_{60} \le 25.995) = P(N_{60} \le 24 + 0.57\cdot 3.5) \approx \Phi(3.5) \approx 0.9998. \tag{2.9}$$

Which number should we choose from the above two? Formally, the question is meaningless. First, we do not know the accuracy of the approximation and cannot trust these numbers "up to the fourth digit." Second, we are approximating the discrete distribution of

N_t by the continuous normal distribution, and this should also be taken into account. Common sense says that it is natural to take an intermediate value and to not use too detailed estimates. Say, we can write

$$P(N_{60} \leq 25) = P(N_{60} \leq 25.5) \approx P(N_{60} \leq 24 + 0.57 \cdot 2.63) \approx \Phi(2.63) \approx 0.996.$$

Similarly,

$$P(N_{60} \leq 22) = P(N_{60} \leq 22.5) \approx P(N_{60} \leq 24 - 0.57 \cdot 2.63) \approx 1 - \Phi(2.63) \approx 0.004.$$

Then we can conclude (more precisely, guess) that

$$P(23 \leq N_{60} \leq 25) \approx 0.996 - 0.004 = 0.992 \approx 0.99.$$

However, until we know the accuracy of approximation, we can consider this merely as an heuristic estimate.

It is also noteworthy that when we estimate a probability which, in any case, is large (as $P(N_{60} \leq 25) \approx 0.99$), we can hope that the accuracy of approximation is essentially less than the probability which we are estimating. However, when estimating a probability which is small (say, $P(N_{60} \leq 22) \approx 0.004$), we should not trust much the estimate obtained: the accuracy of estimation may be larger than what we are estimating.

It also makes sense to note that the bounds for the rate of convergence in the CLT exist (one was considered in Section **9**.2.1), not saying that in practical situations, we can also estimate the accuracy of approximation by computer simulation. □

---

3 SOME PROOFS

We will give heuristic proofs of some of the above assertions, not aiming at complete rigor but rather making these assertions plausible.

Proof of (2.1). The r.v. N_t is the number of renewals that occurred through time t, and T_{N_t} is the time of the last renewal before or at time t. Hence, $T_{N_t} \leq t$, and $t < T_{N_t+1}$. Thus, $T_{N_t} \leq t < T_{N_t+1}$. Dividing this by N_t, we have

$$\frac{T_{N_t}}{N_t} \leq \frac{t}{N_t} < \frac{T_{N_t+1}}{N_t}.$$

This may be rewritten as

$$\frac{T_{N_t}}{N_t} \leq \frac{t}{N_t} < \frac{T_{N_t+1}}{N_t+1} \cdot \frac{N_t+1}{N_t}. \tag{3.1}$$

The number of renewals during an infinite period of time may be finite only if after some time moment, no renewals will happen. This may be true only if a τ_i turns out to be infinite, which is impossible. So, with probability one,

$$N_t \to \infty \quad \text{as} \quad t \to \infty. \tag{3.2}$$

On the other hand, by the law of large numbers, with probability one,

$$\frac{T_n}{n} = \frac{1}{n}\sum_{i=1}^{n}\tau_i \to \mu.$$

Hence, in view of (3.2), one may conjecture that the same will be true if we replace the non-random n with the random N_t; that is, with probability one,

$$\frac{T_{N_t}}{N_t} = \frac{1}{N_t}\sum_{i=1}^{N_t}\tau_i \to \mu \text{ as } t\to\infty. \tag{3.3}$$

This is indeed true but certainly requires a proof, and the point is not only in the fact that the number of terms in $\sum_{i=1}^{N_t}$ is random but also in the fact that N_t depends on the terms τ_i's (the larger τ's, the less the number of renewals during a fixed period $[0,t]$).

We skip a rigorous proof of (3.3) and, taking it for granted, come back to (3.1). By virtue of (3.2), with probability one,

$$\frac{N_t+1}{N_t} \to 1,$$

which together with (3.3) implies that the very left- and right-hand sides of (3.1) converge to μ with probability one. Then, by the squeezing principle of Calculus, so does t/N_t. ■

A more rigorous proof of (2.1), though concerning convergence in probability, is relegated to Exercise 9 provided with detailed advice.

Proof of (2.7). Let the normalized r.v. $N_t^* = \frac{N_t - t/\mu}{\sqrt{t\sigma^2/\mu^3}}$. Let us fix x and consider

$$P(N_t^* \geq x) = P\left(\frac{N_t - t/\mu}{\sqrt{t\sigma^2/\mu^3}} \geq x\right) = P\left(N_t \geq \frac{t}{\mu} + x\sqrt{t\sigma^2/\mu^3}\right) = P\left(N_t \geq n_t\right), \tag{3.4}$$

where

$$n_t = n_t(x) = \frac{t}{\mu} + x\sqrt{t\sigma^2/\mu^3}.$$

(In comparison with (2.4), we have $+$ before x, but x is an arbitrary number.)

The number n_t is not an integer; however, to simplify our reasoning, we will ignore this fact and treat n_t as an integer.

(To be absolutely rigorous, we should consider the integer part $[n_t]$ as well as $[n_t]+1$, and show that, as $t\to\infty$, the replacement n_t by these integers leads to the same limit. We skip these technicalities.)

By virtue of (1.1),

$$P(N_t \geq n_t) = P(T_{n_t} \leq t) = P\left(\frac{T_{n_t} - \mu n_t}{\sigma\sqrt{n_t}} \leq \frac{t-\mu n_t}{\sigma\sqrt{n_t}}\right) = P\left(T_{n_t}^* \leq z_t\right),$$

where $T_{n_t}^* = \frac{T_{n_t} - \mu n_t}{\sigma\sqrt{n_t}}$, and $z_t = \frac{t-\mu n_t}{\sigma\sqrt{n_t}}$. If $t\to\infty$, then $n_t\to\infty$ also, and by the CLT, $P(T_{n_t}^* \leq x) \to \Phi(x)$ for any x. On the other hand,

$$z_t = \frac{t - t - x\mu\sqrt{t\sigma^2/\mu^3}}{\sigma\sqrt{\frac{t}{\mu} + x\sqrt{t\sigma^2/\mu^3}}} = -\frac{x\mu\sqrt{t\sigma^2/\mu^3}}{\sigma\sqrt{\frac{t}{\mu} + x\sqrt{t\sigma^2/\mu^3}}} = -\frac{x\sigma\sqrt{t/\mu}}{\sigma\sqrt{\frac{t}{\mu} + x\sqrt{t\sigma^2/\mu^3}}} \to -x,$$

as $t \to \infty$. Consequently,

$$P(N_t \geq n_t) = P(T_{n_t} \leq t) \to \Phi(-x) = 1 - \Phi(x).$$

(To be absolutely honest, we should take into account that in $P(T^*_{n_t} \leq z_t)$, we are simultaneously considering the limits for $T^*_{n_t}$ and z_t. We can do it because $\Phi(x)$ is a continuous function; we skip the details.)

Together with (3.4), this implies that $P(N^*_t \geq x) \to 1 - \Phi(x)$, or

$$P(N^*_t < x) \to \Phi(x). \tag{3.5}$$

From this, it follows, in particular, that $P(N^*_t = x) \leq P(x \leq N^*_t < x + \delta) \to \Phi(x + \delta) - \Phi(x)$ for any $\delta > 0$. Because $\Phi(x)$ is a continuous function, letting $\delta \to 0$, we get that $P(N^*_t = x) \to 0$, and together with (3.5) this implies that

$$P(N^*_t \leq x) \to \Phi(x). \blacksquare$$

---

4 EXERCISES

1. Does N_t depend on τ_i's? Give a particular example.
2. Which of the following is true? Compare with (1.1). Justify your answers.

   (a) $N_t > n$ if and only if $T_n < t$;

   (b) $N_t \leq n$ if and only if $T_n \geq t$;

   (c) $N_t < n$ if and only if $T_n > t$.

3. Tourists arrive at a historic park (for simplicity, one at a time) according to a Poisson process at a rate of 40 per hour. Each five people take an excursion in a minivan. Let N_t be the number of excursions arranged by time t.

   (a) Show that N_t is a renewal process. What is the distribution of the time between consecutive "renewals" (excursions)? Find the probability that an interarrival time will exceed 10 minutes.

   (b) What is the number of excursions per hour in the long run?

   (c) Estimate the probability that the number of excursions during eight working hours will fall into the interval $[60, 70]$.

4. Let the interarrival times τ_i have a geometric distribution; namely, $P(\tau_i = m) = p(1-p)^{m-1}$ for $m = 1, 2, ...$ and a parameter $p \in (0, 1)$.

   (a) Show that T_n has a negative binomial distribution. With which parameters?

   (b) Show that

   $$P(N_t = n) = \binom{[t]}{n} p^n (1-p)^{[t]-n}$$

   for $n \leq [t]$, where $[t]$ stands for the integer part of t.

   (c) Is the process N_t that with independent increments?

5. We have already considered in Exercise **3**.53 the notion of the coefficient of variation σ/m of a r.v. X with a mean of m and a standard deviation of σ. Find the limit of the coefficient of variation for N_t as $t \to \infty$ and interpret your result.

6. Suppose that for a renewal process N_t, the interarrival times have the Poisson distribution with a mean of λ. Find the distribution of T_n and write a formula for $P(N_1 \geq n)$. For which t an s is it true that $P(N_t \geq n) = P(N_s \geq n)$?

7. In a store, the inter-occurrence times of the replenishment of stock for a particular product are i.i.d. r.v.'s with a mean of five days and a standard deviation of one day. Each replenishment action costs \$1500. Estimate the 0.95-quantile for the total replenishment expenses during a quarter; that is, a number q such that the probability that the total expenses will not exceed q is 0.95.

8. (*An inspection paradox.*) Consider a renewal process N_t with i.i.d. interarrival times τ_i. Say, a piece of equipment serves until it breaks down; upon failure it is instantly replaced, and the renewal process continues.

   (a) Suppose we want to "inspect" the process, *fixing* a time t and observing the length of the interarrival time interval that contains the time t. Say, we observe the lifetime of the equipment that was working at time t. Argue that the length of the time interval mentioned does not have the same distribution as τ_i's and on the average, this length is larger then $E\{\tau_i\}$. (*Hint*: We do not choose an arbitrary r.v. τ_i, but rather choose the τ_i for which $T_{i-1} < t \leq T_i$ for the fixed t.)

   (b) Answer the same question regarding the interarrival time that follows time t; say, the lifetime of the equipment installed after epoch t.

9. Prove that $\frac{N_t}{t} \xrightarrow{P} \frac{1}{\mu}$ proceeding from the following outline. For i.i.d. τ's, set $\mu = E\{\tau_i\} \neq 0$. For $\varepsilon > 0$, write $P\left(\frac{N_t}{t} - \frac{1}{\mu} > \varepsilon\right) = P\left(N_t > t\left(\frac{1}{\mu} + \varepsilon\right)\right) \leq P(N_t \geq n_t)$, where n_t is the integer part of $t\left(\frac{1}{\mu} + \varepsilon\right)$. By (1.1), $P(N_t \geq n_t) = P(T_{n_t} \leq t) = P\left(\frac{T_{n_t}}{n_t} - \mu \leq \frac{t}{n_t} - \mu\right)$. Note that $\frac{t}{n_t} - \mu \to -\frac{\varepsilon\mu^2}{1+\varepsilon\mu}$ as $t \to \infty$. Hence,
$P\left(\frac{T_{n_t}}{n_t} - \mu \leq \frac{t}{n_t} - \mu\right) \leq P\left(\frac{T_{n_t}}{n_t} - \mu \leq -\frac{\varepsilon\mu^2}{2(1+\varepsilon\mu)}\right)$ for large t. Now note that T_n is the sum of i.i.d. r.v.'s, and the last probability converges to zero, as $t \to \infty$, by the LLN: $\frac{T_{n_t}}{n_t} \xrightarrow{P} \mu$.
Thus, $P\left(\frac{N_t}{t} - \frac{1}{\mu} > \varepsilon\right) \to 0$, for any $\varepsilon > 0$. The probability $P\left(\frac{N_t}{t} - \frac{1}{\mu} < -\varepsilon\right)$ is considered similarly.

10. Estimate $P(N_{60} = 25)$ in the situation of Example 2-3. To what extent should we trust the estimate we obtained?

Chapter 15

Martingales in Discrete Time

1,2

In this chapter, we begin to study a new and important nowadays type of process. We assume all r.v.'s under consideration to have finite expectations.

1 DEFINITIONS AND PROPERTIES

1.1 Two formulas of a general nature

Throughout this chapter, we systematically use the notion of conditional expectation $E\{Y\,|\,X\}$ introduced in Section **3**.6 and clarified there and in subsequent chapters. In what follows below, the symbol X in $E\{Y\,|\,X\}$ may stand for a random vector as well as for a random variable, which has been mentioned in (**6**.5.2.10). In particular, we will repeatedly use the formula for total expectation (see, e.g., (**3**.6.2.1))

$$E\{E\{Y\,|\,X\}\} = E\{Y\}. \tag{1.1.1}$$

We will need two more simple relations. In the first reading, the reader may take these relations at a heuristic level.

Consider two r.v.'s or r.vec.'s: X and $\widetilde{X}$.

> If $\widetilde{X} = g(X)$ where $g(\cdot)$ is a one-to-one function, then $E\{Y\,|\,\widetilde{X}\} = E\{Y\,|\,X\}$.

(1.1.2)

To show this, let us recall how we defined the r.v. $E\{Y\,|\,X\}$. First, we have defined the regression function $m(x) = E\{Y\,|\,X = x\}$, and after that, we set $E\{Y\,|\,X\} = m(X)$. So, when considering $\widetilde{X}$, we define the regression function $\widetilde{m}(x) = E\{Y\,|\,\widetilde{X} = x\}$, and we set $E\{Y\,|\,\widetilde{X}\} = \widetilde{m}(\widetilde{X})$.

To make it explicit, let $X, \widetilde{X}$ be r.v.'s, and $g(x) = x^3$. The general case is considered absolutely similarly, and the reader is invited to do this in Exercise 1. So, let $\widetilde{X} = X^3$. We have

$$\widetilde{m}(x) = E\{Y\,|\,\widetilde{X} = x\} = E\{Y\,|\,X^3 = x\} = E\{Y\,|\,X = x^{1/3}\} = m(x^{1/3}).$$

Then,

$$E\{Y\,|\,\widetilde{X}\} = \widetilde{m}(\widetilde{X}) = m(\widetilde{X}^{1/3}) = m(X) = E\{Y\,|\,X\}.$$

If $g(\cdot)$ is not a one-to-one function, then different values of X may correspond to the same value of $\widetilde{X}$, and (1.1.2) may be not true. However, we can proceed as follows.

First, let us look again at (1.1.1). In the l.-h.s., we first compute the expected value of Y given an additional information (about the value of X), and after that, we compute the expected value of the conditional expectation. Such a procedure leads to the unconditional expectation in the r.-h.s.

Since conditional expectations inherit the main properties of "usual" expectations, we can write a counterpart of (1.1.1) for conditional expectations. In particular, we can replace the expectation $E\{\cdot\}$ in (1.1.1) by the conditional expectation $E\{\cdot\,|\widetilde{X}\}$. The only thing we need for such a generalization to be true is that the information on which the interior conditional expectation is based should be either more detailed than, or at least equal to, the information based on values of $\widetilde{X}$.

If $\widetilde{X} = g(X)$, this requirement is fulfilled because given X, we know exactly which value $\widetilde{X}$ has assumed (but perhaps not vice versa). Thus,

If $\widetilde{X} = g(X)$ where $g(\cdot)$ is a function, then $E\{Y\,|\widetilde{X}\} = E\{E\{Y\,|X\}\,|\widetilde{X}\}$. (1.1.3)

(We first condition Y on X and after that, on $\widetilde{X}$.)

If $g(\cdot)$ is a one-to-one function, then (1.1.3) is trivial: in view of (1.1.2), the l.-h.s. in (1.1.3) is equal to $E\{Y\,|X\}$, and the r.-h.s. equals $E\{E\{Y\,|X\}\,|X\} = E\{Y\,|X\}$.

We proceed to random processes.

1.2 Martingales: General properties and examples

In this chapter, we consider processes in discrete time $t = 0, 1, \ldots$.

Beginning to build a general framework, we presuppose the existence of an original basic process ξ_t on which all other processes under consideration depend. We may interpret ξ_t as a global characteristic of the "state of nature" at time t. In general, ξ_t may take on values from an arbitrary space; for example, ξ_t may be a vector process.

We will use the notation ξ^t for the whole trajectory $\xi_0, \xi_1, \ldots, \xi_t$ through time t, and sometimes call it the *history of the process* by time t. So, the conditional expectation of a r.v. given ξ^t is the conditional expectation given the entire history through time t. The exposition below is designed in a way that the notation ξ^t will not cause a confusion with the power symbol.

For *all other processes* X_t to be considered, we assume that for each t, the r.v. X_t is completely determined by the values of ξ^t. In other words, X_t is a function of ξ^t. We say that X_t is *adapted* to ξ_t. Note also that since given ξ^t, the value of X_t is known,

$$E\{X_t\,|\,\xi^t\} = X_t. \tag{1.2.1}$$

EXAMPLE 1. Suppose each ξ_t takes on real values. Let $X_0 = \xi_0$, $X_1 = \xi_0\xi_1$, $X_2 = \xi_0\xi_1\xi_2$, and so on: $X_t = \xi_0 \cdot \ldots \cdot \xi_t$. Clearly, the process X_t is adapted to ξ_t. □

A process X_t is called a *martingale* with respect to the basic process ξ_t if for all $t = 0, 1, 2, \ldots$,

$$E\{X_{t+1}\,|\,\xi^t\} = X_t. \tag{1.2.2}$$

When considering martingales, it is convenient to use the game or investment interpretation and view X_t as the capital of a player or an investor at time t. In this case, definition (1.2.2) means that if t is the present time, then *on the average*, the future capital X_{t+1} is equal to what the investor has already reached at time t. So, if for instance, the investor was lucky before or at time t and got a capital larger than the initial capital, then the *expected* future capital is determined by the capital that has been reached and does not depend on the initial capital anymore.

Since the basic process is fixed, we will often omit the reference to ξ_t, just calling X_t a martingale. However, it is important that if (1.2.2) holds, then the process X_t is a martingale with respect to itself too, that is, for all $t = 0, 1, ...,$

$$E\{X_{t+1} \,|\, X^t\} = X_t, \tag{1.2.3}$$

where $X^t = (X_0, ..., X_t)$, the history of the process by time t. We use the same symbolism as for ξ^t, and will do the same for other processes below.

To justify (1.2.3), we recall that X^t is a function of ξ^t. Hence, by general rule (1.1.3), $E\{X_{t+1} \,|\, X^t\} = E\{E\{X_{t+1} \,|\, \xi^t\} \,|\, X^t\} = E\{X_t \,|\, X^t\} = X_t$.

If we work with just one process X_t having property (1.2.3), we can take as a basic process the process X_t itself. *By convention, we will do it each time when in a particular problem, the basic process is not specified.* However, in general, it is reasonable and convenient to suppose that there is one basic process on which all other processes under consideration depend.

Before turning to examples, it is also convenient to consider the process of the increments of X_t. Set $Z_0 = 0$, and $Z_{t+1} = X_{t+1} - X_t$, the profit in one step (play) after time t if we adopt the game interpretation. Then, in view of (1.2.1) and (1.2.2),

$$E\{Z_{t+1} \,|\, \xi^t\} = E\{X_{t+1} - X_t \,|\, \xi^t\} = E\{X_{t+1} \,|\, \xi^t\} - E\{X_t \,|\, \xi^t\} = X_t - X_t = 0.$$

In the game interpretation, this means that whatever the history of the process is, the expected profit in the next step given this history equals zero. We call such a game *fair*.

R.v.'s Z_t for which

$$E\{Z_{t+1} \,|\, \xi^t\} = 0 \text{ for all } t = 0, 1, ... \tag{1.2.4}$$

are called *martingale-differences* with respect to the process ξ_t.

Note also that in view of rule (1.1.3),

$$E\{Z_{t+1} \,|\, X^t\} = E\{E\{Z_{t+1} \,|\, \xi^t\} \,|\, X^t\} = E\{0 \,|\, X^t\} = 0,$$

that is, Z_t are martingale-differences with respect to X_t also. Furthermore, since there is a one-to-one correspondence between Z^t and X^t, in view of (1.1.2), $E\{Z_{t+1} \,|\, Z^t\} = E\{Z_{t+1} \,|\, X^t\} = 0$; so, Z_t are martingale-differences with respect to themselves also.

At last, note that from (1.2.4) and formula (1.1.1), it follows, in particular, that

$$E\{Z_t\} = E\{E\{Z_t \,|\, \xi^{t-1}\}\} = E\{0\} = 0 \quad \text{for all } t = 0, 1, ... \tag{1.2.5}$$

Furthermore, since $Z_t = X_{t+1} - X_t$,

$$X_t = X_0 + (X_1 - X_0) + (X_2 - X_1) + ... + (X_t - X_{t-1}) = X_0 + Z_1 + ... + Z_t. \qquad (1.2.6)$$

The interpretation is clear: the total capital is equal to the initial capital plus the sum of all profits in the previous plays.

Thus, a martingale may be represented as the sum of martingale-differences plus the initial value. The converse assertion is also true.

Indeed, let X_t satisfy (1.2.6) where Z_t's are martingale-differences. Then,

$$E\{X_{t+1} \mid \xi^t\} = E\{X_t + Z_{t+1} \mid \xi^t\} = E\{X_t \mid \xi^t\} + E\{Z_{t+1} \mid \xi^t\} = X_t + 0 = X_t.$$

Thus,

> A random sequence X_t is a martingale if and only if it may be represented as the sum of martingale-differences plus an initial value.

Let us proceed to examples. The first is trivial and the second is simple but both shed some light on the nature of martingales.

EXAMPLE 2. Let $X_0 = \xi_0 = 0$, $X_t = \xi_1 + ... + \xi_t$, for $t > 0$, where ξ_i's, $i = 1, 2, ...$, are independent r.v.'s and $E\{\xi_i\} = 0$ for all i. Because the r.v. ξ_{t+1} does not depend on $\xi_1, ..., \xi_t$, we have $E\{\xi_{t+1} \mid \xi^t\} = E\{\xi_{t+1}\} = 0$. So, ξ_t's are martingale-differences and, consequently, X_t is a martingale. Thus, the notion of a martingale may be viewed as a generalization of the notion of a sum of independent variables with zero means. The essential difference is that in (1.2.6), we do not require the Z's to be independent, but rather to have zero conditional expectations $E\{Z_{t+1} \mid \xi^t\}$.

Now, suppose that $E\{\xi_i\} = m \neq 0$. Then X_t is not a martingale. However,

$$\text{The process } Y_t = X_t - mt \text{ is a martingale.} \qquad (1.2.7)$$

Indeed, $Y_t - mt = \sum_{i=1}^{t}(\xi_i - m)$, and $E\{\xi_i - m\} = 0$. This simple observation will lead us to non-trivial constructions in Section 1.3.

EXAMPLE 3. Let $X_0 = \xi_0 = 1$, $X_t = \xi_1 \cdot ... \cdot \xi_t$, for $t > 0$, where ξ_i's, $i = 1, 2, ...$, are i.i.d. positive r.v.'s with $E\{\xi_i\} = 1$. We have

$$E\{X_{t+1} \mid \xi^t\} = E\{\xi_1 \cdot ... \cdot \xi_t \cdot \xi_{t+1} \mid \xi^t\} = E\{X_t \cdot \xi_{t+1} \mid \xi^t\} = X_t E\{\xi_{t+1} \mid \xi^t\} = X_t E\{\xi_{t+1}\} = X_t.$$

Thus, X_t is a martingale. Suppose now that ξ's are still positive, but $m = E\{\xi_i\} \neq 1$. How should we normalize X_t to make it a martingale?

Since $m > 0$, we may consider the process

$$Y_t = m^{-t} X_t = \frac{\xi_1}{m} \cdot ... \cdot \frac{\xi_t}{m},$$

and the expected value of the new factors equals one. So, Y_t is a martingale.

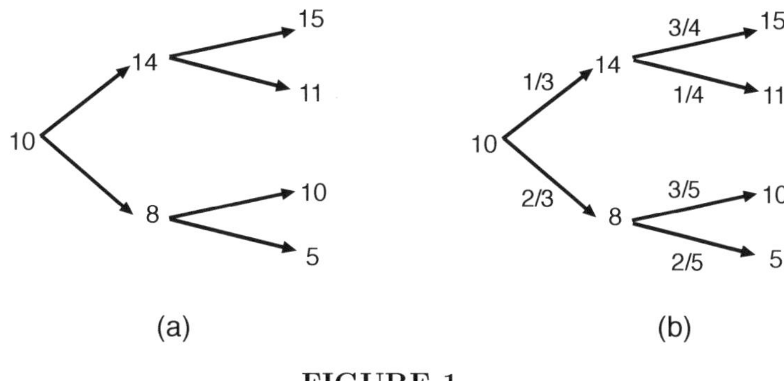

FIGURE 1.

EXAMPLE 4. Let S_t, $t = 0, 1, 2$, be a stock price at three consecutive moments of time. All possible outcomes are shown in Fig. 1a. The initial price is 10. At the end of the first period, the price may be either 14 or 8. If at time $t = 1$ the price occurs to be 14, then its next value may be either 15 or 11, while if S_1 equals 8, then at the end of the second time period, the price may be either 10 or 5. (We already considered such models in Section **2**.1.2.2).

Let us find a probability measure P on paths of the tree, for which the process S_t is a martingale with respect to itself. In Financial Mathematics, such a measure is also called a *risk neutral measure*. It is widely used for pricing various financial products in financial markets, and we will consider this issue in Section 3.2.

For the measure P we are looking for, set $p = P(S_2 = 15 \,|\, S_1 = 14)$, and $p' = P(S_2 = 10 \,|\, S_1 = 8)$ (these are probabilities to "move up" starting from 14 and 8, respectively; see also Fig. 1a). In order to have $E\{S_2 \,|\, S_1\} = S_1$, we need $15 \cdot p + 11 \cdot (1 - p) = 14$, and $10 \cdot p' + 5 \cdot (1 - p') = 8$, which gives $p = 3/4$, and $p' = 3/5$; see Fig. 1b.

For $E\{S_1 \,|\, S_0\} = S_0$, we should have $14 \cdot p'' + 8 \cdot (1 - p'') = 10$, where $p'' = P(S_1 = 14)$, which gives $p'' = 1/3$. The probabilities p, p', p'' *completely specify the probabilities of all four possible paths in the tree:* the probability of the path $10 \to 14 \to 15$ is $\frac{1}{3} \cdot \frac{3}{4} = \frac{1}{4}$; for the path $10 \to 14 \to 11$, it is $\frac{1}{3} \cdot \frac{1}{4} = \frac{1}{12}$; the probabilities for the two paths remaining are $\frac{2}{3} \cdot \frac{3}{5} = \frac{2}{5}$ and $\frac{2}{3} \cdot \frac{2}{5} = \frac{4}{15}$, respectively. See also Fig. 1b.

EXAMPLE 5 (*Pòlya's urn model*). There are r red and b black balls in an urn. We select at random a ball and put it back with c additional balls of the color drawn. After that, we draw another ball and continue sampling in the same fashion. So, this model is opposite, in a sense, to sampling without replacement. Let $X_0 = \frac{r}{r+b}$, the proportion of red balls at the beginning, and X_t is the proportion of red balls in the urn after the tth drawing and replacement of the ball drawn together with c balls. We prove that X_t is a martingale with respect to itself.

Let the r.v.'s N_{rt} and N_{bt} be the respective numbers of red and black balls in the urn after the tth drawing. Then the r.v. $X_t = N_{rt}/(N_{rt} + N_{bt})$. First, let us condition on N_{rt}, N_{bt}. We have

$$\begin{aligned} E\{X_{t+1}\,|\,N_{rt},N_{bt}\} &= \frac{N_{rt}+c}{N_{rt}+N_{bt}+c}\cdot\frac{N_{rt}}{N_{rt}+N_{bt}}+\frac{N_{rt}}{N_{rt}+N_{bt}+c}\cdot\frac{N_{bt}}{N_{rt}+N_{bt}} \\ &= \frac{N_{rt}}{N_{rt}+N_{bt}}\cdot\left(\frac{N_{rt}+c}{N_{rt}+N_{bt}+c}+\frac{N_{bt}}{N_{rt}+N_{bt}+c}\right)=\frac{N_{rt}}{N_{rt}+N_{bt}}\cdot 1 = X_t. \end{aligned}$$

Now, again using (1.1.3), we get

$$E\{X_{t+1}\,|\,X^t\} = E\{E\{X_{t+1}\,|\,N_{rt},N_{bt}\}\,|\,X^t\} = E\{X_t\,|\,X^t\} = X_t.$$

EXAMPLE 6 is important and deep. Let ξ_t be an *arbitrary* process, and V be a r.v. which we view as a global characteristic of the whole process ξ_t. For instance, let ξ_t assume real values, and $V=\max_{0\le t<\infty}\xi_t$ if such a maximum is finite, or V is the first moment when ξ_t will reach a particular level. Let $X_t = E\{V\,|\,\xi^t\}$, the conditional expected value of V given the history of the process ξ_t by time t. The process X_t is a martingale.

To show this, we again use (1.1.3):

$$E\{X_{t+1}\,|\,\xi^t\} = E\{E\{V\,|\,\xi^{t+1}\}\,|\,\xi^t\} = E\{V\,|\,\xi^t\} = X_t. \;\square$$

In the conclusion of this section, we establish a simple but important property of martingales.

Proposition 1 *If X_t is a martingale, then for any $t\ge 0$,*

$$E\{X_t\} = E\{X_0\}. \tag{1.2.8}$$

(The expected values do not change in time.)

Proof. We prove it by induction. For $t=0$, (1.2.8) is trivial. Let (1.2.8) be true for $t=n$. By the formula for total expectation (1.1.1) and the definition of martingales, $E\{X_{n+1}\} = E\{E\{X_{n+1}\,|\,\xi^n\}\} = E\{X_n\} = E\{X_0\}$ by the induction assumption. ■

EXAMPLE 7. Let us come back to the Pòlya urn model from Example 5. The proportions of red balls, i.e., X_t's, are random. Nevertheless, by Proposition 1, the expected proportions $E\{X_t\} = E\{X_0\} = \frac{r}{r+b}$; that is, as at the first draw. See also Exercise 5. □

From now on, in all examples, if the r.v. ξ_0 is not specified, the reader can set $\xi_0 = 0$.

1.3 Predictable sequences. Doob's decomposition

Consider a basic process ξ_t, and a sequence of r.v.'s $\mathcal{Y}=\{Y_t\}$ such that for each t, the r.v. Y_{t+1} is a function of ξ^t. Such a sequence is called *predictable*.

To clarify the significance of this definition, let us view t as the present time and assume ξ^t, that is, the history of the basic process through time t, to be known.

If we had considered an arbitrary process X_t adapted to (based on) the basic process ξ_t, then from the standpoint of time t, the future value of X_{t+1} would have been random, still unknown. However, the value of Y_{t+1}, being a function of ξ^t, is known or "predictable." (At time t, we already know which value the process $\mathcal{Y}$ will assume at the next step.)

EXAMPLE 1. A gambler participates in a sequence of independent plays (turns). Let r.v. $\xi_t, t = 1, 2, \ldots$, take on the value $+1$ if the turn t is successful, and -1 otherwise. In this example, we do not need to specify the corresponding probabilities. The process ξ_t is the basic process for the gambler.

Suppose that after play t is over, the gambler makes a bet of Y_{t+1} at the next play $t+1$. The gambler is free to choose any stake Y_{t+1} depending on *known* results, so Y_{t+1} depends on $\xi_1, \ldots, \xi_t$, and the sequence $\{Y_t\}$ is predictable.

For example, assume that the gambler bets \$1 if she/he lost the last play, and increases the stake by \$1 if she/he won. Then $Y_{t+1} = 1$ if $\xi_t = -1$, and $Y_{t+1} = Y_t + 1$ if $\xi_t = 1$.

As another example, assume that the gambler bets \$1 each time but determines at what moment she/he will quit playing. Suppose that the rule of quitting is such that "to play or not to play" *after* a time t is completely determined by the history of plays, that is, by ξ^t. In other words, the moment of quitting is a r.v. τ, and the occurrence of the event $\{\tau \leq t\}$ (the gambler quits before or at time t) is completely determined by ξ^t. We may describe such a situation setting $Y_{t+1} = 0$ if $\tau \leq t$, and $Y_{t+1} = 1$ otherwise. (To quit means to bet zero.) Thus, whether $Y_{t+1} = 0$ or not depends on ξ^t. □

Route 1 ⇒ *page 364*

2

In particular, the notion of a predictable sequence is connected with the following important question. In the simple Example 2 that concerned a sum of independent r.v.'s, the centering procedure we applied there, transformed the original process into a martingale. Is it possible to generalize the procedure mentioned in a way that it would transform somehow an arbitrary process into a martingale?

Let Y_t, $t = 0, 1, 2, \ldots$, be an *arbitrary* process with finite expectations. Set $\eta_0 = Y_0$, and $\eta_t = Y_t - Y_{t-1}$, the increment over the period $(t-1, t]$. Clearly, $Y_t = \eta_0 + \eta_1 + \ldots + \eta_t$.

Let us view the process η_t as basic: the process Y_t is completely determined by η_t.

Next, set $A_0 = E\{\eta_0\} = E\{Y_0\}$, and $A_{i+1} = E\{\eta_{i+1} \mid \eta^i\}$ for $i = 0, 1, 2, \ldots$. We see that the sequence A_i is predictable (A_{i+1} depends only on η^i).

Set $B_t = A_0 + \ldots + A_t$, and observe that the process B_t is also predictable. Indeed, $B_{t+1} = A_0 + \ldots + A_t + A_{t+1}$, and the last term A_{t+1} is completely determined by η^t. Let

$$M_t = Y_t - B_t = \sum_{i=0}^{t} \eta_i - \sum_{i=0}^{t} A_i = \sum_{i=0}^{t} (\eta_i - A_i). \tag{1.3.9}$$

We show that the process M_t is a martingale with respect to η_t (and hence with respect to Y_t because there is a one-to-one correspondence between the Y_t's and its increments η_t's).

Let $Z_i = \eta_i - A_i$. Then $E\{Z_i \mid \eta^{i-1}\} = E\{\eta_i \mid \eta^{i-1}\} - E\{A_i \mid \eta^{i-1}\} = E\{\eta_i \mid \eta^{i-1}\} - E\{E\{\eta_i \mid \eta^{i-1}\} \mid \eta^{i-1}\} = E\{\eta_i \mid \eta^{i-1}\} - E\{\eta_i \mid \eta^{i-1}\} = 0$; that is, Z_i's are martingale- differences.

So, while in the simple situation of Example 1, to make a sum of independent variables a martingale, we subtracted the corresponding expectations, *in the general case*, we can get a martingale if *we subtract conditional expectations*. Let us put it as

Proposition 2 *(Doob's decomposition). Any process Y_t may be presented as the sum*

$$Y_t = M_t + B_t, \tag{1.3.10}$$

where M_t is a martingale and B_t is a predictable process (with respect to the original process Y_t or its increments, which is the same).

The process B_t is called a *compensator.*

EXAMPLE 2. Let $\xi_1, \xi_2, ...$ be i.i.d. r.v.'s, $Y_0 = 0$, and for $t = 1, 2, ...$, the process $Y_t = \xi_1 + \xi_1\xi_2 + ... + \xi_1\xi_1 \cdot ... \cdot \xi_t$. Then, the increments of Y_t are the r.v.'s $\eta_t = \xi_1 \cdot ... \cdot \xi_t$. There is an one-to-one correspondence between ξ^t and η^t, so it does not matter on which process we will condition. It is convenient to choose the former. We have $E\{\eta_{t+1} \,|\, \xi^t\} = E\{\eta_t\xi_{t+1} \,|\, \xi^t\} = \eta_t E\{\xi_{t+1} \,|\, \xi^t\} = \eta_t E\{\xi_{t+1}\}$. Thus, if $m = E\{\xi_t\} = 0$, then η_t's are martingale-differences, and hence Y_t is a martingale. If $m \neq 0$, we centralize the r.v.'s η_t. However, we do it subtracting conditional mean values rather than unconditional; namely, we consider

$$\begin{aligned} A_t &= E\{\eta_t \,|\, \xi^{t-1}\} = E\{\xi_1\xi_1 \cdot ... \cdot \xi_{t-1}\xi_t \,|\, \xi^{t-1}\} = \xi_1\xi_1 \cdot ... \cdot \xi_{t-1}E\{\xi_t \,|\, \xi^{t-1}\} \\ &= \xi_1\xi_1 \cdot ... \cdot \xi_{t-1}E\{\xi_t\} = \xi_1\xi_1 \cdot ... \cdot \xi_{t-1}m. \end{aligned}$$

The martingale in Doob's decomposition is

$$M_t = \sum_{i=1}^{t}(\eta_i - A_i) = \sum_{i=1}^{t}(\xi_1\xi_1 \cdot ... \cdot \xi_{i-1}\xi_i - \xi_1\xi_1 \cdot ... \cdot \xi_{i-1}m) = \sum_{i=1}^{t}\xi_1\xi_1 \cdot ... \cdot \xi_{i-1}(\xi_i - m).$$

The decomposition itself may be written as

$$Y_t = \sum_{i=1}^{t}\xi_1 \cdot ... \cdot \xi_i = \sum_{i=1}^{t}\xi_1 \cdot ... \cdot \xi_{i-1}(\xi_i - m) + \sum_{i=1}^{t}\xi_1 \cdot ... \cdot \xi_{i-1}m.$$

The first sum is a martingale, the second is a predictable process. □

1.4 Martingale transform

Let $X_t, t = 0, 1, 2, ...$, be a martingale with respect to a basic process ξ_t (and hence, with respect to itself also). Set $Z_{t+1} = X_{t+1} - X_t,\ t = 1, 2, ...$, the corresponding sequence of martingale-differences. Then $X_t = X_0 + Z_1 + ... + Z_t$.

Let Y_t be a sequence predictable with respect to the process ξ_t, and the process

$$W_t = X_0 + Y_1Z_1 + ... + Y_tZ_t. \tag{1.4.1}$$

Such a process is called a *martingale transform.*

EXAMPLE 1. Let us come back to Example 1.3-1 keeping the same notation. Suppose that ξ_t's are independent and take on the values $+1$ or -1 with equal probabilities. Let $X_0 = 0$ and $X_t = \xi_1 + ... + \xi_t$, the difference between the numbers of successful and non-successful plays by time t. Now the process X_t is a martingale, and the corresponding martingale-differences $Z_t = \xi_t$.

As in Example 1.3-1, we denote by Y_t the gambler's stake at play t, and suppose that the sequence $\{Y_t\}$ is predictable; that is, Y_t depends on $\xi_1, ..., \xi_{t-1}$, the results of the previous plays. Then the total capital is given by (1.4.1) with $Z_t = \xi_t$.

Generations of gamblers tried to find a betting strategy Y_t which would transform games from non-favorable or fair into favorable, i.e., into games for which $E\{W_{t+1} \,|\, \xi^t\} > W_t$ and for any given history, the conditional mean profit at each next step $E\{Y_t \xi_t \,|\, \xi^{t-1}\} > 0$. □

Proposition 3 *The transform* W_t *in (1.4.1) is a martingale; that is,* $E\{W_{t+1} \,|\, \xi^t\} = W_t$.

Proof. Since Y_{t+1} depends only on ξ^t, we have $E\{Y_{t+1}Z_{t+1} \,|\, \xi^t\} = Y_{t+1}E\{Z_{t+1} \,|\, \xi^t\} = 0$ because Z_{t+1} is a martingale-difference. Hence, the sequence $\{Y_t Z_t\}$ is that of martingale-differences also and, consequently, W_t is a martingale. ■

Route 1 ⇒ page 388

2

2 OPTIONAL TIME AND SOME APPLICATIONS

2.1 Definitions and stopping property

Consider a basic process ξ_t, $t = 0, 1, ...$, and an integer valued r.v. τ such that

> The occurence of the event $\{\tau = t\}$ is completely determined by the values of the r.v.'s $\xi_0, ..., \xi_t$. (2.1.1)

The r.v. τ with property (2.1.1) is called an *optional time* or a *Markov time*. Sometimes instead of "time," we will say "moment" meaning a time moment.

Condition (2.1.1) means that if t is the present time and if we know the history of the process through time t, then we do know whether the event mentioned occurred before or at time t. (Since we suppose that at time t, we know ξ^t, we know whether one of the events $\{\tau = 1\}, ..., \{\tau = t\}$ has occurred, and if so, which of them has occurred.)

Usually, in particular problems, we are interested in a process X_t adapted to ξ_t rather than in ξ_t itself, and the r.v. τ is the moment when a certain event connected with the evolution of X_t occurs.

EXAMPLE 1. (a) A typical example is a *hitting time*; that is, the first time moment at which the process X_t reaches a level a. In other terms, $\tau = \min\{t : X_t \geq a\}$.

(b) Suppose that X_t assumes only integer values, and $X_0 = 0$. The moment τ when the process takes on an odd value at the first time is also an optional (Markov) time.

EXAMPLE 2. A typical example where a r.v. τ is not an optional time is the moment when the process attains its maximum over a certain period. More precisely, for a fixed T, consider the r.v.

$$\tau = \min\{t \leq T : X_t = \max_{s \leq T} X_s\}. \tag{2.1.2}$$

In order to determine whether a moment t is a point of maximum, we should know the *future* values of the process *after* the time t. Consequently, in general, the event $\{\tau = t\}$ is not determined by ξ^t.

Let, for example, X_t be a stock price at time t. Then τ in (2.1.2) is the time when the price attains its maximum. If τ were an optional moment, the stockholders would have known when to sell their shares to maximize their profit. Clearly, this is not the case. □

If in a particular problem, we deal with one process, say, X_t, then we can choose as a basic process, the process X_t itself, and replace the ξ's in definition (2.1.1) by the X's.

On the other hand, the framework above covers the situations where τ, being still random, does not depend on the process of interest X_t at all. (For example, a gambler when deciding whether to quit or not to quit playing, does not proceed from the results of the plays but, say, just flips a coin each time.)

In this case, to model the random character of τ, we may define a basic process as the process $\xi_t = \{(\eta_t, X_t)\}$, where the process η_t does not depend on the process X_t. (In the example with a gambler, X_t will reflect the results of the plays, while η_t will be connected with flipping a coin.) If we consider a r.v. τ for which the occurrence of the event $\{\tau = t\}$ is determined only by η^t, then τ will not depend on X_t.

Generalizing definition (2.1.1), we also call τ an optional time if condition (2.1.1) holds for each finite t but the probability $P(\tau < \infty) < 1$. To clarify this, suppose that τ is the (random) moment at which a certain event will occur. In the case $P(\tau < \infty) < 1$, this means that with a positive probability, the event mentioned will never occur. In other words, with a positive probability, $\tau = \infty$.

For example, we know that for the random walk with the probability of moving up $p > 1/2$, the ruin probability is $q_u = [(1-p)/p]^u < 1$, where u is the initial value of the process (see Section **2**.2.4). So, if τ is the moment of ruin, then $P(\tau = \infty) = 1 - q_u > 0$.

We will call an optional time τ for which $P(\tau < \infty) = 1$ a *stopping time.*

A r.v. Z for which $P(Z < \infty) < 1$, is called *improper* or *defective.* Thus, an optional but not stopping time is an improper r.v.

Now, let X_t be a martingale. We turn to a theorem allowing to solve in an explicit way many problems on global characteristics of processes. This theorem concerns conditions under which property (1.2.8) in Proposition 1 continues to be true when we replace the certain time t by a random stopping time τ. More precisely, we will establish conditions under which a martingale X_t and a stopping time τ satisfy

$$\text{The martingale stopping property: } E\{X_\tau\} = E\{X_0\}. \tag{2.1.3}$$

First, let us realize that (2.1.3) is not always true.

EXAMPLE 3 (*The doubling strategy)* is classical. A gambler plays a game of chance consisting of a sequence of independent bets with probability $p > 0$ of winning at each bet. As above, let $\xi_t = 1$ if play t is successful, and $\xi_t = -1$, otherwise.

The gambler follows the strategy we already discussed in Example **3**.1.1-13. Having started with a stake of one, the gambler plays until the first win, doubling her/his bet after each loss. After the first win, the gambler quits.

If Y_t is the stake at play t, this corresponds to the bet $Y_1 = 1$, and $Y_{t+1} = 2^t I_{t+1}$ for $t = 1, 2, \dots$, where $I_{t+1} = 1$ if $\xi_1 = \dots = \xi_t = -1$ (there were only losses, and the gambler keeps playing) and $I_{t+1} = 0$ otherwise (the gambler won before or at time t and has quitted). The capital at time 0 is $W_0 = 0$, and the capital at time $t > 0$ is $W_t = Y_1\xi_1 + \dots + Y_t\xi_t$.

If the first success happens at time $k+1$, the gambler's profit is $2^k - (1+2+4+\dots+2^{k-1}) = 1$. Let τ be the moment of the first win, that is, the number of plays to be played. The r.v. τ has a geometric distribution, $P(\tau > k) = (1-p)^k \to 0$, as $k \to \infty$, provided $p > 0$, and $P(\tau < \infty) = 1$

Hence, τ is a stopping time. The doubling strategy allows a gambler to get one unit of money with probability one, though this presupposes that the gambler should, at least theoretically, have an infinite initial capital.

The sequence I_t is predictable, and the process W_t is a particular case of the process (1.4.1).

Set $p = 1/2$. Then $E\{\xi_i\} = 0$ for all i's, and by Proposition 3, the process W_t is a martingale.

Now, the profit at the moment τ is equal to one, that is, $W_\tau = 1$. On the other hand, $W_0 = 0$, which means that for W_t, property (2.1.3) does not hold. □ [1]

Next, we establish some conditions under which (2.1.3) is true.

Theorem 4 *(Optional stopping theorem). Let X_t be a martingale and τ be a stopping time. Then the martingale stopping property (2.1.3) holds if at least one of the following conditions is fulfilled.*

1. *There exists a constant c such that the r.v. $\tau \le c$ with probability one.*
2. *There exists a constant C such that the r.v. $|X_t| \le C$ for all $t \le \tau$ with probability one.*
3. *The expectation $E\{\tau\} < \infty$, and there exists a constant C such that the conditional expectation $E\{|Z_{t+1}| \,|\, \xi^t\} < C$ for all $t = 0, 1, \dots$ with probability one, where the martingale-differences $Z_{t+1} = X_{t+1} - X_t$.*

We skip the proof; it may be found, e.g., in [24], [43], [48], [57].

To comment on the conditions above, let us interpret τ as the time at which the process stops to run. Condition 1 means that the process will stop before or at a finite time c. Under Condition 2, the process $|X_t|$ will not exceed a finite level C before or at the stopping time τ. Condition 3 means that the conditional expected increments of the process are bounded.

[1] Modern dictionaries (see, e.g., [56]) give three meanings of the word "martingale." The first concerns a strap of a horse's harness keeping the horse from rearing its head; the second - a device for keeping a sail in a certain position; the third - a system of betting in which, after a losing wager, the amount bet is doubled (that is, what we considered above). Probably, the use of the word in the third definition came by analogy with the first. Non-mathematical dictionaries do not give the fourth, and nowadays the widespread mathematical meaning of the word. To the author's knowledge, in the mathematical sense and by analogy with the gambling case, the term martingale was first used by J. Ville in [54]. Later, J. L. Doob's book [12] (the 1st edition came out in 1953) made the martingale an important chapter of Probability Theory.

EXAMPLE 4. For illustration, we show that none of these three conditions are satisfied in the situation of Example 3 in which the stopping property does not hold. First, τ has the geometric distribution and assumes any positive integer value with a positive probability. So, τ is not bounded. For $t < \tau$ (the process did not stop yet), the profit $W_t = -(1 + ... + 2^{t-1}) = -2^t + 1$ and, hence, is not bounded. The same concerns the profit in one play. In our example, the role of the martingale-differences Z_t is played by the r.v.'s $Y_t\xi_t$. If $t+1 < \tau$, then $Y_{t+1}\xi_{t+1} = -2^t$, and the r.v. $E\{|Y_{t+1}\xi_{t+1}| \,|\xi^t\}$ assumes the value 2^t. Hence, the r.v. $E\{|Y_{t+1}\xi_{t+1}| \,|\xi^t\}$ is not bounded. □

Now, we turn to examples where Theorem 4 proves to be useful.

2.2 Wald's identity

Proposition 5 *Let $\xi_1, \xi_2, ...$ be i.i.d. r.v.'s, and let a finite $m = E\{\xi_i\}$. Let τ be a stopping time with respect to the process ξ_t and such that $E\{\tau\} < \infty$. Set $S_\tau = \sum_{i=1}^{\tau} \xi_i$. Then*

$$E\{S_\tau\} = mE\{\tau\}. \tag{2.2.4}$$

It is recommended to the reader to look up the result (**3**.6.2.7) where—in a slightly different notation—(2.2.4) was proved in the case when τ does not depend on the ξ's. The result (2.2.4) is much stronger, since now the number of the terms in the sum may depend on the values of separate terms.

Proof. Since m is finite, so is $\mu = E\{|X_i|\}$. Let $X_0 = 0$, $X_t = S_t - mt$, where $S_t = \sum_{i=1}^{t} \xi_i$. In Example 2.1-2, we have shown that X_t is a martingale. The corresponding martingale-differences are $Z_{t+1} = \xi_{t+1} - m$, and since the ξ's are independent, $E\{|Z_{t+1}| \,|X^t\} = E\{|\xi_{t+1} - m|\} \leq E\{|\xi_{t+1}|\} + |m| = \mu + |m|$. So, Condition 3 of Theorem 4 holds, and we can write $0 = E\{X_0\} = E\{X_\tau\} = E\{S_\tau - m\tau\} = E\{S_\tau\} - mE\{\tau\}$, which implies (2.2.4). ■

An alternative direct proof is discussed in Exercise 15.

EXAMPLE 1. Consider the classical random walk as it is described in Section **2**.2.4 but in slightly different notation. Let $S_0 = 0$ (the process starts from zero level); $S_t = \xi_1 + ... + \xi_t$; $\xi_i = \pm 1$ with probabilities p and $q = 1 - p$, respectively; $p > 1/2$. Let a be a positive integer, $\tau_a = \min\{t : S_t \geq a\}$, the time of reaching the level $a > 0$ at the first time.

Since $m = E\{\xi_t\} = 2p - 1 > 0$, by the LLN, $\frac{1}{t}S_t \to 2p - 1 > 0$ as $t \to \infty$ with probability one. Then, with the same probability, $S_t \to \infty$, and hence, starting from 0, the process will reach the level a with probability one. Consequently, τ_a is a stopping time.

Formally, to find $E\{\tau_a\}$ with use of Wald's identity (2.2.4), we should first prove that $E\{\tau_a\} < \infty$. We skip this preliminary step and turn directly to the value of $E\{\tau_a\}$.[2]

[2]The finiteness of $E\{\tau_a\}$ may be proved, say, by the passage time method from Section **5**.2 or, using the exponential bounds from Section **8**.3 and formula $E\{\tau_a\} = \sum_{n=0}^{\infty} P(\tau_a > n)$. For the reader familiar with Mathematical Analysis, the easiest way could be to use Fatou's lemma (see, e.g., [24], [47]) stating that for any sequence $X_n \overset{\text{a.s.}}{\to} X$, it is true that $E\{X\} \leq \liminf E\{X_n\}$. For an integer n, consider the stopping moment $\min\{\tau_a, n\}$ whose mean is finite. By (2.2.4), $E\{S_{\min\{\tau_a,n\}}\} = mE\{\min\{\tau_a, n\}\}$, and since $E\{S_{\min\{\tau_a,n\}}\} \leq a$, we have $E\{\min\{\tau_a, n\}\} \leq a/m$. Then, by Fatou's lemma, $E\{\tau_a\} \leq \liminf E\{\min\{\tau_a, n\}\} \leq a/m < \infty$.

By (2.2.4), we have $E\{S_{\tau_a}\} = mE\{\tau_a\} = (2p-1)E\{\tau_a\}$. In each step, S_t increases or decreases by one. Hence, at time τ, the process will exactly take on the value a (rather than will overshoot the level a). In other words, $S_{\tau_a} = a$. Then $a = (2p-1)E\{\tau_a\}$, and

$$E\{\tau_a\} = \frac{a}{2p-1}. \tag{2.2.5}$$

For example, if $p = 2/3$ and $a = 1$, then starting from zero, the process will reach the (nearest) level one in three steps on the average. See also Exercise 13.

EXAMPLE 2. In the previous example, let $p = 1/2$ and hence $m = 0$. In this case, starting from zero, the process again will reach any level a with probability one. It may be proved, for instance, in the following way.

The probability to reach a starting from zero in a steps (moving only up) is $\left(\frac{1}{2}\right)^a > 0$. We also proved (even twice, in Sections **4**.4.1 and **5**.5.2) that the symmetric random walk starting from zero revisits zero infinitely often. Then the probability that the process will not reach the a-level during n returns to zero is less than $\left(1-\left(\frac{1}{2}\right)^a\right)^n \to 0$ as $n \to \infty$.[3]

Now, let us figure out to what $E\{\tau_a\}$ is equal. Assume that $E\{\tau_a\} < \infty$. Then, by Wald's identity, we would have $a = 0 \cdot E\{\tau_a\} = 0$, which contradicts the assumption $a > 0$. Consequently, $E\{\tau_a\} = \infty$, which is not trivial at all.

This should be understood correctly. The above assertion does not mean that the process will be moving to a infinitely long, however the *mean* time it will take, is infinite. In its turn, it may be understood as follows. Suppose we run independent replicas of the same random walk; that is, we repeat the experiment many times. Let τ_{ai} be the value of the stopping time τ_a in the ith replica. Then, by the LLN, with probability one, $\frac{1}{n}(\tau_{a1} + ... + \tau_{an}) \to \infty$ as $n \to \infty$.

EXAMPLE 3. Let now, in the same scheme, $p = 1/2$ and $a = 0$. Then

$$\tau_0 = \min\{t : S_t = 0,\ t = 1, 2, ...\},$$

the time needed to revisit 0 starting from 0. In Section **4**.4.1, we have proved that $P(\tau_0 < \infty) = 1$ and $E\{\tau_0\} = \infty$. It was not easy. We show that the latter fact practically immediately follows from Example 2, so the martingale technique works here quite efficiently.

Let τ' be the time needed to reach 0 after the first step (regardless of whether the process moved up or down). Then $\tau_0 = 1 + \tau'$, and it suffices to show that $E\{\tau'\} = \infty$.

In accordance with the notation τ_a, let τ_1 be the time of reaching 1 starting from 0. Let τ'_1 be the time needed to reach 0 starting from 1, and τ'_{-1} be the time of reaching 0 starting from -1. Since we consider a symmetric random walk, the r.v.'s τ_1, τ'_1, and τ'_{-1} have the same distribution. Then $P(\tau' = k) = \frac{1}{2}P(\tau'_1 = k) + \frac{1}{2}P(\tau'_{-1} = k) = \frac{1}{2}P(\tau_1 = k) + \frac{1}{2}P(\tau_1 = k) = P(\tau_1 = k)$. Thus, τ' and τ_1 also have the same distribution. As has been proved, $E\{\tau_a\}$ is infinite for any $a \geq 1$. Consequently, $E\{\tau'\}$ is also infinite.

EXAMPLE 4. Consider the general random walk when, as in Example 1, $S_t = \xi_1 + ... + \xi_t$ but the ξ's are arbitrary i.i.d. r.v.'s with a finite mean $m > 0$. Let $a > 0$ and τ_a be defined

[3] Another way to show the same is to come back to Markov chains, recall that in our case all states belong to one recurrence class, and appeal to (**5**.5.2.7).

as above, but we do not assume a to be an integer any more. We again skip the formal preliminary step of proving that $E\{\tau_a\} < \infty$.

In this case, we cannot write that $S_\tau = a$ since the process may overshoot the level a. Nevertheless, we can write $S_{\tau_a} \geq a$, which implies that $mE\{\tau_a\} = E\{S_{\tau_a}\} \geq a$ and hence,

$$E\{\tau_a\} \geq a/m.$$

There is a case, however, when we can write a precise value $E\{\tau_a\}$. Let the ξ's be exponentially distributed. Then, due to the memoryless property, given that the process has exceeded the level a, the overshoot has the same exponential distribution with the same parameter as the original ξ's. Consequently, $E\{S_{\tau_a}\} = a +$ (*the mean value of the overshoot*) $= a + m$. This implies

$$E\{\tau_a\} = \frac{a+m}{m} = \frac{a}{m} + 1. \tag{2.2.6}$$

The case where the ξ's are exponential is strongly relevant to the Poisson process; we discuss it in Exercise 18.

In Exercise 14, the reader is invited to prove that if $m = 0$, then $E\{\tau_a\} = \infty$. □

2.3 The ruin probability for the simple random walk

To demonstrate the power of the martingale method, we come back to the classical random walk and show how quickly one can compute the ruin probability in this case by making use of Theorem 4. It is convenient to use the notation of Section **2.2.4**.

As in that section, let $X_0 = u$, $X_t = u + \xi_1 + ... + \xi_t$ for $t = 1, 2, ...$, where the ξ's are independent and take on values ± 1 with probabilities p and $q = 1 - p$, respectively. We assume that u is natural and $0 \leq u \leq a$ for some fixed natural $a > 0$. See also Section **2.2.4** for detail.

Let $\tau = \min\{t : X_t = 0 \text{ or } a\}$, the moment of reaching the boundaries of the corridor $[0, a]$. We set $p_u = P(X_\tau = a)$, the probability that the process will first reach the level a, and $q_u = P(X_\tau = 0)$, the probability that the process will first reach the level 0, i.e., the ruin probability.

First, let $p = 1/2$. Then X_t is a martingale.

Because $0 \leq X_t \leq a$ for $t \leq \tau$, Condition 2 of Theorem 4 holds. Consequently, by Theorem 4,

$$E\{X_\tau\} = E\{X_0\} = u.$$

On the other hand, $E\{X_\tau\} = ap_u + 0(1 - p_u) = ap_u$. Thus, $p_u = u/a$.

Let $p \neq 1/2$. Then X_t is not a martingale. We set $r = q/p$ and consider the process

$$Y_t = r^{X_t}.$$

The conditional expectation

$$E\{Y_{t+1} \mid \xi^t\} = E\{r^{X_t + \xi_{t+1}} \mid \xi^t\} = r^{X_t} E\{r^{\xi_{t+1}} \mid \xi^t\} = Y_t E\{r^{\xi_{t+1}}\} = Y_t(rp + r^{-1}q).$$

By the choice of r, we have $rp + r^{-1}q = q + p = 1$, and hence, $E\{Y_{t+1} \mid \xi^t\} = Y_t$. Thus, Y_t is a martingale.

For $t \leq \tau$, the values of Y_t lie between $r^0 = 1$ and r^a. (Depending on whether $r > 1$ or not, $r^a > 1$ or < 1.) Hence, Condition 2 of Theorem 4 again holds. Applying Theorem 4, we have

$$E\{Y_\tau\} = E\{Y_0\} = r^u.$$

On the other hand, $E\{Y_\tau\} = r^a p_u + r^0 \cdot (1 - p_u) = 1 + p_u(r^a - 1)$. From this, we easily get $p_u = (r^u - 1)/(r^a - 1)$, which coincides with the result in Section **2**.2.4.

---

3 MARTINGALES AND A FINANCIAL MARKET MODEL

As *a good example of applications of martingales and stopping times*, we show here why the martingale theory plays an essential role in modeling financial markets. A more detailed exposition of financial models may be found, e.g., in [11], [21], [48], [49]), [58].

3.1 A pricing procedure

We begin with a static model distinguishing two moments of time: $t = 0$ and $t = 1$; so to speak, "today" and "tomorrow." From a modeling point of view, any *asset* or *security* may be identified with a r.v., say X, of the asset's future value. The r.v. $X = X(\omega)$ is defined on a sample space $\Omega = \{\omega\}$ of possible outcomes. We will follow the tradition of calling, in this context, outcomes ω states of nature.

For now, suppose that there are a finite number of states of nature, say, $\omega_1, ..., \omega_m$. Among all possible assets, we will distinguish the simplest securities $e_1(\omega), ..., e_m(\omega)$ called *Arrow's* or *state* and such that

$$e_k(\omega) = \begin{cases} 1 & \text{if } \omega = \omega_k, \\ 0 & \text{if } \omega \neq \omega_k. \end{cases} \tag{3.1.1}$$

In other words, $e_k(\omega)$ pays one unit of money (let, for brevity, it be \$1) if ω_k occurs, and pays nothing otherwise.

For a r.v. $X = X(\omega)$, set $x_1 = X(\omega_1), ..., x_m = X(\omega_m)$. Certainly, some x_i's may coincide. We will use the representation

$$X(\omega) = x_1 e_1(\omega) + ... + x_m e_m(\omega) = \sum_{i=1}^{m} x_i e_i(\omega). \tag{3.1.2}$$

Indeed, if $\omega = \omega_k$, then all $e_i(\omega) = 0$ excepting $e_k(\omega)$ which equals 1, and hence $X(\omega)$ in this case equals x_k.

So, the state securities $e_1 = e_1(\omega), ..., e_m = e_m(\omega)$ play the role of a basis, and representation (3.1.2) may be interpreted as if the investor had x_1 units of the security e_1, x_2 units of the security e_2, ... , x_m units of the security e_m.

Our next goal is to define the notion of the price of an asset X, which we will view as a function $C(X)$. It is worth emphasizing that $C(X)$ is the "today" price at time $t = 0$ for having the random capital X at the future time $t = 1$.

Suppose that $C(X)$ is a linear function in the sense that if we buy k_1 units of an asset X_1 and k_2 units of an asset X_2, then for the r.v. $X = k_1X_1 + k_2X_2$, we should pay $C(X) = k_1C(X_1) + k_2C(X_2)$.

Adopting this linearity property, we obtain from (3.1.2) that for any r.v. X, its price

$$C(X) = \sum_{k=1}^{m} x_k C(e_k) = \sum_{k=1}^{m} x_k \psi_k, \tag{3.1.3}$$

where x_k's are the values of X, and $\psi_k = C(e_k)$, the price of the security e_k.

Let a r.v. $X_0 \equiv 1$; that is, all $x_k = 1$, and the asset pays \$1 for sure. Then $C(X_0) = \sum_{k=1}^{m} \psi_k$, which is not surprising. In view of (3.1.2), the purchase of X_0 is equivalent to the purchase of a unit of *each* state security, and this is equivalent to a payment of \$1 at the future time moment for sure. The value of the asset X_0 is certainly not \$1 but rather the *present value* of the obligation of paying \$1 in the future.

Suppose that in the market we consider, there is a possibility to invest \$1 at time $t = 0$ and get $1 + r$ dollars for sure at time $t = 1$. The quantity r is called a *risk-free interest.* Then, investing $\frac{1}{1+r}$, we get \$1, and hence the present value of the future payment of \$1 is $\frac{1}{1+r}$. We call it also a *discount factor*; see the Appendix, Section 3 for detail. Thus,

$$\psi_1 + ... + \psi_m = \frac{1}{1+r}. \tag{3.1.4}$$

Set $\pi_k = (1+r)\psi_k$ and rewrite (3.1.3) as

$$C(X) = \frac{1}{1+r} \sum_{k=1}^{m} x_k \pi_k. \tag{3.1.5}$$

Let us observe that, in view of (3.1.4),

$$\pi_1 + ... + \pi_m = 1, \tag{3.1.6}$$

and denote by $\boldsymbol{\pi}$ the probability measure assigning the probabilities π_k to the respective outcomes ω_k.

The sum $\sum_{k=1}^{m} x_k \pi_k$ in (3.1.5) is the expected value of X with respect to the probability measure $\boldsymbol{\pi}$. It is worth emphasizing that at least formally, the "probabilities" π_k have nothing to do with the "real" probabilities (if any) of the outcomes ω. We will see later that π's rather reflect the beliefs of the market participants regarding the future.

Let the symbol $E_P\{\cdot\}$ denote the expectation operation with respect to a probability measure P. Then, we may rewrite (3.1.5) as

$$C(X) = \frac{1}{1+r} E_{\boldsymbol{\pi}}\{X\}. \tag{3.1.7}$$

Now, we may forget that the space of outcomes Ω was finite and adopt (3.1.7) as a definition of the pricing procedure in the general case, where $\boldsymbol{\pi}$ is an arbitrary probability measure on a space Ω of an arbitrary nature.

Thus, in the model we have built, any pricing procedure is completely determined by a measure $\boldsymbol{\pi}$ and a risk-free interest r. Namely, the price of an asset is the discounted expectation of the future value of this asset with respect to the measure mentioned. Consequently, to *specify* a pricing procedure for all assets, we should *specify* the interest r and the measure $\boldsymbol{\pi}$, and use the definition (3.1.7).

In the examples below, in the case of a finite Ω, we set $\boldsymbol{\psi} = (\psi_1, ..., \psi_m)$ and call it a "pricing vector."

EXAMPLE 1. In a country, all future prices on the next day depend on whether tomorrow it will be sunny, or cloudy, or rainy. Suppose that pricing in the market corresponds to the vector $\boldsymbol{\psi} = (0.5, 0.2, 0.1)$. What is the risk-free interest r in such a model? We have $\frac{1}{1+r} = \psi_1 + \psi_2 + \psi_3 = 0.8$, and hence, $r = (1/0.8) - 1 = 0.25$.

Suppose that the above prices reflect people's belief regarding the likelihood of the tomorrow weather conditions. Which probability do people, perhaps unconsciously, assign to the event that it will be sunny tomorrow? The corresponding probability vector is $\boldsymbol{\pi} = 1.25\,(0.5, 0.2, 0.1) = (0.625, 0.250, 0.125)$, so the probability mentioned is 0.625.

Suppose a company takes an obligation (say, selling insurance against bad weather) to pay \$50 if it is cloudy, and \$200 if it is rainy. How much should such an obligation cost? In accordance with (3.1.3), the price is $0 \cdot 0.5 + 50 \cdot 0.2 + 200 \cdot 0.1 = 30$.

If somebody bets that it will not be raining tomorrow, what are the odds of such a bet if they reflect the price situation in the market? (So, we are talking not about real probabilities but rather about beliefs.) The odds are 0.875 to 0.125, and the odds ratio is 7. □

3.2 A static financial market model and the arbitrage theorem

Next, we consider a so far static market model, again distinguishing two moments of time: $t = 0$ and $t = 1$. Suppose the market consists of n assets. Let S_{0j} be the price of a unit of the jth asset at time 0, and $S_{1j} = S_{1j}(\omega)$ be the future price of the same unit at time 1 if a state of nature ω occurs. Thus, S_{0j} is a number, while S_{1j} is a random variable. Let the price vectors $\mathbf{S}_0 = (S_{01}, ..., S_{0n})$, $\mathbf{S}_1(\omega) = (S_{11}(\omega), ..., S_{1n}(\omega))$.

We suppose that the first asset is not random, and

$$S_{01} = 1, \text{ and } \quad S_{11}(\omega) = 1 + r \tag{3.2.1}$$

for some $r \geq 0$ and all ω. One may view such an asset as a bank account with a risk-free interest r for the time period under consideration. Naturally, the price of a *unit* of this asset at the time of the initial investment is one unit of money.

Consider an investor whose trading strategy (or portfolio) is represented by a vector

$$\boldsymbol{\theta} = (\theta_1, ..., \theta_n),$$

where θ_j is the amount invested into the jth asset at time $t = 0$. We may also think that θ_j is the number of units of the jth asset purchased, but we allow θ_j to be any number, not necessarily an integer.

Moreover, θ_j may be negative, which means borrowing the asset. For the non-random asset 1, this means borrowing money from a bank at an interest rate of r. For random asset j and $\theta_j < 0$, the investor borrows $|\theta_j|$ units of the asset and immediately sells them with the obligation to return the asset borrowed in the future. (So, the investor will benefit if the price drops, and she/he will lose if the price rises.) Such an operation is called *short selling*.

EXAMPLE 1. Let the initial price-vector $\mathbf{S}_0 = (1,4,3)$.

(a) If $\boldsymbol{\theta} = (300,400,50)$, then this means that the investor deposited $\$1 \cdot 300 = \$300$, and bought 400 units of asset 2 and 50 units of asset 3. So, totally, the investor has invested $\$300 + \$4 \cdot 400 + \$3 \cdot 50 = \$2050$.

(b) If $\boldsymbol{\theta} = (-300,400,0)$, then the investor borrowed \$300, and bought 400 units of asset 2 for \$1,600. Since the investor owes \$300, her/his "wealth" at time $t = 0$ equals $1600 - 300 = 1300$ dollars.

(c) If $\boldsymbol{\theta} = (300,-400,0)$, then this means that the investor borrowed 400 units of asset 2, sold them for $\$4 \cdot 400 = \$1600$, and deposited into a free-risk rate account $\$300 + \$1600 = \$1900$. The investor's wealth at time $t = 0$ is $-1600 + 300 = -1300$ dollars. □

In general, the initial investment (or the wealth of the investor at time $t = 0$) equals

$$W_0 = \theta_1 \cdot S_{01} + \theta_2 \cdot S_{02} + ... + \theta_n \cdot S_{0n} = \langle \boldsymbol{\theta}, \mathbf{S}_0 \rangle,$$

where, as usual, $\langle .,. \rangle$ stands for dot (or scalar) product.

The final value of the portfolio (or the final wealth) at time $t = 1$ is the r.v.

$$W_1 = W_1(\omega) = \theta_1 \cdot S_{11}(\omega) + \theta_2 \cdot S_{12}(\omega) + ... + \theta_n \cdot S_{1n}(\omega) = \langle \boldsymbol{\theta}, \mathbf{S}_1(\omega) \rangle.$$

Next, we introduce the notion of *arbitrage*. To simplify formulations, we consider the case of a finite number m of states of nature $(\omega_1, ..., \omega_m)$.

We say that for the market we are describing, there is an arbitrage opportunity if there exists a trading strategy $\boldsymbol{\theta}_{\text{arb}}$ such that

$$W_0^{\text{arb}} = \langle \boldsymbol{\theta}_{\text{arb}}, \mathbf{S}_0 \rangle \leq 0, \tag{3.2.2}$$

while for the final value of the portfolio, for all ω,

$$W_1^{\text{arb}} = W_1^{\text{arb}}(\omega) = \langle \boldsymbol{\theta}_{\text{arb}}, \mathbf{S}_1(\omega) \rangle \geq 0 \text{ and at least for one } \omega, \text{ the wealth } W_1^{\text{arb}}(\omega) > 0. \tag{3.2.3}$$

That is, there exists a strategy such that the investor does not invest anything or even borrows some money or assets, and nevertheless she/he loses nothing and for some outcome gets a positive income. Such a situation is also referred to as "a sure win," "a free lunch," etc.

EXAMPLE 2. Suppose $\mathbf{S}_0 = (S_{01}, S_{02}) = (1,1)$ (there are two assets, and the units of both cost \$1). Let the risk-free interest $r = 0$, and hence $S_{11} = 1$ also. Assume that there are two states of nature, ω_1 and ω_2, and

$$S_{12}(\omega) = \begin{cases} 2 & \text{if } \omega = \omega_1, \\ 1 & \text{if } \omega = \omega_2. \end{cases} \tag{3.2.4}$$

Clearly, one can borrow \$1 at zero interest, and invest it into the second asset. If ω_2 occurs, the investor will get \$1 and will return it to the bank. Whereas, if ω_1 occurs, the investor will get \$2, will return \$1 to the bank, and will have a profit of \$1. Such a strategy corresponds to $\theta = (-1,1)$. In this case, $W_0 = -1 \cdot 1 + 1 \cdot 1 = 0$, $W_1(\omega_2) = -1 \cdot 1 + 1 \cdot 1 = 0$ also, but $W_1(\omega_1) = -1 \cdot 1 + 1 \cdot 2 = 1 > 0$.

On the other hand, as we will see a bit later, if

$$S_{12}(\omega) = \begin{cases} 2 & \text{if } \omega = \omega_1, \\ 1/2 & \text{if } \omega = \omega_2, \end{cases} \tag{3.2.5}$$

then an arbitrage is impossible. □

A market is said to be arbitrage free or *viable* if there is no arbitrage strategy. Not going deeply into details, note that in reality, as a rule, markets with arbitrage opportunities are not stable, and "good" pricing mechanisms eliminate such situations. So, we are interested in viable markets.

Let us come back to the general scheme, and note that the prices $\mathbf{S}_0$ and $\mathbf{S}_1$ correspond to different moments of time. To compare them, we should evaluate the future prices from the stand point of the initial time. Since \$1 to be paid at time $t = 1$ costs $1/(1+r)$ at time $t = 0$, we introduce the discounted future price vector $\widetilde{\mathbf{S}}_1 = \widetilde{\mathbf{S}}_1(\omega) = (\widetilde{S}_{11}(\omega), ..., \widetilde{S}_{1n}(\omega)) = \frac{1}{1+r}\mathbf{S}_1(\omega)$. The initial prices do not have to be discounted, and we set $\widetilde{\mathbf{S}}_0 = \mathbf{S}_0$.

The arbitrage theorem below connects the notions of arbitrage and a martingale. By convention, if a probability measure $\boldsymbol{\pi}$ assigns probabilities π_k to the respective ω_k, we sometimes represent it as the probability vector $\boldsymbol{\pi} = (\pi_1, ..., \pi_m)$. A measure $\boldsymbol{\pi}$ is said to be strictly positive if $\pi_i > 0$ for all i.

Theorem 6 *(The arbitrage theorem). The market described above is viable if and only if there exists a strictly positive probability measure* $\boldsymbol{\pi}$ *such that*

$$E_{\boldsymbol{\pi}}\{\widetilde{\mathbf{S}}_1\} = \widetilde{\mathbf{S}}_0. \tag{3.2.6}$$

First of all, it is again worthwhile emphasizing that the probabilities π_k have nothing to do with the "real" probabilities (if any); the theory just does not involve them.

Secondly, even in this static model, property (3.2.6) may be viewed as a *martingale* one. We have two moments of time. Consider the price process $\widetilde{\mathbf{S}}_0,\ \widetilde{\mathbf{S}}_1$. Since $\widetilde{\mathbf{S}}_0$ is not random, $E_{\boldsymbol{\pi}}\{\widetilde{\mathbf{S}}_1\} = E_{\boldsymbol{\pi}}\{\widetilde{\mathbf{S}}_1 \,|\, \widetilde{\mathbf{S}}_0\}$, and property (3.2.6) may be rewritten as

$$E_{\boldsymbol{\pi}}\{\widetilde{\mathbf{S}}_1 \,|\, \widetilde{\mathbf{S}}_0\} = \widetilde{\mathbf{S}}_0.$$

So, with respect to the probability measure $\boldsymbol{\pi}$, the process is $\{\widetilde{\mathbf{S}}_0,\ \widetilde{\mathbf{S}}_1\}$ is a *martingale*. In the dynamic model in Section 3.3, we consider an unlimited number of time moments.

Note also that, in view of the definitions of $\widetilde{\mathbf{S}}_0,\ \widetilde{\mathbf{S}}_1$, relation (3.2.6) may be rewritten as

$$\mathbf{S}_0 = \frac{1}{1+r} E_{\boldsymbol{\pi}}\{\mathbf{S}_1\}. \tag{3.2.7}$$

A model, like the one above or a more complicated dynamic model below, where the probability measure also has a martingale property, is often referred to as a model of a *risk neutral world*.

Proof of Theorem 6 in the sufficiency part is simple. Suppose that there exists a strictly positive measure $\boldsymbol{\pi}$ such that (3.2.7) is true. Assume that an arbitrage strategy $\boldsymbol{\theta}_{\text{arb}}$ exists. Then

$$W_0^{\text{arb}} = \langle \boldsymbol{\theta}_{\text{arb}}, \mathbf{S}_0 \rangle = \langle \boldsymbol{\theta}_{\text{arb}}, \frac{1}{1+r} E_{\boldsymbol{\pi}}\{\mathbf{S}_1(\omega)\} \rangle = \frac{1}{1+r} E_{\boldsymbol{\pi}}\{\langle \boldsymbol{\theta}_{\text{arb}}, \mathbf{S}_1(\omega) \rangle\} = \frac{1}{1+r} E_{\boldsymbol{\pi}}\{W_1^{\text{arb}}(\omega)\}.$$

All $\pi_k > 0$, the wealth $W_1^{\text{arb}}(\omega) \geq 0$, and at least for one ω, we have $W_1^{\text{arb}}(\omega) > 0$. Consequently, $E_{\boldsymbol{\pi}}\{W_1^{\text{arb}}(\omega)\} > 0$. Hence, $W_0^{\text{arb}} > 0$, which contradicts the definition of arbitrage. ■

The proof of necessity appeals to some facts from Linear Programming; we skip it.

EXAMPLE 3. Let us revisit Example 2. Consider a measure $\boldsymbol{\pi} = (\pi_1, \pi_2)$. In the case (3.2.4), since $r = 0$, the martingale property (3.2.7) would mean that $1 \cdot \pi_1 + 2 \cdot \pi_2 = 1$. Since $\pi_1 + \pi_2 = 1$, this implies that $\pi_2 = 0$. So, there is no positive measure $\boldsymbol{\pi}$ for which (3.2.7) is true, and hence there is an arbitrage opportunity. We saw that it is indeed true.

Consider the case (3.2.5). In this case, property (3.2.7) together with $r = 0$ implies that $2 \cdot \pi_1 + \frac{1}{2} \cdot \pi_2 = 1$. Along with $\pi_1 + \pi_2 = 1$, this leads to $\pi_1 = \frac{1}{3}, \pi_2 = \frac{2}{3}$. Thus, a positive probability measure for which (3.2.7) holds exists, and by Theorem 6, there is no arbitrage opportunity. □

The reason why we consider the arbitrage theorem is based on the conviction that "good" prices are those for which the market is arbitrage free (and hence, stable), and Theorem 6 allows us to determine such prices. In particular, this means the following.

Suppose that proceeding from the prices of the assets of a market, we found a positive and unique measure $\boldsymbol{\pi}$ with respect to which the prices of these assets constitute a martingale. Suppose that we added to the market one more asset whose future value is a r.v. X. Then for the market to remain arbitrage free, the price of X should be equal to

$$C(X) = \frac{1}{1+r} E_{\boldsymbol{\pi}}\{X\}. \tag{3.2.8}$$

EXAMPLE 4. Consider the asset (3.2.5) which we will view as a stock. Let the risk-free interest $r = 0$. Thus, our market consists of two assets: a bank account with zero rate (so, this is just money) and the stock we specified. Suppose that at time $t = 0$, a company issued a ticket allowing at time $t = 1$ to have a dinner for \$200 if the stock price rises, or get a \$2 hot dog if the price drops. Thus, we have added one more asset to the market. How much should such a ticket cost at $t = 0$? The theory gives a simple but non-trivial answer.

In Example 3, we have found that the only positive martingale measure $\boldsymbol{\pi}$ satisfying condition (3.2.7) is $\boldsymbol{\pi} = (\pi_1, \pi_2) = (\frac{1}{3}, \frac{2}{3})$. For the r.v. $X(\omega)$, the value of the ticket at time $t = 1$, we have $X(\omega_1) = 200$, and $X(\omega_2) = 2$. Then, in accordance with (3.2.8), the price $C(X) = \frac{1}{1+0} E_{\boldsymbol{\pi}}\{X\} = 200 \cdot \frac{1}{3} + 2 \cdot \frac{2}{3} = \$68$. □

Assets whose prices are completely determined by the prices of some original assets are called *derivatives*. So, the ticket above is a derivative with respect to the stock. The next examples are more "serious."

EXAMPLE 5. As in Examples 2-3, let $\mathbf{S}_0 = (1,1)$ and S_2 be as in (3.2.5). First, let $r = 0$.

Suppose that at time $t = 0$, the investor bought one unit of the asset 2 for the price $S_{02} = \$1$. The investor takes a risk because at time $t = 1$, the price can drop to \$0.5. The investor wishes to insure her/himself against this, and to this end, she/he buys what is called a *put option*: a security that gives the investor a right to sell the stock in the future for a designated price K which is called an *exercise price*.

Let, say, $K = \$1$. What is a "fair" price for the insurance allowing to sell the stock unit for the same price \$1 (whatever the real price is)? "Fair" means the absence of an arbitrage opportunity. Denote the value of the option in the future by $X = X(\omega)$. If ω_1 occurs, then the investor will not exercise her/his right selling for \$1 what costs \$2. So, $X(\omega_1) = 0$; the put option in this case costs nothing. If ω_2 occurs, then the investor may sell the stock for \$1 while the market price is only \$0.5. Hence, $X(\omega_2) = 1 - 0.5 = 0.5$.

(The value of the option itself does not depend on whether the investor bought a stock at time $t = 0$ or not. The investor may buy only the option not buying the stock. If the price for the stock drops, the investor will buy one unit of the stock for \$0.5, and exercising the option, will immediately sell the stock for \$1.)

Since the martingale measure is $\boldsymbol{\pi} = (\frac{1}{3}, \frac{2}{3}))$, the initial price of X should be equal to $c = E_{\boldsymbol{\pi}}\{X\} = 0 \cdot \frac{1}{3} + 0.5 \cdot \frac{2}{3} = \frac{1}{3} \approx 33$¢.

EXAMPLE 6. Consider now the same stock and the same put option as above, but suppose that $r = 0.1$. Then for the stock, (3.2.7) implies $1 = \frac{1}{1+0.1}\left(2 \cdot \pi_1 + \frac{1}{2} \cdot \pi_2\right)$. It is straightforward to verify that together with $\pi_1 + \pi_2 = 1$, this leads to $\pi_1 = 0.4$ and $\pi_2 = 0.6$.

Then, by (3.2.7), the price of the put option above equals $c = \frac{1}{1+0.1}(0 \cdot 0.4 + 0.5 \cdot 0.6) \approx 0.27 = 27$¢.

EXAMPLE 7. Now, for the situation of Example 6, consider a derivative X that gives the right to *buy* the stock for an exercise price $K = \$1$. Such a derivative is referred to as a *call option.* In this case, the investor will exercise the option only if the price rises. Thus, the value of $X(\omega_1) = 2 - 1 = 1$, and $X(\omega_2) = 0$. Then, the initial price of X should be equal to $\frac{1}{1+0.1}(1 \cdot 0.4 + 0 \cdot 0.6) \approx 0.36 = 36$¢. □.

3.3 A dynamic model. Binomial trees

To build a dynamic model, we extend the sequence $\mathbf{S}_0, \mathbf{S}_1(\omega)$, and introduce a vector-process $\mathbf{S}_0, \mathbf{S}_1(\omega), \mathbf{S}_2(\omega), ..., \mathbf{S}_T(\omega)\}$, where the final moment time T is usually called a time horizon or a maturity date, the price vector $\mathbf{S}_t(\omega) = (S_{t1}(\omega), ..., S_{tn}(\omega))$, and $S_{tj}(\omega)$ is the price of the jth asset at time t if a state of nature ω occurs.

As above, the first asset is assumed to be non-random and may be interpreted as a bank account with a risk-free interest. We assume that this interest is the same over all periods and equals an $r \geq 0$. Thus, the counterpart of (3.2.1) is

$$S_{01} = 1, \;\; S_{11} = 1 + r, \;\; S_{21} = (1+r)^2, \;\; ... \;\; , S_{T1} = (1+r)^T. \tag{3.3.1}$$

Accordingly, the discounted price vector $\widetilde{\mathbf{S}}_t(\omega) = \dfrac{1}{(1+r)^t}\mathbf{S}_t(\omega)$.

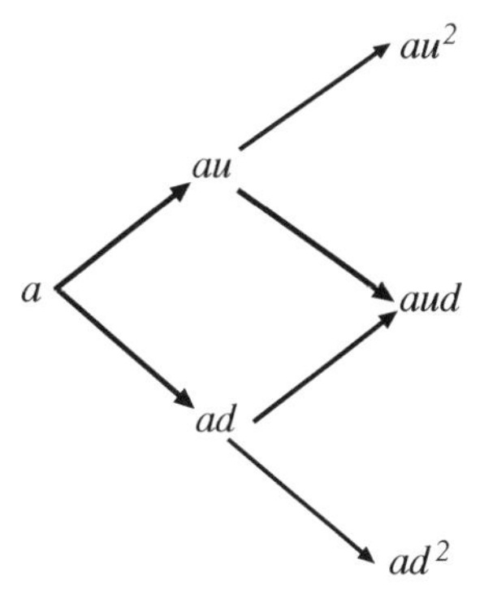

FIGURE 2.

Any outcome ω corresponds to a particular realization of the price process.

As *an example*, consider the simplest dynamic scheme, namely, a *binomial tree* which we have already introduced in Section **2**.1.2.2. In this case, the market consists of a bank account and a stock whose price either increases each period by a factor $u > 1$, or decreases by $d < 1$. (The symbols came after "up" and "down.") Then the evolution of prices may be presented as a tree.

The case $T = 2$ is illustrated in Fig. 2, where the initial stock price $S_{02} = a$, and each vertex is marked by the corresponding value of the price. We see that there are four possible paths and, accordingly, four outcomes (or states of nature) ω corresponding to these paths. In the general case of a horizon of T, there are 2^T different paths (and hence, 2^T outcomes).

Let us come back to a general model. Denote by $\boldsymbol{\theta}_0 = (\theta_{01},...,\theta_{0n})$ the initial portfolio of the investor. Then her/his initial wealth is $W_0 = \langle \boldsymbol{\theta}_0, \mathbf{S}_0 \rangle$. At the next moment $t = 1$, the prices may change, and the new wealth will become $\langle \boldsymbol{\theta}_0, \mathbf{S}_1 \rangle$. The investor may switch to another portfolio, say, $\boldsymbol{\theta}_1 = (\theta_{11},...,\theta_{1n})$, but since we do not presuppose a possibility of a cash flow from outside, the new portfolio must have the same value: $\langle \boldsymbol{\theta}_0, \mathbf{S}_1 \rangle$ must be equal to $\langle \boldsymbol{\theta}_1, \mathbf{S}_1 \rangle$. This is a condition on the choice of a portfolio $\boldsymbol{\theta}_1$. After that, the process continues in the same fashion.

Denote by $\boldsymbol{\theta}_t = (\theta_{t0},...,\theta_{tn})$ the portfolio at time $t = 0,...,T-1$. (At the last moment, the investor does not change the portfolio.) A sequence $\boldsymbol{\theta} = (\boldsymbol{\theta}_0,...,\boldsymbol{\theta}_{T-1})$ is said to be a *strategy* of the investor during the whole period. As above, the portfolios should satisfy the condition

$$\langle \boldsymbol{\theta}_{t-1}, \mathbf{S}_t \rangle = \langle \boldsymbol{\theta}_t, \mathbf{S}_t \rangle.$$

Such a strategy is called *self-financing*.

If the investor adopts a strategy $\boldsymbol{\theta}$, then the r.v. $W_{t\boldsymbol{\theta}} = W_{t\boldsymbol{\theta}}(\omega) = \langle \boldsymbol{\theta}_t, \mathbf{S}_t(\omega) \rangle$ is the wealth of the investor at time t.

We identify paths and outcomes, and again consider the case of a finite number m of outcomes $\omega_1,...,\omega_m$. We say that there is *an arbitrage opportunity* in the market if there exists a strategy $\boldsymbol{\theta}_{\text{arb}}$ such that

$$W_{0\boldsymbol{\theta}_{\text{arb}}} \le 0, \text{ while } W_{T\boldsymbol{\theta}_{\text{arb}}}(\omega) \ge 0 \text{ for all } \omega, \text{ and } W_{T\boldsymbol{\theta}_{\text{arb}}}(\omega) > 0 \text{ for at least one } \omega.$$

As above, for a probability measure $\boldsymbol{\pi}$ assigning probabilities π_k to the respective outcomes (paths) ω_k, the expectation with respect to $\boldsymbol{\pi}$ will be denoted by $E_{\boldsymbol{\pi}}\{\cdot\}$.

As before, we call a market *viable* if there is no arbitrage opportunity.

Theorem 7 *(The arbitrage theorem). The market described above is viable if and only if there exists a strictly positive probability measure $\boldsymbol{\pi}$ such that the discounted price process $\widetilde{\mathbf{S}}_t$ is a martingale with respect to $\boldsymbol{\pi}$; that is, for each $t = 1,...,T-1$,*

$$E_{\boldsymbol{\pi}}\{\widetilde{\mathbf{S}}_{t+1} \,|\, \widetilde{\mathbf{S}}_t,...,\widetilde{\mathbf{S}}_0\} = \widetilde{\mathbf{S}}_t. \tag{3.3.2}$$

We skip the proof. Taking into account the definition of the discounted price-process, we may rewrite (3.3.2) as

$$\mathbf{S}_t = \frac{1}{1+r} E_{\boldsymbol{\pi}}\{\mathbf{S}_{t+1} \,|\, \mathbf{S}_t,...,\mathbf{S}_0\}. \tag{3.3.3}$$

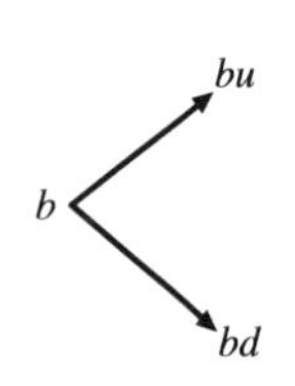

FIGURE 3.

(It does not matter whether we condition on $\widetilde{\mathbf{S}}_t,...,\widetilde{\mathbf{S}}_0$ or $\mathbf{S}_t,...,\mathbf{S}_0$ because there is a one-to-one correspondence between these two vectors.)

Let us return to the binomial tree described above and consider just one vertex and two "edges" starting from this vertex; see Fig. 3. Denote by b the price of an asset corresponding to the vertex chosen. Then the prices at the neighbor vertices are bu and bd.

The martingale measure $\boldsymbol{\pi}$ we are going to specify, is defined on paths. Denote by π the probability (corresponding to $\boldsymbol{\pi}$) that a path going through the vertex we have chosen, will move up; in other words, that given that the asset has a price of b, the next price will be bu. Then $1-\pi$ is the probability that the next price will be bd. By (3.3.3), $b = \frac{1}{1+r}(bu\cdot\pi + bd\cdot(1-\pi))$. So, b may be canceled out, and as is easy to calculate,

$$\pi = \frac{(1+r)-d}{u-d}, \quad 1-\pi = \frac{u-(1+r)}{u-d}. \tag{3.3.4}$$

Thus, π does not depend on the vertex chosen, and hence completely determines the whole measure $\boldsymbol{\pi}$. For example, for $T = 2$ (see also Fig. 2), the probability of the path "up, up" (the price rises twice) is π^2, the probability of the path "up, down" is $\pi(1-\pi)$, and so on.

For both probabilities in (3.3.4) to be positive, we should have

$$d < 1+r < u. \tag{3.3.5}$$

This is a condition for the positivity of $\boldsymbol{\pi}$, and hence, for the viability of the market.

EXAMPLE 1. Consider a two-period model (say, two years) with a risk-free rate of 5%, and a stock with an initial price of \$100. Let $u = 1.1$ and $d = 0.95$. (That is, the stock price may either rise 10% or drop 5%.) The corresponding tree with prices is shown in Fig. 4a. By (3.3.4), $\pi = \frac{1.05-0.95}{1.1-0.95} = \frac{2}{3}$, and $1-\pi = \frac{1}{3}$. One may say that the prices reflect the "belief" or expectation that, "the odds are 2 : 1 in favor of the rise of the price." □

Suppose we add to the market a derivative whose evolution is presented by a price process X_t. In accordance with (3.3.3), for the market to remain viable, we should have

$$X_t = \frac{1}{1+r} E_{\boldsymbol{\pi}}\{X_{t+1} \,|\, \mathbf{S}_t, \ldots, \mathbf{S}_0\}. \tag{3.3.6}$$

This allows us to find the "no-arbitrage" prices for the derivative.

EXAMPLE 2. Let us come back to the situation of Example 1 and consider a put option *on the stock described there* with an exercise price of $K = \$105$. The option may be exercised only at the (maturity) time $T = 2$. What is the price of the option at time $t = 0$?

The corresponding tree is depicted in Fig. 4b; below, we comment on all values there.

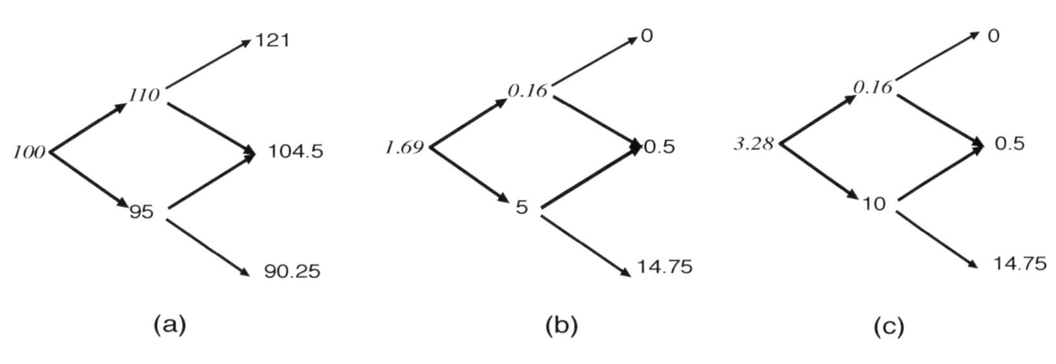

FIGURE 4.

The idea is to move from the end backward. At the final time, the investor will not exercise the put option if the price of the stock is greater then $K = 105$, that is when $S_{22} = 121$. Hence, the value of the option in this case is zero. This corresponds to the very right upper vertex in the tree. If $S_{22} = 104.5$ or 90.25, which is less than $K = 105$, the option will be exercised, and the value of the option is 0.5 or 14.75, respectively. This corresponds to the rest very right vertices.

We move backward by one step and consider, first, the upper vertex among those two that correspond to time $t = 1$. Being at this vertex, the option price process may move either to the value 0 or to the value 0.5. The martingale measure we found in Example 1, is specified by $\pi = 2/3$. Consequently, in accordance with (3.3.6), the option price at this position should be equal to $\frac{1}{1+0.05}\left(0 \cdot \frac{2}{3} + 0.5 \cdot \frac{1}{3}\right) \approx 0.16$, which is shown in Fig. 4b. For the other vertex at $t = 1$, the option price equals $\frac{1}{1+0.05}\left(0.5 \cdot \frac{2}{3} + 14.75 \cdot \frac{1}{3}\right) = 5$.

Let us move to $t = 0$. From the initial point, the process X_t moves either to the value 0.16 (as a matter of fact, this is an approximate value) or to the value 5. Thus, the initial option price is $\frac{1}{1+0.05}\left(0.16 \cdot \frac{2}{3} + 5 \cdot \frac{1}{3}\right) \approx 1.69$. □

Next, we show how the notion of a *stopping time* may work in our model. Derivatives which, as above, may be exercised only at the end of a specified term are called European. More flexible financial instruments that be may be exercised at earlier moments during a designated term are called American.

In a model for such a derivative with values X_t, we determine a stopping time τ, the moment when the derivative will be exercised, and consider the no-arbitrage price which is equal to $X_0 = E_{\boldsymbol{\pi}}\{\widetilde{X}_\tau\}$ where $\boldsymbol{\pi}$ is a martingale measure and $\widetilde{X}$ is a discounted price. Thus, we use the stopping property (2.1.3) where, in our case, X_0 is not random.

EXAMPLE 3. Consider the same option as in Example 2 but allow it to be exercised at the last moment $t = 2$ as well as at the previous moment $t = 1$ (but not at the very beginning upon purchasing). The price tree in this case is depicted in Fig. 4c. The very right vertices are treated as in Example 2: if the option is exercised at the end, its values are the same as for the European option considered above.

Now, consider $t = 1$ and suppose that the situation that occurred corresponds to the upper vertex. The investor should decide whether to exercise the option right away or wait for time $t = 2$. The stock price in this case is \$110 and is larger than the exercise price $K = \$105$. So, the action that corresponds to exercising has zero value. The value of the other action (to wait) has been computed in Example 2: it is \$0.16. Thus, the investor keeps the option not exercised, and the value of the option at this vertex is \$0.16.

Consider the lower vertex at the same $t = 1$. In Example 2, we have computed that in this case, if the option is not exercised at this moment, the value of the option for such a decision is \$5. On the other hand, if the stock that now costs \$95 is sold for \$105, then it will bring a \$10 profit. So, a reasonable decision would be "to exercise," and the value of the option at this point is \$10, which is reflected in Fig. 4c.

Consider $t = 0$. By assumption, the option cannot be exercised immediately, and the process X_t may move either to the value 0.16 or to the value \$10. Then the initial option price is $\frac{1}{1+0.05}\left(0.16 \cdot \frac{2}{3} + 10 \cdot \frac{1}{3}\right) \approx 3.28$; see again Fig. 4c. □

4 LIMIT THEOREMS FOR MARTINGALES

In this section, we consider limit theorems for martingales: the strong LLN and the CLT. In particular, as a corollary from the LLN for martingales, we will derive the LLN for i.i.d. r.v.'s as was promised in Section **3**.5.2. The CLT will be given without a proof. It may be found, e.g., in [19], [30], [34], [42].

4.1 The strong LLN for martingales

Let $X_0 = 0$, and $X_t = Z_1 + ... + Z_t$, where r.v.'s Z_i are martingale-differences with respect to a basic process ξ_t:

$$E\{Z_i \,|\, \xi^{i-1}\} = 0 \quad \text{for all } i = 1, 2, ... \,. \tag{4.1.1}$$

Without loss of generality, the reader can also identify the process ξ_t with X_t or Z_t.

As was already shown (see (1.2.5)), $E\{Z_i\} = 0$ for all i. Suppose that the variance $\sigma_j^2 = Var\{Z_j\} < \infty$, and set $\overline{X}_t = \frac{1}{t}X_t = \frac{1}{t}\,(Z_1 + ... + Z_t)$.

Theorem 8 *(The strong LLN for martingales). If*

$$\sum_{j=1}^{\infty} \frac{\sigma_j^2}{j^2} < \infty, \tag{4.1.2}$$

then

$$P\left(\overline{X}_t \to 0\right) = 1. \tag{4.1.3}$$

EXAMPLE 1. Suppose that all variances $\sigma_j^2 \le \sigma^2$ for a $\sigma \ge 0$. Then (4.1.2) is true, and the LLN takes place. □

As we will see, this theorem follows from the next theorem that has its intrinsic value.

4.2 On convergence of random series

Now, we are interested when the series $\sum_{i=1}^{\infty} Z_i$ converges; that is, with probability one, there exists a r.v. $X_\infty = \lim_{t\to\infty} X_t$.

Theorem 9 *Let (4.1.1) is true and*

$$\sum_{j=1}^{\infty} \sigma_j^2 < \infty. \tag{4.2.1}$$

Then

$$P\left(\left|\sum_{i=1}^{\infty} Z_i\right| < \infty\right) = 1. \tag{4.2.2}$$

EXAMPLE 1. Let the Z's be independent, and Z_j be uniformly distributed on $\left[-\frac{1}{j}, \frac{1}{j}\right]$. Then $E\{Z_j\} = 0$, and $\sigma_j^2 = Var\{Z_j\} = \dfrac{1}{3j^2}$. Hence, the series in (4.2.1) converges, and so does $\sum_{i=1}^{\infty} Z_i$ with probability one. □

4.3 The central limit theorem

As above, let $X_0 = 0$, and $X_t = \sum_{j=1}^{t} Z_j$, where $t = 1,2,\ldots$, and Z_j's are martingale-differences with respect to an original process ξ_t. We know that in this case, $E\{Z_j\} = 0$. Note also that the martingale-differences are uncorrelated. Indeed, for any $j > i$,

$$E\{Z_jZ_i\} = E\{E\{Z_jZ_i\,|\,\xi^{j-1}\}\} = E\{Z_iE\{Z_j\,|\,\xi^{j-1}\}\} = E\{Z_i \cdot 0\} = 0.$$

Hence, $B_t^2 = Var\{X_t\} = \sum_{j=1}^{t} \sigma_j^2$, where σ_j^2 is the variance of Z_j. (See Section **10**.1.1.)

Along with the "usual" variances σ_j^2, we consider the conditional variances $\hat{\sigma}_j^2 = E\{Z_j^2\,|\,\xi^{j-1}\}$, and set $V_t^2 = \sum_{j=1}^{t} \hat{\sigma}_j^2$. The main condition below requires V_t to be asymptotically close to the variance of X_t; namely, we require

$$\frac{V_t}{B_t} \xrightarrow{P} 1 \text{ as } t \to \infty. \tag{4.3.1}$$

This is not a mild condition. To clarify it, consider an example where $\hat{\sigma}_j^2$'s are not random and equal the unconditional variances σ_j^2.

EXAMPLE 1. We begin with a somewhat heuristic reasoning. At the end of the example, we will make all definitions rigorous.

Consider a sequence of arbitrary functions $g_j(x_0, x_1, \ldots, x_{j-1}) \geq 1$, and a sequence of r.v.'s η_j, $j = 0, 1, \ldots$. Set $W_j = g_j(\eta_0, \eta_1, \ldots, \eta_{j-1})$. It is worth emphasizing that W_j is completely determined by $\eta_0, \ldots, \eta_{j-1}$; in terms of Section 1.3, the sequence W_j is predictable with respect to the process η_t. Note also that $W_j \geq 1$.

Now, suppose that for $j = 1, 2, \ldots$, given $\eta_0, \ldots, \eta_{j-1}$,

$$Z_j = \begin{cases} W_j & \text{with probability } \frac{1}{2W_j^2} \\ 0 & \text{with probability } 1 - \frac{1}{W_j^2} \\ -W_j & \text{with probability } \frac{1}{2W_j^2}. \end{cases} \tag{4.3.2}$$

As is easy to see,

$$E\{Z_j\,|\,\eta^{j-1}\} = 0, \text{ and } E\{Z_j^2\,|\,\eta^{j-1}\} = 1.$$

So, with respect to the process η_t, the r.v.'s Z_j are martingale-differences, and the conditional variances are not random.

However, formally speaking, this does not completely fit the framework we built in Section 1.2. The point is that given the values of the η's, the values of the Z's are still random. Consequently, the process η_t cannot serve as a basic process for Z_t or X_t. In (4.3.2), we have determined only the *distribution* of Z_j given η_{j-1}, but did not define the r.v. Z_j itself.

Certainly, this is a formal issue, and we may fix it, for example, as follows.

Let ζ_j, $j = 1, 2, ...$ be uniform on $[0,1]$ r.v.'s, which are independent between themselves, and are independent of the η's. Let us define Z_j as a function of η's and ζ's by

$$Z_j = \begin{cases} W_j & \text{if } \zeta_j \leq \frac{1}{2W_j^2} \\ 0 & \text{if } \frac{1}{2W_j^2} < \zeta_j \leq 1 - \frac{1}{2W_j^2} \\ -W_j & \text{if } \zeta_j > 1 - \frac{1}{2W_j^2}. \end{cases} \tag{4.3.3}$$

Since ζ's are uniform on $[0,1]$, (4.3.3) implies (4.3.2), and the sequence $\xi_t = (\eta_t, \zeta_t)$ may serve as a basic process.

Clearly, X_t is a martingale with respect to ξ_t, and consequently, with respect to itself too. Thus, the conditional variances $\hat{\sigma}_j^2 = 1$. Hence, $V_t^2 = B_t^2 = t$, and (4.3.1) is true. □

As in Section **9**.2, we introduce the Lyapunov fraction

$$L_t = \frac{1}{B_t^3} \sum_{j=1}^{t} E\{|Z_j|^3\},$$

and consider the normalized process $X_t^* = X_t / B_t$.

Theorem 10 *Suppose (4.3.1) holds, and $L_t \to 0$ as $t \to \infty$. Then, for any x,*

$$P(X_t^* \leq x) \to \Phi(x) \quad as \quad t \to \infty. \tag{4.3.4}$$

EXAMPLE 1 revisited. Suppose additionally that for a constant C and all j, we have $E\{W_j\} \leq C$.

Certainly, this is true, for instance, if the functions $g_j(x_0, ..., x_{j-1}) \leq C$. Say, if

$$g_j(x_0, x_1, ..., x_{j-1}) = 1 + \frac{1}{1 + x_0^2 + x_1^2 + ... + x_{j-1}^2},$$

then we may set $C = 2$. However, the boundedness of the functions $g_j(\cdot)$ is not necessary.

Because $E\{|Z_j|^3 \,|\, \xi^{j-1}\} = |W_j|^3 \cdot \dfrac{1}{W_j^2} = |W_j|$,

$$E\{|Z_j|^3\} = E\{E\{|Z_j|^3 \,|\, \xi^{j-1}\}\} = E\{|W_j|\} \leq C.$$

We have shown that $B_t^2 = t$ and (4.3.1) holds. Furthermore,

$$L_t = \frac{1}{t^{3/2}} \sum_{j=1}^{t} E\{|Z_j|^3\} \leq \frac{1}{t^{3/2}} Ct = C \frac{1}{\sqrt{t}} \to 0 \text{ as } t \to \infty.$$

Thus, all conditions of the theorem hold, and X_t^* is asymptotically standard normal.

EXAMPLE 2. In the previous example, given η^{j-1}, the r.v. Z_j assumes only three values. It will make the example more sophisticated if we set

$$Z_j = \begin{cases} Y_j W_j & \text{with probability } \frac{1}{2W_j^2} \\ 0 & \text{with probability } 1 - \frac{1}{W_j^2} \\ -Y_j W_j & \text{with probability } \frac{1}{2W_j^2}. \end{cases} \tag{4.3.5}$$

where positive r.v.'s Y_j are independent between themselves, and of all other r.v.'s in the scheme, and $E\{Y_j^2\} = 1$, $E\{|Y_j|^3\} \le C_1$ for a constant C_1. In Exercise 26, the reader is asked to make sure that Theorem 10 applies in the case. □

Next, we proceed to proofs and start with a famous inequality.

4.4 Kolmogorov's inequality

We keep the above notation. As was shown, the martingale-differences are uncorrelated, and $B_t^2 = Var\{X_t\} = \sigma_1^2 + ... + \sigma_t^2$.

By Chebyshev's inequality (see (**3**.3.3.4)), for any $\varepsilon > 0$,

$$P(|X_t| > \varepsilon) \le B_t^2/\varepsilon^2. \tag{4.4.1}$$

It proves that this inequality may be essentially strengthened. Let $M_t = \max_{k \le t} |X_k|$, the largest value among $|X_1|, ..., |X_t|$.

Theorem 11 *(Kolmogorov's inequality). For any* $\varepsilon > 0$,

$$P(M_t > \varepsilon) \le B_t^2/\varepsilon^2. \tag{4.4.2}$$

(Compare this with (4.4.1).)

Proof. Let us fix t and $\varepsilon > 0$. Let the event $A = \{M_t > \varepsilon\}$, and the events $A_k = \{|X_1| \le \varepsilon, ..., |X_{k-1}| \le \varepsilon, |X_k| > \varepsilon\}$, $k = 1, ..., t$. Clearly, A_k's are disjoint, and

$$A = \bigcup_{k=1}^{t} A_k. \tag{4.4.3}$$

As usual, for any event C, an indicator r.v. $I_C = 1$ if C occurs, and $I_C = 0$ otherwise. Since A_k's are disjoint, among the indicator r.v.'s $I_{A_1}, ..., I_{A_t}$, at most one r.v. equals one, and the others vanish. Then, by (4.4.3),

$$I_A = \sum_{k=1}^{t} I_{A_k}. \tag{4.4.4}$$

From (4.4.4), conditioning on $\xi_1, ..., \xi_k$, we get that

$$\begin{aligned} B_t^2 &= E\{X_t^2\} \ge E\{I_A X_t^2\} = E\left\{\sum_{k=1}^{t} I_{A_k} X_t^2\right\} = \sum_{k=1}^{t} E\{I_{A_k} X_t^2\} \\ &= \sum_{k=1}^{t} E\{E\{I_{A_k} X_t^2 \,|\, \xi_1, ..., \xi_k\}\} = \sum_{k=1}^{t} E\{I_{A_k} E\{X_t^2 \,|\, \xi_1, ..., \xi_k\}\}. \end{aligned} \tag{4.4.5}$$

Next, we use the inequality $E\{Y^2\} \ge (E\{|Y|\})^2$ which is true for any r.v. Y, and any probability distributions, including conditional distributions. (See also Exercise **3**-54.) So,

$$E\{X_t^2 \,|\, \xi_1, ..., \xi_k\} \ge (E\{|X_t| \,|\, \xi_1, ..., \xi_k\})^2 \ge (E\{X_t \,|\, \xi_1, ..., \xi_k\})^2. \tag{4.4.6}$$

Now, note that by the martingale property and (1.1.3), for $k = 1, ..., t-1$,

$$\begin{aligned} E\{X_t \,|\, \xi_1, ..., \xi_k\} &= E\{E\{X_t \,|\, \xi_1, ..., \xi_{t-1}\} \,|\, \xi_1, ..., \xi_k\} = E\{X_{t-1} \,|\, \xi_1, ..., \xi_k\} \\ &= E\{X_{t-2} \,|\, \xi_1, ..., \xi_k\} = ... = E\{X_{k+1} \,|\, \xi_1, ..., \xi_k\} = X_k. \end{aligned} \tag{4.4.7}$$

For $k = t$, (4.4.7) holds automatically. Hence, from (4.4.6), it follows that

$$E\{X_t^2 \,|\, \xi_1, ..., \xi_k\} \geq (X_k)^2. \tag{4.4.8}$$

Eventually, from (4.4.8) and (4.4.5), we get that

$$B_t^2 \geq \sum_{k=1}^{t} E\{I_{A_k} X_k^2\}. \tag{4.4.9}$$

Now, $I_{A_k} \neq 0$ only if A_k occurs, and in this case $|X_k| > \varepsilon$. Hence, (4.4.9) and (4.4.4) imply that $B_t^2 \geq \sum_{k=1}^{t} E\{\varepsilon^2 I_{A_k}\} = \varepsilon^2 \sum_{k=1}^{t} E\{I_{A_k}\} = \varepsilon^2 E\left\{\sum_{k=1}^{t} I_{A_k}\right\} = \varepsilon^2 E\{I_A\} = \varepsilon^2 P(M_t > \varepsilon)$. ■

4.5 Proof of Theorem 9

By Kolmogorov's inequality, for any integers m and l, and $\varepsilon > 0$,

$$P\left(\max_{k=1,...,l} \left|\sum_{i=m+1}^{m+k} Z_i\right| > \varepsilon\right) \leq \frac{1}{\varepsilon^2} \sum_{i=m+1}^{m+l} \sigma_i^2.$$

Because the series $\sum_{i=1}^{\infty} \sigma_i^2$ converges, we can let $l \to \infty$, which implies

$$P\left(\sup_{k\geq 1} \left|\sum_{i=m+1}^{m+k} Z_i\right| > \varepsilon\right) \leq \frac{1}{\varepsilon^2} \sum_{i=m+1}^{\infty} \sigma_i^2. \tag{4.5.1}$$

(The reader who is not familiar with the notion of *supremum* may again replace the sign sup by max.)

Set $Y_m = \sup_{k\geq 1} \left|\sum_{i=m+1}^{m+k} Z_i\right|$. Due to the convergence of $\sum_{i=1}^{\infty} \sigma_i^2$, the r.-h.s. of (4.5.1) converges to zero as $m \to \infty$, which means that $Y_m \xrightarrow{P} 0$ as $m \to \infty$. Then

$$\begin{aligned}
\widetilde{Y}_n &= \sup_{m\geq n} Y_m = \sup_{m\geq n} \sup_{k\geq 1} \left|\sum_{i=m+1}^{m+k} Z_i\right| = \sup_{m\geq n} \sup_{k\geq 1} \left|\sum_{i=n+1}^{m+k} Z_i - \sum_{i=n+1}^{m} Z_i\right| \\
&\leq \sup_{m\geq n} \sup_{k\geq 1} \left|\sum_{i=n+1}^{m+k} Z_i\right| + \sup_{m\geq n} \sup_{k\geq 1} \left|\sum_{i=n+1}^{m} Z_i\right| = 2Y_n \xrightarrow{P} 0 \quad \text{as} \quad n \to \infty.
\end{aligned}$$

Consequently, by Lemma **7**.19, $Y_n \xrightarrow{\text{a.s.}} 0$ as $n \to \infty$. The last assertion is equivalent to the assertion that the series $\sum_{i=1}^{\infty} Z_i$ converges with probability one. ■

4.6 Proof of Theorem 8

From (4.1.2), it follows that $\sum_{j=1}^{\infty} Var\{X_j/j\} < \infty$. Then, by Theorem 9,

$$P\left(\left|\sum_{j=1}^{\infty} (X_j/j)\right| < \infty\right) = 1. \tag{4.6.1}$$

It remains to use the following fact from Calculus, which is referred to as Kronecker's lemma[4]: if a (non-random) series $\sum_{j=1}^{\infty}(a_j/j)$ converges, then the average $\frac{1}{n}\sum_{i=1}^{n} a_i \to 0$ as $n \to \infty$.

For the completeness of the picture and because this lemma may not be found in any Calculus text, we give a proof. It follows from the well known fact that if a sequence $b_n \to b$, then $\frac{1}{n}\sum_{k=1}^{n} b_k \to b$. What we do below is a sort of integration by parts applied to sequences.

Set $b_1 = 0$, and $b_n = \sum_{j=1}^{n-1}(a_j/j)$ for $n > 1$. So, in our case, $b = \sum_{j=1}^{\infty}(a_j/j) < \infty$. We have

$$\sum_{i=1}^{n} a_i = \sum_{i=1}^{n} i(b_{i+1} - b_i) = \sum_{i=1}^{n}((i+1)b_{i+1} - ib_i) - \sum_{i=1}^{n} b_{i+1}.$$

In the first sum in the r.-.s., all terms cancel save $(n+1)b_{n+1} - b_1 = (n+1)b_{n+1}$. The last sum equals $b_{n+1} + \sum_{k=1}^{n} b_k$. Then $\sum_{i=1}^{n} a_i = nb_{n+1} - \sum_{k=1}^{n} b_k$. Hence,

$$\frac{1}{n}\sum_{i=1}^{n} a_i = b_{n+1} - \frac{1}{n}\sum_{i=1}^{n} b_i \to b - b = 0. \blacksquare$$

4.7 Proof of the LLN for i.i.d. r.v.'s

Now, we will prove the strong LLN for i.i.d. r.v.'s., i.e., Theorem **3**.15. It is reasonable to switch to notations of Chapter 3.

Let $S_n = X_1 + ... + X_n$, where the r.v.'s X_j are i.i.d. Without loss of generality, we set $E\{X_j\} = 0$. Clearly, X_j's are martingale differences, and S_n is a martingale.

4.7.1 Sufficiency

If the variance of the X's (the same for all) had been finite, the LLN in its sufficiency part would have immediately followed from Theorem 8 since in this case σ_j equals some constant σ, and hence the series in (4.1.2) converges. To cover the general case, we use what is called *truncation* technique, introducing the r.v.'s

$$Y_j = \begin{cases} X_j & \text{if } |X_j| \le j, \\ 0 & \text{if } |X_j| > j. \end{cases} \tag{4.7.1}$$

We have

$$\sum_{j=1}^{\infty} P(Y_j \ne X_j) = \sum_{j=1}^{\infty} P(|X_j| > j) = \sum_{j=1}^{\infty} P(|X_1| > j)$$

because X_j are identically distributed.

Next, we use the general formula (**3**.2.2.1). Writing $[x]$ for the integer part of x, we get

$$\sum_{j=1}^{\infty} P(|X_1| > j) \le \sum_{j=1}^{\infty} P([|X_1|] > j) \le \sum_{j=0}^{\infty} P([|X_1|] > j) = E\{[|X_1|]\} \le E\{|X_1|\} < \infty$$

[4]More precisely, we consider here a version of this lemma.

by assumption. Hence, $\sum_{j=1}^{\infty} P(Y_j \neq X_j) < \infty$, and by Borel-Cantelli's theorem **2**.1, $P(Y_j \neq X_j$ infinitely often $) = 0$. Consequently, it suffices to prove that $\frac{1}{n}\sum_{j=1}^{n} Y_j \overset{\text{a.s.}}{\to} 0$.

We have

$$\frac{1}{n}\sum_{j=1}^{n} Y_j = \frac{1}{n}\sum_{j=1}^{n}(Y_j - E\{Y_j\}) + \frac{1}{n}\sum_{j=1}^{n} E\{Y_j\} = \zeta_{n1} + \zeta_{n2},$$

where ζ_{n1}, ζ_{n2} stand for the first and second averages, respectively. To evaluate the second sum, we observe that

$$E\{Y_j\} = \int_{-j}^{j} x dP(X_j \leq x) = \int_{-j}^{j} x dP(X_1 \leq x) \to E\{X_1\} = 0 \text{ as } j \to \infty.$$

From Calculus, we know that, if $x_k \to 0$ as $k \to \infty$, so does the average $\frac{1}{n}\sum_{k=1}^{n} x_k$ as $n \to \infty$. All of this implies that $\zeta_{n2} \to 0$. It remains to prove that $\zeta_{n1} \overset{\text{a.s.}}{\to} 0$.

By Theorem 8, it suffices to prove that $\sum_{j=1}^{\infty} \frac{1}{j^2} Var\{Y_j\} < \infty$.

Let $Z_j = [|X_j|]$. Since $|Y_j| \leq j$,

$$Var\{Y_j\} \leq E\{Y_j^2\} \leq \sum_{k=0}^{j-1}(k+1)^2 P(Z_j = k) = \sum_{k=0}^{j-1}(k+1)^2 P(Z_1 = k) = \sum_{m=1}^{j} m^2 P(Z_1 = m-1)$$

In the next inequality, we use the fact that $\sum_{j=m}^{\infty} \frac{1}{j^2} \leq \frac{2}{m^2}$. (It may be proved by comparing the sum with the integral $\int_{m-1}^{\infty} \frac{1}{x^2} dx$.) We have

$$\begin{aligned}
\sum_{j=1}^{\infty} \frac{1}{j^2} Var\{Y_j\} &\leq \sum_{j=1}^{\infty} \frac{1}{j^2} \sum_{m=1}^{j} m^2 P(Z_1 = m-1) = \sum_{m=1}^{\infty} m^2 P(Z_1 = m-1) \sum_{j=m}^{\infty} \frac{1}{j^2} \\
&\leq \sum_{m=1}^{\infty} m^2 P(Z_1 = m-1)\frac{2}{m} = 2\sum_{m=1}^{\infty} m P(Z_1 = m-1) = 2\sum_{k=0}^{\infty}(k+1)P(Z_1 = k) \\
&= 2E\{Z_1 + 1\} \leq 2E\{|X_1|\} + 2 < \infty. \blacksquare
\end{aligned}$$

4.7.2 Necessity

Again using (**3**.2.2.1), we get

$$\begin{aligned}
E\{|X_1|\} &\leq 1 + E\{[|X_1|]\} = 1 + \sum_{j=0}^{\infty} P([|X_1|] > j) \leq 1 + \sum_{j=0}^{\infty} P(|X_1| > j) \\
&\leq 2 + \sum_{j=1}^{\infty} P(|X_1| > j) = 2 + \sum_{j=1}^{\infty} P(|X_j| > j) = 2 + \sum_{j=1}^{\infty} P(|\xi_j| > 1), \qquad (4.7.2)
\end{aligned}$$

where $\xi_j = X_j/j$. Now, since $\frac{1}{n}S_n \overset{\text{a.s.}}{\to} c$,

$$\xi_n = \frac{S_n}{n} - \frac{S_{n-1}}{n} = \frac{S_n}{n} - \frac{n-1}{n} \cdot \frac{S_{n-1}}{n-1} \overset{\text{a.s.}}{\to} c - 1 \cdot c = 0.$$

Then, by Proposition **3**.18, $\sum_{j=1}^{\infty} P(|\xi_j| > 1) < \infty$, which together with (4.7.2) implies that $E\{|X_1|\} < \infty$. Then, by the sufficiency part of the theorem (which has been already proved), $\frac{1}{n}S_n \overset{\text{a.s.}}{\to} m$, where $m = E\{X_j\}$. Hence, $c = m$. (In the current proof, we have assumed $m = 0$.)

$\blacksquare$

5 EXERCISES

Section 1

In problems below, if the r.v. ξ_0 is not specified, the reader can set $\xi_0 = 0$, or just $\xi^t = (\xi_1, ..., \xi_t)$.

1. Prove (1.1.2) for the general mapping $g(\cdot)$.
2. Let $\xi_1, \xi_2, ...$ be i.i.d. r.v.'s with zero means. Set $X_0 = X_1 = 0$, $X_t = \sum_{1 \le i < j \le t} \xi_i \xi_j$, $t = 2, 3, ...$.

   Show that X_t is a martingale with respect to ξ_t.
3. Let $\xi_1, \xi_2, ...$ be independent r.v.'s uniformly distributed on $[0, a]$. Let $X_0 = 1$ and $X_t = C_t \xi_1 \cdot ... \cdot \xi_t$. For which constant C_t is X_t a martingale?
4. Let $\xi_1, \xi_2, ...$ be independent r.v.'s having an exponential distribution, and $E\{\xi_i\} = 1$. Let $S_0 = 0$, $S_t = \xi_1 + ... + \xi_t$, and $X_t = C_t \exp\{-S_t\}$, where C_t is a constant such that X_t is a martingale. (a) Find C_t. (b) Find $\lim_{t \to \infty} X_t$. How fast is such a convergence? (*Advice*: You may use the fact that, by the LLN, $\frac{1}{t}(\xi_1 + ... + \xi_t) \to 1$ with probability one.)
5. Show directly that in the urn model from Example 1.2-5, the probability of selecting a red ball at the second drawing is the same as at the first.
6. Let $\xi_1, \xi_2, ...$ be i.i.d. r.v.'s with zero mean and variance $\sigma^2 > 0$, $S_0 = 0$, $S_t = \xi_1 + ... + \xi_t$ for $t = 0, 1, ...$ Show that $X_t = S_t^2 - \sigma^2 t$ is a martingale with respect to ξ_t.
7. Let $X_t = 2^t \xi_1 \cdot ... \cdot \xi_t$, where ξ's are independent and take on values 0 and 1 with equal probabilities. Show that X_t is a martingale. What is $E\{X_t\}$? What is $\lim_{t \to \infty} X_t$? (*Hint*: Realize how long there will be no zeros among ξ's.)
8.* (a) Let $\xi_1, \xi_2, ...$ be independent standard normal r.v.'s, $S_0 = 0$, $S_t = \xi_1 + ... + \xi_t$, $X_0 = 1$, and $X_t = C_t \exp\{S_t\}$ for $t = 1, ...$ Find a constant C_t for which X_t is a martingale. Find $\lim_{t \to \infty} X_t$. How will the problem change if we consider $X_t = C_t \exp\{-S_t\}$?

   (b) Do the same if ξ's are normal with a mean of m and a variance of σ^2.
9. (a) Let $\xi_0, \xi_1, \xi_2, ...$ be independent r.v.'s and $E\{\xi_i\} = 0$ for all i. Let $X_0 = 0$ and $X_t = Z_1 + ... + Z_t$ for $t \ge 1$ where $Z_k = \xi_{k-1} \xi_k$. Are Z's independent? Show that X_t is a martingale.

   (b)* Let $E\{\xi_i\} = m \neq 0$. Write Doob's decomposition.
10.* Let X_t be a branching process as it was defined in Sections **4**.1, 3. Suppose that μ, the mean number of the offsprings of a particular particle, is positive. (a) Show that the process $Y_t = \frac{1}{\mu^t} X_t$ is a martingale with respect to itself. (b) Show the same for the process $W_t = u^{X_t}$ where u is the extinction probability. (*Hint*: Look at (**4**.3.4).)
11.* Let $\xi_1, \xi_2, ...$ be i.i.d. r.v.'s with a m.g.f. $M(z)$, and $S_t = \xi_1 + ... + \xi_t$. Show that $X_t = M^{-t}(z) \exp\{z S_t\}$ is a martingale with respect to ξ_t.

Route 1 ⇒ page 435

*Section 2**

12. Let τ be an optional time as it was defined in (2.1.1). (a) Is the occurrence of event $\{\tau > t\}$ completely determined by the values of the r.v.'s $\xi_0, ..., \xi_t$? (b) Explain why $P(\tau < \infty) = \sum_{t=0}^{\infty} P(\tau = t)$.

13. In the situation of Example 2.2-1, explain why it takes considerably long to reach the level 1 starting from the *neighbor* level 0. Show that $E\{\tau_a\} > a$ if $p < 1$, not appealing to formula (2.2.5). What will happen if p is getting close to $1/2$?

14. In the situation of Example 2.2-4, prove that $E\{\tau_a\} = \infty$ if $m = 0$.

15. An alternative and direct proof of Wald's identity (2.2.4) uses indicator r.v.'s $I_t = 1$ if $\tau \geq k$, and $= 0$ otherwise. Show that, in the notation of Proposition 5, $S_\tau = \sum_{k=1}^{\infty} \xi_k I_k$, and that the r.v.'s ξ_k and I_k are independent. Derive from this (2.2.4). (*Hint*: Whether the stopping time $\tau \geq k$ or not (that is, whether the process "stopped" after the moment $k-1$) is completely determined by the r.v.'s $\xi_1, ..., \xi_{k-1}$. In other terms, the sequence $\{I_k\}$ is predictable.)

16. A gambler plays a game of chance favorable for him, winning in each play \$5 with a probability of 0.55 and losing the same amount with probability 0.45. The gambler decides to play until the total gain will reach \$100. Assuming that the gambler has sufficiently enough money at his disposal, find how long it will take on the average. What is the answer for the case of equal probabilities?

17. Let us revisit Exercise 6. Let $S_0 = 0$, and a stopping time $\tau < c$ for a constant c with probability one. Using the optional stopping theorem, prove that $E\{S_\tau^2\} = \sigma^2 E\{\tau\}$. Compare it with the standard formula $Var\{S_t\} = \sigma^2 t$.

18. Consider a homogeneous Poisson process N_t with parameter λ. Show that the formula $E\{N_t\} = \lambda t$ is consistent with (2.2.6).

19. Consider the doubling strategy from Example 2.1-3 in the case of a game of roulette. Assume that a gambler always bets on red. Then at each play, the probability of success is $p = 9/19$ (there are 18 red, 18 black, and 2 green cells).

    (a) Is the profit W_t a martingale?

    (b) What will happen if after each failure, the gambler bets the amount c times as large as in the previous play? Consider the cases $1 < c < 2$ and $c > 2$ separately.

    (c) Let us consider the situation of Example **3**.1.1-13. Suppose that once the gambler quits (either winning one or losing being not able to apply the doubling strategy), he starts over with a stake of one and doubling the bet after each loss; that is, he repeats the whole game (that is, the whole sequence of plays up to quitting). Using the LLN, analyze the behavior of the total profit for a long sequence of such (independent) games. (To make your result consistent with the answer to the next problem, recalculate the probability of losing with a larger accuracy.)

    (d) Solve Problem 19c for the general case where the probability of winning in a separate try equals a (given) $p > 0$, and the gambler is not able to apply the doubling strategy if the maximal bet is 2^m.

Sections 3* and 4*

20. (a) Does the dimension of the price vector $\boldsymbol{\psi}$ equal the number of the assets in the market or the number of future states of nature? (b) Does the dimension of the trading strategy vector $\boldsymbol{\theta}$ equal the number of assets in the market or the number of future states of nature?

21. Assume that in a country, there is a law which prohibits from selling short risky assets but allows one to borrow money with a fixed rate. When modeling such a market, what restrictions on the model of Section 3.2 should be imposed?

22. Assume that in a market, the pricing procedure (3.1.3) corresponds to the vector $\boldsymbol{\psi} = (0.2, 0.03, 0.07, 0.1, 0.2, 0.2)$. (a) How many states of nature can an investor face in this market? (b) Which states of nature do people consider in this case most likely, least likely? (c) If an investor borrows \$1000 today, how much should she/he return at the end of the period?

23. The price of a stock is currently \$50. It is expected that at the end of a time period, the price will be either \$75 or \$25. The risk-free interest over this period is zero. (a) Prove that there is no arbitrage opportunity. (b) Write the prices of call and put options with an exercise price of \$50. (c) Write the price of the state security that corresponds to the increase of the stock price. (d) Solve the problems above for $r = 0.1$. (e) Give an example of r for which there is an arbitrage opportunity. Point out how a sure win may be accomplished.

24. Proceeding from a common-sense point of view, interpret the fact that the price of the European option in Example 3.3-2 is larger than the price of the American option in Example 3.3-3.

25. Consider a stock whose initial price is \$100, and each year the price either increases by 5% or drops by 6%. Let a risk-free interest be 4%.

    (a) Build a two-year tree for stock prices and determine a martingale measure. Is it unique? Find the probability corresponding to this measure ("the probability in the risk neutral world") that the price will drop exactly one time.
    (b) Find the price of a European call option with an exercise price of \$98.
    (c) Find the price of a European put option with an exercise price of \$98.
    (d) Find the price of an American put option with an exercise price of \$98. Compare with the price of the European put.
    (e) Provide an Excel sheet for solving Problems 25b and 25c. Play a bit with input parameters.
    (f) You buy a ticket which gives you the following options. You can exercise your ticket at the end of each year during a two-year period but just one time. If at this moment the price of the stock is higher than the initial price, then you are paid \$100 if you exercise the ticket at the end of the first year, and \$200 if you exercise it at the end of the second year. At the end of the first year, you may also just quit (losing the money you paid for the ticket). However, if you did not quit in the first year, and the current price at the end of the second year is not higher than the initial price, you pay a fee of \$100. (i) How much should such a ticket cost? (ii) Find the optimal strategy for a buyer of the ticket.
    (g) Using software, say Excel, solve Problems 25b and 25c for a three-year period.
    (h) Using software, say Excel, solve Problem 25f for a three-year period for the exercise payments \$100, \$200, and \$300, and a fee of \$400.

26. Show that Theorem 10 applies in the case of Example 4.3-2.

27. Let X_t, $t = 0, 1, \dots$, be a non-negative martingale, $M_t = \max_{k \le t} X_k$. Show that for any t and $x > 0$,

$$P(M_t \ge x) \le \frac{E\{X_0\}}{x}. \tag{5.1}$$

    (*Hint*: First, do not forget (1.2.8). Secondly, the proof of (5.1) may be similar to that of (4.4.2) and easier.)

28. Consider a symmetric random walk: $X_0 = u$ and $X_t = u + \xi_1 + \dots + \xi_t$, for $t = 1, 2, \dots$, where an integer $u > 0$ and ξ's are independent and assume the values ± 1 with equal probabilities. Let τ be the moment of ruin, that is, $\tau = \min\{t : X_t = 0\}$. Let $Y_t = X_t$ if $t \le \tau$, and $Y_t = 0$ if $t \ge \tau$. We may think about a gambler who starts with the initial capital u, and quits at the moment of ruin. (We know that $P(\tau < \infty) = 1$; see Sections **4**.4.1 and **5**.5.2.) Show that Y_t is a martingale. Using (5.1), estimate the probability that before ruin, the gambler will reach a level $a \ge u$.

Chapter 16

Brownian Motion and Martingales in Continuous Time

1 BROWNIAN MOTION AND ITS GENERALIZATIONS

In this section, we continue to explore Brownian motion or the Wiener process w_t defined in Section **12**.3.

1.1 Further properties of the standard Brownian motion

1.1.1 Non-differentiability of trajectories

The definition of w_t in Section **12**.3 requires the trajectories of the process to be continuous. Let us turn to differentiability.

As before, let w_Δ stand for the increment of w_t and $\Phi(x)$ denote the standard normal d.f.

To determine whether w_t has a derivative at a point t, consider a time interval $\Delta = (t, t+\delta]$ and explore the behavior of the r.v. $\eta_\Delta = w_\Delta/\delta$ as $\delta \to 0$.

By definition, the r.v. w_Δ is normal with zero mean and a standard deviation of $\sqrt{\delta}$. Then for $x \geq 0$, we have $P(|\eta_\Delta| > x) = P(|w_\Delta| > x\delta) = 2(1 - \Phi(x\delta/\sqrt{\delta})) = 2(1 - \Phi(x\sqrt{\delta})) \to 2(1 - \Phi(0)) = 1$ as $\delta \to 0$. Since the last relation is true for an arbitrary large x, this means that when δ is approaching zero, the r.v. $|\eta_\Delta|$ takes on arbitrary large values with a probability close to one. Rigorously, $|\eta_\Delta| \to \infty$ as $\delta \to 0$ in probability (for a definition of this type of convergence see Section **3**.5.1).

As a matter of fact, an even stronger property is true. Namely, with probability one, trajectories of Brownian motion (that is, w_t as a function of t) are nowhere differentiable; i.e., the derivative does not exist at any point. (See, e.g., [37, p.32] or the outline of a proof in [24, p.268].)

This is an amazing property. Trajectories are continuous but not smooth, and the process fluctuates infinitely frequently in any arbitrary small time interval. However, this is not an obstacle for applications. If we are interested in the increments of the process over intervals that perhaps are small but not infinitesimally small, then we are dealing with r.v.'s w_Δ which are normal in the mathematical and usual sense as well, and hence are tractable.

When in 1872, K. Weierstrass constructed a function that was continuous but non-differentiable at any point, it was a significant mathematical achievement. Some people considered this function pathological, others—a mathematical masterpiece, but regardless, this function looked exotic. Nowadays, the Wiener process whose trajectories are functions with the same property, serves as a good model for many applied problems.

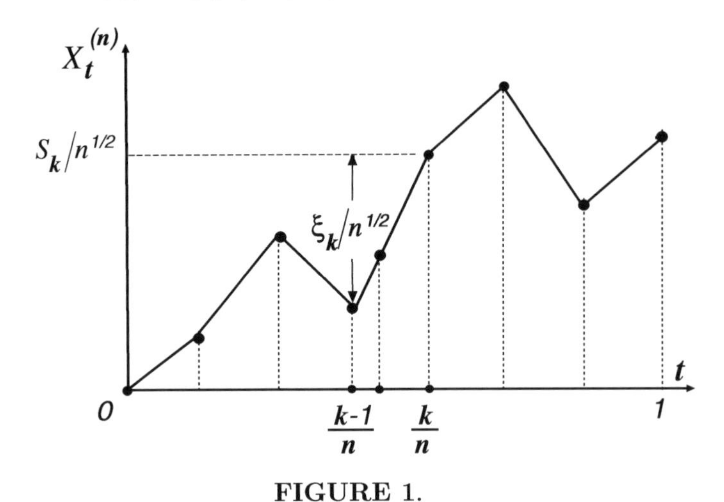

FIGURE 1.

1.1.2 Brownian motion as an approximation: The Donsker–Prokhorov invariance principle

Let $\xi_1, \xi_2, ...$ be i.i.d. r.v.'s having zero means and unit variances. Let $S_0 = 0$ and $S_k = \xi_1 + ... + \xi_k$.

Let us consider the time interval $[0,1]$ and for each $n = 1, 2, ...$, construct a piecewise linear (or polygonal) random process $X_t^{(n)}$ on $[0,1]$ as follows.

We divide $[0,1]$ into n intervals $\Delta_k = (\frac{k-1}{n}, \frac{k}{n}]$, $k = 1, ..., n$, of the same length $\frac{1}{n}$. The end points of these intervals are the points $t_k = t_{kn} = k/n$. At the points t_k, we set (see also Fig. 1)

$$X_{t_k}^{(n)} = \frac{1}{\sqrt{n}} S_k,$$

and for $t_{k-1} \leq t \leq t_k$, we define $X_t^{(n)}$ as a linear function whose graph connects points $(t_{k-1}, X_{t_{k-1}}^{(n)})$ and $(t_k, X_{t_k}^{(n)})$; see again Fig. 1.

The process so constructed is called a *partial sum process*. We may view it as the sequence of partial sums $S_1, ..., S_n$, compressed in a way that it runs in the interval $[0,1]$. Since ξ_k's are independent, the process $X_t^{(n)}$ is that with independent increments. We will see that for large n, the fluctuations of the piecewise linear process $X_t^{(n)}$ are approaching those of Brownian motion.

Proposition 1 *For any* t,

$$X_t^{(n)} \xrightarrow{d} w_t \quad as \quad n \to \infty, \tag{1.1.1}$$

where the convergence $\xrightarrow{d}$ *means the convergence of the distributions of the corresponding r.v.'s. (see also Section 7.3).*

We prove it at the end of this section, but first note that, as a matter of fact, an essentially stronger assertion is true. Namely, not only the marginal distributions (for separate t's) of the process $X_t^{(n)}$ converge to the corresponding marginal distributions of w_t, but the probability distribution of the process $X_t^{(n)}$ as a whole (that is, the joint distribution of the values of the process at different time moments t) converges to the distribution of the standard Wiener process. This fact is referred to as the *Donsker–Prokhorov invariance principle*. A rigorous statement may be found, for example, in [16], [24]. By virtue of this principle, the Wiener process may be viewed as a continuous approximation of the partial sum process.

EXAMPLE 1. Let ξ_i take on values ± 1 with equal probabilities. Then the process of partial sums corresponds to the symmetric random walk considered in Section **2**.2.4. We have shown there that, starting from a level u, the symmetric random walk will reach a level a before hitting zero level with the probability $p_u = u/a$. The corresponding ruin probability is $q_u = (a-u)/a$. Consider the process $X_t = u + w_t$. This is Brownian motion starting from level u. In view of the invariance principle, we may conjecture that the corresponding ruin probability for X_t will be the same as for the symmetric random walk. In Section 2.3.2, we will show that this is indeed true. □

Proof of Proposition 1. For a fixed $t \in (0,1]$, let $k = k(n)$ be the smallest integer which is not less than tn. Formally, $k = tn$ if tn is an integer, and $k(n) = [tn] + 1$ otherwise. (As usual, $[a]$ denotes the integer part of a.) Because $t > 0$, we have $k(n) \to \infty$ as $n \to \infty$.

For k so defined, $t \in \Delta_k$. Indeed, if tn is an integer, then $t = \frac{k}{n} \in \Delta_k$. If tn is not an integer, we have $tn < k < tn + 1$, which implies $\frac{k-1}{n} < t < \frac{k}{n}$.

Since the ξ's have zero means and unit variances, $E\{S_{k(n)}\} = 0$ and $Var\{S_{k(n)}\} = k(n)$. Because $X_t^{(n)}$ is a linear on $[t_{k-1}, t_k]$,

$$X_t^{(n)} = X_{t_{k(n)}}^{(n)} - \frac{t_k - t}{t_k - t_{k-1}} \cdot \frac{\xi_k}{\sqrt{n}} = \sqrt{\frac{k(n)}{n}} \cdot \frac{1}{\sqrt{k(n)}} S_{k(n)} - \frac{t_k - t}{t_k - t_{k-1}} \cdot \frac{\xi_k}{\sqrt{n}}$$

(See also Fig. 1). By construction, $\frac{k(n)}{n} \to t$ as $n \to \infty$. By the CLT,

$$\frac{1}{\sqrt{k(n)}} S_{k(n)} \xrightarrow{d} Z,$$

where Z is a standard normal r.v. Also, $\left| \frac{t_k - t}{t_k - t_{k-1}} \cdot \frac{\xi_k}{\sqrt{n}} \right| < 1 \cdot \frac{|\xi_k|}{\sqrt{n}} \xrightarrow{d} 0$.

Hence, $X_t^{(n)} \xrightarrow{d} \sqrt{t} Z$. The r.v. $\sqrt{t} Z$ is normal with zero mean and variance t, that is, it has the same distribution as w_t. ■

1.1.3 The distribution of w_t, hitting times, and the maximum value of Brownian motion

First, note that since w_t is normal with zero mean and variance t, we can explicitly write its density and the d.f. In accordance with (**6**.2.4.3), for all $t > 0$, the density of w_t is

$$f_t(x) = \frac{1}{\sqrt{2\pi t}} \exp\left\{ -\frac{x^2}{2t} \right\}, \tag{1.1.2}$$

and the d.f. is

$$F_t(x) = \Phi\left(\frac{x}{\sqrt{t}} \right). \tag{1.1.3}$$

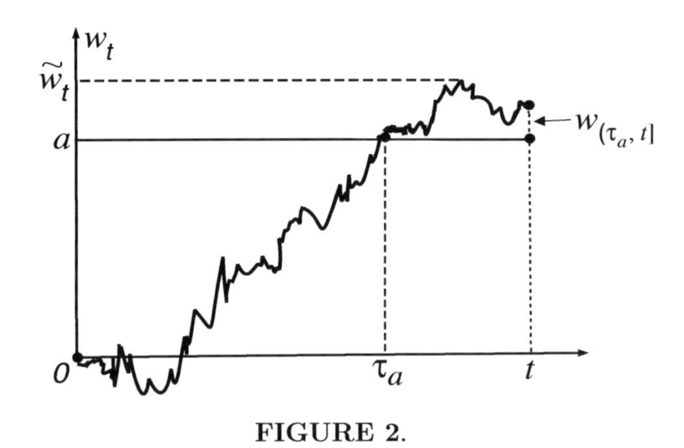

FIGURE 2.

For $t = 0$, the r.v. $w_0 = 0$.

The next two definitions are illustrated in Fig. 2. For $a > 0$, we set $\tau_a = \min\{t \geq 0; w_t = a\}$, the time at which the process, starting from zero, first hits a level a.

We define also $\widetilde{w}_t = \max\limits_{0 \leq s \leq t} w_s$, the maximal value of the process over the interval $[0,t]$. Since trajectories w_t are continuous, the maximum exists.

The characteristics τ_a and $\widetilde{w}_t$ are connected with each other:

$$\tau_a \leq t \quad \text{if and only if} \quad \widetilde{w}_t \geq a. \tag{1.1.4}$$

Indeed, if the maximum value of the process over the period $[0,t]$ was greater than or equal to a, then the process "had to cross or hit" the level a. Since the process is continuous, it could not overshoot a and was equal to a at the moment of crossing or hitting.

Furthermore, by the formula for total probability,

$$P(w_t \geq a) = P(w_t \geq a \,|\, \tau_a > t)P(\tau_a > t) + P(w_t \geq a \,|\, \tau_a \leq t)P(\tau_a \leq t). \tag{1.1.5}$$

The first conditional probability clearly equals zero because if a continuous process reached the a-level at the first time after time t, the process cannot be larger than or equal to a at time t.

The second conditional probability

$$P(w_t \geq a \,|\, \tau_a \leq t) = 1/2. \tag{1.1.6}$$

To show this, one may reason as follows. It is given that the random moment τ_a is less than or equal to t; see also Fig. 2. The value of the process at the moment τ_a is exactly equal to a. Hence, if $t > \tau_a$, the r.v.

$$w_t = a + w_{(\tau_a,t]},$$

where $w_{(\tau_a,t]}$ is the increment of the process over the time interval $(\tau_a,t]$; see Fig. 2. Then, w_t will be greater than or equal to a if and only if $w_{(\tau_a,t]} \geq 0$. Since w_t is a process with independent increments, after the moment τ_a, the process starts over, and its increments do not depend on the history of the process before τ_a. On the other hand, all increments w_Δ are symmetric r.v.'s, and hence, $w_{(\tau_a,t]}$ is equally likely to be positive or negative. This implies (1.1.6).

Thus, from (1.1.5) it follows that $P(w_t \geq a) = \frac{1}{2}P(\tau_a \leq t)$. Combining it with (1.1.3), we eventually obtain that

$$P(\tau_a \leq t) = 2\left(1 - \Phi\left(a/\sqrt{t}\right)\right). \tag{1.1.7}$$

In view of (1.1.4), $P(\widetilde{w}_t < a) = 1 - P(\widetilde{w}_t \geq a) = 1 - P(\tau_a \leq t) = 2\Phi\left(a/\sqrt{t}\right) - 1$. This is a continuous function in a. Therefore, $P(\widetilde{w}_t = a) = 0$, and

$$P(\widetilde{w}_t \leq a) = 2\Phi\left(a/\sqrt{t}\right) - 1. \tag{1.1.8}$$

It is worth noting that in (1.1.7)–(1.1.8), we consider the distribution functions of two r.v.'s: τ_a and $\widetilde{w}_t$, respectively. In (1.1.7), the argument of the d.f. is t and a is a parameter, while in (1.1.8), the roles of these two quantities switch: a is the argument of the d.f. and t is a parameter.

EXAMPLE 1. You own a stock whose current price per share is $S_0 = 100$. The price changes in time as the process $S_t = S_0 \exp\{\sigma w_t\}$, where a number $\sigma > 0$. In Sections 1.2 and 2.3.4 below, we discuss in detail why such a model may be adequate.

Let σ equal, say, 0.15. You have decided to sell your shares when the price increases by 10%. What is the probability that this will not happen within the first year?

You are going to sell your stock at the *first time* when $S_t \geq 1.1S_0$. This is equivalent to the inequality $\exp\{\sigma w_t\} \geq 1.1$, or $w_t \geq \frac{1}{\sigma}\ln(1.1) \approx 0.63$. (Note that the answer does not depend on S_0.) So, you will not sell your shares if w_t does not reach the level 0.63 during the time period $[0,1]$. The probability of this event is $P(\widetilde{w}_1 < 0.63)$ and equals $2\Phi(0.63) - 1 \approx 0.47$. As we have seen, $\widetilde{w}_1$ is a continuous r.v., so it does not matter whether to write $P(\widetilde{w}_1 < x)$ or $P(\widetilde{w}_1 \leq x)$. □

Formula (1.1.8) leads to an unexpected, at first glance, conclusion. Consider the probability that during a fixed time interval $[0,T]$ the process will take on only non-positive values. In this case, the graph of the realization will be under the t-axis. The probability in hand is $P(\widetilde{w}_T \leq 0)$. In accordance with (1.1.8), it is equal to $2\Phi\left(\frac{0}{\sqrt{T}}\right) - 1 = 0$ for any arbitrary small $T > 0$. This means that starting from zero, the process cannot move down assuming for a while only negative values. On the contrary, during any arbitrary small interval $[0,T]$, the process will cross zero level with probability one. It may be proved that before "leaving zero," the process fluctuates around zero infinitely often, so to speak, rapidly oscillating around zero. Since the state of the process at any time may be viewed as the initial state with respect to the future evolution, the same conclusion concerns the behavior of the process in a neighborhood of any point. The evolution of the process is by no means smooth but consists of an infinite number of small but frequent movements up and down.

1.2 Brownian motion with drift and geometric Brownian motion

In Chapter 6, we defined a normal r.v. with mean m and variance σ^2 as a r.v. $m + \sigma X$, where the r.v. X is standard normal.

We define a *Brownian motion with a drift parameter* μ and *a variance parameter* σ^2 as the process

$$X_t = \mu t + \sigma w_t, \quad t \geq 0, \tag{1.2.1}$$

where w_t is Brownian motion.

Since w_t is a process with independent increments, so is X_t, and for any time interval Δ, the increment X_Δ is normal with mean $\mu|\Delta|$ and variance $\sigma^2|\Delta|$, where $|\Delta|$ stands for the length of Δ. In particular, the r.v. X_t is $(\mu t, \sigma^2 t)$-normal, and hence for any $t > 0$, the density of X_t is

$$f_t(x) = \frac{1}{\sqrt{2\pi t}\sigma} \exp\left\{-\frac{(x-\mu t)^2}{2\sigma^2 t}\right\}. \tag{1.2.2}$$

Next, we consider a process $Y_t = Y_0 \exp\{X_t\}$, where Y_0 is a certain positive number, and $X_t = \mu t + \sigma w_t$, a Brownian motion with drift. The process Y_t is called a *geometric Brownian motion*. Since $X_0 = 0$, the number Y_0 is the initial value of the process Y_t. Because $\ln(Y_t) = \ln(Y_0) + X_t$, and X_t is normally distributed, so is the r.v. $\ln(Y_t)$. The distributions of the r.v.'s whose logarithms are normal are said to be *log-normal distributions*; we already considered them in Section **9**.1.1.

The geometric Brownian motion is widely used in many areas; for example, for modeling investment processes. In Example **9**.1.1-5, we already discussed why the future value of an asset, for instance, a future stock price, may be closely approximated by a log-normal r.v.

Now, we show how the problem may be treated in continuous time. The approach we will use is based on an infinitesimal argument. Let B_0 be an investment into a risk-free asset. Suppose that the interest is compounded continuously at a rate δ, and let B_t be the result of investment at time t.[1] In this case, the relative growth over an infinitesimally small interval $[t, t+dt]$ is

$$\frac{dB_t}{B_t} = \delta dt, \tag{1.2.3}$$

which leads to the solution $B_t = B_0 e^{\delta t}$. (See also the Appendix, Section 3.)

Now, let S_0 be an investment into a risky asset and S_t be the corresponding result at time t. We may also view S_0 as an initial price (per share) of a security and S_t as the price at moment t. By analogy with (1.2.3), assume that

$$\frac{dS_t}{S_t} = m dt + \sigma dw_t, \tag{1.2.4}$$

where m, σ are parameters, and dw_t is the infinitesimal increment of Brownian motion over the infinitesimal interval $[t, t+dt]$. The difference between (1.2.3) and (1.2.4) is that in (1.2.4), we have added a random component.

Since the length of the interval $[t, t+dt]$ is dt, the variance of dw_t is equal to dt. So, for the r.v. dS_t / S_t, we have

$$E\{dS_t / S_t\} = m dt, \; Var\{dS_t / S_t\} = \sigma^2 dt.$$

It is natural to call m the *expected return* (per unit of time). The quantity σ is called a *volatility* and in this context, it is considered a measure of riskiness.

[1]The notation B is traditional in this context and comes after "bond." Regarding a risk-free rate, another traditional (and even more frequently used in Finance models) notation is r. We keep the notation δ we used in previous chapters and the Appendix, Section 3.

Solving (1.2.4) is not as easy as solving (1.2.3) since we cannot integrate (1.2.4) directly: as we know, w_t is not differentiable. The corresponding theory leads to the following solution:

$$S_t = S_0 \exp\{(m - \sigma^2/2)t + \sigma w_t\}, \tag{1.2.5}$$

that is, to the geometric Brownian motion with $\mu = m - \sigma^2/2$.

Derivations of (1.2.5) at different levels of rigor may be found in almost any textbook on Financial Mathematics (see, e.g., [11], [21], [48], [49]), [58]). All derivations are based on the famous differentiation formula obtained first by K. Ito. We omit the proof of (1.2.5) but will use it in examples below.

---

2 MARTINGALES IN CONTINUOUS TIME

2.1 General properties and examples

Definitions and properties below are very similar to those in Chapter 15; so we will be, to some degree, brief.

First, we presuppose the existence of a basic process ξ_t to which all other processes we will consider are adapted. That is, for any process X_t we consider below, and for any t, the trajectory $\{X_u;\ 0 \le u \le t\}$ is completely determined by the corresponding trajectory of the basic process $\{\xi_u;\ 0 \le u \le t\}$.

In practically all schemes below, the role of the basic process is played by a Wiener process w_t.

As in Chapter 15, we use the symbol ξ^t for the whole trajectory $\{\xi_u;\ 0 \le u \le t\}$, and the same symbolism applies to all other processes.

A process X_t in continuous time is called a *martingale* with respect to a basic process ξ_t if for any $t, s \ge 0$,

$$E\{X_{t+s} \mid \xi^t\} = X_t. \tag{2.1.1}$$

Similarly to what we did in Chapter 15, one may show that the martingale so defined is a martingale with respect to itself also. Below, if we say that a process is a martingale and do not refer to a basic process, we will mean that the process is a martingale with respect to itself.

As in Chapter 15, we will often use the game or investment interpretation and view X_t as the total capital of a player or investor by time t. In this case, definition (2.1.1) means that if t is the present time, then on the average, the future capital X_{t+s} is equal to what the player has already reached at time t.

EXAMPLE 1. Let ξ_t be a process with independent increments. Let $m_t = E\{\xi_t\}$ and $X_t = \xi_t - m_t$. Then X_t is a martingale. For instance, if N_t is a homogeneous Poisson process with parameter λ, then $X_t = N_t - \lambda t$ is a martingale (with respect to itself). Another example: since $E\{w_t\} = 0$, Brownian motion is a martingale.

Indeed, since $E\{X_t\} = 0$ for any $t \ge 0$, we have $E\{X_{(t,t+s]}\} = E\{X_{t+s}\} - E\{X_t\} = 0$. Then $E\{X_{t+s} \mid \xi^t\} = E\{X_t + X_{(t,t+s]} \mid \xi^t\} = X_t + E\{X_{(t,t+s]} \mid \xi^t\}$. Since the increment $X_{(t,t+s]}$ does not depend on ξ^t, we have $E\{X_{(t,t+s]}\,|\,\xi^t\} = E\{X_{(t,t+s]}\} = 0$, and hence $E\{X_{t+s}\,|\,\xi^t\} = X_t$. □

In the next proposition which is used in a number of problems below, the basic process is w_t.

Proposition 2 *Let a geometric Brownian motion $Y_t = Y_0 \exp\{\mu t + \sigma w_t\}$ where a positive Y_0 is certain, and let $\alpha = \mu + \sigma^2/2$. Then the process $Z_t = e^{-\alpha t} Y_t$ is a martingale with respect to w_t.*

Note that since there is a one-to-one correspondence between the processes Y_t, Z_t and w_t, it does not matter whether to condition on Y^t, or Z^t, or w^t, and hence Z_t is also a martingale with respect Y_t and itself. We will see that it is more convenient to condition on w_t.

Furthermore, $Z_t = Y_0 \exp\{-\mu t - (\sigma^2 t/2) + \mu t + \sigma w_t\} = Y_0 \exp\{-\sigma^2 t/2 + \sigma w_t\}$; that is, the μt terms cancel out. We use this fact in the proof below, but for future references, it is convenient to have Proposition 2 as it is stated above.

Proof. Since w_t is a process with independent increments,

$$\begin{aligned} E\{Z_{t+s}\,|\,w^t\} &= Y_0 \exp\{-\sigma^2(t+s)/2\} \cdot E\{\exp\{\sigma w_{t+s}\}\,|\,w^t\} \\ &= Y_0 \exp\{-\sigma^2(t+s)/2\} \cdot E\{\exp\{\sigma(w_t + w_{(t,t+s]})\}\,|\,w^t\} \\ &= Y_0 \exp\{-\sigma^2(t+s)/2\} \cdot \exp\{\sigma w_t\} \cdot E\{\exp\{\sigma w_{(t,t+s]}\}\,|\,w^t\} \\ &= Y_0 \exp\{-\sigma^2 t/2 + \sigma w_t\} \cdot \exp\{-\sigma^2 s/2\} \cdot E\{\exp\{\sigma w_{(t,t+s]}\}\} \\ &= Z_t \cdot \exp\{-\sigma^2 s/2\} \cdot E\{\exp\{\sigma w_{(t,t+s]}\}\}. \end{aligned} \tag{2.1.2}$$

The last expectation is equal to the value of the moment generating function of the r.v. $w_{(t,t+s]}$ at point σ. The r.v. $w_{(t,t+s]}$ is normal with zero mean and variance s. Hence [see (**8**.1.2.7), or the Appendix, Table 2], $E\{\exp\{\sigma w_{(t,t+s]}\}\} = \exp\{\sigma^2 s/2\}$, and (2.1.2) implies the assertion of the proposition. ■

EXAMPLE 2. Consider the stock price process S_t from (1.2.5). Suppose that there exists also a risk-free asset with the interest compounded continuously at a risk-free rate of δ. Then from the standpoint of time 0, the present value of the stock price at time t is $\widetilde{S}_t = e^{-\delta t} S_t$. (See the Appendix, Section 3.)

On the other hand, for the process S_t, the α from Proposition 2 equals $(m - \frac{1}{2}\sigma^2) + \frac{1}{2}\sigma^2 = m$. Consequently, because we multiply S_t by $e^{-\delta t}$, the process $\widetilde{S}_t$ is a martingale if and only if $m = \delta$.

Assume that it is so. Such a situation is often referred to as a "risk neutral world" for the following reason. Since $\widetilde{S}_t$ is a martingale,

$$E\{\widetilde{S}_{t+u}\,|\,\widetilde{S}^t\} = \widetilde{S}_t. \tag{2.1.3}$$

Suppose that the present time is the initial time $t = 0$, and we are speculating about possible values of the future price. When comparing the possible prices at two different future moments of time, t and $t+u$, we should not compare the prices themselves (S_t and S_{t+u}) but rather their present values from the standpoint of the initial time: $\widetilde{S}_t$ and $\widetilde{S}_{t+s}$.

The relation (2.1.3) means that in the risk neutral world, on the average, the present value keeps the level it has already reached. □

Next, we establish the counterpart of Proposition 1 from Chapter 15.

Proposition 3 *If X_t is a martingale, then for any $t \geq 0$,*

$$E\{X_t\} = E\{X_0\}. \tag{2.1.4}$$

Proof. Setting $t = 0$ in (2.1.1) and taking into account that $\xi^0 = \xi_0$, we have $E\{X_s \mid \xi_0\} = X_0$. Computing the expected values of both sides, we get that $E\{X_s\} = E\{X_0\}$. It remains to replace s by t. ■

EXAMPLE 3. Let us revisit Example 2. By Proposition 3, from (2.1.3) it follows that $E\{\widetilde{S}_t\} = E\{\widetilde{S}_0\}$. On the other hand, $\widetilde{S}_0 = S_0$, the initial price, and $\widetilde{S}_0$ is not random. Eventually, $E\{\widetilde{S}_t\} = S_0$. Thus, in the situation of Example 2, the prices themselves change even on the average, but the mean *present values* of the future prices do not change. □

2.2 Risk neutral evaluation. Black–Scholes formula

We will not go too deep into the model below. Our goal is to demonstrate possibilities of Martingale Theory. Let us consider again the stock process (1.2.5):

$$S_t = S_0 \exp\{(m - \sigma^2/2)t + \sigma w_t\}.$$

As was already told, the characteristic m is called an expected return and σ is called a volatility. The latter reflects the level of "uncertainty."

Let δ be a risk-free rate, and the present value process $\widetilde{S}_t = e^{-\delta t} S_t$. As was shown above, if $m = \delta$, then the process $\widetilde{S}_t$ is a martingale, and hence the corresponding probability measure on the trajectories of the process is a martingale measure. It is worth emphasizing that this measure may differ from the "real" probability measure (if any) characterizing the market. The reader may also look up the corresponding theory for discrete time in Section **15**.3.

For the sake of simplicity, we skip a rigorous statement of the *arbitrage theorem* in the continuous time case. Loosely put, it says that for the market to be arbitrage free, the prices of assets should be equal to the discounted expected values of the prices of these assets in *the risk neutral world.*

More precisely, let a process X_t present the evolution of the prices of an asset with an original price X_0, and suppose that X_t is adapted to the original price process S_t. This means that for any t, the evolution of the process X_t during the time interval $[0,t]$ is completely determined by the evolution of S_t during the same period. Then, for the market to be arbitrage free, it is necessary that for any t,

$$X_0 = e^{-\delta t} \hat{E}\{X_t\}, \tag{2.2.1}$$

where $\hat{E}\{\cdot\}$ denotes expectation in the risk neutral world; in our case, for $m = \delta$.

EXAMPLE 1 is similar to Example **15**.3.2-4. Now we will see how its continuous time counterpart looks. Consider the stock above. Suppose that a company issued tickets that allowed the buyers of these tickets to have—at a maturity date T—a \$200 dinner if the price $S_T > S_0$, and a \$2 hot dog otherwise. How much should such tickets cost at $t = 0$?

The r.v. X_T, the price of the ticket at time T, equals either 200 or 2. Let $\hat{S}_t = S_0 \exp\{(\delta - \sigma^2/2)t + \sigma w_t\}$, the stock price for the case $m = \delta$. By (2.2.1),

$$X_0 = e^{-\delta T}\left(200 \cdot P(\hat{S}_T > S_0) + 2 \cdot P(\hat{S}_T \leq S_0)\right) = e^{-\delta T}\left(200 - 198 \cdot P(\hat{S}_T \leq S_0)\right). \quad (2.2.2)$$

The inequality $\hat{S}_T \leq S_0$ is equivalent to $\exp\{(\delta - \sigma^2/2)T + \sigma w_T\} \leq 1$, or $(\delta - \sigma^2/2)T + \sigma w_T \leq 0$. Hence, $P(\hat{S}_T \leq S_0) = P\left(w_T \leq \frac{((\sigma^2/2) - \delta)T}{\sigma}\right)$. Since w_T is normal with zero mean and a variance of T, eventually,

$$P(\hat{S}_T \leq S_0) = \Phi\left(\frac{1}{\sqrt{T}} \cdot \frac{((\sigma^2/2) - \delta)T}{\sigma}\right) = \Phi\left(\frac{((\sigma^2/2) - \delta)\sqrt{T}}{\sigma}\right),$$

where $\Phi(\cdot)$ is the standard normal d.f. Then by (2.2.2),

$$X_0 = e^{-\delta T}\left(200 - 198 \cdot \Phi\left(\frac{((\sigma^2/2) - \delta)\sqrt{T}}{\sigma}\right)\right).$$

Assume that $T = 1$ year, the annual risk-free interest rate is 3%, and the annual volatility is 5%. Substituting $\delta = 0.03$ and $\sigma = 0.05$, we get $X_0 \approx \$139.8$. In Exercise 21, the reader is invited to consider the case $\delta = 0$. $\square$

Next, we derive the famous Black–Scholes formula for the prices of options with an exercise price of K and a maturity time of T. (For definitions of options, see Section **15**.3.)

In the case under consideration, at the maturity time T, the value of a call option is $X_T = S_T - K$ if $S_T > K$, and $S_T = 0$ otherwise. We write this as $X_T = (S_T - K)^+$, where the symbol x^+ stands for $\max\{x, 0\}$. Then, in accordance with (2.2.1), the option price is $c = e^{-\delta T} E\{(\hat{S}_T - K)^+\}$. For a put option with the same exercise price K, its price $p = e^{-\delta T} E\{(K - \hat{S}_T)^+\}$.

Proposition 4 *For the options above,*

$$c = S_0 \Phi(d_+) - K e^{-\delta T} \Phi(d_-), \quad (2.2.3)$$

$$p = K e^{-\delta T} \Phi(-d_-) - S_0 \Phi(-d_+), \quad (2.2.4)$$

where

$$d_{\pm} = \frac{\ln(S_0/K) + (\delta \pm \sigma^2/2)T}{\sigma\sqrt{T}}. \quad (2.2.5)$$

If $K = S_0$, then (2.2.5) becomes simpler: $d_{\pm} = \frac{1}{\sigma}(\delta \pm \sigma^2/2)\sqrt{T}$. We discuss some properties of the Black–Scholes formulas and consider examples in Exercises 22 and 23.

Proof of Proposition 4. The main idea is to use the process $\hat{S}_t$. In the rest of the proof, we just enjoy integration. As above, set $\hat{S}_T = S_0 \exp\{\mu T + \sigma w_T\}$, where $\mu = \delta - \sigma^2/2$. Then the r.v. $Z_T = \ln(\hat{S}_t/S_0))$ is normal with a mean of μT and a variance of $\sigma^2 T$. Furthermore,

$\hat{S}_T = S_0 e^{Z_T}$, and

$$\begin{aligned} e^{\delta T} \cdot c &= E\{(\hat{S}_T - K)^+\} = E\{(S_0 e^{Z_T} - K)^+\} \\ &= \int_{-\infty}^{\infty} (S_0 e^z - K)^+ \frac{1}{\sqrt{2\pi T}\sigma} \exp\{-(z-\mu T)^2/(2T\sigma^2)\} dz \\ &= \frac{1}{\sqrt{2\pi T}\sigma} \int_{\ln(K/S_0)}^{\infty} (S_0 e^z - K) \exp\{-(z-\mu T)^2/(2T\sigma^2)\} dz \end{aligned}$$

because $S_0 e^z - K \geq 0$ if $z \geq \ln(K/S_0)$. We use the change of variable $y = (z - \mu T)/(\sigma\sqrt{T})$. It is straightforward to calculate that the lower integration limit for y's is $-d_-$. Then,

$$e^{\delta T} \cdot c = \frac{1}{\sqrt{2\pi}} S_0 e^{\mu T} \int_{-d_-}^{\infty} e^{y\sigma\sqrt{T}} e^{-y^2/2} dy - K \frac{1}{\sqrt{2\pi}} \int_{-d_-}^{\infty} e^{-y^2/2} dy.$$

The second term equals $K(1 - \Phi(-d_-)) = K\Phi(d_-)$. Furthermore, completing the square in the exponent of the integrand below and making the change variable $u = y - \sigma\sqrt{T}$, we have

$$\begin{aligned} &\frac{1}{\sqrt{2\pi}} \int_{-d_-}^{\infty} e^{y\sigma\sqrt{T}} e^{-y^2/2} dy = e^{T\sigma^2/2} \frac{1}{\sqrt{2\pi}} \int_{-d_-}^{\infty} e^{-(y-\sigma\sqrt{T})^2/2} dy \\ &= e^{T\sigma^2/2} \frac{1}{\sqrt{2\pi}} \int_{-d_- - \sigma\sqrt{T}}^{\infty} e^{-u^2/2} du = e^{T\sigma^2/2} (1 - \Phi(-d_- - \sigma\sqrt{T}) \\ &= e^{T\sigma^2/2} \Phi(d_- + \sigma\sqrt{T}) = e^{T\sigma^2/2} \Phi(d_+), \end{aligned}$$

because, as is straightforward to verify, $d_- + \sigma\sqrt{T} = d_+$.

Combining all of this, and recalling that $\mu = \delta - \sigma^2/2$, we get

$$c = e^{-\delta T} \left(S_0 e^{(\delta - \sigma^2/2)T} e^{T\sigma^2/2} \Phi(d_+) - K\Phi(d_-) \right) = S_0 \Phi(d_+) - K e^{-\delta T} \Phi(d_-).$$

The relation (2.2.4) is proved similarly. ■

2.3 Optional stopping time and some applications

2.3.1 Definitions and examples

The ideas and results below are similar to those in Chapter 15.

First, we presuppose the existence of a basic process ξ_t to which all other processes under consideration are adapted.

Regarding optional (or Markov) moments, the continuous time case is somewhat more complicated because for each time moment t, we should take into account the behavior of the process in an infinitesimal neighborhood of t. Without going too deeply into the theory, we just state that in this case, it is reasonable to define an *optional time* τ as a r.v. such that

The event $\{\tau < t\}$ is completely determined by the values of ξ^t, i.e., the trajectory ξ_s for all $s \in [0, t]$. (2.3.1)

The fact that the point t is not included in the event $\{\tau < t\}$ allows us to avoid some technical difficulties. However, this circumstance is not essential. For a "good process" in continuous time, the probability that a certain event will happen *exactly* at a fixed time t, as a rule, is zero. For example, for the Poisson process, the probability that a new arrival will occur exactly at a fixed time t, is zero (a point is an interval of zero length).

The definition of stopping time is the same as in the discrete time case. The same concerns

$$\text{The martingale stopping property: } E\{X_\tau\} = E\{X_0\}. \tag{2.3.2}$$

Theorem 5 *Let X_t be a martingale and τ be a stopping time. Then the martingale stopping property (2.3.2) holds if at least one of the following conditions is true.*

1. *There exists a constant c such that the r.v. $\tau \leq c$ with probability one.*
2. *There exists a constant C such that the r.v. $|X_t| \leq C$ for all $t \leq \tau$ with probability one.*

See, e.g., [24], [43], [48], [57]. As an example, consider

2.3.2 The ruin probability for the Brownian motion with drift

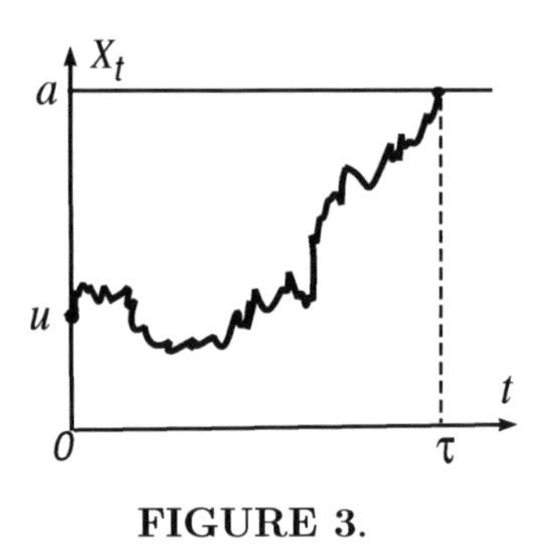

FIGURE 3.

Set $X_t = u + \mu t + \sigma w_t$, where w_t is the standard Brownian motion, $\sigma > 0$. So, the initial value $X_0 = u$.

Assume $u \in [0, a]$, where $a > 0$, and suppose that the process, starting from the point $(0, u)$, evolves until the first moment when it reaches the boundary of the "corridor" $[0, a]$. See Fig. 3; for the particular trajectory in this figure, the process first reaches the upper level a. Denote by p_u and q_u the probabilities that the process will reach the boundary, and it will occur, respectively, at the upper level a or at the lower level 0.

One may conjecture that the process will reach the boundary with unit probability:

$$p_u + q_u = 1. \tag{2.3.3}$$

This is indeed true and is stated in Proposition 6 below.

Let $\tau = \min\{t : X_t = 0 \text{ or } a\}$, the first moment when the process reaches the boundary of the "corridor"; see Fig. 3. If (2.3.3) had not held, then τ would have been equal to infinity with a positive probability, but as was told, it is not true.

By the definition of τ, the probabilities $p_u = P(X_\tau = a)$, and $q_u = P(X_\tau = 0)$.

EXAMPLE 1. As we did when dealing with random walk, let us view X_t as a model of the evolution of the capital of a company. The company runs until either its capital reaches a certain amount a, or the company is ruined. Then τ is the lifetime of the company and q_u is the probability that the company will be ruined before it reaches the level planned. □

Proposition 6 *For $\mu = 0$,*

$$p_u = u/a. \tag{2.3.4}$$

For $\mu \neq 0$,

$$p_u = \frac{1 - e^{-\gamma u}}{1 - e^{-\gamma a}}, \text{ where } \gamma = \frac{2\mu}{\sigma^2}. \tag{2.3.5}$$

In both cases, $q_u = 1 - p_u$. In the former, $q_u = 1 - u/a$, in the latter, $q_u = \frac{e^{-\gamma u} - e^{-\gamma a}}{1 - e^{-\gamma a}}$.

It makes sense to note that μ in (2.3.5) may be positive or negative as well. Keeping in mind that $0 \leq u \leq a$, the reader may double-check that, nevertheless, in both cases, $0 \leq p_u \leq 1$.

It is also noteworthy that the expression in (2.3.5) converges to that in (2.3.4) as $\mu \to 0$ (one may use L'Hópital rule).

In the case where there is no upper barrier, that is, when $a = \infty$, the probability p_u is the probability that the process will never reach zero, and, accordingly, q_u is the ruin probability.

Corollary 7 *If $a \to \infty$, then $p_u \to 0$, $q_u \to 1$ for $\mu \leq 0$; and $p_u \to 1 - e^{-\gamma u}$ for $\mu > 0$. Thus, for $\mu > 0$, and $a = \infty$, the ruin probability*

$$q_u = 1 - p_u = e^{-\gamma u} = \exp\{-2u\mu/\sigma^2\}. \tag{2.3.6}$$

The reader is invited to derive Corollary 7 from Proposition 6 on her/his own.

EXAMPLE 1 revisited. Suppose that the growth of the company per unit time (say, a year) is a r.v. with a mean of \$150,000 and a standard deviation of \$15,000. If Δ is a unit time interval, then the increment $X_\Delta = \mu + \sigma w_\Delta$, and w_Δ has a standard normal distribution. Thus, $\mu = 150\text{K}$, and $\sigma = 15\text{K}$. Then, by (2.3.6), as is easy to compute, $q_u = \exp\{-4u/3\}$. Assume that the company wishes the ruin probability to be less than or equal to 1%. It will be the case if $\exp\{-4u/3\} \leq 0.01$, which leads approximately to $u \geq 3.46$ or \$3,460. □

Proof of Proposition 6. First, let $\mu = 0$. In this case, $X_t = u + \sigma w_t$, and the probability that during a time interval $[0, T]$, the process X_t will never reach the level a equals

$$P(\max_{t \leq T} X_t \leq a) = P(\max_{t \leq T} w_t \leq (a-u)/\sigma) = 2\Phi\left(\frac{a-u}{\sigma\sqrt{T}}\right) - 1$$

in accordance with (1.1.8). This probability converges to $2\Phi(0) - 1 = 0$ as $T \to \infty$, which means that the process will reach the level a with probability one. Thus, τ is a stopping time, which in particular implies (2.3.3).

The process X_t is a martingale (see Example 2.1-1 and Exercise 16), and $0 \leq X_t \leq a$ for $t \leq \tau$. Hence, Condition 2 of Theorem 5 holds and $E\{X_\tau\} = E\{X_0\} = u$. On the other hand, $E\{X_\tau\} = ap_u + 0(1 - p_u) = ap_u$, where $p_u = P(X_\tau = a)$. This leads to $p_u = u/a$.

Let $\mu \neq 0$. By the LLN, with probability one, $X_t \to \infty$ if $\mu > 0$, and $X_t \to -\infty$ if $\mu < 0$. This means that with probability one, the process will exit the "corridor" $[0, a]$, and hence τ is a stopping time.

Next, we use the following nice technique. Let $W_t = \exp\{s(\mu t + \sigma w_t)\}$, where s is a number. The exponent is the Brownian motion $\widetilde{\mu} t + \widetilde{\sigma} w_t$, where the drift $\widetilde{\mu} = s\mu$ and the parameter $\widetilde{\sigma} = s\sigma$. We apply to this process Proposition 2. In our case, the characteristic α from this proposition is equal to $\widetilde{\alpha} = \widetilde{\mu} + \widetilde{\sigma}^2/2 = s(\mu + s\sigma^2/2)$. Now, we choose s for which $\widetilde{\alpha} = 0$, that is, we set $s = -\gamma$, where $\gamma = 2\mu/\sigma^2$. For the s chosen, by Proposition 2, the process W_t is a martingale.

If we multiply a martingale by a constant, the martingale property continues to hold (why?). Hence, the process $Y_t = e^{-\gamma u}W_t = \exp\{-\gamma(u + \mu t + \sigma w_t)\} = \exp\{-\gamma X_t\}$ is also a martingale.

Since $0 \leq X_t \leq a$ for $t \leq \tau$, the values of Y_t lie between 1 and $e^{-\gamma a}$, and Condition 2 of Theorem 5 holds. Finally, we have

$$E\{Y_\tau\} = E\{Y_0\} = e^{-\gamma u}.$$

On the other hand, $E\{Y_\tau\} = e^{-\gamma a}p_u + 1 \cdot (1 - p_u) = 1 + p_u(e^{-\gamma a} - 1)$. This readily leads to (2.3.5). ■

2.3.3 The distribution of the ruin time in the case of Brownian motion

Next, we consider the model of the previous section in the particular case $a = \infty$. So, again $X_t = u + \mu t + \sigma w_t$, $\sigma > 0$, the initial point $u \in [0, \infty)$, and $\tau = \min\{t : X_t = 0\}$. However, now we solve a more difficult problem; namely, we will find the distribution of the r.v. τ. We have already seen in Corollary 7 that for $\mu \leq 0$, this r.v. is a stopping time, i.e., $P(\tau < \infty) = 1$, while for $\mu > 0$, this is not the case because $\tau = \infty$ with a positive probability. As was already said, such r.v.'s are called *improper* or *defective*.

Proposition 8 *For $\sigma > 0$, all μ, and $t > 0$,*

$$P(\tau \leq t) = \Phi\left(-\frac{u}{\sigma\sqrt{t}} - \frac{\mu\sqrt{t}}{\sigma}\right) + \exp\left(-\frac{2\mu}{\sigma^2}u\right)\Phi\left(-\frac{u}{\sigma\sqrt{t}} + \frac{\mu\sqrt{t}}{\sigma}\right), \tag{2.3.7}$$

where $\Phi(x)$ is the standard normal d.f.

Some comments. Certainly, $P(\tau \leq 0)$ should be equal to 0 if $u > 0$. The reader is invited to check that, indeed, the r.-h.s. of (2.3.7) converges to zero as $t \to 0$ for any positive u.

The reader can also verify that the r.-h.s. of (2.3.7) converges to one as $t \to \infty$ for $\mu \leq 0$, and the same limit equals $\exp\left(-\frac{2\mu}{\sigma^2}u\right)$ for $\mu > 0$. As should be expected, the last formula coincides with (2.3.6).

Now, let us set $\mu = 0$, and replace the symbol u by a. So, $X_t = a + w_t$. The process w_t is symmetric, and the distribution of the time it takes to reach zero starting at a, is equal to the distribution of the time it takes to reach a starting from zero. Hence, in our case, the r.v. τ has the same distribution as τ_a in (1.1.7). So, (2.3.7) must imply (1.1.7). This is indeed true. In the case in hand, $P(\tau \leq t) = \Phi\left(-\frac{a}{\sigma\sqrt{t}}\right) + \Phi\left(-\frac{a}{\sigma\sqrt{t}}\right) = 2\Phi\left(-\frac{a}{\sigma\sqrt{t}}\right) = 2\left(1 - \Phi\left(\frac{a}{\sigma\sqrt{t}}\right)\right)$.

Examples of direct applications of this theorem are given in Exercise 27. An example concerning a hitting time is considered in the next section.

Proof of Proposition 8. We restrict ourselves to a short but somewhat non-rigorous proof in order to demonstrate again the usefulness of the martingale stopping property (2.3.2).

As in Section 2.3.2, consider the process $W_t = \exp\{s(\mu t + \sigma w_t)\}$, where s plays the role of a free parameter. We have seen in Section 2.3.2 that the characteristic α from Proposition

2 in this case is $\widetilde{\alpha} = s\mu + s^2\sigma^2/2$. So, the process $\exp\{-\widetilde{\alpha}t + s(\mu t + \sigma w_t)\}$ is a martingale. Multiplying it by e^{su}, which does not change the martingale property, we come to the process $Y_t = \exp\{-\widetilde{\alpha}t + su + s(\mu t + \sigma w_t)\} = \exp\{-\widetilde{\alpha}t + sX_t\}$ which is also a martingale.

The lack of rigor in the next step is due to the fact that we apply property (2.3.2) to the optional time τ which for $\mu > 0$ is an improper (defective) r.v., that is, not a stopping time. The justification of this step requires some work and we omit it here. (One way is to provide calculations for the barrier $a < \infty$ and let $a \to \infty$ at the very end.)

Since $X_0 = u$, making use of the martingale stopping time property, we have $E\{Y_\tau\} = E\{Y_0\} = e^{su}$. On the other hand, $E\{Y_\tau\} = E\{\exp\{-\widetilde{\alpha}\tau\}\}$, since $X_\tau = 0$ by the definition of τ. Thus, $E\{\exp\{-\widetilde{\alpha}\tau\}\} = e^{su}$, or

$$E\{\exp\{-(s\mu + s^2\sigma^2/2)\tau\}\} = \exp\{su\}. \tag{2.3.8}$$

Consider a $z \geq 0$, set $s\mu + s^2\sigma^2/2 = z$, and solve this as an equation for s. We choose the negative root

$$s = -\frac{1}{\sigma^2}\left(\mu + \sqrt{\mu^2 + 2z\sigma^2}\right). \tag{2.3.9}$$

(The expression in the parentheses is positive whatever μ is: positive or negative. If we had chosen the positive root, the r.-h.s. of (2.3.8) would have been larger than one, while the l.-h.s. would have been less than one.) Then we can rewrite (2.3.8) as

$$E\{e^{-z\tau}\} = \exp\left\{-\frac{u}{\sigma^2}\left(\mu + \sqrt{\mu^2 + 2z\sigma^2}\right)\right\}. \tag{2.3.10}$$

The l.-h.s. is the m.g.f. of τ; more precisely, $M_u(-z)$ if $M(z)$ is the m.g.f. of τ. We should only remember that if τ is improper, then $M_u(0) = P(\tau < \infty) < 1$.

The verification that (2.3.10) is indeed the m.g.f. of the distribution (2.3.7) is lengthy but consists in pure integration, so we can turn to tables of integrals, for example, in [1], [18]. At least, what we have done completes the probabilistic part of the problem. ■

2.3.4 Hitting time for the Brownian motion with drift

Next, we connect Proposition 8 with the problem of hitting time. Let $X_t = \mu t + \sigma w_t$, and let $\tau_a = \min\{t : X_t = a\}$, the time of hitting a level $a > 0$. It makes sense to emphasize that now we consider only one barrier. Assume $\mu > 0$. Then X_t is a process with a positive drift, and it will reach the level a with probability one. (Proceeding from the LLN, show that $P(X_t > a) \to 1$ as $t \to \infty$ for *any a*.)

The event $\{X_t = a\} = \{\mu t + \sigma w_t = a\} = \{a - \mu t - \sigma w_t = 0\}$. Since w_t is a symmetric r.v., the distribution of the process $-\sigma w_t$ coincides with the distribution of σw_t. Hence, the probabilities of all events we consider will not change if we replace the process $-\sigma w_t$ by σw_t.

Thus, we may consider the first time t when the process $\widetilde{X}_t = a - \mu t + \sigma w_t$ will reach zero level. This is the ruin time for the process $\widetilde{X}_t$. In order to use the results of the previous section, in these results, we should replace μ by $-\mu$ and set $u = a$. So, we have arrived at

Corollary 9 *For the process $X_t = \mu t + \sigma w_t$ above,*

$$P(\tau_a \leq t) = \Phi\left(-\frac{a}{\sigma\sqrt{t}} + \frac{\mu\sqrt{t}}{\sigma}\right) + \exp\left(\frac{2\mu}{\sigma^2}a\right)\Phi\left(-\frac{a}{\sigma\sqrt{t}} - \frac{\mu\sqrt{t}}{\sigma}\right). \tag{2.3.11}$$

EXAMPLE 1. We generalize Example 1.1.3-1. As in this example, suppose that an investor owns a stock whose current price is $S_0 = 100$, and the price changes in time as the process $S_t = S_0 \exp\{\mu t + \sigma w_t\}$, where the expected return $\mu = 0.1$ and the volatility $\sigma = 0.15$. The difference is that now we consider a non-zero drift. The investor decides to sell her/his shares when the price increases by 10%. Find the probability that this will not happen within the first year.

The investor will sell the stock at the first time when $S_t \geq 1.1S_0$. The last inequality is equivalent to $\exp\{\mu t + \sigma w_t\} \geq 1.1$, or $\mu t + \sigma w_t \geq a = \ln(1.1)$. It remains to use (2.3.11) with $t = 1$, $\mu = 0.1, \sigma = 0.15$. Calculations give $P(\tau_a \leq 1) \approx 0.74$, and $P(\tau_a > 1) \approx 0.26$. □

---

3 EXERCISES

Section 1

1. Provide an Excel worksheet illustrating the invariance principle of Section 1.1.2. To do this, simulate values of some independent r.v.'s, for example, exponential or uniform, construct sums S_k, and then the process $X_t^{(n)}$ (consider only points $t = k/n$). Provide charts with the graphs of particular realizations of $X_t^{(n)}$. Pay attention how the graphs look for large n.

2. Let w_t be Brownian motion. (a) Are w_1 and w_5 independent r.v.'s? (b) Answer the same question for w_1 and $w_{(1,5]}$. (c) Find $P(w_1 + w_5 < 4)$.

3. Let w_{t1} and w_{t2} be independent Brownian motions. Find σ for which the process $w_{t1} + 2w_{t2}$ has the same distribution as σw_{t1}.

4. Continuing Exercise 3, consider the process $x_t = \sigma_1 w_{t1} + \sigma_2 w_{t2}$, where σ_1, σ_2 are numbers. Let $\sigma^2 = \sigma_1^2 + \sigma_2^2$. Show that the process x_t/σ is a standard Brownian motion, and hence x_t may be represented as σw_t, where w_t is a standard Brownian motion. Next, consider the case $\sigma_1 = -1, \sigma_2 = 0$. Explain why the fact that $-w_t$ is Brownian motion is almost obvious. (*Advice*: First, show that x_t is the process with independent increments. Secondly, consider the distribution of x_Δ.)

5. A physical process evolves as w_t. You decided to measure time in different units setting $t = as$, where a is a fixed scale parameter and s is time measured in the new units. Argue that you cannot model the original process by w_s. Find c for which the process cw_s would be an adequate model for the same physical process.

6. Assume that in the situation of Example 1.1.3-1, you decided to sell the stock not when the price increases by 10%, but when it drops by 10%. Will the answer change? If yes, find it. If no, justify the answer.

7. The prices for two stocks evolve as the processes $S_{t1} = 10\exp\{\sigma_1 w_{t1}\}$ and $S_{t2} = 11\exp\{\sigma_2 w_{t2}\}$, where w_{t1} and w_{t2} are independent Brownian motions, $\sigma_1 = 0.1$, and $\sigma_2 = 0.2$. What are the initial prices for the stocks? Find the probability that during a year the price S_{t1} will meet the price S_{t2}. (*Advice*: Use the result of Exercise 4.)

8. Find $Corr\{w_t, w_s\}$ and show that it vanishes as $t \to \infty$ for any fixed s.

9. Show that $\widetilde{w}_t$ and $|w_t|$ have the same distribution.

10. (a) Compute the probability density of the r.v. τ_a from Section 1.1.3. Show that $E\{\tau_a\} = \infty$.

    (b) Proceeding from the invariance principle, connect heuristically this fact with what we obtained in Example **15**.2.2-2: for the symmetric random walk, the mean time of reaching a level a equals infinity.

11. Prove that $E\{\widetilde{w}_t\} = \sqrt{2t/\pi}$. Show that to obtain this answer, it suffices to consider the case $t = 1$. Is the expected value of the maximum value of Brownian motion on the interval $[0,2]$ twice as large as that on the interval $[0,1]$?

12. For any non-random interval Δ, the r.v. w_Δ is normal. Is it true if Δ is random? Give an example. Is the r.v. $w_{(\tau_a,\, t]}$ normal? Look also at the proof of (1.1.6).

13. Customers are calling to a company, and each customer is equally likely to be a man or a woman. Using the Brownian motion approximation, estimate the probability that during the first $n = 100$ calls, the difference between the numbers of men and women will not exceed $a = 10$. (*Advice*: When applying the invariance principle from Section 1.1.2, set $\xi_k = \pm 1$. Note also that the inequality $\max_{k \le n} S_k \le b$ may be rewritten as $\max_{k \le n}(S_k/\sqrt{n}) \le b/\sqrt{n}$, and it is reasonable to set $b = a\sqrt{n}$.)

14. Find $E\{Y_t\}$ and $Var\{Y_t\}$ for the geometric Brownian motion Y_t defined in Section 1.2. (*Advice*: Use the formula for the m.g.f. of the normal distribution.)

15. Suppose that the "chaotic" motion of a particle along a line is depicted by a Markov process Y_t with continuous trajectories. Let $p(y\,|\,x,t)$ be the density of Y_t under the initial-time condition $Y_0 = x$. A. Einstein has shown that the function $p(y\,|\,x,t)$ should satisfy the equation

$$\frac{\partial p}{\partial t} = \frac{1}{2}\sigma^2 \frac{\partial^2 p}{\partial x^2} \tag{3.1}$$

which is called the diffusion equation with a diffusion coefficient σ^2. Show that the density of the r.v. $Y_t = x + w_t$ satisfies (3.1) for $\sigma = 1$. Modify Y_t to make this true for $\sigma \neq 1$.

Section 2

16. Show that any process ξ_t with independent increments such that $E\{\xi_\Delta\} = 0$ for any interval Δ, is a martingale. Consider, as an example, w_t. Explain why the assertion of this exercise is formally more general than the assertion of Example 2.1-1 (though it is very close).

17. Let N_t be a non-homogeneous Poisson process. Is N_t a martingale? Is the process $X_t = N_t - E\{N_t\}$ a martingale (with respect tow_t)?

18. Show that any process X_t such that $E\{X_{(t,t+\delta]}\,|\,X^t\} = 0$ for any t and $\delta \ge 0$ is a martingale.

19. Let $Y_t = C_t e^{4w_t}$. What should the constant C_t be for Y_t to be a martingale?

20. Let a stock price process S_t be a geometric Brownian motion with an annual volatility of $\sigma = 0.1$, and an expected annual return of $m = 0.15$ per annum. Let $S_0 = 100$.

    (a) Write a basic differential equation for S_t and a solution to this equation.

    (b) Find the probability that at time $T = 6$ months, the price will be 5% larger than the original price. Does the answer depend on the original price?

    (c) Find the probability that a call option with an exercise price of \$101 will be exercised at $T = 6$ months.

    (d) Do the same for a put option with the same exercise price.

(e) Assume now that the *current* time is $t < T = 6$ months, and we do know the price S_t. We are thinking about the future price S_T. Write a representation for S_T.

(f) Consider a call option with an exercise price of K and the maturity time T. The *current* time is $t < T$. What is the probability that (*from the standpoint of the time t*) the option will be exercised? (*Hint*: Certainly, the answer is not a number but an expression involving t and S_t, K, and T.)

21. Make sure that the answer in Example 2.2-1 becomes much simpler if $\delta = 0$.

22. This exercise concerns some properties of the Black-Scholes formulas (2.2.3) and (2.2.4).

(a) Consider a call option. If T is very small, the option will be ready to be exercised practically immediately, and hence its price should be close to $(S_0 - K)^+$. Show that, indeed, the price in (2.2.3) converges to this limit as $T \to 0$.

(b) If $\sigma \to 0$, the stock is becoming riskless, and in the risk neutral world, its price will grow to $S_0 e^{\delta T}$. Then, at time T, the value of the call option will be $(S_0 e^{\delta T} - K)^+$, and the present value is $e^{-\delta T}(S_0 e^{\delta T} - K)^+ = (S_0 - Ke^{-\delta T})^+$. Show that the Black-Scholes formula is consistent with this reasoning.

(c) Carry out the reasoning analogous to those in Exercises 22a and 22b for a put option and find the corresponding limits.

(d) If S_0 is very large while K is not, the put option is unlikely to be exercised, so the present value of the put option is close to zero. Show that this is consistent with the Black-Scholes formula.

23. Consider the stock described in Exercise 20 and assume the risk-free rate $\delta = 0.05$. (a) Find the prices for a put and call option with an exercise price of $K = \$100$ and maturity time $T = 6$ months. Show why the general formula becomes simpler in this case. (b) Richard has an \$100 bet with John that at time $T = 6$ months the stock price will be 5% larger than the original. Who to whom and how much should pay at time $t = 0$, for the bet to be fair?

24. Find the limit of (2.3.5) as $\mu \to 0$ and σ fixed. Interpret the answer.

25. Find the probability that the Wiener process w_t will ever hits the line $c + bt$ where c and b are positive numbers.

26. A capital growth process is well approximated by the Brownian motion with drift $X_t = u + \mu t + \sigma dw_t$ from Section 2.3.2. The ruin probability for a certain choice of parameters occurs to be $\frac{1}{16}$. How will this probability change if (a) the parameter σ is doubled; (b) the process X_t is multiplied by 2; (c) the initial surplus is doubled ?

27. The level of a water reservoir changes, in appropriate units, accordingly to the process $X_t = 1 + 2t + 3w_t$, where w_t is a standard Brownian motion. (Certainly, X_t may be negative, so this is just an approximation.) Find the probability that the reservoir (a) will never be empty, (b) will not be empty during the first two units of time.

28. Assume that in the situation of Example 2.3.4-1, you decided to sell the stock not when the price increases by 10% but when it decreases by 10%. Will the answer change? If yes, find it. If no, justify the answer. Is the selling time proper (non-defective) in this case? Find the probabilities that the price will never drop by 10%, and will not drop by 10% during the first year.

29. The prices for two stocks change as the processes $S_{t1} = 10\exp\{\mu_1 t + \sigma_1 w_{t1}\}$ and $S_{t2} = 11\exp\{\mu_2 t + \sigma_2 w_{t2}\}$, where w_{t1} and w_{t2} are independent Brownian motions, $\mu_1 = 0.15$, $\mu_2 = 0.11$, $\sigma_1 = 0.15$, and $\sigma_2 = 0.1$. What are the initial prices for the stocks? Find the probability that during a year the price S_{t1} will meet the price S_{t2}.

Chapter 17

More on Dependency Structures: Classification and Limit Theorems

2

There exists the opinion that the main difference between the general Measure Theory and Probability Theory is that the latter involves the notion of independency. Actually, it is disputable that this is the main difference: along with purely mathematical schemes, Probability Theory is concerned with a great many models of real phenomena, and the "common-sense significance" of abstract schemes, the physical or economic interpretation of results, to a large extent distinguish Probability from other mathematical disciplines. Nevertheless, the notion of independency is indeed a distinctive feature.

Another matter is that this does not mean that we should concentrate on models involving independent variables. On the contrary, such models, though important, represent ideal simplified situations, and one of important tasks of Probability Theory is the study of various types of dependency.

In this chapter, we continue considering dependency structures, providing a classification and presenting some dependency characteristics.

In the case of not just two but many r.v.'s, one may distinguish two types of dependencies.

In the first, different groups of r.v.'s—say, different pairs of r.v.'s—may have essentially different joint distributions. For example, for some pairs, the dependency may be strong, while for others—weak.

As an example, consider a homogeneous ergodic Markov chain $X_0, X_1, X_2, \ldots$. In Section 5.4, we have shown that the dependence of X_t on the initial state X_0 is vanishing as $t \to \infty$. Since for a homogeneous chain, any time moment s may be viewed as initial, the dependence of X_t on X_s is also vanishing as $t - s \to \infty$. In other words, the dependence between the r.v.'s X_t and X_s may be strong if the time moments t and s are close, and it is weak when t and s are far away from each other.

Let us also recall that in the case of a homogeneous Markov chain, the probability distribution of the process is completely determined by the initial distribution and the one-step transition probabilities. Suppose that we somehow permuted the r.v.'s; say, instead of the sequence X_0, X_1, X_2, X_3, X_4, we consider the sequence X_2, X_1, X_4, X_3, X_0. The r.v.'s are the same, just being considered in a different order, but now the Markov structure is distorted, and a formal presentation of the joint distribution of the latter sequence may not reveal the main pattern of the dependency.

So, the order in which we consider the r.v.'s, or how they are arranged, matters: it allows one to present the distribution in an explicit way.

Let us turn to the second type of dependency structures. In this case, the joint distribution of the r.v.'s is symmetric (or close to that) in the sense that the change of the order in which the r.v's are being considered, does not change the joint distribution of the r.v.'s (or changes it to a small extent). In particular, all pairs of the r.v.'s have the same (or almost the same) joint distribution. For example, one can watch such a property when the dependency between the r.v.'s is specified by one common factor having the same impact on all r.v.'s.

A good example is the scheme of Example **6**.5.2-2 where we considered a sequence of trials with the same but random probability of success. Namely, we select at random a number θ from an interval $[0,1]$ and arrange a sequence of n independent (after the θ has been selected) trials with the probability of success equal to θ.

Let $X_j = 1$ or 0, $j = 1,...,n$, be the indicator of success in the jth trial. The r.v.'s $X_1,...,X_n$ are dependent (since all of them depend on the choice of θ), but their joint distribution is symmetric. In particular, the joint distribution of any pair (X_i, X_j), $i \neq j$, are the same; say, (X_1, X_2) and (X_1, X_n) have the same joint distribution. See Example **6**.5.2-2 for more detail.

In Sections 1 and 2, we consider the first type mentioned, and in Section 3— the versions of the central limit theorem (CLT) for some dependencies of this type. Section 4 is concerned with the second type of dependency and the corresponding limit theorems.

Our goal is to give a preliminary idea about possible dependency structures, and mostly we will restrict ourselves to typical examples. So, the exposition is somewhat fragmentary.

---

1 ARRANGEMENT STRUCTURES AND THE CORRESPONDING DEPENDENCIES

Besides Markov chains, we considered a number of other collections of r.v.'s $\{X_t\}$ where t might be interpreted as a time parameter, and the description of the dependency was strongly connected with the corresponding arrangement of the r.v.'s. Cases in point are martingales, birth-death processes, etc.

One of the ways to generalize such schemes is to consider the collection of r.v.'s $\{X_{\boldsymbol{v}}\}$, where the parameter $\boldsymbol{v}$ is of a general nature and, in particular, is not interpreted any more as a time parameter. Say, we are interested in the tomorrow highest air temperature at different locations of a region, $\boldsymbol{v} = (v_1, v_2)$ stands for the coordinates of locations, and $\{X_{\boldsymbol{v}}\}$ is the (future) temperature at the location with the coordinates (v_1, v_2).

In another typical scheme, $\boldsymbol{v}$ denotes a vertex in a graph, and $\{X_{\boldsymbol{v}}\}$ is a r.v. associated with the corresponding vertex.

Suppose, for instance, a model of an economy deals with a fixed number of economic agents (companies, individuals), and the relations between the agents are specified by a graph. Namely, we identify the agents with the vertices, and each edge of the graph designates a "connection" between the corresponding agents. The r.v.'s or r.vec.'s $X_{\boldsymbol{v}}$ in this case may represent (random) characteristics of the agents, like the amounts of commodities the agents possess, their demands, etc.

Coming back to the general scheme of r.v.'s defined on a graph, note that when we connect the dependency structure with the arrangement of the r.v.'s, it may be natural to assume that the dependence between the r.v.'s corresponding to close vertices is stronger, in a sense, than that between the r.v.'s associated with "distant" from each other vertices. In the next section, we discuss one of typical examples—what is called Markov fields.

1.1 Markov random fields

Consider a graph characterized by a set of vertices and edges. We restrict ourselves to undirected graphs (edges do not have directions). A particular graph of this type is presented in Fig. 1.

In an undirected graph, two vertices are said to be neighbors if there is an edge connecting them. For example, in Fig. 1, the vertex $\boldsymbol{v}_1$ has two neighbors: $\boldsymbol{v}_2$ and $\boldsymbol{v}_4$, while the vertex $\boldsymbol{v}_9$ has five neighbors: $\boldsymbol{v}_5$, $\boldsymbol{v}_6$, $\boldsymbol{v}_8$, $\boldsymbol{v}_{10}$, and $\boldsymbol{v}_{11}$.

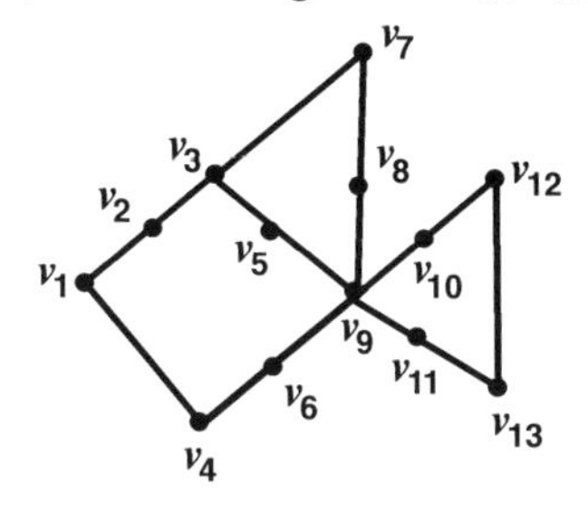

FIGURE 1.

Let us come back to the general scheme. To each vertex $\boldsymbol{v}$, we assign a r.v. $X_{\boldsymbol{v}}$, and call the collection $\{X_{\boldsymbol{v}}\}$ a *random field* defined on a graph.

Denote by $\mathcal{N}_{\boldsymbol{v}}$ the set of all neighbors of a vertex $\boldsymbol{v}$.

A random field $\{X_{\boldsymbol{v}}\}$ is said to be a *Markov random field* if for any vertex $\boldsymbol{v}$, *given* the values of the r.v.'s corresponding to the neighbor vertices from $\mathcal{N}_{\boldsymbol{v}}$, the probability distribution of $X_{\boldsymbol{v}}$ does not depend on the values of the r.v.'s corresponding to the other vertices. Formally, for any set B,

$$P(X_{\boldsymbol{v}} \in B \,|\, X_{\boldsymbol{u}} \text{ for all } \boldsymbol{u} \neq \boldsymbol{v}) = P(X_{\boldsymbol{v}} \in B \,|\, X_{\boldsymbol{u}} \text{ for all } \boldsymbol{u} \in \mathcal{N}_{\boldsymbol{v}}). \tag{1.1.1}$$

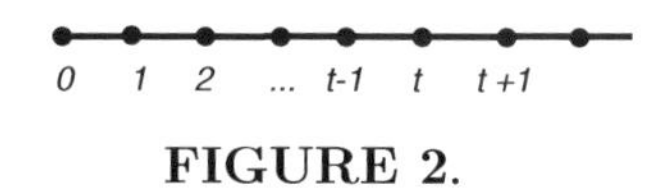

FIGURE 2.

Note that a "usual" Markov chain X_t in discrete time $t = 0, 1, \ldots$ may be viewed as a Markov chain defined on a graph whose vertices are integer points on a line; see Fig. 2. Since in this case any vertex, save the first, has exactly two neighbors, condition (1.1.1) may be rewritten as (see also Fig. 2)

$$P(X_t \in B \,|\, X_u \text{ for all } u \neq t) = P(X_t \in B \,|\, X_{t-1}, X_{t+1}). \tag{1.1.2}$$

Formally, this looks different in comparison with the classical definition of the Markov property in Chapter 5, but it may be proved that the latter definition is equivalent to requirement (1.1.2).

Let us consider an example of a Markov random field where the parameter $\boldsymbol{v}$ essentially differs from a time parameter.

EXAMPLE 1 (*Ising's model*). Originally, the famous scheme we consider below, was built as a model of the ferromagnetic material structure. Consider a graph, and suppose that at each vertex, there is a small dipole or "spin" which may be in one of two positions, or in other words, may have one of two orientations. We indicate these two positions by the symbols $+$ and $-$.

As a particular example, one may consider a two dimensional lattice where each vertex is a point (i, j) with positive integer coordinates. Suppose the lattice is finite; $i, j = 1, \ldots, M$. Then, say, for $M = 5$, a particular possible configuration of dipoles may look as in Fig. 3: at each vertex, either $+$ or $-$ "is sitting." Note also that in this case, interior vertices have four "neighbors" while vertices on the boundary have two or three neighbors.

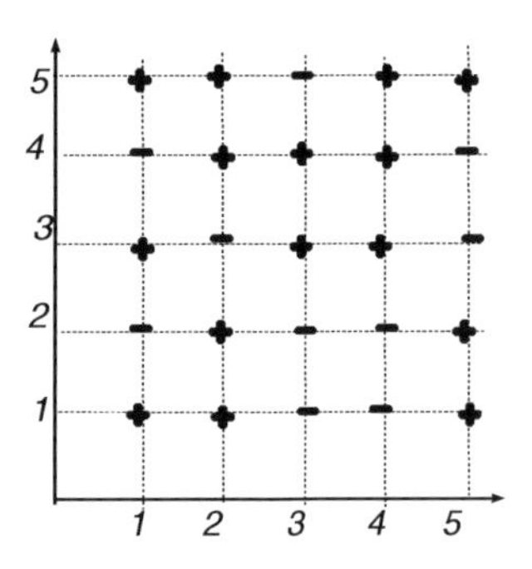

FIGURE 3.

Now, let us consider a general graph with a finite number M of vertices. Each outcome ω in such a scheme corresponds to one of possible configurations of pluses and minuses. The number of all possible outcomes ω is 2^M, and to define a probability measure on the sample space $\Omega = \{\omega\}$, it suffices to define the probability $p(\omega)$ for each ω.

Let a r.v. $\xi_{\boldsymbol{v}} = \xi_{\boldsymbol{v}}(\omega) = 1$ or -1 depending on whether the orientation of the dipole in the vertex $\boldsymbol{v}$ is positive or negative.

In Ising's model, the energy of the system is defined as the function

$$U(\omega) = k_1 \sum\nolimits'_{(\boldsymbol{v},\boldsymbol{u})} \xi_{\boldsymbol{v}}(\omega)\xi_{\boldsymbol{u}}(\omega) + k_2 \sum\nolimits_{\boldsymbol{v}} \xi_{\boldsymbol{v}}(\omega), \tag{1.1.3}$$

where in $\sum'_{(\boldsymbol{v},\boldsymbol{u})}$, the summation is over all pairs of vertices $(\boldsymbol{v},\boldsymbol{u})$ which are neighbors of each other, and in $\sum_{\boldsymbol{v}}$, the summation is over all vertices.

The first sum in (1.1.3) represents the energy connected with the interactions of dipoles, and we assume that only the interactions between neighboring dipoles matter. The constant k_1 is specified by properties of the material.

The second sum in (1.1.3) characterizes the influence of an external magnetic field, and the constant k_2 is specified by the intensity of this field.

Ising's model presupposes that

$$p(\omega) = \frac{1}{S}\exp\left\{-\frac{1}{kT}U(\omega)\right\}, \tag{1.1.4}$$

where T is the temperature, k is a universal constant, and the normalizing quantity

$$S = \sum_{\omega} \exp\left\{-\frac{1}{kT}U(\omega)\right\}$$

with the summation over all ω. So, $\sum_{\omega} p(\omega) = 1$.

The model above leads to many interesting conclusions (see, e.g., [25]) and proved to be applicable in many other situations in Physics, Biology, and even Sociology. We will just state (and show) that if the probability measure $p(\omega)$ is defined as in (1.1.4), then the collection of the r.v.'s $\{\xi_{\boldsymbol{v}}\}$ is a Markov random field. The proof is technical.

▶ Since each $\xi_{\boldsymbol{v}}$ assumes only two values, it suffices to prove that for any vertex $\boldsymbol{v}$,

$$P(\xi_{\boldsymbol{v}} = 1 \,|\, \xi_{\boldsymbol{u}} = a_{\boldsymbol{u}} \text{ for } \boldsymbol{u} \neq \boldsymbol{v}) = P(\xi_{\boldsymbol{v}} = 1 \,|\, \xi_{\boldsymbol{u}} = a_{\boldsymbol{u}} \text{ for } \boldsymbol{u} \in \mathcal{N}_{\boldsymbol{v}}), \tag{1.1.5}$$

where $\mathcal{N}_{\boldsymbol{v}}$ is the set of the neighbors of $\boldsymbol{v}$, and each number $a_{\boldsymbol{u}}$ equals either 1 or -1.

First of all, note that (1.1.4) may be rewritten as

$$p(\omega) = \frac{1}{S}\exp\{-V(\omega)\}, \tag{1.1.6}$$

where

$$V(\omega) = c_1 \sum\nolimits'_{(\boldsymbol{v},\boldsymbol{u})} \xi_{\boldsymbol{v}}(\omega)\xi_{\boldsymbol{u}}(\omega) + c_2 \sum\nolimits_{\boldsymbol{u}} \xi_{\boldsymbol{u}}(\omega),$$

and c_1, c_2 are other constants ($c_1 = k_1/(kT)$, $c_2 = k_2/(kT)$).

Consider the l.-h.s. of (1.1.5), and observe that

$$P(\xi_{\boldsymbol{v}}=1,\xi_{\boldsymbol{u}}=a_{\boldsymbol{u}} \text{ for } \boldsymbol{u}\neq\boldsymbol{v})=\frac{1}{S}\exp\left\{-c_1\sum_{\boldsymbol{u}\in\mathcal{N}_{\boldsymbol{v}}}a_{\boldsymbol{u}}-c_1\sum_{(\boldsymbol{u},\boldsymbol{s})}{}^{\prime,\neq\boldsymbol{v}}a_{\boldsymbol{u}}a_{\boldsymbol{s}}-c_2\left(1+\sum_{\boldsymbol{u}\neq\boldsymbol{v}}a_{\boldsymbol{u}}\right)\right\}, \tag{1.1.7}$$

where in $\sum_{(\boldsymbol{u},\boldsymbol{s})}^{\prime,\neq\boldsymbol{v}}$ the summation is over all $\boldsymbol{u},\boldsymbol{s}$ that are not equal to $\boldsymbol{v}$ and are neighbors of each other. (When writing the expression for the r.-h.s. of (1.1.7), we set $\xi_{\boldsymbol{v}} = 1$.)

To find $P(\xi_{\boldsymbol{u}} = a_{\boldsymbol{u}} \text{ for } \boldsymbol{u}\neq\boldsymbol{v})$, we should add up $P(\xi_{\boldsymbol{v}} = 1, \xi_{\boldsymbol{u}} = a_{\boldsymbol{u}} \text{ for } \boldsymbol{u}\neq\boldsymbol{v})$ and $P(\xi_{\boldsymbol{v}} = -1, \xi_{\boldsymbol{u}} = a_{\boldsymbol{u}} \text{ for } \boldsymbol{u}\neq\boldsymbol{v})$. This leads to

$$P(\xi_{\boldsymbol{u}}=a_{\boldsymbol{u}} \text{ for } \boldsymbol{u}\neq\boldsymbol{v})=\frac{1}{S}\left[\exp\left\{-c_1\sum_{\boldsymbol{u}\in\mathcal{N}_{\boldsymbol{v}}}a_{\boldsymbol{u}}-c_1\sum_{(\boldsymbol{u},\boldsymbol{s})}{}^{\prime,\neq\boldsymbol{v}}a_{\boldsymbol{u}}a_{\boldsymbol{s}}-c_2\left(1+\sum_{\boldsymbol{u}\neq\boldsymbol{v}}a_{\boldsymbol{u}}\right)\right\}\right.$$
$$\left.+\exp\left\{+c_1\sum_{\boldsymbol{u}\in\mathcal{N}_{\boldsymbol{v}}}a_{\boldsymbol{u}}-c_1\sum_{(\boldsymbol{u},\boldsymbol{s})}{}^{\prime,\neq\boldsymbol{v}}a_{\boldsymbol{u}}a_{\boldsymbol{s}}-c_2\left(-1+\sum_{\boldsymbol{u}\neq\boldsymbol{v}}a_{\boldsymbol{u}}\right)\right\}\right]. \tag{1.1.8}$$

To find $P(\xi_{\boldsymbol{v}} = 1 \,|\, \xi_{\boldsymbol{u}} = a_{\boldsymbol{u}} \text{ for } \boldsymbol{u}\neq\boldsymbol{v})$, we should divide the r.-h.s. of (1.1.7) by the r.-h.s. of (1.1.8). A number of terms cancel, and

$$P(\xi_{\boldsymbol{v}}=1\,|\,\xi_{\boldsymbol{u}}=a_{\boldsymbol{u}} \text{ for } \boldsymbol{u}\neq\boldsymbol{v})=\frac{\exp\left\{-c_1\sum_{\boldsymbol{u}\in\mathcal{N}_{\boldsymbol{v}}}a_{\boldsymbol{u}}-c_2\right\}}{\exp\left\{-c_1\sum_{\boldsymbol{u}\in\mathcal{N}_{\boldsymbol{v}}}a_{\boldsymbol{u}}-c_2\right\}+\exp\left\{c_1\sum_{\boldsymbol{u}\in\mathcal{N}_{\boldsymbol{v}}}a_{\boldsymbol{u}}+c_2\right\}}. \tag{1.1.9}$$

We see that the above conditional probability is specified only by the values of $a_{\boldsymbol{u}}$ for $\boldsymbol{u} \in \mathcal{N}_{\boldsymbol{v}}$. Hence, this probability must be equal to the r.-h.s. of (1.1.5), and the validity of the Markov property is established. ◀ □

1.2 Local dependency

To introduce the next dependency type, first, consider a process X_t where the univariate discrete parameter $t = 0, 1, ...$ and may be interpreted as time. The r.v.'s $\{X_t\}$ are said to be *m-dependent* if for any t, the r.v. X_t does not depend on the collection $\{X_s : |s-t| > m\}$. In other words, X_t does not depend on r.v.'s X_s when the time-moments t and s are m time-units apart.

EXAMPLE 1. Consider a sequence of independent r.v.'s ξ_s, where $s = 0, 1, 2, ...$, and a function $\varphi(x_0, x_1, ..., x_m)$. For $t = 0, 1, 2, ...$, set

$$X_t = \varphi(\xi_t, \xi_{t+1}, ..., \xi_{t+m}). \tag{1.2.1}$$

For instance, if $m = 1$ and $\varphi(x_0, x_1) = x_0 x_1$, then

$$X_0 = \xi_0\xi_1,\ X_1 = \xi_1\xi_2,\ X_2 = \xi_2\xi_3, ... \tag{1.2.2}$$

The reader is invited to make sure that the r.v.'s in (1.2.1) are m-dependent, and regarding the particular case (1.2.2), for example, X_0 does depend on X_1, but does not depend on $X_2, X_3, ...$ □

Consider a more general scheme; namely, a collection of r.v.'s $\{X_{\boldsymbol{v}}\}$, where as above, $\boldsymbol{v}$ denotes a vertex in a graph.

For a vertex $\boldsymbol{v}$, we call a collection of vertices $\mathcal{D}_{\boldsymbol{v}}$ a *dependency neighborhood* of $\boldsymbol{v}$ if the r.v. $X_{\boldsymbol{v}}$ does not depend on the collection of r.v.'s $\{X_{\boldsymbol{u}}, \boldsymbol{u} \notin \mathcal{D}_{\boldsymbol{v}}\}$. Loosely put, $X_{\boldsymbol{v}}$ may depend only on r.v.'s belonging to its dependency neighborhood.

So, we distinguish two terms: a neighborhood of a vertex (which is specified only by the graph and has nothing to do with a dependency structure), and a dependency neighborhood (which is specified by the probability distribution of the r.v.'s defined on the graph).

Certainly, if we add to a dependency neighborhood one more vertex, the new set will also be a dependency neighborhood (so to speak, with a leeway). Below, when talking about dependency neighborhoods, we will mean the minimal sets with this property.

Two more things are noteworthy. First, since any r.v. depends on itself, any $\mathcal{D}_{\boldsymbol{v}}$ contains $\boldsymbol{v}$ itself. Secondly, if any $\mathcal{D}_{\boldsymbol{v}}$ contains only $\boldsymbol{v}$, then the r.v.'s $X_{\boldsymbol{v}}$ are independent.

EXAMPLE 2. Let us view m-dependent r.v.'s X_t as a sequence of r.v.'s defined on a graph whose vertices are integer points on a line; see Fig. 2. Then, for any "time"-moment t, its dependency neighborhood $\mathcal{D}_t = \{t-m, t-m+1, ..., t-1, t, t+1, ..., t+m-1, t+m\}$; that is, $\mathcal{D}_t = \{s : |s-t| \leq m\}$.

EXAMPLE 3. Consider an arbitrary graph with vertices $\{\boldsymbol{v}\}$, and a collection of *independent* r.v.'s $\xi_{\boldsymbol{v}}$. Suppose $X_{\boldsymbol{v}}$ is a function of $\xi_{\boldsymbol{v}}$ and the r.v.'s $\xi_{\boldsymbol{u}}$ for $\boldsymbol{u} \in \mathcal{N}_{\boldsymbol{v}}$. In other words, the value of $X_{\boldsymbol{v}}$ is completely determined by the values of $\xi_{\boldsymbol{v}}$ and all $\xi_{\boldsymbol{u}}$ for $\boldsymbol{u}$ from the neighborhood of $\boldsymbol{v}$ with respect to the graph.

(Say, for the economy model discussed in p.410, we may interpret this as follows. The sites where the economic agents are situated are identified with the vertices of the graph. The situation in a site $\boldsymbol{v}$ is characterized by a r.v. $\xi_{\boldsymbol{v}}$. We suppose that in each site $\boldsymbol{v}$, there is only one agent, and its "behavior" depends only on the situation in its site and in the neighboring—with respect to the graph—sites.)

The reader is invited to realize that in this case, the dependency neighborhood of $\boldsymbol{v}$ is

$$\mathcal{D}_{\boldsymbol{v}} = [\boldsymbol{v}] \cup \mathcal{N}_{\boldsymbol{v}} \cup_{\boldsymbol{u} \in \mathcal{N}_{\boldsymbol{v}}} \mathcal{N}_{\boldsymbol{u}}, \tag{1.2.3}$$

where $[\boldsymbol{v}]$ is the set containing only $\boldsymbol{v}$. Thus, $\mathcal{D}_{\boldsymbol{v}}$ consists of $\boldsymbol{v}$, all its neighbors, and all neighbors of the neighbors.

Let us look, for instance, at Fig. 1. For the scheme of the current example, the dependency neighborhood of $\boldsymbol{v}_1$ is $\{\boldsymbol{v}_1, \boldsymbol{v}_2, \boldsymbol{v}_3, \boldsymbol{v}_4, \boldsymbol{v}_6\}$, and the dependency neighborhood of $\boldsymbol{v}_7$ is $\{\boldsymbol{v}_7, \boldsymbol{v}_2, \boldsymbol{v}_3, \boldsymbol{v}_5, \boldsymbol{v}_8, \boldsymbol{v}_9\}$. $\square$

1.3 Point processes

The next type of dependency we consider in this chapter, is connected with a random distribution of points in a space. In Chapter 13, we considered models where the points were time moments in a time axis, and these moments corresponded to arrival times (of customers, messages, etc.). In particular, we considered a counting process N_t, the number of arrivals before or at time t.

Here, we introduce a "spatial" generalization of a counting process, where points under consideration are those from a multidimensional space. Say, we try to model the distribution of the locations of fires in a forest area, earthquakes, trees in a forest, or stars in the universe.

When modeling the distribution of points in a multidimensional space, we should realize that if, for example, $\boldsymbol{x}$ is a point on a plane, unlike in the univariate case, the very expression "before $\boldsymbol{x}$" does not make much sense. Hence, we cannot consider a literal generalization of a counting process N_t, and should state the problem somewhat differently.

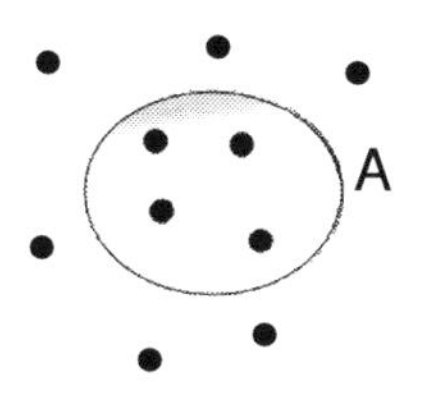

FIGURE 4. $N_A = 4$.

Consider a collection of d-dimensional r.vec.'s (or points) $\{\mathbf{X}_j, j = 1, 2, ...\}$, and for a set A in $\mathbb{R}^d$, define the r.v. N_A equal to the number of points in A; see, for instance, Fig. 4 illustrating the two-dimensional case.

Formally, the quantity N_A may be also defined as follows. For a point $\boldsymbol{x} \in \mathbb{R}^d$, we define the indicator $I_A(\boldsymbol{x}) = 1$ if $\boldsymbol{x} \in A$, and $= 0$ otherwise.

Then the number of random points $\mathbf{X}_j$ in A may be represented as

$$N_A = \sum_j I_A(\mathbf{X}_j).$$

We suppose that the joint distribution of $\{\mathbf{X}_j, j = 1, 2, ...\}$ is such that for any bounded A, the number of points in A is finite with probability one; that is, $P(N_A < \infty) = 1$ for any bounded A.

In general, for any *non-random* distribution of points in a space, a function $Q(A)$ equal to the number of points in A, is called a *counting measure*. Since for any A, the quantity N_A defined above is a random variable, N_A as a function of A, is a *random counting measure*. Another widely used term for N_A is a *point process*.

In the framework under consideration, the counterpart of the independence-of-increments property may be stated as follows:

$$\text{For any } \textit{disjoint} \text{ sets } A_1, A_2, ..., \text{ the r.v.'s } N_{A_1}, N_{A_2}, ... \text{ are independent.} \tag{1.3.1}$$

Sometimes, this property is called "*complete randomness*."

We call the function

$$\mu(A) = E\{N_A\}$$

the *intensity* of the point process (random measure) N_A. We assume $\mu(A)$ to be finite for any bounded A. If A is not bounded, then $\mu(A)$ may be infinite.

A process (or a random measure) is called a *Poisson point process* (or a *Poisson random measure*) if (1.3.1) holds, and for any bounded A, the r.v. N_A has the Poisson distribution with a mean of $\mu(A)$. That is, for a bounded A,

$$P(N_A = k) = e^{-\mu(A)} \frac{(\mu(A))^k}{k!}. \tag{1.3.2}$$

EXAMPLE 1. Let random points represent the locations of the origin points of wildfires in a forest region during a certain period. So, $d = 2$.

Suppose the distribution of the origin points is well modeled by a Poisson counting measure with a rate of $\beta = 0.05$ per square mile. Since the rate is constant, we set $\mu(A) = \beta s(A)$, where $s(A)$ is the area of A.

Let Y be the distance from a particular point, say, a visitor center, to the nearest fire origin point. Let O_r be a disk of radius r centered at the visitor center. Then, in accordance with (1.3.2),

$$\begin{aligned} P(Y > r) &= P(N_{O_r} = 0) = \exp\{-\mu(O_r)\} = \exp\{-\beta s(O_r)\} \\ &= \exp\{-\beta\pi r^2\} \approx \exp\{-0.157 \cdot r^2\}. \end{aligned}$$

For the mean distance from the center to the nearest fire, using (**6**.1.3.2), we have

$$E\{Y\} = \int_0^{\infty} P(Y > r)dr = \int_0^{\infty} \exp\{-\beta\pi r^2\}dr \approx \int_0^{\infty} \exp\{-0.157 \cdot r^2\}dr \approx 2.24. \quad (1.3.3)$$

(The numerical answer may be obtained either with use of software, or appealing to the normal distribution; see Exercise 5.)

EXAMPLE 2. Let us add to the model of Example 1 one more parameter, namely, time. Suppose that the counting measure N_A is defined on sets A from a three dimensional space of points (x_1, x_2, t), where the first two coordinates are spatial coordinates designating locations in the region, while the third coordinate is a time parameter.

Let B be a set in a plane, and $A = B \times [t_1, t_2]$. The last symbol means that if a random point fell in A, then a fire occurred in the region B between time moments t_1 and t_2. We suppose that for a fixed region B, the mean number of fire occurrences is proportional to the length of the time period, and for A specified above, we set

$$\mu(A) = \lambda s(B)(t_2 - t_1),$$

where λ is the mean number of occurrences per unit of time (say, a year) and unit of area (say, a sq. mile).

Consider the event $\{N_{O_r \times [0,t]} = 0\}$ that during t years, no fire occurs within a distance of r from the center. Then

$$P(N_{O_r \times [0,t]} = 0) = \exp\{-\lambda\pi r^2 t\}. \ \square$$

---

2 MEASURES OF DEPENDENCY

In this section, we touch on some characteristics which measure the "degree of dependency" between collections of r.v.'s.

2.1 Covariance functions

For a collection of variables $\{X_{\boldsymbol{u}}\}$, where $\boldsymbol{u}$ is a parameter of an arbitrary nature, the *covariance function*

$$C(\boldsymbol{u}, \boldsymbol{v}) = Cov\{X_{\boldsymbol{u}}, X_{\boldsymbol{v}}\}, \quad (2.1.1)$$

where, as usual, Cov stands for covariance. We also define the *correlation function*

$$R(\boldsymbol{u}, \boldsymbol{v}) = Corr\{X_{\boldsymbol{u}}, X_{\boldsymbol{v}}\}, \quad (2.1.2)$$

where *Corr* stands for correlation.

EXAMPLE 1. Let N_t be a homogeneous Poisson process with intensity λ. Since N_t is a process with independent increments, and $N_{t+s} = N_t + N_{(t,t+s]}$, for the covariance function $C(\cdot,\cdot)$, we have

$$\begin{aligned} C(t,t+s) &= Cov\{N_t, N_{t+s}\} = Cov\{N_t, N_t + N_{(t,t+s]}\} = Cov\{N_t, N_t\} + Cov\{N_t, N_{(t,t+s]}\} \\ &= Var\{N_t\} + 0 = \lambda t. \end{aligned}$$

Then, for the correlation function,

$$R(t,t+s) = Corr\{N_t, N_{t+s}\} = \frac{Cov\{N_t, N_{t+s}\}}{\sqrt{Var\{N_t\}}\sqrt{Var\{N_{t+s}\}}} = \frac{\lambda t}{\sqrt{\lambda t}\sqrt{\lambda(t+s)}} = \sqrt{\frac{t}{t+s}}.$$

In particular, for time moments that are far apart from each other, formally for $s \to \infty$, the correlation function $R(t,t+s) \to 0$, which reflects the fact that the dependence between N_t and N_{t+s} is vanishing for large s.

EXAMPLE 2 (*An autoregressive model*). We consider a *linear autoregressive* model that deals with a sequence of r.v.'s $X_0, X_1, X_2, \ldots$ such that for $t = 1, 2, \ldots$,

$$X_t = c_1 X_{t-1} + c_2 X_{t-2} + \ldots + c_{t-k} X_{t-k} + \varepsilon_t, \tag{2.1.3}$$

where ε_t is a r.v. (In the case of $t < k$, we set $X_{t-k} = 0$.) Relation (2.1.3) is usually interpreted as an attempt to predict the value of a characteristic at time t (that is, X_t), given the values of this characteristic at previous moments $t-1, t-2, \ldots$. In a simple model (2.1.3), the predictor is a linear combination $c_1 X_{t-1} + \ldots + c_{t-k} X_{t-k}$ of k previous values (so, k may be called an order or depth of prediction), and the r.v. ε_t is a random error of prediction.

To make the model more illustrative (though simplified), let us restrict ourselves to the case $k = 1$ and rewrite (2.1.3) as

$$X_t = r X_{t-1} + \varepsilon_t \tag{2.1.4}$$

(replacing c_1 by the symbol r). We suppose $r \geq 0$.

One may conjecture that if $r < 1$, then the dependence of X_t on the values of the process in a "remote past" is vanishing. Let us look at it carefully. We have

$$\begin{aligned} X_t &= r X_{t-1} + \varepsilon_t = r(r X_{t-2} + \varepsilon_{t-1}) + \varepsilon_t = r^2 X_{t-2} + r\varepsilon_{t-1} + \varepsilon_t \\ &= r^2 (r X_{t-3} + \varepsilon_{t-2}) + r\varepsilon_{t-1} + \varepsilon_t = r^3 X_{t-3} + r^2 \varepsilon_{t-2} + r\varepsilon_{t-1} + \varepsilon_t \\ &= \ldots = r^t X_0 + r^{t-1}\varepsilon_1 + r^{t-2}\varepsilon_2 + \ldots + r\varepsilon_{t-1} + \varepsilon_t. \end{aligned} \tag{2.1.5}$$

We assume the r.v.'s $X_0, \varepsilon_1, \varepsilon_2, \ldots$ to be mutually independent, and without loss of generality, set $E\{X_0\} = E\{\varepsilon_i\} = 0$, $i = 1, 2, \ldots$. If it is not so, we apply centering.

Let $Var\{\varepsilon_i\} = \sigma^2$ for all i. To make the formulas below more explicit, we assume that X_0 has the same variance σ^2. The case where $Var\{X_0\}$ is different, is treated similarly, and—qualitatively—leads to the same conclusions.

Along with this assumption, (2.1.5) implies

$$E\{X_t\} = 0, \;\; Var\{X_t\} = r^{2t}\sigma^2 + \sigma^2(r^{2(t-1)} + \ldots r^2 + 1) = \sigma^2 \frac{1 - r^{2(t+1)}}{1 - r^2}. \tag{2.1.6}$$

Now, let us view t as an initial point, and consider a time moment $t+s$. Similarly to (2.1.5), we can write $X_{t+s} = r^s X_t + r^{s-1}\varepsilon_{t+1} + ... + r\varepsilon_{t+s-1} + \varepsilon_{t+s}$, and

$$Cov\{X_{t+s}, X_t\} = E\{X_{t+s}X_t\} = r^s E\{X_t^2\} + E\{X_t(r^{s-1}\varepsilon_{t+1} + ... + r\varepsilon_{t+s-1} + \varepsilon_{t+s})\}.$$

Since X_t does not depend on $\{\varepsilon_{t+1}, ..., \varepsilon_{t+s}\}$, the second expectation equals $E\{X_t\}E\{(r^{s-1}\varepsilon_{t+1} + ... + r\varepsilon_{t+s-1} + \varepsilon_{t+s})\} = 0$, and

$$Cov\{X_{t+s}, X_t\} = r^s Var\{X_t\}. \tag{2.1.7}$$

From this and (2.1.6), we come to the covariance function

$$C(t, t+s) = r^s \sigma^2 \frac{1 - r^{2(t+1)}}{1 - r^2}.$$

We see that, if $0 \le r < 1$, then for a fixed t, the covariance function $C(t, t+s) \to 0$ as $s \to \infty$, and the rate at which the covariance function is vanishing, is exponential ($r^s = e^{\ln rs} = e^{-|\ln r|s}$).

Now, let us fix s and consider large t's; formally, let $t \to \infty$. In other words, we consider the system in the long run, approaching a stationary regime. In this case,

$$C(t, t+s) \to r^s \frac{\sigma^2}{1 - r^2}. \tag{2.1.8}$$

Next, consider the correlation function. From (2.1.7), we get

$$R(t, t+s) = r^s \frac{Var\{X_t\}}{\sqrt{Var\{X_t\}}\sqrt{Var\{X_{t+s}\}}} = r^s \sqrt{\frac{Var\{X_t\}}{Var\{X_{t+s}\}}}.$$

From this and (2.1.6), after simple algebra, we come to

$$R(t, t+s) = r^s \sqrt{\frac{1 - r^{2(t+1)}}{1 - r^{2(t+s+1)}}}. \tag{2.1.9}$$

In the long run, formally, as $t \to \infty$, we come to a very simple relation:

$$R(t, t+s) \to r^s \text{ for any fixed } s \text{ and } t \to \infty. \tag{2.1.10}$$

EXAMPLE 3. Consider a collection of r.v.'s $\{X_{\boldsymbol{v}}\}$ defined on vertices of a graph. As above, we denote by $\mathcal{D}_{\boldsymbol{v}}$ the dependency neighborhood of $\boldsymbol{v}$. The reader is invited to realize that the covariance function

$$C(\boldsymbol{v}, \boldsymbol{u}) = 0 \text{ for } \boldsymbol{u} \notin \mathcal{D}_{\boldsymbol{v}}. \ \square$$

It remains to note that, as was already mentioned in Section **10**.1, covariance is though a convenient but simple characteristic; sometimes, too simple. We know that uncorrelated r.v.'s may be dependent, and the correlation coefficient is not informative enough in the case of non-linear regression (unlike the linear case in Example 2).

The next characteristic is more complicated but proves to be more efficient in describing dependency types.

2.2 Mixing

Below, we use the symbol sup for supremum. The reader who is not familiar with this notion, may view it as maximum.

Consider two collections of r.v.'s, $\mathcal{X} = \{X_{\boldsymbol{u}}\}$ and $\mathcal{Y} = \{Y_{\boldsymbol{u}}\}$, where $\boldsymbol{u}$ stands for a running parameter of an arbitrary nature. Let $\mathcal{F}_{\mathcal{X}}$ be the collection of all events whose occurrences are completely determined by the values of the r.v.'s from $\mathcal{X}$. Let $\mathcal{F}_{\mathcal{Y}}$ be the corresponding collection of events for $\mathcal{Y}$. Let

$$\alpha(\mathcal{X},\mathcal{Y}) = \sup\{|P(AB) - P(A)P(B)|,\ A \in \mathcal{F}_{\mathcal{X}}, B \in \mathcal{F}_{\mathcal{Y}}\}.$$

We will call it a *mixing coefficient* for $\mathcal{X}$ and $\mathcal{Y}$. Clearly, the collections $\mathcal{X}$ and $\mathcal{Y}$ are independent if and only if $\alpha(\mathcal{X},\mathcal{Y}) = 0$. In the case of dependency, the mixing coefficient "measures the degree of dependency."

Consider, for example, a random process X_t in discrete time $t = 0,1,2,...$, and set $\mathcal{X}_t = \{X_1,...,X_t\}$, and $\widetilde{\mathcal{X}}_t = \{X_t, X_{t+1},...\}$. So, if we view t as the present time moment, then the collection $\mathcal{X}_t$ concerns "the present and the past," while $\widetilde{\mathcal{X}}_t$ deals with the present and future values of the process.

Now, consider the collections $\mathcal{X}_t$ and $\widetilde{\mathcal{X}}_{t+n}$. In this case, we deal with the past and future values of the process with a gap of n time units. Set

$$\alpha(n) = \sup_t \alpha(\mathcal{X}_t, \widetilde{\mathcal{X}}_{t+n}). \tag{2.2.1}$$

Note that the larger n, the less the collection of events $\mathcal{F}_{\widetilde{\mathcal{X}}_{t+n}}$. Therefore, the coefficient $\alpha(\mathcal{X}_t, \widetilde{\mathcal{X}}_{t+n})$ is non-increasing in n, and hence, so is $\alpha(n)$.

If $\alpha(n)$ is a decreasing function, then "the dependence between the past and the future is getting weaker" when the gap between them is increasing.

We say that the process satisfies the *strong mixing* or α*-mixing* condition if

$$\alpha(n) \to 0 \text{ as } n \to \infty. \tag{2.2.2}$$

EXAMPLE 1. Let X_t be a finite homogeneous Markov chain as it was described in Chapter 5 whose notation we will follow. Suppose that the chain is ergodic; that is, the transition probabilities $p_{ij}^{(t)} \to \pi_j$ as $t \to \infty$ for any i, and the vector $\boldsymbol{\pi} = (\pi_1, \pi_2, ...)$ represents a stationary distribution.

Moreover, suppose that this convergence is uniform; namely, there exists a decreasing function $q(t) \to 0$, as $t \to \infty$, such that

$$\max_{i,j} |p_{ij}^{(t)} - \pi_j| \leq q(t). \tag{2.2.3}$$

For instance, we have shown in Section **5**.4.4, that under Doeblin's condition of Theorem **5**.4 , (2.2.3) is true with

$$q(t) = C\rho^t$$

for a positive C and ρ such that $0 \leq \rho < 1$. (See (**5**.4.4.11).)

In the case (2.2.3), *strong mixing does hold.*

We demonstrate how to show this, restricting ourselves to events $A = \{X_0 = i_0, X_1 = i_1, ..., X_t = i_t\}$ and $B = \{X_{t+n} = i_{t+n}, X_{t+n+1} = i_{t+n+1}, ..., X_{t+n+s} = i_{t+n+s}\}$. Here, s is a positive integer, and all i_k's denote arbitrary states of the chain.

By the multiplication rule,

$$\begin{aligned} P(A) &= \pi_{0\,i_0} p_{i_0\,i_1} \cdot ... \cdot p_{i_{t-1}\,i_t}, \; P(B) = \pi_{t+n\,i_{t+n}} p_{i_{t+n}\,i_{t+n+1}} \cdot ... \cdot p_{i_{t+n+s-1}\,i_{t+n+s}}, \\ P(AB) &= \pi_{0\,i_0} p_{i_0\,i_1} \cdot ... \cdot p_{i_{t-1}\,i_t} p^{(n)}_{i_t\,i_{t+n}} p_{i_{t+n}\,i_{t+n+1}} \cdot ... \cdot p_{i_{t+n+s-1}\,i_{t+n+s}}. \end{aligned}$$

There are common factors, and

$$\begin{aligned} |P(AB) - P(A)P(B)| &= |\pi_{0\,i_0} p_{i_0\,i_1} \cdot ... \cdot p_{i_{t-1}\,i_t} \cdot p_{i_{t+n}\,i_{t+n+1}} \cdot ... \cdot p_{i_{t+n+s-1}\,i_{t+n+s}} \cdot (p^{(n)}_{i_t\,i_{t+n}} - \pi_{t+n\,i_{t+n}})| \\ &\leq |p^{(n)}_{i_t\,i_{t+n}} - \pi_{t+n\,i_{t+n}}| = |(p^{(n)}_{i_t\,i_{t+n}} - \pi_{i_{t+n}}) - (\pi_{t+n\,i_{t+n}} - \pi_{i_{t+n}})| \\ &\leq |p^{(n)}_{i_t\,i_{t+n}} - \pi_{i_{t+n}}| + |\pi_{t+n\,i_{t+n}} - \pi_{i_{t+n}}|. \end{aligned} \tag{2.2.4}$$

Furthermore,

$$\begin{aligned} |\pi_{t+n\,i_{t+n}} - \pi_{i_{t+n}}| &= |\sum_i \pi_{0i} p^{(t+n)}_{i,i_{t+n}} - \pi_{i_{t+n}} \sum_i \pi_{0i}| = |\sum_i \pi_{0i}(p^{(t+n)}_{i,i_{t+n}} - \pi_{i_{t+n}})| \\ &\leq \max_{i,j} |p^{(t+n)}_{ij} - \pi_j| \sum_i \pi_{0i} = \max_{i,j} |p^{(t+n)}_{ij} - \pi_j|. \end{aligned}$$

From this, (2.2.4), and (2.2.3), it follows that

$$|P(AB) - P(A)P(B)| \leq q(n) + q(t+n) \leq 2q(n). \tag{2.2.5}$$

Note also that the r.-h.s. of (2.2.5) does not depend on A and B as well as on t and s.

More complicated sets A and B are treated similarly, and eventually, we will get

$$\alpha(n) \leq C_1 q(n) \quad \text{for a constant } C_1. \; \square \tag{2.2.6}$$

---

3 LIMIT THEOREMS FOR DEPENDENT RANDOM VARIABLES

We saw above a number of examples where for a collection of r.v's, the dependence between r.v.'s that are, in a sense, distant from each other, is weak. We will see now that such a property may ensure the validity of the Central Limit Theorem (CLT); that is, the approximate normality of the sums of r.v.'s. We have already mentioned such an opportunity in Chapter 9 and considered a CLT for martingales in Section **15**.15. Now, we state a CLT for two more schemes.

We do not aim to present the most developed and general results, restricting ourselves to simple but useful sufficient conditions of normal approximation, and we will skip proofs. Some of them may be found, e.g., in [8], [9], [23], [29], [42], and [51]. Our goal is again to give an introductory presentation, to illustrate what one can expect when dealing with dependent r.v.'s.

3.1 A limit theorem for stationary sequences with strong mixing

In this section, we consider random sequences, or processes in discrete time, $X_0, X_1, X_2, ...$ such that for any "time moments" $t_1, ..., t_k$,

$$\text{The joint distribution of the r.v.'s } X_{t_1+s}, ..., X_{t_k+s} \text{ is the same for all } s = 0, 1, ... \quad (3.1.1)$$

In other words, the joint distribution does not change upon a shift in time. In particular, the r.v.'s X_t are identically distributed. Indeed, set $k = 1$ and $t_1 = 0$. Then (3.1.1) deals with one r.v. X_s and claims that its distribution is the same for all s.

Such processes are called *stationary*.

EXAMPLE 1. Let r.v.'s $\xi_0, \xi_1, \xi_2, ...$ be independent and identically distributed (i.i.d.), and let the one-dependent sequence $X_0 = \xi_0\xi_1$, $X_1 = \xi_1\xi_2, X_2 = \xi_2\xi_3, ...$ The reader is invited to realize that this sequence is stationary.

EXAMPLE 2. In this example, we keep the notation of Chapter 5. Let X_t, $t = 0, 1, ...$ be a homogeneous ergodic Markov chain whose initial distribution $\boldsymbol{\pi}_0$ equals a stationary distribution $\boldsymbol{\pi}$. We know that in this case, all distribution $\boldsymbol{\pi}_t$ of the r.v.'s X_t are equal to the same $\boldsymbol{\pi}$. Also, since for a homogeneous Markov chain the transitional matrix $\mathcal{P}$ does not depend on time, for a chain in a stationary regime, at each time moment, the process starts over as from the beginning. So, the chain is stationary. □

Without loss of generality, we assume also that

$$E\{X_0\} = 0 \quad (3.1.2)$$

(and because the X's are identically distributed, $E\{X_t\} = 0$ for all t).

Let $S_n = X_1 + ... + X_n$. We might include into the sum the initial r.v. X_0, and the results would be practically the same. However, for the former definition, some formulas below will be simpler.

Let $B_n^2 = Var\{S_n\}$. Since we consider dependent r.v.'s, B_n^2 does not have to be equal to the sum of the variances of the X's. Moreover, B_n^2 does not have to grow when n is increasing.

EXAMPLE 3. Let $\xi_0, \xi_1, ...$ be i.i.d., and $X_1 = \xi_1 - \xi_0$, $X_2 = \xi_2 - \xi_1, ..., X_t = \xi_t - \xi_{t-1}, ...$ Then $S_n = (\xi_1 - \xi_0) + (\xi_2 - \xi_1) + ... + (\xi_{n-1} - \xi_{n-2}) + (\xi_n - \xi_{n-1}) = \xi_n - \xi_0$. So, neither the variance $Var\{S_n\} = Var\{\xi_n\} + Var\{\xi_0\} = 2Var\{\xi_0\}$, nor the sum S_n itself grows. □

Thus, we should consider $B_n^2 = Var\{S_n\}$ more carefully. In the general case,

$$Var\{S_n\} = \sum_{j=1}^{n} Var\{X_j\} + 2 \sum_{1 \le i < j \le n} Cov\{X_i, X_j\} \quad (3.1.3)$$

(see Section **10**.1.1). If (3.1.2) is true, this may be rewritten as

$$Var\{S_n\} = \sum_{j=1}^{n} E\{X_j^2\} + 2 \sum_{1 \le i < j \le n} E\{X_i X_j\}. \quad (3.1.4)$$

Let $\sigma^2 = Var\{X_j\} = E\{X_j^2\}$.

Since the process is stationary, the distribution of a pair (X_s, X_{s+j}) does not depend on s, and hence, $E\{X_sX_{s+j}\} = E\{X_0X_j\}$, for all s. The reader is invited to realize that in this case, (3.1.4) may be rewritten as

$$\begin{aligned} B_n^2 &= \sum_{j=1}^{n} E\{X_j^2\} + 2\sum_{j=1}^{n-1}(n-j)E\{X_0X_j\} = n\sigma^2 + 2\sum_{j=1}^{n-1}(n-j)E\{X_0X_j\} \\ &= n\left(\sigma^2 + 2\sum_{j=1}^{n-1} E\{X_0X_j\}\right) - 2\sum_{j=1}^{n-1} jE\{X_0X_j\}. \end{aligned} \quad (3.1.5)$$

Suppose that the sum $\sum_{j=1}^{\infty} jE\{X_0X_j\}$ absolutely converges; that is,

$$\sum_{j=1}^{\infty} j|E\{X_0X_j\}| < \infty. \quad (3.1.6)$$

Then the sum $\sum_{j=1}^{\infty} |E\{X_0X_j\}|$ which is not greater than the sum above, converges also, and from (3.1.5), it follows that

$$\frac{1}{n}B_n^2 \to \sigma_1^2, \quad (3.1.7)$$

where

$$\sigma_1^2 = \sigma^2 + 2\sum_{j=1}^{\infty} E\{X_0X_j\}. \quad (3.1.8)$$

Thus, $B_n \sim \sigma_1\sqrt{n}$, and if $\sigma_1 > 0$, then we may normalize S_n by $\sigma_1\sqrt{n}$ as well as by B_n. Since σ_1 admits an explicit representation, the normalization by $\sigma_1\sqrt{n}$ may turn out to be preferable.

Theorem 1 *Suppose that for a stationary sequence $X_0, X_1, X_2, \ldots$, we have $E\{X_0\} = 0$, $E\{|X_0|^3\} < \infty$, and for the strong mixing coefficients $\alpha(n)$ defined in (2.2.1),*

$$\sum_{n=1}^{\infty} \alpha^{1/3}(n) < \infty. \quad (3.1.9)$$

Then, first, the sum in the definition (3.1.8) absolutely converges (and hence σ_1 is finite). Secondly, if $\sigma_1 > 0$, then for any x,

$$\lim_{n\to\infty} P\left(\frac{S_n}{B_n} \le x\right) = \lim_{n\to\infty} P\left(\frac{S_n}{\sigma_1\sqrt{n}} \le x\right) = \Phi(x), \quad (3.1.10)$$

where, as usual, $\Phi(x)$ is the standard normal distribution function.

First, note that because the r.v.'s X_t are identically distributed, under the conditions of the theorem, $E\{|X_t|^3\} < \infty$ for all t. From (3.1.9), it follows that $\alpha(n) \to 0$, so the sequence X_t satisfies the strong mixing condition. However, (3.1.9) presupposes a certain rate of convergence. For example, if $\alpha(n) \sim \frac{1}{n^p}$, then for (3.1.9) to hold, we should have $p > 3$.

EXAMPLE 4. Let us come back to the scheme of Example 1, and assume $E\{\xi_i\} = 0$, $E\{|\xi_i|^3\} < \infty$. Let $Var\{\xi_i\} = v^2 \neq 0$.

Since the collections $\{X_0, ..., X_t\}$ and $\{X_{t+2}, X_{t+3}, ...,\}$ are independent, the mixing coefficient $\alpha(n) = 0$ for $n \geq 2$, so (3.1.9) obviously holds.

We have $E\{|X_0|^3\} = E\{|\xi_0\xi_1|^3\} = E\{|\xi_0|^3\}E\{|\xi_1|^3\} < \infty$.

Since ξ's are independent, $E\{X_0\} = E\{\xi_0\xi_1\} = E\{\xi_0\}E\{\xi_1\} = 0$, and hence, (3.1.2) is true.

Furthermore, if $s > t+1$, the r.v.'s X_t and X_s are independent, and hence, are not correlated. If $s = t+1$, then

$$E\{X_tX_s\} = E\{X_tX_{t+1}\} = E\{\xi_t\xi_{t+1}^2\xi_{t+2}\} = E\{\xi_t\}E\{\xi_{t+1}^2\}E\{\xi_{t+2}\} = 0,$$

and, X_t and X_{t+1} are non-correlated too. Then, $B_n^2 = \sum_{j=1}^n Var\{X_j\} = nVar\{X_1\} = nE\{X_1^2\} = nE\{\xi_1^2\xi_2^2\} = nE\{\xi_1^2\}E\{\xi_2^2\} = nv^4$. Thus, $\sigma_1 = v^2$, and

$$P\left(\frac{S_n}{v^2\sqrt{n}} \leq x\right) \to \Phi(x).$$

EXAMPLE 5. Suppose that in Example 4, $E\{\xi_i\} = m \neq 0$, and consider the centered r.v.'s $X_t = \xi_t\xi_{t+1} - E\{\xi_t\xi_{t+1}\} = \xi_t\xi_{t+1} - m^2$. Let us set $v^2 = E\{\xi_0^2\}$. Then

$$Var\{X_0\} = E\{(\xi_0\xi_1)^2\} - (E\{\xi_0\xi_1\})^2 = E\{\xi_0^2\}E\{\xi_1^2\} - (E\{\xi_0\}E\{\xi_1\})^2 = v^4 - m^4.$$

Furthermore, while for $s > t+1$, we still have $Cov\{X_t, X_s\} = 0$, for $s = t+1$,

$$\begin{aligned} Cov\{X_t, X_{t+1}\} &= Cov\{X_0, X_1\} = Cov\{\xi_0\xi_1, \xi_1\xi_2\} \\ &= E\{\xi_0\xi_1 \cdot \xi_1\xi_2\} - E\{\xi_0\xi_1\}E\{\xi_1\xi_2\} = m \cdot v^2 \cdot m - m^4 = m^2(v^2 - m^2). \end{aligned}$$

Then, replacing in (3.1.8) expectations of products by the covariances, we have

$$\sigma_1^2 = v^4 - m^4 + 2m^2(v^2 - m^2) = (v^2 - m^2)(v^2 + 3m^2). \qquad (3.1.11)$$

We see that the normalization differs from that in Example 4. Assume $Var\{\xi_i\} \neq 0$, and hence, $v > m$. Then $\sigma_1 > 0$, and by Theorem 1,

$$P\left(\frac{S_n}{\sigma_1\sqrt{n}} \leq x\right) \to \Phi(x).$$

EXAMPLE 6. Consider a stationary ergodic Markov chain X_t, as it is described in Example 2. Suppose that $E\{|X_0^3|\} < \infty$, and the chain satisfies Doeblin's condition of Theorem **5**.4. As has been shown in Example 2.2-1, in this case, for some $C_1 \geq 0$, and a positive $\rho < 1$,

$$\alpha(n) = C_1\rho^n.$$

This implies the convergence of the sum in (3.1.9).

Set $m = E\{X_t\}$, and $Y_t = X_t - m$. Clearly, the process $\{Y_t\}$ is also stationary. By definition, the characteristic $\alpha(n)$ for X's and Y's are the same. (For example, the set of all events determined by values of X_1 coincides with the set of all events determined by values of $X_1 - m$.)

So, excluding the case of zero variances, Theorem 1 applies to the centered sequence Y_t. □

In the next section, we do not impose the stationarity condition, and the r.v.'s do not have to be identically distributed. On the other hand, the dependency structure we will consider, is simpler than in Theorem 1.

3.2 A limit theorem for the case of local dependency

Let us come back to the scheme of Section 1.2. Let $\{X_{\mathbf{v}}\}$ be a collection of r.v.'s, where the running parameter $\mathbf{v}$ corresponds to vertices of a graph. It is again worth emphasizing that this scheme covers the case of a "usual" sequence of r.v.'s $X_1, X_2, \ldots$.

Suppose that the graph contains an infinite number of vertices, and consider an infinite increasing sequence of sets of vertices $\mathcal{A}_1 \subset \mathcal{A}_2 \subset \mathcal{A}_3 \subset \ldots$ such that the number of vertices in $\mathcal{A}_n$ converges to infinity as $n \to \infty$. Set

$$S_n = \sum_{\mathbf{v} \in \mathcal{A}_n} X_{\mathbf{v}}.$$

If we consider just a sequence $X_1, X_2, \ldots$, then we may choose $\mathcal{A}_n = \{1, \ldots, n\}$, and in this case, $S_n = X_1 + \ldots + X_n$.

Let $B_n^2 = Var\{S_n\}$, and as usual, the normalized sum $S_n^* = \dfrac{1}{B_n}(S_n - E\{S_n\})$.

Let $\mathcal{D}_{\mathbf{v}}$ be the dependency neighborhood of $\mathbf{v}$, and let for any set of vertices $\mathcal{A}$, the symbol $|\mathcal{A}|$ denote the number of points in $\mathcal{A}$.

Theorem 2 *Suppose there exist positive constants k, C and c such that*

$$|\mathcal{D}_{\mathbf{v}}| \leq k, \quad E\{|X_{\mathbf{v}}|^3\} \leq C \ \text{for all} \ \mathbf{v}, \quad \text{and} \quad B_n^2 \geq c|\mathcal{A}_n| \ \text{for all} \ n.$$

Then, for any x,

$$P(S_n^* \leq x) \to \Phi(x) \quad as \quad n \to \infty.$$

We consider an example in Exercise 9.

---

4 CONDITIONAL INDEPENDENCY. SYMMETRIC DISTRIBUTIONS. DE FINETTI'S THEOREM

Now, we consider an essentially different dependency type.

4.1 Conditional independency

Consider a r.vec. $\mathbf{X} = (X_1, \ldots, X_n)$. Its probability distribution (i.e., all probabilities $P(\mathbf{X} \in B)$ for $B \subseteq \mathbb{R}^n$) is completely determined by the joint d.f. $F(x_1, \ldots, x_n) = P(X_1 \leq x_1, \ldots, X_2 \leq x_n)$. We discussed this in detail in Section **7**.1 for the one-dimensional case; the multidimensional framework is treated similarly. So, we will deal with the d.f. F.

We will call the r.v.'s $X_1, \ldots, X_n$ *conditionally independent* if there exists a r.v. or a r.vec. Θ such that for any $x_1, \ldots, x_n$, the conditional d.f. given $\Theta = \theta$, that is,

$$P(X_1 \leq x_1, \ldots, X_n \leq x_n \,|\, \Theta = \theta) = P(X_1 \leq x_1 \,|\, \Theta = \theta) \cdot \ldots \cdot P(X_n \leq x_n \,|\, \Theta = \theta). \tag{4.1.1}$$

One may say that, given Θ, the r.v.'s $X_1, \ldots, X_n$ are independent. Note also that in a non-trivial case, the dimension of Θ is less than n since otherwise we can set $\Theta = (X_1, \ldots, X_n)$, and the definition will become trivial.

In the case of an infinite sequence $X_1, X_2, \ldots$, we call the r.v.'s X_i conditionally independent if (4.1.1) holds for any, arbitrary large, n with the *same* Θ.

EXAMPLE 1. Let Θ, $\xi_1, \xi_2, \ldots$ be independent r.v.'s, and $X_j = g(\Theta, \xi_j)$ for a function $g(x,y)$.

For instance, let X_j be the amount of ice cream to be consumed tomorrow by the jth citizen of a city. X_j may depend on two factors: a factor ξ_j reflecting particular circumstances connected with the jth citizen, and a common factor Θ characterizing tomorrow's weather conditions, say, the temperature.

Once we know that Θ has assumed a particular value θ, we are dealing with the r.v.'s $g(\theta, \xi_1), g(\theta, \xi_2), \ldots$ which are independent.

For example, given today's weather conditions, the sizes of ice cream consumption for different people are independent, but this is not the case for the future consumption since it depends on the common factor: weather. □

To make the model below more substantial, we also assume that given Θ, all X's are identically distributed; namely, for any j and x,

$$P(X_j \leq x \,|\, \Theta = \theta) = F_\theta(x), \tag{4.1.2}$$

where for each θ, the function $F_\theta(x)$ is a d.f., the same for all j. In this case, (4.1.1) may be rewritten as

$$P(X_1 \leq x_1, \ldots, X_n \leq x_n \,|\, \Theta = \theta) = F_\theta(x_1) \cdot \ldots \cdot F_\theta(x_n).$$

Such a scheme may be also interpreted as follows. Suppose that we have at our disposal a family of probability distributions $\{F_\theta\}$ marked by a parameter θ which may be univariate or multivariate as well. First, we select a parameter θ as a value of a r.v. (or a r.vec.) Θ, which leads to the choice of the distribution F_θ. We may view it as a selection of a probability distribution in accordance with some random mechanism.

Once the choice of a probability distribution has been performed, we consider independent r.v.'s $X_1, X_2, \ldots$ having the same (conditional, given $\Theta = \theta$) distribution F_θ.

Replacing in (4.1.2) a particular value θ by the r.v. Θ itself, we may also write that for each X_j,

$$P(X_j \leq x \,|\, \Theta) = F_\Theta(x).$$

This is a "random" d.f. marked by the r.v. (or r.vec.) Θ. Then, by the formula for total probability, the unconditional d.f. of each X_j is $P(X_j \leq x) = E\{F_\Theta(x)\}$, and for the whole sequence $X_1, \ldots, X_n$,

$$F(x_1, \ldots, x_n) = E\{P(X_1 \leq x_1, \ldots, X_n \leq x_n \,|\, \Theta)\} = E\{F_\Theta(x_1) \cdot \ldots \cdot F_\Theta(x_n)\}. \tag{4.1.3}$$

We call the r.v.'s $X_1, \ldots, X_n$ so defined *conditionally i.i.d r.v.'s.*

EXAMPLE 2 (*The Pòlya model*). Consider Poisson r.v.'s $X_1, X_2, \ldots$ with a common random (!) parameter Λ. For instance, the X's are daily numbers of traffic accidents in a region, and Λ is the *mean* daily number of accidents during the current season. (Say, Λ reflects to what extent the current winter is severe, and in the next winter, the value of Λ may turn out to be different.)

If we suppose that given Λ, the daily numbers of accidents are independent, we will come to the scheme above, and Λ will play the role of a common factor.

To consider an interesting particular case, assume Λ to have the exponential distribution with a parameter a. We will see that in this case, each X has the second version of the geometric distribution with parameter $p = \frac{a}{1+a}$.

In the proof below, we use moment generating functions (m.g.f.'s).

Given $\Lambda = \lambda$, each X has the Poisson distribution with parameter λ. The corresponding m.g.f. $E\left\{e^{zX} \,|\, \Lambda = \lambda\right\} = \exp\{\lambda(e^z - 1)\}$.

Then the m.g.f. of X is $E\left\{e^{zX}\right\} = E\left\{E\left\{e^{zX} \,|\, \Lambda\right\}\right\} = E\left\{\exp\{\Lambda(e^z - 1)\}\right\}$.

The very right expression may be viewed as the m.g.f. of Λ at the point $e^z - 1$. The m.g.f. of the exponential distribution with parameter a is $1/(1 - z/a)$. Replacing in the last expression z by $e^z - 1$, we get that $E\left\{e^{zX}\right\} = 1/\left(1 - (e^z - 1)/a\right)$.

The reader is invited to check that after simple algebra the last expression may be rewritten as $p/\left(1 - qe^z\right)$, where $p = \frac{a}{1+a}$ and $q = 1 - p$. This is the m.g.f. of the geometric distribution; see, e.g., Section **8**.1.2.2. We consider a generalization of this example in Exercise 11. □

Let us return to the general scheme and consider a separate X omitting the index. Set $m_\Theta = E\{X \,|\, \Theta\}$, which we view as the (random!) mean value of X's given Θ. Let $\sigma^2_\Theta = Var\{X \,|\, \Theta\}$, the (random) variance given Θ; that is,

$$\sigma^2_\Theta = E\{(X - m_\Theta)^2 \,|\, \Theta\} = E\{X^2 \,|\, \Theta\} - m^2_\Theta. \tag{4.1.4}$$

First of all, note that

$$E\{m_\Theta\} = E\{E\{X \,|\, \Theta\}\} = E\{X\}. \tag{4.1.5}$$

However, it would be a mistake to think that $E\{\sigma^2_\Theta\} = Var\{X\}$. As a matter of fact, in view of (4.1.4),

$$E\{\sigma^2_\Theta\} = E\{X^2\} - E\{m^2_\Theta\}. \tag{4.1.6}$$

EXAMPLE 3 deals with a simplified but illustrative scheme. Let $\xi_1, \xi_2, ...$ be i.i.d. r.v.'s with zero mean and unit variance. Let r.v.'s α and $\beta > 0$ do not depend on ξ's. For $t = 1, 2, ...$, set

$$X_t = \beta\xi_t + \alpha. \tag{4.1.7}$$

The common factor is presented by the r.vec. $\Theta = (\alpha, \beta)$, and given Θ, the r.v.'s X_t are i.i.d.

Now, since $E\{\xi_i\} = 0$, and due to the independency assumption,

$$m_\Theta = E\{X_t \,|\, \Theta\} = \beta E\{\xi_t \,|\, \Theta\} + \alpha = \beta E\{\xi_t\} + \alpha = \beta \cdot 0 + \alpha = \alpha.$$

Similarly, since $E\{\xi_i^2\} = 1$,

$$\begin{aligned} \sigma^2_\Theta &= E\{(X_t - m_\Theta)^2 \,|\, \Theta\} = E\{(X_t - \alpha)^2 \,|\, \Theta\} = E\{(\beta\xi_t)^2 \,|\, \Theta\} \\ &= \beta^2 E\{\xi_t^2 \,|\, \Theta\} = \beta^2 E\{\xi_t^2\} = \beta^2 \cdot 1 = \beta^2. \end{aligned}$$

Thus, α in (4.1.7) is the (random) mean given the common factor Θ, and β is the (random) standard deviation. Actually, it could be seen from the very beginning, but for us it was useful to watch how the general scheme works in this simple case. □

The next fact we establish is not only important but looks nice.

Proposition 3 *For $i \neq j$,*

$$Cov\{X_i, X_j\} = Var\{m_\Theta\}. \tag{4.1.8}$$

In particular,

> The r.v.'s X_j above are always non-negatively correlated, and they are uncorrelated if and only if the conditional expectation m_Θ is non-random (that is, assumes only one value). (4.1.9)

From this and (4.1.6), it follows that if the r.v.'s X_j are not correlated, then $m_\Theta = m$, and hence $E\{\sigma_\Theta^2\}$ does equal to $Var\{X\}$.

Proof of Proposition 3. Let $m = E\{X_t\}$. Clearly, it suffices to consider X_1 and X_2. In view of (4.1.5), and because, given Θ, the r.v.'s X_1 and X_2 are conditionally independent and have the same conditional expectation,

$$\begin{aligned} Cov\{X_1, X_2\} &= E\{(X_1 - m)(X_2 - m)\} = E\{E\{(X_1 - m)(X_2 - m) \mid \Theta\}\} \\ &= E\{E\{X_1 - m \mid \Theta\}E\{X_2 - m \mid \Theta\}\} = E\{(m_\Theta - m)(m_\Theta - m)\} \\ &= E\{(m_\Theta - m)^2\} = Var\{m_\Theta\}. \ \blacksquare \end{aligned}$$

4.2 Symmetrically dependent or exchangeable r.v.'s. De Finetti's theorem

R.v.'s $X_1, ..., X_n$ are called *symmetrically dependent* or *exchangeable* if their joint distribution does not depend on the order in which these r.v.'s are arranged.

For example, X_1, X_2 are exchangeable if the joint distribution of (X_1, X_2) is the same as for (X_2, X_1).

If a sequence $X_1, X_2, ...$ is infinite, we call the r.v.'s $X_1, X_2, ...$ exchangeable, if any finite sequence $X_1, ..., X_n$ is exchangeable.

Since the joint distribution of $X_1, ..., X_n$ is completely determined by the corresponding distribution function (d.f.) $F(x_1, ..., x_n) = P(X_1 \leq x_1, ..., X_n \leq x_n)$, the r.v.'s $X_1, ..., X_n$ are symmetrically dependent if and only if the function $F(x_1, ..., x_n)$ is symmetric; that is, does not change its value under any rearrangement of its arguments.

Say, the d.f. $\frac{1}{2}(x_1^2 x_2 + x_1 x_2^2)$, $0 \leq x_1 \leq 1, 0 \leq x_2 \leq 1$, is symmetric, while the d.f. $\frac{1}{2}(x_1^3 x_2 + x_1 x_2^2)$ is not. (The reader may double-check that both functions are d.f.'s.)

In particular, symmetrically dependent r.v.'s are identically distributed. Indeed, let for simplicity, $n = 2$. Then $P(X_1 < x) = F(x, \infty) = F(\infty, x) = P(X_2 \leq x)$.

EXAMPLE 1. Let $Y_1, ..., Y_n$ be an arbitrary sequence of r.v.'s. Let us permute $Y_1, ..., Y_n$ at random; that is, in a way that all permutations are equally likely. Then, after such a random permutation, the joint distribution will become symmetric.

To show this rigorously, denote by $Q(x_1, ..., x_n)$ the joint d.f. of $Y_1, ..., Y_n$, and by $(m_1, ..., m_n)$ a particular permutation of the indices $1, ..., n$. If the r.v.'s are rearranged in accordance with this permutation, then the new d.f. will be equal to $Q(x_{m_1},...,x_{m_n})$. For any particular permutation, the probability that it will be performed is $1/n!$. Then the result of the random permutation will lead to a sequence of r.v.'s $X_1,...X_n$ whose d.f. is the function

$$R(x_1,...,x_n) = \frac{1}{n!} \sum_{(m_1,...,m_n)} Q(x_{m_1},...,x_{m_n})$$

which is clearly symmetric. (Say, $Q(x_1,x_2)$ may be non-symmetric, but the function $\frac{1}{2}(Q(x_1,x_2)+Q(x_2,x_1))$ is symmetric.)

In particular, when considering a sum $X_1 + ... + X_n$, we may think that the terms are symmetrically dependent, since otherwise we can permute them at random, which will not change the sum.

(Another matter is that it may be not efficient since it may distort the presentation of the distribution in a non-desirable way.)

EXAMPLE 2 (*Pòlya's urn model*). This popular example has been already considered in Example **15**.1.2-5, and has nothing in common with Pòlya's model in Example 4.1-2.

There are r red and b black balls in an urn. We select at random a ball and put it back with c additional balls of the color drawn. After that, we draw another ball and continue sampling in the same fashion.

Let $X_j = 1$ if the jth ball selected is red, and $X_j = 0$ otherwise. We show that the infinite sequence $X_1, X_2, ...$ is that of exchangeable r.v.'s.

Let $p(c_1,...,c_n) = P(X_1 = c_1,,...,X_n = c_n)$, where c_i's are equal to either 1 or 0. To prove that the X's are exchangeable, it suffices to prove that $p(c_1,...,c_n)$ does not change after any permutation of $c_1,...,c_n$. Since any permutation may be performed by a sequence of permutations of adjacent c's, it suffices to consider only permutations of this type.

Consider the realizations $c_1,...,c_j,c_{j+1},c_{j+2}$ and $c_1,...,c_j,c_{j+2},c_{j+1}$. In both cases, the numbers of red balls after draw $j+2$ *are the same,* and this is the only information we should know to compute the probabilities of realizations after the draw $j+2$. Consequently, it suffices to show that $p(c_1,...,c_j,c_{j+1},c_{j+2}) = p(c_1,...,c_j,c_{j+2},c_{j+1})$. Dividing this by $P(X_1 = c_1,...,X_j = c_j)$, we come to the equivalent relation

$$\begin{aligned} & P(X_{j+1} = c_{j+1}, X_{j+2} = c_{j+2}, \,|X_1 = c_1,...,X_j = c_j) \\ = \; & P(X_{j+1} = c_{j+2}, X_{j+2} = c_{j+1}, \,|X_1 = c_1,...,X_j = c_j). \end{aligned}$$

We restrict ourselves to proving that

$$P(X_{j+1} = 0, X_{j+2} = 1\,|\,X_1 = c_1,...,X_j = c_j) = P(X_{j+1} = 1, X_{j+2} = 0\,|\,X_1 = c_1,...,X_j = c_j). \tag{4.2.1}$$

Suppose that after the jth draw, given $X_1 = c_1,...,X_j = c_j$, there are n_r red and n_b black balls. Then the l.-h.s.of (4.2.1) is equal to $\frac{n_b}{n_r+n_b} \cdot \frac{n_r}{n_r+n_b+c}$, while the r.-h.s. is $\frac{n_r}{n_r+n_b} \cdot \frac{n_b}{n_r+n_b+c}$. These expressions are identical.

Thus, the r.v.'s $X_1, X_2, ...$ are exchangeable. Then, in particular, they are identically distributed, and $P(X_j = 1)$, the probability to select a red ball at draw j, is the same as for the first draw, that is, $r/(b+r)$. □

Now, let us observe that the d.f. in (4.1.3) is symmetric, and hence, *any conditionally independent and identically distributed r.v.'s $X_1, ..., X_n$ are exchangeable.* A natural question is whether the converse assertion is true. Without additional conditions, the answer is negative.

EXAMPLE 3. Let $X_1 = \pm 1$ with equal probabilities, and $X_2 = -X_1$. The reader may verify that X_1, X_2 are exchangeable, and $E\{X_j\} = 0$. On the other hand, $E\{X_1X_2\} = -E\{X_1^2\} = -1$, which is inconsistent with (4.1.9). □

The remarkable theorem below says that if a sequence of exchangeable r.v.'s is infinite, then these r.v.'s are conditionally independent. We may also understand this as follows. Suppose that a finite sequence of r.v.'s $X_1, ..., X_n$ has a symmetric joint distribution. Suppose also that we can somehow continue (extend) this sequence to an infinite sequence $X_1, ..., X_n, X_{n+1}, X_{n+2}, ...$ keeping the exchangeability property for the whole infinite sequence. Then the original n r.v.'s are conditionally independent (certainly, as well as all r.v.'s in the infinite sequence).

Theorem 4 *(De Finetti). Let $X_1, X_2, ...$ be an infinite sequence of symmetrically dependent r.v.'s. Then there exist a r.v. or a r.vec. Θ and a family of d.f.'s $\{F_\Theta(x)\}$ such that for any $n = 1, 2, ...,$ the representation (4.1.3) is true.*[1]

EXAMPLE 4. Let us come back to Example 2. The infinite sequence $X_1, X_2, ...$ in this example is that of exchangeable r.v.'s. By Theorem 4, these r.v.'s are conditionally independent. On the other hand, each r.v. X_j takes on only two values: 1 and 0. Hence, its distribution, conditional or unconditional, is determined by one parameter: the probability that X_j will assume the value 1. Therefore, the common factor Θ may be identified with the probability that $X_j = 1$, and the distribution of the r.v.'s X_j may be modeled as follows.

We view Θ as a r.v. with values from $[0, 1]$, and given $\Theta = \theta$, we view $X_1, X_2, ...$ as independent r.v.'s assuming values 1 and 0 with probabilities θ and $1 - \theta$ respectively. In particular, $P(X_1 = 1, ..., X_k = 1 \mid \Theta = \theta) = \theta^k$. Hence,

$$P(X_1 = 1, ..., X_k = 1) = E\{\Theta^k\} = \int_0^1 \theta^k dG(\theta), \tag{4.2.2}$$

where $G(\theta)$ is the d.f. of Θ.

Relation (4.2.2) gives a way to find $G(\theta)$. Clearly,

$$P(X_1 = 1, ..., X_k = 1) = \frac{r}{r+b} \cdot \frac{r+c}{r+b+c} \cdot ... \cdot \frac{r+(k-1)c}{r+b+(k-1)c}, \tag{4.2.3}$$

and it remains to find the distribution of a r.v. whose kth moment equals the r.-h.s. of (4.2.3). In Exercise 13, we give detailed advice how to show that the probability density of such a distribution is

$$g(\theta) = G'(\theta) = \frac{1}{B(u, v)} \theta^{u-1}(1-\theta)^{v-1}, \text{ where } 0 \le \theta \le 1, \tag{4.2.4}$$

[1]It makes sense to note that in the context of this theorem, the random element Θ in (4.1.3) may have a complicated nature, and we should understand (4.1.3) rather in terms of distributions than in terms of random variables. As a rule, in concrete schemes, this circumstance does not matter. More detailed statements and references may be found in [10], [13], [42].

the Beta-function $B(u,v) = \frac{\Gamma(u)\Gamma(v)}{\Gamma(u+v)}$ and $u = r/c$ and $v = b/c$.

Regarding the Beta-function, see Section **6**.4.2 for detail. Distribution (4.2.4) is called the *Beta-distribution* with parameters u, v; we already considered it in Exercise **6**.51.

4.3 Limit theorems for exchangeable r.v.'s

Consider an infinite sequence of exchangeable r.v.'s $X_1, X_2, \dots$. Since these r.v.'s are also conditionally i.i.d., we will keep the notation from Section 4.1.

4.3.1 The law of large numbers

Suppose $E\{X_j\}$ is finite (and the same for all j since X's are identically distributed). Then without loss of generality, we may assume $E\{X_j\} = 0$, which implies $E\{m_\Theta\} = 0$.

We will see that in the case of conditionally independent r.v.'s, we should not expect the validity of the "usual" LLN, which in the case of $E\{X_j\} = 0$ would mean that

$$\overline{X}_n = \frac{X_1 + \dots + X_n}{n} \xrightarrow{P} 0. \tag{4.3.1}$$

As a matter of fact, the distribution of $\overline{X}_n$ approaches the distribution of the r.v. m_Θ which may be non-degenerate, that is, may assume more than one value, being, so to speak, "really random." More precisely, the following is true.

Theorem 5 *As $n \to \infty$, the d.f. of $\overline{X}_n$, that is,*

$$P(\overline{X}_n \le x) \to P(m_\Theta \le x) \tag{4.3.2}$$

at least for all x's at which $P(m_\Theta \le x)$, as a function of x, is continuous.

At points where $P(m_\Theta \le x)$ jumps, the convergence may not take place, but it does not matter. We discussed this type of convergence in Section 7.3. Theorem 5 will be proved in Section 4.3.3.

EXAMPLE 1. For the Pòlya urn model from Examples 4.2-2 and 4, $m_\Theta = E\{X_j \,|\, \Theta\} = P(X_j = 1 \,|\, \Theta) = \Theta$, where the r.v. Θ has the Beta-distribution with the Beta-density $g(\theta)$ given in (4.2.4). So, the distribution of m_Θ is continuous, and for any x,

$$P(\overline{X}_n \le x) \to \int_0^x g(\theta)d\theta \;\text{ for }\; 0 \le x \le 1.$$

Thus, though the probability of selecting a red ball is the same for all draws and equal $\frac{r}{r+b}$, the r.v. $\overline{X}_n$, the proportion of red balls selected after n draws does not approach this probability, and $\overline{X}_n$ is not getting "less random" as n is increasing. As a matter of fact, the distribution of $\overline{X}_n$ approaches the Beta-distribution mentioned.

Another matter is that the expected value of this distribution, that is, $E\{\Theta\} = P(X_1 = 1) = \frac{r}{r+b}$ (see also (4.2.2) and (4.2.3)). So, the mean proportion of red balls does equal $\frac{r}{r+b}$, but the proportion itself does not approach this probability. □

Now, consider the case when the "usual" LLN, i.e., (4.3.1), does hold.

Corollary 6 *Relation (4.3.1) is true if and only if $P(m_\Theta = 0) = 1$, which in the case of finite variances is equivalent to the uncorrelatedness of the r.v.'s X_j.*

The first assertion of the corollary immediately follows from (4.3.2), the second from (4.1.9).

EXAMPLE 2. Let $\xi_1, \xi_2, ...$ be i.i.d. r.v.'s with zero mean, and Θ be a r.v. independent of ξ's. Let $X_j = \Theta \cdot \xi_j$. Clearly, $X_1, X_2, ...$ is an infinite sequence of exchangeable r.v.'s. This is a simple scheme, and the LLN may be established in a direct way. Indeed, because ξ's are i.i.d., to them does the classical LLN apply. Then, with probability one,

$$\frac{X_1 + ... + X_n}{n} = \Theta \cdot \frac{\xi_1 + ... + \xi_n}{n} \to \Theta \cdot 0 = 0.$$

However, for us it is interesting how Corollary 6 works in this case. We have

$$E\{X_j \,|\, \Theta\} = E\{\Theta \cdot \xi_j \,|\, \Theta\} = \Theta E\{\xi_j \,|\, \Theta\} = \Theta E\{\xi_j\} = \Theta \cdot 0 = 0.$$

Thus, $m_\Theta = 0$ with probability one, the X's are not correlated, and (4.3.1) does follow from Corollary 6. □

4.3.2 The central limit theorem

Let again $E\{X_j\} = 0$, and let $Var\{X_j\} = \sigma^2 \neq 0$. Set $S_n = X_1 + ... + X_n$. Since S_n/n has a distribution close to that of m_Θ, the sum S_n has an order of $m_\Theta \cdot n$, and m_Θ with a positive probability is not equal to zero if the X's are correlated.

Now, suppose X's are uncorrelated, and hence $P(m_\Theta = 0) = 1$. Then, in particular, $E\{\sigma^2_\Theta\} = \sigma^2$.

We will see that in this scheme, it is more convenient to normalize S_n just by $\sqrt{n}$, setting $S^*_n = S_n/\sqrt{n}$.

Let $Q(v) = P(\sigma_\Theta \leq v)$, the d.f. of the r.v. σ_Θ, and let Y denote a standard normal r.v. independent of all other r.v.'s involved. To clarify the CLT below, let us consider first the r.v. $\sigma_\Theta Y$. We may view it as a normal r.v. with the "random standard deviation" σ_Θ. Conditioning on σ_Θ, we have

$$\begin{aligned} P(\sigma_\Theta Y \leq x) &= \int_0^\infty P(\sigma_\Theta Y \leq x \,|\, \sigma_\Theta = v) dP(\sigma_\Theta \leq v) = \int_0^\infty P(vY \leq x \,|\, \sigma_\Theta = v) dQ(v) \\ &= \int_0^\infty P(vY \leq x) dQ(v) = \int_0^\infty \Phi(x/v) dQ(v). \end{aligned} \tag{4.3.3}$$

The distribution in the r.-h.s.of (4.3.3) is called *weighted normal*: we are considering the weighted mixture of normal distributions with different variances.

Theorem 7 *For the scheme under consideration,*

$$P(S^*_n \leq x) \to P(\sigma_\Theta Y \leq x) = \int_0^\infty \Phi(x/v) dQ(v) \;\; as \;\; n \to \infty. \tag{4.3.4}$$

EXAMPLE 1. Let us come back to Example 4.3.1-2. Assume $Var\{\xi_j\} = 1$, and set $\sigma^2 = E\{\Theta^2\}$. Then $Var\{X_j\} = E\{X_j^2\} = E\{\Theta^2\}E\{\xi_j^2\} = \sigma^2$.

Furthermore, since $m_\Theta = 0$ and $E\{\xi_j\} = 0$, we have $\sigma^2_\Theta = E\{X^2_j \,|\, \Theta\} = E\{\Theta^2 \cdot \xi^2_j \,|\, \Theta\} = \Theta^2 E\{\xi^2_j \,|\, \Theta\} = \Theta^2 E\{\xi^2_j\} = \Theta^2$. So, $\sigma^2_\Theta = \Theta^2$.

The scheme is simple, and we may treat it directly:

$$S^*_n = \frac{1}{\sqrt{n}} \cdot \Theta \cdot (\xi_1 + ... + \xi_n) = \Theta \cdot \frac{1}{\sqrt{n}} (\xi_1 + ... + \xi_n).$$

Since ξ's are i.i.d. with zero mean and unit variance, the distribution of $\frac{1}{\sqrt{n}}(\xi_1 + ... + \xi_n)$ is asymptotically standard normal. Then the distribution of S^*_n is close to the distribution of the r.v. $\Theta \cdot Y$, which in our case, leads to (4.3.4).

Let, for example, Θ assume values 2 and 4 with probabilities $1/2$. Then

$$P(S^*_n \le x) \to \frac{1}{2} P(2Y \le x) + \frac{1}{2} P(4Y \le x) = \frac{1}{2} \Phi\left(\frac{x}{2}\right) + \frac{1}{2} \Phi\left(\frac{x}{4}\right).$$

EXAMPLE 2, to some extent, is contrived but illustrative. Consider the general situation of Theorem 7, and suppose that σ_Θ has the d.f. $Q(v) = 1 - e^{-v^2/2}$ for $v \ge 0$.

We will show that in this case, the limiting distribution has the density $f(x) = \frac{1}{2} e^{-|x|}$; rather simple and quite different from the normal. This distribution, called *two-sided exponential*, has been already considered in Exercises **6**.41 and **8**.4. The reader is recommended to graph $f(x)$.

It is easier to use m.g.f.'s. Since the m.g.f. of the standard normal r.v. Y is $e^{z^2/2}$, the m.g.f. of the r.v. $\sigma_\Theta Y$ is

$$\begin{aligned} E\{e^{z\sigma_\Theta Y}\} &= = E\{E\{e^{z\sigma_\Theta Y} \,|\, \Theta\}\} = E\{e^{z^2 \sigma^2_\Theta/2}\} = \int_0^\infty e^{z^2 v^2/2} dQ(v) \\ &= \int_0^\infty e^{z^2 v^2/2} e^{-v^2/2} d(v^2/2) = \int_0^\infty e^{-v^2(1-z^2)/2} d(v^2/2) = \int_0^\infty e^{-t(1-z^2)} dt. \end{aligned}$$

(In the last step, we used the variable change $t = v^2/2$.) The last integral is finite for $|z| < 1$ and is equal to $1/(1-z^2)$. Thus, the m.g.f. of the limiting distribution is

$$E\{e^{z\sigma_\Theta Y}\} = \frac{1}{1-z^2} = \frac{1}{1-z} \cdot \frac{1}{1+z}. \tag{4.3.5}$$

The first factor is the m.g.f. of a r.v. η having the standard exponential distribution, and the second factor is the m.g.f. of $-\eta$. Hence, if η_1 and η_2 are independent standard exponential r.v.'s, then the m.g.f. of the r.v. $\eta_1 - \eta_2$ is equal to the r.-h.s. of (4.3.5). Computing the corresponding convolution, it is straightforward to show that the density of $\eta_1 - \eta_2$ is the $f(x)$ above. (This is exactly what we did in Exercise **6**.41.) □

Now, let us observe that the r.-h.s of (4.3.4) is equal to a normal distribution if the distribution Q is concentrated at just one point. Since $E\{\sigma^2_\Theta\} = \sigma^2$, this point should be σ, and $P(\sigma_\Theta = \sigma)$ should equal one. Note also that, because $m_\Theta = 0$ with probability one, $\sigma^2_\Theta = E\{X^2_\Theta \,|\, \Theta\}$.

On the other hand, the r.v.'s $X^2_1, X^2_2, ...$ are also exchangeable and given Θ, they are independent. Then, by (4.1.9), $E\{X^2_\Theta \,|\, \Theta\}$ (and hence σ^2_Θ) assume only one value if and only if $X^2_1, X^2_2, ...$ are not correlated. This leads to

Corollary 8 *In the situation of Theorem 7,*

$$P(S_n^* \leq x) \to \Phi(x/\sigma) \tag{4.3.6}$$

(the normal distribution with a variance of σ^2) if and only if the r.v.'s $X_1, X_2, \ldots$ are uncorrelated, and the r.v.'s $X_1^2, X_2^2, \ldots$ are uncorrelated also.

It is worth emphasizing that the uncorrelatedness of $X_1, X_2, \ldots$ together with the uncorrelatedness of $X_1^2, X_2^2, \ldots$ do not imply the independence of the X's.

EXAMPLE 3. Consider two sequences of independent r.v.'s: $\alpha_1, \alpha_2, \ldots$ and $\beta_1, \beta_2, \ldots$. Assume that

$$\alpha_j = \begin{cases} 1 & \text{with probability } 1/2 \\ -1 & \text{with probability } 1/2 \end{cases}, \text{ and } \beta_j = \begin{cases} 2 & \text{with probability } 1/4 \\ 0 & \text{with probability } 1/2 \\ -2 & \text{with probability } 1/4 \end{cases}.$$

Clearly, $E\{\alpha_j\} = E\{\beta_j\} = 0$, and $Var\{\alpha_j\} = Var\{\beta_j\} = 1$.

Let a r.v. Θ assume two values: 1 and 2 with equal probabilities, and the whole sequence $\{X_1, X_2, \ldots\}$ is equal to the sequence $\{\alpha_1, \alpha_2, \ldots\}$ if $\Theta = 1$, and $\{X_1, X_2, \ldots\}$ is equal to the sequence $\{\beta_1, \beta_2, \ldots\}$ if $\Theta = 2$.

The reader is invited to make sure that in this case, $m_\Theta = 0$ and $\sigma_\Theta = 1$ with probability one. Furthermore, $P(S_n^* \leq x) = \frac{1}{2}P\left(\frac{1}{\sqrt{n}}(\alpha_1 + \ldots + \alpha_n) \leq x\right) + \frac{1}{2}P\left(\frac{1}{\sqrt{n}}(\beta_1 + \ldots + \beta_n) \leq x\right)$ $\to \frac{1}{2}\Phi(x) + \frac{1}{2}\Phi(x) = \Phi(x)$.

On the other hand, the r.v. X_j are strongly dependent: if, for instance, X_1 has assumed the value 1, all other X's cannot be equal to 2. □

4.3.3 Proofs

We will sketch proofs of Theorems 5 and 7, skipping formalities connected with "moving limits inside expectations."

Proof of Theorem 5. Given Θ, the r.v.'s X_j are i.i.d., and by the "usual" LLN, the conditional d.f. $P(\overline{X}_n \leq x \mid \Theta = \theta)$ converges to the d.f. of a r.v. assuming only one value m_θ. Hence, given $\Theta = \theta$,

$$P(\overline{X}_n \leq x \mid \Theta = \theta) \to 0 \text{ if } x < m_\theta, \text{ and } P(\overline{X}_n \leq x \mid \Theta = \theta) \to 1 \text{ if } x > m_\theta. \tag{4.3.7}$$

Furthermore, by the formula for total expectation,

$$P(\overline{X}_n \leq x) = E\left\{P(\overline{X}_n \leq x \mid \Theta)\right\} = E\left\{P(\overline{X}_n \leq x \mid \Theta) \mid m_\Theta < x\right\} P(m_\Theta < x)$$
$$+ E\left\{P(\overline{X}_n \leq x \mid \Theta) \mid m_\Theta = x\right\} P(m_\Theta = x) + E\left\{P(\overline{X}_n \leq x \mid \Theta) \mid m_\Theta > x\right\} P(m_\Theta > x).$$

Let x be a continuity point of the d.f. $P(m_\Theta \leq x)$. Then $P(m_\Theta = x) = 0$, and by (4.3.7), $P(\overline{X}_n \leq x) \to E\{1\} P(m_\Theta < x) + 0 + E\{0\} P(m_\Theta > x) = P(m_\Theta < x) = P(m_\Theta \leq x)$. ■

Proof of Theorem 7. Since $P(m_\Theta = 0) = 1$, given $\Theta = \theta$, the r.v.'s X_j are i.i.d. with zero mean and a variance of σ_θ^2. Hence, by the "usual" CLT, the conditional d.f. $P(S_n^* \leq x \mid \Theta = \theta)$ $\to \Phi(x/\sigma_\theta)$. (In the case $\sigma_\theta = 0$, we understand the r.-h.s. as the d.f. of a r.v. assuming only zero value.) Consequently, $P(S_n^* \leq x) = E\{P(S_n^* \leq x \mid \Theta)\} \to E\{\Phi(x/\sigma_\Theta)\}$. The last expression is just another representation of the r.-h.s. of (4.3.4). ■

5 EXERCISES

1. In Example 1.1-1, comment on how (1.1.9) would look for the lattice in Fig. 3.
2. For the scheme of Example 1.2-3, (a) find the dependency neighborhood of the vertex $\mathbf{v}_6$ in Fig. 1; (b) describe dependency neighborhoods for the lattice in Fig. 3 (certainly, ignoring pluses and minuses there).
3. Regarding the point process from Section 1.3, is $\mu(A)$ a counting measure?
4. For a point Poisson process N_A, find $Corr\{N_A, N_B\}$.
5. In Example 1.3-1, (a) realize that the integral in (1.3.3) may be computed with use of a table for the standard normal distribution; (b) model the case where the intensity of fires is growing proportionally to the distance from the center, and find $P(Y > r)$.
6. (*A binomial point process.*) Suppose n points are selected at random and independently from a unit cube. Let N_A be the number of points fallen in a region A in the cube. (a) Does this point process satisfy the complete randomness condition (1.3.1)? (b) Find the intensity $\mu(A)$, and the probability that all points will fall in the unit ball inscribed in the cube.
7. Suppose that in the region within ten miles from a hospital, the locations of emergency calls during a particular day are distributed accordingly to a Poisson point process. The region consists of two subregions with equal areas. In the first, the intensity equals 0.03 per sq. mile, in the second it is 0.01. Find the probability that there will be not more than 3 emergency calls in the region.
8. State the CLT for the scheme of Example 3.1-1 in the case where ξ's are uniform (a) on $[-1,1]$; (b) on $[0,1]$.
9. Consider a two-dimensional lattice with vertices $\mathbf{v} = (i, j)$; $i, j = 1, 2, \ldots$. To each vertex $\mathbf{v}$, we assign a r.v. $\xi_{\mathbf{v}}$ and assume ξ's to be i.i.d. A neighborhood $\mathcal{N}_{\mathbf{v}}$ consists of the nearest vertices. Let $X_{\mathbf{v}} = \xi_{\mathbf{v}} \cdot \prod_{\mathbf{u} \in \mathcal{N}_{\mathbf{v}}} \xi_{\mathbf{u}}$ (the product of the r.v.'s in the neighborhood). State the CLT for the case where ξ's are uniform on $[-1,1]$.
10. Let $Y, Y_1, Y_2, \ldots$ be i.i.d. r.v.'s, $E\{Y\} = 0$, $Var\{Y\} = 1$. Let the random vector $(X_{1n}, \ldots, X_{nn}) = (Y_1, \ldots, Y_n)$ with probability $1 - 1/n$, and $(X_{1n}, \ldots, X_{nn}) = (Y, \ldots, Y)$ with probability $1/n$. Let $S_n = X_{1n} + \ldots + X_{nn}$. Show that $P\left(\frac{1}{\sqrt{n}} S_n \leq x\right) \to \Phi(x)$ while $Var\{S_n\} = 2n - 1 \sim 2n$, which says that the standard deviation of the sum of r.v.'s is not always a good normalization even in the case of normal convergence.
11. In the situation of Example 4.1-2, prove that if Λ has the Γ-distribution with parameters (a, ν), then X's are negative binomial with parameters (p, ν), where $p = \frac{a}{1+a}$.
12. State the LLN for the scheme of Example 4.1-2. Interpret your result.
13. Consider the Pòlya urn model (Examples 4.2-2 and 4). (a) Show that the density of Θ is indeed the Beta-density from (4.2.4). In particular, make sure that $E\{\Theta\}$ indeed equals $\frac{r}{b+r}$. (*Advice*: Proceeding from the formula $\Gamma(t+1) = t\Gamma(t)$, show that the r.-h.s. of (4.2.3) equals $\frac{\Gamma(k+r/c)\Gamma((r+b)/c)}{\Gamma(k+(r+b)/c)\Gamma(r/c)}$. Using the formula $B(u, v) = \frac{\Gamma(u)\Gamma(v)}{\Gamma(u+v)}$, find the moments of the Beta-distribution. Compare what you got.) (b) Let W_n be the proportion of red balls in the urn after the nth drawing. Compare the limiting distributions of W_n and $\overline{X}_n$. (c) Find these distributions for the case $r = b = c$; for instance, when in the beginning, there are one red and one black ball, and each time one ball is added.
14. Is the limiting distribution in Theorem 7 always symmetric?

Chapter 18

Comparison of Random Variables. Risk Evaluation

1,2

1 PREFERENCE ORDER. FIRST SIMPLE CRITERIA

1.1 Preference order

We begin with

EXAMPLE 1 that is not quite serious but explicit. Suppose you are asked to choose one of two ways of payment for a job. In the first case, you will get \$50; in the second, a coin will be tossed. If it comes up heads, you will get \$100, otherwise—nothing. So, you are asked to compare two r.v.'s of the future income:

$$X = 50, \text{ and } Y = \begin{cases} 100 & \text{with probability } 1/2 \\ 0 & \text{with probability } 1/2 \end{cases}. \tag{1.1.1}$$

A decision may depend on a person and on a situation. The mean values are identical, and numerous observations show that in such cases, most people prefer the certain amount; in our example, it is X. We may call such people *risk averters*. On the other hand, \$50 is not a big money to be afraid of losing it. So, some people may prefer an alternative involving a risk. These people would exhibit a *risk lover* behavior.

Assume that you have preferred X to Y above, and after that you are asked to compare the r.v. Y with the certain amount $X_1 = 49$. Though now $E\{Y\} > E\{X_1\}$, you may still prefer the certain amount; that is, X_1. Then you are asked to compare Y with $X_2 = 48$, and so on, until you start to hesitate. Suppose this happened at $X_{10} = 40$. This may be interpreted as if you view the random income Y and the certain \$40 as equivalent. Loosely speaking, you are ready to pay \$10 for stability. □

This chapter addresses a theory of comparison of risky alternatives. As a rule, we will talk about possible values of a future profit (which takes on negative values in the case of losses). While comparison criteria may vary, they usually have one feature in common. When choosing one out of possible alternatives, we have competing interests: we want the profit to be large, and we want the risk to be low. Usually, we can reach a certain level of stability only by sacrificing a part of the profit—we should pay for stability. So, our decision is a result of a trade-off between the possible gain and stability.

Formally, in each situation, we specify a class of r.v.'s under consideration, and try to establish a rule allowing, for *each* pair (X,Y) of r.v.'s from the class mentioned, to determine whether X is better than Y, or X is worse than Y, or these two random variables are equivalent. We will denote it, respectively, as $X \succ Y$, $X \prec Y$ and $X \simeq Y$. The symbolism

$X \succsim Y$ will mean that X "is not worse than Y"; that is, X is either better than or equivalent to Y. We will call the rule of comparison itself a *preference order* and denote it by $\succsim$.

An order $\succsim$ is said to be preserved, or completely determined, by a function $V(X)$ if for any X, Y from the class under consideration,

$$X \succsim Y \text{ if and only if } V(X) \geq V(Y). \tag{1.1.2}$$

The function $V(X)$ may be viewed as a measure of "quality" of X: the larger $V(X)$, the better X, and X is not worse than Y if and only if $V(X) \geq V(Y)$.

We call a preference order that is preserved by such a function *calculable* and restrict ourselves to such orders.

EXAMPLE 2 (*The mean-value criterion*). Suppose we compare r.v.'s proceeding just from their mean values. Then $V(X) = E\{X\}$. Certainly, this is a non-flexible rule of comparison. For instance, for such a rule, the r.v.'s in (1.1.1) are equivalent, which may not reflect real preferences. □

When we deal with r.v.'s of a future profit, we follow the rule "the larger, the better"; so, it is natural to consider only preference orders having the following *monotonicity property*: if $X \geq Y$ with probability one, then X is not worse than Y.

In the case of (1.1.2), the *monotonicity property* is equivalent to the requirement

$$\text{If } P(X \geq Y) = 1, \text{ then } V(X) \geq V(Y). \tag{1.1.3}$$

EXAMPLE 3. The mean criterion is certainly monotone. Indeed, if $P(X - Y \geq 0) = 1$, then $E\{Y - X\} \geq 0$, which implies $E\{X\} \geq E\{Y\}$. □

1.2 Several simple criteria

The simplest criterion has been considered in Examples 2-3 above. The next popular criterion concerns

1.2.1 Value-at-Risk (VaR)

We use the characteristic $q_\gamma(X)$, the γ-quantile of a r.v. X, which has been defined in Section **7**.1.2. Loosely put, it is the largest number x for which $P(X \leq x) \leq \gamma$. For the sake of convenience, we repeat this definition here and in a bit different fashion.

Let $F(x)$ be the d.f. of X, and $q_\gamma = q_\gamma(X)$. If the r.v. X is continuous and its d.f. is strictly increasing, then q_γ is the unique number q for which $F(q) = \gamma$; see Fig. 1a.

If there are many numbers q for which $F(q) = \gamma$, the definition we adopt, chooses the right end point of the interval where $F(x) = q$; see Fig. 1b,c. In the literature, one can find definitions where the γ-quantile is the left end point, or the middle, or even any point from this interval. The difference is not essential.

If the r.v. takes on some values with positive probabilities (and hence the d.f. has "jumps"), it may happen that there is no number q such that $F(q) = \gamma$; see Fig. 1d. Then we choose the point at which $F(x)$ "jumps" over the level γ.

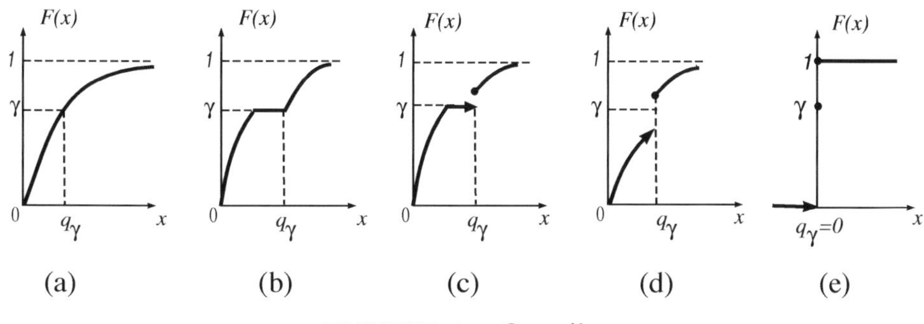

FIGURE 1. Quantiles

In particular, if $X = 0$ with probability one, the point 0 is the γ-quantile for all $\gamma \in [0,1)$ (see Fig. 1e).

As above, we view r.v.'s as those of a profit, and for brevity, for an individual or a company whose future profit we consider, we use the term "investor."

Let γ be a fixed level of probability, viewed as sufficiently small. Assume that an investor with a random profit of X does not take into account events whose probabilities are less than γ. Then for such an investor the worst, smallest conceivable level of the profit is $q_\gamma(X)$.

Let, for instance, $\gamma = 0.05$. Then the probability that the income will be less than $q_{0.05}$ is not larger than 5%, and non-formally speaking, $q_{0.05}$ is the smallest (worst) value of the income among all values which may occur with 95% probability. One may say that $q_{0.05}$ is the value at 5% risk. Certainly, q_γ may be negative, which would correspond to losses.

We define the *VaR criterion* as the criterion that is preserved by the function $V(X) = q_\gamma(X)$; so X is "not worse" than Y if and only if $q_\gamma(X) \geq q_\gamma(Y)$.

In applications, for the γ-quantile of X, the notation $VaR_\gamma(X)$ is frequently used; we will keep the notation $q_\gamma(X)$.

The VaR criterion is monotone. Indeed, if $X \geq Y$ with probability one, than $P(X \leq x) \leq P(Y \leq x)$, so in this case $q_\gamma(X) \geq q_\gamma(Y)$.

The criterion under consideration may be called cautious: the decision maker compares "pessimistic" scenarios for alternatives under consideration. The particular choice of $\gamma = 0.05$ is very common, but it has rather a psychological explanation: 0.01 is "too small," while 0.1 is "too large."

As a matter of fact, whether a particular value of γ should be viewed as small or not depends on the situation.

We can view a probability of 0.05 as small if it is the probability that it will be rainy tomorrow. However, the same number should be considered very large if it is the probability of being involved in a traffic accident: it would mean that on the average, we are likely to be involved in an accident one out of twenty times we are in traffic.

EXAMPLE 1. Consider two r.v.'s of future income: X that is uniform on $[0,2]$ and Y having the standard exponential distribution. So, the r.v.'s have the same means.

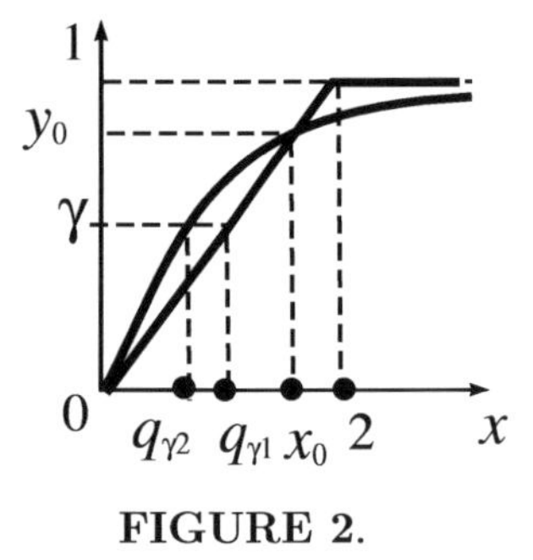

FIGURE 2.

The respective d.f.'s, $F_X(x) = x/2$ for $x \in [0,2]$ and $F_Y(x) = 1 - e^{-x}$ for $x \geq 0$, are depicted in Fig. 2. They intersect at a point $x_0 \approx 1.59$; the corresponding value of the d.f.'s is $y_0 = F_X(x_0) = F_Y(x_0) \approx 0.79$. For $\gamma < y_0$, the respective γ-quantiles, $q_{\gamma 1} = q_\gamma(X)$ and $q_{\gamma 2} = q_\gamma(Y)$, are also shown in Fig. 2. We see that, in this case, $q_{\gamma 1} > q_{\gamma 2}$, and hence $X \succ Y$. The reader is invited to realize that for $\gamma > y_0$, the conclusion will be the opposite.

EXAMPLE 2. Let a r.v. X take on values $0, 10, 20$ with probabilities $0.1, 0.1, 0.8$, respectively, and let a r.v. Y assume the same values with probabilities 0.07, 0.14, 0.79. The parts of the graphs of the d.f.'s $F_X(x)$ and $F_Y(x)$ are given in Fig. 3. For $\gamma = 0.05$, we have $q_\gamma(X) = q_\gamma(Y) = 0$ (see Fig. 3), and the r.v.'s X and Y are equivalent under the VaR criterion. For $\gamma = 0.08$, we have $q_\gamma(X) = 0$, while $q_\gamma(Y) = 10$, that is, X is worse than Y. So, the result of comparison again depends on γ.

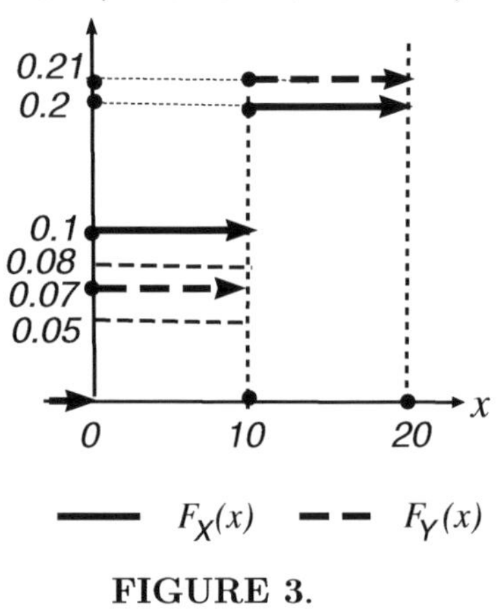

FIGURE 3.

EXAMPLE 3. Let X be normal with mean m and variance σ^2. Since the d.f. of X is $\Phi\left(\frac{x-m}{\sigma}\right)$, where $\Phi(x)$ is the standard normal d.f., the γ-quantile of X is a solution to the equation $\Phi\left(\frac{q-m}{\sigma}\right) = \gamma$. Denote by $q_{\gamma s}$ the γ-quantile of the *standard* normal distribution, i.e., $\Phi(q_{\gamma s}) = \gamma$. Then we can rewrite the equation mentioned as $\frac{q-m}{\sigma} = q_{\gamma s}$, and hence,

$$q_\gamma(X) = m + q_{\gamma s}\sigma.$$

The coefficient $q_{\gamma s}$ depends only on γ. As a rule, people choose $\gamma < 0.5$, and in this case $q_{\gamma s} < 0$. For example, if $\gamma = 0.05$, then $q_{\gamma s} \approx -1.64$ (see the Appendix, Table 4), and the VaR criterion is preserved by the function $q_\gamma(X) \approx m - 1.64\sigma$. Criteria of the type

$$V(X) = m - k\sigma, \tag{1.2.1}$$

where k is a positive number, are frequently used in practice, and not only for normal r.v.'s—as we will see in Section 1.2.4, maybe too frequently. The expression in (1.2.1) can be interpreted as follows. If X as a future profit, and variance (or standard deviation) is a measure of riskiness, then we want the mean m to be as large as possible and σ as small as possible. This is reflected by the minus sign in (1.2.1). The number k may be viewed as a weight assigned to the standard deviation.

EXAMPLE 4. There are n assets with random *returns* $X_1, ..., X_n$. As was already mentioned in previous chapters, the term "return" means that, if you invest \$1 in, say, the first asset, you will get an income of X_1 dollars. For example, if the today price of a stock is \$11, while the yesterday price was \$10, the return for this period is 1.1. Certainly, the returns X_i may be less than one.

Assume $X_1, ..., X_n$ to be independent and their distributions to be closely approximated by the normal distribution with mean m and variance σ^2.

Let us compare two strategies of investing n million dollars: either investing the whole sum in one asset, for example, in the first, or distributing the investment sum equally between n assets. We proceed from the VaR criterion with $\gamma = 0.05$.

For the first strategy, the income will be the r.v. $Y_1 = nX_1$. The mean $E\{Y_1\} = nm$, and $Var\{Y_1\} = n^2\sigma^2$. So, to compute $q_\gamma(Y_1)$, we should replace in (1.2.1) m by nm, and σ by $n\sigma$. Replacing $q_{\gamma s}$ by its approximate value -1.64, we have

$$q_\gamma(Y_1) = mn - 1.64n\sigma = 10m - 16.4\sigma$$

if, say, $n = 10$. For the second strategy, the income is the r.v. $Y_2 = X_1 + ... + X_n$. Hence, $E\{Y_2\} = nm$, $Var\{Y_2\} = n\sigma^2$, and

$$q_\gamma(Y_2) = mn - 1.64\sqrt{n}\sigma \approx 10m - 5.2\sigma.$$

Thus, the second strategy is preferable, which might be expected from the very beginning. Nevertheless, in the next example, we will see that if the X_i's have a distribution different from normal, we may jump to a different conclusion.

EXAMPLE 5[1]. There are ten *independent* assets such that investment in each with 99% probability gives 4% profit, and with 1% probability, the investor loses the whole investment sum. Assume that we invest \$10 million and compare the same two strategies as in Example 4. Let us again apply the VaR criterion with $\gamma = 0.05$.

If we invest all \$10 million in the first asset, we will get \$10.4 million with probability 0.99, and in the notation of the previous example, $q_\gamma(10X_1) = 10.4$.

For the second strategy, the number of successful investments has the binomial distribution with parameters $p = 0.99$, $n = 10$. If the number of successes is k, then the income is $x_k = k \times 1\text{million} \times 1.04$. So, if X is the total income, then $P(X \leq x_k) = P(N \leq k)$, where N is the corresponding binomial r.v. The values of the income and the corresponding cumulative probabilities computed with use of Excel, are given in the table below.

k	≤ 6	7	8	9	10
The income x_k	≤ 6.24	7.28	8.32	9.36	10.4
$P(X \leq x_k)$	$\leq 2.002 \cdot 10^{-6}$	0.000114	0.004266	0.095618	1

The 0.05-quantile of this distribution is $9.36 < 10.4$. Therefore, following VaR, we should choose the first investment strategy.

However, if we choose as γ, a number slightly smaller than 0.01—for example $\gamma = 0.0095$, then the result will be different. In this case, $q_\gamma(10X_1) = 0$, while the 0.0095-quantile of the distribution presented in the table, is still 9.36.

Certainly, the results of the comparison above should not be considered real recommendations. On the contrary, the last example indicates a limitation of the application of VaR, and shows that this criterion is sensitive to the choice of γ. □

[1]This example is close to an example from [2] presented also in [4, p.14] with the corresponding reference.

1.2.2 An important remark: Rather risk measures than criteria

This simple but important remark concerns the two criteria above and practically all other criteria we will consider in this chapter. The point is that we do not have to limit ourselves to using only one criteria each time. On the contrary, we can combine them.

For example, when considering a random income X, we may compute its expectation $E\{X\}$ and its quantile $q_\gamma(X)$. In this case, we will know what we can expect on the average, and what is the worst conceivable outcome. When comparing two r.v.'s, we certainly may take into account both characteristics. How we will do this depends on our preferences. The simplest way is to consider the linear combination $\alpha E\{X\}+\beta q_\gamma(X)$, where α and β play the role of weights we assign to the mean and to the quantile. The larger β, the more pessimistic we are.

Under such an approach to risk assessment, various functions $V(X)$ present not criteria but rather characteristics of the random income X. In this context, criteria $V(X)$ are frequently called *risk measures*.

Route 1 ⇒ *page 441*

2

1.2.3 Tail conditional expectation or Tail-Value-at-Risk

We slightly touch on a modification of VaR. First, consider

EXAMPLE 1. Two r.v.'s, X and Y, of future income are distributed as follows:

X takes on	values	-1	10	20
	with probabilities	0.03	0.47	0.5
Y takes on	values	-600	10	20
	with probabilities	0.01	0.49	0.5

The probabilities that the income will be negative in both cases are small: 3% and 1%, though the former probability is three times larger than the latter. For $\gamma=0.025$, we would have $q_\gamma(X)=-1$ and $q_\gamma(Y)=10$. So, under the VaR criterion, Y is preferable, which does not look natural. While we may neglect negative values of the income in the first case, this may be unreasonable in the second: a loss of 600, while the maximal possible gain is just 20, can be too serious to ignore, even if such an event occurs with a small probability of 1% which is a third of the probability for the first case. □

In situations as above, we speak about the possibility of *large deviations*, or a *heavy tail* of the distribution. One of ways to take the possibility of large deviations into account is to consider the mean values of large deviations; more precisely, the function

$$V_{\text{tail}}(X)=E\{X\,|\,X\le q_\gamma(X)\}.$$

EXAMPLE 1 revisited. Consider the r.v.'s from Example 1, and $\gamma=0.025$. As we already saw, in this case, $q_\gamma(X)=-1$ and $q_\gamma(Y)=10$. If the r.v. X took on a value less or equal -1, then it can take on only one value -1. Hence,

$$V_{\text{tail}}(X)=E\{X\,|\,X\le -1\}=-1.$$

For Y, we have

$$V_{\text{tail}}(Y) = E\{Y \,|\, Y \leq 10\} = -600 \cdot \frac{0.01}{0.5} + 10 \cdot \frac{0.49}{0.5} = -2.2 < -1.$$

So, $X \succ Y$.

On the other hand, as can be easily computed, if we replace 600 by 500, then the conclusion will be opposite. So, the criterion is flexible in taking into account the relations between values and probabilities. □

1,2

1.2.4 The mean-and-standard-deviation criterion

This criterion is the same as (1.2.1), but it concerns not only normal r.v.'s, and the motivation and derivation are different. Consider an investor expecting a random income X. Set $m_X = E\{X\}$, $\sigma_X^2 = Var\{X\}$. Suppose the investor identifies the riskiness of X with its standard deviation (or variance), and wishes the mean income m_X to be as large as possible, and the standard deviation σ_X—as small as possible. The quality of the r.v. X for such an investor is determined by a function of m_X and σ_X. In the simplest case, it is a linear function, and we can write the preserving function $V(X)$ as

$$V(X) = m_X - k\sigma_X, \tag{1.2.2}$$

where a weight $k \geq 0$.

In Section **10**.2.3, we already considered a similar criterion $\widetilde{V}(x) = \tau m_X - \sigma_X^2$. The fact that the latter criterion involves variance rather than standard deviation does not matter much: both characteristics are of the same nature, and they uniquely determine each other. The choice of variance in Chapter 10 was connected with the convenience of calculations.

The fact that, in Section **10**.2.3, we assigned a weight to mean value rather than to variance is entirely a tribute to tradition, and does not matter at all. Indeed, we may write $\widetilde{V}(X) = \tau(m_X - \frac{1}{\tau}\sigma_X^2)$. Now, a weight is assigned to variance, while the front factor τ does not change the comparison rule: if, comparing r.v.'s X and Y, we check whether $\widetilde{V}(X) > \widetilde{V}(Y)$, then the factor τ will cancel.

In this chapter, we consider (1.2.2). In a certain sense, it is more convenient because both terms in (1.2.2) have the same units of measurement (say, dollars if we deal with income, rather than dollars in square).

When introducing (1.2.2), we did *not* assume the r.v.'s under consideration to be normal, which we did when deriving similar criterion (1.2.1). We will see that the former approach may cause problems. The mean-standard-deviation or mean-variance criteria are very popular and at first glance look quite natural. *However, in some situations the choice of such criteria may contradict common sense.*

EXAMPLE 1. Let $X = 0$ and for a number $a \geq 1$,

$$Y = \begin{cases} a & \text{with probability } \frac{1}{a}, \\ 0 & \text{with probability } 1 - \frac{1}{a}. \end{cases}$$

Clearly, $E\{Y\} = 1$ and $Var\{Y\} = E\{Y^2\} - (E\{Y\})^2 = a - 1$. Then,

$$V(Y) = 1 - k\sqrt{a-1}, \quad \text{while } V(X) = 0 - k \cdot 0 = 0. \tag{1.2.3}$$

So, whatever positive k is, we can choose a sufficiently large a for which $V(Y) < 0$. On the other hand, $V(X) = 0$, and under the mean-and-standard deviation criterion, Y is worse than X, whereas $P(X \leq Y) = 1$. Clearly, if we replace in (1.2.3) the standard deviation $\sqrt{a-1}$ by the variance $a-1$, then the difference will be even more dramatic. $\square$

Note also that it would be a mistake to think that the example above is contrived, and in practice problems, we do not watch such cases. In Exercise 7, the reader is invited to compare a r.v. having the Pareto distribution (frequently used in many applications) and a uniform variable, and to watch a similar phenomenon.

It is also worth noting that the linearity of the function in the r.-h.s. of (1.2.2) is not an essential circumstance, neither is the choice of particular r.v.'s X and Y.

> One may observe the same phenomenon for $V(X)$ equal to almost *any* function $g(m_X, \sigma_X)$ of the mean and standard deviation. Moreover, for any r.v. X, we may point out a r.v. Y such that $P(Y \geq X) = 1$ whereas $V(Y) < V(X)$.

More precisely, this means the following. Let $V(X) = g(m_X, \sigma_X)$. To avoid cumbersome formulations, assume that $g(x,y)$ is smooth. Since we want the mean to be large and the variance to be small, it is natural to assume that the partial derivatives $g_1(x,y) = \frac{\partial}{\partial x} g(x,y) > 0$, $g_2(x,y) = \frac{\partial}{\partial y} g(x,y) < 0$.

Proposition 1 *Assume, in addition, that the partial derivatives $g_1(x,y)$ and $g_2(x,y)$ are continuous functions. Then for any r.v. X with a finite variance, there exists a r.v. Y such that $P(Y \geq X) = 1$, while*

$$g(m_X, \sigma_X) > g(m_Y, \sigma_Y).$$

We will prove it in the end of this section.

Proposition 1 is a strong argument against using variance as a measure of risk. However, *if we restrict ourselves to a sufficiently narrow class of r.v.'s, the monotonicity property may hold.*

In particular, this is true if we consider only normal r.v.'s because there are no two normal r.v.'s, X and Y with different variances and such that $X \leq Y$ with probability one.

To show it rigorously, assume that the normal r.v.'s X and Y mentioned exist. We have $P(X \leq x) = \Phi((x - m_X)/\sigma_X)$ and $P(Y \leq x) = \Phi((x - m_Y)/\sigma_Y)$. Since $P(Y \geq X) = 1$, it is true that $P(Y \leq x) \leq P(X \leq x)$, and hence $\Phi((x - m_Y)/\sigma_Y) \leq \Phi((x - m_X)/\sigma_X)$ for any x.

The function $\Phi(x)$ is strictly increasing. Therefore, from the last inequality, it follows that $\frac{x - m_Y}{\sigma_Y} \leq \frac{x - m_X}{\sigma_X}$ for *all* x. Certainly, this cannot be true if $\sigma_Y \neq \sigma_X$ because two lines with different slopes intersect and at only one point.

On the other hand, if $\sigma_Y = \sigma_X$, the comparison is trivial: $Y \succsim X$ if $m_Y \geq m_X$.

The case of normal r.v.'s is simple because the normal distribution is characterized only by two parameters: mean m and variance σ^2. Each normal distribution may be identified with a point (m, σ) in a plane, and the rule of comparison will be equivalent to a rule of comparison of points in this plane.

If we consider a family of distributions with three or more parameters but still compare these distributions basing on their means and variances, we may come to paradoxes similar to what we saw above. Again, *this does not mean that we should not use the mean-variance criterion* (we did it, for example, in Section **10**.2.3), *but it does mean that we should be cautious.*

Proof of Proposition 1 uses the Taylor expansion for functions of two variables. Let $E\{X\} = m$, $Var\{X\} = \sigma^2$, and a number $\varepsilon \in (0,1)$. Set $Y = X + \xi_\varepsilon$, where the r.v. ξ_ε is independent of X, and

$$\xi_\varepsilon = \begin{cases} \varepsilon^{-1} & \text{with probability } \varepsilon^3, \\ 0 & \text{with probability } 1-\varepsilon^3. \end{cases}$$

Obviously, $P(Y \geq X) = 1$. Furthermore, $E\{\xi_\varepsilon\} = \varepsilon^2$, $Var\{\xi_\varepsilon\} = \varepsilon - \varepsilon^4$, and hence, $E\{Y\} = m + \varepsilon^2$, and $Var\{Y\} = \sigma^2 + \varepsilon - \varepsilon^4$.

Then $\sigma_Y = \sqrt{\sigma^2 + \varepsilon - \varepsilon^4}$. Applying Taylor's expansion for this function of ε (see the Appendix, (2.2.1)), and assuming $\sigma \neq 0$, we get that $\sigma_Y = \sigma + \frac{1}{2\sigma}\varepsilon + o(\varepsilon)$, where here and below $o(\varepsilon)$ stands for a remainder negligible as $\varepsilon \to 0$. (See also the Appendix, Section 2.1.)

By the Taylor expansion for $g(x,y)$, we have $g(m_Y, \sigma_Y) = g\left(m + \varepsilon^2, \sigma + \dfrac{1}{2\sigma}\varepsilon + o(\varepsilon)\right) =$

$$g(m,\sigma) + g_1(m,\sigma)\varepsilon^2 + g_2(m,\sigma)\left(\frac{1}{2\sigma}\varepsilon + o(\varepsilon)\right) + o(\varepsilon) = g(m,\sigma) + \frac{1}{2\sigma}g_2(m,\sigma)\varepsilon + o(\varepsilon).$$

Because $g_2(m,\sigma^2) < 0$ and the remainder $o(\varepsilon)$ is negligible for small ε, there exists $\varepsilon > 0$ such that $\frac{1}{2\sigma}g_2(m,\sigma^2)\varepsilon + o(\varepsilon) < 0$.

For such an ε, we have $g(m_Y, \sigma_Y^2) < g(m, \sigma^2) = g(m_X, \sigma_X^2)$.

In the case $\sigma = 0$, we have $\sigma_Y = \sqrt{\varepsilon - \varepsilon^4} = \sqrt{\varepsilon} + o(\varepsilon)$, and the proof is similar. ■

2 EXPECTED UTILITY

2.1 Expected utility maximization (EUM)

EXAMPLE 1 (*The Saint Petersburg paradox*). The problem below was first investigated by the Swiss mathematician Daniel Bernoulli in a paper of 1738 when D. Bernoulli worked in St. Petersburg. Let us come back to the game of chance that was already considered in Example **3**.2.3-1. The game consists of tossing a regular coin until a head appears. If the first head appears at the kth toss, the payment equals 2^k, say, dollars. So, the payment is a r.v. X taking on values $2, 4, 8, ..., 2^k, ...$ with probabilities $\frac{1}{2}, \frac{1}{4}, \frac{1}{8}, ..., \frac{1}{2^k}, ...$, respectively, and $E\{X\} = 2 \cdot \frac{1}{2} + 4 \cdot \frac{1}{4} + 8 \cdot \frac{1}{8} + ... = 1 + 1 + 1 + ... = \infty$

By the LLN, this means that, if the game is played repeatedly, and X_j is the payment in the jth play, then with probability one, $\frac{1}{n}(X_1 + .. + X_n) \to \infty$ as $n \to \infty$.

So, in the long run, the average payment will be greater than *an arbitrary large* number.

Then if a player had proceeded from the LLN, she/he would have agreed to pay any, arbitrary large, entry price for participating in each play. Certainly, it does not reflect preferences of real people: most would not agree to pay each time, for example, \$100 if even they are guaranteed to participate in a large number of plays. (Would the reader agree to pay \$100 each time?) □

There exists a purely mathematical solution to this paradox based on the fact that in this particular case, $\frac{X_1 + ... + X_n}{n \log_2 n} \to 1$ with probability one. A not very short proof may be found, e.g., in [13], [42]. Thus, if the entry price for each play depends on the number of plays n and equals $c = \log_2 n$, then for large n, the total payment for participating in n plays will be close to the total gain: the price c would be "fair."

This solution is strongly connected with the particular problem under consideration. Fortunately, D. Bernoulli did not know the fact mentioned, and the general solution he suggested proved to be very fruitful and, in the twentieth century, had led to a developed theory.

D. Bernoulli proceeded from the simple observation that the "degree of satisfaction" of having capital, or in other words, the "utility of capital," depended on the particular amount of capital in a nonlinear way. For example, if we give \$1000 to a person with a wealth of \$1,000,000, and the same \$1000 to a person with zero wealth, the former will feel much less satisfied than the latter.

To model it, D. Bernoulli assumed that the satisfaction of possessing a capital x, or the "utility" of x, may be measured by a function $u(x)$ that as a rule is not linear. Such a function is called a *utility function*, or a *utility of money function*. The word "satisfaction" would possibly reflect the significance of this concept better, but the term "utility" has already been adopted.

In the above framework, the utility function can be viewed as a characteristic of the individual, as if the individual is endowed by this function; so to speak, it is "built into the mind." To some extent, we can talk about the utility function of a company too. In this case, it reflects the preferences of the company.

D. Bernoulli himself suggested as a good candidate for a "natural" utility function the function $u(x) = \ln x$, assuming that the increment of the utility is proportional to the relative growth of the capital rather than to the absolute growth. More specifically, if a capital x is increased by a small dx, then the increment of the utility, $du(x)$, is proportional to dx/x, that is, $du = k\frac{dx}{x}$ for a constant k. A solution is $u(x) = k \ln x + C$, where C is another constant. These constants may be completely specified by the choice of units in which we measure utility. So, without loss of generality, we can set $k = 1$, $C = 0$, and come to $u(x) = \ln x$.

Consider now a random income X. In this case, the utility of the income is the r.v. $u(X)$. Bernoulli's suggestion was to proceed from the *expected utility* $E\{u(X)\}$.

EXAMPLE 1 revisited. Assume that the utility function of the player in St. Petersburg's paradox is $u(x) = \ln x$. Then the expected utility

$$E\{u(X)\} = \sum_{k=1}^{\infty} u(2^k)2^{-k} = \sum_{k=1}^{\infty} \ln(2^k)2^{-k} = (\ln 2)\sum_{k=1}^{\infty} k2^{-k} = 2\ln 2,$$

and, unlike $E\{X\}$, the expected utility is finite. (To realize that $\sum_{k=1}^{\infty} k2^{-k} = 2$, it suffices to recall that this is the expected value of the geometric r.v. with the parameter $p = 1/2$; see Example **3**.2.1-3.)

Let c be a price for participating in a play. We have seen that the mean-value criterion had led to a non-realistic conclusion that c could be arbitrary large. Proceeding from the expected utility concept, one may accept a price c whose utility is equal to the expected utility of X; that is, $u(c) = E\{u(X)\}$. We have $\ln c = 2\ln 2$, and $c = 4$. □

The concept of expected utility leads to the *expected utility maximization* (EUM) criterion: for a given $u(x)$ and for each two r.v.'s, X and Y, of a future income, we say that

$$X \succsim Y \quad \text{if and only if} \quad E\{u(X)\} \geq E\{u(Y)\}. \tag{2.1.1}$$

This is equivalent to choosing, as the preserving function in (1.1.2), the function $V(X) = E\{u(X)\}$. If $E\{u(X)\} = E\{u(Y)\}$, we say that X and Y are *equivalent.*

We will call an individual whose decisions are based on the rule (2.1.1) for some $u(x)$, an *expected utility maximizer* (EU maximizer). However, it is worth emphasizing that when talking about an EU maximizer, we mean that the person's preferences *may be described* by (2.1.1), but this does not imply in any way that calculations are really going in the mind.

A good image illustrating this was suggested in [31]. A thrown ball exhibits a trajectory described as a solution to a certain equation, but no one thinks that the ball "has and itself solves" this equation. People do not get confused about the ball but they sure do about models of other people.

Next, note that *the preference order (2.1.1) does not change if $u(x)$ is replaced by any function $u^*(x) = bu(x) + a$, where b is positive and a is an arbitrary number.* Indeed, if we replace in (2.1.1) u by u^*, then b and a will cancel.

Thus, u may be defined up to a linear transformation, and the scale in which we measure utility may be chosen at our convenience. In particular, there is nothing wrong or strange if u assumes negative values.

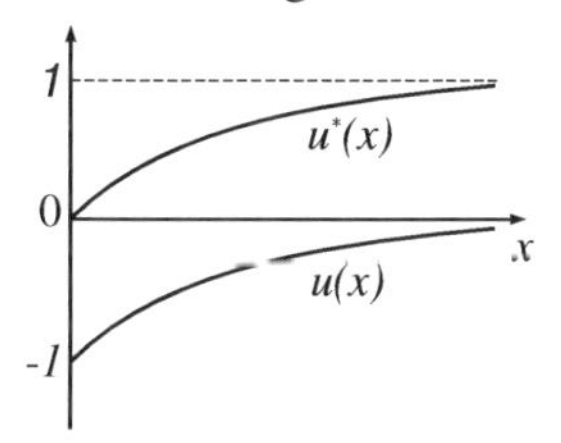

FIGURE 4.

EXAMPLE 2. As was already mentioned, when considering $u(x) = k\ln x + C$, we may set $k = 1$ and $C = 0$, restricting ourselves to $u(x) = \ln x$.

EXAMPLE 3. Let $u(x) = -1/(1+x)$, $x \geq 0$; see Fig. 4. Should the fact that $u(x)$ is negative for all x's make us uncomfortable? Not at all. Consider $u^*(x) = u(x) + 1 = x/(1+x)$; see again Fig. 4. The new function is positive but generates the same preference order. The sign of $u(x)$ does not matter; what matters when we compare X and Y is whether $E\{u(X)\}$ is larger than $E\{u(Y)\}$ or not. □

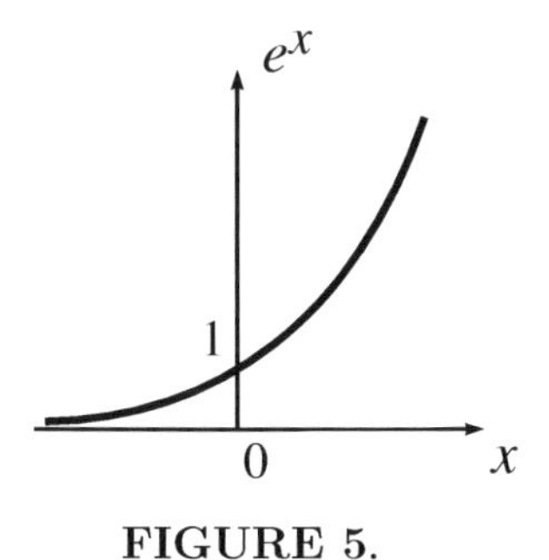

FIGURE 5.

EXAMPLE 4. (a) *A reckless gambler.* Let a gambler's utility function $u(x) = e^x$. Negative x's correspond to losses, and positive—to gains. The values of $u(x)$ for negative x's with large absolute values practically do not differ, while in the region of positive x's, the function $u(x)$ grows fast; see Fig. 5. We may interpret it as if the gambler is not concerned about possible losses and is highly enthusiastic about large gains. Note that the function

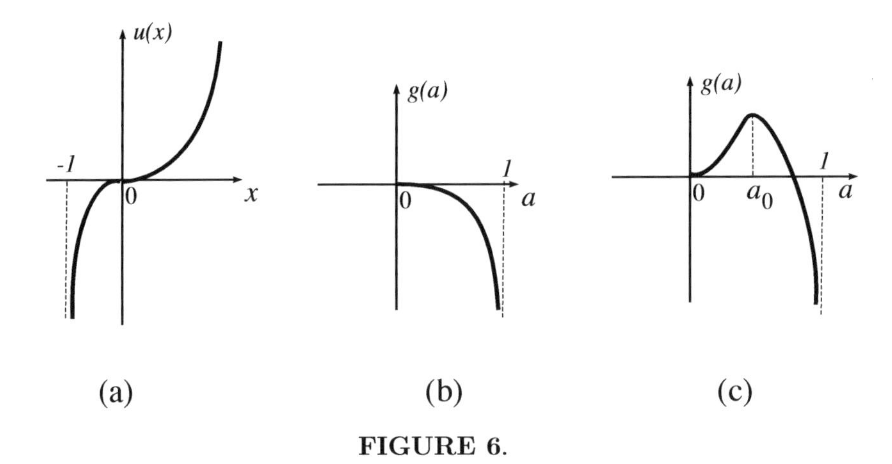

FIGURE 6.

e^x is convex, and as we will see later, the convexity of the utility function is relevant to the inclination to risk.

Consider a game in which the gambler wins a dollars with a probability of p, and loses the same amount a with the probability $q = 1 - p$. So, we deal with $X = \pm a$ with the mentioned probabilities. Assume $p < q$.

In our case, $E\{u(X)\} = e^a p + e^{-a} q$. The gambler will participate in such a game if X is better than a r.v. $Y \equiv 0$, which amounts to $E\{u(X)\} > u(0) = 1$. This is equivalent to $e^a p + e^{-a} q > 1$.

With the variable change $e^a = y$, the last inequality may be reduced to the quadratic inequality $py^2 - y + q > 0$; the reader is invited to check it on her/his own. One root of the corresponding quadratic equality is one, the other is $\frac{q}{p}$. (Since $p+q = 1$, the discriminant of the equation is $1 - 4pq = (p+q)^2 - 4pq = p^2 + q^2 - 2pq = (p-q)^2$.)

Since $y \geq 1$, the solution to the inequality is $y > q/p$, and consequently, $a > \ln(q/p)$. Thus, the gambler is inclined to bet large stakes, and will participate in the game only if $a > \ln(q/p)$. For instance, if $p = \frac{1}{4}$, the lowest stake acceptable for the gambler is $\ln 3 \approx 1.1$.

(b) *A cautious gambler.* Consider now a gambler who by no means wants to lose a unit of money. What this unit is equal to, \$1,000,000 or just \$100, depends on the gambler. On the other hand, the gambler does not mind taking some risk and participating in a game with a moderate stake. The utility function of such a gambler may look as in Fig. 6a: $u(x) \to -\infty$ as $x \to -1$, and $u(x)$ is growing as a convex function for positive x's. For example, the function

$$u(x) = \begin{cases} kx^2 & \text{for } x \geq 0, \\ \ln(1-x^2) & \text{for } -1 < x < 0, \end{cases}$$

has a similar graph. The parameter k indicates the gambler's inclination to risk. The larger k, the steeper $u(x)$ for positive x's. We wrote x^2 in $\ln(1-x^2)$ to make the function smooth at the origin.

Consider the same r.v. X as in the previous example. Then $E\{u(X)\} = ka^2 \cdot p + \ln(1 - a^2) \cdot q$. Denote the r.-h.s. by $g(a)$. The gambler will participate in the game if $E\{u(X)\} > u(0) = 0$, which amounts to $g(a) > 0$. The reader can readily verify that, if $k \leq q/p$, then

$g'(a) \le 0$, and hence, the graph of $g(a)$ looks as in Fig. 6b. So, $g(a)$ does not assume positive values. In this case, the gambler will refuse to play.

The reader may also double check that for $k > q/p$, the graph of $g(a)$ looks as in Fig. 6c. Now, $g(a)$ is positive in a neighborhood of zero. The maximum of $g(a)$ is attained at $a_0 = \sqrt{1-q/(kp)}$ (check by differentiation). For example, for $p = \frac{1}{4}$, and $k = 4$, we get $a_0 = \frac{1}{2}$, so the gambler's optimal behavior is to bet a half unit of money.

EXAMPLE 5 (*Expected utility and insurance*). Suppose that a potential customer of an insurance company has a wealth of w, and she/he faces a possible random loss ξ. Assume that the customer is an EU maximizer with a utility function $u(x)$. What premium G would the customer be willing to pay to insure the risk?

The customer's wealth after paying the premium becomes $X = w - G$, while if she/he does not buy the insurance, the wealth will equal the r.v. $Y = w - \xi$.

Then, in accordance with the principle (2.1.1), a premium G will be acceptable for the customer only if $u(w-G) \ge E\{u(w-\xi)\}$. (The l.-h.s. is not random.) In the boundary case, for the maximal acceptable premium $G_{\max}$,

$$u(w - G_{\max}) = E\{u(w-\xi)\}. \tag{2.1.2}$$

Let, for instance, $u(x) = \sqrt{x}$. As we will see a bit later, such a function may serve as a good example of a utility function.

Let $w = 1$, and let ξ take on values 1 and 0 with respective probabilities p and $q = 1-p$. For example, the customer insures her/his home against fire, and she/he presupposes that in the case of a fire, the home will be completely destroyed. We view the home value as a unit of money. If, when thinking about insurance, the customer takes into account only the value of the home itself, then the wealth w that is taken into account equals one.

By (2.1.2),

$$\sqrt{1-G_{\max}} = E\{\sqrt{1-\xi}\}.$$

On the other hand, $E\{\sqrt{1-\xi}\} = (\sqrt{1-1})p + (\sqrt{1-0})q = q$, and hence, $\sqrt{1-G_{\max}} = q$. Thus, $G_{\max} = 1-q^2$.

Assume, for instance, that the probability of a fire during a year is $p = 0.001$. Then, as is easy to compute, $G_{\max} = 0.001999$. Let the unit of money, that is, the home value, be \$500,000. Then the *maximum* accepted annual premium $G_{\max} = \$500000 \cdot 0.001999 = \$999.5$. □

The next notion we introduce is a *certainty equivalent*. First, note that any number c may be viewed as a r.v. taking only one value c. Consider a preference order $\succsim$ not necessarily connected with expected utility maximization. Assume that for a r.v. X, we can find a number $c = c(X)$ such that $c \simeq X$ with respect to the order $\succsim$. In other words, the decision maker is indifferent whether to choose c or X. The number $c(X)$ so defined is called the *certainty equivalent* of X.

For an EU maximizer with a utility function $u(x)$, the relation $c \simeq X$ is equivalent to $E\{u(X)\} = E\{u(c)\}$, and since c is not random, $E\{u(X)\} = u(c)$. If u is a one-to-one function, then there exists the inverse function $u^{-1}(y)$, and

$$c(X) = u^{-1}(E\{u(X)\}).$$

EXAMPLE 6. Let us come back to (1.1.1), and suppose $u(x) = \sqrt{x}$. Then $c(Y) = (E\{\sqrt{Y}\})^2 = \left(\sqrt{100}\cdot\frac{1}{2} + \sqrt{0}\cdot\frac{1}{2}\right)^2 = 25$. So, for such a person, Y is equivalent to just \$25, which says that the person is a strong risk averter.

EXAMPLE 7. (a) In the situation of Example 4a, $u^{-1}(x) = \ln x$. So, the certainty equivalent $c(X) = \ln(e^a p + e^{-a} q)$. For example, for $p = \frac{1}{4}$ and $a = 10$, we would have $c(X) = \ln(\frac{1}{4}e^{10} + \frac{3}{4}e^{-10}) \approx 8.614$, which is fairly close to 10. It is not surprising: the gambler does not care much about losses.

(b) Consider Example 4b for $p = \frac{1}{4}$ and $k = 4$. Now, the situation is quite different. The gambler bets $a = a_0 = \frac{1}{2}$. In this case, $E\{u(X)\} = g\left(\frac{1}{2}\right) = \frac{1}{4}\cdot 4\cdot\frac{1}{4} + \frac{3}{4}\cdot\ln(\frac{3}{4}) \approx 0.034$. On the other hand, $u^{-1}(y) = \sqrt{y/k}$ for positive y's. So, in our case the certainty equivalent $c(X) \approx \sqrt{0.034/4} \approx 0.0922$. □

Note that the certainty equivalent of a certain number a is, of course, this number: $c(a) = u^{-1}(E\{u(a)\}) = u^{-1}(u(a)) = a$.

2.2 Some particular utility functions. How to estimate utility functions

First, we consider some "classical" examples.

1. *Positive-power functions.* Let $u(x) = x^\alpha$ for all $x \geq 0$ and some $\alpha > 0$; see Fig. 7a. The expected utility in this case is considered only for positive r.v.'s, and $E\{u(X)\} = E\{X^\alpha\}$, the moment of X of the order α.

If $\alpha = 1$, then $E\{u(X)\} = E\{X\}$, and the EUM criterion coincides with the mean-value criterion. For $\alpha < 1$, the function $u(x)$ is concave, for $\alpha > 1$ - convex. We will see soon that this is strongly connected with the attitude to risk. If $\alpha < 1$, then the individual exhibits a risk averter's behavior; if $\alpha > 1$, we deal with a risk lover.

The certainty equivalent of a r.v. X is $c(X) = (E\{X^\alpha\})^{1/\alpha}$. In the simplest case $\alpha = 1$, the certainty equivalent $c(X) = E\{X\}$.

EXAMPLE 1. Let X be uniform on $[0, b]$. Then $c(X) = \left(\int_0^b x^\alpha \frac{1}{b} dx\right)^{1/\alpha} = \left(\frac{1}{1+\alpha} b^\alpha\right)^{1/\alpha}$ $= \left(\frac{1}{1+\alpha}\right)^{1/\alpha} b$. Because $(1+\alpha)^{1/\alpha}$ is decreasing in α, the smaller α, the smaller the certainty equivalent. □

2. *Negative-power functions.* Let $u(x) = -1/x^\alpha$ for all $x > 0$ and some $\alpha > 0$; see Fig. 7b. We again deal only with positive r.v.'s, and $E\{u(X)\} = -E\{X^{-\alpha}\}$. The fact that $u(x)$ is negative does not matter, but the fact that $u(x) \to -\infty$, as $x \to 0$, is meaningful: now the investor is much more afraid of being ruined than in the previous case when $u(x) \to 0$ as $x \to 0$. The fact that $u(x) \to 0$ as $x \to +\infty$, indicates the saturation effect: the investor does not distinguish much large values of the capital (so to say, not very greedy). Compare it with the previous case of the positive power where $u(x) \to +\infty$ as $x \to +\infty$.

3. *The logarithmic utility function,* $u(x) = \ln x$, $x > 0$, is in a sense intermediate between the two cases above and has been already discussed.

4. *Quadratic utility functions.* Consider $u(x) = 2ax - x^2$, where parameter $a > 0$; the multiplier 2 is written for convenience. The reader is invited to graph this parabola. Cer-

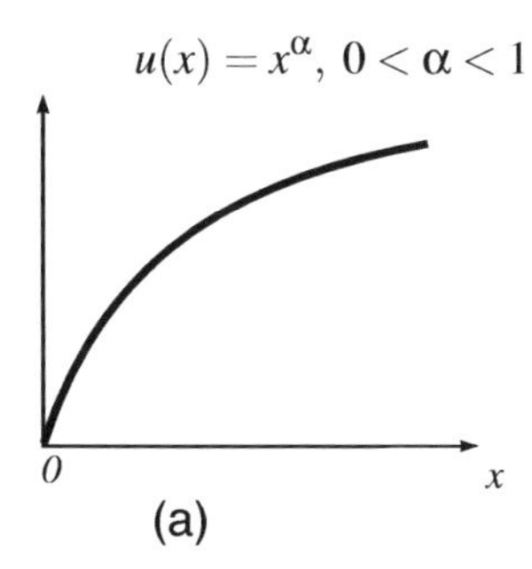

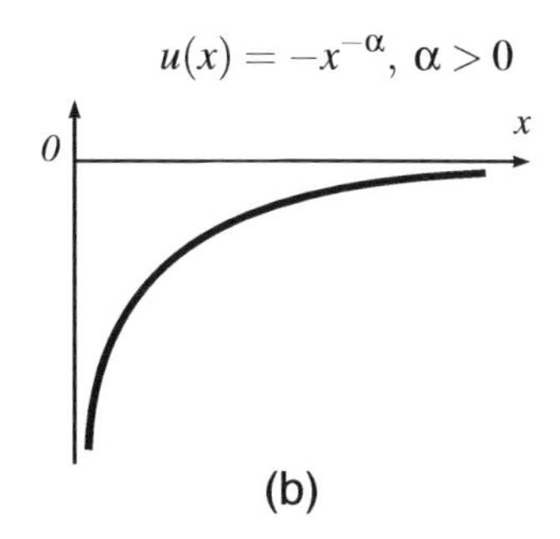

FIGURE 7. Positive- and negative-power utility functions.

tainly, such a utility function is meaningful only for $x \leq a$ when the function is increasing. Hence, in this case, we consider only r.v.'s X such that $P(X \leq a) = 1$. Negative values of X are interpreted as the case where the investor loses or owes money. We have $E\{u(X)\} = 2aE\{X\} - E\{X^2\} = 2aE\{X\} + (E\{X\})^2 - Var\{X\}$. Thus, the expected utility is a quadratic function of the mean and variance.

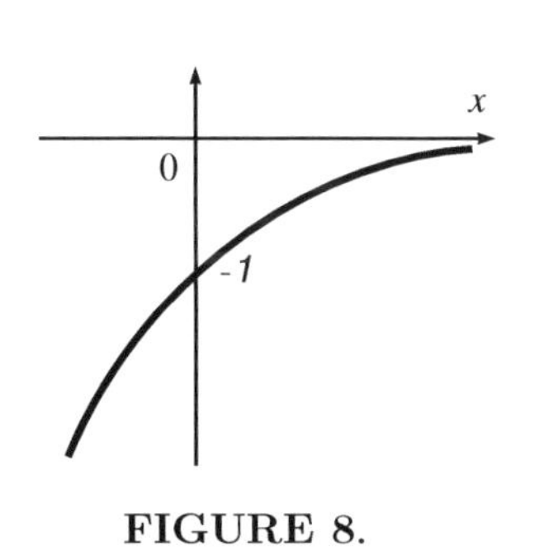

FIGURE 8.

5. *Exponential utility functions.* Let $u(x) = -e^{-\beta x}$, where parameter $\beta > 0$, and the function is considered for all x's. The graph is given in Fig. 8. Since $u(x) \to 0$, as $x \to \infty$, faster than any power function, the saturation effect in this case is stronger than in Case 2. The expected utility $E\{u(X)\} = -E\{e^{-\beta X}\} = -M(-\beta)$, where $M(z) = E\{e^{zX}\}$, the *moment generating function* of X.[2]

EXAMPLE 2. Let X be distributed exponentially with parameter a. Then $M_X(z) = a/(a-z)$. Simple calculations lead to $c(X) = \frac{1}{\beta}[\ln(a+\beta) - \ln a)]$. □

In conclusion, we consider a (somewhat naive) example illustrating how one may estimate utility functions in particular cases.

EXAMPLE 3. We believe that Chris is an EU maximizer, and we try to determine his utility function $u(x)$. Since the scale in which utility is measured does not matter, we can set, say, $u(0) = 0$ and $u(100) = 1$, where money is measured in convenient units.

You invite Chris to compare a game with a prize $X = 100$ or 0 with equal probabilities and a payment of 50 for sure. Chris finds X worse than a certain payment of 50. Then you reduce 50 to $49, 48, ...,$ and so on, up to the moment when Chris starts to hesitate. Assume that it happens at $c = 40$. Then we can view c as the certainty equivalent of X. This means that $u(c) = E\{u(X)\} = \frac{1}{2}u(100) + \frac{1}{2}u(0) = \frac{1}{2} \cdot 1 + \frac{1}{2} \cdot 0 = \frac{1}{2}$. Hence $u(40) = 0.5$, and we know the value of $u(x)$ at one more point. You can continue such a process, for example, figuring out how much Chris values a r.v. $X_1 = 100$ and 40 with equal probabilities. Assume that Chris's answer is 60. Then $u(60) = \frac{1}{2}u(100) + \frac{1}{2}u(40) = \frac{1}{2} \cdot 1 + \frac{1}{2} \cdot \frac{1}{2} = 0.75$. □

[2]The reader who skipped Chapter 8 may take $M(z)$ as it is defined here and ignore the term. If X has a density $f(x)$, then $M(z) = \int_{-\infty}^{\infty} e^{zx} f(x)dx$. If X is exponential with parameter a (see Example 2 below), then as is easy to calculate, $M(z) = \int_{-\infty}^{\infty} e^{zx} ae^{-ax}dx = a\int_{-\infty}^{\infty} e^{x(z-a)}dx = a/(a-z)$ for $z < a$.

In real life, it works not so well as in nice theoretical examples. The problem is not in mathematical modeling but in making results of such an inquiry reliable, reflecting the real preferences of the individual. This is a psychological rather than mathematical question. The difficulty is that answers depend on the situation, on the form in which the questions are asked, whether the questioning involves real money or the experiment is virtual, and on many other psychological, economic and social issues. These problems are beyond the scope of a book on mathematical modeling.

2.3 Risk aversion

Below, by the symbol Z_ε, we will denote a r.v.

$$Z_\varepsilon = \begin{cases} \varepsilon & \text{with probability } 1/2, \\ -\varepsilon & \text{with probability } 1/2, \end{cases}$$

where $\varepsilon > 0$. We will talk about the *risk aversion* of an individual with a preference order $\succsim$ if the following condition holds.

Condition Z: For any r.v. X, any $\varepsilon > 0$, and any r.v. Z_ε independent of X, it is true that $X \succsim X + Z_\varepsilon$.

Condition Z reflects the rule "the less stable, the worse." An investor with preferences satisfying this property would not accept an offer resulting in either an additional income with probability $1/2$ or a loss of the same amount and with the same probability.

It is important to emphasize that *Condition Z concerns an arbitrary preference order*, not only the EUM criterion.

An individual whose preference order satisfies Condition Z is called a *risk averter*. If $X \precsim X + Z_\varepsilon$ for any X, any $\varepsilon > 0$, and any Z_ε independent of X, then we call such an individual a *risk lover*.

Certainly, a person may be neither a risk averter nor a risk lover. For instance, it may happen that for some particular X and ε, it is true that $X \succsim X + Z_\varepsilon$, and for another r.v., say, X^*, it may turn out that $X^* \precsim X^* + Z_\varepsilon$.

The fact that we consider in Condition Z a non-strict relation $\succsim$ is not essential. We do it to avoid below some superfluous constructions. Formally, the above definition does not exclude the case when an individual is simultaneously a risk averter and a risk lover, that is, $X \simeq X + Z_\varepsilon$ for all X and ε. In this case, we say that the individual is *risk neutral*. For example, this is the case if the individual proceeds just from expected values $E\{X\}$.

Next, we consider the EUM criterion and figure out when *this particular criterion* satisfies Condition Z.

Proposition 2 *Let $\succsim$ be an EUM order defined in (2.1.1). Then Condition Z holds if and only if $u(x)$ is concave.*

We prove it at the end of this section.

Usually we deal with smooth utility functions, so to check whether an EU maximizer with a utility function $u(x)$ is a risk averter, it suffices to check the second derivative $u''(x)$.

For example, for $u(x) = x^{\alpha}$, we have $u''(x) = \alpha(\alpha-1)x^{\alpha-2}$. Thus, $u''(x) < 0$ for $\alpha < 1$, which corresponds to the risk aversion case, and for $\alpha > 1$ we deal with a risk lover. The case $\alpha = 1$ when $E\{u(X)\} = E\{X\}$ may be assigned to both types: the person is *risk neutral*. Other utility functions are considered in Exercise 20.

Whether a person is a risk averter or a risk lover (or neither) depends, of course, not only on her/his personality but on the particular situation. You may be a risk averter in routine life but if you have decided to spend some time (and money) in a casino, during this time you are definitely a risk lover.

There is also strong evidence based on experiments that many people incline to behave as risk averters when concerned with future gains (positive values of X), and as risk lovers when facing losses.

For example, a person may choose \$500 for sure rather than \$1,000 with probability 1/2. However, the same person may prefer to take a risk of losing \$1,000 with probability 1/2 rather than to lose \$500 for sure. A utility function in this case may look as in Fig. 9.

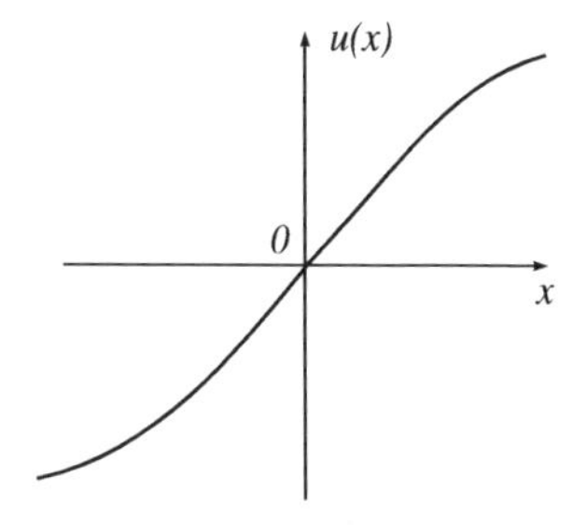

FIGURE 9.

Certainly, the utility function may be more complicated or—better to say—more sophisticated. *For example*, in the region of moderate x's, the function may be concave and in the region of large income values—convex.

The fact that the concavity of utility functions is relevant to risk aversion is also strongly connected with Jensen's inequality we proved in Section **3**.2.4. We showed there that for any concave $u(x)$,

$$E\{u(X)\} \le u(E\{X\}). \tag{2.3.1}$$

From a utility theory point of view, this means that an EU maximizer with a concave utility function would prefer the certain mean value $E\{X\}$ to the r.v. X.

EXAMPLE 1. (*Jensen's inequality and certainty equivalent*). Let $u(x)$ be a concave one-to-one increasing utility function, and $c = c(X)$ be a certainty equivalent of a r.v. X. By the definition of certainty equivalent and (2.3.1), $u(c) = E\{u(X)\} \le u(E\{X\})$. So, $u(c) \le u(E\{X\})$. If we view X as a r.v. of future income, it is natural to assume $u(x)$ to be increasing. Then, $c \le E\{X\}$. Thus,

> In the case of risk aversion, the certainty equivalent $c(X) \le E\{X\}$.

For the risk lover, a similar argument leads to $c(X) \ge E\{X\}$.

Consider a particular example. Let X be exponential with parameter a, and $u(x) = -e^{-\beta x}$, $\beta > 0$. The function is concave, and the person with such a utility function is a risk averter. Continuing the calculations from Example 2.2-2, we have

$$c(X) = \frac{1}{\beta}[\ln(a+\beta) - \ln a)] = -\frac{1}{\beta}(\ln a)[1 - \frac{\ln(a+\beta)}{\ln a}] = \frac{1}{\beta}\left(\ln\frac{1}{a}\right)\left[1 - \frac{\ln(a+\beta)}{\ln a}\right]. \tag{2.3.2}$$

Let X be "large"; formally let $a \to 0$. Then $E\{X\} = \frac{1}{a} \to \infty$. Since the third factor in (2.3.2) converges to one, $c(X) \sim \frac{1}{\beta}\ln\left(\frac{1}{a}\right)$. (The relation $u \sim v$ means $\frac{u}{v} \to 1$.) Thus, in our case,

$$c(X) \sim \frac{1}{\beta}\ln(E\{X\}).$$

Since $\ln x$ is much smaller than x for large x's, the certainty equivalent is much smaller than the mean value $E\{X\}$. We interpret this as saying that the individual is a strong risk averter.

EXAMPLE 2 (*Jensen's inequality and insurance*). The same inequality is relevant to the basic question of insurance: *why is it possible?*

Assume that an insured is an EU maximizer and consider (2.1.2). If the client is a risk averter (which is natural to assume since the client is willing to pay to insure the risk), then $u(x)$ is concave, and by Jensen's inequality, $u(w - G_{\max}) = E\{u(w-\xi)\} \le u(E\{w-\xi\}) = u(w - E\{\xi\})$. Since u is increasing, this implies $w - G_{\max} \le w - E\{\xi\}$, or $G_{\max} \ge E\{\xi\}$.

Thus, the maximum premium the client agrees to pay is larger than (or, in the boundary case, equals) the average coverage of the risk, $E\{\xi\}$.

So, *the company will get on the average more than it will pay,* which means that the company can function.

To the contrary, if the client had been a risk lover, from Jensen's inequality it would have followed that $G_{\max} \le E\{\xi\}$, and insurance would have been impossible.

Let us come back, for instance, to Example 2.1-5. We computed $G_{\max}$=\$999.5, while $E\{\xi\}$=\$500. □

Proof of Proposition 2. *Sufficiency.* If $u(x)$ is concave, then for any x_1, x_2 from the domain of $u(x)$, we have (see the Appendix, Section 2.3)

$$\frac{u(x_1)+u(x_2)}{2} \le u\left(\frac{x_1+x_2}{2}\right). \tag{2.3.3}$$

Hence, conditioning on X, we may write

$$E\{u(X+Z_\varepsilon)\} = E\{E\{u(X+Z_\varepsilon)\,|\,X\}\} = E\left\{\frac{1}{2}u(X+\varepsilon)+\frac{1}{2}u(X-\varepsilon)\right\} \le E\{u(X)\}.$$

Necessity. Let x_1, x_2 be two numbers such that $x_1 < x_2$. Set

$$x_0 = \frac{x_1+x_2}{2},\ \varepsilon = \frac{x_2-x_1}{2},$$

and $X \equiv x_0$. By definition of risk aversion, $E\{u(X+Z_\varepsilon)\} \le E\{u(X)\}$. For the particular X above, it means that $\frac{1}{2}u(x_0+\varepsilon)+\frac{1}{2}u(x_0-\varepsilon) \le u(x_0)$. By the choice of x_0 and ε, this implies (2.3.3). The last property is called midconcavity, and formally, in some exotic cases, it does not imply concavity; that is, (2.3.1) in the Appendix for all $\lambda \in [0,1]$. However, the utility function u is non-decreasing, and hence it is bounded on any closed interval where it is defined. It is known that in this case midconcavity implies concavity. See, e.g., [20]. ■

Route 1 ⇒ *page 457*

3 GENERALIZATIONS OF THE EUM CRITERION

The EUM approach may be considered a first approximation to the description of people's preferences. Over the years, there has been a great deal of discussion about the adequacy of the EUM criterion. Many experiments have been provided and a number of examples have been suggested, showing that the EUM approach is far from being efficient in all situations.

The existence of such examples is not surprising. On the contrary, it would have been surprising if the behavior of such sophisticated (and sometimes not quite understandable) creatures as human beings had been always well described by such a simple scheme.

We will briefly touch on some generalizations of the expected utility theory. It is noteworthy that this theory as well as what we consider below are often classified as chapters of Mathematical Psychology.

As above, we will view r.v.'s as those of future income. First, consider a counterexample to the expected utility approach.

3.1 Allais' paradox

The following example is probably the most famous. Though being contrived, it is very illustrative. Consider r.v.'s X_1, X_2, X_3, X_4 that all take on values \$0, \$10 millions, or \$30 millions. The corresponding probabilities are given in the following table.

	\$0	\$10 million	\$30 million
X_1	0	1	0
X_2	0.01	0.89	0.1
X_3	0.9	0	0.1
X_4	0.89	0.11	0

Apparently, a majority of people would prefer X_1 to X_2, reasoning as follows. Ten million dollars is a lot of money, for ordinary people as inconceivable as thirty million. So, it is better to get ten for sure than to go in pursuit of thirty million at the risk of receiving nothing (even if the probability of this is very small). Thus, $X_1 \succ X_2$.

The situation with X_3 and X_4 is different. Now, the probabilities of receiving nothing are large—and hence one should be ready to lose—and these probabilities are practically the same. Then it is reasonable to choose the variant with the larger prize. So, $X_3 \succ X_4$.

Now, suppose that a person is an EU maximizer with a utility function $u(x)$. Suppose that she/he compares the following two alternatives.

For the first, she/he chooses with probability $\frac{1}{2}$ the r.v. X_1 and with the same probability X_3. Let Y_1 be the resulting r.v. Then $E\{u(Y_1)\} = \frac{1}{2}E\{u(X_1)\} + \frac{1}{2}E\{u(X_3)\}$.

For the second alternative, the person chooses with equal probabilities either X_2 or X_4. If Y_2 is the resulting r.v. in this case, then $E\{u(Y_2)\} = \frac{1}{2}E\{u(X_2)\} + \frac{1}{2}E\{u(X_4)\}$.

If the person follows the natural logic above, then for her/him, $E\{u(X_1)\} > E\{u(X_2)\}$ and $E\{u(X_3)\} > E\{u(X_4)\}$. Hence,

$$E\{u(Y_1)\} > E\{u(Y_2)\}. \tag{3.1.1}$$

On the other hand, the r.v. Y_1 takes on values 0, 10, and 30 with the respective probabilities $\frac{1}{2}(0+0.9) = 0.45$, $\frac{1}{2}(1+0) = 0.5$, and $\frac{1}{2}(0+0.1) = 0.05$, and the r.v. Y_2 assumes the same values with the same (!) probabilities $\frac{1}{2}(0.01+0.89) = 0.45$, $\frac{1}{2}(0.89+0.11) = 0.5$, and $\frac{1}{2}(0.1+0) = 0.05$.

Thus, Y_1 and Y_2 have the same distribution, which contradicts (3.1.1).

Next, we address two directions (of many) in which the EUM criterion can be generalized.

3.2 Implicit or comparative utility

In this section, instead of a utility function $u(x)$, we introduce a function $v(x,y)$ which we will call an *implicit* or *comparative utility* function, and interpret it as a function indicating to what extent income x is preferable to income y. In other words, $v(x,y)$ is the comparative utility of x with respect to y. In light of this, we assume $v(x,x) = 0$, $v(x,y) \geq 0$ if $x \geq y$, and $v(x,y) \leq 0$ if $x \leq y$. Sometimes, one can assume $v(x,y) = -v(y,x)$ but in the general case, it may be false: x may be "better" than y to a smaller extent than y is "worse" than x; see also Example 3 below.

It is natural to assume $v(x,y)$ to be non-decreasing in x and non-increasing in y, which reflects the property "the larger, the better."

EXAMPLE 1. Let

$$v(x,y) = \frac{x-y}{1+|x|+|y|}. \tag{3.2.1}$$

In this case, for small x and y, the comparative utility almost equals the difference $x-y$, while for large x's and y's, the function $v(x,y)$ is close to the relative difference: $x-y$ is divided by $1+|x|+|y|$. □

We define the certainty equivalent of a r.v. X as a solution to the equation

$$E\{v(X,c)\} = 0, \tag{3.2.2}$$

provided that this solution exists and is unique. Note that such a solution is a function of X. The interpretation is clear: $c = c(X)$ is the certain amount whose comparative utility with respect to X equals zero on the average.

EXAMPLE 2. Let $v(x,y)$ be given by (3.2.1), and let $X = 1$ or 0 with equal probabilities. Then (3.2.2) is equivalent to $\frac{1}{2}v(1,c) + \frac{1}{2}v(0,c) = 0$. Obviously, c should be between 0 and 1. So, $c \geq 0$ and the equation may be written as

$$\frac{1}{2} \cdot \frac{1-c}{1+1+c} + \frac{1}{2} \cdot \frac{0-c}{1+0+c} = 0.$$

This is a quadratic equation. Its positive solution is

$$c = \frac{\sqrt{3}-1}{2} \approx 0.366.$$

So, $c < E\{X\} = 0.5$, and we face a sort of risk aversion. □

Once we have defined what is certainty equivalent in this case, we can define the corresponding preference order by the rule

$$X \succsim Y \ \text{ if and only if } \ c(X) \geq c(Y).$$

First of all, note that this scheme includes the classical EU maximization as a particular case. Indeed, let $v(x,y) = u(x) - u(y)$, where u is a utility function. In this case, (3.2.2) implies $E\{u(X)\} - E\{u(c)\} = 0$, and since c is certain, we have $E\{u(X)\} = u(c)$. Assume that u is increasing, so its inverse u^{-1} exists. Then, $c(X) = u^{-1}(E\{u(X)\})$, as in the classical case. Because $u^{-1}(x)$ is increasing, the relation $c(X) \geq c(Y)$ is equivalent to the relation $E\{u(X)\} \geq E\{u(Y)\}$.

Let now $v(x,y)$ be concave with respect to x. Consider a r.v. X and set $m = E\{X\}$ and $c = c(X)$. Note that, once X is fixed, c is a number. By definition, $v(m,m) = 0$. Then, by Jensen's inequality, $v(m,m) = 0 = E\{v(X,c)\} \leq v(E\{X\},c) = v(m,c)$. Thus, $v(m,m) \leq v(m,c)$. Since $v(x,y)$ is non-increasing in y, this implies that $c(X) \leq E\{X\}$.

A good example for the function $v(x,y)$ above is $v(x,y) = g(x-y)$, where $g(s)$ is a concave increasing function such that $g(0) = 0$. Note that in this case we should not expect $v(x,y) = -v(x,y)$.

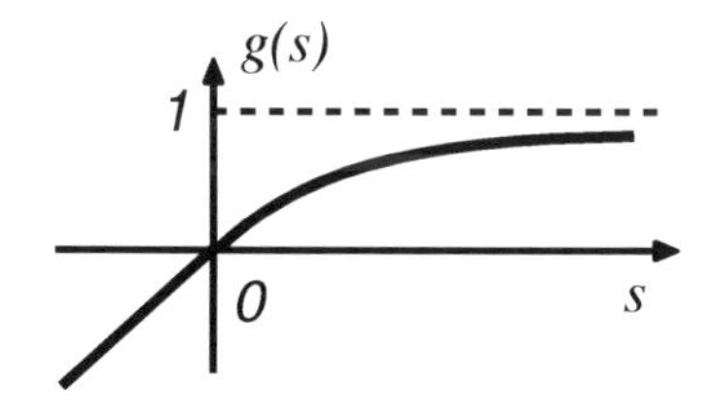

FIGURE 10.

EXAMPLE 3 (*A jealous person*). Let Mr. J.'s implicit utility function $v(x,y) = g(x-y)$, where

$$g(s) = \begin{cases} \dfrac{s}{1+s} & \text{if } s \geq 0, \\ s & \text{if } s < 0. \end{cases}$$

The function $g(s)$ is concave, its graph is given in Fig. 10.

Suppose Mr. J.'s wealth is x, and he compares it with a Mr. A.'s wealth y. If x is much larger than y, the comparative utility $v(x,y)$ is, nevertheless, not large, and $v(x,y) \to 1$ as $x-y \to \infty$. (Mr. J. does not think that his wealth is much more valuable than that of Mr. A.)

On the other hand, if x is much smaller than y, the comparative utility is negative with a large absolute value, and $v(x,y) \to -\infty$, as $x - y \to -\infty$. (Now Mr. J. considers himself much less happy than Mr. A.)

It looks like Mr. J. is pretty jealous and not very thoughtful.

Let again a r.v. $X = 1$ or 0 with equal probabilities. In this case, equation (3.2.2) is reduced to $\frac{1}{2}g(1-c) + \frac{1}{2}g(-c) = 0$, which leads to

$$\frac{1-c}{1+1-c} = c.$$

This is a quadratic equation. The solution that lies between 0 and 1 is $c = c(X) = \frac{2}{3+\sqrt{5}} \approx 0.382$. □

3.3 Rank dependent expected utility

The next approach essentially differs from that above. For simplicity, we restrict ourselves to non-negative r.v.'s. Consider a non-negative r.v. X with a d.f. $F(x)$, and two functions: $u(x)$ viewed as a *utility* function, and a function $\Psi(p)$ defined on $[0,1]$ which we call a *transformation* or *weighting* function.

We assume $\Psi(p)$ to have all properties of a d.f.: it is non-decreasing, right continuous, $\Psi(0)=0$, $\Psi(1)=1$. Assume also that $\Psi(p)$ is continuous at $p=1$.

We consider an individual whose preferences over distributions F are preserved by the function

$$V(F) = \int_0^\infty u(x)d\Psi(F(x)). \tag{3.3.1}$$

The transformation function Ψ reflects the attitude of the individual to different probabilities. The individual, when perceiving information about the distribution F, "transforms in her/his mind" the actual distribution function $F(x)$ into another one, $F_\Psi(x) = \Psi(F(x))$, underestimating or overestimating real probabilities.

► Let us show that $F_\Psi(x)$ is indeed a distribution function. First, since we consider non-negative r.v.'s, we have $F(x)=0$ for $x<0$. Second, due to properties of Ψ, the function $F_\Psi(x) = \Psi(F(x))$ is non-decreasing and right continuous. Moreover, for $x<0$, we have $F_\Psi(x){=}\Psi(F(x)) = \Psi(0) = 0$. Also, since $\Psi(p)$ is continuous at $p=1$, $F_\Psi(\infty) = \lim_{x\to\infty}\Psi(F(x)) = \Psi(1) = 1$. ◄

A typical question is why we transform distribution functions rather than densities. For example, because a transformation $\Psi(f(x))$ of a density f may be not a density, since $\int_0^\infty \Psi(f(x))dx$ may not equal one; so we would no longer deal with a probability distribution.

The quantity (3.3.1) is referred to as a *Rank Dependent Expected Utility (RDEU)*. Another term in use is a *distorted expectation.*

The corresponding preference order $\succsim$ is preserved by the function $V(F)$, that is, $F \succsim G$ if and only if $V(F) \geq V(G)$.

A simple example is $\Psi(p) = p^\beta$. If $\beta = 1$, the subject perceives F as is, and deals with the "usual" expected utility. If $\beta < 1$, the individual overestimates the probability that the income will be less than a fixed value: $(F(x))^\beta > F(x)$ for all x such that $F(x) < 1$. The individual is "security-minded." In the case $\beta > 1$, the individual underestimates the probability mentioned, being "potential-minded" (or "opportunity-minded").

EXAMPLE 1. Let X be uniformly distributed on $[0,1]$, and $\Psi(p) = p^\beta$. We have

$$F(x) = \begin{cases} 0 & \text{if } x<0, \\ x & \text{if } x\in[0,1], \\ 1 & \text{if } x>1, \end{cases} \quad \text{and} \quad F_\Psi(x) = \begin{cases} 0 & \text{if } x<0, \\ x^\beta & \text{if } x\in[0,1], \\ 1 & \text{if } x>1. \end{cases}$$

Then the density of the transformed distribution,

$$f_\Psi(x) = F'_\Psi(x) = \begin{cases} 0 & \text{if } x<0, \\ \beta x^{\beta-1} & \text{if } x\in[0,1], \\ 0 & \text{if } x>1. \end{cases}$$

For example, if $\beta > 1$, then the density $f_\Psi(x)$ is increasing, and while for the original distribution all values of X are equally likely, in "the individual's mind" it is not so: smaller values are considered less likely.

To the contrary, if $\beta < 1$, then the density $f_\Psi(x) \to \infty$ (!) as $x \to 0$, that is, the individual strongly overestimates the probability to get nothing.

The case $\beta = 0$ corresponds to an "absolutely pessimistic" individual: $F_\Psi(x) = 0$ for $x < 0$, and $= 1$ for $x \geq 0$, that is, F_Ψ is the distribution of a r.v. $X \equiv 0$. In this case, the individual expects that she/he will get nothing for sure.

EXAMPLE 2. Let F be the distribution of a r.v. taking only two values, say, a and $b > a$ with respective probabilities p and $1-p$. Then $F(x) = 0$ if $x \in [0,a)$, $F(x) = p$ if $x \in [a,b)$, and $F(x) = 1$ if $x \in [b,\infty)$. Consequently, $\Psi(F(x)) = 0$ if $x \in [0,a)$, $\Psi(F(x)) = \Psi(p)$ if $x \in [a,b)$, and $\Psi(F(x)) = 1$ if $x \in [b,\infty)$. Then, by (3.3.1),

$$V(F) = u(a)\Psi(p) + u(b)[1-\Psi(p)]. \tag{3.3.2}$$

In this case, $\Psi(\cdot)$ "transforms" just one probability p.

Note also that if a r.v. $X \equiv c$, then its d.f. $Q_c(x) = 1$ or 0 for $x \geq c$, and $x < c$, respectively. Hence, $\Psi(Q_c(x))$ also equals 1 or 0 for $x \geq c$, and $x < c$, respectively, and by (3.3.1) or (3.3.2),

$$V(X) = u(c). \tag{3.3.3}$$

EXAMPLE 3. This is merely a study example. It is known that Mr. K's utility function $u(x) = \sqrt{x}$. When comparing two annual income plans: $X = \$100{,}000$ or a r.v. $Y = \$50{,}000$ or $\$200{,}000$ with probabilities $1/2$, Mr. K chooses the former. However, the EU criterion leads to a slight preference for the latter plan: $E\{u(X)\} \approx 316$ while $E\{u(Y)\} \approx 335$. Assume that Mr. K's preferences correspond to the RDEU preference order.

By (3.3.3), we have $V(X) = u(10^5)$ and, by (3.3.2), $V(Y) = u(5\cdot 10^4)\Psi(1/2) + u(2\cdot 10^5)[1-\Psi(1/2)]$.

It is easy to calculate that $V(X) > V(Y)$ if $\Psi(1/2) > 0.59$. This means that Mr. K overestimates the probability of the unlucky event to get \$50,000, perceiving it as, say, 0.6 or higher while the real probability equals $1/2$. □

A more developed theory may be found, e.g., in [32], [36], and [43].

---

4 EXERCISES

1,2

Section 1

1. This exercise concerns the VaR criterion with a parameter γ. R.v.'s X below—with or without indices—correspond to an income.

   (a) Let a r.v. X_1 take on values $0, 1, 2, 3$ with probabilities $0.1, 0.3, 0.4, 0.2$, respectively, and a r.v. X_2 take on the same values with probabilities $0.1, 0.4, 0.2, 0.3$. Find all γ's for which $X_1 \succsim X_2$.

(b) Let X_1 be uniform on $[0,1]$, and X_2 be exponential with $E\{X_2\} = m$. When is the relation $X_2 \succsim X_1$ true for all γ's? Let $m = 0.98$. Find all γ's for which $X_2 \succsim X_1$.

(c) Let r.v.'s $X_1 = 1 - \xi_1$, $X_2 = 1 - \xi_2$, where the loss ξ_1 is uniform on $[0,1]$, and ξ_2 is exponential with $E\{\xi_2\} = m$. When is the relation $X_1 \succsim X_2$ true for all γ's? Let $m = 0.98$. Find all γ's for which $X_1 \succsim X_2$. (*Advice*: Use Exercise **7**.5.)

(d) In Exercise 1c, let ξ_1 be uniform on $[0,3]$, and let ξ_2 be uniform on $[1,2]$. Find all γ's for which $X_1 \succsim X_2$.

2.* For the case of the TailVaR criterion, solve (a) Exercise 1a; (b) Exercise 1d.

3. Let r.v.'s $X_1,...,X_{10}$ be independent exponential r.v.'s with unit means. Suppose that $Y_i = 1.1 \cdot X_i$ for $i = 1,...,10$, represent the returns for the investments in 10 independent assets. (Thus, since $E\{X_i\} = 1$, an investment in each asset gives 10% profit on the average.) Proceeding from the VaR criterion, figure out what is more profitable: to invest \$10 in one asset, or split it between 10 assets. It is recommended to use Excel or another software. (*Hint*: We know that $S_n = X_1 + ... + X_n$ has the Γ-distribution with parameters $(1,n)$; see Section **6**.4.2. The corresponding command in Excel for quantiles is =GAMMAINV$(p,\nu,1/a)$ where a is the scale parameter. The answer may depend on γ.)

4.* Consider two assets with returns $1+\xi_i$, $i = 1,2$, where ξ's have a joint normal distribution, $E\{\xi_i\} = 0$, $Corr\{\xi_1,\xi_2\} = \rho$. Let K_i be the *profit* of an investment in the ith asset, and K be the total *profit*. Prove that $q_\gamma(K) = \sqrt{q_\gamma^2(K_1) + q_\gamma^2(K_2) + 2\rho q_\gamma(K_1)q_\gamma(K_2)}$.

5. Take real data on the daily stock prices for the stocks of two companies for one year from, say, http://finance.yahoo.com or another similar site. Using the VaR criterion, for different values of γ, compare the performance of the companies. (The analysis should be based on returns, that is, on the ratios of the prices on the current and previous days.) Estimate the mean return for each company. Characterize and compare the performance of the companies, taking into account both characteristics.

6. Let X_1 and X_2 be the returns of two securities with respective means m_1 and m_2, and variances σ_1^2 and σ_2^2. We invest one unit of money: α in the first security and $1-\alpha$ in the second. So, the investment return is the r.v. $X = \alpha X_1 + (1-\alpha)X_2$. Assume X_1 and X_2 to be independent. (a) Find an α optimal under the mean-variance criterion (1.2.2). (b) Consider two particular cases: $m_1 = m_2$ and $k \to \infty$. (c) Find an α minimizing the variance of the investment return. Compare it with what you got before and the results of Exercise **3**.61.

7. Let X be uniformly distributed on $[0,1]$, and let Y take on values from $[1,\infty)$, and $P(Y > x) = 1/x^\alpha$ for all $x \geq 1$ and some $\alpha > 2$. This is a version of the Pareto distribution we discussed in Example **6**.1.3-3 and Exercise **6**.9. Verify that $m_Y = \frac{\alpha}{\alpha-1}$ and $\sigma_Y^2 = \frac{\alpha}{(\alpha-2)(\alpha-1)^2}$. Show that though $P(Y \geq X) = 1$, for α sufficiently close to 2, Y is worse than X under criterion (1.2.2).

8. Regarding the phenomenon we watched in Section 1.2.4, discuss to what extent the situation changes if we replace in (1.2.2) the standard deviation by the variance. In particular, consider the situation of Exercise 7.

Sections 2 and 3

9. Let $u_1(x)$ and $u_2(x)$ be John's and Mary's utility functions, respectively. How do John's and Mary's preferences differ if (i) $u_1(x) = 2u_2(x) + 3$; (ii) $u_1(x) = -2u_2(x) + 3$?

10. Consider an EU maximizer with utility function $u(x) = x^\alpha$; $0 \leq \alpha < 1$. (a) Find a certainty equivalent of a r.v. $X_1 = b > 0$ or 0 with respective probabilities p and $q = 1-p$. (b) Compare it with the certainty equivalent of a r.v. X_2 uniform on $[0,b]$, which is found in Example 2.2-1.

Which certainty equivalent is larger for $p = 1/2$? (The means are the same.) Interpret the answer.

11. Consider a r.v. X such that $P(X > x) = x^{-1}$ for $x \geq 1$. (This is a particular case of the Pareto distribution.) Does X have a finite expected value? Does X have a finite certainty equivalent for $u(x) = \sqrt{x}$? If so, find it.

12. Consider the EUM criterion with the exponential utility function $u(x) = -e^{-\beta x}$.

    (a) Show that if for some X and Y, we have $X \succsim Y$, then $w + X \succsim w + Y$ for any number w. (The number w may be interpreted as the initial wealth, and X and Y as random incomes corresponding to two investment strategies. The above assertion claims that *in the exponential utility case, the preference relation between X and Y does not depend on the initial wealth.*)

    (b) (*Additivity property.*) Show that for any two independent r.v.'s X and Y, the certainty equivalent $c(X_1 + X_2) = c(X_1) + c(X_2)$. Interpret this, viewing X_1, X_2 as the results of two independent investments.

    (c) Show that for, say, $u(x) = \sqrt{x}$ the above two assertions are not true. (*Hint*: In examples, one of two variables may take only one value, and the other—just two.)

13. Let X be exponential, and $u(x) = -e^{-\beta x}$. Show that for the certainty equivalent $c(X)$, we have $\frac{c(X)}{E\{X\}} \to 1$ as $E\{X\} \to 0$, that is, $c(X) \approx E\{X\}$ if $E\{X\}$ is small. Interpret it considering $u(x)$ for small x's. (*Hint*: Use Example 2.2-2.)

14. Provide calculations similar to those in Example 2.1-5 for $w = 1$ and ξ uniform on $[0, 1]$.

15. In Example 2.2-1, compare the expected values and certainty equivalents for different values of α including the case $\alpha \to 0$ and $\alpha \to \infty$. Interpret results.

16. A customer of an insurance company is an EU maximizer with a total wealth of 100. She/he is facing a random loss ξ for which $P(\xi = 0) = 0.9$, $P(\xi = 50) = 0.05$, $P(\xi = 100) = 0.05$.

    (a) Let the utility function of the customer be $u(x) = x - 0.005x^2$ for $0 \leq x \leq 100$. Graph it. Why do we consider this range of x's? Is the customer a risk averter?

    (b) What would you say in the case $u(x) = x + 0.005x^2$?

    (c) For the case 16a, find the maximal premium the customer would be willing to pay. Is the premium you found greater or less than $E\{X\}$? Might you predict it from the very beginning?

    (d) Solve Exercise 16c for $u(x) = 200x - x^2 + 349$. (*Advice*: Look at this function carefully before starting calculations.)

17. Take real data on the daily stock prices for the stocks of two companies for one year from, say, http://finance.yahoo.com or another similar site. Considering a particular utility function, for instance, $u(x) = -e^{-0.001 \cdot x}$, determine which company is better for an EU maximizer with this utility function. (*Advice:* Look at the comment in Exercise 5. To estimate the expected value $E\{u(X)\}$, where X is a random return, we can use the usual estimate $\frac{1}{n}[E\{u(X_1)\} + ... + E\{u(X_n)\}]$, where $\{X_1, ..., X_n\}$ is the time series based on the data. Excel is convenient for such calculations.) Add to your analysis the characteristics considered in Exercise 5. Try to describe the performance of the companies, taking into account all characteristics you computed.

18. (a) Is Condition Z from Section 2.3 a requirement on (i) distribution functions, or (ii) random variables, or (iii) the preference order under consideration?

(b) Is Condition Z based on the concept of expected utility?

19. Is it true that for an expected utility maximizer to be a risk averter, his/her utility function should have a negative second derivative? Let $u(x) = x$ for $x \in [0,1]$, and $u(x) = \frac{1}{2} + \frac{1}{2}x$ for $x \geq 1$. Graph this function. Is an EU maximizer with this utility function a risk averter? (*Advice*: Look up the definition of concavity.)

20. Check for risk aversion the criteria with utility functions from Section 2.2.

21. Suppose that in the situation of Example 1.1-1, a person has preferred Y. Would you expect the same answer if we change units of money and compare, say, \$50,000 and $Y = \$100,000$ with probability $1/2$? How may the utility function look in this case?

22. Let the utility function of a person be $u(x) = e^{ax}$, $a > 0$. Graph it. Is the person a risk averter or a risk lover? Show that in this case, the results of the comparison of risky alternatives does not depend on the initial wealth and has the additivity property for certain equivalents (see Exercise 12a for a precise definition).

23. Let X be exponential, and $u(x) = -e^{-\beta x}$. Show that the certainty equivalent $c(X) \to 0$ as $\beta \to \infty$. Interpret it in terms of risk aversion.

24. Let Michael be an EU maximizer with utility function $u(x) = -\exp\{-\beta x\}$ and $\beta = 0.001$. (For this value of β, the values of expected utility in this problem will be in a natural scale.) Michael compares stocks of two mutual funds. The current price for each is \$100 per share. Michael believes that in a year the price for the first stock will be on the average either 10% higher or 10% lower with equal probabilities, while for the second stock 10% up or down should be replaced by a slightly higher figure, say, 11%. (a) Which mutual fund is "better" for Michael? Do we need to calculate something to answer this question? (b) Now, assume that the second fund invites all people who buy 100 shares to a dinner valued at \$$k$. Which k would make a difference?

25. Consider two r.v.'s both taking on values $1,2,3,4$. For the first r.v., the respective probabilities are $0.1, 0.2, 0.5, 0.2$, for the second, $0.1, 0.3, 0.3, 0.3$. Which r.v. is better in the EUM-risk-aversion case? Justify the answer.

26. This is a generalization of Exercise 25. Consider a r.v. taking values $x_1, x_2, \ldots$ with probabilities $p_1, p_2, \ldots$, respectively. Let the x_i's be equally spaced, that is, $x_{i+1} - x_i$ equals the same number for all i. Assume that for a particular i, we replaced probabilities p_{i-1}, p_i, p_{i+1} by probabilities $p_{i-1} + \Delta$, $p_i - 2\Delta$, $p_{i+1} + \Delta$ where a positive $\Delta \leq p_i/2$. Has the new distribution become worse or better in the EUM-risk-aversion case? Justify the answer.

27.* In Example 3.2-3, find the certainty equivalent of a r.v. $X = d > 0$ or 0 with probabilities p and q, respectively. Analyze and interpret the case of p close to one.

28.* Let $g(s)$ defined in Section 3.2 equal $1 - e^{-s}$ for $s > 0$, and s for $s \leq 0$. Graph $g(s)$. Let c be the certainty equivalent of a r.v. X. Is it true that $c \leq E\{X\}$? Estimate the certainty equivalent for X equal to 1 or 0 with equal probabilities.

29.* Let an individual's preferences be described by the RDEU criterion with $\Psi(p) = 1 - (1 - p)^{\beta}$. In other words, $\Psi(p)$ is close to a power function for p close to one, that is, for p's corresponding to large values of r.v.'s. Let X be an exponential r.v. with a distribution F. Show that in this case, the transformation of F corresponds to dividing X by β.

30.* Consider the RDEU scheme. Let F be the uniform distribution on $[0,b]$, $u(x) = x^{\alpha}$, and $\Psi(p) = p^{\beta}$. Find a certainty equivalent and show that for $\beta > 1$ it is larger than for $\beta = 1$ (the expected utility case). Interpret this. Consider the case $\beta < 1$.

Appendix

Tables. Some Facts from Calculus and the Theory of Interest

1 TABLES

TABLE 1. Some basic distributions

Distributions		cdf	mean	variance	Some properties
DISCRETE:	probabilities				
Binomial	$\binom{n}{m}p^m q^{n-m}$, $m=0,1,...,n$	A step function	np	npq	The distribution of "the number of successes"
Geometric: 1st version	pq^{m-1}, $m=1,2,...$	A step function	$1/p$	q/p^2	1. The distribution of "the 1st success" 2. The lack of memory 3. $P(X>m)=q^m$
Geometric: 2nd version	pq^m, $m=0,1,2,...$	A step function	q/p	q/p^2	$P(X>m)=q^{m+1}$
Negative binomial: 1st version	$\binom{m-1}{\nu-1}p^\nu q^{m-\nu}$, $m=\nu,\nu+1,...$	A step function	ν/p	$\nu q/p^2$	The distribution of "the νth success"
Negative binomial: 2nd version	$\binom{\nu+m-1}{m}p^\nu q^m$, $m=0,1,2,...$	A step function	$\nu q/p$	$\nu q/p^2$	
Poisson	$e^{-\lambda}\lambda^m/m!$, $m=0,1,2,...$	A step function	λ	λ	The sum of two independent Poisson r.v.'s with parameters λ_1 and λ_2 have the Poisson distribution with the parameter $\lambda_1+\lambda_2$
CONTINUOUS:	pdf				
Uniform	$1/(b-a)$, $a\le x\le b$	$\frac{x-a}{b-a}$, $a\le x\le b$	$\frac{a+b}{2}$	$\frac{(b-a)^2}{12}$	All values from $[a,b]$ are equally likely
Exponential	ae^{-ax}, $x\ge 0$	$1-e^{-ax}$, $x\ge 0$	$1/a$	$1/a^2$	1. The lack of memory 2. $P(X>z)=e^{-az}$

TABLE 1. (Continued)

Distributions		cdf	mean	variance	Some properties
Γ-distribution	$a^{\nu}x^{\nu-1}e^{-ax}/\Gamma(\nu)$, $x \geq 0$	—	ν/a	ν/a^2	
Normal	$\frac{1}{\sqrt{2\pi}\sigma}e^{-\frac{(x-m)^2}{2\sigma^2}}$	$\Phi\left(\frac{x-m}{\sigma}\right)$	m	σ^2	The sum of independent (m_1, σ_1^2)- and (m_2, σ_2^2)-normal r.v.'s is $(m_1+m_2, \sigma_1^2+\sigma_2^2)$-normal

TABLE 2. Some basic m.g.f.'s

Distributions		Moment generating function $M(z)$
DISCRETE:	probabilities	
Binomial	$\binom{n}{m}p^m q^{n-m}$, $m = 0, 1, ..., n$	$(pe^z + q)^n$
Geometric: 1st version	pq^{m-1}, $m = 1, 2, ...$	$e^z p/(1-qe^z)$, exists for $z < \ln(1/q)$
Geometric: 2nd version	pq^m $m = 0, 1, 2, ...$	$p/(1-qe^z)$, exists for $z < \ln(1/q)$
Negative binomial: 1st version	$\binom{m-1}{\nu-1}p^{\nu}q^{m-\nu}$, $m = \nu, \nu+1, ...$	$[e^z p/(1-qe^z)]^{\nu}$, exists for $z < \ln(1/q)$
Negative binomial: 2nd version	$\binom{\nu+m-1}{m}p^{\nu}q^m$, $m = 0, 1, 2, ...$	$[p/(1-qe^z)]^{\nu}$, exists for $z < \ln(1/q)$
Poisson	$e^{-\lambda}\lambda^m/m!$, $m = 0, 1, 2, ...$	$\exp\{\lambda(e^z - 1)\}$
CONTINUOUS:	pdf	
Uniform	$1/(b-a)$, $a \leq x \leq b$	$\left(e^{zb} - e^{za}\right)/[z(b-a)]$
Exponential	ae^{-ax}, $x \geq 0$	$1/(1-z/a)$, exists for $z < a$
Γ-distribution	$a^{\nu}x^{\nu-1}e^{-ax}/\Gamma(\nu)$ $x \geq 0$	$[1/(1-z/a)]^{\nu}$, exists for $z < a$
Normal	$\frac{1}{\sqrt{2\pi}\sigma}\exp\left\{-\frac{(x-m)^2}{2\sigma^2}\right\}$	$\exp\left\{mz + \sigma^2 z^2/2\right\}$

TABLE 3. The standard normal distribution function $\Phi(x)$.

x	0	0.01	0.02	0.03	0.04	0.05	0.06	0.07	0.08	0.09
0	0.5	0.504	0.508	0.512	0.516	0.5199	0.5239	0.5279	0.5319	0.5359
0.1	0.5398	0.5438	0.5478	0.5517	0.5557	0.5596	0.5636	0.5675	0.5714	0.5753
0.2	0.5793	0.5832	0.5871	0.591	0.5948	0.5987	0.6026	0.6064	0.6103	0.6141
0.3	0.6179	0.6217	0.6255	0.6293	0.6331	0.6368	0.6406	0.6443	0.648	0.6517
0.4	0.6554	0.6591	0.6628	0.6664	0.67	0.6736	0.6772	0.6808	0.6844	0.6879
0.5	0.6915	0.695	0.6985	0.7019	0.7054	0.7088	0.7123	0.7157	0.719	0.7224
0.6	0.7257	0.7291	0.7324	0.7357	0.7389	0.7422	0.7454	0.7486	0.7517	0.7549
0.7	0.758	0.7611	0.7642	0.7673	0.7704	0.7734	0.7764	0.7794	0.7823	0.7852
0.8	0.7881	0.791	0.7939	0.7967	0.7995	0.8023	0.8051	0.8078	0.8106	0.8133
0.9	0.8159	0.8186	0.8212	0.8238	0.8264	0.8289	0.8315	0.834	0.8365	0.8389
1	0.8413	0.8438	0.8461	0.8485	0.8508	0.8531	0.8554	0.8577	0.8599	0.8621
1.1	0.8643	0.8665	0.8686	0.8708	0.8729	0.8749	0.877	0.879	0.881	0.883
1.2	0.8849	0.8869	0.8888	0.8907	0.8925	0.8944	0.8962	0.898	0.8997	0.9015
1.3	0.9032	0.9049	0.9066	0.9082	0.9099	0.9115	0.9131	0.9147	0.9162	0.9177
1.4	0.9192	0.9207	0.9222	0.9236	0.9251	0.9265	0.9279	0.9292	0.9306	0.9319
1.5	0.9332	0.9345	0.9357	0.937	0.9382	0.9394	0.9406	0.9418	0.9429	0.9441
1.6	0.9452	0.9463	0.9474	0.9484	0.9495	0.9505	0.9515	0.9525	0.9535	0.9545
1.7	0.9554	0.9564	0.9573	0.9582	0.9591	0.9599	0.9608	0.9616	0.9625	0.9633
1.8	0.9641	0.9649	0.9656	0.9664	0.9671	0.9678	0.9686	0.9693	0.9699	0.9706
1.9	0.9713	0.9719	0.9726	0.9732	0.9738	0.9744	0.975	0.9756	0.9761	0.9767
2	0.9772	0.9778	0.9783	0.9788	0.9793	0.9798	0.9803	0.9808	0.9812	0.9817
2.1	0.9821	0.9826	0.983	0.9834	0.9838	0.9842	0.9846	0.985	0.9854	0.9857
2.2	0.9861	0.9864	0.9868	0.9871	0.9875	0.9878	0.9881	0.9884	0.9887	0.989
2.3	0.9893	0.9896	0.9898	0.9901	0.9904	0.9906	0.9909	0.9911	0.9913	0.9916
2.4	0.9918	0.992	0.9922	0.9925	0.9927	0.9929	0.9931	0.9932	0.9934	0.9936
2.5	0.9938	0.994	0.9941	0.9943	0.9945	0.9946	0.9948	0.9949	0.9951	0.9952
2.6	0.9953	0.9955	0.9956	0.9957	0.9959	0.996	0.9961	0.9962	0.9963	0.9964
2.7	0.9965	0.9966	0.9967	0.9968	0.9969	0.997	0.9971	0.9972	0.9973	0.9974
2.8	0.9974	0.9975	0.9976	0.9977	0.9977	0.9978	0.9979	0.9979	0.998	0.9981
2.9	0.9981	0.9982	0.9982	0.9983	0.9984	0.9984	0.9985	0.9985	0.9986	0.9986
3	0.9987	0.9987	0.9987	0.9988	0.9988	0.9989	0.9989	0.9989	0.999	0.999

TABLE 4. The quantiles of the standard normal distribution: $\Phi^{-1}(y)$.

y	0.8	0.85	0.9	0.91	0.92	0.93	0.94	0.95	0.96	0.97	0.98	0.99	1
$\Phi^{-1}(y)$	0.842	1.036	1.2816	1.341	1.4051	1.476	1.555	1.645	1.751	1.88079	2.05	2.33	infinity

2 SOME FACTS FROM CALCULUS

2.1 The "little o" notation

What we discuss below is not a notion but rather a notation which turns out to be useful in many calculations. We begin with a particular definition.

Denote by the symbol $o(x)$ (little-o of x) *any* function $o(x) = \varepsilon(x)x$, where the function $\varepsilon(x) \to 0$ as $x \to 0$. We may say that $o(x) \to 0$ faster than x as $x \to 0$. (The term "small-o" is also in use but much rarer.)

Another way to define $o(x)$ is to say that

$$\frac{o(x)}{x} \to 0 \quad \text{as} \quad x \to 0,$$

which is equivalent to what was said above.

For example, $x^2 = o(x)$, and $x^{3/2} = o(x)$, while $\sqrt{x}$ is not $o(x)$.

Heuristically, the formula $x^2 = o(x)$ means that "x^2 is much smaller than x" for small x's.

Certainly, we can replace x by, say, x^2 and in this case, the expression $o(x^2)$ denotes any function such that $[o(x^2)/x^2] \to 0$. For instance, $x^3 = o(x^2)$.

EXAMPLE 1. If

$$f(x) = 1 + 2x^2 + o(x^2) \quad \text{as} \quad x \to 0, \tag{2.1.1}$$

then we can say that $f(x)$ converges to one, as $x \to 0$, at a rate of $2x^2$ up to a remainder $o(x^2)$ which is negligible in comparison with the main term $2x^2$ for small x's. Formally, (2.1.1) means that $f(x) - 1 - 2x^2 = o(x^2)$, that is,

$$\frac{f(x) - 1 - 2x^2}{x^2} \to 0 \quad \text{as} \quad x \to 0.$$

EXAMPLE 2. At what rate does the function $f(x) = (1 + 4x^2)^3$ converge to one as $x \to 0$? By the formula $(1 + a)^3 = 1 + 3a + 3a^2 + a^3$ (which is a particular case of the binomial formula (**1**.3.4.7)),

$$(1 + 4x^2)^3 = 1 + 3 \cdot 4x^2 + A(x),$$

where $A(x)$ is the sum of the terms containing higher powers of $4x^2$, that is, $(4x^2)^2$ and $(4x^2)^3$. The sum of such terms converges to zero faster than x^2. Hence, the remainder $A(x)$ is $o(x^2)$, and we can write

$$(1 + 4x^2)^3 = 1 + 12x^2 + o(x^2).$$

The term $o(x^2)$ is a negligible remainder.

Certainly, all of this is very simple, and we just want to demonstrate that if the accuracy of an order of x^2 is enough for us, then we can avoid superfluous calculations. □

It is also worth emphasizing that $o(x)$ denotes not a particular function but a function, or *some* function, with the above property. Therefore, although we can write, for example, the expression $2o(x)$, it will not make much sense since multiplying $o(x)$ by 2 we get a function with the same property: it converges to zero faster than x. Thus, we can write $2o(x) = o(x)$. For the same reason, we may write that $o(x) + o(x) = o(x)$, etc.

EXAMPLE 3. Let us approximate the function $f(x) = (1+x)^4 + (1+3x^2)^3$ for small x's by a quadratic function. Similar to what we did in Example 2, and recalling that $(1+a)^4 = 1+4a+6a^2+4a^3+a^4 = 1+4a+6a^2+o(a^2)$, we have

$$f(x) = 1+4x+6x^2+o(x^2)+1+3\cdot 3x^2+o(x^2) = 2+4x+15x^2+o(x^2).$$

So, $f(x) = 2+4x+15x^2+$ the term which is negligible in comparison with the main term as $x \to 0$. □

The same relations may be established for x's tending to any point, or for $x \to \infty$. For example, if we write $o(x)$ as $x \to \infty$, then we mean a function $o(x)$ such that

$$\frac{o(x)}{x} \to 0 \text{ as } x \to \infty.$$

For instance, $x^{-2} = o(x^{-1})$ as $x \to \infty$, or $(x-1)^3 = o((x-1)^2)$ as $x \to 1$.

EXAMPLE 4. Let $f(x) = \frac{1}{1+x^2}$. Clearly, $f(x) \to 0$ as $x \to \infty$, and we may write this as $f(x) = o(1)$ because the latter relation means $\frac{f(x)}{1} \to 0$. Moreover, since

$$\frac{1}{1+x^2} - \frac{1}{x^2} = \frac{-1}{x^2(1+x^2)} = o\left(\frac{1}{x^2}\right),$$

we may write

$$\frac{1}{1+x^2} = \frac{1}{x^2} + o\left(\frac{1}{x^2}\right),$$

that is, as $x \to \infty$, the function $\frac{1}{1+x^2}$ equals $\frac{1}{x^2}$ up to a negligible remainder. □

2.2 Taylor's expansions

2.2.1 General expansion

The *Taylor expansion* concerns approximations of sufficiently smooth functions $f(x)$ by polynomials in a neighborhood of a point x_0. Without loss of generality, we can set $x_0 = 0$. If it is not so, it suffices to translate the origin to x_0, or more precisely, to consider instead of $f(x)$ the function $f_0(x) = f(x+x_0)$, and apply the Taylor expansion to the latter function.

Let $f(x)$ be n times differentiable in a neighborhood of zero, that is, for $x \in \Delta = (-d, d)$ for some $d > 0$. Suppose also that the nth derivative $f^{(n)}(x)$ is continuous in the same neighborhood. Then, the *Taylor formula* states that for all $x \in \Delta$,

$$f(x) = \sum_{k=0}^{n} \frac{f^{(k)}(0)}{k!} x^k + o(x^n), \tag{2.2.1}$$

where the remainder $o(x^n)$ is negligible with respect to x^n as $x \to 0$. More precisely,

$$\frac{o(x^n)}{x^n} \to 0 \text{ as } x \to 0. \tag{2.2.2}$$

If f has all derivatives, then under some additional conditions, we can set $n = \infty$, and write

$$f(x) = \sum_{k=0}^{\infty} \frac{f^{(k)}(0)}{k!} x^k. \tag{2.2.3}$$

For example, this is true if $|f^{(n)}(x)| \leq M^n$ for some M and all n in the neighborhood where we consider the expansion (2.2.3).

2.2.2 Some particular expansions

All expansions below are verified by making use of the general formulas (2.2.1) and (2.2.3).

The exponential function: for all x's,

$$e^x = 1 + x + \frac{x^2}{2!} + \ldots = \sum_{k=0}^{\infty} \frac{x^k}{k!}. \tag{2.2.4}$$

For $x \to 0$, it often suffices to consider the first two or three terms, that is, the approximations

$$e^x = 1 + x + o(x), \tag{2.2.5}$$

or

$$e^x = 1 + x + \frac{x^2}{2} + o(x^2). \tag{2.2.6}$$

The logarithmic function. For $\ln(1+x)$, the expansion

$$\ln(1+x) = \sum_{k=1}^{\infty} (-1)^{k-1} \frac{x^k}{k!} \tag{2.2.7}$$

is true for $-1 < x \leq 1$. In particular,

$$\ln(1+x) = x - \frac{x^2}{2} + o(x^2) \quad \text{as} \quad x \to 0. \tag{2.2.8}$$

From the last relation, it follows that

$$\ln(1+x) \sim x \text{ as } x \to 0,$$

where, as usual, the symbol $a(x) \sim b(x)$ means that $\frac{a(x)}{b(x)} \to 1$.

The power function. Next, we consider the function $(1-x)^{-\alpha}$ for $\alpha > 0$. The Taylor expansion here is true for $|x| < 1$, and is given by the formula

$$(1-x)^{-\alpha} = \sum_{m=0}^{\infty} \binom{-\alpha}{m} (-x)^m = \sum_{m=0}^{\infty} \binom{\alpha + m - 1}{m} x^m, \tag{2.2.9}$$

where for any real r

$$\binom{r}{k} = \frac{r(r-1)\cdots(r-k+1)}{k!}. \tag{2.2.10}$$

The second equality in (2.2.9) is based on the formula

$$\binom{-\alpha}{m} = (-1)^m \binom{\alpha+m-1}{m},$$

which is true since

$$\begin{aligned}\binom{-\alpha}{m} &= \frac{(-\alpha)(-\alpha-1)\cdots(-\alpha-m+1)}{m!} = \frac{(-1)^m(\alpha)(\alpha+1)\cdots(\alpha+m-1)}{m!} \\ &= (-1)^m \binom{\alpha+m-1}{m}.\end{aligned}$$

For $\alpha = 1$, the coefficient $\binom{\alpha+m-1}{m} = \binom{m}{m} = 1$, and for $|x| < 1$ we have

$$\frac{1}{1-x} = 1 + x + x^2 + \ldots = \sum_{k=0}^{\infty} x^k. \tag{2.2.11}$$

We see that this is just the formula for the geometric series, which can certainly be derived without Taylor's expansion from the well known formula

$$1 + x + x^2 + \ldots + x^n = \frac{1 - x^{n+1}}{1-x}. \tag{2.2.12}$$

The last formula is true for all $x \neq 1$.

2.3 Concavity

In introductory Calculus courses, a concave function $u(x)$ is often defined as a function for which the second derivative $u''(x) \leq 0$. For us, this definition is somewhat restrictive. The definition below does not contradict the above definition if u is twice differentiable.

Definition. We say that a function $u(x)$ defined on an interval I is *concave* if for any $x_1, x_2 \in I$ and any $\lambda \in [0,1]$,

$$\lambda u(x_1) + (1-\lambda)u(x_2) \leq u(\lambda x_1 + (1-\lambda)x_2). \tag{2.3.1}$$

See, as an illustration, Fig.1.

It is known that $u(x)$ so defined is continuous at any interior point of I; see, e.g., [55]. The following proposition may be viewed as another definition of concavity.

Proposition 1 *A function $u(x)$ is concave on an interval I if and only if for any interior point $x_0 \in I$, there exists a number c, perhaps depending on x_0, such that for any $x \in I$,*

$$u(x) - u(x_0) \leq c(x - x_0). \tag{2.3.2}$$

Note that Proposition 1 does not presuppose that the number c is unique. If $u(x)$ is differentiable at x_0, then (2.3.2) is true for $c = u'(x_0)$; see Fig.2i. However, if $u(x)$ is not

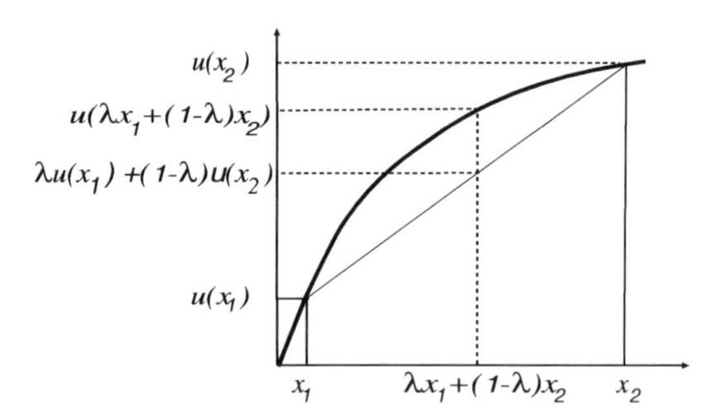

FIGURE 1. The first definition of a concave function.

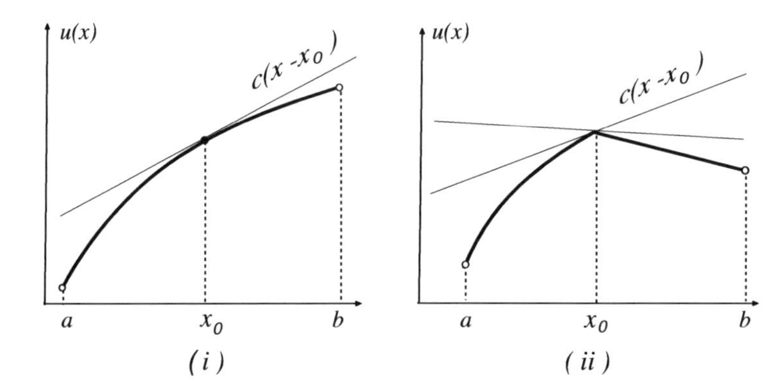

FIGURE 2. The second definition of a concave function.

smooth at x_0, then there may be many c's for which (2.3.2) is true; see Fig.2ii. A proof may be also found in [55]. The line $c(x-x_0)$ is called a *support* of $u(x)$ at x_0.

We call a function $u(x)$ *convex* if (2.3.1) (or (2.3.2)) is true with the replacement of the inequality sign $\leq$ by the sign $\geq$. Note that there is no need to explore convex function separately because if a function $u(x)$ is convex, then the function $-u(x)$ is concave.

---

3 SOME FACTS FROM THE THEORY OF INTEREST

In this book, we consider a number of examples that address stochastic models involving cash flows and capital growth. It is convenient to present facts from the "usual" non-stochastic theory of interest in one place.

3.1 Compounding interest

Assume that we deposit an amount C_0 into a bank account, and the bank credits (or compounds) interest monthly proceeding from an annual *rate* δ equal, say, to 5%. The *crediting* (or *compounding*) procedure will be carried out in the following way.

The bank will divide the rate δ by 12, and at the end of the first month, the interest $C_0\frac{\delta}{12}$

will be credited. So, the initial capital C_0 will increase to $C_0(1+\delta/12)$. At the end of the next month, this amount will be again multiplied by $(1+\delta/12)$, the total amount will equal $C_0(1+\delta/12)^2$, and so on, up to the end of the year when the final amount will be equal to $C_0(1+\delta/12)^{12} = C_0(1+0.05/12)^{12} \approx C_0 \cdot 1.0512$.

We see that the real annual profit per one unit of money will be equal to $i = (1+\delta/12)^{12} - 1 \approx 0.0512$ or 5.12%, which is a bit larger than the rate 5%. The quantity $(1+\delta/12)^{12} - 1$ (in our case, 5.12%) is called an *effective interest rate* or *yield*, and the reader can find both words, rate and yield, in her/his bank statement and compare the numbers there.

If, *keeping the same annual rate* δ, the bank compounds interest not twelve but n times a year, it will lead to the amount $C_0(1+\delta/n)^n$. As is known, the function $(1+\delta/n)^n$ is increasing in n. So, the more often the interest is compounded, the better for the investor.

Furthermore, $(1+\delta/n)^n \to e^\delta$ as $n \to \infty$. If $n = \infty$, we say that the interest is *compounded continuously*. The annual growth factor in this case is e^δ; at the end of the year, the capital invested becomes equal to $C_1 = C_0 e^\delta$; and the effective interest rate or yield is equal to $e^\delta - 1$. Certainly, with respect to real investment, it is just an approximation.

At the end of t years, the capital will become equal to

$$C_t = C_0 e^{\delta t}. \tag{3.1.1}$$

One may show this similarly to what we did above, but a better and more general way is to consider a continuous time model and proceed as follows.

Assume that during an infinitesimally small interval $[t, t+dt]$, the relative capital growth (that is, dC_t/C_t) is proportional to the length of the interval (that is, dt). Identifying the proportionality coefficient with the rate δ above, we come to the equation

$$\frac{dC_t}{C_t} = \delta dt. \tag{3.1.2}$$

A solution to this equation is (3.1.1), which the reader may double-check by a straightforward substitution.

3.2 Present value and discount

For certainty and brevity, we take a year as a unit of time, and \$1 as a unit of money.

For understandable reasons, the value of \$1 to be paid tomorrow is not \$1; it is less than that. We define a *discount* factor v_t as the value of \$1 to be paid at time t, if evaluation is carried out from the standpoint of the present time $t = 0$. In another terminology, v_t is the *present value* of \$1 to be paid at time t.

In the models of this book, we understand v_t as the amount of money which one should invest into a risk-free security at time $t = 0$, in order to have a unit of money at time t. (So, we presuppose that such a security does exist. As an example, we may think about term investments with a fixed guaranteed rate.)

Consider a unit time interval (say, a calendar year), and assume that the risk-free effective interest rate or yield for this period is equal to r. That is, investing \$1 at the beginning of the period, the investor gets $1+r$ dollars at the end. Then, investing $\frac{1}{1+r}$ dollars, the investor will obtain $\frac{1}{1+r} \cdot (1+r) = 1$ dollar at the end of the period. Thus, to get one unit at the end, one should invest $\frac{1}{1+r}$ at the beginning. Hence,

$$v_1 = v = \frac{1}{1+r}. \tag{3.2.1}$$

Let time be discrete, that is, $t = 0, 1, \ldots$, and let the risk-free interest in each period $[t-1, t]$ be the same and equal r. Then, reasoning in the same fashion, we conclude that from the stand point of the initial time $t = 0$, the present value of \$1 to be paid at time t, is

$$v_t = \left(\frac{1}{1+r}\right)^t = v^t, \tag{3.2.2}$$

where t is an integer.

In the continuous time model, we proceed from (3.1.1). For C_t to be equal to one (dollar), the initial investment C_0 should be equal to $e^{-\delta t}$. Thus, the present value of \$1 to be paid at time t, is

$$v_t = e^{-\delta t}. \tag{3.2.3}$$

The last formula is consistent with (3.2.2): we can write

$$v_t = e^{-\delta t} = \left(e^{-\delta}\right)^t = v^t$$

if we set $v = e^{-\delta}$, the discount for one year.

Next, assume that an investor expects a future cash flow during n periods of time with payments at the beginning of each period. Such a sequence of payments is called an *annuity-due*, or simply *annuity*. Denote by c_t the cash at time t or, more precisely, at the beginning of the period $[t, t+1)$. The present value of this payment is $v^t c_t$. The total cash flow is the sequence $c_0, c_1, c_2, \ldots, c_{n-1}$, and the present value of this flow is the number

$$a_n = c_0 + vc_1 + v^2 c_2 + \ldots + v^{n-1} c_{n-1} = \sum_{t=0}^{n-1} v^t c_t.$$

In particular, if all $c_t = 1$, then, by virtue of (2.2.12),

$$a_n = 1 + v + \ldots + v^{n-1} = \frac{1-v^n}{1-v}, \tag{3.2.4}$$

provided $v \neq 1$. For $n \to \infty$, and $v < 1$,

$$a_n \to \frac{1}{1-v}.$$

EXAMPLE 1. How much money should we deposit into a risk-free account, for the bank to pay \$1 to our descendants at the beginning of each year during the next one billion years? Certainly, the answer to this question depends on from which discount v we will proceed. Suppose that we believe that on the average, the annual risk free interest will be about 4%. So, $r = 0.04$. As is easy to derive from (3.2.1), $\frac{1}{1-v} = 1 + \frac{1}{r}$ which in our case amounts to just \$26. $\square$

References

The list below should not be considered a bibliography. It is a list of books and papers to which we refer in this textbook.

[1] Abramowitz, M. and Stegun, I., eds., *Handbook of Mathematical Functions with Formulas, Graphs, and Mathematical Tables*, New York, 1972.

[2] Albanese, C., Credit Exposure, Diversification Risk and Coherent VaR, *Working Paper, Department of Mathematics, University of Toronto,* September, 1997.

[3] Arias, E., United States Life Tables, 2007, *Monthly vital statistics report,* vol. 59, no. 9, National Vital Statistics Reports, September 28, 2011.

[4] Artzner, P., Delbaen, F., Eber, J.-M. and Heath, D., Coherent measures of Risk, *Math. Finance,* 9, 3, 1999.

[5] Bening, V.E. and Korolev, V.Yu., *Generalized Poisson Models and Their Applications in Insurance and Finance*, VSP, Utrecht, 2002.

[6] Bhattacharya, R.N. and Rao, R.R., *Normal Approximation and Asymptotic Expansion*, SIAM, 2010.

[7] Bhattacharya, R.N., and Waymire, E.C., *Stochastic Processes With Applications*, Society for Industrial & Applied Mathematics, 2009.

[8] Billingsley, *Convergence of Probability Measures*, 2nd edition, Wiley-Interscience, 1999.

[9] Bradley, R. C., *Introduction to Strong Mixing*, Kendrick Press, Heber City, Utah, 2007.

[10] Chow, Y.S. and Teicher, H., *Probability Theory, Independence, Interchangeability, Martingales*, Springer-Verlag, New York, 1978.

[11] Cvitanic, J. and Zapatero, F., *Introduction to the Economics and Mathematics of Financial Markets*, MIT Press, Cambridge, Massachusetts, 2004.

[12] Doob, J.L., *Stochastic Processes*, Wiley-Interscience, New York, 1990; the first edition, 1953.

[13] Feller, W., *Introduction to Probability Theory*, Volumes I-II, 2, Wiley, New York, 1971.

[14] Fishburn, P.C., *Non-linear Preference and Utility Theory*, Johns Hopkins University Press, Baltimore, 1988.

[15] Fishburn, P.C., *The Foundations of Expected Utility*, Springer, 2010.

[16] Gikhman, I.I. and Skorohod, A. V., *The Theory of Stochastic Processes,* Springer-Verlag, New York, 1974-1979.

[17] Girko, V.L., *Theory of Random Determinants*, Springer, 1990.

[18] Gradshteyn, I.S. and Ryzhik, I.M., *Table of Integrals, Series, and Products*, Academic Press, San Diego, 2000.

[19] Hall, P. and Heyde, C.C., *Martingale Limit Theory and its Application*, Academic Press, New York, 1980.

[20] Hardy, G.H, Littlewood, J.E. and Polya, G. *Inequalities*, 2nd ed., Cambridge University Press, Cambridge, 1959.

[21] Hull, J.C., *Options, Futures, and Other Derivatives*, Prentice Hall, New Jersey, 2006.

[22] Horn, R.A. and Johnson, C.A., *Matrix Analysis*, Cambridge University Press, Cambridge, 1985.

[23] Ibragimov, I.A. and Linnik, Yu. V., *Independent and stationary sequences of random variables*, Wolters-Noordhoff, Groningen, 1971.

[24] Kallenberg, O., *Foundations of Modern Probability*, Springer, New York, 2000.

[25] Kindermann, R. and Snell, J.L, *Markov Random Fields and Their Applications,* American Mathematical Society, 1980.

[26] Korolev, V., and Shevtsova I., An improvement of the BerryEsseen inequality with applications to Poisson and mixed Poisson random sums, *Scandinavian Actuarial Journal*, 2010

[27] Lay, D., *Linear Algebra and its Applications,* Pearson, Addison-Wesley, 2006.

[28] Lesigne, E., *Heads or Tails: An Introduction to Limit Theorems in Probability*, AMS, 2005.

[29] Lin, Z and Lu, C., *Limit theory for Mixing Dependent Random Variables*, Kluwer, 1996.

[30] Liptser, R.Sh., and Shiryayev, A.N., *Theory of Martingales (Mathematics and its Applications)*, Springer, 1989.

[31] Luce, R.D. and Raifa, H., *Games and Decisions: Introduction and Critical Survey,* John Wiley, New York, Wiley 1957; the last edition, Dover, 1989.

[32] Luce, R.D., *Utility of Gains and Losses*, Erlbaum, Mahwah, New Jersey, 2000.

[33] Minc, H. , *Nonnegative Matrices,* Wiley, New York, 1998.

[34] Neveu, J., *Discrete-parameter martingales*, Amsterdam: North-Holland; New York, American Elsevier, 1975

[35] Petrov, V.V., *Limit Theorems of Probability Theory. Sequences of Independent Random Variables*, Clarendon Press, Oxford, 1995.

[36] Quiggin, J., *Generalized Expected Utility Theory*, Kluwer, Boston, 1993.

[37] Revuz, D. and Yor, M., *Continuous Martingales and Brownian motion*, Springer-Verlag, Berlin, New York, 1991.

[38] Rogozin, B.A., A remark on the paper "A moment inequality with an application to the central limit theorem" by C. G. Esseen, *The Theory of Probability and its Applications*, 5, 1960.

[39] Ross, S.M., *A First Course in Probability*, 6th ed., Prentice Hall, Upper Saddle River, New Jersey, 2002.

[40] Ross, S.M., *Introduction to Probability Models*, 9th edition, 2006.

[41] Rotar, V.I., On Summation of RandomVariables in Non-Classical Situation, Russian Mathematical Survey, 37:6 (1982), 151-175.

[42] Rotar, V.I., *Probability Theory*, World Scientific, 1997.

[43] Rotar, V.I., *Actuarial Models: The Mathematics of Insurance*, Chapman & Hall/ CRC, 2006.

[44] Selvin, Steve (1975). "A problem in probability" (letter to the editor). *American Statistician*, 29(1): 67, 1975.

[45] Senatov, V.V., *Normal Approximation: New Results, Methods And Problems*, V.S.P. Intl Science, 1998.

[46] Shevtsova I., A refinement of bounds and Lypunov's theorem, *Doklady Mathematics RAN*, 2010, Vol. 435, No. 1, pp. 26-28.

[47] Shiryaev, A.N., *Probability*, Springer, New York, New York, 1996.

[48] Shiryaev, A.N., *Essentials of Stochastic Finance: Facts, Models, Theory*, World Scientific, River Edge, New Jersey, 1999.

[49] Stampfli, J. and Goodman, V., *The Mathematics of Finance, Modeling and Hedging*, Brooks/Cole, Pacific Grove, CA, 2001.

[50] Taylor H.M., and Karlin, S., *An Introduction to Stochastic Modeling*, 3rd edition, Academic Press, 1998.

[51] Tuyen, D.Q., *Convergence of Dependent Random Variables: Central Limit Theorems, Berry-Esseen Bounds, Martingale-like Sequences, C-sequences, Strong Laws*, Lap Lambert Academic Publishing, 2010.

[52] Tyurin, I. A refiniment of the bounds of the constants in Lyapunov's theorem. *Russian Math. Survey*, 2010, v.65, #3, p.201-202.

[53] Tyurin, I. New estimates of the convergence rate in the Lyapunov theorem // arXiv:0912.0726v1, 3 December, 2009.

[54] Ville, J., *Étude Critique de la Notion de Collectif*, Gauthier-Villars, 1939.

[55] Webster, R., *Convexity*, Oxford University Press, Cambridge [Eng.], 1994.

[56] *Webster's New Universal Unabridged Dictionary*, Barnes and Noble Books, 2003.

[57] Williams, D., *Probability with Martingales*, Cambridge University Press, 1991.

[58] Williams, R.J., *Introduction to the Mathematics of Finance*, American Mathematical Society, 2006.

[59] Zolotarev, V.M., *Modern Theory of Summation of Random Variables*, V.S.P. Intl. Science, Utrecht, the Netherlands,1997.

Answers to Exercises

Possible additional remarks and errata will be posted in *http://actuarialtextrotar.sdsu.edu*

CHAPTER 1

1. F, F, F, T.
2. (a) 4^n. (b) 2^{20}. (c) $10!$. (d) Ω may be identified with the collections of all pairs (m,n) for $m = 0,1,\dots, n = 0,1,\dots; |\Omega| = \infty$.
3. For any pair, the events are not disjoint.
8. (a) 0.9; (b) 0; (c) 0.4.
9. $0.1, 0.15, 0.15, 0.6; 0.4$.
10. (a) $\frac{1}{2}, \frac{3}{4}$, (b) $\frac{1}{3}, 0$.
12. $0.1, 0.1, 0.5, 0.3$.
13. $P(C) = P(A) + P(B) - 2P(AB)$.
14. $P(AB) \geq 0.1$.
16. $8 \cdot 10^6$, $\frac{\binom{6}{3} \cdot 8 \cdot 9^3}{8 \cdot 10^6} = 0.01458$, $\frac{\binom{6}{2} 9^4 + \binom{6}{3} \cdot 7 \cdot 9^3}{8 \cdot 10^6} \approx 0.025$.
17. $2/3$.
19. $4.71974 \cdot 10^{-7}, \approx 0.0046725$.
22. (a) $4/\binom{52}{13}$; (b) $12/[52!/((13!)^2 26!)]$.
23. (a) $1/19$; (b) ≈ 0.30650.
25. ≈ 0.3398.
27. 45.
28. ≈ 0.0005, ≈ 0.025; the answers do not depend on n.
29. (a) ≈ 0.275, (b) 0. ≈ 0.659.
31. (a) $1/12^4$; (b) ≈ 0.382; (c) ≈ 0.0796.
32. $\frac{100!}{(20!)^5 5^{100}}$.
33. (a) ≈ 0.000113. (b) ≈ 0.00553. (c) ≈ 0.114. (d) ≈ 0.0020. (e) ≈ 0.0569.
34. (b) ≈ 0.5275; (c) $(51 \quad k)/1326$.
35. (a) ≈ 0.0011. (b) ≈ 0.1260.
36. (a) 1; (c) $\frac{(n+1)!}{2n^n}$.
37. $n > 6$.
38. (a) 0.192.
41. ≈ 0.476.
43. (a) The former.
47. $\approx \frac{2.5}{e}$.

CHAPTER 2

1. 0.68.
3. (b) Yes, if their probabilities are positive. (c) Yes.
4. The events are dependent.
6. The events are dependent.
10. $p_4 + p_1p_2p_3 - p_1p_2p_3p_4$. (b) $p_5(p_1 + p_2 - p_1p_2)(p_3 + p_4 - p_3p_4) + (1 - p_5)(p_1p_3 + p_2p_4 - p_1p_2p_3p_4)$.
13. ≈ 0.396

16. The probability in question is ≈ 0.539.
18. (a) ≈ 0.00823; (b) ≈ 0.0823; (c) ≈ 0.00274; (d) ≈ 0.004; (e) ≈ 0.004; (f) ≈ 0.039.
20. $2/3$.
21. $1/12$.
22. $\binom{26}{13}/\binom{39}{13}$.
25. (b) $(1-p_1)/(1-p_1+p_2)$.
26. $6/11$.
27. 0.9, 1/9.
30. $p_5(p_1+p_2-p_1p_2)(p_3+p_4-p_3p_4)+(1-p_5)(p_1p_3+p_2p_4-p_1p_2p_3p_4)$.

item[**34.**] ≈ 0.150.
35. (a) $368/693$; (b) $99/368$.
36. ≈ 0.587.
37. (a) 0.0613.... (b) 0.032. (d) ≈ 0.489.
40. (a) $u > 4.19$. (c) $8/9$, $p \geq 100/105$.
41. After the fourth loss.
42. No.
44. No.

CHAPTER 3

2. (b) $1, 35/36, 35/36, 32/36, 1/36$; (c) $1/6$.
4. $\begin{matrix} 0 & 1 & 2 \\ 21/38 & 15/38 & 1/19 \end{matrix}$.
6. $\begin{matrix} 0 & 1 & 2 & 3 & 4 \\ 5/15 & 4/15 & 3/15 & 2/15 & 1/15 \end{matrix}$.
7. ≈ 0.762.
9. $1/3$.
10. $P(S=n) = \frac{\min\{n-1, 13-n\}}{36}$, $n = 2, ..., 12$.
16. (a) $-n, -n+2, ..., n-2, n$. (b) $P(W_{2k}=0) = \binom{2k}{k}4^{-k}$, $P(W_{2k+1}=0) = 0$.
18. Yes.
19. $P(S=3) = 1/15, P(S=4) = 6/15, P(S=5) = 8/15$.
21. 0.1.
26. $-7/9$ and $-8/9$.
27. 1.
28, 29. $E\{X\} = 2^{n-1} + \frac{1}{2}$, $(n-1)2^{-n}$, 0.
32. $E\{X\} = 26$, $E\{Y\} = 25$.
34. ≈ 5.031.
35. 2.
36. $\frac{1}{\lambda}(1-e^{-\lambda})$.
37. 12.25.
38. No.
43. Underestimated.
46. $13/6$.
48. (a) $35/12$; (b) $\frac{11515}{144} \approx 79.97$.
49. $4/3$; $14/9$.
51. $\frac{302}{3}\lambda + \frac{2}{3}\lambda^2$, where $\lambda = 60,000$.
52. (b) $\frac{2}{3} - \frac{1}{2^n} + \frac{1}{3 \cdot 4^n}$; $2/3$.
54. $E\{X^2\}$.
56. $8 \cdot \frac{1}{4} = 2$. No. No.
57. 350, $\frac{875}{3}$.
58. the standard deviation is $\sqrt{n-1}$ and is large for large n.
61. If all $\sigma_i \neq 0$, then the share of the investment to the ith security equals σ_i^{-2}/A, where $A = \sum_{i=1}^{n} \sigma_i^{-2}$. The minimal variance equals $1/A$.
63. ≈ 0.782.

64. 7.5, if we count the day of the third drop.
66. $\binom{5+k}{k}2^{-5-k}$.
70. The mean is 2, the variance is $2(1-\frac{1}{750}) \approx 1.9973....$ An estimate of the probability is ≈ 0.947.
71. An estimate is 0.758.
72. $\sqrt{q}$, $\sqrt{q/(pn)}$, $1/\sqrt{\lambda}$.
73. ≈ 0.5503.
76. (a) Yes. (b) No; converges.
77. No. Yes. $P(S=100) \approx 0.055$, $P(S=200) \approx 0.082$, $P(S=350) \approx 0.082$.
79. (a) ≈ 0.544, (b) ≈ 0.544, (c) ≈ 0.113.
80. 0.9775.
81. 0; the standard deviation is $\approx \$2.92$; Chebyshev's inequality gives ≈ 0.081, ≈ 0.036. More precise estimates are given in Example **8**.3.2-1.
84. (a) $0.17n$; (b) $2n/5$; (c) $n(m+1)/(2m)$.
85. (a) 6.5; (b) 4; (4) $\sum_{k=1}^{m}(1/k) \sim \ln m$.
88. (a) $\mu\lambda$, No; (b) $(\mu+\mu^2)\lambda$.

CHAPTER 4

1. $(pz/(1-qz))^{\nu}$.
6. p/q if $p<q$, and 1 otherwise.
8. 1.
10. ≈ 0.427.
11. 2.
15. 15/16.
17. ≈ 80.
18. ≈ 0.128.

CHAPTER 5

1. (b) 0.000049.
2. For $n=4$, the transition matrix $\mathcal{P} = \begin{Vmatrix} 0 & 1 & 0 & 0 & 0 \\ 0.25 & 0 & 0.75 & 0 & 0 \\ 0 & 0.5 & 0 & 0.5 & 0 \\ 0 & 0 & 0.75 & 0 & 0.25 \\ 0 & 0 & 0 & 1 & 0 \end{Vmatrix}$.

4. $$\mathcal{P} = \begin{Vmatrix} 1-p_1 & p_1 & 0 \\ (1-p_1)p_2 & p_1p_2+(1-p_1)(1-p_2) & p_1(1-p_2) \\ (1-p_1)p_2^2 & p_1p_2^2+2(1-p_1)p_2(1-p_2) & 2p_1p_2(1-p_2)+(1-p_2)^2 \end{Vmatrix}.$$

5. $$\mathcal{P} = \begin{Vmatrix} f_0 & f_1 & f_2 & f_3 & \cdots \\ 0 & f_0+f_1 & f_2 & f_3 & \cdots \\ 0 & 0 & f_0+f_1+f_2 & f_3 & \cdots \\ 0 & 0 & 0 & f_0+f_1+f_2+f_3 & \cdots \\ 0 & 0 & 0 & 0 & \cdots \\ \vdots & \vdots & \vdots & \vdots & \cdots \end{Vmatrix}.$$

8. 0.18, 0.4.
9. 9, 0.
10. No, it suffices to assume that the migration of each citizen is governed by the same transition matrix.
11. $\approx (0.220, 0.641, 0.139)$.
13. 0, 3, 4, 3, 0.
14. ≈ 12.8.
16. (a) ≈ 58.2; (b) $(1/0.15) \approx 6.7$; (c) ≈ 23.0.
17. 9.5; (b) ≈ 1.66.

18. ≈ 15.98.
19. $\approx (0.41, 0.59)$; precisely $(\frac{7}{17}, \frac{10}{17})$.
20. (a) $3/7$ and $4/7$; (b) $2/7$.
21. $(0.5, 0.5)$.
23. $\frac{5}{17}, \frac{8}{17}, \frac{4}{17}$.
29. ≈ 2.67.
30. ≈ 0.57.
32. (a) ≈ 2.11. (b) ≈ 0.307.
34. (a) The limiting distribution is Poisson with parameter λ. (b) $P(T_0 > m) = \lambda^m/m!$, and $E\{T_0\} = e^{\lambda}$.
35. Three classes: $\{0,1\}, \{2\}, \{3\}$. The first two states are transient, the last two—absorbing.
38. All states communicate, and hence, are recurrent.
39. $\mathcal{P}$, 2.

CHAPTER 6

1. (a) $3/32, \approx 0.015, 0, 0, 0$; (b) $F(x) = x^6/64$ for $0 \le x \le 2$; (c) $12/7, \approx 59.08$; (d) $3/49$.
3. (b) $\frac{e-\sqrt{e}}{e-1}$; (c) $e^x/(e-1)$.
4. $F(x) = x^2/64$ for $0 \le x \le 8$.
5. $F_Y(x) = \frac{1}{2}(\sin x + 1)$ and $f_Y(x) = \frac{1}{2}\cos x$ for $-\frac{1}{2}\pi \le x \le \frac{1}{2}\pi$; and $F_Z(x) = \sin x$ and $f_Z(x) = \cos x$ for $0 \le x \le \frac{1}{2}\pi$.
6. $P(\frac{\pi}{6} \le X \le \frac{\pi}{3}) = (\sqrt{3}-1)/2$.
7. (a) $3/4$ and $1/4$.
8. $P(k \le X \le k+1) = 1/2^{k+1}$; $P(k \le X \le k+\frac{1}{2} \,|\, k \le X \le k+1) = 1/2$; $E\{X\} = 3/2$; $Var\{X\} = 25/12$.
9. $\theta/(\alpha-1)$ for $\alpha > 1$; $\theta^2\alpha/[(\alpha-1)^2(\alpha-2)]$ for $\alpha > 2$.
10. (b) $\delta_{\text{optimal},1} = 2m(1-\delta)$ and $\delta_{\text{optimal},2} = m\ln(1/\delta)$; $\delta_{\text{optimal},1} > \delta_{\text{optimal},2}$ for $\delta > \delta_0 \approx 0.20$.
13. (b) Say, the distribution from Exercise 7c.
14. (b) $(\ln 2)/a$; ≈ 1.78.
15. m, 0, $(\nu-1)/a$ for $\nu > 1$.
18. $0.008, \approx 0.0894$.
19. $1/3, 1/3$.
21. The probabilities in hand are $1-e^{-3/2}$, $e^{-1}-e^{-3/2}$.

22. $\approx 0.144, \approx 6.952, \approx 6.952$.
25. (c) $1/\sqrt{\nu}$.

item[**26.**] (d) ≈ 0.668.
28. $\Phi(3/4) - \Phi(-5/4) \approx 0.668$.
29. ≈ 10.32.
30. $\pi/4$.
31. x^3, $3/4$, $3/80$.
32. $1/5, 0$.
34. Independent; $2x$, $3y^2$.
36. $c = 3/2$; $\frac{1}{2} + \frac{3}{2}x^2$ for $0 \le x \le 1$, $\frac{1}{2} + \frac{3}{2}y^2$ for $0 \le y \le 1$.
37. The answer to the last question: geometric with $p = P(X_0 > x_0)$.
40. (a) $2(e^{-x} - e^{-2x})$; (b) $f(x) = x/2$ for $0 \le x \le 1$, $f(x) = 1/2$ for $1 \le x \le 2$, $f(x) = (3-x)/2$ for $2 \le x \le 3$.
42. ≈ 0.954.
43. ≈ 0.504.
52. $f(y\,|\,x) = (x+y)/(x+0.5)$ for $0 \le y \le 1$; $E\{Y\,|\,X\} = 0.5 + 1/(12X+6)$.
53. $f(y\,|\,x) = 2y/x^2$ for $0 \le y \le x$; $E\{Y\,|\,X\} = \frac{2}{3}X$.

55. The conditional density is $f(x\,|\,t) = 2x/t^2$ for $0 \le x \le t$. If T is the total waiting time, then $E\{X\,|\,T\} = \frac{2}{3}T$.
57. The *Pareto density* $f(x) = 1/(1+x)^2$ for $x \ge 0$.
60. $E\{N\} = 3/2$; $Var\{N\} = 9/4$.

CHAPTER 7

1. (a) 0.5, 0.4, 0.4, 0.3, 0.1, 0, 0, 0, 0.56..., 0.083..., 0.083.... (b) 0.5, 1, 0.4. (c) Decomposition (**7**.1.3.3) is true with $\alpha = 0.35$; the discrete component F_d is the distribution of a r.v. X_d assuming values 1, 3, 4 with respective probabilities $\frac{2}{7}, \frac{3}{7}, \frac{2}{7}$; the continuous component has the density

$$f_c(x) = \begin{cases} 5/13 & \text{for } x \in [0,1], \\ 3/13 & \text{for } x \in [3,4], \\ 10/39 & \text{for } x \in (4,5.5], \\ 0 & \text{otherwise.} \end{cases} \qquad E\{X\} = 2.7875.$$

2. 3.
3. (a) $a + y(b-a)$; (b) $-m\ln(1-y)$.
4. $m + \sigma z$, where z takes on respective values $0, \approx -0.524, \approx 1.23$.
6. Say, $X \equiv 0$.
8. Decomposition (**7**.1.3.3) is true with $\alpha = 2/3$; the discrete component F_d is the distribution of a r.v. X_d assuming values 1, 2 with respective probabilities $\frac{1}{5}, \frac{4}{5}$; the continuous component is exponential with parameter $a = 2$; the mean equals $41/30$, the variance equals ≈ 0.566.
9. Decomposition (**7**.1.3.3) is true with $\alpha = 0.6$; the discrete component F_d is the geometric distribution with $p = 1/3$, the continuous component is uniform on $[0,2]$; $a = 0.2$. The mean is 2.2, the variance equals 4.693....

CHAPTER 8

5. (b).
6. If $M(z) \sim 0.2e^{3z}$ as $z \to \infty$, then $P(X \le 3) = 1$ and $P(X = 3) = 0.2$.
8. Yes. No.
9. $M''(0) = \sigma^2 + m^2$.
10. The means are the same, the variance for M_2 are larger.
16. $\exp\left\{50\left(\frac{1}{1-100z} - 1\right)\right\}$.
17. ≈ 0.889.
18. $\exp\left\{\lambda_1\left(e^{\lambda_2(e^z-1)} - 1\right)\right\}$.
19. $\frac{2}{3e} \approx 0.245$.
20. The distribution of S is that of $Y_1 + ... + Y_\nu$, where Y_i's are independent r.v.'s, and each has the distribution (**8**.2.3.7).
21. $A = \mu/(\mu+\delta)$, $a = 1/(\mu+\delta)$, the premium is μ.
22. $\exp\{-x^2/2\}$.
23. $\exp\{-x^2/(4n)\}$ for $x \le n$, and $\exp\{-x/4\}$ for $x > n$.
24. For $p \le 0.5$, we have $P(\hat{p} - p \ge 0.05) < 0.024$. So, the confidence level is larger than 0.976.

CHAPTER 9

1. ≈ 3 h.10 min.
2. ≈ 0.988.
4. (a) $q(a) = 2\Phi\left(\frac{a}{\sigma}\right) - 1$, where $\sigma = \sqrt{\frac{35}{12}} \approx 1.7079$. (b) $q(a) = 0.9$ for $a \approx 1.645\sigma \approx 2.809$. (c) $z \approx 28.09$.
5. (c) $2(1 - \Phi(c))$.
6. $2(1 - \Phi(c))$.
7. $\Phi(1/2) \approx 0.692$.
8. (a) ≈ 0.694.

15. Close to $1/2$.
16. ≈ 0.51.
18. No condition is required.
19. Let C stands for a constant. If $a > \frac{1}{2}$, then $L_n \not\to 0$. If $a = \frac{1}{2}$, then $L_n \sim \frac{C}{(\ln n)^{3/2}}$.

If $\frac{1}{3} < a < \frac{1}{2}$, then $L_n \sim C\frac{1}{n^{3(1-2a)/2}}$. The power index $\frac{3(1-2a)}{2} < \frac{1}{2}$.

If $a = \frac{1}{3}$, then $L_n \sim C\frac{\ln n}{\sqrt{n}}$. If $a < \frac{1}{3}$, then $L_n \sim \frac{C}{\sqrt{n}}$.
21. 1.

CHAPTER 10

1. T, T, T, F, F, T, T, T, F.
4. 0. No.
5. Yes, because the Cauchy-Schwartz inequality is true for *all* r.v.'s including $|\xi|, |\eta|$.
6. $-1/\sqrt{2}$. It suffices to assume that X and Y are uncorrelated and have the same variances.
8. $1/\sqrt{1+k}$.
10. $[(1-2p)^n - (1-p)^{2n}]/[(1-p)^n - (1-p)^{2n}]$, where $p = 1/k$.
11. $5/7$.
12. Yes.
14. $\alpha = (1-\rho k)/(1+k^2-2\rho k)$, where $k = \sigma_1/\sigma_2$.
15. (a) $\frac{\sqrt{2}}{3}$. (b) 0.
17. (a) $\rho = \sqrt{21}/5$.
18. $\beta = \rho\sigma_Y/\sigma_X$.
19. $0.73\ldots$.
20. $\rho \neq \pm 1$.
21. Sum up all elements of $\mathcal{C}$.
24. $\det(\mathcal{C}) = 0$.
27. (a) 1.07 or 7% of profit; 0.00018. (b) $\approx (0.74, 0.26)$, $\approx (60\%, 29.6\%, 10.4\%)$.
30. $\frac{1}{\sqrt{2\pi(\sigma_1^2+\sigma_2^2+2\rho\sigma_1\sigma_2)}}\exp\left\{-\frac{x^2}{2(\sigma_1^2+\sigma_2^2+2\rho\sigma_1\sigma_2)}\right\}$.
31. (b) (i) $-5, 276, \frac{1}{\sqrt{552\pi}}\exp\{-(x+5)^2/552\}$; (ii) $-1, 11.04, \frac{1}{\sqrt{22.08\pi}}\exp\{-(x+1)^2/22.08)\}$.
33. (a) Yes. The distribution with the density $f(r) = re^{-r^2/2}$, and the distribution uniform on $[0, 2\pi]$.
35. $f_{\chi_k}(x) = \frac{2^{1-k/2}}{\Gamma(k/2)}x^{k-1}e^{-x^2/2}$.
38. ≈ 0.9997, ≈ 0.3935.
41. If, say, $\mathcal{C} = \left\| \begin{array}{cc} \sigma_1^2 & 0 \\ 0 & \sigma_2^2 \end{array} \right\|$, then $\mathcal{D} = \left\| \begin{array}{cc} \sigma_1^{-1} & 0 \\ 0 & \sigma_2^{-1} \end{array} \right\|$.
42. $M_{\mathbf{X}}(\mathbf{z}) = \prod_{i=1}^k M_{X_i}(z_i)$.
43. $1 - 1/\sqrt{e}$.
44. *Hint*: The sets presented are combinations of convex sets.

CHAPTER 11

1. (a) $\overline{F}_1(n-1)\overline{F}_2(n-1)$; (b) $(1-\overline{F}_1(n))\overline{F}_2(n) + \overline{F}_1(n)(1-\overline{F}_2(n))$; (c) $(1-\overline{F}_1(n))(1-\overline{F}_2(n))$; (d) $\overline{F}_1(n)+\overline{F}_2(n)-\overline{F}_1(n)\overline{F}_2(n)$; (e) $(\overline{F}_1(n-1)-\overline{F}_1(n))(\overline{F}_2(n-1)-\overline{F}_2(n))$.
2. (a) The distribution of a r.v. $a+Y$, where $F_Y(x) = (x/d)^n$ for $0 \leq x \leq d$, and $E\{Y\} = d\frac{n}{n+1}$.
(b) $F_Y(x) = \exp\{\lambda((x/d)-1)\}$ for $0 \leq x \leq d$, and $E\{Y\} = d(1-\frac{1}{\lambda}(1-e^{-\lambda}))$. In Part (b), the price is less in the mean and in the sense of the FSD.
3. 30.125; ≈ 30.318.
4. $1-e^{-x}$.
6. $8n/(4n-1) \to 2$.

7. $z+2z^2-3z^3+z^4$, where $z=F(x)$; $23/60$.

8. ≈ 0.253.

9. $G(x)(1-(1-F(x))^2)^2+(1-G(x)(1-(1-F^2(x))^2)$.

12. $\alpha=1$, and $\underset{\sim}{W}_n^*$ has exactly the exponential distribution with the same parameter as for the X's for all n.

13. (a) $\underset{\sim}{W}_n=\frac{1}{\sqrt{n}}\underset{\sim}{W}_n^*$, the limiting d.f. for $\underset{\sim}{W}_n^*$ is $1-e^{-x^2}$. (b) $\underset{\sim}{W}_n=\frac{1}{n}\underset{\sim}{W}_n^*$, the limiting d.f. for $\underset{\sim}{W}_n^*$ is standard exponential.

14. $\underset{\sim}{W}_n=1+\frac{1}{\sqrt{n}}\underset{\sim}{W}_{Yn}^*$, the limiting d.f. for $\underset{\sim}{W}_{Yn}^*$ is $1-e^{-x^2}$.

15. (a) $\widetilde{W}_n=2-\frac{1}{n}\widetilde{W}_n^*$; the limiting d.f. for $\widetilde{W}_n^*$ is $1-e^{-2x}$ (that is, exponential with parameter 2). (b) $\widetilde{W}_n=2-\frac{1}{n^{1/3}}\widetilde{W}_n^*$; the limiting d.f. for $\widetilde{W}_n^*$ is $1-e^{-x^3}$. (c) $\widetilde{W}_n=-\frac{1}{n}\widetilde{W}_n^*$; and $\widetilde{W}_n^*$ is standard exponential for all n.

21. $\widetilde{W}_n=n^{1/4}\widetilde{W}_n^*$; the limiting d.f. for $\widetilde{W}_n^*$ is e^{-16/x^4}.

22. $\widetilde{W}_n=n\widetilde{W}_n^*$, the limiting d.f. for $\widetilde{W}_n^*$ is $e^{-1/(\pi x)}$; $\underset{\sim}{W}_n=n\underset{\sim}{W}_n^*$, the limiting d.f. for $\underset{\sim}{W}_n^*$ is $1-e^{-1/(\pi|x|)}$.

29. Yes.

30. $\mu(x)=\frac{\alpha}{100-x}$ for $x\in[0,100]$; $s(120)=0$.

31. $\mu(x)=\frac{1}{a-x}$ for $x\in[0,a]$

34. The new force of mortality should be $\ln 2/k$ less.

35. (a) 0.125, (b) ≈ 0.674.

39. 1, 1.06.

42. Yes. If w_1,w_2 are the proportions of the new heaters ($w_1=0.3, w_2=0.7$), and x is the age of a heater selected at random, then denoting by μ_i the respective hazard rates ($\mu_1=1/10$ and $\mu_2=1/8$), we have $P(T(x)>t)=w_1(x)\exp\{-\mu_1 t\}+w_2(x)\exp\{-\mu_2 t\}$, where the proportions at time x are given by

$$w_i(x)=\frac{w_i\exp\{-\mu_i x\}}{w_1\exp\{-\mu_1 x\}+w_2\exp\{-\mu_2 x\}},\ i=1,2.$$

43. $\mu(x)\sim x$ as $x\to\infty$.

CHAPTER 13

2. (a) $\approx 0, \approx 0.208$. (b) $\approx 0.082, 0, \approx 0.082$. (c) ≈ 0.265. (d) $\frac{4}{5}=0.8$, $\sqrt{\frac{4}{25}}=0.4$.

3. Follow the scheme of Exercise 2 with $\lambda=2$.

4. (a) ≈ 0.368. (b) ≈ 0.59, 1h., ≈ 0.73h.

5. (a) $\lambda t+\lambda^2 t(t+s)$. (b)t+ m/λ, m/λ^2. (c) 0, $\sqrt{t/(t+s)}$.

6. 1/27, 19/27.

7. $20\lambda, 20\lambda+75\lambda^2$. In general, $\lambda(a+b/2), \lambda(a+b/2)+\lambda^2b^2/12$.

8. $t<c\approx 0.206$.

10. $80/3, 4\sqrt{15}/3; \approx 0.671$.

11. (c) $730/\pi$, $\approx 2(1-\Phi(1))\approx 0.317$.

14. ≈ 0.135.

16. (a) 24/49. (b) ≈ 0.186. (c) 2/7.

17. 0.1, ≈ 0.316.

18. $\rho/(1-\rho)^2$.

22. (a) $\mu/(\lambda+\mu)$. (b) The mean is ρ which is less than one if $\lambda<\mu$.

23. (b) 1/2.

26. (a) The second particle will arrive at random time T_2 having the Γ-distribution with parameters $(\lambda, 2)$, and after that there will be no arrivals neither departures. (b) It may be viewed as a system with two servers and no queue: once both servers are busy, no particles enter the system.
27. (b) π_k is decreasing if $\rho < 1$. (d) $\pi_k \to 0$ as $\rho \to \infty$ for $k = 0, ..., a-1$; while $\pi_a \to 1$.
28. (a) $\frac{4}{7}, \frac{2}{7}, \frac{1}{7}$ for $\rho = 1/2$, and $\frac{1}{7}, \frac{2}{7}, \frac{4}{7}$ for $\rho = 2$.
29. $1/\sqrt{\rho}$.
30. $k \geq 9$.
31. The limiting distribution is Poisson with parameter $\rho = \lambda/\mu$.
34. $\pi_0 = 0, \pi_1 = 2\mu/(2\mu+\lambda), \pi_2 = \lambda/(2\mu+\lambda)$.

CHAPTER 14

1. For example, compare $P(N_1 \geq 1 \,|\, \tau_1 > 1)$ and $P(N_1 =\geq \,|\tau_1 < 1)$.
2. F, F, T.
3. (a) The τ's have the Γ-distribution with parameters $a = 40$ and $\nu = 5$; $P(\tau > \frac{1}{6}) \approx 0.21$. (b) 8. (c) ≈ 0.821.
6. T_n is Poisson with parameter $n\lambda$. For t and s such that $[t] = [s]$.
7. $29,094.

CHAPTER 15

3. $C_t = (2/a)^t$.
4. (a) $C_t = 2^t$. (b) $X_t \to 0$ with probability one.
7. $E\{X_t\} = 1$; $X_t \to 0$ with probability one.
8. (a) $C_t = e^{-t/2}$; $X_t \xrightarrow{P} 0$. (b) $C_t = e^{-mt-t\sigma^2/2}$; $X_t \xrightarrow{P} 0$.
16. 200.
19. (a) No. (c) In each game (a sequence of plays up to quitting), the gambler loses on the average ≈ 53¢. In accordance with the LLN, if the gambler keeps applying the doubling strategy n times, then her/his loss is close to $0.507n$ dollars with probability close to one. (d) The mean profit for one game equals $1-(2q)^{m+1}$, where $q = 1-p$. So, the mean profit is negative if $p < \frac{1}{2}$.
20. The latter. The former.
21. $\theta_i \geq 0$ for $i \geq 2$.
22. (a) 6. (c) $1,250.
23. (a) $\pi_1 = 0.5 > 0, \pi_2 = 0.5 > 0$. (b) $12.5, $12.5. (c) 50$c$. (d) $\pi_1 = 0.6 > 0, \pi_2 = 0.4 > 0$; $\approx$ $9.09, $\approx$ $13.64; ≈ 55¢. (d) $r > 0.5$.
24. The price of an American derivative is not smaller than that of its European counterpart because the former has all options of the latter and additional options.
25. (a) $\pi \approx 0.909, \approx 0.165$. (b) ≈ 9.47. (c) ≈ 0.07. (d) ≈ 0.36; see Exercise **15**.24. (f) ≈ 145.18; one should quit at the first step if the price drops, and keep waiting otherwise.
28. $\leq u/a$.

CHAPTER 16

2. (a) No. (b) Yes. (c) $\Phi(\sqrt{2}) \approx 0.921$.
3. $\sigma = \sqrt{5}$.
5. $c = \sqrt{a}$.
6. The answer is different: the probability in hand is ≈ 0.52.
7. ≈ 0.670.
8. For $t \geq s$, we have $Corr\{w_t, w_s\} = \sqrt{\frac{s}{t}} \to 0$ as $t \to \infty$.

10. (a) $f_{\tau_a}(t) = \frac{a}{\sqrt{2\pi}t^{3/2}} \exp\left\{-\frac{a^2}{2t}\right\}$. Thus, $f_{\tau_a}(t) \sim \frac{a}{\sqrt{2\pi}t^{3/2}}$ as $t \to \infty$, and hence, $\int_0^\infty t f_{\tau_a}(t)dt$ diverges.
12. No.
13. 0.683.
14. $E\{Y_t\} = Y_0 \exp\left\{\mu t + \frac{1}{2}\sigma^2 t\right\}$; $Var\{Y_t\} = Y_0^2 \exp\{2\mu t + \sigma^2 t\}\left(\exp\left\{\sigma^2 t\right\} - 1\right)$.
17. No. Yes.
19. $C_t = e^{-8t}$.
20. (a) $\frac{dS_t}{S_t} = 0.15dt + 0.1dw_t$; $S_t = 100\exp\{0.145t + 0.1w_t\}$. (b) ≈ 0.63. (c) ≈ 0.812. (d) ≈ 0.188. (e) $S_T = S_t \exp\{0.145(T-t) + 0.1w_{T-t}\}$, where $T = 0.5$. (f) $1 - \Phi\left(\frac{\ln(K/S_t) - 0.145(T-t)}{0.1\sqrt{T-t}}\right)$.
21. The price equals $200 - 198\Phi(\sigma\sqrt{T}/2) \approx 99.03$.
23. (a) $c \approx 4.19$ and $p \approx 1.72$. (b) For the bet to be fair on the average, at time $t = 0$, Richard should pay John \$25 61¢.
24. u/a.
25. $\exp\{-2bc\}$.
26. The probability will become (a) $1/2$, (b) the same, (c) $1/256$.
27. (a) ≈ 0.359. (b) ≈ 0.393.
28. The answer is different: the probability of interest ≈ 0.723. The selling time is improper, and the probability that the prices will never drop by %10 is 0.607.
29. 10 and 11. The probability of interest is ≈ 0.667.

CHAPTER 17

2. (a) $\boldsymbol{v}_1, \boldsymbol{v}_4, \boldsymbol{v}_5, \boldsymbol{v}_6, \boldsymbol{v}_8, \boldsymbol{v}_9, \boldsymbol{v}_{10}, \boldsymbol{v}_{11}$.
3. No.
4. $\frac{\mu(AB)}{\sqrt{\mu(A)}\sqrt{\mu(B)}}$.
5. (a) The integrand is of the type $\exp\{-cx^2\}$. (b) $\mu(d\boldsymbol{x}) = \beta|\boldsymbol{x}|d\boldsymbol{x}$, $\exp\{-2\beta\pi r^3/3\}$.
6. (a) No. (b) $\mu(A) = nV(A)$, where $V(A)$ is the volume of A, $\left(\frac{\pi}{6}\right)^n$.
7. ≈ 0.128.
8. (a) $P\left(\frac{3S_n}{\sqrt{n}} \leq x\right) \to \Phi(x)$. (b) $P\left(\frac{12S_n - 3n}{\sqrt{13n}} \leq x\right) \to \Phi(x)$.
9. $P\left(\frac{S_n}{B_n} \leq x\right) \to \Phi(x)$, where $B_n^2 = 3^{-5}|A_n|$, and $|A_n|$ is the number of points in A_n.
12. $P(X_n \leq x) \to P(\Lambda \leq x)$. In the particular case of Example **17**.4.1-2, $P(\Lambda \leq x) = 1 - e^{-ax}$.
13. (b) W_n and X_n have the same limiting distribution. (c) The distribution is uniform on $[0, 1]$.
14. Yes.

CHAPTER 18

1. (a) $\gamma \in [0, 0.7) \cup [0.8, 1]$. (b) $m \geq 1$; for $\gamma \geq \gamma_0 \approx 0.04$. (c) $m \geq 1$; $\gamma \leq 0.96$. (d) $\gamma \geq 0.5$.
2. (a) $\gamma \in [0, 01) \cup [0.4, 0.7) \cup [0.8, 1]$; (b) $\gamma = 1$.
3. The investment into one asset is more profitable for $\gamma > \gamma_0 \approx 0.68$.
6. (a) $\alpha = \frac{m_1 - m_2 + 2k\sigma_2^2}{2k(\sigma_1^2 + \sigma_2^2)}$; (b) $\alpha = \frac{\sigma_2^2}{\sigma_1^2 + \sigma_2^2}$.
10. $c(X_1) < c(X_2)$ for $p < \frac{1}{1+\alpha}$. If $p = 1/2$, then $c(X_1) < c(X_2)$ because $\alpha < 1$. In this case, the values of X_1 are "more dispersed."
11. $E\{X\} = \infty$ while $c(X) = 4$.
14. $5/9$.
16. (c) 25. (d) The same.
18. (a) On the preference order. (b) No.
24. $k \geq \$79$.
25. The former.
27. $c = \frac{2rd}{r + 1 + d + \sqrt{(r+1+d)^2 - 4rd}}$, where $r = p/q$. If $p \to 1$, then $c \to d$.
28. $c \approx 0.433$.
30. $c = b\left(\frac{\beta}{\beta + \alpha}\right)^{1/\alpha}$.

Index